# ENGINEERING GRAPHICS

William P. Spence

# ENGINEERING GRAPHICS

Prentice-Hall, Inc., Englewood Cliffs, NJ 07632

*Library of Congress Cataloging in Publication Data*

SPENCE, WILLIAM PERKINS.  (date)
    Engineering graphics.

    Includes index.
    1. Engineering graphics.  I. Title.
T353.S573 1984      604.2      83-17686
ISBN 0-13-278879-9

Editorial/Production Supervision: Karen Skrable
Interior and Cover Design: Maureen Eide
Manufacturing Buyer: Anthony Caruso
Cover Photo: NASA

Printed in the United States of America

10  9  8  7  6  5  4  3  2  1

ISBN 0-13-278879-9

Prentice-Hall International, Inc., *London*
Prentice-Hall of Australia Pty. Limited, *Sydney*
Editora Prentice-Hall do Brasil, Ltda., *Rio de Janeiro*
Prentice-Hall Canada Inc., *Toronto*
Prentice-Hall of India Private Limited, *New Delhi*
Prentice-Hall of Japan, Inc., *Tokyo*
Prentice-Hall of Southeast Asia Pte. Ltd., *Singapore*
Whitehall Books Limited, *Wellington, New Zealand*

This book is dedicated to
Bettye Margaret Spence
in acknowledgment of her
support, encouragement,
and productive assistance.

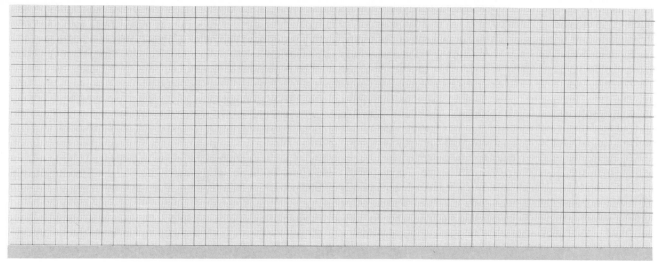

# CONTENTS

# 5

## Geometric Constructions 82

# 6

## The Principles of Orthographic Projection 109

# 7

## Technical Sketching 141

# 8

## Descriptive Geometry: Spatial Relationships 159

# 9

## Auxiliary Views 182

# 16

## Mechanical Fasteners: Bolts, Keys, Nuts, Screws, Springs, and Washers  392

# 17

## Permanent Fasteners: Welding and Riveting  418

# 18

## Preparation of Production Drawings  439

# 19

## Developments  479

# 20

## Intersections  507

# 21

## Vector Analysis  531

# 22

## Graphical Kinematics  556

# 23

## Graphical Presentation and Analysis  590

# 24

## Computers in Engineering and Graphics  632

# 25

## Engineering Design Problems  670

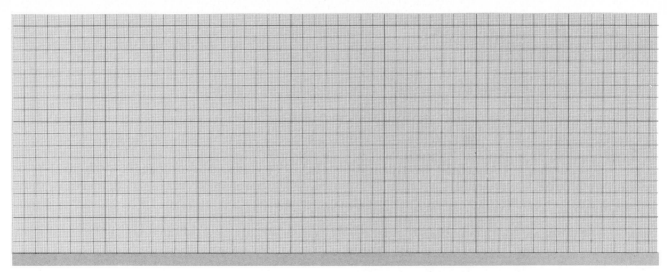

# PREFACE

This text has been designed to provide basic and advanced instruction in engineering graphics. The basic content was developed from the results of an analysis of course outlines of introductory graphics courses being offered in accredited engineering schools. Additional material was included to reflect the rapidly changing techniques and procedures in engineering graphics and design.

The text is of sufficient breadth and depth to meet the needs of instructors of both one- and two-semester courses. Additional materials are included for those who wish to go beyond this level.

The coverage of engineering graphics is handled in such a way that a student with no prior experience can successfully comprehend the material and progress through the subject with a minimum of instruction. The basic principles have been explained as clearly as possible and have been carefully illustrated in order to increase student interest and encourage self-study. The instructor is therefore freed from teaching every minute detail and can devote teaching efforts to the major concepts in engineering graphics. A second color has been used to emphasize important parts of each illustration. Steps of procedure have been placed on the same page as the illustrations whenever possible. This makes it much easier for the student to follow the instructions. An extensive appendix provides additional technical data needed for design problems.

Some engineering graphics instructors emphasize the techniques of drafting and technical information; this type of material is extensively covered. Other instructors prefer to reduce emphasis on these and introduce the basic concepts of engineering design; *Engineering Graphics* permits this approach to be used while including extensive technical data. One chapter is devoted to the engineering design process, and references to it have been woven into various other chapters. A second chapter is devoted to student design projects and how they might be handled in an academic environment. Therefore, there is sufficient design information to offer a basic experience. Computers are standard tools of the engineer, and their basic principles are covered. The use of computers in graphics and design, including hardware, software, and languages, are introduced. Sample computer drafting programs are shown with a complete explanation.

Metrics are thoroughly covered. The text includes an explanation of the metric system, symbols, pronunciation, and applications to engineering graphics. Metric standards are used when available.

An extensive array of problems are available at the ends of chapters. They cover the areas discussed in the chapter and give the instructor a wide choice of types of problems as well as various levels of difficulty.

The text begins by introducing the use of

graphics in various engineering fields. The concepts of engineering design are then covered, followed by a study of the graphic language. The discussion of tools used in the production and reproduction of graphics are followed by the techniques of geometric construction. A considerable effort has been made to present the principles of orthographic projection in an easily understood form. Since engineering ideas often start with sketches, techniques for producing engineering sketches are presented.

The key to successful engineering design is an understanding of the principles and applications of descriptive geometry. Two chapters have been devoted to this important area. Related to these chapters are chapters on auxiliary views and revolutions.

Other means of presenting engineering ideas and designs, such as pictorial drawings and the use of sectional views, are given emphasis.

Considerable space is devoted to dimensioning, tolerancing, and surface finish. Careful attention has been given to standards and inch and metric requirements. Geometric tolerancing material includes the use of the various symbols, feature control principles, position, symmetry, form, runout cylindricity, straightness, flatness, profile angularity, parallelism tolerances, and the control of surface quality.

Two chapters are devoted to mechanical and permanent fasteners, including the details of inch and metric threads and fasteners.

Production drawings are the end product of the design effort. An entire chapter has been devoted to this topic and includes many problems for the students to solve.

Surface development and intersections occur on every product and are covered by two chapters. Since analysis is a function of the engineer, a chapter on vector analysis is included. Design involves the application of kinematic principles that are covered in a complete chapter that includes linkages, cams, and gears.

The presentation and graphic analysis of data are necessary parts of the engineering design process. Techniques for presenting data and making engineering analyses of experimental data and physical relationships are explained. Charts of various types and the use of graphical algebra and calculus are included.

No publication of this size and depth can be completed without the help of many individuals and companies. These are acknowledged in the credits, and my thanks is again extended to each. Special acknowledgment is due to Dr. A. O. Brown, Dr. Joe Porter, and Professor Gene Chambers for their assistance with several of the chapters. And finally thanks is due to the reviewers who, while anonymous, read and commented on the chapters as they were developed.

WILLIAM P. SPENCE

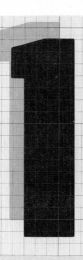

# Graphics in Engineering

Many young people have an early interest in a career in engineering. Often they are not certain what an engineer does, but they want to be one. Young people do not generally have the opportunity to get into the plants and laboratories of industry and see what actually takes place. Even if their parents are engineers, they may not be certain what tasks they perform on the job. A study of the early chapters of this text plus a well-rounded experience in an engineering graphics course will give a brief look into some of the problems and opportunities of engineering.

## WHAT ARE THE MAJOR AREAS OF WORK?

The work of an engineer covers a wide range of activities. Some work in areas—such as manufacturing management—where they draw upon a wide variety of experiences and knowledge. Others work in areas—such as nuclear research—where they draw upon a narrow but intensive area of specialization. Following are the broad areas in which engineers work and illustrations of how engineering graphics is used.

## Engineers in Research

Engineers involved in research are attempting to apply known principles to solve existing problems or to discover new knowledge. This is usually a practical type of research that can involve many things, such as the use of materials or the development of a new process. This type of work is often slow to show results. While scientific knowledge is the basis for research, many attempts at solutions are by trial and error. Research engineers learn from failure as well as success. They add to the knowledge that can be used by others (Fig. 1–1).

Most of the time, research engineers work on research teams. The team can have a variety of people besides engineers. If the project involves chemical and biological factors, chemists and biologists may be on the team. Technologists and technicians also are a part of the team. They do much of the practical work, such as building prototypes or running routine tests.

Possible solutions and test procedures and equipment require the use of engineering drawings.

The research engineer must be able to read engineering drawings and produce engineering sketches. Engineering drafters work from these sketches to produce the needed finished drawings (Fig. 1–2).

*Research engineers* must be thoroughly prepared in mathematics and the sciences. They must have a strong curiosity and a desire to question, try, reason, and speculate. Usually, advanced degrees are necessary for employment.

## Engineers in Development

The area of development involves the actual use of discoveries. Development follows the research activity and sometimes occurs along with it. Because the two activities overlap, many companies combine them into a single research and development department.

Much of the work in engineering development involves improving and redesigning existing products. Engineers study other solutions. They find out why the product is not totally satisfactory and propose a new design. This could involve actual field testing of the product after the laboratory model has proven satisfactory. The final key to success is to have the solution perform satisfactorily under the actual conditions for which it was designed (Fig. 1–3).

Engineers in development must keep up to date on new discoveries and materials. This requires that they have a regular time for reading professional journals and attend seminars and other educational programs.

Development involves a great deal of drafting, including both sketches and finished engineering drawings. From these are developed the models or prototypes that are used in testing and manufacturing planning, which usually results in revisions, drawing changes, and more testing. If the project is successful, final drawings are used for production.

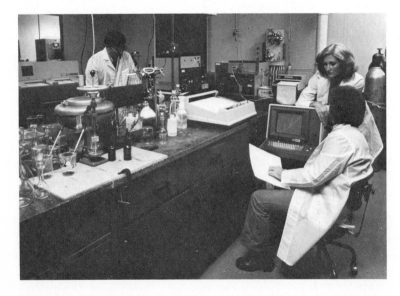

FIGURE 1–1 Engineering research involves the use of computer graphics for engineering testing and data analysis. (Courtesy of Tektronix, Inc.)

FIGURE 1–2 Research engineers must be able to produce engineering sketches and drawings.

## Engineers in Design

The design engineer works on both mass-produced and single-item products. For example, a typical *mass-produced product* is the radio. Its various components provide a variety of possible design solutions. These include AM and FM reception, a variety of electronic circuits and components, the inclusion of cassette recording and playing capabilities, shortwave bands, a clock, and other possibilities. The components are housed in a variety of packages, from a basic rectangle to a square to a sphere. They can be designed for use in a home, car, or boat or as portable units. The design engineer is involved with designing a salable product that will function in the described environment and will meet the needs of the purchaser at a price people will pay.

An example of a *single-item design* project would be a multistory office building or a satellite. These are designed for a specific situation and are not likely to be built again (Fig. 1–4).

The design engineer works with many materials and must be thoroughly prepared in the basic

FIGURE 1–3 Development engineers evaluate designs using laboratory and field tests. (Courtesy of Cessna Aircraft Company.)

FIGURE 1–4 Design engineers working with a satellite. (Courtesy of National Aeronautics and Space Administration.)

FIGURE 1–5 Engineers in design use drawings to record ideas and check possible solutions. (Courtesy of Bendix.)

principles of engineering. Often, the engineer will call on specialists in other disciplines for advice and assistance. The design engineer must know the thousands of stock parts available and have access to catalogs listing them. These are used because they are cheaper, and final cost is a primary aspect of the design solution. A product made with stock 6-mm bolts will be cheaper than if special bolts of an unusual size have to be made especially for that product. The engineer in construction will certainly consider the use of stock 4 × 8-ft sheets of plywood rather than requiring special sizes such as 4 ft 6 in. × 8 ft 6 in. be manufactured.

As a product is developed, several acceptable solutions might be reached. In making decisions, the engineer will consider the cost of manufacture and the ease of acquiring materials. Marketing staff are often consulted to see which solution might be best for sales promotion. In the end the design engineer must consider all factors and select the final solution.

An important part of the design process is the production of engineering drawings (Fig. 1–5). Possible solutions can be first tested by making appropriate drawings. The location and sizing of parts can be checked with a drawing. Various solutions can be analyzed and revised before expensive models or prototypes are built. Changes developing from the construction of models or prototypes are recorded on drawings (Fig. 1–6). The engineer must be able to read and produce design drawings and supervise the engineering drafting staff.

The design engineer must have a good general engineering background. Interest in the practical and a concern for cost are essential. Since design engineers often direct the work of others, preparation in management is helpful.

## Engineers in Manufacturing

The engineer in manufacturing is closely involved with the production of the product. This includes aspects such as fabrication and assembly of parts, establishing the work flow in the plant, quality control, safety, equipment selection, standards of workmanship, and production schedules (Fig. 1–7). A knowledge of how materials of all kinds are pro-

FIGURE 1-6 Engineers conducting tests with a scale model of a dam to predetermine water-flow characteristics. (Courtesy of Tennessee Valley Association.)

FIGURE 1-7 Engineers in manufacturing are responsible for supervising a wide range of activities in the manufacturing facility. (Courtesy of Unimation Inc.)

FIGURE 1-8 Engineers in manufacturing use many different types of drawings.

cessed, the means of joining materials, and the best equipment to do the work is needed. Since the engineer in manufacturing often supervises large numbers of people, a knowledge of psychology and management skills are vital.

The engineer in manufacturing constantly uses drawings. These include plant layouts, electrical, hydraulic, and pneumatic circuitry, engineering drawings of the products being produced, drawings relating to the equipment in the plant, and design-change drawings (Fig. 1-8).

## Engineers in Construction

Engineers in construction may be involved in the design phase, as discussed earlier, or in the supervision of the construction project. They may work in estimating and figuring construction costs for bidding purposes. They may supervise the construction process and make certain the product is built as indicated by the drawings and specifications (Fig. 1-

FIGURE 1-9 Many engineers in construction are employed as field engineers who supervise the day-to-day construction activities. (Courtesy of Armstrong World Industries.)

9). They must be certain the needed materials arrive on the site when needed, so scheduling is a big factor. They are often responsible for hiring various subcontractors, such as the plumbers, electricians, and concrete workers. They have to keep extensive records, including payroll, materials ordered and received, and an equipment inventory.

Engineers in construction supervision must be able to read construction drawings. They must know the symbols used for all the trades involved and where to look for items on the drawings. Construction drawings are large, very detailed, and run to many pages. They are coordinated with the specifications, and the engineer must be able to read and relate these to each other. Construction engineers should be able to sketch or produce drawings when special circumstances require them.

## Engineers in Plant Maintenance

A plant engineer is in charge of the overall maintenance of the buildings, equipment within these buildings, and the grounds surrounding the plant. Control of utilities is often a part of the responsibility. This operation requires a large staff, so the engineer is a manager as well as an engineer.

In this position the engineer works in a wide range of areas. Basic plant operation requires a knowledge of machinery, how it should be installed, operated, and maintained. A knowledge of electricity, electronics, hydraulics, and pneumatics is vital, because these perform the control functions on machines. An understanding of building construction, the materials of construction, and the various trades, such as plumbing, electrical, and masonry work, is necessary (Fig. 1-10).

The special features of a plant will require the

FIGURE 1-10 Plant engineers are responsible for the overall operation of the buildings, utilities, and equipment located in the building. (Courtesy of Atomic Energy Commission.)

engineer to get specialized knowledge. For example, in a chemical plant there are many unusual situations to handle; in a nuclear generating plant, the plant engineer faces an entirely different set of circumstances. The plant engineer must work closely with the engineers in manufacturing. They must work together to select equipment, locate it, install it, and keep it running. The problems of handling materials before, during, and after the manufacturing phase must be continually assessed.

Engineers in plant maintenance regularly use drawings of all kinds: architectural drawings relating to the building, utilities, or remodeling, for example. Engineering drawings related to the machinery, its installation, and maintenance are vital to the successful operation of the plant. Sometimes the plant engineer has to generate drawings for some aspect of the operation, such as moving utilities or revamping the production line. Plant maintenance could not function without extensive use of drawings.

Engineers in plant maintenance need mechanical, civil, and electrical engineering experience plus specialized knowledge related to the industry in which they are working.

## Engineers in Sales

An engineer who gets on well with people and enjoys discussing products in technical terms can be very successful in sales. Many of the products of industry cannot be sold without an assessment of the needs of a possible purchaser, a clear technical analysis of the problem, and a demonstration of how a specific product will function in a given situation. Engineers in sales not only represent a product but also serve as consultants to prospective purchasers and help them solve their manufacturing problems. They present materials to individuals and groups (Fig. 1–11). They must be ready to answer any questions about their product and about the general manufacturing field in which they work. For example, in plastics the sales engineer should be able to explain the workings of the extruders, injection molders, and other equipment being sold. The engineer should also be able to discuss the various thermoset and thermoplastic resins and their composition and characteristics. Which resin is best for a particular use in a particular machine? The sales engineer must be able to respond.

Sales engineers travel a great deal. They must have a knowledge of the plants of all potential customers in their region. This comes from regular visits to each plant, whether the company is actively looking for new equipment or not.

FIGURE 1–11 Engineers in sales present information about their product to potential customers. (Courtesy of Bendix.)

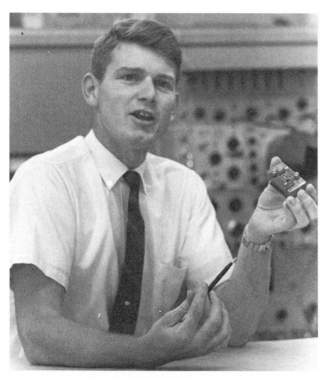

After equipment is in place, the sales engineer must be certain that plant personnel know how to operate and maintain it. This might mean doing some in-plant teaching and regular follow-up to see that everything is operating properly.

Sales engineers use drawings rather frequently. Sometimes they work from drawings prepared by the customers. Other times they are expected to have drawings prepared to present to the customer as part of their sales proposal. In either case, drawings of all kinds are part of the life of a sales engineer.

A sales engineer must have a broad engineering background plus actual work experience with the equipment being sold. This means several years of preparation before actually entering the sales field.

## Engineers in Management

Most companies have rather common elements. These include a plant with equipment, a labor force, a marketing staff, and a financial organization. Industrial managers work with these units regardless of the products being produced. Because the products of industry are becoming increasingly technical and complex, engineers are finding success in upper-level management positions. Here their engineering preparation becomes a background resource upon which they can draw in making decisions concerning the overall operation of the company. Management engineers do not usually get involved directly in the engineering decisions made by staff engineers. Instead they are concerned with the financial, human relations, and marketing aspects of the business. Engineers in management positions are expected to review a wide variety of engineering drawings. They have the responsibility for the ultimate success of the company, and intensive examination and approval of product designs and revisions are part of this success.

Engineers entering management need a good understanding of the principles of engineering. Work experience is essential. Preparation in business and management is important.

## Engineers as Consultants

Some engineers work alone or form a company and sell their services to any who need help in solving engineering problems. These persons will spend time at the company employing them and bring their full resources together to solve a particular problem. Consulting engineering companies will have a variety of different types of engineers on their staff. They can have them work as a team, using the expertise of the various disciplines to solve

FIGURE 1–12 Engineers are employed at university facilities.

the problem. Some companies tend to specialize in a narrow field rather than offering a full range of engineering services.

Consultants use and make a wide variety of engineering drawings. The types vary with the industry. In metal manufacturing, for example, orthographic drawings of the parts of a product are used. Dimensioning and tolerancing are critical. In a petroleum plant, piping drawings assume importance. These can be orthographic or pictorial.

## Engineers as Teachers

After engineers have earned a master's or doctorate and have had actual experience in industry, they can join university engineering faculties. Here they have the opportunity to work with young people interested in the field. They can use their education and experience to prepare the next generation of engineers. Often engineering faculty become involved in research and consulting.

In engineering schools, the faculty present courses in engineering graphics and other engineering subjects (Fig. 1–12). This provides the student with a thorough preparation in the theory of orthographic projection and other forms of drawing. Engineers make and use drawings constantly and must have a detailed understanding of the language of graphic communication.

Drafting and sketching skills are used in many of the courses taken in engineering. Reports must be prepared and lettered. Design solutions must be presented. Drawings in class and in textbooks become part of the learning process.

# The Engineering Design Process

## DESIGN

The term *design* means different things to different people. The various meanings found in dictionaries include: to conceive, contrive, form a plan for; to draw a sketch of; to have a goal or purpose. Design can refer to artistic creation using color, form, proportion, and texture. Design includes diverse areas such as abstract painting or sculpture, the function and appearance of industrial products, and the generation of advertising layouts. It includes the development of new three-dimensional physical objects or the redevelopment of an object.

Design involves creation, planning, and drawing to meet a specific need. It is a proposal for producing something but does not usually get involved in the actual production or manufacture of the item. In some areas, as in art, the designer (artist) has great freedom of action, working with few limitations. In other areas, as in engineering, rather severe limitations, such as size, cost, or ease of manufacture, are often imposed.

Design can be classified into several general categories. These include abstract, aesthetic, and functional. In actual practice these may overlap and reinforce each other.

*Abstract design* is intended to create a sense of visual interest. It has no relationship to a functional or useful purpose (Fig. 2–1). Such designs do provide a source of inspiration for those involved in product design (Fig. 2–2).

*Aesthetic design* is the application of the principles of form, proportion, color, and texture to an object with the goal of improving its overall appearance. It does not serve to make the product function better but does make it more pleasing in appearance (Fig. 2–3). The shell enclosing the garden tractor does not add to its power or operating efficiency but does contribute to its overall appeal to the consumer.

*Functional design* focuses on how a unit operates, whereas the attractiveness or decorative features are of little or no importance (Fig. 2–4). Automotive

FIGURE 2–2  This product reflects the influence of abstract design as applied to useful products. (Courtesy of Herman Miller, Inc.)

parts are not seen as the auto is used; they are designed to function most effectively regardless of their appearance.

Some blending of these design categories often occurs. For example, in Fig. 2–5 are several electric erasers used by engineering drafters. Notice that

FIGURE 2–1  This abstract sculpture, *The First Voyage of Ulysses*, creates a sense of visual interest. (Artist, Robert Blunk.)

FIGURE 2–3  The appearance of this product is improved by the application of aesthetic design. (Courtesy of Sears, Roebuck and Co.)

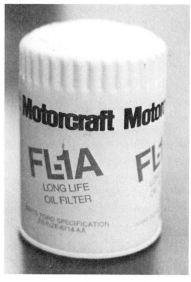

FIGURE 2–5 This product is functional yet is aesthetically pleasing. (Courtesy Teledyne Post.)

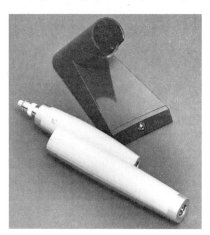

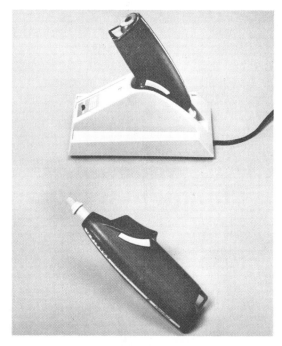

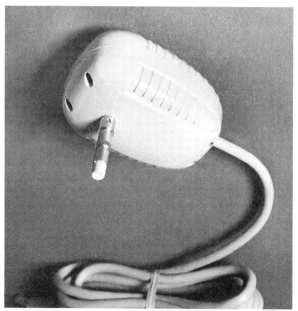

they were designed to be lightweight, easily and comfortably held, yet pleasing in overall appearance. Notice that the example shows four solutions to the same design problem, all of which are in production.

## PRECEPTS OF DESIGN

The precepts of contemporary design have evolved from centuries of work of artists, architects, designers, craftsmen, and engineers all over the world. Creative solutions that developed first for handmade products and eventually for factory-produced products forged present-day design concepts. Following is a summary of design concepts evolved from developments in the past 100 years.

1. Modern design should fulfill the practical needs of modern life.
2. Modern design should express the spirit of our times.
3. Modern design should benefit by contemporary advances in the fine arts and pure sciences.
4. Modern design should take advantage of new materials and techniques and develop familiar ones.
5. Modern design should develop the forms, textures, and colors that spring from the direct fulfillment of requirements in appropriate materials and techniques.
6. Modern design should express the purpose of an object, never making it seem to be what it is not.
7. Modern design should express the qualities and beauties of the materials used, never making the materials seem to be what they are not.
8. Modern design should express the methods used to make an object, not disguising mass production as handicraft or simulating a technique not used.
9. Modern design should blend the expression of utility, materials, and process into a visually satisfactory whole.
10. Modern design should be simple, its structure evident in its appearance, avoiding extraneous enrichment.
11. Modern design should master the machine for the service of man.
12. Modern design should serve as wide a public as possible, considering modest needs and limited costs no less challenging than the requirements of pomp and luxury.*

## THE ENGINEERING DESIGNER

The engineering designer, to be totally effective, must have a thorough preparation in the area in which work is to be performed. This includes both

formal education and on-the-job experience. For example, an electrical engineer would have a difficult time designing a plastic blow-molding machine. If the electrical engineer works with people who are knowledgeable in this area, all can contribute to the design of the controls. This is why much design is done by a team and why consultants are widely used. As the engineer gets field experience, the scope of possible work tends to broaden. If an engineer is employed in an industry in which his or her background is weak, additional formal education in that area would be beneficial.

Engineers often have limited knowledge about the materials of industry and processes of production. Again, actual work experience in production situations will help; however, it may take years to become competent. Engineering technologists are prepared to be very knowledgeable in all aspects of production and should form an important part of the engineering team.

In all cases, a knowledge of graphics is important. This applies to research and development as well as production. The use of computers in design does not reduce the need for graphics knowledge; in some cases, it increases its importance. A design involves written specifications, engineering test reports, and computer data, but most important to complete communication of an idea are the sketches and design drawings. No other form of communication can describe a design as quickly and completely as some form of a drawing.

## RESEARCH AND DEVELOPMENT

Contemporary industry utilizes a carefully structured, systematic approach to the design and production of products. Among the many subdivisions of the process to produce useful products is research and development. This area is responsible for the generation of creative ideas, the refinement of these ideas, and the utilization of market research to determine the needs of society. Research and development can be subdivided into several areas (Fig. 2–6). Engineers and technicians are involved in all of these activities.

1. *Product research and development* involves the creation of new products and the redevelopment and improvement of existing products.
2. *Process research and development* involves improving existing manufacturing processes such as cutting, forming, machining, finishing, and joining. It includes experimentation to develop new processes and the utilization of new technology, as computers, in the manufacturing operations.
3. *Materials research and development* involves improving the commonly used materials of industry

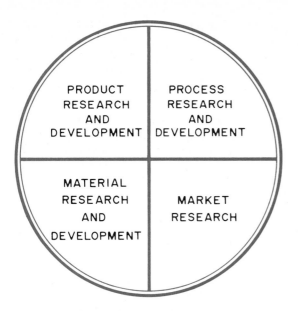

FIGURE 2–6 Research and development is often divided into four distinct areas.

and finding new uses for them. It includes experimentation to develop new materials. The development of plastics is a classic example.

4. *Market research* involves determining the need for new products as well as new uses and markets for existing products. It may point out how existing products might be altered to meet changes occurring in the marketplace. As an example, the high cost of gasoline dramatically changed the products of the automotive industry. Market research should note these changes and recommend that they be reflected in the products.

## ENGINEERING DESIGN

Engineering design is the process of arriving at a practical and economical solution to a problem. It may involve developing a *new product* or improving an existing product. It may be the development of a *new system,* such as a means to remove pollution from the air in a manufacturing situation. This problem-solving process can include a wide range of activities, including the use of computers, mathematics, drafting, consulting, testing, and any other activity needed to bring the project to a satisfactory conclusion.

The design process usually involves many people. Some work together as a design team, each person contributing according to academic preparation and industrial experience (Fig. 2–7). Usually, those who begin a project do not finish it. The development process passes on to others who are better prepared for each segment of the work. For example, those who have the original, creative ideas are not necessarily the best prepared to select materials or make decisions that would simplify the product to make production easier and less costly.

Engineering design begins with the recognition of a need or problem. This need or problem is isolated and carefully defined. Possible solutions must be developed. Materials to be used must be analyzed. Ultimately, after all preliminary information is gathered, possible solutions are proposed. One solution is accepted as the best. It is developed into a

FIGURE 2–7 The engineering-design process usually involves several people. (Courtesy of Boeing Co.)

final product, which is tested, and revisions are made based on the tests. Finally, the resulting product or system is put into production and delivered to the consumer.

All this involves the very best thinking and efforts of everyone involved. Each problem or need studied has certain aspects that make it unique. Ingenuity and creative thinking are essential to success. *Creativity* is a mental process that recognizes a problem and formulates an imaginative solution. What was successful on one project may not be acceptable on another. Engineering designers must not fit themselves into a routine pattern but must remain flexible to follow any plan of action that holds the promise of success.

The engineering designer must be a creative person but must also realize that any final solution has some restrictions. Among these are the limitations of scientific and technical knowledge, the materials available, and the need to produce a useful solution at a reasonable cost. Creative designers must have persistence. When one idea fails or is rejected, they must be able to come up with several more. New thoughts, combinations, and approaches must be recognized rapidly and acted upon. The willingness to hitchhike (take somebody else's idea and work with it) often produces outstanding results.

Following is a generalized discussion of steps often used in the engineering design process. These can be varied to suit individual situations. In all cases there is need for feedback. *Feedback* occurs when at a particular step the solution runs into problems that oblige the designers to go back to an earlier step and repeat the study at that point. The revised solution then proceeds through the steps again.

## THE ENGINEERING DESIGN PROCESS

There are a variety of approaches to engineering design; different designers and companies use different approaches, and certain problems may require different approaches than others. In all cases it is important that an organized plan be followed. A plan will clearly mark the procedure and set times for decisions to be made. This will bring the process to an orderly conclusion in the time available.

A general approach could be divided into seven parts: identify and define the problem, develop a work plan, prepare solutions and alternatives, select a solution, design the product, test the solution, and implement the final design. Each of these parts is broken down into substeps (Fig. 2–8).

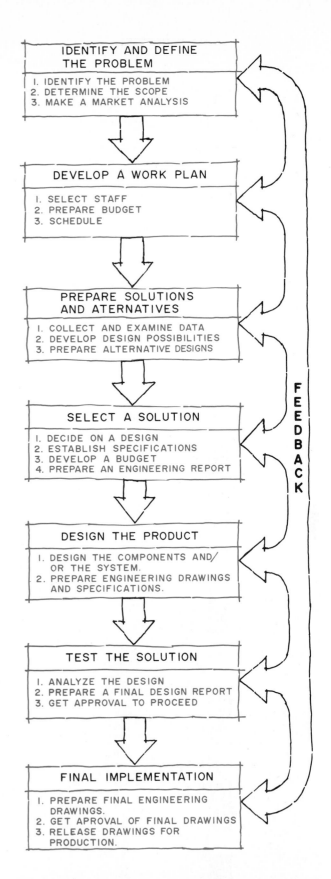

FIGURE 2–8 A typical approach to the engineering design process.

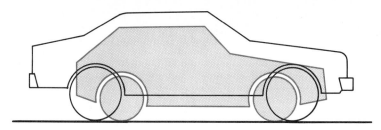

FIGURE 2–9 The demands of the marketplace often influence product design decisions.

## Identify and Define the Problem

The first step leading into a design situation is to identify and define a problem.

*Identify the problem* Somewhere within a company, problems develop that lead to the need for so-

Realize that the steps shown are typical and may change slightly with the problem being studied or the particular engineering staff organization.

lutions by the engineering design staff. The marketing staff may recognize the need for a new product to help meet competition from other companies. There may be changes in the marketplace that must be met. The downsizing of automobiles is a classic example (Fig. 2–9). An existing product may be obsolete or not functioning satisfactorily and need design attention. These and other factors will generate engineering design problems (Fig. 2–10).

The staff in the area where a problem is recog-

FIGURE 2–10 Obsolescence and new technical developments lead to constant improvement and redesign of products. (Courtesy of General Electric Co.)

nized usually prepares a written proposal. This identifies the problem and may contain suggestions for potential solutions. Top management reviews the proposal and decides whether to proceed with an engineering design study or reject the proposal. If the decision is to proceed, the designers must determine the scope of the problem.

The problem often changes as more people review it. It should be rewritten as often as necessary until those responsible accept it.

*Determine the scope* The designer will visit with the people presenting the problem and try to get a full understanding of the situation. It is important to keep accurate notes of all information gathered. The limitations of the problem are recorded. For example, a small automobile might be limited to a four-cylinder engine. The specific criteria that must be met by a satisfactory solution are also listed. For example, the auto must carry four adults. If there are things that are not clear, they should be recorded as questions to be answered as the study progresses. For example, the desired weight of the auto might not have been established. It might be recorded as unknown or placed in a range of weights or given a maximum. The desired cost of the finished product is a very important limitation that must be constantly kept in mind. The scope statement will often include sketches or diagrams.

*Make a market analysis* The market for the product is studied during the initial planning stages and continues as the product is refined and developed. This generally begins with a market survey to find if there is a need for the product and how the consumer reacts to the initial plans for the proposed product. As the product is developed, various aspects of it need to be tested on the market. For ex-

ample, the original survey may show an interest in a product, such as an electric grass clipper. The consumer may like the proposed design. As the engineering design team progresses they decide to change the design to a cordless unit that has to be recharged after one hour of use. Marketing needs to make a new analysis of this product. Possibly marketing will continue to check consumer reaction to both the cord and cordless grass clipper. After prototypes have been developed, they will get them to selected consumers for additional analysis.

The marketing staff will examine other aspects of the use of the product. For example, does the product have only regional use? Who is the typical consumer? Does it have uses other than the original? Are there social, cultural, or religious stigmas?

Marketing and product-development staffs must communicate continuously as the product is developed and tested prior to its entering the actual mass-production stage. When the design has been finalized and accepted by management, the advertising plan is developed, using the data gathered in the marketing studies and from engineering design. Graphics techniques are useful in presenting the findings from marketing surveys. Charts and graphics are among those widely used (Fig. 2–11).

## Develop a Work Plan

After the problem is clearly identified, an organized approach to solving it must be developed. While this can involve many factors, the basic questions are: (1) Who is to do the work? (2) What facilities are needed? (3) What will be the cost to develop a solution? and (4) How much time is to be allowed for each segment of the study? (Fig. 2–12).

The engineer in charge of design will usually

FIGURE 2–11 Marketing staff members are expected to make reports to engineering and management. (Courtesy of Caterpillar Tractor Co.)

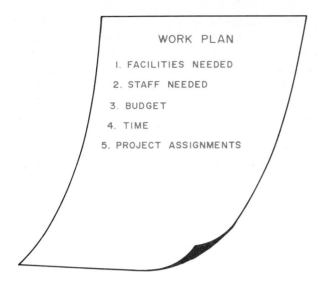

WORK PLAN

1. FACILITIES NEEDED

2. STAFF NEEDED

3. BUDGET

4. TIME

5. PROJECT ASSIGNMENTS

FIGURE 2–12 The parts that commonly make up a work plan.

make project assignments to those on the staff. The persons selected depend upon the project, its technical requirements, and the academic and design experience of the staff. Often, several people are assigned to a project as a team. They then subdivide the work, each being responsible for a particular part of the investigation. The company must provide facilities in which the staff can work. These needs can vary a great deal.

Time is always a major factor. Its allocation can be based on a variety of situations. If the design project is a reaction to an immediate emergency, a solution may be needed as soon as it is possible to generate one.

There may be a target date when the design, as for a new product, must be finished so that manufacturing can begin. In this case, the allocation of time to the various phases of the design process works backwards from the date the product is needed. The division of time into the various phases of the design process is usually based on previous experience with similar projects. The following hypothetical case will illustrate the problems in planning the time allocation.

Assume that a company has decided to manufacture a garden cultivator for the home gardener market. The marketing staff wants to begin advertising and release the first units to wholesale and retail outlets in January. This means that the members of the manufacturing staff have to decide how long it will take to begin producing finished products. If they decide it will take six months to set up and begin manufacturing the cultivator, they must have the completed, tested product design by the previous July 1. It is from this date that the design team begins to plan backwards to see how much time it has to do its work. If it is already February 1 when the

decision to produce a design for a cultivator was made, the team has only five months to do the total design and product testing. If the chief designer believes it can be done, the project will continue. If it cannot be done that fast, marketing will most likely have to change the date of the initial sales campaign. The entire process of design, manufacturing, and marketing becomes flexible and depends upon the success of the people handling the project before them.

Another part of the work plan is an estimate of the funds needed to bring in a solution. This budget will include funds for items such as staff salaries, facilities, materials, computer time, and consultants.

Regardless of the design situation, a work plan is developed to guide the process. If unexpected problems occur, the plan may have to be altered, but this should be avoided whenever possible.

## Prepare Solutions and Alternatives

This procedure includes the collection and examination of data, the development of design possibilities, the preparation of alternate designs, and the checking of the designs for feasibility.

*Collect and examine data* When starting into a design project, it is necessary to review the literature to see what has been published relating to the problem. Often the work of others will be of great help when deciding what approach to use. You can find things that did not work, so there is little need to try them again. You can find research and reports of approaches that were successful. These need to be considered as you develop your solution.

There are hundreds of books and magazines that report research and other findings. The key to data collection is to have a way to review these rapidly and read only those that relate to your product.

Large companies have technical libraries where the card catalog will direct the researcher to books by topics (Fig. 2–13). Usually the catalog lists books by title, author, and subject. Material in books does not reflect the very latest developments, because it takes several years to write and publish a book.

Professional magazines contain more current developments. These can be found through the reference section of a library. Here they will have the *Engineering Index* and the *Applied Science and Technology Index*. These contain abstracts of important articles in engineering. They are organized so that all articles on a certain subject are listed together. The abstract will tell you if the article might be of value. It also identifies the magazine in which the complete article was published. You must secure this magazine to read the article. Most libraries have past issues of magazines on microfiche. A microfiche is a negative that holds 60 pages of printed material. A

FIGURE 2–13 Many companies maintain technical libraries. (Courtesy of Cessna Aircraft Company.)

FIGURE 2–14 An engineer reviewing technical literature on microfiche. (Courtesy of 3M.)

reader-printer is used to enlarge the page on a screen so that it can be read (Fig. 2–14). The reader-printer will also produce a printed copy of each page.

Another way to find out about and keep up with technical developments is to attend local and national meetings in your discipline. You will also get to know the leading researchers in the country. When you are personally involved in a design problem, these people can be contacted for advice and assistance. Many work as consultants on a fee basis and can make significant contributions to your design decisions.

Above all else, do not work in a vacuum. Seek advice and technical data as you pursue a design solution.

Be certain to analyze critically everything you read. The fact that it has been published does not mean that it is accurate or that it fits your situation.

Compare results from several different studies, or try it out yourself and reach your own conclusions. Be a critical reader of research and technical reports.

*Develop design possibilities*  After a problem has been defined and a work plan developed, preliminary ideas are recorded for use in reaching a solution. It is at this time that the engineer is the most creative. The ideas developed at this point do not have to be evaluated for feasibility. They may be practicable or impracticable, but in either case they might stimulate additional solutions (Fig. 2–15).

The first attempts are recorded by sketches and notes. The goal is to record the idea before it is forgotten. These sketches may or may not be in detail. Refinements can be added later as decisions are made (Fig. 2–16).

Sometimes the first ideas are generated by brainstorming. *Brainstorming* is a term used to describe a session in which several people attempt to solve a problem by offering spontaneous ideas for solutions regardless of their practicality. Even the most unusual or sometimes ridiculous ideas are recorded. The group tries to produce as large a number of solutions as possible. No one is criticized for an idea. Criticism would inhibit the free thinking needed for the generation of original, creative, unusual ideas. Often an idea suggested by one person leads another person to add on to it or offer alternate solutions (Fig. 2–17).

*Prepare alternate designs*  The engineer will have an array of suggested solutions or partial solutions. From these an attempt is made to develop several acceptable solutions. It is possible to combine elements from various ideas into a single solution. As the project moves toward a final decision, the inferior alternatives will be weeded out. The favorable solutions will be examined, changed, and evaluated until one or more possible final solutions exist.

These designs will be reduced to scale drawings that set sizes and limits and describe shape and movement. Since they are to scale, the concepts can be shown very accurately. Angles, lengths, and sizes

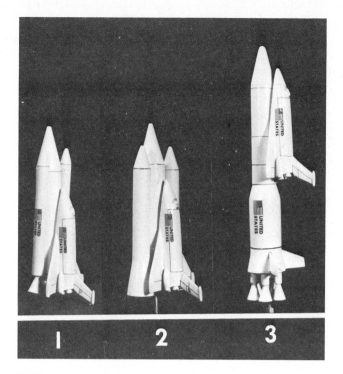

FIGURE 2–15 Several proposed design solutions for the space shuttle. (Courtesy of National Aeronautics and Space Administration.)

FIGURE 2–16 The first design ideas are recorded rapidly with freehand sketches. These sketches show various proposed configurations of Spacelab modules and pallets fitted into the cargo bay of the space shuttle. (Courtesy of National Aeronautics and Space Administration.)

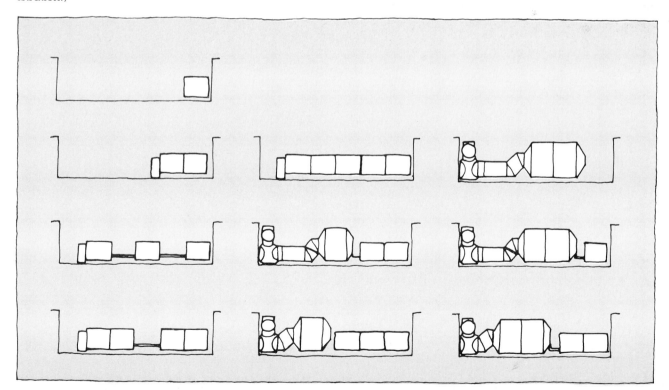

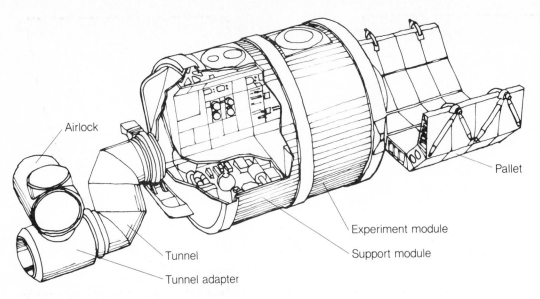

Airlock

Pallet

Experiment module

Support module

Tunnel

Tunnel adapter

FIGURE 2–17 A freehand design sketch of the Space-lab support module and experiment module with the connecting pressurized tunnel. (Courtesy of National Aeronautics and Space Administration.)

can be described with orthographic drawings and the use of descriptive geometry. Known dimensions are added to show the size and locations of the existing design. These are often changed as the design is refined toward a single final solution.

Some companies have computer systems that aid in the design and drafting process. The design is developed in the computer and displayed on a CRT (cathode ray tube). The design can be changed rapidly, and alternate ideas can quickly be put in the computer and displayed on the screen. This has revolutionized the design and decision-making processes (Fig. 2–18).

## Select a Solution

Of the many possible solutions, the one that appears the most feasible is chosen for development. Sometimes several appear as equal in value, and they all may proceed toward refinement before the final one is chosen.

*Decide on a design* A decision on the design to use must be made. Those responsible can accept the design as presented, accept parts but reject other aspects of the design, or reject the entire proposal. Everything that is known about the project to this point is studied. This includes things such as the

FIGURE 2–18 Computer drafting has greatly changed the engineering design process. (Courtesy of Tektronix, Inc.)

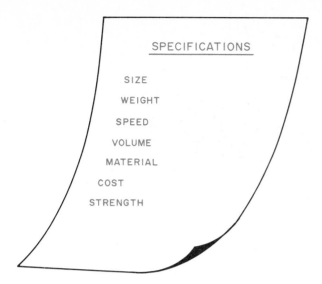

FIGURE 2–19 Design specifications produce a realistic matrix within which the final solution must fall.

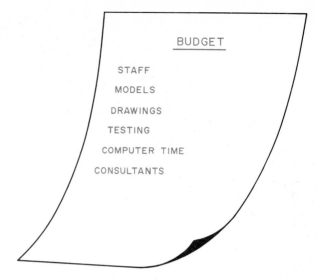

FIGURE 2–20 The design team must have an operating budget.

market survey, engineering drawings and reports, cost estimates, and manufacturing problems. Of great importance are the opinions of experienced engineers. They can often point out factors that, based on their experience, work for or against the proposed design.

*Establish specifications* Once a proposed design solution has been accepted, complete and realistic specifications for the product must be written to finalize decisions on matters such as size, weight, speed, volume, and materials. The ultimate final design will be developed according to these specifications (Fig. 2–19).

*Develop a budget* Once a proposed design has been accepted, the engineering team can proceed to produce a final design, develop models and prototypes, prepare the drawings needed, and perform the tests necessary to confirm that the product is ready for production. A development budget is necessary so that the team has funds to continue the design work (Fig. 2–20).

*Prepare an engineering report for a design review* When the preliminary design process is complete, the engineering staff usually prepares a report on the project. This details factual data related to the project and includes recommendations for action on the project. Often this report is presented orally to those responsible for the final decision. The well-prepared engineer will use charts, slides, and other visual aids to present the design concepts, justification, and the recommended solution. The goal is to explain accurately and clearly the results of the design study. It is not an attempt to cover up design problems but to present the truth. If an engineer does not do a good job of presenting the design re-

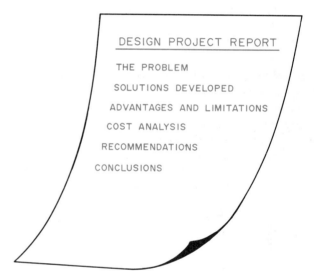

FIGURE 2–21 The results of the preliminary design work are reported to management in a design report.

port, a satisfactory design could be rejected (Fig. 2–21). If the project is approved and the funds are made available, the final design work begins.

## Design the Product

At this point some members of the original design team may leave the project. New people with knowledge in areas such as materials and manufacturing may join the design team. The goal is to resolve all design aspects. This could include readjusting the size, shape, motion, or other aspects of the original design. The functional specifications are constantly kept in mind.

*Design the components of the system* In this phase of design, the creative aspects of design are restricted by the realities of the specifications and limits imposed by materials and manufacturing techniques. Cost is another limiting factor. The design team must face the practical realities limiting it and even at times alter the product design to accommodate these limitations.

Scale drawings are used to formulate the final concept and check proposed solutions. Some design problems, such as angles or oblique lines or surfaces, are developed with the techniques of descriptive geometry. Dimensions are used to record size and location decisions.

Engineers need to calculate loads and stresses and to size parts accordingly. The design of gears, cams, fasteners, and other features must be finalized. Sometimes three-dimensional models and mock-ups are used to assist in making decisions.

A product is made up of many parts. Each of these poses a design problem. Each part must serve its function and relate to the other parts within the product.

Some design projects involve systems engineering. *Systems engineering* involves the design of a total functioning, operational unit that has a number of subsystems, each of which contains a number of individual components. To be effective, each component must function properly within its subsystem. Each subsystem must perform its function within the total system before it is operational.

The design process for a systems engineering problem is similar to that used when designing a product or component. Following is a typical design procedure for a systems engineering problem.

1. Recognize the need for the system.
2. Identify the system required to meet the need.
3. Identify the subsystems of the design.
4. Design the subsystems.
5. Construct models of the subsystems and combine them to form the total system.
6. Identify problems and develop solutions.
7. Test the model and evaluate performance.
8. Put the system into extensive use, evaluate the design, and redesign where necessary.

To illustrate systems engineering, examine an automobile electrical system. The system is needed to provide electrical power to run the engine, accessories such as a heater, provide lights, and operate the on-board computer (Fig. 2–22). Each function becomes a subsystem. A battery is the source of storage for electrical power to activate the system. To keep it charged, an alternator is used. A computer keeps control over engine functions. The various lights require sockets and bulbs. The heater and air conditioner use electric motors, which become part of the electrical system. A means of controlling directional signals and stop lights must be developed. Other accessories, such as a clock, radio, tape deck, and cigarette lighter, must be developed. Each becomes a design problem in itself, but when assembled with the others each becomes a part of the total electrical system.

When one component is changed, it may or may not affect the other components. For example, if larger motors and lights that require more electricity are used, this could possibly affect the battery to

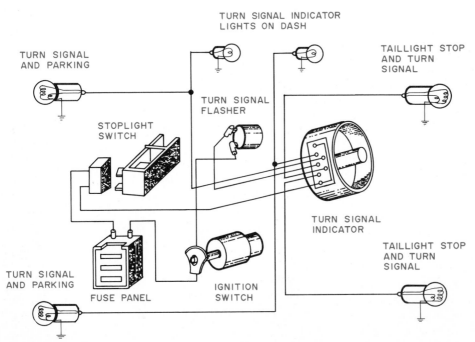

FIGURE 2–22 Systems engineering involves the design of an operating unit made up of several subunits.

FIGURE 2–23 Engineering drawings are made of the final solution.

be used, which in turn may affect the alternator specifications.

A more complex system would be the design of a nuclear generating plant. The total product is so large and complex that the major subsystems will themselves have subsystems.

*Prepare engineering drawings and specifications* The final result of the design process is a set of detailed engineering drawings and a revised set of manufacturing specifications (Fig. 2–23). These together are the result of the best thinking, planning, calculating, and experience of the design team. The solution must now be tested.

## Test the Solution

Before presenting the solution to the decision makers for approval to begin production, the solution must be tested. Some of the following engineering analyses take place as the product is being developed. Some occur once the design solution is firm. The timing of each will vary with the nature of the product.

*Analyze the design* The chosen design solution must be carefully analyzed before a final decision is reached. Often, the analysis reveals features that need additional study, and the project is returned to an earlier step in the design process for further consideration.

Analysis reveals the feasibility of the recommended design. At this point, the limitations set by the materials available, manufacturing processes, cost, usefulness of the product, its durability, and possible acceptance by the consumer become impor-

AREAS OF ANALYSIS

HUMAN ENGINEERING ANALYSIS

PERFORMANCE ANALYSIS

PHYSICAL ANALYSIS

ENGINEERING ANALYSIS

COST ANALYSIS

ENVIRONMENTAL ANALYSIS

FIGURE 2–24 The final design is subjected to a detailed analysis.

tant. The realities of the technology and marketplace now are applied to the proposed design. Sometimes an imaginative concept must be altered in the face of reality.

There are many areas to consider in analyzing the design. The major areas of analysis are human engineering analysis, performance analysis, physical analysis, engineering analysis, cost analysis, and environmental analysis (Fig. 2–24).

*Human engineering analysis* involves relating the product to the human body. This includes all aspects of a human's actions and functions, such as sight, hearing, length of body members, and comfortable body positions. For example, the handle on an electric grass clipper should be sized and shaped so it can be comfortably held by the human hand. The height of shelves in storage areas is related to the maximum comfortable reach of a human (Fig. 2–25).

FIGURE 2–25 Human engineering involves relating the final design to the limitations prescribed by the human body. (Courtesy of Boeing Co.)

FIGURE 2–26 Many designs require careful analysis of the limitations of human vision. The relationship between the gauges and those who operate them is critical for the safe operation of the aircraft. (Courtesy of Boeing Co.)

The driver and passenger seats in an automobile must provide support for the seated body and room for the legs. The controls must be easily seen and reached by the driver.

Human vision will control the size of many things. Gauges must be large enough to be read quickly. The use of color and lights can increase visibility. The cockpit of a commercial airliner is a classic example of the use of many gauges (Fig. 2–26).

Sound generated by the product must be controlled. The human ear can safely accept noises up to 90 decibels for periods of less than 8 hours. If sound is used as a warning, the type and level of sound are important. An example is the piercing, pulsating warning given by a piece of heavy construction equipment when it is running in reverse (Fig. 2–27). The stall warning on small private aircraft is designed to get the attention of the most inattentive pilot.

As these and other human factors are analyzed, the designer realizes that the human body varies in size a great deal. Studies of the human body have revealed typical body sizes for various percentages of the population. Henry Dreyfuss produced body measurements that describe various percentages of the adult population in the United States (Fig. 2–28).*

---

*Other sources of design data of the human form include R. W. Bailey, *Human Performance Engineering: A Guide for System Designers* (Englewood Cliffs, N.J.: Prentice-Hall, 1982) and E. J. McCormick, *Human Factors in Engineering and Design* (New York: McGraw-Hill, 1976)

FIGURE 2–27 This wheel loader with a coil carrier produces a loud warning sound when it is backing. (Courtesy of Caterpillar Tractor Co.)

*Performance analysis* involves a study to see if the design solution will perform as intended by those who developed the original concept (Fig. 2–29). For example, a thermostat that will not control the temperature of your house within the temperature range desired is not satisfactory. Most original designs are not complete failures but may fail to reach the level of performance expected. In effect they have failed to be totally functional, and the design must be reworked. Sometimes several possible designs are carried to the testing stage. The one that

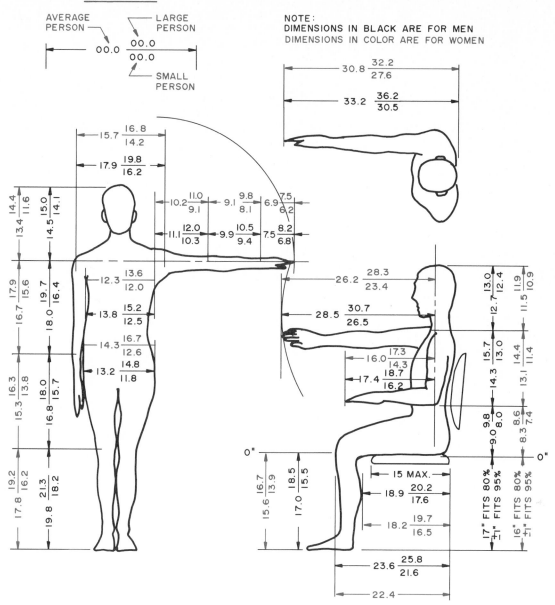

proves the most satisfactory will be the one that goes into production (Fig. 2–30).

Models are used to test products for performance before they are put into production. A model is a three-dimensional replica of the product. It may be to scale or full-size.

A model can be used to get an idea of what the product will look like from various locations. It can be used to check proportions, clearances, fit, and other design specifications. Some are actual working models and can be tested under actual working conditions (Fig. 2–31).

There are a variety of models that may be used. The designer may make a model to help visualize the design. It may have only the features that are of immediate concern. It can be built of wood, cardboard, clay, or any other useful material.

Once a design is relatively firm, a scale model may be built. The scale selected depends on the product and the use of the model. It should be large enough to permit the design to be tested. For example, a scale model of an airplane is tested in a wind tunnel. It must be of a size to permit the

FIGURE 2–29 A lunar module pilot is checking the bending and stretching movement of the pressure garment assembly of a lunar module pilot space suit. (Courtesy of National Aeronautics and Space Administration.)

FIGURE 2–30 A production analysis team making a study of the production capabilities of an earth-moving product. (Courtesy of Caterpillar Tractor Co.)

FIGURE 2–31 This research scale model lets engineers try out new concepts and designs before full-size prototypes are built. (Courtesy of Caterpillar Tractor Co.)

needed tests to be made. It must be accurate enough to reflect the design of the exterior surfaces.

Sometimes a mock-up is used. A *mock-up* is a full-size replica that gives an overall appearance of the product. Mock-ups help designers to visualize the relationships between components and are used to check for size, space, color, and other aspects (Fig. 2–32).

FIGURE 2–32 A mock-up of a commercial aircraft permits a close check of all design details. (Courtesy of Boeing Co.)

When the design is firm and needs thorough testing, a prototype is built. A *prototype* is a full-size working model built to the design specifications. It is used to test the product under actual working conditions (Fig. 2–33). Some products require extensive field testing of prototypes before production starts. The results of the field testing often lead to design changes or the substitution of a different ma-

terial. Before a manufacturer brings a product on the market, its components are extensively field-tested (Fig. 2–34).

Other types of models are variations of those just described. When a complex system, such as a refinery, is being designed, the total plant is laid out as a scale model. It shows each system and its relationships with the others (Fig. 2–35). Architectural

FIGURE 2–33 Astronauts testing space suits by performing duties (picking up small rock samples) expected upon landing on the surface of the moon. (Courtesy of National Aeronautics and Space Administration.)

FIGURE 2–34 Prototypes are field-tested before production begins. (Courtesy of Caterpillar Tractor Co.)

FIGURE 2–35 Engineers working with a scale model of a proposed anhydride facility. (Courtesy of Monsanto Chemical Intermediates Co.)

models are scale models of buildings that are used mainly to visualize a building and check the design for mass and exterior detail.

Often, models help the designer make decisions that are needed to produce the finished design. They are an important part of the design process and are a check on the performance of the design.

*Physical analysis* involves a study of the physical properties, such as weight, volume, and strength, of the materials used. For example, the completed product usually has a maximum design weight. An electric grass clipper must be very light. If it is cordless, the maximum weight would include the design feature used to store the electricity. The electric iron is an example of a product that over the years has continually had its weight reduced. The materials selected must be analyzed. Does a part have to be metal, or can a lighter plastic be used? If electrical shock is to be avoided, nonconductors are used in critical areas. Do some materials provide a more pleasing exterior? These and other questions are studied during physical analysis.

The design is checked repeatedly against the specifications. Does it measure up to the expectations of those who developed the original concept? What can be done to bring the design within the specifications and still produce a useful product?

*Engineering analysis* involves the application of scientific and engineering principles to the proposed design. The preliminary engineering drawings establish the shape, sizes, paths of motion, and other features of the product. The engineer must use the

principles of chemistry, physics, mathematics, kinematics, strength of materials, dynamics, and other engineering areas to analyze the design. The computer is a tool that permits complex analyses to be made rapidly. It can simulate operating conditions, loads, and other stresses on the product.

The strength of materials used is critical to the proper functioning and the life of the product. This relates to the size of the part and the material used. If a part is "overengineered" (made stronger than actually required), the cost of the product becomes unnecessarily high; so may the weight (Fig. 2–36).

FIGURE 2–36 Actual components, such as the wing shown here, are extensively tested. (Courtesy of Cessna Aircraft Co.)

Engineering analysis involves studying the data gathered from laboratory and field tests of the product (Fig. 2–37). This can include mathematical manipulations or graphical analysis of empirical data. Mathematical evaluation of data can also be handled graphically by using graphical algebra and calculus (Fig. 2–38). A design solution can also be checked by graphical means. For example, clearances can be calculated and checked by drawing the parts involved to scale (Fig. 2–39).

*Cost analysis* is a very important part of the overall analysis process. The product must be able to be produced at a cost that enables the manufacturer, wholesaler, and retailer to sell it at a profit. Regardless of how well the product works or how attractive it is, if it cannot compete with similar products in selling price at each level, it has a handicap. As the engineer alters the design, cost is one

FIGURE 2–37 These starters are being tested on a life-test program. They will be operated continuously for an extended period of time. Careful records will be kept during the entire test period. (Courtesy of Square D Company.)

FIGURE 2–38 Data showing brake performance for an off-highway truck. (Courtesy of Caterpillar Tractor Co.)

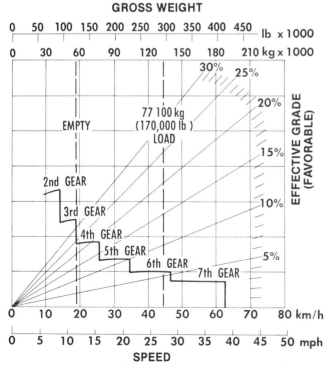

FIGURE 2–39 The digging envelope for an excavator can be checked graphically. (Courtesy of Caterpillar Tractor Co.)

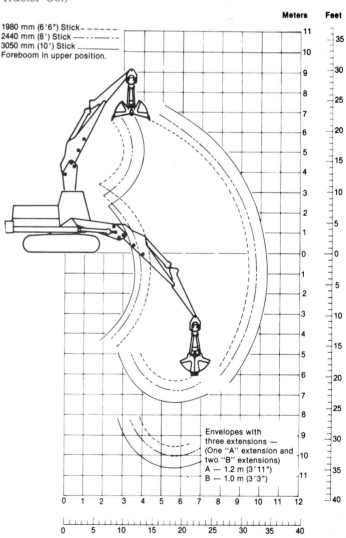

factor always to consider. This analysis must cover the total cost, which includes items such as materials, labor, manufacturing, distribution, inventory, and financing.

*Environmental analysis* has assumed an increasing importance. What is the impact of the product on the environment? Some conditions are regulated by the federal government, and products cannot be sold unless they meet environmental standards. Others are conditions expected by the consumer. Examples include possible radiation danger from microwave ovens or television sets and health dangers from auto emissions and ureaformaldehyde foam insulation. The engineering team must be constantly alert for unfavorable environmental effects and overcome them as part of the overall design process.

Another environmental consideration is working conditions in the plant while the product is being manufactured. Will the materials used produce noxious fumes that harm the workers and pollute the neighborhood? Will harmful vapors, smoke, or liquids be produced that require special disposal facilities? Can the materials used in the product be changed to reduce or eliminate these hazards to the environment? The engineer must be always alert for such possibilities.

*Prepare a final design report*  Before the project is released to manufacturing, the design team prepares a final report that details all aspects of the project. Engineering drawings and specifications are an important part of the report. Results of tests, prototype experiences, recent marketing analyses, and manufacturing data are included.

*Get approval to proceed*  The final design report is submitted to those responsible for making the decision to proceed with the manufacture of the product. Additional suggestions may be made at these hearings. Part of the design may be rejected, and revised material may be requested. Once the final approval is received, the product is ready for implementation.

## Implement the Final Design

After approval to proceed with the manufacture of the product is received, several steps remain. These include the preparation of the final engineering drawings, approval of these drawings, and their release for production.

All design changes are incorporated into a final set of engineering drawings. These will be used for the actual manufacture of the product and by companies supplying parts and materials. The final drawings must then be approved by those responsible for the final decision. When all has been approved, the drawings are released to those responsible for getting the product into production. This involves activities such as ordering materials, building jigs and fixtures, locating machines in the plant in needed locations, and preparing the workers who will build and assemble the product. Often, representatives of the engineering staff are called upon during this phase to assist with the process.

# Problems

Problems are to be recorded on 8½ × 11 in. grid or plain paper. Sketches are to be made in a neat, professional manner. Lettering should follow recommended engineering style and be ⅛ in. high.

1. Bring to class products that exhibit the characteristics of abstract, aesthetic, and functional design. Be prepared to explain to the class why you chose each product.

2. Select one product that you believe represents a high level of creative engineering design. Prepare a report of not over two pages explaining why you selected it.

3. Study the various motorcycles on the market. Prepare a report citing the improvements you would like to see made. Explain what you would recommend be done to replace the elements you have criticized.

4. List the systems that exist in your home. Prepare a list of others that would make the home more pleasant or more efficient.

5. Assume that you have been asked to develop a device to replace the conventional doorbell's push button, transformer, and chimes. Prepare a report explaining each of the steps you would follow to come up with a final design proposal.

6. Study the current automobiles and prepare a report citing the significant design improvements made in the last three years.

7. Assume that you are going to make a marketing study for a new type of camping stove. Explain in detail how you approach the study, whom you would contact, and how you would analyze the findings.

8. Check your school library and list as many sources as you can find that would be of possible use as technical references in studying one of the following: pacemaker, home sound system, automotive pollution, electrical power generation, and computers.

9. Pick a problem, such as the need for warmer but lighter weight clothing, and lead the class in a brainstorming session to produce as many ideas for new clothing designs as possible in a 30-minute period.

10. Visit the school computer center and have the person in charge explain how it can be used in engineering design and analysis.

11. Pick a product that represents a system and break it down into subsystems. An automobile would be a good choice.

12. Study the chairs used in your classroom. Make a human engineering analysis of the chair and report its shortcomings. Then redesign the chair to reflect what you have learned.

13. Select rods of identical size made of several different materials, such as oak, steel, aluminum, and plastic. Subject 10 samples of each to compression and tensile tests, record the data, and report it on a chart.

14. A product is to have a casting that could be brass, aluminum, cast iron, or steel. It is to weigh ½ lb. Make a cost analysis for each material and report your findings on a chart.

15. Look around your school for possible situations where there is some pollution of the atmosphere. Report the cause of the situation and prepare a preliminary proposal with sketches for a solution.

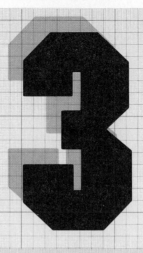

# Graphics Language, Measurements, and Standards

In order to communicate design ideas, suggestions, decisions, or revisions the engineer uses graphics of various types. These drawings may circulate in an engineering office or within a local manufacturing plant or be transmitted to other sites many miles away. Generally, the person who originates the drawing is not available to explain what is meant. Drawings must be easily read and understood. To enable these communications to occur, the graphics system used has been standarized. The drawings mean the same thing to anyone trained in preparing and reading them. This chapter reviews the major features of graphics language and the standards and measurement systems used.

## LETTERING

Drawings contain dimensions and notes. These must be lettered so that they are clear and there is no chance for misreading a size or a note. Errors of this kind are inexcusable and are not tolerated in engineering offices.

Most companies establish standards for lettering used on their drawings. This includes the size of letters and the style. Two types of lettering are used,

## Forming the Letters

Good lettering begins with letters properly formed. Recommended proportions for vertical and inclined lettering are given in Figs. 3–2 and 3–3. Some letters are as wide as they are high. There is a tendency to make letters tall and narrow. This reduces legibility and damages the overall appearance of a drawing. Other letters are slightly taller than they are wide. The *W* is an exception in that it is wider than it is tall.

The properly formed letter must be stable. If you draw a square and divide it horizontally in half, it will appear top-heavy. The same illusion applies to letters and numbers.

Letters such as *B*, *E*, *F*, and *H* are drawn with the central horizontal element slightly above center. *K*, *S*, *X*, and *Z* and the numbers *2*, *3*, and *8* are drawn with the width of the top half smaller than the bottom half. These techniques provide the appearance of stability. Examine the proportions on the numbers *4*, *5*, *6*, and *9* and see how stability is achieved by varying the proportions.

To develop good lettering requires not only studying proportion but also developing a planned way to form each letter. A suggested order of strokes for forming letters is also shown in Figs. 3–2 and 3–3. Develop the habit of forming each letter with the same order of strokes each time.

A third factor in good lettering is practice. The ability to form each letter and number will never be developed without a conscious effort to form each letter properly and reduce this to an automatic skill. Then practice and practice some more.

## Spacing Lettering

The space between letters must appear to be equal. Because the letters are different shapes, the space cannot actually be equal. When letters with tall vertical parts, such as *H* and *I*, come next to each other, more space is allowed than when letters having open space around them, such as *A* and *V*, fall together. The spacing is done by eye. Through experience letters can be placed so that they appear to be equally spaced (Fig. 3–4).

The space between words equals the size of the letter *O* (Fig. 3–5).

## Lettering Guidelines

In order to keep lettering the proper height and the lines straight, light, horizontal guidelines are always drawn. They are spaced according to the height of the letters to be drawn.

Guidelines are drawn very lightly. They should be dark enough so that they serve their purpose but

vertical and inclined. All-capital, vertical Gothic lettering is the most common. Lowercase letters are used only when required for abbreviations. Inclined lettering is drawn on angle of 68°. Never mix vertical and inclined lettering on a drawing. When drawings are revised, the lettering should match the original style as closely as possible. Words and sentences must be arranged in a pleasing and legible format. The space between words and letters must appear to be uniform. Lettering is not usually underlined.

On inch drawings up to 17 × 22 in., lettering height is 0.125 in. On metric drawings up to 297 × 420 mm, lettering height is 3.5 mm. Larger drawings use the larger letters. Recommended sizes are shown in Fig. 3–1.

If drawings are to be photographed and stored on microfilm, lettering size is important. It must be large enough so it will reduce and be able to be enlarged again. The sizes shown in Fig. 3–1 will usually be large enough for use on drawings to be microfilmed.

FIGURE 3–1 Recommended heights for lettering on engineering drawings.

| METRIC (mm) | | CUSTOMARY (in.) | |
|---|---|---|---|
| Size of Drawing | Letter Height | Size of Drawing | Letter Height |
| A4, A3 | 3.5 | 8.5 × 11 (A) | .125 |
| A2 | 4.0 | 11 × 17 (B) | .125 |
| A1, A0 | 4.5 | 17 × 22 (C) | .125 |
| Titles | 7.0 | 22 × 34 (D) | .156 |
| Drawing number in title block | 8.0 | 34 × 44 (E) | .156 |
| | | Titles | .250 |
| | | Drawing number in title block | .250 (.312 over 17 × 22 in.) |

THESE LETTERS ARE $\frac{5}{6}$ AS WIDE AS THEY ARE HIGH.

THESE LETTERS ARE AS WIDE AS THEY ARE HIGH.

THIS W IS WIDER THAN IT IS HIGH.

NUMERALS ARE $\frac{5}{6}$ AS WIDE AS THEY ARE HIGH.

VERTICAL CAPITAL GOTHIC LETTERS AND NUMBERS.

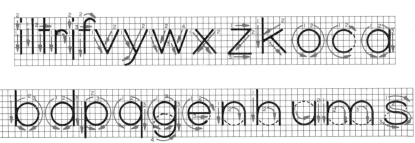

VERTICAL LOWERCASE GOTHIC LETTERS.

FIGURE 3–2  Vertical capital and lowercase Gothic lettering.

FIGURE 3–3  Inclined capital and lowercase Gothic lettering.

THESE LETTERS ARE $\frac{5}{6}$ AS WIDE AS THEY ARE HIGH.

THESE LETTERS ARE AS WIDE AS THEY ARE HIGH.

THIS W IS WIDER THAN IT IS HIGH.

NUMERALS ARE $\frac{5}{6}$ AS WIDE AS THEY ARE HIGH.

INCLINED CAPITAL GOTHIC LETTERS AND NUMBERS.

INCLINED LOWERCASE GOTHIC LETTERS.

SPACES EQUAL

SPACING BETWEEN LETTERS IS

SPACES NOT EQUAL

SPACING BETWEEN LETTERS IS

FIGURE 3–4 Space the letters in each word so that they appear to be an equal distance apart.

ONE LETTER SPACE
BETWEEN WORDS.

KEEP○THE○SPACE○BETWEEN
WORDS○EQUAL○TO○ONE○LETTER.

FIGURE 3–5 The space between words equals the width of one letter.

not so dark that they reproduce when prints are made from the drawing. If they are light enough, they need not be erased (Fig. 3–6).

If it is difficult to keep the lettering vertical or on a uniform incline (for inclined lettering), vertical or inclined guidelines can be drawn (Fig. 3–6). Inclined guidelines are drawn at a 68° angle. They are spaced at random. When fractions are lettered, guidelines are drawn for the numeral and the fraction (Fig. 3–7). Total height of the fraction is equal to two normal letters. Since the numbers must not touch the bar between them, they are drawn slightly smaller. Always draw guidelines for fractions.

Guidelines can be drawn using a lettering guide (Fig. 3–8). It spaces the guidelines in both inches and millimeters. To draw metric guidelines, find the row of holes marked metric (mm). These holes and the numbers on the left side of the disc relate to metric heights for guidelines. The column of six holes enables guidelines to be spaced equally if the holes marked with the right-side brackets are used. They may be half-spaced if those with the left brackets are used. Set the size of lettering wanted opposite the mark *M*. In Fig. 3–8, this was set on 8, giving guidelines for letters 8 mm high.

The numbers at the bottom right side of the wheel space guidelines in inches, from 2 to 10. They show the height of lettering in thirty-seconds of an inch. In Fig. 3–8, the number 8 is opposite the vertical mark. This means the guidelines will be $8/32$, or ¼ in., high.

FIGURE 3–6 Always use guidelines when lettering.

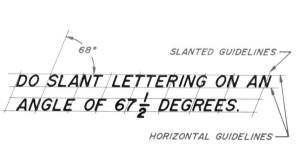

FIGURE 3–7 Lettering common fractions.

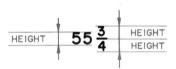

FIGURE 3–8 A lettering guide can be used to locate horizontal, vertical, and inclined guidelines in inches and millimeters. (Courtesy of Olson Manufacturing Co.)

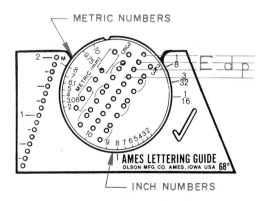

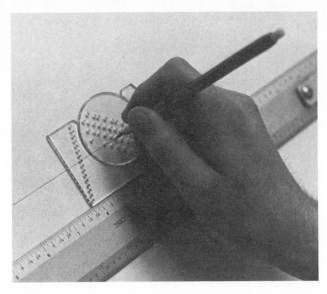

FIGURE 3–9 Place the lettering guide on a straight-edge to draw horizontal guidelines.

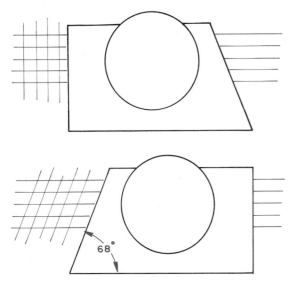

FIGURE 3–10 Using a lettering guide to draw vertical and inclined guidelines.

The lettering guide is placed against a straightedge. A pencil is placed in the proper holes. The pencil and the guide are slid along the straightedge to draw a guideline (Fig. 3–9).

Vertical guidelines are drawn using the vertical edges of the guide. One edge is 90° and the other is 68° (Fig 3–10).

## Lettering Techniques

Select the proper pencil for lettering. A 2H pencil is generally preferred, though some drafters prefer something softer. Softer pencils tend to smudge a

drawing, so care must be taken to blow away excess carbon. Some cover completed lettering with a piece of paper to try to reduce smearing and smudging.

Keep the pencil sharp. Rotate it in your fingers after every letter or two. This keeps the point uniformly rounded. If the pencil is permitted to get dull, the thickness of the lines forming the lettering will increase, and the result will not appear uniform.

When lettering, hold the pencil in a comfortable, relaxed manner. Do not grip it too tightly. Place your arm and body in a comfortable position. The paper may be placed at any convenient angle.

## Lettering with Ink

There are a variety of mechanical lettering devices used to letter with ink. One type uses a plastic template with the letters recessed in it. The ink pen is held in a scriber. A pin on the scriber slides into the recessed letter form on the template. This guides the ink pen, which is attached to it (Fig. 3–11). These templates and pens are available in a variety of sizes.

The template is held in place on a straightedge. As the words are written, it is slid right and left on the straightedge as each letter is located and moved into position.

Ink dries very rapidly but it is still necessary to be careful not to move over an inked area before it is dry.

Another method uses a plastic template with the letters cut all the way through the plastic. The letters are inked with a technical drafting pen. This pen has a point through which the ink flows. Ink is stored in the handle in a plastic tube. Keep the cap on the point when it is not in use to keep the ink from hardening in the point (Fig 3–12).

FIGURE 3–11 The template guides the scriber to form each letter. The scriber can be adjusted to form vertical and inclined letters. (Courtesy of Keuffel and Esser Co.)

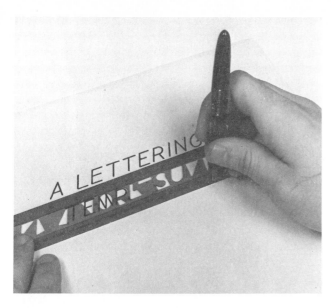

FIGURE 3-12 This lettering guide has the letters cut out like a stencil. The technical fountain pen is used to form the letters.

These templates are available in a range of letter sizes. The pens also have a variety of tips to produce lines of different widths.

Ink images are also placed on drawings with various types of typewriters and computer printers. These use cloth or carbon ribbon as the image producer (Fig 3-13). Some computer printers use an ink-filled pen much like a felt-tip or ballpoint (Fig. 3-14).

## Pressure-Sensitive Letters

Large letters such as titles are sometimes placed on drawings with pressure-sensitive letters. These letters are available in sheets. They have an adhesive on the back. They are cut from the sheet, placed on the drawing, and rubbed until they adhere (Fig. 3-15). They are available in many sizes and styles (Fig. 3-16).

FIGURE 3-13 A long-carriage open-end typewriter prepared for typing on engineering drawings. (Courtesy of Diagram Corporation.)

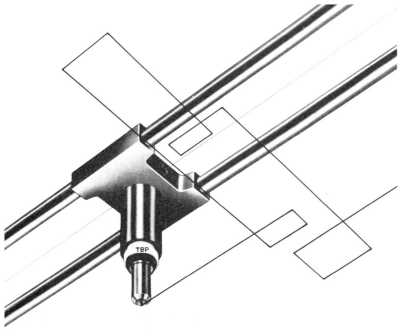

FIGURE 3-14 Computer plotters produce drawings using special ink pens. (Courtesy of Koh-I-Noor Rapidograph, Inc.)

FIGURE 3–15   The steps for applying transfer-type letters to drawings. (Courtesy of Artype, Inc.)

FIGURE 3–16   A few of the many letter forms available in transfer-type materials. (Courtesy of Artype, Inc.)

**ABCDEFG
NOPQRSTU
1234567890**

abcdefghijklmno
ABCDEFGHIJKLMN
1234567890&

**abcdefghijklmn
ABCDEFGHIJKLMN
1234567890**

abcdefghijklmnopqrstuv
ABCDEFGHIJKL

Another system uses a disc containing letters of the alphabet. The words are composed by dialing each letter wanted and printing it on a transparent tape. After the word or words needed are composed, the machine cuts the tape. The pressure-sensitive transparent tape with the letters printed on it are peeled from a protective backing and pressed into place on the drawing (Fig. 3–17).

## LINE SYMBOLS

Drawings are made of a variety of lines. They vary in thickness and shape. Line symbols help people reading the drawing to interpret what is meant.

Accepted standard line symbols are published by the American National Standards Association in publication ANSI Y14.2M-1979 (Fig. 3–18).

Lines can be thick or thin. Ink lines are usually drawn thicker than pencil lines because of the difficulty of drawing uniformly thick lines with pencil. The thin line measures about 0.35 mm or 0.015 in., and the thick line about 0.7 mm or 0.03 in. It can be seen that the thick line is twice as thick as the thin line. This helps when reading a drawing because the major outlines are thick and interior details are usually thin. All lines are drawn absolutely black. A thin line is as black as a thick line. This is necessary if the drawing is to reproduce properly.

The lines that are drawn thick are the visible, cutting-plane, and break lines. Thin lines include the hidden, section, extension, dimension, break, phantom, and stitch lines. These are shown in Fig. 3–19.

*Visible lines*   are used for all visible edges that are seen when looking at an object. The symbol is a thick, solid line.

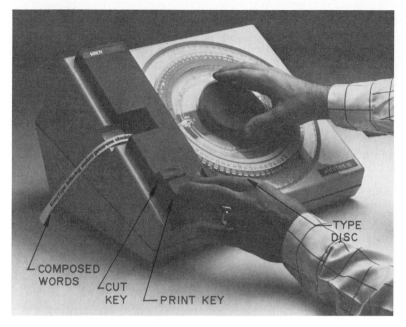

COMPOSED WORDS
CUT KEY
PRINT KEY
TYPE DISC

FIGURE 3–17 This unit produces words on pressure-sensitive tape. The tape is pressed into place on the drawing. (Courtesy of Kroy Inc.)

## LINE THICKNESSES

THICK LINE      .032" OR 0.7 mm

THIN LINE      .016" OR 0.35 mm

## LINE SYMBOLS

VISIBLE LINE — THICK

HIDDEN LINE — THIN
DASHES .10" TO .20" OR 3 TO 5 mm
SPACES .03" TO .06" OR I TO 2 mm

CENTER LINE — THIN
LONG DASHES .75" TO 1.50" OR 20 TO 40 mm
SHORT DASHES .10" OR 3 mm
SPACES .03" TO .06" OR I TO 2 mm

DIMENSION LINE — THIN
EXTENSION LINE — THIN

LEADER — THIN

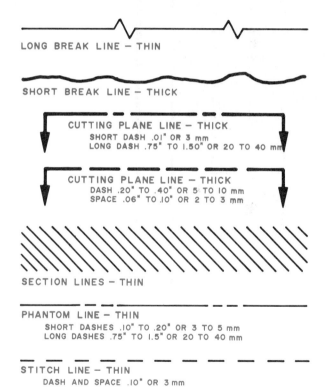

LONG BREAK LINE — THIN

SHORT BREAK LINE — THICK

CUTTING PLANE LINE — THICK
SHORT DASH .01" OR 3 mm
LONG DASH .75" TO 1.50" OR 20 TO 40 mm

CUTTING PLANE LINE — THICK
DASH .20" TO .40" OR 5 TO 10 mm
SPACE .06" TO .10" OR 2 TO 3 mm

SECTION LINES — THIN

PHANTOM LINE — THIN
SHORT DASHES .10" TO .20" OR 3 TO 5 mm
LONG DASHES .75" TO 1.5" OR 20 TO 40 mm

STITCH LINE — THIN
DASH AND SPACE .10" OR 3 mm

FIGURE 3–18 Standard line symbols.

*Hidden lines* show surfaces that are not seen when you look at an object. They are shown with a thin dashed line. The dashes are about 0.3 mm or 0.02 in. long. The spaces between the dashes measure about 1 mm or 0.04 in. They are drawn by eye. Do not measure each dash.

*Dimension lines* are used to show the extent of a dimension or a particular size. They are thin solid lines with an arrowhead on each end. They feature a number reporting the distance indicated.

A *leader* is a special form of dimension line. It connects a dimension or note to the drawing.

*Extension lines* are used with dimension lines. They extend a point out from a view to which the dimension line refers. They are thin solid lines.

*Center lines* are used to locate geometric axes. They are thin lines made of long and short dashes. The long dashes range from 20 to 35 mm or ¾ to 1½ in. The short dashes measure about 3 mm or ⅛ in. The length of the long dash will vary with the size of the

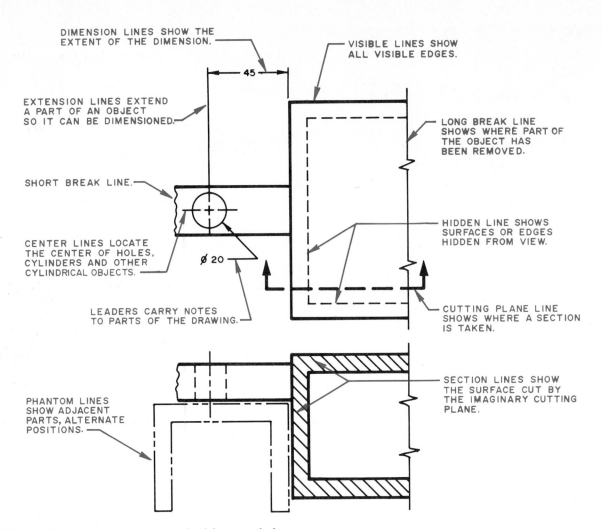

DIMENSION LINES SHOW THE EXTENT OF THE DIMENSION.

EXTENSION LINES EXTEND A PART OF AN OBJECT SO IT CAN BE DIMENSIONED.

SHORT BREAK LINE.

CENTER LINES LOCATE THE CENTER OF HOLES, CYLINDERS AND OTHER CYLINDRICAL OBJECTS.

LEADERS CARRY NOTES TO PARTS OF THE DRAWING.

PHANTOM LINES SHOW ADJACENT PARTS, ALTERNATE POSITIONS.

VISIBLE LINES SHOW ALL VISIBLE EDGES.

LONG BREAK LINE SHOWS WHERE PART OF THE OBJECT HAS BEEN REMOVED.

HIDDEN LINE SHOWS SURFACES OR EDGES HIDDEN FROM VIEW.

CUTTING PLANE LINE SHOWS WHERE A SECTION IS TAKEN.

SECTION LINES SHOW THE SURFACE CUT BY THE IMAGINARY CUTTING PLANE.

45

Ø 20

FIGURE 3-19 Applications for standard line symbols.

part illustrated. The long dash extends 1.5 mm or ¹⁄₁₆ in. beyond the edge of the circular element. If a center line is very long, it can be a series of long and short dashes.

*Section lines* are drawn on the surface that has been cut when drawing a section view. They are thin lines spaced 1.5 to 3 mm or ¹⁄₁₆ to ⅛ in. apart. The larger the surface to section, the wider the spacing between the lines. They are spaced by eye. Section lines are placed on an angle to the edges of the surface, usually 45°, 30°, and 60°.

*Cutting plane* lines are used on sectioned drawings to show where the section was taken. They are thick dashed lines. Two kinds are in use. One uses only short dashes, while the other uses one long and two short dashes of equal length. They usually end with an arrow pointing in the direction of sight used to make the section.

*Break lines* are used to show where a part of an object has been theoretically broken away. It may be a thick or thin line, depending upon the break. Short breaks are drawn freehand with a solid irregular line. Long breaks are drawn with a thin straight line with Z shapes inserted in several places.

*Phantom lines* are used to show the location of a part that moves. The part is drawn in one position

with visible lines and in its rotated position with phantom lines. The symbol is a long dash followed by two short dashes. It is a thin line.

*Stitch lines* are used to show stitching or the stitching process. They are thin dashed lines with dashes and spaces of equal length.

*Arrowheads* are small points on the ends of dimension lines. They are used to show the points from which dimensions are taken. They are usually about 3 mm or ⅛ in. long. They are one third as wide as they are long (Fig. 3–20).

## ABBREVIATIONS

Part of the language of graphics is the extensive list of abbreviations used on engineering drawings. Complete details are available in "Abbreviations for Use on Drawings," ANSI Y1.1-1972, published by the American National Standards Institute. Since it is difficult to remember all of these, the standard should be part of a designer's working library. Selected abbreviations are given in Appendix A.

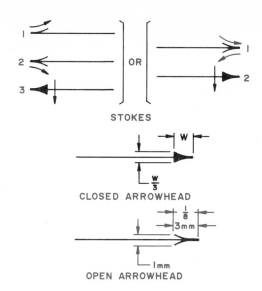

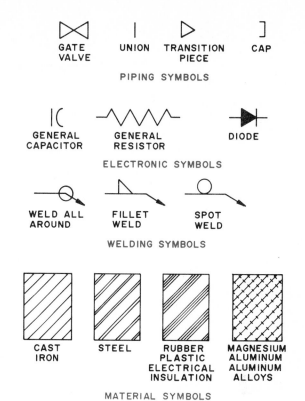

FIGURE 3–20  How to draw arrowheads.

FIGURE 3–21  Examples of the symbols used on engineering drawings.

## SYMBOLS

In many cases it is not possible actually to draw a specific item as part of a drawing. For example, an electrical drawing for a power plant cannot detail each switch or motor as it really appears. Instead, a symbol is used. A *symbol* is a simple drawing that represents a more complex item.

Symbols for various industries are standardized and available from the American National Standards Institute. A listing of their publications is in Appendix M. Symbols are shown in detail in this text when appropriate. Some examples appear in Fig. 3–21.

## BORDERS AND TITLE BLOCKS

The actual layout of the border and title block depends on the drafting standards of each company. As a general practice, borders like those shown in Fig. 3–22 are recommended. If sheets are to be bound, leave 25 mm or 1 in. on the edge to be bound.

Title blocks give information about the company and the drawing. Typical of the information are the company name, company location, finishes, drawing number, part name, part number, data, scale, tolerances, and the names of the drafter and checker. Recommended designs are found in ANSI Y14.1-1980, "Drawing Sheet Size and Format." A typical block is shown in Fig. 3–23. Title blocks are usually located in the lower right-hand corner. If it is a metric drawing, this information is often given in or just above the title block.

| SHEET SIZE | BORDER |
|---|---|
| Metric | |
| A4, A3 | 6 mm all sides |
| A2 | 10 mm all sides |
| A1, A0 | 12 mm all sides |
| Customary | |
| A, B | ¼ in. all sides |
| C | ⅜ in. all sides |
| D, E | ½ in. all sides |

FIGURE 3–22  Recommended border widths for standard-size engineering drawings.

FIGURE 3–23  A typical title block.

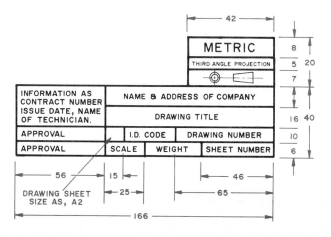

FIGURE 3–24 Typical printed title block and drawing revision schedule. (Courtesy of Stanpat Products Inc.)

Many companies purchase their drawing papers with the border and title block printed on them. Title blocks are also available printed on translucent cellulose acetate film with an adhesive back. They are stuck to the drawing. The front surface has a matte finish that accepts pencil, ink, and typing (Fig. 3–24).

# DRAFTING AND INDUSTRIAL STANDARDS

Standards are concerned with uniformity of communication, manufacture, or use of a product. The U.S. National Bureau of Standards defines engineering standards as "a document to assure dimensional compatibility, quality, and performance and uniformity of language." Standards may typically describe screw-thread dimensions, clothing sizes, chemical composition and mechanical properties of steel, methods of testing for sulfur in oil, or a code for highway signs.

Engineers must be aware and utilize the standards related to the project in which they are involved. Such standards are detailed in the various chapters of this book wherever appropriate.

The industrial countries of the world each have their own organization to establish internal standards. In the United States this is the American National Standards Institute. In Canada it is the Canadian Standards Association.

## American National Standards Institute (ANSI)

Five engineering associations joined together in 1918 to develop national drafting standards. From this initial effort grew the American Standards Association (ASA). In 1966, the name was changed to United States of America Standards Institute (USASI). In 1969, the name became the American National Standards Institute (ANSI).

ANSI identifies industrial and public needs for national consensus standards and coordinates their development. It resolves national problems related to standards and represents the United States in the establishment of international standards. The standards are developed in cooperation with various societies and associations. For example, the American Petroleum Institute (API) developed the standard "Flanged Steel Safety Relief Valves." When a standard is accepted nationally, it becomes a part of the American National Standards Institute standards.

Standards are designated by a code system that was adopted in 1977. The code consists of the ANSI initials plus an abbreviation of the sponsoring organization, such as API. This is followed by catalog code numbers and the date the standard was accepted. The Flange Steel Safety Relief Valve Standard is coded ANSI/API 526-1969. Not all standards have been changed to this new coding system, but they will be as they are revised.

Of the approximately 10,500 ANSI standards, some are used more frequently than others in engineering design. Of critical importance is the ANSI Drafting Manual. It is made up of 16 separate publications. The newer standards are designated by the code letter Y. The older ones, using the Z designation, will be changed to Y as they are revised. These and other related standards are given in the appendix.

## International Standards Organizations

The International Organization for Standardization (ISO) is a worldwide body that coordinates the development of engineering and product standards for all nations. The International Electrotechnical Commission (IEC) is involved with the development of electrical and electronics standards. Both organizations have members in all the major industrialized nations of the world. They establish and publish internationally accepted standards. The American National Standards Institute is the representative of the United States to these groups. Some ISO standards of interest and value to engineers appear in Appendix Z. Both ANSI and ISO standards can be purchased from the American National Standards Institute, 1430 Broadway, New York, N.Y. 10018.

# MEASUREMENTS

The engineer must thoroughly understand both the customary and the metric system of measurement.

| DECIMAL FRACTION | NAME |
|---|---|
| 1.0 | one inch |
| 0.1 | one-tenth inch |
| 0.01 | one-hundredth inch |
| 0.001 | one-thousandth inch |
| 0.0001 | one ten-thousandth inch |
| 0.00001 | one hundred-thousandth inch |
| 0.000001 | one-millionth inch |

FIGURE 3–25 Inch decimal fractions.

Everything an engineer does relates in some way to some form of measurement.

## The Customary System

The basic linear unit is the *yard*. It is divided into 3 *feet*. Each 1-ft unit is divided into *common fractions* or *decimal fractions.* Common fractions are specified in units divisible by 2, such as ½, ¼, ⅛, ⅒, ¹⁄₃₂, or ¹⁄₆₄. Decimal fractions are specified in one or more decimal parts of an inch, such as 0.1, 0.5, or 0.75. The proper way to express these is shown in Fig. 3–25.

Occasionally it is necessary to convert from a common fraction to a decimal fraction or vice versa. A table giving decimal equivalents is in Appendix B.

## The Metric System (SI)

The metric system is used by almost every country in the world. The first attempt to develop a world-wide system of measurement was made in France in 1790. The French developed a basic unit of length called the meter. Measurements of capacity (volume) and mass (weight) were derived from this unit of length. In this way length, volume, and weight were related. It was designed as a decimal system based on the unit 10. Over the years, various adjustments have been made in the original system.

In 1866, Congress passed a bill permitting the use of the metric system in the United States. In 1876, the United States established the Metric Bureau. It supplied metric materials to all who wanted them.

In 1890, the United States joined the International Bureau of Weights and Measures and received international prototype meters and kilograms. They were deposited at the U.S. Office of Weights and Measures. Since that time they have become the standards used for determining the yard and the pound.

In 1901, the National Bureau of Standards replaced the Office of Weights and Measures. Today it preserves the prototype meters and kilograms.

Bills have been introduced in Congress recommending the adoption of the metric system, but none was passed until the Metric Conversion Act of 1975. The Act declares that the policy of the United States is to coordinate and plan for increasing use of the metric system and provide for the establishment of a United States Metric Board to coordinate voluntary conversion to the metric system.

The International Bureau of Weights and Measures, which is located in France, establishes standards and changes in the metric system that are accepted over the world. In 1960, it adopted a modern metric system, the Système International d'Unités (International System of Units, abbreviated SI). Some changes have been made since 1960. This is the system that is replacing the customary system in the United States. The move is to use SI metrics for all product design.

## SI Metric Base Units

The SI metric system has seven base units. These are shown in Fig. 3–26. Other measurement units are derived from these seven base units.

*Meter*    A *meter* (m) is equal in length to 1 650 763.73 wavelengths of the orange-red light given off by the atom krypton-86. This definition is such that it can be commonly understood around the world.

*Kilogram*   The *kilogram* (kg) is the unit to measure mass. Mass is often referred to as weight. Actually, the weight of an object is dependent upon gravitational pull, so weight varies around the world.

Mass is based on a standard kilogram cylinder located at the International Bureau of Weights and Measures in France.

*Second*   A *second* (s) is a unit of time. One second is equal to 9 192 631 770 cycles of movement of the cesium atom in an atomic clock. This is the unit we are accustomed to using.

FIGURE 3–26 The basic units of SI metrics.

| QUANTITY | LENGTH | MASS | TIME | ELECTRIC CURRENT | TEMPERATURE | LUMINOUS INTENSITY | AMOUNT OF SUBSTANCE |
|---|---|---|---|---|---|---|---|
| Unit | meter | kilogram | second | ampere | kelvin | candela | mole |
| SI symbol | m | kg | s | A | K | cd | mol |

*Kelvin*  A *kelvin* (K) is an absolute scale of temperature having a zero point of $-273.16°C$. One kelvin is equal to $\frac{1}{273.16}$ of the thermodynamic temperature of the triple point of water. This is the temperature at which water exists as solid, liquid, and vapor.

For nonscientific temperature measurements, the Celsius (°C) temperature scale is used. This scale is based on the use of 0° as the freezing point, and 100° as the boiling point, of water. A degree celsius is $\frac{1}{100}$ of this scale. Celsius has replaced the centigrade scale.

*Ampere*  An *ampere* (A) is the base unit for the measurement of electrical current. It is defined as the amount of current in two parallel wires one meter apart that will result in a specific force between the wires.

*Candela*  A *candela* (cd) is a measure of luminous intensity (the amount of light given off by an object). It is equal to the amount of light given off by platinum at its freezing point under pressure. At its freezing point platinum is glowing hot.

*Mole*  The *mole* (mol) is the unit of substance. It is equal to the number of particles contained in a specific amount and type of carbon. It is used in special scientific measurements.

*Supplementary units*  The *radian* and the *steradian* are two supplementary units in the SI metric system. They are used to measure plane and solid angles. The *radian* (rad) is the plane angle between two radii of a circle when those radii cut off an arc along the circumference equal in length to the radius.

A *steradian* (sr) is the solid angle that, with its vertex in the center of a sphere, cuts off an area on the surface of the sphere equal to the radius squared (Fig. 3–27).

Steradians and radians are used in the scientific angular measurement. In engineering design projects, the angles are measured in *degrees* (°), *minutes* ('), and *seconds* ("). A circle has 360 degrees, each degree is divided into 60 minutes, and each minute into 60 seconds.

## SI Metric Unit Prefixes

A prefix is a word part added to the beginning of a word that changes the word's meaning. Metric prefixes are added to SI metric unit names to represent multiples or submultiples of that unit. As an example, 1000 meters is called a *kilometer*, the prefix *kilo-* meaning "thousand." Likewise, $\frac{1}{1000}$ of a meter is a *millimeter*, the prefix *milli-* meaning "one thousandth."

The common prefixes used in the metric system are shown in Fig. 3–28.

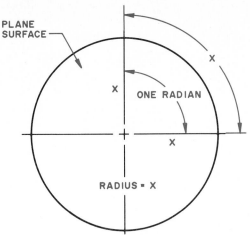

A RADIAN IS A PLANE ANGLE GENERATED WHEN THE LENGTH OF THE ARC CUT BY TWO RADII EQUALS THE LENGTH OF THE RADIUS.

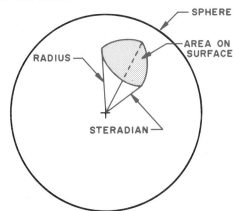

A STERADIAN IS THE SOLID ANGLE WITH ITS VERTEX AT THE CENTER OF A SPHERE WHICH CUTS OFF AN AREA ON THE SURFACE OF THE SPHERE EQUAL TO THE RADIUS SQUARED.

FIGURE 3–27  The radian and steradian are supplementary units in SI metrics.

## The Derived Units

Derived units are combinations of base, supplementary or other derived units. They are generated by multiplying or dividing. For example, the square meter is the derived unit for area. It is obtained by multiplying the length in meters by the width in meters. This gives the derived unit, square meter ($m^2$).

Some of the derived units have special names and symbols. They are used to show some derived units in a simpler way than using the base units. For example, the newton (N) is the special derived unit for force. If it were expressed in base units, it would appear as $kg \cdot m/s^2$. This is difficult to say and write. It is easier to say *newton* and write *N*. Other derived units with special names appear in Fig. 3–29.

Derived units used frequently in engineering design include units of area, volume, velocity, acceleration, force, pressure, work, and electrical power.

| MULTIPLICATION FACTOR | PREFIX | SYMBOL | PRONUNCIATION* | MEANING |
|---|---|---|---|---|
| $1\ 000\ 000\ 000 = 10^9$ | giga | G† | jig'a (a as in about) | one billion times |
| $1\ 000\ 000 = 10^6$ | mega | M† | as in megaphone | one million times |
| $1000 = 10^3$ | kilo | k† | as in kilowatt | one thousand times |
| $100 = 10^2$ | hecto | h | heck'toe | one hundred times |
| $10 = 10^1$ | deka | da | deck'a (a as in about) | ten times |
| BASE UNIT $1 = 10^0$ | | | | |
| $0.1 = 10^{-1}$ | deci | d | as in decimal | one tenth of |
| $0.01 = 10^{-2}$ | centi | c† | as in sentiment | one hundredth of |
| $0.001 = 10^{-3}$ | milli | m† | as in military | one thousandth of |
| $0.000\ 001 = 10^{-6}$ | micro | μ† | as in microphone | one millionth of |
| $0.000\ 000\ 001 = 10^{-9}$ | nano | n† | nan'oh (an as in ant) | one billionth of |

*The first syllable of every prefix is accented. For example, the preferred pronunciation of *kilometer* places the accent on the first syllable, not the second.
†Most commonly used and preferred prefixes. *Centimeter* is used mainly for measuring the body, clothing, sporting goods, and some household articles.

FIGURE 3–28 Metric prefixes are used with the metric units to indicate a multiple or sub-multiple of the unit.

| MEASUREMENT | NAME | SYMBOL |
|---|---|---|
| frequency | hertz | Hz |
| force | newton | N |
| pressure or stress | pascal | Pa |
| energy | joule | J |
| power | watt | W |
| electric charge | coulomb | C |
| electric potential | volt | V |
| capacitance | farad | F |
| electrical resistance | ohm | Ω |
| conductance | siemens | S |
| magnetic flux | weber | Wb |
| magnetic flux density | tesla | T |
| inductance | henry | H |
| luminous flux | lumen | lm |
| illuminance | lux | lx |
| radioactive source | becquerel | Bq |
| absorbed dose | gray | Gy |

FIGURE 3–29 The derived units that have special names. There is only one name, symbol, and formula for each measure.

*Area* Area, as described above, is found by multiplying the width by the length. The basic unit is the square meter ($m^2$). The derived unit depends upon the prefix. The area can be expressed in terms of *square meters* ($m^2$), *square centimeters* ($cm^2$) or *square millimeters* ($mm^2$).

Land area is expressed in square hectometers ($hm^2$) or square kilometers ($km^2$).

*Volume* Volume is derived by multiplying the length by the width by the height of rectangular objects. A cube that measures one meter on each side has a volume of one *cubic meter* ($m^3$). Other derived units of volume are *cubic decimeters* ($dm^3$), *cubic centimeters* ($cm^3$), and *cubic millimeters* ($mm^3$). The cubic decimeter is also known as a *liter*.

*Velocity* Velocity is the speed at which an object is moving. It is measured in the number of meters an object is moving each second, or *meters per second* (m/s). Certain speeds, as of machine tools, might be expressed as *meters per minute* (m/min).

*Acceleration* Acceleration is the rate at which the velocity is changing. It is measured in *meters per second squared* ($m/s^2$).

*Force* Force is power made operative against a resistance. The derived unit of force is the *newton* (N). The base unit designation is $kg \cdot m/s^2$ (kilograms per meter per second squared).

*Pressure* Pressure is measured in *pascals* (Pa). One pascal is one newton of force acting over one square meter of area ($N/m^2$).

*Work* The unit of work is the *joule* (J), the force of one newton acting through a distance of one meter ($N \cdot m$).

*Electrical power* The measurement of electrical power is the *watt* (W). Power is the rate of doing work. One watt is equal to the energy of one joule per second (J/s).

## SI Metric Units of Length

Units of length are the most frequently used when designing and manufacturing a product. The meter is the basic metric unit of length. Since the metric

```
            1 meter is the basic unit
   10 decimeters (dm) = 1 meter (m)
  100 centimeters (cm) = 1 meter (m)
 1000 millimeters (mm) = 1 meter (m)
            also remember
     1 decimeter (dm) = 10 centimeters (cm)
     1 centimeter (cm) = 10 millimeters (mm)
```

FIGURE 3-30  The subdivisions of the meter.

system is based on the unit 10, all subdivisions of the meter are divided by 10. The subdivisions of a meter are given in Fig. 3–30. A meter is divided into decimeters (dm, tenths of a meter). A decimeter is divided into centimeters (cm, tenths of a decimeter). A centimeter is divided into millimeters (mm, tenths of a centimeter).

A metric unit can be changed from one unit to another by moving the decimal point. For example, 25 m equals 250 dm or 2 500 cm or 25 000 mm. By moving the decimal to the left, you can express the figure 350 mm as 35 cm, 3.5 dm, or 0.35 m.

Measurements on engineering drawings are always specified in millimeters or decimal parts of a millimeter (Fig. 3–31). Centimeters and decimeters are not used. Meters are used on large drawings such as land surveys. Surface finish on machine parts is stated in micrometers (μm). A micrometer is one millionth of a meter.

A scale marked in millimeters compared to an inch scale appears in Fig. 3–32.

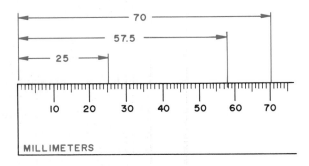

FIGURE 3-31  How to measure distances in millimeters.

FIGURE 3-32  A comparison of a metric scale with common fractions and decimal fractions of an inch.

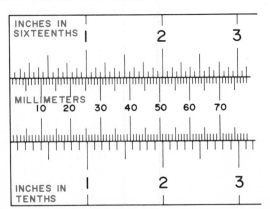

# METRIC DRAFTING STANDARDS

Since metric units are not yet totally familiar in our society, the designer must make a special effort to see that those used are correct. This includes the proper prefixes, spelling, abbreviations, and use of properly derived units. A metric practices manual should be standard equipment for every engineer. Following are some of the frequently used metric practices.

## Spelling

The accepted spelling of the units of length and volume is still undecided. Some people use *meter* and *liter*, while others write *metre* and *litre*. Until this is resolved, either spelling can be considered correct. The engineer should be consistent in using the *er* or *re* form and be sure not to mix them on a drawing or in a report.

The correct use, spelling, and pronunciation of prefixes is important. An error could lead to a costly mistake.

## Plurals

When a unit name is written out in full, it is made plural by adding an *s*. The unit name without the *s* is singular.

| CORRECT PLURAL | INCORRECT PLURAL |
|---|---|
| 350 meters | 350 meter |
| 75 newtons | 75 newton |

Metric symbols, as *m* or *kg*, represent both the singular and plural form. An *s* is never added to a symbol. The lowercase *s* is the symbol for *second*, and its use as a plural would be confusing.

| CORRECT PLURAL | INCORRECT PLURAL |
|---|---|
| 75 mm | 75 mms |
| 15 kg | 15 kgs |

## Capitalization

Metric unit names and their prefixes are not capitalized unless one is the first word in a sentence. An exception is degrees Celsius. The word *degrees* is lowercase, but *Celsius* is always capitalized.

Symbols for units are always lowercase except those named after a person. That is, units derived from a proper name are lowercase when spelled out, as *newton*, but the symbol is capitalized, as *N*. The exception to this is the liter. The symbol is capital *L*, simply to keep a lowercase *l* from being confused with the numeral 1.

| NAME | SYMBOL |
|---|---|
| millimeter | mm |
| newton | N |
| liter | L |

No space is left between the prefix and the unit name. Some prefixes are capitalized so they cannot be confused with other symbols.

| UNIT | SYMBOL |
|---|---|
| kilogram | kg |
| megameter | Mm |
| centimeter | cm |

## Symbols

The proper symbols must always be used. For example, the symbol *A* is used for *ampere*, not the old abbreviation *amp*.

## Spacing

A space is left between a number and the unit name or symbol.

| CORRECT SPACING | INCORRECT SPACING |
|---|---|
| 25 m | 25m |
| 15 kg | 15kg |

The exception is when writing a temperature. A space is not left between the number and the Celsius symbol, °C.

| CORRECT SPACING | INCORRECT SPACING |
|---|---|
| 70°C | 70° C |

Spaces are used instead of commas between large series of numbers. When writing numbers of five or more digits, separate the digits in groups of three, starting from the decimal marker and counting either right or left. Use a space instead of the comma. A four-digit number does not use a space or a comma.

| CORRECT SPACING | INCORRECT SPACING |
|---|---|
| 2571 | 2 571 or 2,571 |
| 15 069 | 15069 or 15,069 |
| 19 172.341 62 | 9172.34162 or 9,172.34162 |

## Decimal Fractions

The metric system is a decimal system. All dimensions use a decimal marker (period) to separate whole numbers from fractions. Common fractions are never used.

| CORRECT FRACTIONS | INCORRECT FRACTIONS |
|---|---|
| 0.5  mm | ½ mm |
| 0.75 kg | ¾ kg |

When writing whole numbers and fractions, both are expressed in the same unit. Never mix metric units.

| CORRECT UNITS | INCORRECT UNITS |
|---|---|
| 5.3 cm | 5 cm 3 mm |
| 8.5 m | 8 m 5 dm |

In some countries the decimal marker used is a comma or a raised decimal point. Since these may cause confusion, the United States has selected to use the baseline decimal marker.

| CORRECT DECIMAL MARKER | NOT USED IN U.S. |
|---|---|
| 50.5 ml | 50,5 ml |
| 75.3 mm | 75·3 mm |

Numbers with a value of less than 1 have a 0 placed to the left of the decimal marker. This prevents overlooking the decimal marker.

| CORRECT USE | INCORRECT USE |
|---|---|
| 0.7  cm | .7 cm |
| 0.35 g | .35 g |

## Punctuation

The symbols used to indicate metric units are not abbreviations. Therefore, they are not followed by a period.

| CORRECT USE | INCORRECT USE |
|---|---|
| 10 m | 10 m. |
| 5 kg | 5 kg. |

Area and volume are stated in this order: numerical size, metric symbol, superscript (indicating squared or cubed). The superscript for area is 2, as in $m^2$ (square meter), and for volume it is 3, as in $m^3$ (cubic meter).

| CORRECT USE | INCORRECT USE |
|---|---|
| 10 $m^2$ | 10 square m or 10 sq m |
| 5 $cm^3$ | 5 cubic cm or 5 cu cm or 5 cc. |

## PRONUNCIATION

The recommended way to pronounce SI metric prefixes and other units in the United States is to accent the first syllable of each prefix so as to ensure that the prefix will retain its identity. For example, the term *kilometer* is accented on the first syllable: *kilometer* (Fig. 3–33).

| PREFIX | PRONUNCIATION* |
|---|---|
| exa | ex'a (*a* as in *about*) |
| peta | pet'a (*e* as in *pet*, *a* as in *about*) |
| tera | as in *terrace* |
| giga | jig'a (*i* as in *jig*, *a* as in *about*) |
| mega | as in *megaphone* |
| kilo | kill'oh |
| hecto | heck'toe |
| deka | deck'a (*a* as in *about*) |
| deci | as in *decimal* |
| centi | as in *centipede* |
| milli | as in *military* |
| micro | as in *microphone* |
| nano | nan'oh (*an* as in *ant*) |
| pico | peek'oh |
| femto | fem'toe (*fem* as in *feminine*) |
| atto | as in *anatomy* |

| SELECTED UNITS | PRONUNCIATION |
|---|---|
| candela | can *dell*'a |
| joule | rhymes with *tool* |
| kilometer | *kill*'oh meter |
| pascal | rhymes with *rascal* |
| siemens | same as *seamen's* |

*The first syllable of every prefix is accented to assure that the prefix will retain its identity. Therefore, the preferred pronunciation of *kilometer* places the accent on the *first* syllable, not the second.

FIGURE 3–33 Aids to pronunciation of metric prefixes and selected units.

FIGURE 3–34 Selected conversion factors.

| WHEN YOU KNOW | | YOU CAN FIND | IF YOU MULTIPLY BY |
|---|---|---|---|
| LENGTH | inches | millimeters | 25.4 |
| | feet | millimeters | 300.48 |
| | yards | meters | 0.91 |
| | miles | kilometers | 1.61 |
| | millimeters | inches | 0.04 |
| | meters | yards | 1.1 |
| | kilometers | miles | 0.6 |
| AREA | square inches | square centimeters | 6.45 |
| | square feet | square meters | 0.09 |
| | square yards | square meters | 0.83 |
| | square miles | square kilometers | 2.6 |
| | acres | square hectometers (hectares) | 0.4 |
| | square centimeters | square inches | 0.16 |
| | square meters | square yards | 1.2 |
| | square kilometers | square miles | 0.4 |
| | hectares | acres | 2.5 |
| MASS (Weight) | ounces | grams | 28.0 |
| | pounds | kilograms | 0.45 |
| | tons (short) | metric tons | 0.9 |
| | grams | ounces | 0.04 |
| | kilograms | pounds | 2.2 |
| | metric tons | tons (short) | 1.1 |

Fig. 3–34 cont.

| WHEN YOU KNOW | | YOU CAN FIND | IF YOU MULTIPLY BY |
|---|---|---|---|
| VOLUME | bushels | cubic meters | 0.04 |
| | cubic feet | cubic meters | 0.03 |
| | cubic inches | cubic centimeters | 16.4 |
| | cubic yards | cubic meters | 0.8 |
| FLUID VOLUME | ounces | milliliters | 30.0 |
| | pints | liters | 0.47 |
| | quarts | liters | 0.95 |
| | gallons | liters | 3.8 |
| | milliliters | ounces | 0.03 |
| | liters | pints | 2.1 |
| | liters | quarts | 1.06 |
| | liters | gallons | 0.26 |
| TEMPERATURE | degrees Fahrenheit | degrees Celsius | 0.6 (after subtracting 32) |
| | degrees Celsius | degrees Fahrenheit | 1.8 (then add 32) |
| POWER | horsepower | kilowatts | 0.75 |
| | kilowatts | horsepower | 1.34 |
| PRESSURE | pounds per square inch (psi) | kilopascals | 6.9 |
| | kilopascals | pounds per square inch | 0.15 |
| VELOCITY (Speed) | miles per hour | meters per second | 0.45 |
| | miles per hour | kilometers per hour | 1.6 |
| | kilometers per hour | miles per hour | 0.6 |

## METRIC CONVERSION

Whenever possible, a designer working on a product to be manufactured to metric sizes will design it initially using accepted metric units. The drawings will be dimensioned in metric units with no consideration of inch sizes. If a product is being redesigned and converted to metric, it should have the sizes changed to preferred metric design sizes. For example a part that measures 1 in. (25.4 mm) would be redesigned as an even 25 mm. This type of conversion is called *hard conversion.* It produces a truly metric product and is less costly to manufacture.

If a designer simply converts the inch sizes to equivalent metric sizes, the product is not changed and is not truly metric. This type of simple conversion is called *soft conversion.* It gives very unusual metric sizes, which are expensive to manufacture, are unacceptable, and are to be avoided.

If there is need to convert inch sizes to metric sizes, tables or conversion units can be used. Direct metric conversion tables for linear measurement are in Appendix B. Selected conversion factors for other metric units are in Fig. 3–34.

## Problems

1. Using the engineers' grid paper with eight squares to the inch, practice lettering by copying the paragraphs pertaining to lettering. Letter on every other line. Use an H or 2H pencil.

2. Go to the school library and check out a technical report that is on microfilm. Examine it carefully, then put it on a microfilm reader. As you view the report on the screen, record factors about the image that you think could be improved. Especially note the style and size of letters used.

3. Write a definition for the following technical terms.

| | |
|---|---|
| guidelines | kilogram |
| mm | kelvin |
| ANSI | ampere |
| leader | candela |
| extension line | mole |
| center line | radian |
| section line | steradian |
| cutting plane | $m^2$ |
| break line | m/s |
| phantom line | N |
| symbol | Pa |
| engineering standards | hard conversion |
| ISO | soft conversion |
| meter | |

# Drafting Instruments and Techniques

An extensive array of drafting tools is available. Selection of the proper tool for a particular job is important. The technique of using each tool must be mastered to produce acceptable engineering drawings. It is also important to use the highest-quality tools available. Time is money, and good tools save time.

## DRAFTING TABLES

Drafting tables are available in many styles. Some are four-legged tables with hinged tops. These permit the drafter to give the top some slope. Other tables are built with two legs. Some have controls to assist in raising and lowering the height of the top as well as tilting it to various angles (Fig. 4–1).

Generally, drafters prefer to work seated in a chair with a comfortable back rather than on the old type of drafting stool. This reduces fatigue. Since the height of the tops of many tables can be changed, the relationship between the chair and the table top can be easily adjusted (Fig. 4–2).

Drawing-table tops are usually wood or a wood product such as hardboard over a cellular core. They are usually covered with a vinyl drawing-board covering material.

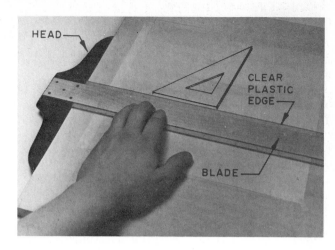

FIGURE 4–3 The T-square serves as a base for the triangles.

## THE T-SQUARE

The T-square is an inexpensive tool used to draw horizontal lines. It serves as a base to hold other tools such as drafting triangles (Fig. 4–3). Most have clear plastic edges. This enables the drafter to see any lines that are beneath the edge of the T-square. Usually the plastic is thinner than the wood portion of the blade. It does not actually touch the paper. This prevents ink from flowing under it when drawings are being inked (Fig. 4–4). T-squares are available in lengths from 18 in. (457 mm) to 60 in. (1524 mm).

They must be stored flat or hanging so they do not develop a bow. Care must be taken to protect the plastic edges from nicks. The head must be kept tightly joined to the blade. With good care, the head will seldom become loose.

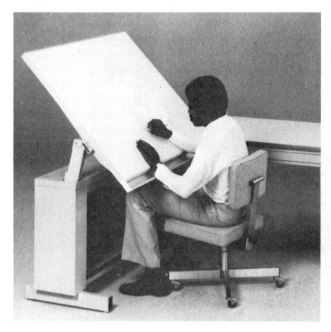

FIGURE 4–1 A drafting table that can be adjusted to suit the drafter. (Courtesy of American Hamilton.)

FIGURE 4–2 Drafters can work seated or standing. (Courtesy of American Hamilton.)

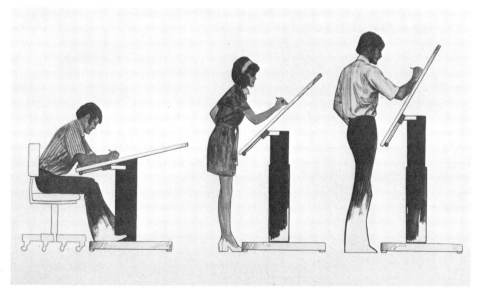

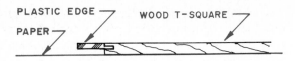

FIGURE 4–4 The raised edge on a T-square prevents ink from flowing under the blade.

## THE PARALLEL STRAIGHTEDGE

A parallel straightedge is used for the same purpose as the T-square. It is a horizontal wood blade with plastic edges. It runs on a series of wires with a set of pulleys. The parallel straightedge can be moved up and down the board on the wires and always remains parallel with the top of the drawing board (Fig. 4–5).

## DRAFTING MACHINES

The two basic types of drafting machines are the arm type and the track type.

The arm type fastens to the top edge of the drafting table. The two arms pivot on each other. The head of the machine is located at the end of the lower arm. As the head is moved up or down or left or right, the scales fastened to the head remain in positions parallel to their original setting (Fig. 4–6).

The track type moves on two tracks. One track is fastened to the top edge of the drafting table. The other rides on this first track and is perpendicular to it. The head is fastened to the vertical track and

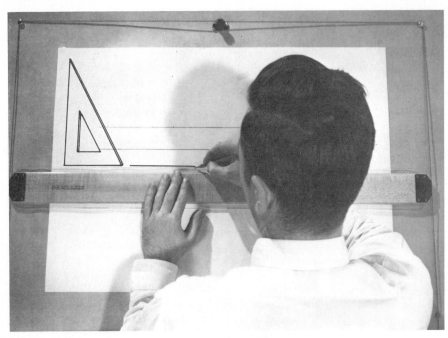

FIGURE 4–5 The parallel straightedge is used to draw horizontal lines and serves as a base for the triangles. (Courtesy of Keuffel and Esser Co.)

FIGURE 4–6 An arm-type drafting machine. (Courtesy of Keuffel and Esser Co.)

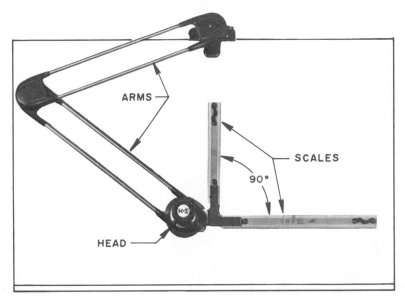

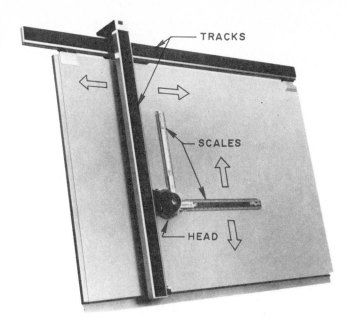

FIGURE 4–7 A track-type drafting machine. (Courtesy of Keuffel and Esser Co.)

slides up and down on it. The vertical track slides left and right on the top track (Fig. 4–7).

These drafting machines are a combination of several drafting tools. The horizontal scale serves the same purpose as the T-square or parallel straightedge. The vertical scale replaces the triangles used with the T-square and parallel straightedge. Each blade has linear measurements marked on it, so it replaces the drafting scale. The head contains a scale in degrees, replacing the protractor.

The advantage to drafting machines is that they speed up drafting time, since all these tools are assembled into one machine.

FIGURE 4–8 The head of a drafting machine. (Courtesy of Keuffel and Esser Co.)

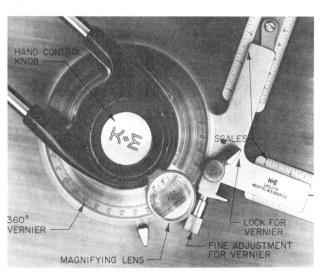

Machines are made for right-handed and left-handed people. Right-handed people hold the lead with the left hand and the pencil in their right hand. A left-handed machine would have the scales facing opposite those in Figs. 4–6 and 4–7.

The scales are moved to various angles by releasing a lock, pressing a button, and turning the head. Some frequently used angles, especially 30°, 45°, and 60°, are set with notches cut into the head. Others are held in place by tightening the locking handle (Fig 4–8).

The scales are made in a variety of lengths. They are available with different standard scales on their edges. When it is necessary to change to a different scale, the old scale is simply removed and a new scale snapped into place. Scales are usually plastic or aluminum.

## TRIANGLES

The two triangles commonly used with the T-square or parallel straightedge are the 45° and the 30°-60° triangle. They are sold in different sizes. The size given is the length of the longest side of the right angle (Fig. 4–9). Triangles are available in clear and colored plastic. On some, the edge is recessed to help keep ink from running under it. Since triangles are easily damaged, they must be used and stored with care.

## SCALES

Scales are tools used to lay out linear distances. On the edges of the tool are markings, such as inches or millimeters. These are in various spacings that permit the enlargement or reduction of a drawing. Each series of graduations is also called a scale.

Scales are made in flat, oval, and triangular shapes (Fig. 4–10). The scales may be open-divided

FIGURE 4–9 Triangles used in drafting. (Courtesy of J. S. Staedtler, Inc.)

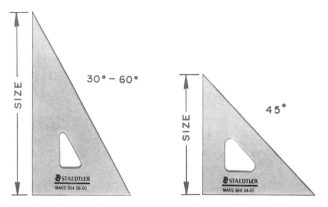

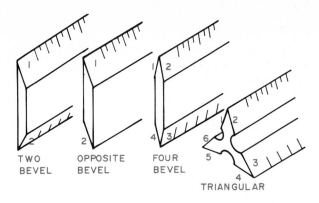

TWO BEVEL   OPPOSITE BEVEL   FOUR BEVEL   TRIANGULAR

FIGURE 4-10 Shapes of drafting scales.

or fully divided. *Open-divided scales* have a section of the scale on the end that is fully subdivided into small units. The remainder of the scale's length is marked with only the main units of measure. *Fully divided scales* have the entire length subdivided. Metric scales are usually fully divided (Fig. 4–11).

Several types of scales are available. They are used for various types of drafting. These include the architect's scale, engineer's chain scale, mechanical draftsman's scale, and metric scale. Each has different types of subdivisions (Fig. 4–11).

The *architect's scale* is used to lay out drawings of buildings and related things such as cabinets, plumbing, and electrical systems. It is used to lay out distances in feet and inches. It uses customary inches and feet as its units of measure, so it can only be used in nonmetric design.

All architect's scales are based on customary measurements. Each major division represents 1 ft. Subdivisions represent inches and fractions of an inch. Therefore, a major unit on this scale is divided into 12 subdivisions (inches). Each of these can be subdivided into fractions (½ in., ¼ in., and so on).

The scales commonly used are designated ³⁄₃₂, ⅛, ³⁄₁₆, ¼, ⅜, ½, ¾, 1, 1½, and 3. Each of these represents a foot on the scale. For example, on the ¼ scale, ¼ in. on the scale represents 1 ft on the drawing.

To read an architect's scale, begin at the 0 point. Count off the number of feet along the length of the scale. Count off the inches on the open-divided end (Fig. 4–12).

The *engineer's chain scale* (sometimes called the civil engineer's scale) is divided into decimal parts of an inch. The divisions are 10, 20, 30, 40, 50, and 60 parts of an inch. The scale marked 20 means that each subdivision on the inch is equal to ¹⁄₂₀ of an inch. It is used by engineers when making maps using customary measurements. It is also used for machine drawings in decimal fractions of an inch (Fig. 4–13). For example, the divisions on the 10 scale are each 0.10 in., while on the 50 scale they are 0.02 in.

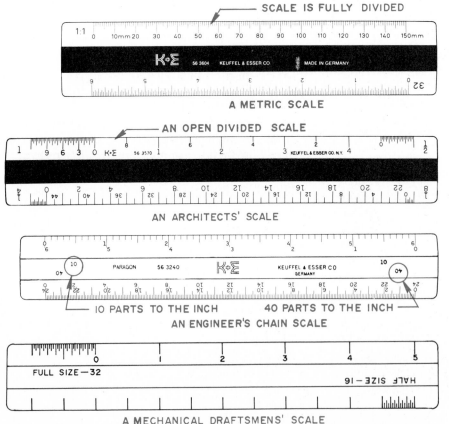

SCALE IS FULLY DIVIDED

A METRIC SCALE

AN OPEN DIVIDED SCALE

AN ARCHITECTS' SCALE

10 PARTS TO THE INCH   40 PARTS TO THE INCH
AN ENGINEER'S CHAIN SCALE

FULL SIZE—32   HALF SIZE—16
A MECHANICAL DRAFTSMENS' SCALE

FIGURE 4–11 The types of scales used in drafting. (Courtesy of Keuffel and Esser Co.)

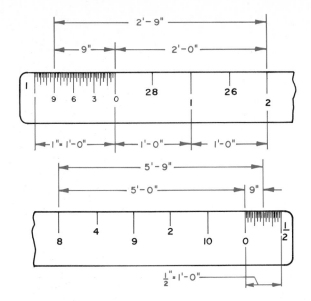

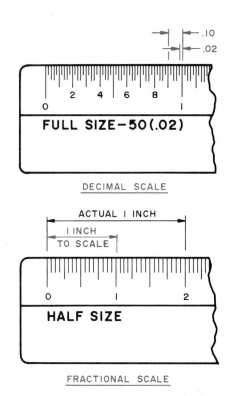

FIGURE 4–12 How to read an architect's scale.

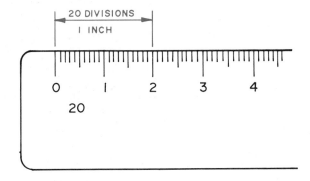

FIGURE 4–13 How to read an engineer's chain scale.

FIGURE 4–14 How to read a mechanical draftsman's scale.

FIGURE 4–15 Mechanical draftsman's scales are used to draw things at a reduced size.

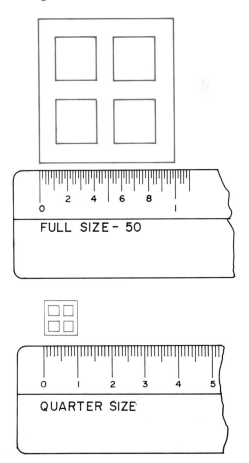

The *mechanical draftsman's scale* (sometimes called the mechanical engineer's scale) is available with fractional and decimal dimensions. The fractional scales are full, half, quarter, and eighth size. These are divided into ⅟₃₂, ⅟₁₆, ⅛, and ¼ divisions. The decimal scales are available in full, half, and quarter size. They are divided into ⅟₅₀ in. (0.02 in.) and ⅟₁₀ in. (0.10 in.) (Fig. 4–14).

These scales are used to draw products full size or to a reduced size. For example, if the quarter scale is used, the product is drawn one-fourth actual size (Fig. 4–15). A summary of the customary scales is in Fig. 4–16.

*Metric scales* are used to design products in metric units. Since engineering drawings use only millimeters, this is the basic unit of the metric scale. Metric scales are based on a ratio such as 1:10. This means that 1 mm on the scale represents 10 mm on the object. The most frequently used metric scale ratios for reduction of a drawing are 1:2, 1:3, 1:5, and 1:10. Those used to enlarge a drawing are 2:1 and 5:1. Whenever possible, engineers prefer to draw products full size (1:1) (Fig. 4–17).

| ARCHITECT'S SCALES | ENGINEER'S CHAIN SCALES | MECHANICAL DRAFTSMAN'S SCALES | |
|---|---|---|---|
| 3/32″ = 1′-0″ | 10 Scale: 1″ = 1′, 10′, 100′, 1000′ | Fractional Scales | |
| 1/8″ = 1′-0″ | 20 Scale: 1″ = 2′, 20′, 200′, 2000′ | Full | 1″ = 1″ |
| 3/16″ = 1′-0″ | 30 Scale: 1″ = 3′, 30′, 300′, 3000′ | Half | 1/2″ = 1″ |
| 1/4″ = 1′-0″ | 40 Scale: 1″ = 4′, 40′, 400′, 4000′ | Quarter | 1/4″ = 1″ |
| 3/8″ = 1′-0″ | 50 Scale: 1″ = 5′, 50′, 500′, 5000′ | Eighth | 1/8″ = 1″ |
| 1/2″ = 1′-0″ | 60 Scale: 1″ = 6′, 60′, 600′, 6000′ | Decimal Scales | |
| 3/4″ = 1′-0″ | | Full size | 1.00″ = 1.00″ |
| 1″ = 1′-0″ | | Half size | 0.50″ = 1.00″ |
| 1 1/2″ = 1′-0″ | | Quarter size | 0.25″ = 1.00″ |
| 3″ = 1′-0″ | | | |

FIGURE 4–16 Scales used in various types of engineering.

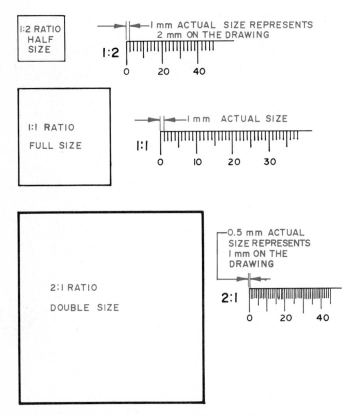

FIGURE 4–17 Metric scale ratios are used to draw things enlarged or reduced from actual size.

Metric scales may contain a single ratio, as 1:1, or two ratios on one edge (Fig. 4–18). When buying metric scales, you must select those that have the ratios you are most likely to use. They are available in 150-mm and 300-mm lengths.

A single-ratio metric scale can be used to lay out a variety of ratios. Because the metric system is on base 10, a scale ratio can provide several other ratios. For example, using a 1:1 scale, the 1 mm marking could equal 1 mm, 10 mm, 100 mm, and so on. Therefore, in Fig. 4–18 the 100-mm distance could also represent 1000 mm or 10 000 mm.

The same technique is used for other scales. For example, on the 1:2 scale, 1 mm could represent 20 mm, 200 mm, and so forth (Fig. 4–19).

Typical uses for commonly used metric scales are given in Fig. 4–20. The 1:3 scale is not widely accepted in other countries but is being used in the United States. It gives a size between the 1:2 and the 1:5 and is often useful.

## TEMPLATES

Templates are thin, flat, plastic tools that have pierced openings of various shapes and sizes. They are used to guide the pencil when drawing these shapes (Fig. 4–21). There are hundreds of different

FIGURE 4–18 A metric scale can be used to measure in several different ratios.

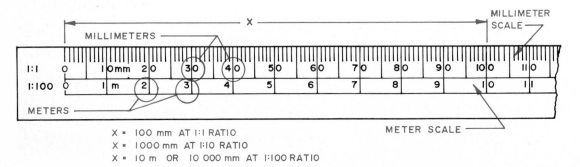

| METRIC SCALE RATIO | MEANING | OTHER APPLICATIONS |
|---|---|---|
| 1:1 | 1 mm = 1 mm (full size) | 1 mm = 10 mm, 100 mm, 1000 mm |
| 1:2 | 1 mm = 2 mm (half size) | 1 mm = 20 mm, 200 mm, 2000 mm |
| 1:3 | 1 mm = 3 mm (one-third size) | 1 mm = 30 mm, 300 mm, 3000 mm |
| 1:5 | 1 mm = 5 mm (one-fifth size) | 1 mm = 50 mm, 500 mm, 5000 mm |
| 1:10 | 1 mm = 10 mm (one-tenth size) | 1 mm = 100 mm, 1000 mm, 10 000 mm |
| 2:1 | 2 mm = 1 mm (double size) | |
| 5:1 | 5 mm = 1 mm (enlarged 5 times) | |

FIGURE 4–19 The meaning of the various metric scale ratios.

| METRIC (mm:mm or m:m) | APPLICATION |
|---|---|
| 1:1 | Machine drawings |
| 1:2 | Machine drawings |
| 1:3 | Machine drawings |
| 1:5 | Machine drawings |
| 1:10 | Machine drawings and architectural details |
| 1:20 | Architectural construction details |
| 1:25 | Architectural construction details |
| 1:50 | Architectural plans and elevations |
| 1:100 | Architectural plot plans |
| 1:200 | Architectural plot plans |
| 1:500 | Architectural site plans |
| 1:1250 | Architectural site maps |
| 1:2500 | Architectural site maps |
| 1:10 000 | Location maps |
| 1:50 000 | Location maps |

FIGURE 4–20 Uses for the various metric scale ratios.

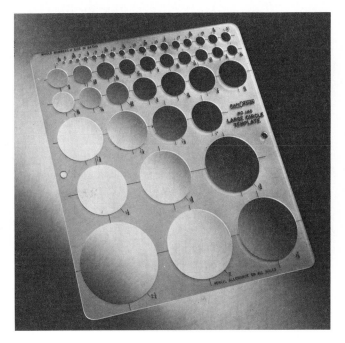

FIGURE 4–21 A typical template. This is a circle template. (Berol USA/RapiDesign.)

kinds available. Some feature standard forms such as circles, squares, and ellipses. Others make arrowheads, brackets, electrical symbols, lettering, plants, human figures, and a host of other forms. To a large extent they have replaced the drafting compass as a major tool for drawing circles.

## INSTRUMENT SETS

Typical sets of drawing instruments contain one or more compasses, a divider, and a circle-extension bar (Fig. 4–22). Compasses are used to draw circles. They have a pin in one end and a lead or inking device in the other. Compasses may be of the friction or center-wheel type. The radius on the friction type is adjusted by pulling on the legs. The center-wheel type is adjusted by rotating the center wheel (Fig. 4–23).

A divider has pins in both legs. They are usually of the friction type. Dividers are used to transfer distances from one part of a drawing to another (Fig. 4–24).

FIGURE 4–22 A typical set of drafting instruments. (Courtesy of J. S. Staedtler, Inc.)

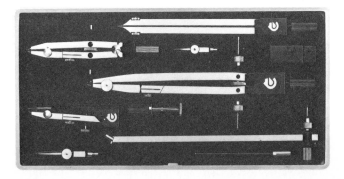

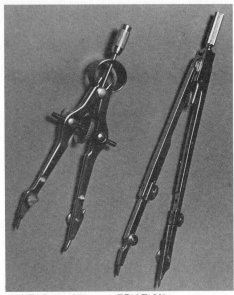

CENTER WHEEL     FRICTION
ADJUSTMENT      ADJUSTMENT

FIGURE 4–23 Center-wheel and friction-type compasses.

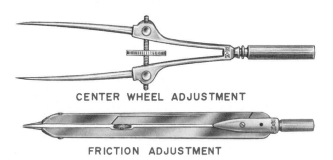

CENTER WHEEL ADJUSTMENT

FRICTION ADJUSTMENT

FIGURE 4–24 Center-wheel and friction-type dividers. (Courtesy of Keuffel and Esser Co.)

## Using a Compass

The best way to set the radius of the circle on a compass is to mark the distance on a piece of paper. Then set the pin on one mark and open the compass until the point on the other leg touches the other mark.

To draw a circle, place the pin on the drawing at the center of the circle. Do not push the pin into the paper. It should be in only far enough to keep it from slipping. Avoid punching holes in the drawing. Hold the compass at the top and rotate it clockwise. Lean the compass in the direction of rotation. Press hard enough to get a dark line (Fig. 4–25). Keep the compass moving and try to complete the circle in one rotation. If it is difficult to get a dark line, put a softer lead in the compass.

The lead in a compass is sharpened to a wedge-shaped point. Sand the point to a wedge shape. Then touch each side lightly on the sandpa-

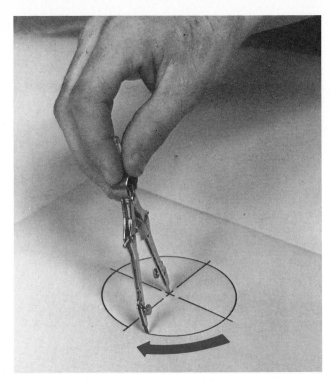

FIGURE 4–25 Hold the compass at the top and lean it in the direction it is rotated.

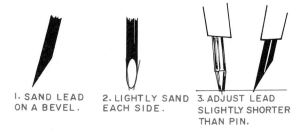

1. SAND LEAD ON A BEVEL.    2. LIGHTLY SAND EACH SIDE.    3. ADJUST LEAD SLIGHTLY SHORTER THAN PIN.

FIGURE 4–26 How to sharpen a compass lead.

per to flatten the edges (Fig. 4–26). Always sand the lead over a wastebasket. Never sand at the drawing table—the loose graphite quickly smudges a drawing. Adjust the lead so it is a little shorter than the pin. This is necessary because the pin penetrates the paper a little.

## Using a Beam Compass

A beam compass has a long horizontal extension arm. It holds a pin holder and a lead holder or ink pen. The leg with the pin is placed on the center of the circle and held with one hand. It is rotated as described for a regular compass (Fig. 4–27).

Another device for drawing large-radius curves is the adjustable ruler. This tool is used to draw arcs with radii from 6¾ to 200 in. The metric model has a range from 170 to 5000 m (Fig. 4–28).

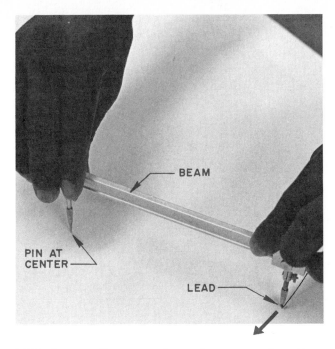

FIGURE 4–27 Drawing a large-diameter circle with a beam compass.

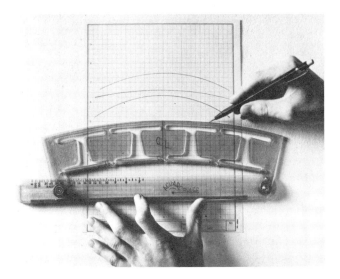

FIGURE 4–28 An adjustable ruler used to draw large-diameter arcs. (Courtesy of Hoyle Engineering Co.)

## Using a Divider

The distance to be transferred by a divider is measured on the drawing. If it is to be taken from a scale, mark the distance from the scale on a piece of paper. Then set the divider points using the marks on the paper.

The divider is held at the top with one hand. It can be used to step off equal distances. This is done by locating one point on the paper. Swing the divider and locate a second point. Holding that point

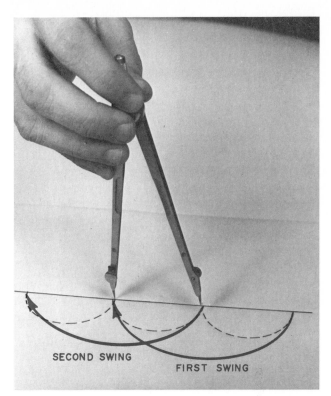

FIGURE 4–29 Using a divider to step off equal distances.

in place, swing the divider so the first point marks the third location (Fig. 4–29). Repeat this across the paper to obtain as many equal spaces as you need.

## IRREGULAR CURVES

Irregular curves are thin, plastic tools made in a wide variety of shapes and consisting of curves of various types (Fig. 4–30). They are used to draw noncircular curves. Noncircular curves are curves that do not have a center. They consist of tangent arcs that have varying radii.

FIGURE 4–30 A few of the many styles of irregular curves. (Courtesy of Keuffel and Esser Co.)

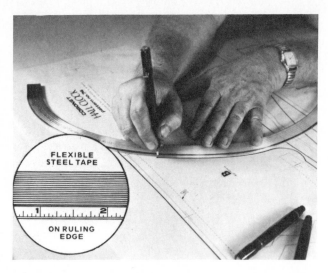

FIGURE 4–31 Drawing an irregular curve with a flexible curve. (Courtesy of Hoyle Engineering Co.)

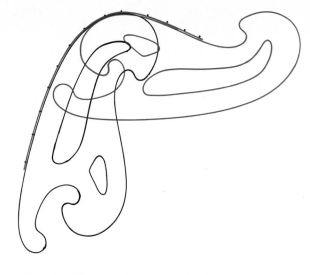

FIGURE 4–32 How to draw an irregular curve.

Some sets of irregular curves are designed for general use, while others are for special use. Special-use curves include ship's curves, railroad curves, mechanical engineering curves, and teardrop-radius curves.

Another type of curve is the flexible curve. Flexible curves are made from metal, plastic, or rubber and can be bent to fit an irregular curve (Fig. 4–31).

A curved line on a drawing that is irregular is usually located by a series of points. These are connected by using the irregular curve tool. Study the path formed by the points. Examine the curves available and select those having curves near to the path of the points. Place the curve next to the points. Try to connect as many points as possible, but never less than three. Draw a line through these points. Move the tool along the points and connect additional points (Fig. 4–32).

## DRAWING PENCILS

There are three types of drawing pencils in use: the wood-cased pencil, a mechanical lead holder that holds leads of the same diameter as those used in wood-cased pencils, and a fine-line lead holder (Fig. 4–33).

Wood-cased pencils come in various degrees of hardness, from 9H, very hard, to 6B, very soft (Fig. 4–34). These leads are graphite and are used on all types of drafting papers, vellum, and tracing paper. The diameter of the lead gets larger as the lead gets softer.

Leads in the B to 4B range are used for shading and pencil renderings. F and HB leads are commonly used for freehand sketching. H and 2H leads are used for finished lines on drawings. They find the greatest use. The harder leads, 3H and harder, are used for layout work that must be extremely light.

Leads used on polyester drafting film are made of a special plastic substance. They are water-resistant and do not smear easily. A special vinyl eraser is used to remove them. Plastic leads are available in five degrees of hardness. These are E1, soft; E2, medium; E3, hard; E4, extra hard; and E5, super hard.

The mechanical lead holder uses leads having the same degrees of hardness as wood-cased pencils. The leads are slid into the pencil. They are sharpened just like wood-cased pencils (Fig. 4–35).

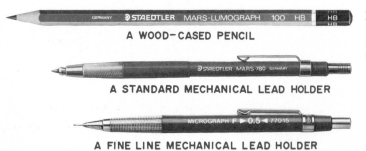

**A WOOD-CASED PENCIL**

**A STANDARD MECHANICAL LEAD HOLDER**

**A FINE LINE MECHANICAL LEAD HOLDER**

FIGURE 4–33 The types of pencils commonly used in engineering drafting. (Courtesy of J. S. Staedtler, Inc.)

PLASTIC LEADED PENCILS

GRAPHITE LEAD PENCILS

FIGURE 4–34 The grades of graphite and plastic-leaded pencils.

FIGURE 4–35 Leads for mechanical drafting pencils are available in packets. (Courtesy of J. S. Staedtler, Inc.)

The fine-line automatic lead holder uses leads that are very small in diameter. The sizes commonly used are 0.3 mm, 0.5 mm, and 0.7 mm. They use the same degrees of hardness as wood-cased pencils. These pencils do not have to be pointed. Since the lead is so small, it always produces the desired line size. As the lead is used, more lead is pushed out by pressing a button on the end of the lead holder.

## Sharpening Drafting Pencils

Wood-cased pencils have the wood cut away with a pencil sharpener that has draftsmen's cutters. The cutters remove the wood but do not cut the lead (Fig. 4–36). The lead is pointed with some type of lead pointer (Fig. 4–37). The pointer has a fine abrasive-paper lining or steel cutters. The point is placed in the pencil holder and the top of the pointer is rotated. This rubs the lead against the abrasive paper, forming a cone-shaped point (Fig. 4–38). For some jobs drafters like to use a wedge point. Wedge points are formed using sandpaper.

Do not form conical points with sandpaper.

FIGURE 4–36 Wood-cased pencils are sharpened by first cutting away only the wood.

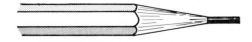

FIGURE 4–37 Pencil leads are pointed in a lead pointer. (Courtesy of J. S. Staedtler, Inc.)

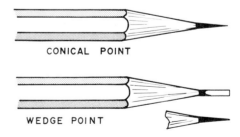

CONICAL POINT

WEDGE POINT

FIGURE 4–38 The two ways to point drafting pencil leads.

This allows graphite to get on the board, on the floor, in the air, and on your hands. It becomes very difficult to keep a drawing clean.

## Making Quality Lines

It is important that pencil lines be the proper width and as black as possible. The width is important because it is part of the correct line symbol. Thin lines are drawn 0.015 to 0.022 in. (about 0.4 to 0.6 mm) thick. Thick lines are drawn 0.030 to 0.038 in. (about 0.8 to 1.0 mm) thick. Pencil drawings tend to be made close to the thinner dimension. Ink drawings approach the thicker sizes. Lines must be black so that they reproduce easily when copies are made. All lines, thick or thin, are drawn as black as possible. Just because a line is thin does not mean it is drawn light. It must be made very dark.

Most people prefer an H or 2H pencil for finished lines. If it is difficult to get dark lines, a softer pencil, such as an F, could be used. Seldom should anything softer be used because the lines would smear.

The forming of the line symbols must be carefully observed. The length of dashes and spaces between dashes should be the proper size and uniform in length. Dashes must start and stop with a uniform width.

It is considered best technique to draw each line to its proper width with the first stroke. If this cannot be done, go back over them again. Be careful when redrawing a line to get the second try directly on top of the first. Keep the width consistent as you try for increased darkness.

Rotate your pencil as you draw. This keeps the cone-shaped point round. It prevents lines from starting thin and ending up thick because of a flat spot forming on the bottom of the lead (Fig. 4–39). Rotation is not necessary when using fine-line mechanical pencils because the lead is very small and uniform in diameter.

To draw thin lines, sharpen the point. Then wear it down a little on a piece of scrap paper until it produces the line width desired. To draw thick lines, dull the point even more until the desired width is reached.

Keep the pencil close to the edge of the straightedge. Hold the pencil in a comfortable position about 25 mm from the end. Keep it at right angles to the paper and slant it about 60° in the direction the line is being drawn (Fig. 4–40). Fine-line

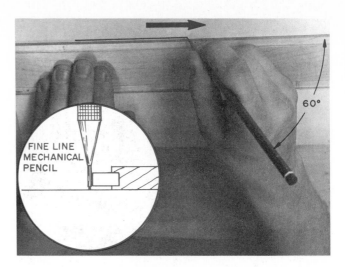

FIGURE 4–40 Slant the pencil to about 60°.

mechanical pencils are held almost perpendicular to the paper. If slanted, their metal tip will score the paper.

Do not press down too hard to get a dark line. Use a softer pencil. Excess pressure will produce grooves in the paper that cannot be removed if a change is needed.

## INKING INSTRUMENTS

Considerable inking is done in certain industries. It is fast and permanent, especially on plastic drafting film. The ink used is a black, waterproof drawing ink. There are several kinds available, including one for use on plastic drafting film. Inks with different drying times are also available. Waterproof drawing inks come in a variety of colors.

Lines are inked using a technical fountain pen (Fig. 4–41). This pen has a variety of point sizes. The sizes correspond to ANSI line-width designations. They are also coordinated in metric sizes (Fig. 4–42).

The ink is held in a cartridge in the handle. It can be used a long time before it needs to be refilled. When the pen is not in use, keep the cap screwed over the tip to prevent the ink from hardening (Fig. 4–43). If it hardens, a pen-cleaning fluid can be forced through the point with a pressure bulb, or the point can be placed in an ultrasonic cleaner (Fig. 4–44). It is best if the point is not taken apart, as the fine feed wire can be damaged.

To start the flow of ink, hold the point down

FIGURE 4–41 A technical fountain pen. (Courtesy of J. S. Staedtler, Inc.)

FIGURE 4–39 When drawing lines, rotate the pencil to get a uniform line width.

ROTATE PENCIL

UNIFORM WIDTH LINE

.13/5x0  .18/4x0  .25/3x0  .30/00  .35/0  .45/1  .50/2  .70/2½  .80/3  1.0/3½  1.2/4  1.4/5  2.0/6

● ISO LINE WIDTHS

LINE WIDTHS CODE: .13/5X0 MEANS LINE WIDTH
IS .13 mm OR AMERICAN STANDARD SIZE 00000.
METRIC WIDTHS FROM .13 TO 2.0 mm
AMERICAN STANDARD WIDTHS FROM 000000 TO 6.

FIGURE 4–42  Standard ink-line widths. (Courtesy of
J. S. Staedtler, Inc.)

CAP          POINT          GRIP          INK CARTRIDGE          HANDLE

FIGURE 4–43  The parts of a technical fountain pen.
(Courtesy of J. S. Staedtler, Inc.)

FIGURE 4–44  An ultrasonic cleaner for cleaning ink-
pen points. (Courtesy of Keuffel and Esser Co.)

over a piece of scrap paper. If the ink does not
flow in this position, shake the pen up and down
gently. The clicking noise you will hear is the flow-
regulating device sliding up and down. This causes
ink to flow if the pen is slightly clogged.

### Inking Drawings

Ink will tend to run under triangles and straightedges. To prevent this, use tools that have a recessed edge, or tape thin pieces of cardboard to the
bottom of them. This raises them above the surface.

If the ink runs under and smears, allow it to
dry. Then erase with erasers designed for that purpose. Do not use the typical gritty ink eraser. It
damages the drafting paper and makes it impossible
to ink over the spot.

After inking a line, let it dry normally. Do not
use a blotter. This removes much of the ink from the
line. If an ink line is shiny, it is still wet; if it is dull,
it is dry.

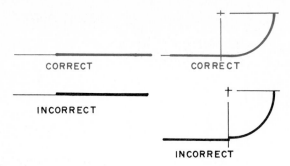

FIGURE 4–45 Ink lines should be centered on the pencil construction lines.

FIGURE 4–46 The proper way to hold a technical fountain pen.

If inking over pencil construction lines, center the ink line on the pencil line. This will enable ink lines to meet smoothly at corners or curves (Fig. 4–45).

Keep a piece of scrap paper handy. Before placing the pen on the paper, touch it lightly to the scrap paper. This will show you if it is flowing correctly.

The pens operate best if kept reasonably full of ink. Hold the pen perpendicular to the paper. When it touches the paper, move it immediately in the direction of the line. When the line is finished, lift it vertically off the paper (Fig. 4–46).

### Order of Inking

Before starting to ink, plan how you are going to proceed. Here is a suggested order.

1. Ink all center lines. Do the circular lines first, then the horizontal lines, then the vertical straight lines.
2. Ink all arcs and circles.
3. Ink all horizontal visible lines.
4. Ink all vertical visible lines.

5. Ink all inclined visible lines.
6. Ink all hidden lines.
7. Ink section lines, extension lines, and dimension lines.
8. Ink arrowheads.
9. Ink all dimensions, notes, and titles.

## ERASERS

Erasers are designed for particular types of leads or ink and different drafting surfaces. Soft rubber erasers are used on drafting papers and vellums. Vinyl erasers are used to remove plastic lead lines from plastic sheets. Ink can be removed from plastic sheets with several erasers designed for this purpose. One type has an ink solvent incorporated into the eraser.

Erasers are available in block and stick form (Fig. 4–47). Electric erasers may be cordless and run on rechargeable batteries or connected to a 120-volt outlet. Various kinds of erasing materials are made in long, round strips. They are inserted inside the eraser shaft (Fig. 4–48).

FIGURE 4–47 Block and stick erasers. (Courtesy of J. S. Staedtler, Inc.)

FIGURE 4–48 A cordless electric eraser. (Courtesy of Keuffel and Esser Co.)

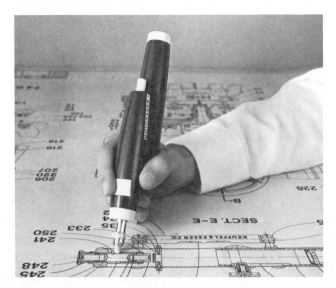

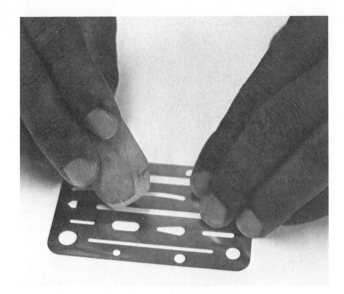

FIGURE 4–49 Using an erasing shield.

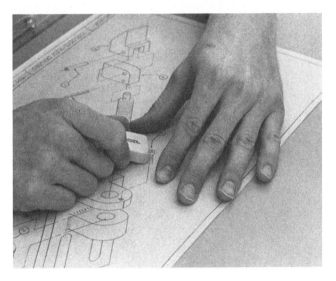

FIGURE 4–50 When erasing, hold the paper tightly between two fingers so that it will not wrinkle.

An erasing shield is used when erasing a small area. It protects the lines around the area being erased (Fig. 4–49). It helps if you hold the paper tight between two fingers in the area being erased. This prevents the sheet from buckling and creasing (Fig. 4–50).

## CLEANLINESS

Cleanliness is important in drafting. Never sharpen a pencil over a table. Wipe the pencil point clean after it is sharpened. Keep a soft, clean cloth available at all times. Wipe your triangles, straightedges, drafting machine scales, irregular curves, and other tools frequently. They pick up graphite as they are moved across drawings. Occasionally, wash them

FIGURE 4–51 A drafting brush. (Courtesy of J. S. Staedtler, Inc.)

with a mild soap and lukewarm water and dry them thoroughly.

Some drafters like to use a drafting brush. It is used to remove eraser crumbs from the surface of a drawing without smearing the lines (Fig. 4–51).

You can cover the completed part of a drawing with a clean piece of paper. This prevents instruments from smearing finished lines.

Some people use eraser crumbs to clean a drawing. These are available in small bags. The crumbs are powdered over the drawing and lightly rubbed. They are brushed away with a drafting brush. While this cleans a drawing, it does reduce the darkness of the lines at the same time.

## DRAWING LINES

Drawings are made up of horizontal, vertical, and inclined lines as well as regular and irregular curves.

### Drawing Straight Lines

Horizontal lines are drawn with the top edge of a T-square or parallel straightedge or the scales on drafting machines.

To use a T-square, place the head on the left side of the board. Left-handed people prefer to place it on the right side. A right-handed person holds the head against the table with the left hand. Left-handed persons hold the head with the right hand. The T-square is slid up and down the board with this hand. Use the same hand to move a parallel straightedge up and down the table top.

Right-handed persons draw horizontal lines from left to right. Left-handed persons draw them from right to left (Fig. 4–52). Slant the pencil about 60° in the direction the line will be drawn (a fine-line automatic pencil is held upright).

On a drafting machine, the horizontal scale is used to draw horizontal lines (Fig. 4–53). The head is held in one hand as the line is drawn.

### Drawing Vertical Lines

When using a T-square or parallel straightedge, it is necessary to use triangles to draw vertical lines. Hold the triangle to the straightedge with one hand. Slant the pencil about 60° in the direction the line is to be drawn (unless using a fine-line automatic pen-

FIGURE 4–52 Right-handed people draw horizontal lines from left to right.

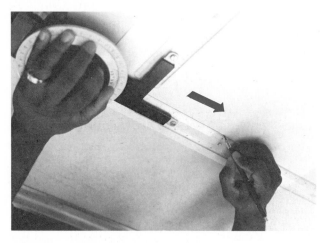

FIGURE 4–53 Drawing a horizontal line with a drafting machine.

FIGURE 4–54 Right-handed people draw vertical lines from bottom to top.

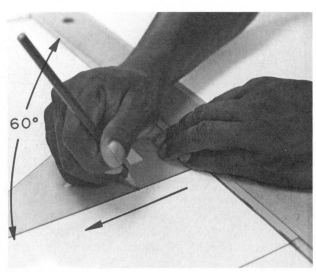

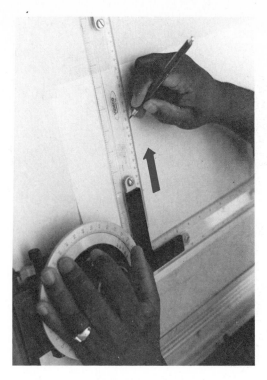

FIGURE 4–55 Vertical lines are drawn using the vertical scale on the drafting machine.

cil). Right-handed persons draw vertical lines from bottom to top. Left-handed persons often prefer to draw them from top to bottom (Fig. 4–54).

The vertical scale on a drafting machine is used to draw vertical lines. The procedure is the same as when using triangles (Fig. 4–55).

## Drawing Inclined Lines

Inclined lines are drawn with the triangle against the T-square or parallel straightedge or by rotating the scales on a drafting machine.

Lines sloping down to the right are drawn in that direction. If sloping down to the left, they are drawn up the slope (Fig. 4–56). The same procedure is used with the drafting machine (Fig. 4–57).

## Drawing Angles

Angles are located and drawn with triangles, a protractor, or the vernier scale on a drafting machine. Triangles can be combined in a variety of ways to locate some angles (Fig. 4–58). With the drafting machine, simply rotate the head while reading the angle indicator on the scale. Lock the head in place and draw the angle.

A protractor is marked in degrees (Fig. 4–59). Place the center point at the bottom of the tool on the point from which the angle is to run. Read the angle on the scale. Mark it with a pencil. Remove the protractor and draw the angle with a triangle.

FIGURE 4–56  How to draw inclined lines.

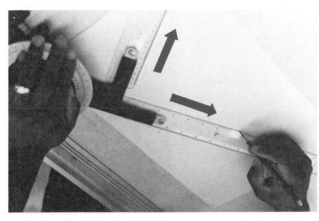

FIGURE 4–57  To draw inclined lines with a drafting machine, rotate the head so that the scales are on the desired angle.

FIGURE 4–58  How to draw commonly used angles by combining the triangles.

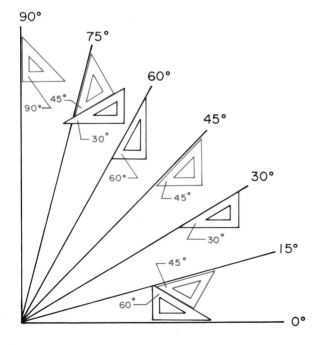

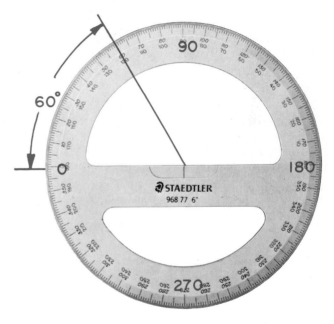

FIGURE 4–59  This protractor contains 360°. Some contain only 180°. (Courtesy of J. S. Staedtler, Inc.)

## Drawing Lines Perpendicular to Each Other

A triangle locates a line perpendicular to a horizontal line drawn with a T-square or parallel straightedge (Fig. 4–60). The scales on a drafting machine are at 90° to each other. To draw a line perpendicular to a horizontal line, move the head until the vertical scale crosses the horizontal line (Fig. 4–61).

To draw a line perpendicular to an inclined line, place a triangle along the inclined line. Place another triangle on the edge of the first triangle. The 90° corner will produce a perpendicular (Fig. 4–62). To do this with a drafting machine, place one scale parallel with the inclined line. Lock the head. Move the head so that the other scale crosses the inclined line. Draw the perpendicular (Fig. 4–63).

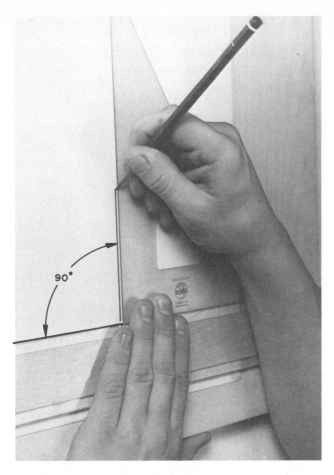

FIGURE 4–60 How to draw a line perpendicular to a horizontal line with a triangle and T-square.

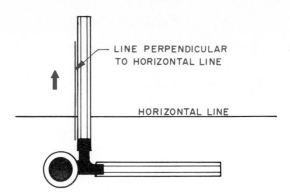

FIGURE 4–61 Drawing perpendicular lines with a drafting machine.

FIGURE 4–62 A perpendicular can be drawn to an inclined line using two triangles.

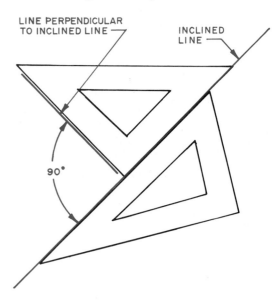

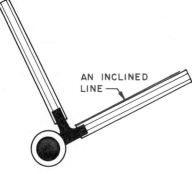

FIRST PLACE A SCALE PARALLEL TO THE INCLINED LINE.

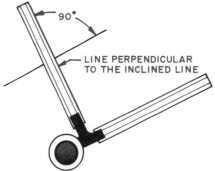

THEN MOVE THE DRAFTING MACHINE UNTIL THE OTHER SCALE CROSSES THE INCLINED LINE. IT IS PERPENDICULAR TO IT.

FIGURE 4–63 Drawing a line perpendicular to an inclined line with a drafting machine.

## Drawing Parallel Lines

Place a triangle along the line. Place a second triangle or a straightedge below this one. Slide the first triangle along the second until the location of the parallel line is reached. Draw the line (Fig. 4–64).

Using a drafting machine, rotate the head until a scale is parallel with the line. Lock the head. Wherever you move the head, you will produce parallel lines (Fig. 4–65).

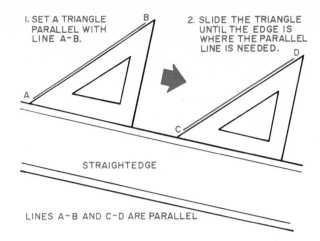

FIGURE 4–64 Drawing parallel lines with triangles and a straightedge.

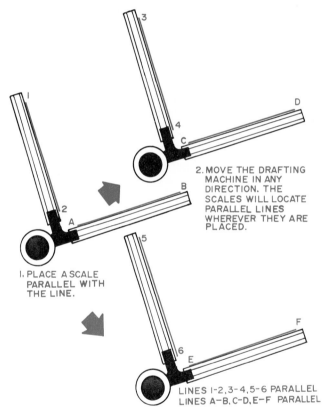

FIGURE 4–65 Drawing parallel lines with a drafting machine.

## DRAFTING PAPERS, FILM, AND GRIDS

Drafting and tracing materials are available in three forms: paper, cloth, and film.

### Papers

Drafting papers are available in a wide range of qualities in terms of strength, erasability, permanence, translucency, cost, and stability.

Opaque papers are referred to as drawing paper. They are usually a heavy paper, white, cream, or light green in color. They are not used a great deal in industry. Drawings made on opaque papers cannot be reproduced. Some papers are smooth on one side and a little rougher on the other. The smooth side is best for inking. The rough side is best for pencil drawings. Papers are sold in sheets.

The translucent papers are of two types: natural tracing paper and vellum or treated tracing paper.

Natural tracing paper is a thin, untreated, natural, translucent paper. If it is made for strength, the translucent properties suffer; if made to emphasize translucency, the strength suffers.

Vellum tracing papers are treated papers that have strength and good translucency. They have a high rag content for good strength. They withstand erasing without leaving marks. Vellum tracing papers can be used for both pencil and ink drawings.

Tracing papers are sold in sheet and roll form.

### Cloths

Tracing cloth is a fabric that has been treated to be translucent. It often has a slight tint. Some types are best for inking, while others are best for pencil work. Usually, the working side is dull and the other side is glazed. Tracing cloths are strong and will last for years. They are sold in sheet and roll form.

### Films

Polyester drafting film is a tough, translucent plastic drafting material. Sheet thicknesses include 0.002, 0.003, 0.004, 0.005, and 0.0075 in. or 0.05, 0.08, 0.10, 0.14, and 0.19 mm. The sheets may be single or double matte. The dull or matte side is used for drawing. A single-matte sheet has one dull side for drawing and one glossy side; double-matte film is dull on both sides.

The film is used for pencil and ink drawings. Plastic-leaded pencils are used. A special ink is also available. Special erasers are required.

This material is very strong and does not shrink or stretch. It is sold in sheet and roll form.

### Drafting Sheet Sizes

Drafting papers, tracing paper, and vellum and polyester films are sold in sheets and rolls. The standard sizes are given in Fig. 4–66, and the most commonly used roll sizes are in Fig. 4–67. Drafting cloth is sold in 20-yd (18 288 mm) rolls in 30-, 36-, and 42-in. widths. If sheets are wanted, they are cut to size on order.

| U.S. CUSTOMARY | | | ISO METRIC | |
| --- | --- | --- | --- | --- |
| | Size (inches) | | | |
| Type | Engineer's | Architect's | Type | Size (millimeters) |
| A | 8½ × 11 | 9 × 12 | A4 | 210 × 297 |
| B | 11 × 17 | 12 × 18 | A3 | 297 × 420 |
| C | 17 × 22 | 18 × 24 | A2 | 420 × 594 |
| D | 22 × 34 | 24 × 36 | A1 | 594 × 841 |
| E | 34 × 44 | 36 × 48 | A0 | 841 × 1189 |

FIGURE 4–66    Sizes of drafting sheets.

| | |
| --- | --- |
| Width | 30″ (762 mm), 36″ (914 mm), 42″ (1067 mm) |
| Length* | 20 yd (18 288 mm), 50 yd (45 720 mm) |

*All widths are available in both lengths.

FIGURE 4–67    Sizes of rolls of drafting media.

## REPRODUCTION OF DRAWINGS

After the drawings have received approval of the checker and anyone else responsible for the final decision, copies are made for use in production. Often, the parts produced are made in several plants in cities or countries other than where the design was completed. This means that clear, durable copies are needed. It is possible to produce copies of the original drawings from which prints can be made. Following is a discussion of some of the most commonly used products.

### Whiteprints

Whiteprints are copies of the original on heavy white paper with the drawing reproduced in blue or black lines. The reproduction paper is coated with a diazo-compound coating that is light-sensitive. This process is often called the *diazo process.* There are two diazo processes, dry and moist (Fig. 4–68).

To produce a copy, the original drawing is placed image side up on the top of the diazo coating. Together they are passed through a whiteprint machine, which exposes them to ultraviolet light (Fig. 4–69). The light penetrates the translucent paper of the original drawing, hitting the diazo coating. This destroys the coating. Wherever there is an image on the drawing, the light does not strike the coating, so it remains.

In the *dry process,* the exposed diazo paper is then fed into an atmosphere of heated ammonia vapor. The vapor develops the diazo coating, producing lines that form the image on the reproduction

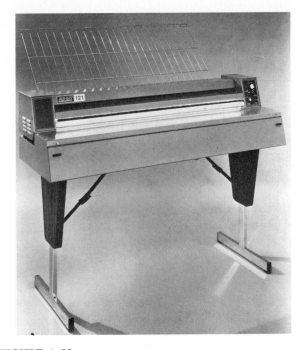

FIGURE 4–68    A whiteprinter.

FIGURE 4–69    How a whiteprinter produces copies of engineering drawings.

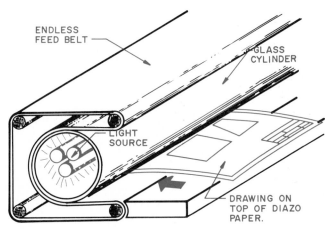

1. EXPOSE THE ORIGINAL DRAWING AND THE DIAZO PAPER TO A LIGHT SOURCE.

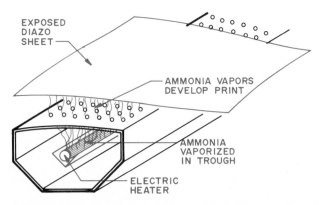

2. THE EXPOSED DIAZO SHEET IS DEVELOPED BY PASSING IT THROUGH AMMONIA VAPORS.

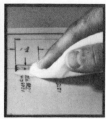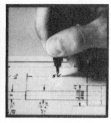

**Just apply TeleGel, dab away image, and redraw!**

FIGURE 4-70 This polyester drafting film is used in ammonia whiteprint equipment to produce copies of original drawings. Corrections can be made with a special fluid. (Courtesy of Teledyne Post.)

(Fig. 4–69). The colored lines appear on the white background.

The *moist process* exposes the original to the diazo-coated paper in the same manner as the dry process. The exposed paper is then fed through a series of rollers in an alkaline coupler, which develops the diazo coating. The moist print is dried when it passes over electric heaters as it leaves the machine.

Another product, TeleTex, is a polyester film upon which the original drawing is reproduced using standard ammonia whiteprint equipment. The image is on a matte-finish surface that will also accept ink and pencil. If a change is desired, the image is removed with a special fluid, TeleGel (Fig. 4–70). Corrections can then be drawn. This material produces a durable reproduction of the original that can be changed with little difficulty.

## Intermediates

An intermediate is a duplicate of the original drawing. The intermediate copy, being translucent, is used to produce additional copies. This preserves the original, which can be filed as a permanent record.

Diazo intermediates are formed on polyester film and developed with ammonia in the same manner as dry diazo prints. Changes can be made on the intermediate without changing the original. Lines can be drawn on it in ink or pencil and erased, just as the original drawing. A special image-release fluid is used when removing large portions of the drawing. The image is produced in black, sepia (dark brown), or red-orange, all of which are excellent light-blocking colors. A custom design using the original design as a beginning can be accomplished by altering the intermediate. Large sections can be cut away and a revised section taped in place, or another intermediate can be run and changes drawn on the cleared area. Intermediates are used to im-

prove old, faded, and worn originals. The intermediate is superior in quality to the original drawing.

Intermediates are also produced using photographic techniques. The original drawing is photographed, and the negative or microfilm copy is used to produce the intermediate. The *negative* is exposed to a special film using an enlarger to produce a full-size copy. This copy can be used to produce additional prints by the diazo process, Xerox, or other common reproduction methods. Corrections or design changes can be made on it, so it can be altered and then used to produce copies. One system removes parts of the drawing by opaquing them out. Another uses a special eradicator fluid that removes the image. A third system requires that the portion to be changed be cut away, a new print run with the changed area blank, and corrections drawn on it.

## Electrostatic Reproductions

Electrostatic reproductions of engineering drawings produce black-line images on a white background. This is a dry process that produces no fumes or odor. The process uses a selenium-coated plate that is given a positive electrical charge. The image on the drawing is projected through a lens onto the plate, where it is retained as a positive electrical charge. The charge is drained away from the rest of the plate. A negatively charged powder, called toner, is spread over the selenium-coated plate. It sticks to the positively charged image. A sheet of paper with a positive charge is placed over the plate. The powder image is attracted to the positively charged paper. It is heated, fusing the powder to the paper to form a reproduction of the engineering drawing.

A high-speed reproduction unit is shown in Fig. 4–71. It can be used to create a new drawing using scissors drafting techniques. It can produce reverse images, produce a drawing with unwanted images removed, reformat a drawing, create a finished drawing from composite overlays, and produce enlarged or reduced copies of existing drawings.

Engineering drawings and reports can be transmitted over long-distance xerography networks using special telephone lines, coaxial cable, or microwave transmission. High-speed transmitting and receiving equipment enables images to be sent from plant to plant with little delay. The transmissions are made and received with a xerographic transceiver. It converts the image on the original drawing to video transmission signals. The printer receives these signals and produces a black line on white background copy (Fig. 4–72).

Another product uses electrostatic reproduc-

FIGURE 4–71  This electrostatic printer produces copies of drawings up to 24 in. wide and 25 ft long. It will reproduce on bond, colored papers, vellum, and other transparent papers. It requires no special venting or plumbing. (Courtesy of Xerox Corporation.)

FIGURE 4–72  This transceiver operates without an attendant, receiving or sending documents by telephone automatically at the rate of one letter-size page every 2 min. It records the image on plain paper. (Courtesy of Xerox Corporation.)

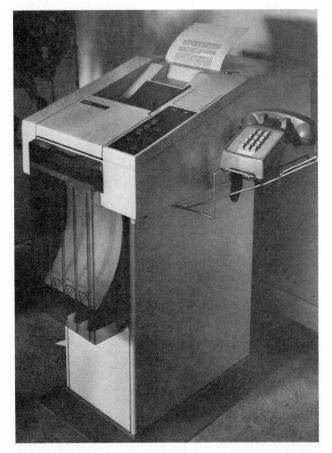

tion to produce drawings on pressure-sensitive appliqué sheets. A detail that is frequently used on many drawings is placed on the platen of the electrostatic machine. A Stanpat® polyester sheet is placed in the paper tray. The electrostatic machine reproduces the image of the drawing on the polyester sheet. A paper cover over the adhesive back is peeled away and the reproduction is attached to the new drawing being developed (Fig. 4–73).

## Adhesive-backed Reproductions

Much drafting time can be saved by reproducing symbols, screws, and other standard parts of drawings on a diazo-sensitized polyester medium with an adhesive back. The symbols are developed on the polyester material, the back is peeled off, and the reproduction is stuck to the original drawing. This greatly speeds drafting time.

## Microfilm

A microfilm is a small negative containing a drawing or other information such as design data or specifications. It is produced by photographing the original drawing or data sheets with a microfilm camera. The exposed film is developed in a film processor (Fig. 4–74). This produces the negative containing the image of the drawing. The developed negative is inspected and mounted in an aperture card. An aperture card is a computer card with a window cutout in which the negative is mounted. This card is about $83 \times 188$ mm and is called a *tab card* (Fig. 4–75).

## STEP 1

Fan the drafting sheets before placing them in the paper tray. If there is a slight curl in the sheets, bend the sheets in the opposite direction to flatten them.

## STEP 2

Place the sheets in the tray with two precautions: The free edge (the edge without adhesive) should be the trailing edge, and the sheets should be placed in the tray so that the image is fused to the white side of the material (not the blue).

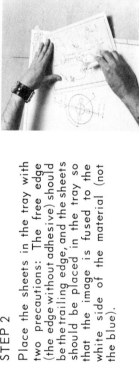

## STEP 3

Raise the platen cover and put the original face down on the document glass. Close the cover. Set the Print Selector on 1 and push the PRINT button.

## STEP 4

Using the free edge, peel the backing off. To apply the item to the drawing, place one side down first and then roll the item down onto the drawing.

## STEP 5

Using a SOFT CLOTH, burnish the STANPAT down rubbing towards the edges.

**FIGURE 4–73** This product will provide reproductions of drawing details on a polyester sheet with an adhesive backing. (Courtesy of Stanpat Products Inc.)

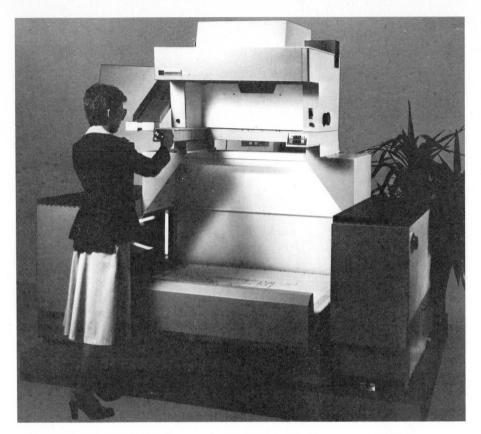

FIGURE 4–74 A microfilm processor camera. It has three reduction ratios: 16×, 24×, and 30×. The copy board permits up to fifteen 8½ × 11 in. single-sided documents to be filmed simultaneously. (Courtesy of 3M.)

FIGURE 4–75 A tab-size aperture card. (Courtesy of Microseal Corporation.)

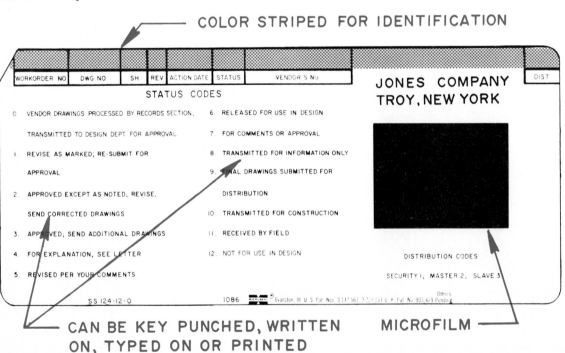

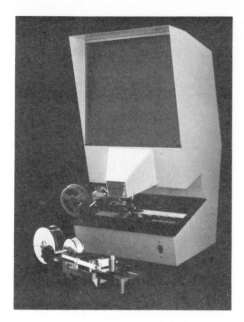

FIGURE 4–76 The microfilm negative and the aperture card are joined with this viewer-inserter. (Courtesy of Microseal Corporation.)

Before the negative is mounted, the aperture card is keypunched with indexing data for automated filing, sorting, and retrieval. It also has printed information for visual identification. This enables a user to read the facts about the material on the card without enlarging the image on a viewer (Fig. 4–76).

Microfilm can also be mounted in *fiche-size* aperture cards. They measure 105 × 148 mm. Fiche-size cards makes the microfilm system compatible with high-speed batch duplicators, fiche readers, and reader-printers (Fig. 4–77).

FIGURE 4–77 Fiche-size aperture cards are larger than tab-size. (Courtesy of Microseal Corporation.)

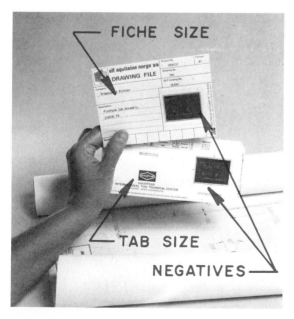

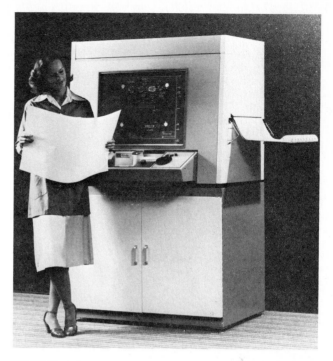

FIGURE 4–78 A reader-printer creates an enlarged view of the drawing on a screen and can print a hard copy. This unit produces full-size 18 × 26 in. and half-size 18 × 13 in. prints. (Courtesy of 3M.)

There are several ways to retrieve the information from the microfilm. The aperture card could be placed in a microfilm reader. This projects on a screen an enlarged image that can be easily read. The image can be enlarged and printed as a hard copy with a reader-printer (Fig. 4–78). A card duplicator can be used to produce additional aperture cards. These can be mailed to other locations, where engineers can read or reproduce the image on the negative.

Microfilm systems are available in 16-, 35-, and 105-mm sizes. It is possible to combine 35-mm and 16-mm negatives on a single aperture card. The 35-mm negative carries the drawing and the 16-mm negative contains the related documents such as change orders or bills of material.

One big advantage to a microfilm system is that a large number of drawings can be stored in a small file drawer (Fig. 4–79).

*Drawing for microfilm reproduction* When a drawing is prepared for microfilming, special attention is given to line uniformity, line density, and lettering.

It is important that all lines of a particular kind be of a standard width. Inked lines produce the best microfilm because the ink produces a line with uniform density.

Lettering must be large enough to be reduced photographically and enlarged with no loss of clar-

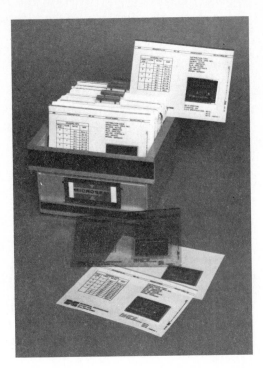

FIGURE 4–79 Microfilm aperture cards are stored in small file drawers. One drawer can hold hundreds of large drawings. (Courtesy of Microseal Corporation.)

FIGURE 4–80 A microfiche contains many pages of material.

ity. Recommended lettering sizes are given in the unit on lettering.

All letters must be carefully formed. The open spaces in a letter must be clear. Letters should not be crowded, or they may blend together when reduced. The space between lines of lettering should be at least half the height of the letter. Leave even more space between paragraphs.

If dimensions and notes are to be typed, use a ribbon that produces a dense, uniform image. Beware of ribbons that apply too much ink to the paper. Carbon ribbons are usually the best. The typewriter must be adjusted so that each image is perfectly clear.

Be careful when erasing. All smudges must be removed because they will photograph. Try to keep the background uniform. If an area has been heavily erased, it might be cleaner than the rest. Some drafters attempt to dirty this area slightly so that it matches the rest of the drawing.

Always store drawings to be microfilmed flat. Draw on flat sheet stock, not on sheets cut from rolled stock.

ED 069 919    MICROFICHE COLLECTION OF CLEARINGHOUSE DOCUMENTS REPORTED IN ABSTRACTS OF INSTRUCTIONAL MATERIALS IN VOCATIONAL AND TECHNICAL EDUCATION (AIM), VOLUME 6, NUMBER 1. OHIO STATE UNIV., COLUMBUS. CENTER FOR VOCATIONAL AND TECHNICAL EDUCATION 72    20699P.

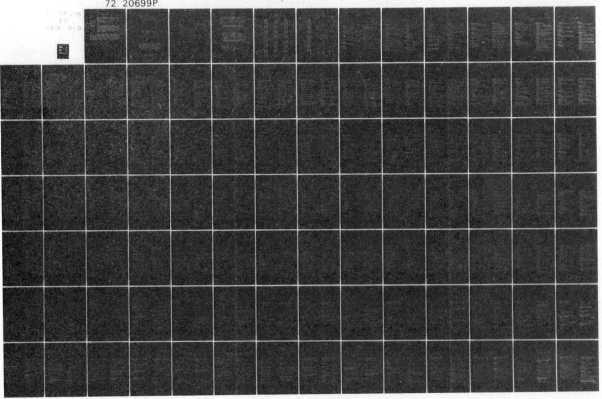

FIGURE 4–81 A microfiche reader-printer permits each page to be read on a screen. It also produces a hard copy of each page. (Courtesy of 3M.)

## Microfiche

A microfiche is a 4 × 6 in. microfilm negative that contains ninety-eight 8½ × 11 in. pages (Fig. 4–80). These are used for reading and storing specifications, drawings, instructions, and other data. They are easily stored in small file drawers. Thousands of drawings and related documents can be stored in one file drawer.

The pages on a microfiche are enlarged to original size with a microfiche reader-printer. The image can be read on the screen, or a hard copy can be produced (Fig. 4–81).

## FILING DRAWINGS

It is necessary to store original drawings and reproduced copies. The common type of files in use are flat drawers, hanging files, and roll files (Fig. 4–82). A system for coding files and indexing them and the storage units is important. It should be possible to find any set of drawings with little searching.

Microfilm aperture cards are usually stored in drawers.

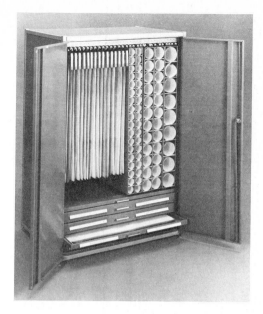

FIGURE 4–82 A typical drawing storage unit. It shows the three commonly used methods: hanging, flat drawers, and tubes. (Courtesy of Plan Hold.)

FIGURE 4–83 Recommended ways to fold engineering drawings.

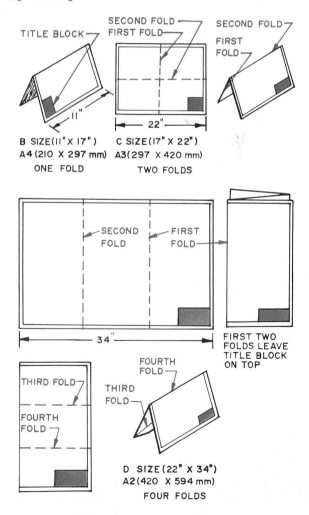

## FOLDING DRAWING REPRODUCTIONS

The original drawing is never folded. This would seriously impair its reproduction capabilities. Reproductions are folded to aid in mailing, filing, or general handling. The various size sheets are folded in different ways. The title block should always be on the front of the folded drawing. Methods of folding are shown in Fig. 4–83. These produce a folded drawing 8½ × 11 in. or, in metric paper, 140 × 210 mm.

## Problems

1. Describe the scales used on the following: architect's scale, engineer's chain scale, mechanical draftsman's scale, metric scale.

2. What are the metric scales in current use?

3. List the various hardnesses of drafting lead, from the hardest to the softest.

4. List the steps for inking a drawing in the proper order.

5. List the three materials upon which drawings are made.

6. List the standard inch and metric paper sizes.

7. Explain the difference between a whiteprint and an intermediate.

8. Describe a microfiche.

9. Draw the following problems on 9 × 12 in. vellum, including the border and title block as shown. Work as carefully and as accurately as possible. Lay out the work with a hard, sharp pencil, such as a 4H. Then darken with a softer pencil. Pay special attention to the condition of the pencil point so that the various line widths are correctly drawn. Part of a satisfactory solution includes locating the views correctly on the paper. They should have approximately equal space on all sides next to the border. Be certain to leave space for the dimensions. If your instructor wishes, you may include the dimensions.

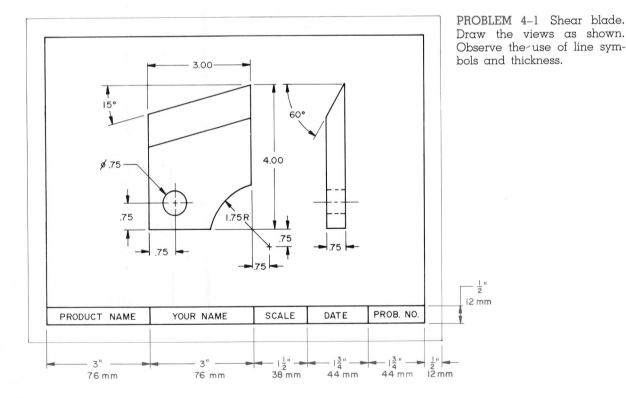

PROBLEM 4–1 Shear blade. Draw the views as shown. Observe the use of line symbols and thickness.

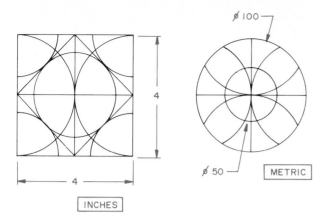

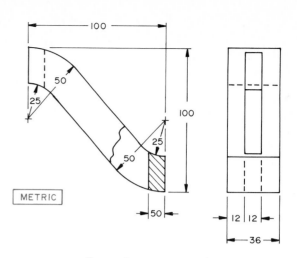

PROBLEM 4–2 Draw these two geometric exercises on one 9 × 12 in. sheet of vellum. Be especially careful to make all the intersections meet perfectly. Draw it full size.

PROBLEM 4–3 Draw this two-view drawing on 9 × 12 in. vellum. Notice it has a partial section showing the lower cross member. Draw it full size.

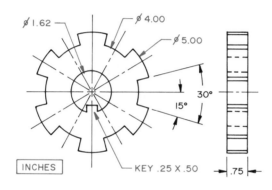

PROBLEM 4–4 Winch brake plate. Draw this problem full size on 9 × 12 in. vellum.

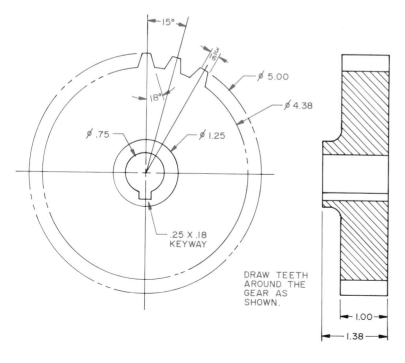

PROBLEM 4–5 Gear. Draw this gear full size on 9 × 12 in. vellum. Draw each tooth around the perimeter of the gear.

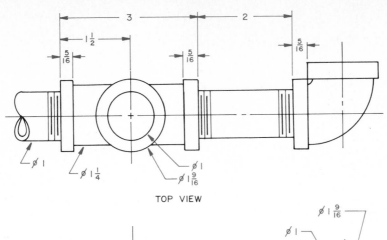

TOP VIEW

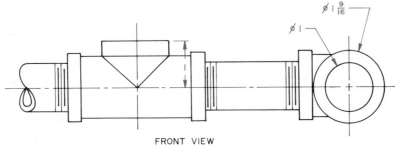

FRONT VIEW

NOTE: MEASUREMENTS ARE
FROM CENTER LINE OF
MEMBERS.

**PROBLEM 4–6** Malleable iron fittings. Draw these fittings full size on 9 × 12 in. vellum.

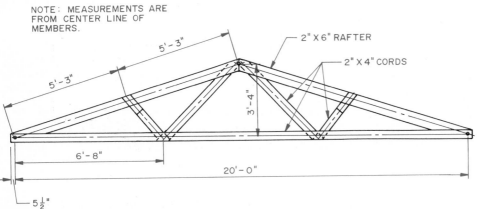

**PROBLEM 4–7** Fink wood framed roof truss. Draw this truss to the scale ½ in. = 1 ft 0 in. if you use 9 × 12 in. vellum. Use scale ¾ in. = 1 ft 0 in. if you use 12 × 18 in. vellum.

**PROBLEM 4–8** Structural steel bracket detail. Draw to the scale ⅜ in. = 1 ft 0 in. on 9 × 12 in. vellum.

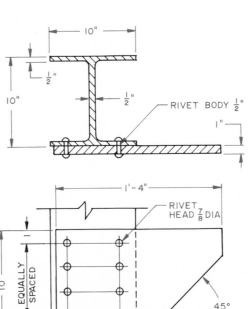

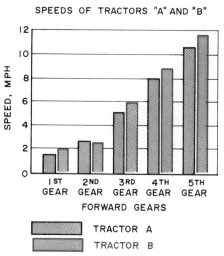

SPEEDS OF TRACTORS "A" AND "B"

TRACTOR A

TRACTOR B

DRAWING NOTE: BARS AND SPACES
BETWEEN THEM 12 mm VERTICAL
SCALE 20 mm.

**PROBLEM 4–9** Chart. Draw this chart on 9 × 12 in. vellum. Draw the bars with colored pencil and shade them in some manner.

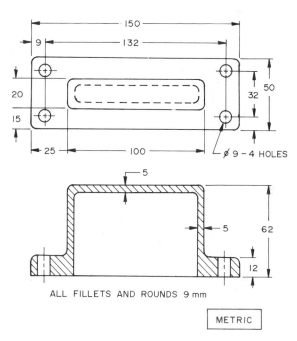

ALL FILLETS AND ROUNDS 9 mm

METRIC

**PROBLEM 4–11** Pressure cap. Draw this part full size on 9 × 12 in. vellum.

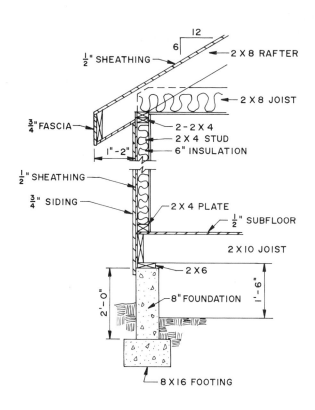

**PROBLEM 4–10** Typical wall section. Draw this wall section on 9 × 12 in. vellum to the scale 1 in. = 1 ft 0 in.

# Geometric Constructions

If you look at the products we use every day, you will notice that they are a configuration of geometric shapes. Most of the designs use one or more of the basic geometric shapes, either as a plane or a solid. When engineering designers begin the problem-solving process by sketching possible solutions, they make frequent use of the basic geometric shapes. As engineering drafters prepare the finished production drawings, they are required to draw these shapes. A knowledge of the structure of geometric forms and how they can be recorded on drawings is one of the everyday tools of the engineer and drafter. In addition, as designers discuss possible solutions, these forms and variations of them become a part of the engineering vocabulary. If a discussion involves a chord or a truncated right square pyramid, the engineer should immediately know what is being discussed (Fig. 5–1).

This chapter presents the most important geometric forms and explains how to draw them. In addition to the drafting procedures, engineering drafters have tools of various types that they use when drawing these shapes. For example, small hexagons can be rapidly drawn with a template. There are times, however, when mechanical aids do not meet the need. It is then that drafters use the construction techniques presented in this chapter. For example, a

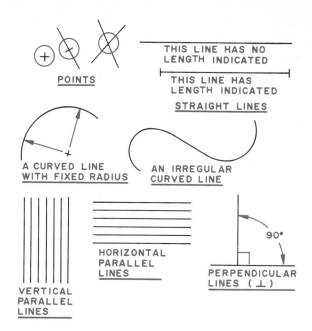

FIGURE 5–2 Points and lines.

If lines are drawn an equal distance apart, they are called parallel. The symbol indicating parallelism is ∥. If they meet at 90°, they are perpendicular. The symbol used for this is ⊥. Sometimes a square is sketched at the intersection to indicate perpendicularity.

A *plane* is a surface in such a position that when a straight line connects two points on the surface, the line lies in that surface. Planes can be generated by connecting three points, two intersecting lines, a point and a line, or two parallel lines (Fig. 5–3).

## ANGLES

Angles are formed when two straight lines intersect. An *acute angle* is one that is less than 90°. An *obtuse*

FIGURE 5–3 Ways to indicate a plane.

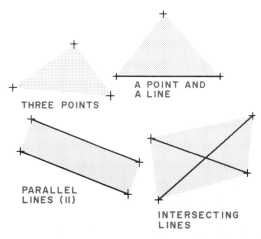

FIGURE 5–1 This version of the space shuttle and a space station clearly shows several of the basic geometric shapes. (Courtesy of National Aeronautics and Space Administration.)

product may require a hexagonal part that is 12 in. across the flats. Templates are not available in these large sizes. The drawing will be developed using basic construction techniques.

## POINTS, LINES, AND PLANES

A *point* is used to indicate a location. It has no physical size (width, depth, or height). It is shown on a drawing by a small cross, a short line crossing a longer line, or the intersection of two lines (Fig. 5–2).

A *line* is a mark on a drawing indicating a length but having no width or depth. A straight line is formed by connecting two points. A curved line is formed by moving a point through space at a predetermined radius or a varying distance from another point (Fig. 5–2).

A line can theoretically extend in length to infinity. If it has a specific length, it can be marked with short crosslines.

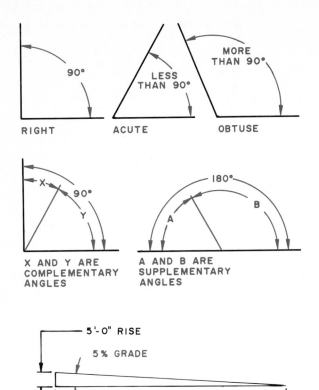

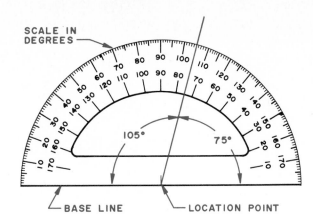

FIGURE 5-5 A protractor is used to lay out angles.

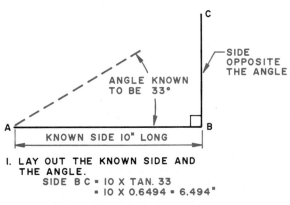

FIGURE 5-4 Angles formed by the intersection of straight lines.

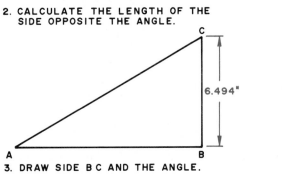

FIGURE 5-6 Angles can be drawn using trigonometry to calculate the length of the side opposite the known angle.

angle is more than 90°. A *right angle* is 90° (Fig. 5–4). The symbol commonly used for angle is ∠.

When two angles combined total 90°, they are called *complementary*. If they total 180°, they are called *supplementary*. A straight line represents an angle of 180°.

Angles are laid out in degrees, minutes, and seconds. A circle has 360°. A degree has 60 minutes. A minute has 60 seconds. The symbols used are degree (°), minute (') and second ("). An angle of 25°15'30" is read "25 degrees, 15 minutes, and 30 seconds." Angles can also be recorded in decimal fractions of degrees, such as 35.5°, and in percent of grade. Percent grade is a ratio between the rise in feet per 100 ft of run.

## How to Draw Angles

Angles are often laid out using a protractor. Place the baseline of the protractor on a line that is one side of the angle. Place the location point on the place where the angle will be formed. Read the angle in degrees on the scale. Mark it on the paper. Connect the mark with the location point (Fig. 5–5).

If greater accuracy is desired when drawing angles with long sides, the tangent method can be used. This involves simple trigonometry. It is based

on the fact that the length of the side opposite the angle is equal to the tangent of the angle multiplied by the length of the other side (Fig. 5–6).

## POLYGONS

A polygon is any closed figure whose sides are formed of straight lines. Those with sides of equal length are called regular polygons. Regular polygons can be inscribed in or circumscribed by a circle (Fig. 5–7).

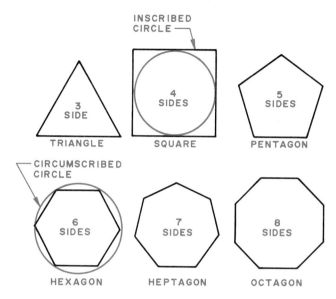

FIGURE 5–7 Commonly used polygons.

## Triangles

A triangle is a three-sided polygon. The sum of the three angles always equals 180°. A *right triangle* has one 90° angle. An *isosceles triangle* has two equal sides and equal angles. An *equilaterial triangle* has three equal sides and three equal angles. A *scalene triangle* has no equal sides or angles.

The length of the sides of right triangles can be found by using the Pythagorean theorem. This states that the square of the hypotenuse (longest side) is equal to the sum of the squares of the other

FIGURE 5–8 The length of the sides of a right triangle is found using the Pythagorean theorem.

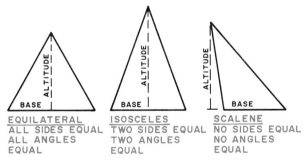

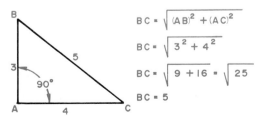

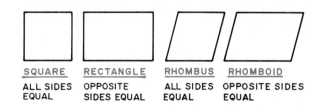

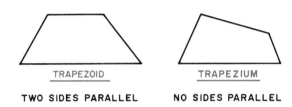

FIGURE 5–9 The opposite sides of a parallelogram are parallel.

FIGURE 5–10 The trapezoid and trapezium are polygons.

two sides (Fig. 5–8); stated another way, the hypotenuse equals the square root of the sum of the squares of the other two sides.

## Parallelograms

Parallelograms are polygons whose opposite sides are parallel and equal in length. The *square* has four equal sides and angles. The *rectangle* has parallel sides of two different lengths, but all angles are equal. The *rhombus* has parallel sides all the same length and no 90° angles. A *rhomboid* has parallel sides of two different lengths and no 90° angles (Fig. 5–9).

## Quadrilaterals

Quadrilaterals are polygons that have four straight sides. The parallelograms in Fig. 5–9 are quadrilaterals. In addition, the trapezoid and trapezium are quadrilaterals. The trapezoid has two parallel sides and the trapezium has no parallel sides (Fig. 5–10).

## CIRCLES AND ARCS

A circle is a closed curve that has all points an equal distance from a center point. The *circumference* of a circle is the distance around the circle. It can be found by multiplying the diameter by 3.1416. The *diameter* is the distance of a straight line from one side to the other that passes through the center. The *radius* is the distance from the center to the circle. A *chord* is a straight line that crosses the circle in two places. An *arc* is a part of the circumference. A *quadrant* is one fourth of the circle. A *segment* is an area of the circle bounded by a chord and an arc. A *sector* is an area bounded by an arc and two radii. A *semi-*

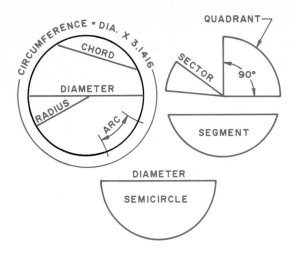

FIGURE 5–11  The parts of circles are arcs.

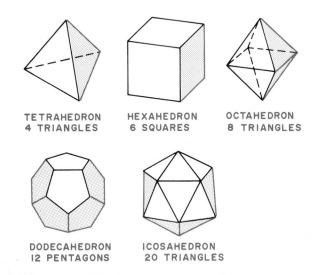

TETRAHEDRON
4 TRIANGLES

HEXAHEDRON
6 SQUARES

OCTAHEDRON
8 TRIANGLES

DODECAHEDRON
12 PENTAGONS

ICOSAHEDRON
20 TRIANGLES

FIGURE 5–12  The basic geometric solids.

circle is half the circle, one side of which is a diameter (Fig. 5–11).

## GEOMETRIC SOLIDS

Solids have width, depth, and height. The five regular solids are the tetrahedron, hexahedron, octahedron, dodecahedron, and icosahedron (Fig. 5–12).

### Prisms

Prisms have two parallel bases that are identical polygons. They are identified by the shape of their base. If the bases are parallelograms, the prisms are *parallelopipeds*. A *right prism* has the sides perpendicular to the base. An *oblique prism* has the sides oblique to the base. If one end of the prism is cut on an angle not parallel with the other base, it is a truncated prism (Fig. 5–13). Truncated means shortened

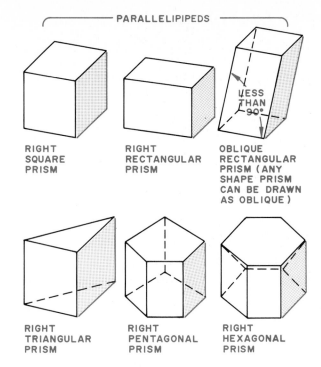

PARALLELIPIPEDS

RIGHT
SQUARE
PRISM

RIGHT
RECTANGULAR
PRISM

OBLIQUE
RECTANGULAR
PRISM (ANY
SHAPE PRISM
CAN BE DRAWN
AS OBLIQUE)

RIGHT
TRIANGULAR
PRISM

RIGHT
PENTAGONAL
PRISM

RIGHT
HEXAGONAL
PRISM

FIGURE 5–13  Typical prisms.

by cutting off the top. If the plane of the cut is parallel with the base, the solid is called a frustum.

### Cylinders

A cylinder is formed by moving a line, called the generatrix, in a circle around a central axis (Fig. 5–14). Each position of the generatrix on the surface of the cylinder is called an element. A cylinder may be

FIGURE 5–14  A cylinder is formed by moving a straight line in a circle about a central axis.

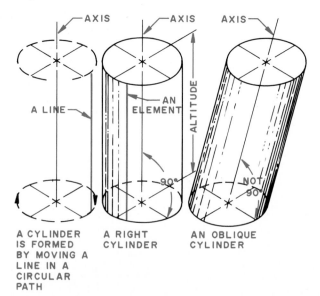

A CYLINDER
IS FORMED
BY MOVING A
LINE IN A
CIRCULAR
PATH

A RIGHT
CYLINDER

AN OBLIQUE
CYLINDER

FIGURE 5–15 The body of the satellite is a cylinder with a cone-shaped unit on top. (Courtesy of National Aeronautics and Space Administration.)

a right or oblique cylinder. The length of the axis is called the altitude (Fig. 5–15).

## Pyramids

A pyramid has a polygon for a base. It may be any type of polygon, such as a triangle or a square. The triangular faces of the pyramid meet at the vertex. The height is called the altitude. Pyramids can be right or oblique and may be truncated (Fig. 5–16).

FIGURE 5–16 Pyramids have a polygon for a base.

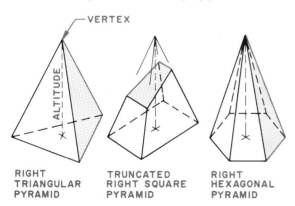

RIGHT TRIANGULAR PYRAMID     TRUNCATED RIGHT SQUARE PYRAMID     RIGHT HEXAGONAL PYRAMID

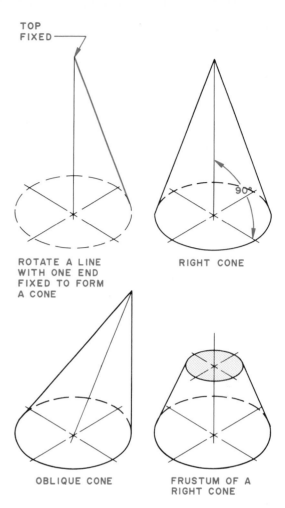

FIGURE 5–17 A cone is formed by rotating a straight line in a circular path, keeping one end fixed at a vertex.

## Cones

A cone is formed by holding one end of a straight line fixed at a vertex and rotating the other in a circle. Cones may be right or oblique and may be truncated or a frustum (Fig. 5–17).

## Spheres

A sphere is a double curved surface of revolution generated by revolving a circle about an axis (Fig. 5–18). The center of the sphere is on the axis of the revolved circle. The ends of the axis of the sphere are called the poles. A double curved surface is one formed by revolving a curved line, producing a surface containing no straight lines.

## Tori

A torus is a double curved surface of revolution generated by revolving a circle around a central axis

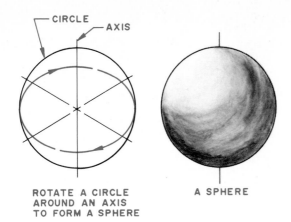

ROTATE A CIRCLE
AROUND AN AXIS
TO FORM A SPHERE

A SPHERE

FIGURE 5–18 A sphere is formed by rotating a circle about its axis.

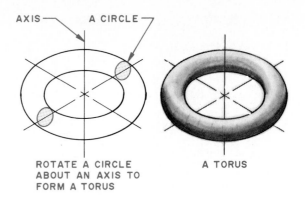

ROTATE A CIRCLE
ABOUT AN AXIS TO
FORM A TORUS

A TORUS

FIGURE 5–19 A torus is formed by rotating a circle around a central axis.

(Fig. 5–19). The center of the circle generating the torus is off the axis of the torus.

## BASIC GEOMETRIC CONSTRUCTIONS

The following material shows how to draw commonly used geometric forms and perform a variety of geometric constructions.

### To Bisect a Line

Two commonly used techniques for bisecting a line include using a compass (Fig. 5–20) or a triangle (Fig. 5–21).

### To Bisect an Arc

An arc is bisected in the same manner as a straight line. Connect the ends of the arc and bisect the straight line (Fig. 5–22).

### To Bisect an Angle

To bisect an angle, each side is marked to the same length. Then the arc is bisected (Fig. 5–23).

### To Transfer an Angle

A known angle can be transferred to a new location using a compass. The steps are shown in Fig. 5–24.

### To Draw a Line Parallel to a Curved Line

Set the radius of a compass equal to the distance between the lines. Draw a series of arcs from points along the given curved line. Use enough points to produce a smooth curve (Fig. 5–25).

### To Divide a Line into Equal Parts

A line can be divided into equal parts using any scale available (Fig. 5–26).

FIGURE 5–20 How to bisect a line.

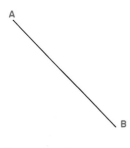

Assignment: Bisect line AB.

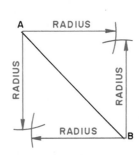

1. First, draw arcs that cross from each end of the line. The radius of the arc must be greater than half the length of the line.

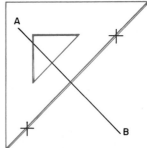

2. Then, connect the intersecting arcs with a straightedge.

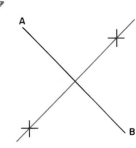

3. Draw the bisecting line.

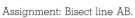

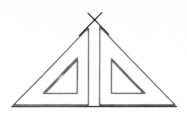

A           B

Assignment: Bisect line AB.

1. Using any triangle, draw lines from each end that intersect.

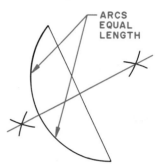

2. Place the triangle on the line and draw a perpendicular.

FIGURE 5–21   How to bisect a line using a triangle.

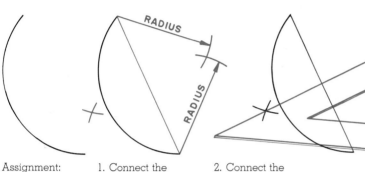

Assignment: Bisect the arc.

1. Connect the ends of the arc and bisect the straight line.

2. Connect the arcs with a straightedge.

3. Draw the bisecting line.

FIGURE 5–22   How to bisect an arc.

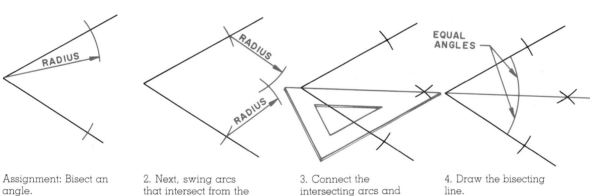

Assignment: Bisect an angle.

1. First, swing an arc of any length locating points on the sides of the arc.

2. Next, swing arcs that intersect from the points on the sides of angle.

3. Connect the intersecting arcs and the vertex of the angle with a straightedge.

4. Draw the bisecting line.

FIGURE 5–23   How to bisect an angle.

FIGURE 5–24   How to transfer an angle to a new location.

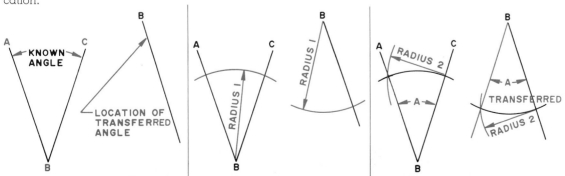

Assignment: Transfer angle ABC to the new location.

1. Draw an arc at any radius on both angles.

2. Set compass to radius 2 and mark it on the transferred angle. Connect B' with the intersection.

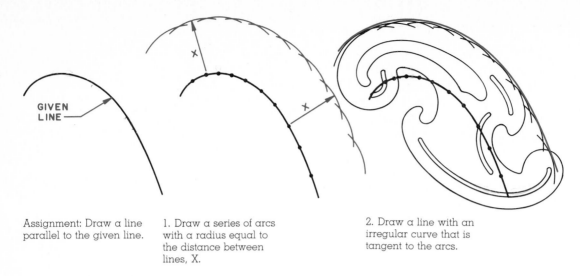

Assignment: Draw a line parallel to the given line.

1. Draw a series of arcs with a radius equal to the distance between lines, X.

2. Draw a line with an irregular curve that is tangent to the arcs.

**FIGURE 5–25** How to draw a line parallel to a given curved line.

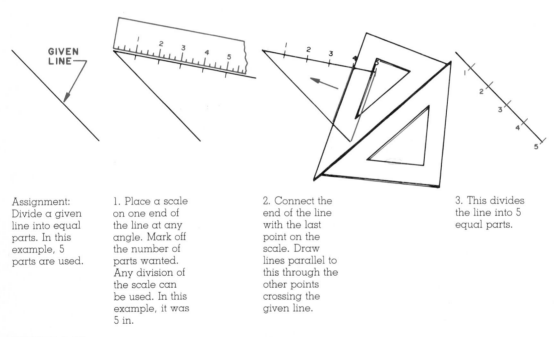

Assignment: Divide a given line into equal parts. In this example, 5 parts are used.

1. Place a scale on one end of the line at any angle. Mark off the number of parts wanted. Any division of the scale can be used. In this example, it was 5 in.

2. Connect the end of the line with the last point on the scale. Draw lines parallel to this through the other points crossing the given line.

3. This divides the line into 5 equal parts.

**FIGURE 5–26** How to divide a line into equal parts.

## To Divide a Line into Proportional Parts

A line can be divided into proportional parts using the same technique as dividing it into equal parts except for the scale used. The divisions on the scale are made proportional (Fig. 5–27).

## To Draw an Isosceles or Equilateral Triangle

The procedure in Fig. 5–28 is used for isosceles and equilateral triangles. The only difference is the length of the base.

## To Draw a Line Through a Point Perpendicular to a Line

A line can be drawn through a point perpendicular to a line using a compass and locating points or with triangles using a drafter's method (Fig. 5–29).

## To Draw a Triangle When the Three Sides Are Known

When a triangle has three sides of unequal but known length, it can be drawn as shown in Fig. 5–30.

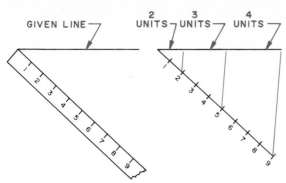

GIVEN LINE

2 UNITS 3 UNITS 4 UNITS

Assignment: Divide a given line into proportional units. In this example, the proportions are 2, 3, and 4 units. This totals 9 units. Locate the units with a scale.

Connect the end of the line with the last point on the scale. Draw lines parallel with it through the points, providing the number of units desired.

FIGURE 5–27 How to divide a line into proportional units.

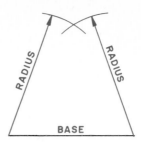

RADIUS    RADIUS

BASE

Assignment: Construct an isosceles or equilateral triangle.

1. From each end of the base, draw arcs having a radius equal to the sides of the triangle.

2. Connect the ends of the base to the intersection of the arcs.

FIGURE 5–28 How to construct an isosceles or equilateral triangle.

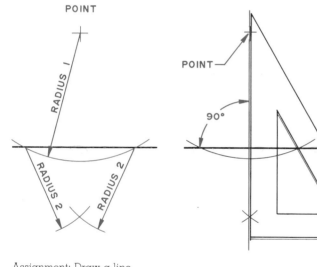

POINT

RADIUS 1

RADIUS 2    RADIUS 2

POINT

90°

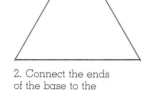

POINT

Assignment: Draw a line through the point perpendicular to the line.

1. Swing an arc, radius 1, from the point so that it crosses the line in two places. From these intersections, swing arcs that cross below the line.

2. Connect the point and the intersection to produce a perpendicular.

FIGURE 5–29 How to draw a line through a point perpendicular to a line.

1. Place triangles as shown. Slide the one parallel with the line until the vertical side crosses the point. Draw the perpendicular line.

FIGURE 5–30 How to construct a triangle when the lengths of the sides are known.

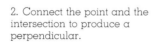

SIDE 1

SIDE 2

SIDE 3

Assignment: Construct a triangle when the length of the sides is known.

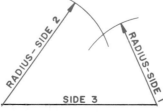

RADIUS - SIDE 2    RADIUS - SIDE 1

SIDE 3

1. From each end of one side, draw arcs having a radius equal to the sides of the triangle.

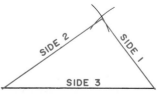

SIDE 2    SIDE 1

SIDE 3

2. Connect the ends of the base side with the intersection of the arcs.

91

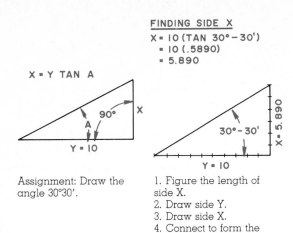

$$X = 10 \, (\text{TAN } 30° - 30')$$
$$= 10 \, (.5890)$$
$$= 5.890$$

$$X = Y \, \text{TAN} \, A$$

90°

A

Y = 10

30° - 30'

X = 5.890

Y = 10

Assignment: Draw the angle 30°30'.

1. Figure the length of side X.
2. Draw side Y.
3. Draw side X.
4. Connect to form the angle.

FIGURE 5–31 How to draw a triangle using the tangent method.

## To Draw an Angle Using the Tangent Method

In addition to laying out angles with a protractor, basic trigonometry can be used accurately to find the lengths of the sides. These lengths are used to draw the angle. It is especially useful when angles are expressed in minutes and seconds (Fig. 5–31). The tangent of angle A is $\frac{x}{y}$. Side X = Y tan A. Assume a convenient size for side Y, as 10. Find the tangent of angle A in a table of natural tangents and multiply it by 10. Draw this distance for side X. This produces the desired angle.

FIGURE 5–32 How to draw a square when one side is known.

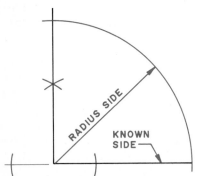

RADIUS SIDE

KNOWN SIDE

RADIUS SIDE

RADIUS SIDE

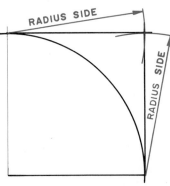

6

3

2

4

5

1

1. First, draw a perpendicular at one end of the known side. Then, draw an arc with a radius the length of the side crossing the perpendicular.

2. Next, draw arcs the length of the known side from the ends of the two sides already drawn. Connect their intersection with the ends of the sides.

1. Using a 45° triangle and a straightedge or a drafting machine, draw the lines as shown above. Notice the suggested order of strokes.

**USING A COMPASS**

**USING A TRIANGLE OR DRAFTING MACHINE**

**USING A CIRCUMSCRIBED CIRCLE**

**USING AN INSCRIBED CIRCLE**

ACROSS THE CORNERS

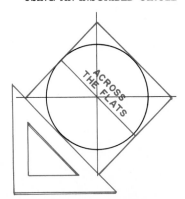

ACROSS THE FLATS

1. Draw a circle having a diameter equal to the across-the-corners distance. Connect the ends of the center lines.

1. Draw a circle having a diameter equal to the across-the-flats distance. Draw lines at 45° tangent to the circle as shown above.

92   Chapter 5

## To Draw a Square

A square can be drawn using a compass or triangle when one side is known. Frequently the across-the-corners distance is given. When this is known, the square can be drawn using a circumscribed circle (Fig. 5–32).

## To Draw a Pentagon

A pentagon can be drawn by finding the length of one side and stepping it off on the circle circumscribing it (Fig. 5–33).

## To Draw a Hexagon

A hexagon can be drawn using the across-the-corners or the across-the-flats distances (Fig. 5–34).

## To Draw an Octagon

An octagon can be drawn with an inscribed circle and a 45° angle or with a compass (Fig. 5–35).

## To Draw a Polygon with Any Given Number of Sides

Regular polygons with any number of sides can be constructed if the length of one side is known (Fig. 5–36).

## To Move a Polygon

It is sometimes necessary to move a known polygon to a new location. The steps for doing this are shown in Fig. 5–37.

## To Find the Center of a Circle

Sometimes the center of a circle is lost. The way to relocate it is shown in Fig. 5–38.

## To Draw a Circle Through Three Points

When it is necessary to pass a circle through three points, use the same procedure as shown in Fig. 5–38. Connect the points with straight lines. Bisect each line. Where the bisectors cross is the center of the circle.

## TANGENTS

A tangent point is the exact point at which one of two meeting lines stops and the other begins. Tangents occur when curved and straight lines or two curved lines meet. Tangent points are very important to quality drafting. If they are not located and used by the drafter, irregular lines will occur (Fig. 5–39).

## To Draw an Arc Tangent to Two Straight Lines at 90°

There are many instances when an arc must be drawn tangent to straight lines meeting at 90°. This is shown in Fig. 5–40. The tangent point is actually a line drawn from the center of the arc perpendicular to the straight line.

FIGURE 5–33 How to draw a pentagon.

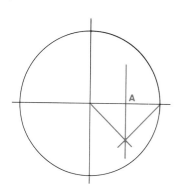

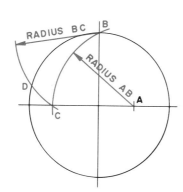

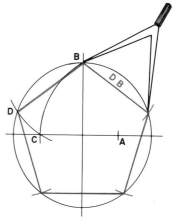

1. Draw a circle that will circumscribe the pentagon. Bisect the radius, locating point A.

2. With point A as a center, draw radius AB until it crosses the center line at C. With B as a center and BC as a radius, draw an arc to cross the circle at D.

3. Connect points D and B. Set distance DB on a divider and step off the other sides around the circle.

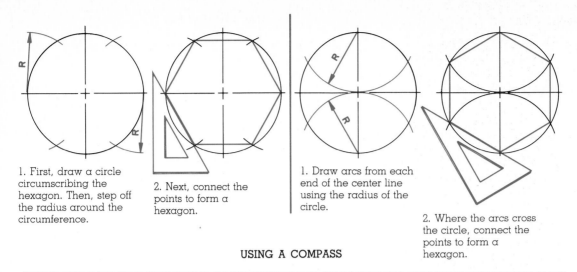

1. First, draw a circle circumscribing the hexagon. Then, step off the radius around the circumference.

2. Next, connect the points to form a hexagon.

1. Draw arcs from each end of the center line using the radius of the circle.

2. Where the arcs cross the circle, connect the points to form a hexagon.

**USING A COMPASS**

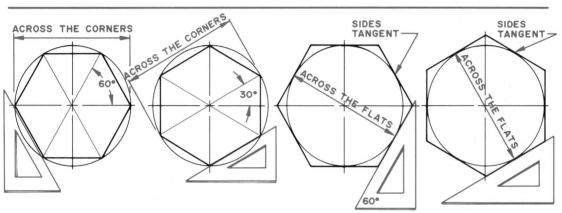

Drawing circumscribed hexagons when the across-the-corners distance is known.

Drawing inscribed hexagons when the across-the-flats distance is known.

**DRAFTING TECHNIQUES**

FIGURE 5–34   How to draw a hexagon.

FIGURE 5–35   How to draw a octagon.

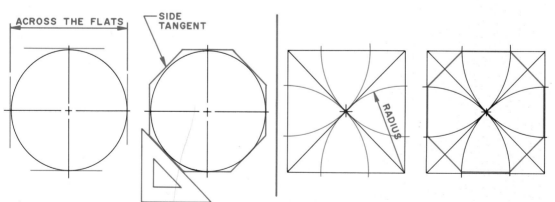

1. Draw a circle with a diameter equal to the across-the-flats dimension. Draw perpendiculars at the ends of each center line tangent to the circle.

2. Draw 45° sides tangent to the circle forming the octagon.

1. Draw a square with sides equal to the across-the-flats dimension. Draw diagonals to find the center. From each corner, draw arcs through the center and cross the square.

2. Connect the intersections on the square forming the octagon.

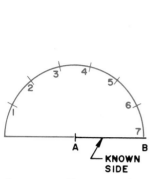

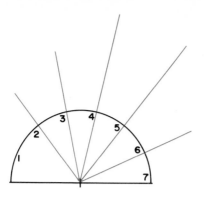

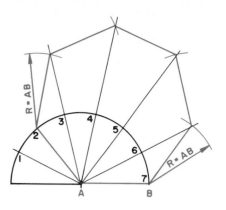

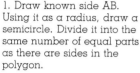

Assignment: Draw a regular polygon with seven sides having length AB.

1. Draw known side AB. Using it as a radius, draw a semicircle. Divide it into the same number of equal parts as there are sides in the polygon.

2. Draw radial lines through the divisions, starting with the second division from the left.

3. Using given side AB as a radius, draw arcs crossing the radial lines. This locates the ends of the sides of the polygon. Connect these points to form the polygon.

FIGURE 5–36  How to draw a polygon with any number of sides.

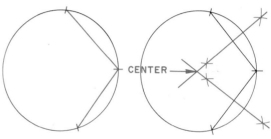

Assignment: Find the center of a circle.

1. Locate three points at random on the circle. Connect these with chords.

2. Bisect each chord. Their intersection is the center of the circle.

FIGURE 5–38  How to find the center of a circle.

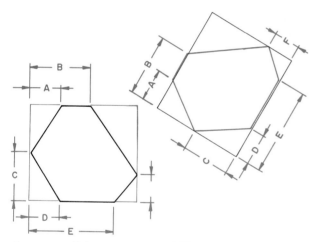

Assignment: Relocate a polygon.

1. First, draw a rectangle around the polygon.

2. Then, draw the rectangle in the new location. Locate each corner of the polygon from a side of the rectangle. Connect the corners to form the polygon.

FIGURE 5–37  How to relocate a polygon.

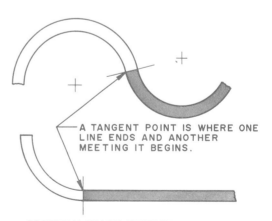

A TANGENT POINT IS WHERE ONE LINE ENDS AND ANOTHER MEETING IT BEGINS.

PROPERLY DRAWN INTERSECTIONS USING TANGENT POINTS TO JOIN TWO LINES.

FIGURE 5–39  A tangent point occurs where two lines meet. One line stops at this point and the other begins there.

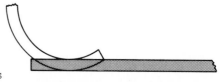

IF TANGENT POINTS ARE NOT LOCATED IRREGULARITIES IN THE LINES OCCUR.

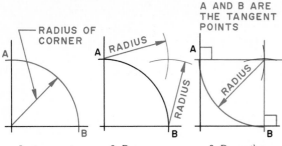

RADIUS OF CORNER

A AND B ARE THE TANGENT POINTS

RADIUS

RADIUS

RADIUS

Assignment: Draw an arc tangent to straight lines.

1. Draw an arc that is the radius of the round corner to be drawn. Points A and B are the tangent points.

2. Draw arcs from points A and B. This locates the center of the round corner.

3. Draw the corner from the center. Points A and B are the tangent points. A and B can also be located by drawing a perpendicular from the center to the straight line when the center is known.

FIGURE 5–40 How to draw an arc tangent to a straight line.

## To Draw an Arc Tangent to Two Straight Lines Not at 90°

The procedure involves locating the center of the arc and drawing perpendiculars from it to both lines. The perpendiculars locate the tangent points (Fig. 5–41).

## To Draw an Arc Tangent to Another Arc and a Straight Line

This is an application of the principles just illustrated. The steps are shown in Fig. 5–42.

## To Draw an Arc Tangent to Two Arcs

Two situations can occur. The arc can be drawn tangent to arcs from the convex or concave side, depending upon how they are situated. When approached from the convex side, the radii of the two arcs are added together. When approached from the concave side, the radii of the two arcs are subtracted.

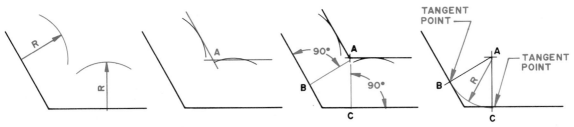

Assignment: Draw an arc tangent to two straight lines not at 90°.

1. Draw an arc having a radius equal to the arc to form the corner.

2. Draw lines parallel with the straight lines and tangent to the arcs. Their intersection, A, is the center.

3. Draw perpendiculars from the center, A, to each straight line. This locates the tangent points, B and C.

4. With A as the center, draw the arc forming the corner. Stop at the tangent points.

FIGURE 5–41 How to draw an arc tangent to two straight lines not at 90° to each other.

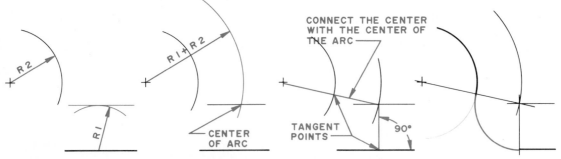

Assignment: Draw an arc tangent to another arc and a straight line.

1. Draw a line parallel with the straight line at a distance equal to the radius of the connecting arc, R₁.

2. Draw an arc from the center of the arc to be connected with the straight line. The radius is equal to its radius, R1, plus the radius of the connecting arc, R2. The point at which it crosses the straight line is the center.

3. Locate the tangent points. Draw a perpendicular from the center to the straight line. Connect the center with the center of the first arc.

4. Draw the arc connecting the first arc and the straight line. Stop it at the tangent points.

FIGURE 5–42 How to draw an arc tangent to another arc and a straight line.

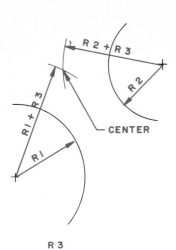
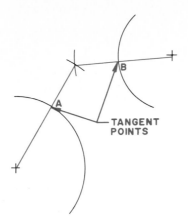
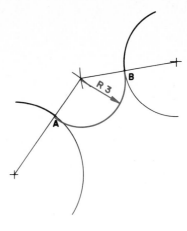

R 3

Assignment: Draw an arc with radius R3 tangent to arcs R1 and R2.

1. From the center of each arc, draw a radius equal to its radius plus the radius of the arc to be drawn tangent, R3. The point at which they cross locates the center.

2. Connect the centers to locate the points of tangency, A and B, on each arc.

3. Set a compass on the radius R3 and draw an arc connecting the two given arcs, R1 and R2. Stop at the tangent points.

FIGURE 5-43 How to draw an arc tangent to two arcs.

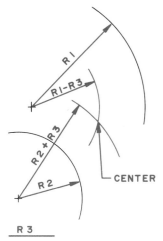
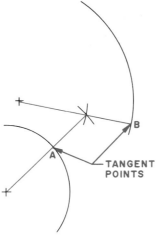
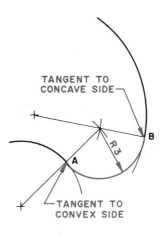

R 3

Assignment: Draw an arc tangent to two arcs when one is concave and the other is convex.

1. Subtract the radius of the connecting arc, R3, from the radius of the concave arc. Add radius, R3, to the convex arc.
From the centers of the concave and convex arcs, swing these radii.

2. Connect the intersection of the two arcs just drawn with the centers of the concave and convex arcs. This locates the points of tangency.

3. Set a compass on the desired radius, R3, and draw the arc. Stop at the tangent points.

FIGURE 5-44 How to draw an arc tangent to two arcs, one from the concave side and the other from the convex side.

The tangent point between arcs when approached from the convex side is found by connecting their center lines. The procedure for constructing an arc tangent with two other arcs is shown in Fig. 5-43. The procedure for finding the tangent points when an arc is approached from the concave side is shown in Fig. 5-44.

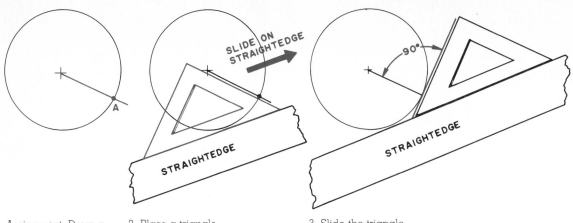

Assignment: Draw a line tangent to the circle at point A.

1. Connect the center and point A.

2. Place a triangle along the line. Rest it on a straightedge.

3. Slide the triangle along the straightedge until the other side reaches the point. Draw the tangent line.

FIGURE 5–45  How to draw a line tangent to a circle at a given point on the circle.

## To Draw a Line Tangent to a Circle at a Known Point on the Circle

The construction follows the principle that the point of tangency between a line and an arc is a line perpendicular to the line. This construction is shown in Fig. 5–45.

## To Draw a Line Tangent to a Circle from a Point Outside the Circle

The first step is to find the point of tangency on the circle (Fig. 5–46). Then draw the tangent line through this point as explained in Fig. 5–45.

## CONIC SECTIONS

Conic sections are formed when a right cone is cut by planes at different angles. This produces four curves of intersection. These are a circle, ellipse, parabola, and hyperbola (Fig. 5–47).

> A *circle* is formed when the intersecting plane is perpendicular to the axis.
>
> An *ellipse* is formed when the intersecting plane is at a greater angle with the axis than the elements on the surface of the cone.
>
> A *parabola* is formed when the intersecting plane is on the same angle with the axis as the elements of the cone.
>
> A *hyperbola* is formed when the intersecting plane makes an angle with the axis smaller than the angle of the element or is parallel with the axis.

### Ellipse

An ellipse is a plane curve in such a form that the sums of the distances of each point on the periphery

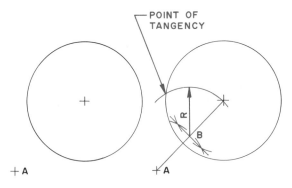

Assignment: Draw a tangent from point A outside the circle to the circle.

1. Connect A and the center of the circle. Bisect this line. At the point of bisection, B, place a compass with a radius to reach the center of the circle. Draw an arc. The point at which it crosses the circle is the tangent point. Draw the tangent line as shown in Fig. 5–44.

FIGURE 5–46  How to draw a line tangent to a circle from a point outside the circle.

from two fixed points on the major axis, the foci, are equal (Fig. 5–48). It has two axes, major and minor.

### To Draw an Ellipse Using the Foci Method

The foci method produces a rather accurate ellipse. It is based on the definition of an ellipse as shown in Fig. 5–48. The more points used to draw the ellipse, the greater will be its accuracy. This is especially true for the ends of the major axis.

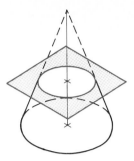

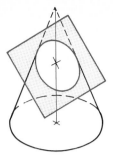

A circle is formed by a plane perpendicular to the axis.

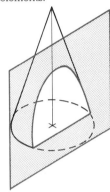

An ellipse is formed by a plane making an angle with the axis greater than the elements.

A parabola is formed by a plane making the same angle with the axis as the elements.

A hyperbola is formed by a plane making an angle with the axis smaller than the elements.

FIGURE 5–47 Conic sections formed when a right cone is cut by a plane at various angles.

FIGURE 5–49 How to draw an ellipse using the foci method.

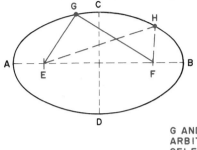

AB MAJOR AXIS
CD MINOR AXIS
FOCI POINTS E AND F
EG + GF = HE + HF

G AND H ARE ARBITRARY POINTS SELECTED ON THE PERIPHERY

FIGURE 5–48 The foci method of drawing an ellipse is based on a relationship between the focus points and points on the circumference of the ellipse.

The steps for drawing a foci ellipse are shown in Fig. 5–49.

### To Draw an Ellipse Using the Four-Center Method

The four-center method is a rapid way to draw an ellipse. It is not as accurate as the foci method but is adequate for many situations (Fig. 5–50).

### Drawing with Ellipse Templates

Ellipse templates are used to draw ellipses accurately and rapidly (Fig. 5–51). They are made in various ellipse angles. The ellipse angle is the angle at which

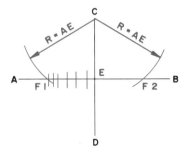

Assignment: Draw an ellipse using the foci method when the axes are known.
1. Draw the major and minor axes. Locate the foci points, F1 and F2, by swinging an arc from the end of the minor axis. The radius equals half the diameter, AE. Decide on the number of points needed to draw one quadrant. The more points, the more accurate the ellipse. Mark these on the major axis between the foci, F1, and the center, E.

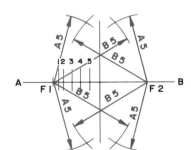

2. Locate the points on the periphery of the ellipse. Use distance A5 as a radius. Place the compass on the foci and draw arcs. Use distance B5 as a radius. Place the compass on the foci and draw arcs. The points at which they cross are points on the ellipse. Repeat for A4, A3, A2, A1 and B4, B3, B2, B1.

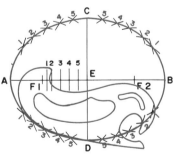

3. Connect the points with an irregular curve.

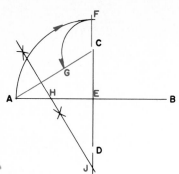

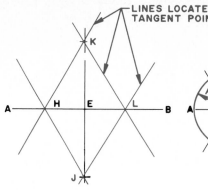

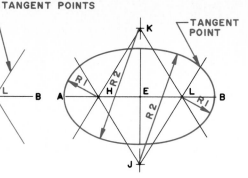

**LINES LOCATE TANGENT POINTS**

**TANGENT POINT**

Assignment: Draw a four-center ellipse.

1. Draw the major and minor axes. Connect the end of the major axis, A, with the end of the minor axis, C. Draw radius AE crossing the minor axis at F. Draw radius CF crossing AC at G. Draw a perpendicular bisector of AG. Extend it until it crosses the major axis at H and the minor axis at J. Extend the minor axis if necessary.

2. Using dividers, transfer EJ to the upper half of the minor axis locating K. Transfer EH to the other half of the major axis, locating L. Connect points H, J, K, and L. They are the center points for drawing the ellipse. The extended lines locate the tangent points.

3. Using the radii developed, R1 and R2, and the tangent points, draw the ellipse.

FIGURE 5–50   How to draw an ellipse using the four-center method.

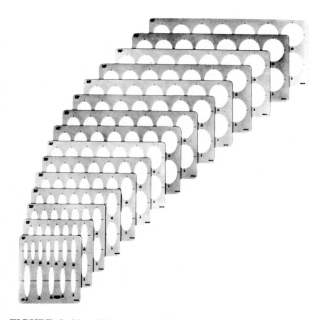

FIGURE 5–51   Ellipse templates.

you would look at a circle to see it as an ellipse. If the line of sight is perpendicular to the circle, it is round. As the circle leans, one axis becomes shorter. This forms an ellipse (Fig. 5–52).

Ellipse templates are sold in sets with angles from 10° to 80°, with a template every 5°.

## To Draw a Parabola

A parabola is a curve having points equidistant from a fixed line, the directrix, and from a fixed point not on the line, the focus (Fig. 5–53). There are several

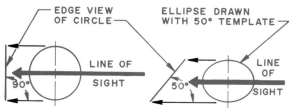

**EDGE VIEW OF CIRCLE**     **ELLIPSE DRAWN WITH 50° TEMPLATE**

LINE OF SIGHT     LINE OF SIGHT

90°     50°

A circle perpendicular to the line of sight projects as a circle.

A circle on an angle to the line of sight projects as an ellipse.

FIGURE 5–52   A circle on an angle to a line of sight appears as an ellipse.

FIGURE 5–53   A parabola is a curve having points equidistant from a fixed line (the directrix) and from a fixed point (the focus) not on the directrix.

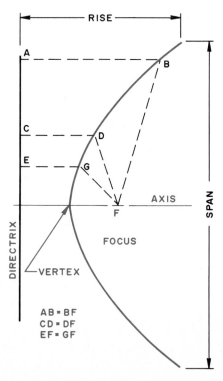

AB = BF
CD = DF
EF = GF

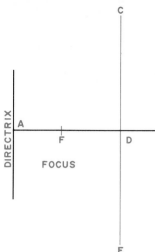

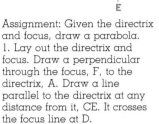

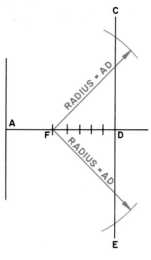

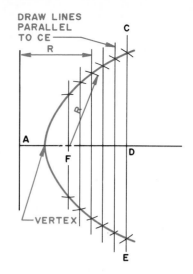

Assignment: Given the directrix and focus, draw a parabola.
1. Lay out the directrix and focus. Draw a perpendicular through the focus, F, to the directrix, A. Draw a line parallel to the directrix at any distance from it, CE. It crosses the focus line at D.

2. Using AD as a radius, draw an arc from the focus, F, above and below the focus line. This locates two points on the parabola.

3. Draw several equally spaced lines parallel with CE in the distance between CE and the focus. Use each of these to find a radius and locate other points on the parabola. The vertex is halfway between A and F. Connect the points with an irregular curve.

FIGURE 5–54  How to draw a parabola when the focus and directrix are known.

ways to draw parabolas. The procedure used depends upon what is known. The two procedures shown are for instances when the focus and directrix are known (Fig. 5–54) and when the rise and span are known (Fig. 5–55).

## To Draw a Hyperbola with the Foci and Vertices Given

A hyperbola is the path made by a point that moves so that the difference of its distances from two fixed points, the foci, is a constant and is equal to the transverse axis of the hyperbola (Fig. 5–56). The transverse axis is the distance between the vertices.

As the hyperbolic curves extend toward infinity, they approach two straight lines, called *asymptotes*. These are found by drawing a circle having the distance between the foci, F1 and F2, as a diameter. Then, construct perpendiculars to the transverse axis at the vertices, X and Y. Where these intersect the circle, locate a point on the asymptote. Connect these points with the center of the circle to locate the asymptotes (Fig. 5–57).

The procedure for constructing a hyperbola when the foci and vertices are known is shown in Fig. 5–58.

A hyperbola can be drawn when the asymptotes are known and one point on the hyperbola is located. When the asymptotes are at right angles

FIGURE 5–55  How to draw a parabola when the rise and span are known.

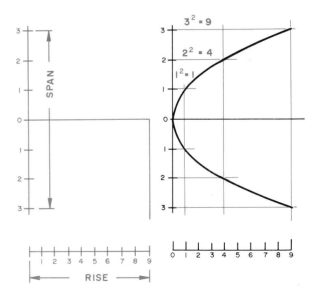

Assignment: Draw a parabola when the rise and span are known.

1. Divide half the span into any number of parts. Divide the rise into the square of the number of spaces on half the span.

2. Connect the number on the span with its square on the rise. This locates points on the parabola.

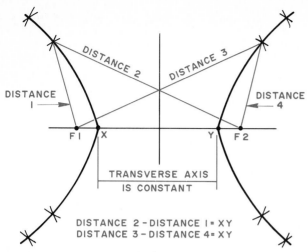

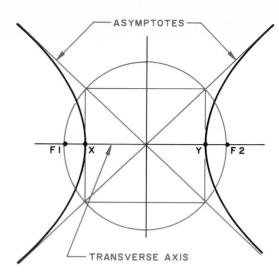

The difference of the distance from the foci, F1 and F2, to each point on the hyperbola is equal to the transverse axis constant, XY. This is true for any point selected on the hyperbola.

FIGURE 5-56 The form of a hyperbola depends on the foci and the transverse distance.

FIGURE 5-57 How to locate the asymptotes of the hyperbola.

to each other, it is an equilateral or rectangular hyperbola. The steps for drawing these are shown in Fig. 5-59.

## OTHER GEOMETRIC CONSTRUCTIONS

### To Draw a Helix

A cylindrical helix is generated by a point moving around the surface of a cylinder and also moving parallel with the axis at a constant speed. The distance the point moves parallel to the axis in one revolution is called the lead (Fig. 5-60).

A helix may be right- or left-handed. A right-handed helix slopes up to the right. A left-handed helix slopes up to the left.

The steps for drawing a cylindrical helix are shown in Fig. 5-61, which also shows a left-handed helix.

A helix can also be formed on a cone. It is formed by a point moving on the surface of the cone. The point moves at a constant rate around the conical surface and parallel with the axis. It is drawn in the same manner as a cylindrical helix. The lead is measured parallel to the axis of the cone. The di-

FIGURE 5-58 How to draw a hyperbola when the foci and vertices are known.

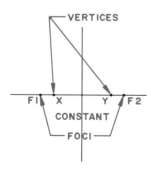

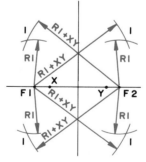

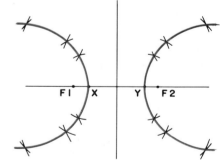

Assignment: Draw a hyperbola when the vertices and foci are known.
1. Draw the axis and locate the vertices, X and Y, and foci, F1 and F2, equal distances from the perpendicular.

2. Find the first point on the hyperbola by drawing radius R1 (any length) from F1 and F2. Then, draw radius R1 + XY from F1 and F2. This locates four points. Change the radius and find additional points.

3. Connect the points with an irregular curve to form the hyperbola.

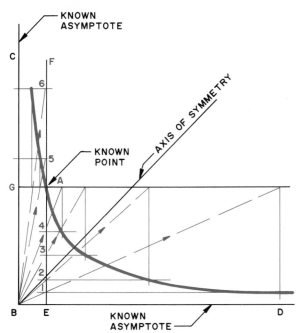

Assignment: Draw a hyperbola when the asymptotes and a point on the hyperbola are known.

1. Draw lines from the point, A, parallel with the asymptotes, BC and BD.

2. Divide EF into any number of parts. Draw lines through them parallel with the asymptotes, BD.

3. Draw lines from B through the points located on EF. Extend them until they cross GA extended.

4. From the point of intersection on GA, draw perpendicular lines, extending them until they cross the horizontal lines drawn through each numbered point. Each such intersection is a point on the hyperbola.

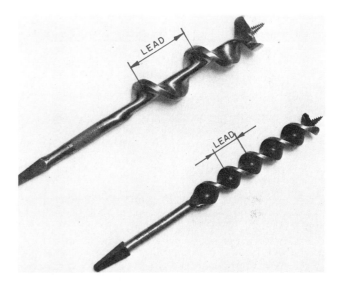

FIGURE 5–60 The threads on these auger bits are in the form of a helix.

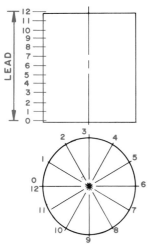

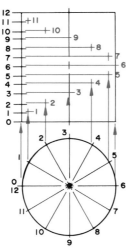

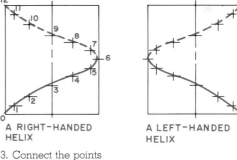

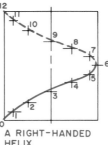

Assignment: Draw a right-hand cylindrical helix.

1. Draw the side and top views of the cylinder. Mark off one lead and divide it into a number of equal parts. Divide the circular view into the same number of parts. The more divisions used, the more accurate the helix will be. Number these divisions.

2. Project the points in the side view horizontally across the cylinder. Project the points in the circular view up to the side view. The points at which they cross the line of the same number are points on the helix. For example, point 1 in the top view intersects point 1 in the side view.

3. Connect the points with an irregular curve to form the helix. Remember that half of each lead will be hidden behind the cylinder.

A HELIX IS A FORM OF AN INCLINED PLANE

FIGURE 5–61 How to draw a helix.

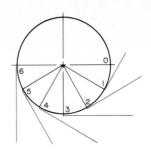

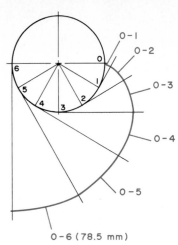

0–6 (78.5 mm)

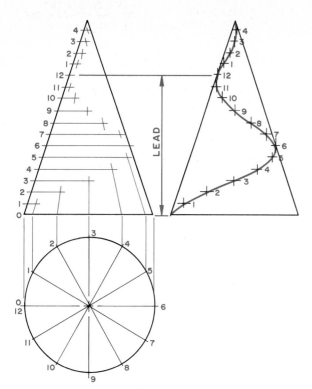

Assignment: Draw a conical helix.

1. Divide the lead into a number of equal parts along the sloped face. Project these horizontally. Divide the circular view into the same number of equal parts. Project the points from the circular view perpendicular to the base. From the base, project each point to the vertex of the cone. The point at which each number line crosses the line of the same number in the side view is a point on the helix.
2. Connect the points with an irregular curve to form the conical helix.

FIGURE 5–62   How to draw a helix on the surface of a cone.

visions are projected to the base of the cone and from there connected with the vertex (Fig. 5–62).

## To Draw an Involute

An involute is a plane curve that is formed by a point on a string when the string is unwound from around a circle or polygon. The involute of a polygon is drawn by adding its sides as it unwinds, using the corners of the polygon as centers for drawing the arcs. A circle may be considered a polygon with an infinite number of sides. It is usually divided into a number of equal parts. Procedures for drawing involutes are shown in Figs. 5–63 and 5–64.

## To Draw a Cycloid

A cycloid is the path of a point fixed on the circumference of a rolling circle. When the circle rolls on a straight line, the path of the point is a cycloid (Fig. 5–65). When the circle rolls on the outside of another circle, the path of the point is a epicycloid

Assignment: Draw an involute of a circle (∅ 50 mm).

1. Divide the circle into a number of equal parts. Draw tangents to each division (90° to division line). Calculate the circumference: $\pi \times \varnothing = 3.1416 \times 50$ mm = 157 mm.

2. Divide the circumference into the number of equal parts (12 in this example): $^{50}/_{12} = 4.16$ mm.
 Starting at point 1, step off along the tangent line a distance equal to arc 0–1 (4.16 mm). At point 2, step off a distance equal to arc 0–2 (8.32 mm). Continue until all desired points are located. Connect these points with an irregular curve to form the involute of the circle.

FIGURE 5–63   How to draw an involute of a circle.

FIGURE 5–64   How to draw an involute of a square and a triangle.

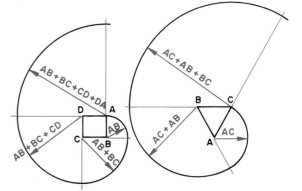

Assignment: Draw the involute of a square.
1. Draw the square and extend the sides.
2. Drawn an arc from B with a radius of AB.
3. Draw an arc from C with a radius equal to AB + BC.
4. Continue increasing the radius of each corner by the length of the side.
5. When the desired number of points have been found, connect them to form the involute.

Assignment: Draw the involute of a triangle.
1. Draw the triangle and extend the sides.
2. Draw an arc from A with a radius of AC.
3. Continue increasing the radius at each corner the length of one side.
4. When the desired number of points have been found, connect them to form the involute.

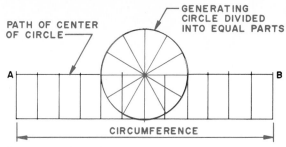

PATH OF CENTER OF CIRCLE

GENERATING CIRCLE DIVIDED INTO EQUAL PARTS

CIRCUMFERENCE

Assignment: Draw the cycloid generated by the above circle moving in a straight line.

1. Draw the circumference tangent to the generating circle.
2. Divide the circle into a number of equal parts.
3. Divide the circumference into the same number of parts.
4. Draw a horizontal line, AB, through the center of the circle. This is the path of the center of the circle as it rolls.

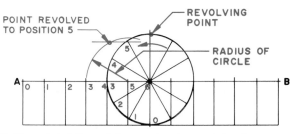

POINT REVOLVED TO POSITION 5

REVOLVING POINT

RADIUS OF CIRCLE

5. Use points 1 through 6 as centers for revolving the circle. Draw a partial circle from position 5 and draw a horizontal line from the point on the circle. This is a partial revolution of the point to the left.

Repeat this for the other positions.

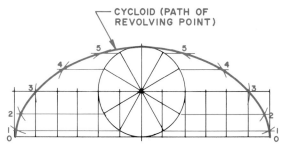

CYCLOID (PATH OF REVOLVING POINT)

6. Locate the revolving point at each of the divisions. Connect the points to form the cycloid.

FIGURE 5–65 How to draw a cycloid generated by a circle moving in a straight line.

(Fig. 5–66). When the circle rolls on the inside of another circle, the path of the point is a hypocycloid (Fig. 5–67).

One use for cycloids is when designing a cycloidic system of gear teeth.

The method for drawing the epicycloid and hypocycloid are similar to that used for the cycloid.

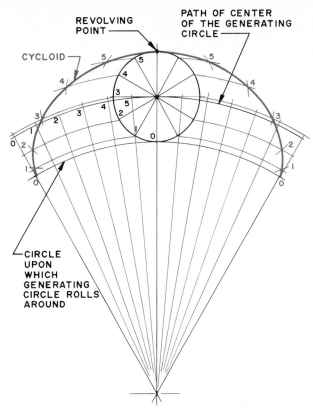

REVOLVING POINT

PATH OF CENTER OF THE GENERATING CIRCLE

CYCLOID

CIRCLE UPON WHICH GENERATING CIRCLE ROLLS AROUND

Assignment: Draw an epicycloid (with the generating circle outside a second circle).

1. Follow the procedure for a cycloid. Notice that the divisions of the large circle run from its center. The horizontal projections of the points are arcs running parallel with the path of the center of the generating circle.

FIGURE 5–66 How to draw an epicycloid with the generating circle on the outside of a second circle.

FIGURE 5–67 How to draw a hypocycloid with the generating circle inside a second circle.

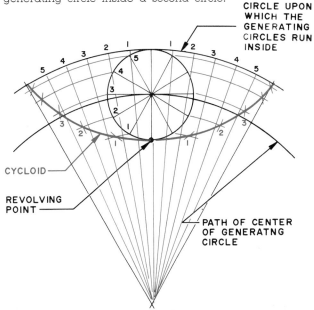

CIRCLE UPON WHICH THE GENERATING CIRCLES RUN INSIDE

CYCLOID

REVOLVING POINT

PATH OF CENTER OF GENERATNG CIRCLE

Assignment: Draw a hypocycloid (with the generating circle inside a second circle).

1. The procedure is the same as for the cycloid and epicycloid.

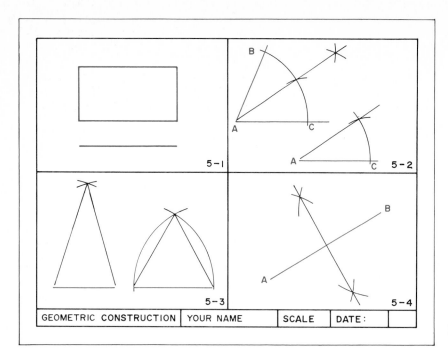

FIGURE 5–68  Suggested problem layout.

# Problems

Solve the following problems using 9 × 12 in. vellum. Draw the border and title block as shown in the example in Fig. 5–68. Since accuracy is important, use a pencil with a sharp point and a hard lead. Be certain the lead in the compass is properly pointed. The required lines should be drawn to a normal degree of darkness and the construction lines much lighter. Leave the construction lines on the finished drawing.

Examine the problems and plan your use of the paper so that several can be drawn on one sheet. In many cases, four can easily fit on one sheet. The first four problems are solved in Fig. 5–68, which shows a good layout.

1. Draw the top and side view of a rectangular plane of any convenient size.

2. Draw a 67° angle and bisect it. Draw half the angle in the same space.

3. Draw an equilateral triangle and an isosceles triangle.

4. Draw a line 75 mm long on a 30° angle and bisect it.

5. Draw an arc having a radius of 6 in. Bisect the arc.

6. Draw a line on a 60° angle. Construct a second line parallel with it and 38 mm away from it.

7. Draw an arc having a radius of 6 in. Draw a second arc parallel with it and 1 in. away from it.

8. Draw a line 4 in. long on a 30° angle. Divide the line into five equal parts.

9. Draw a line 2 in. long. Locate a point 2½ in. above it. Construct a perpendicular from the point to the line.

10. Draw a triangle with sides equal to 50, 75, and 90 mm. Bisect each angle to find the center and use it to draw an inscribed circle.

11. Draw an angle of 42°15′ using the tangent method.

12. Draw a square that has an across-the-corners distance of 2 in.

13. Draw a pentagon that is circumscribed by a 2½-in. circle.

14. Draw a hexagon with an across-the-corners distance of 50 mm.

15. Draw a hexagon with an across-the-flats distance of 2½ in.

16. Draw a square with sides that measure 70 mm. Inscribe a circle and circumscribe a circle about it.

17. Draw a square with 60-mm sides and inscribe an octagon.

18. Draw a regular polygon having seven sides, each 1 in. long.

19. Draw a 3-in. circle. Show how you would find the center if it were not known.

20. Draw an octagon having an inscribed circle with a diameter of 50 mm.

21. Place at random three points approximately 1½ to 2 in. apart. Then construct a circle that will pass through these three points and inscribe a hexagon.

22. Draw two straight lines at 90° to each other. Then draw an arc with a 1½-in. radius tangent to both lines. Mark the tangent point with a short line.

23. Draw arcs having the radii indicated tangent to the arcs in Prob. 5–23. Mark each tangent point with a short line.

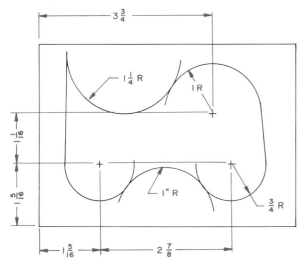

PROBLEM 5–23  Tangency problem.

24. Lay out the object in Prob. 5–24. Locate each tangent point with a short line.

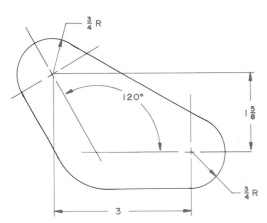

PROBLEM 5–24  Tangency problem.

25. Draw a line tangent to a 2-in.-diameter circle from a point that is 3 in. from the center of the circle.

26. Using the foci method, construct an ellipse that has a major axis of 75 mm and a minor axis of 50 mm.

27. Using the four-center method, draw the ellipse detailed in Problem 5–26. Lay the ellipse in problem 26 over this one to compare the difference in shape.

28. Draw a parabola with the focus 1¼ in. from the directrix and point D 1½ in. from the focus point.

29. Draw a parabola with a span of 4 in. and a rise of 2½ in.

30. Draw a hyperbola that has vertices 1¼ in. from the center line and foci 2 in. from the center line.

31. Draw a hyperbola that has known asymptotes at right angles to each other and a known point on the hyperbola that is 1½ in. from both.

32. Draw a right-hand helix that has a diameter of 50 mm and a lead of 35 mm. Draw it through two complete revolutions.

33. Draw a helix on a right cone having a base diameter of 2¼ in., an altitude of 3 in., and a lead of 2¼ in.

34. Draw an involute of a circle having a diameter of 75 mm.

35. Draw an involute of a ½-in. square.

36. Draw an involute of an equilateral triangle with sides ¾ in. long.

37. Draw a cycloid formed by a point on a circle revolving on a straight line. The diameter of the circle is 2 in. (This solution will require half the 9 × 12 in. vellum sheet used for these problems.)

38. Draw an epicycloid formed by a point on a circle with a diameter of 1¾ in. rolling on an arc with a radius of 5¼ in. (This solution will require a full 9 × 12 in. sheet.)

39. Draw a hypocycloid formed by a point on a circle with a diameter of 1½ in. rolling on an arc with a radius of 5 in. (This solution will require a full 9 × 12 in. sheet.)

40. Draw the sprocket in Prob. 5–40 full size. Leave all construction lines on the drawing.

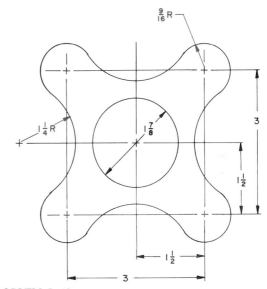

PROBLEM 5–40  Sprocket.

41. Draw the mast step in Prob. 5–41 twice full size. Leave all construction lines on the drawing.

42. Draw the cam lever in Prob. 5–42 full size. Leave all construction lines on the drawing.

43. Draw the latch in Prob. 5–43. Latch A revolves clockwise until it rests on pin B. Draw the latch as shown and then in its revolved position. Show it with phantom lines in the revolved position. Draw it twice full size, and leave all construction lines on the drawing.

44. Draw the bracket in Prob. 5–44 twice full size. Leave all construction lines on the drawing.

45. Draw the pinion in Prob. 5–45 full size. Leave all construction lines on the drawing.

| | MAJOR AXIS | MINOR AXIS |
|---|---|---|
| A | 4.87 | 2.80 |
| B | 3.68 | 2.12 |
| C | 2.68 | 1.56 |

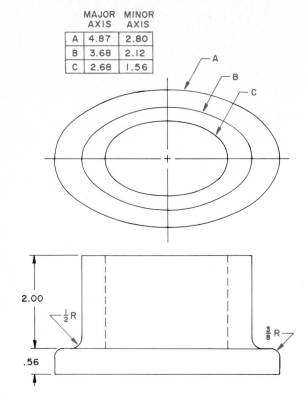

PROBLEM 5–41   Mast step.

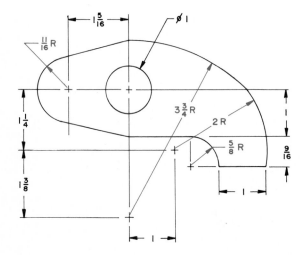

PROBLEM 5–42   Cam lever.

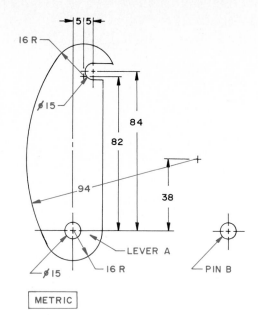

METRIC

PROBLEM 5–43   Latch.

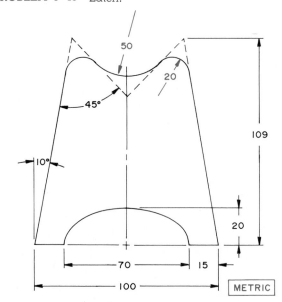

METRIC

PROBLEM 5–44   Bracket.

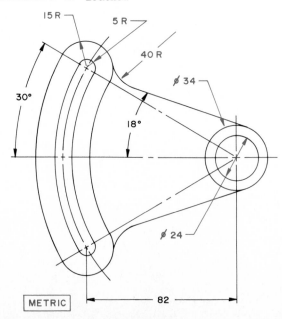

PROBLEM 5–45   Pinion.

METRIC

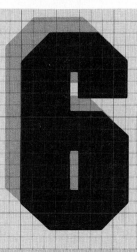

# The Principles of Orthographic Projection

Engineers use different kinds of drawings to describe the products they are designing and manufacturing. Among these, the engineering drawing or multiview drawing is a major means of communicating the design concept.

Engineering drawings are composed of views of the product using a standard system of planes of projection. The drawings are composed of points, lines and planes.

## POINTS

A point is a dimensionless geometric object having no property but location. It indicates a position on a drawing (Fig. 6–1). When points move in space, they form lines, surfaces, and volumes. A point moving in a straight path forms a straight line. A point moving in a circular path with a fixed radius from a center forms a circle (Fig. 6–2).

## LINES

A line is formed when two planes intersect. Theoretically, a line has no thickness and extends to infinity (Fig. 6–3). Lines may be straight or curved.

On engineering drawings, lengths are established when sizes are decided. This gives the line a finite value.

## PLANES

Objects are made up of plane and curved surfaces. A plane is a surface on which a straight line drawn between any two points on the surface lies on that surface. If two straight lines intersect, they form a plane. Two parallel straight lines form a plane. When three points not on a line are connected, they form a plane. When a point and a line are connected, a plane is formed. The top of your drafting table is a plane (Fig. 6–4). Theoretically, planes have no limit. They project to infinity (Fig. 6–5). On engineering drawings, limits are established when sizes are decided.

## ORTHOGRAPHIC PROJECTION

Orthographic projection is the system of projection used when preparing engineering drawings. It is a standardized method of showing three-dimensional

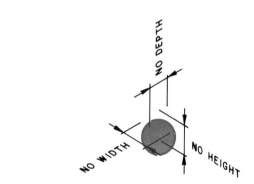

FIGURE 6–1  A point is dimensionless.

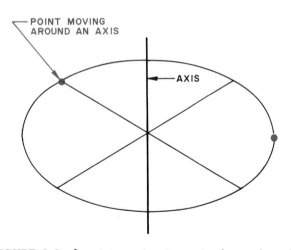

FIGURE 6–2  A point moving in a circular path with a fixed radius from a center forms a circle.

FIGURE 6–3  A line has no thickness and extends to infinity.

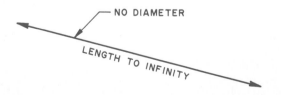

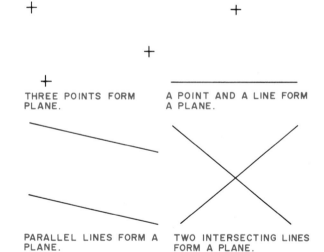

THREE POINTS FORM PLANE.

A POINT AND A LINE FORM A PLANE.

PARALLEL LINES FORM A PLANE.

TWO INTERSECTING LINES FORM A PLANE.

FIGURE 6–4  Various ways to identify a plane.

FIGURE 6–5  Theoretically, a plane has no limits.

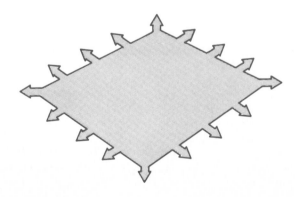

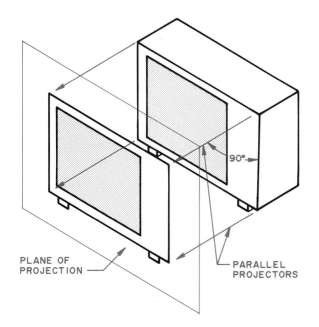

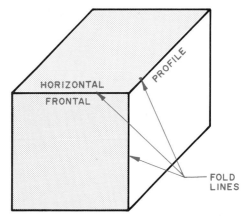

I. THE PRINCIPAL PLANES OF PROJECTION.

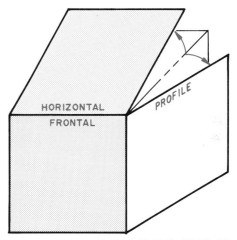

2. THE PLANES UNFOLD LIKE THE SIDES OF A BOX.

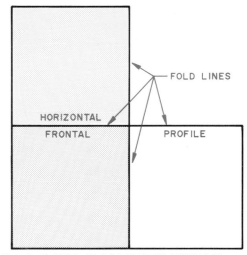

3. THE PLANES OF PROJECTION UNFOLDED FORMING A SINGLE PLANE.

FIGURE 6–6 Orthographic projection involves projecting views perpendicular to planes of projection.

FIGURE 6–7 The three principal planes of projection.

products by drawing views that are projected onto planes of projection using parallel projection lines that are perpendicular to the planes of projection (Fig. 6–6). The projection lines are parallel and perpendicular to the planes of projection because the purpose of the projection is to record the actual size of the product being drawn. If it were not projected perpendicularly and parallelly, the view drawn would be distorted.

The three principal planes of projection are perpendicular to each other. They are identified as *horizontal, frontal,* and *profile* (Fig. 6–7). They form a boxlike shape.

The plane upon which an engineer draws is a sheet of paper. It is necessary to get the principal planes of projection, which are three-dimensional, onto a single plane (the drawing paper). This is done by unfolding the three planes (just like a box) until they all lie in a common plane (the sheet of drafting paper) (Fig. 6–7).

The planes of projection are infinite in size. Sometimes boundaries are arbitrarily drawn for the convenience of the engineer. The intersections between the planes of projection are sometimes called *reference lines.* They are used as points of reference for locating views on the planes of projection.

## Planes of Projection

The planes of projection join and form quadrants. The quadrants are identified as first, second, third, and fourth. The first and third quadrants are used for orthographically projected drawings (Fig. 6–8).

*Third-angle projection* means that the object to be drawn is placed in the third quadrant. It is projected up to the horizontal plane, forward to the frontal plane, and to the right to the profile plane (Fig. 6–9). When these planes are unfolded to form a single

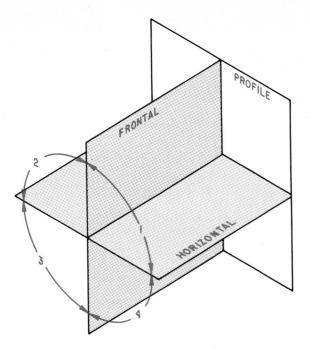

FIGURE 6–8 The planes of projection form four quadrants. Orthographically projected drawings fall in the first or third quadrants.

plane, the views appear as in Fig. 6–10. Notice the symbol indicating that this drawing uses third-angle projection.

The view drawn on the horizontal plane is often called the top view. The view on the frontal plane is called the front view. The view on the profile plane is called the side view. In this case, it is the right-side view of the product.

Third-angle projection is used in the United States. *First-angle projection* is used in some other countries. The product is placed in the first quadrant. The horizontal plane is below the product. The top of the product is projected down to the horizontal plane. The front view projects back to the frontal plane. The side view projects to the profile plane (Fig. 6–11). As these planes are unfolded to form a single plane, the views appear as in Fig. 6–12. Notice the symbol placed on the drawing to indicate that first-angle projection is used.

### Space Locations

When a product is positioned in the third quadrant, the various corners and sides must be located. These

FIGURE 6–9 Third-angle projection places the object in the third quadrant.

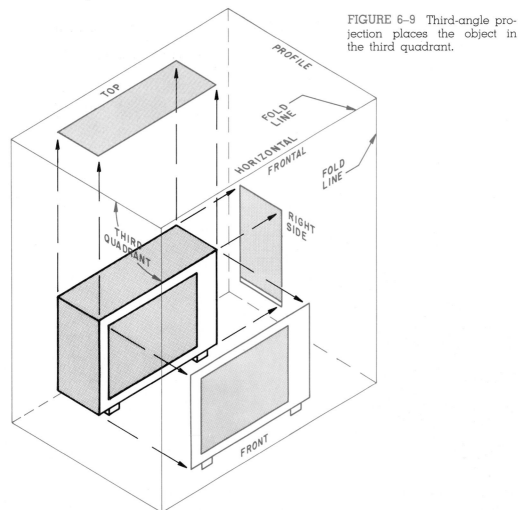

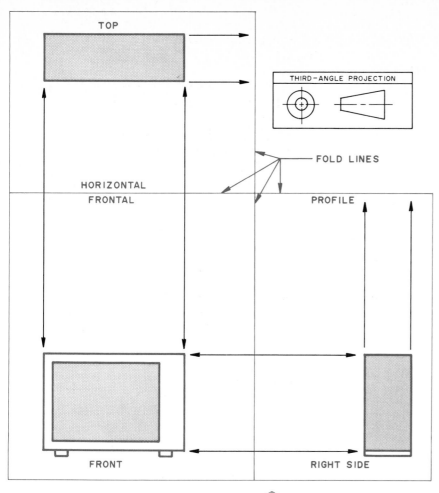

**FIGURE 6-10** The planes of projection from the third quadrant unfolded into a single plane.

**FIGURE 6-11** First-angle projection places the object in the first quadrant.

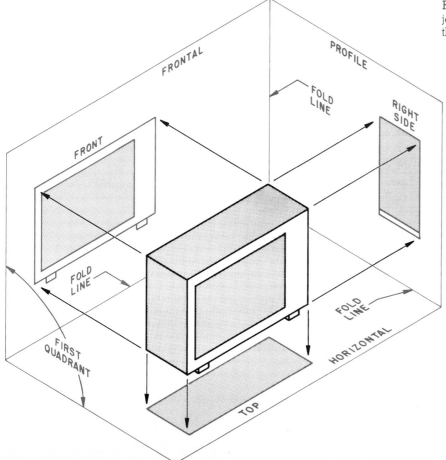

113

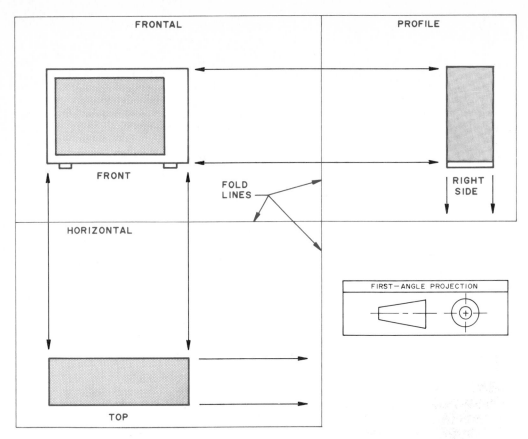

FIGURE 6–12 The planes of projection from the first quadrant unfolded into a single plane.

are measured from the planes of projection. The system for doing this is shown in Fig. 6–13. The location of a point is described as forward or back, right or left, or up or down from the plane to which it is being related. The left and right directions are perpendicular to the profile plane. They appear in the

FIGURE 6–13 A system for describing direction in relation to the planes of projection in orthographic projections.

horizontal and frontal planes (Fig. 6–14). The forward and back directions are perpendicular to the frontal plane. They appear in the horizontal and profile plane (Fig. 6–15). The up and down direc-

FIGURE 6–14 Right and left directions are perpendicular to the profile plane.

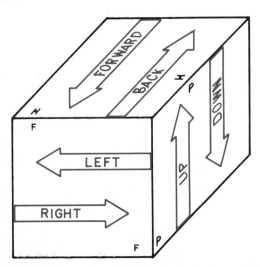

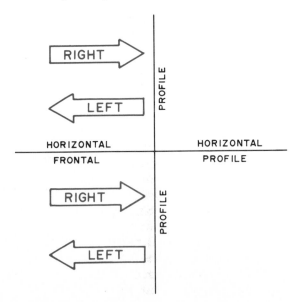

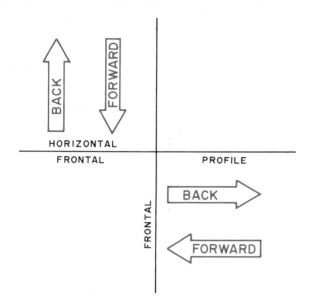

FIGURE 6–15  Forward and backward directions are perpendicular to the frontal plane.

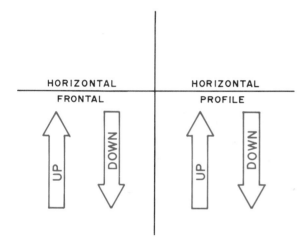

FIGURE 6–16  Up and down directions are perpendicular to the horizontal plane.

tions are perpendicular to the horizontal plane. They appear in the frontal and profile planes (Fig. 6–16). An application of these directions is shown in Fig. 6–17. Here a point is located 3 units back, 2 units left, and 4 units down.

### Height, Width, and Depth

When discussing the size of a product, the terms *height, width,* and *depth* are used. Height is the perpendicular distance between horizontal planes. Width is the perpendicular distance between profile planes. Depth is the perpendicular distance between frontal planes (Fig. 6–18).

When looking at a product, the front view shows the width and height. The side view shows the height and depth. The top view shows the width and depth (Fig. 6–19).

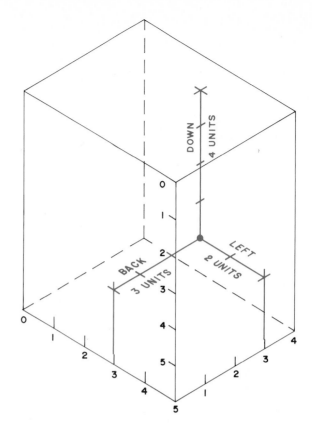

FIGURE 6–17  Directions for locating a point in an orthographic projection.

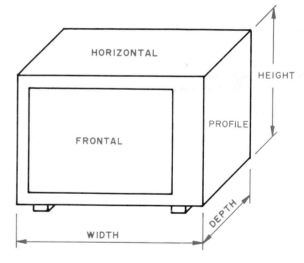

FIGURE 6–18  Terms used to designate overall sizes.

## PROJECTING VIEWS

The three principal views are perpendicular to each other. They form the top, front, and side views of a product. There are six possible normal views. Imagine a product placed in a glass box (Fig. 6–20). If the product is projected to all six sides of the box, three views in addition to the three principal views are

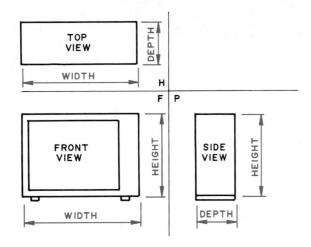

FIGURE 6–19  Size designations on normal views.

FIGURE 6–20  Views can be projected to the six principal planes of projection.

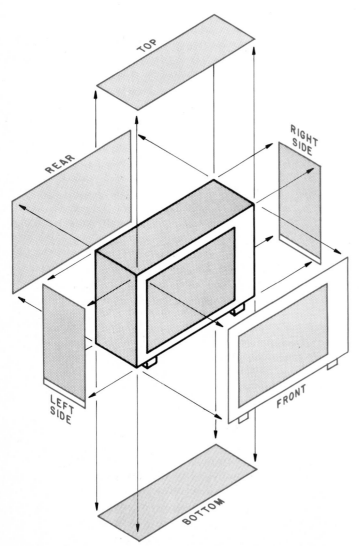

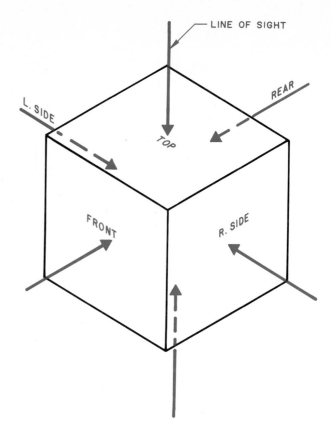

FIGURE 6–21  The lines of sight are perpendicular to the planes of projection.

formed. The lines of sight are perpendicular to the planes of projection (Fig. 6–21).

To represent these views on an engineering drawing, they must all be placed on a plane (the drawing paper). The box with the views projected

FIGURE 6–22  The principal planes unfold to form a single plane.

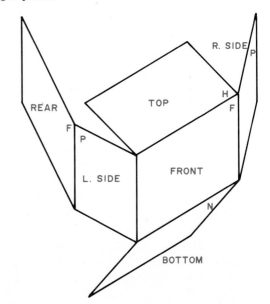

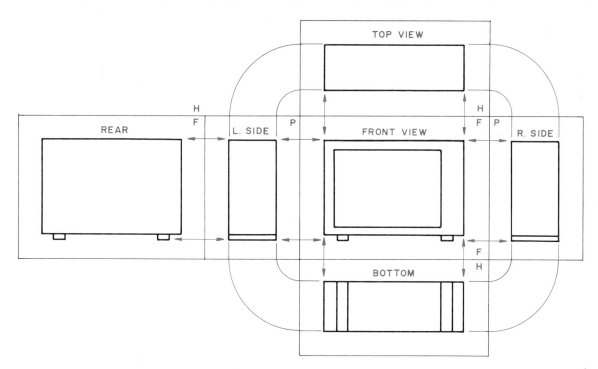

FIGURE 6–23 The six planes of projection unfolded into a single plane (the drawing paper).

FIGURE 6–24 The edges of the planes of projection form the reference lines.

orthographically upon them is unfolded to form a plane (Fig. 6–22). The system has established standard positions for each side of the box. This locates each of the six views in a standard position (Fig. 6–23).

## Arrangement of Views

The front view is the most important view and forms the core of the system. The top view folds up and the bottom view folds down from the front view. The right and left sides fold off each edge of the front view. The rear view normally folds off the left side view.

## Reference Lines

The edges of the imaginary glass box enclosing a product to be drawn are called reference lines or fold lines. When the box is unfolded, they separate the views. Reference lines are really the edge views of the planes of projection. They are used to locate the views on the drawing. They are the starting point from which location dimensions are taken. When you look perpendicular to the horizontal plane, you see the frontal and profile planes in edge view. When you look perpendicular to a profile plane, you see the horizontal and frontal plane in edge view. When you look perpendicular to a frontal plane, you see the horizontal and profile planes in edge view (Fig. 6–24).

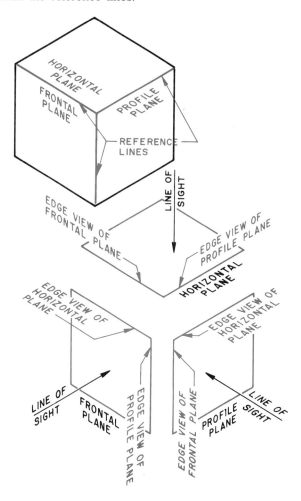

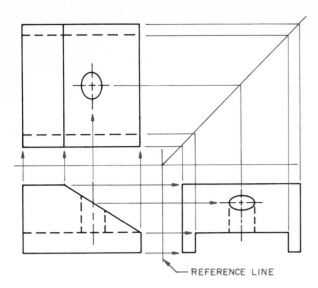

FIGURE 6–25 Adjacent views, as the front and side, share the same reference line.

## Locating and Projecting the Views

Two adjacent views share the same reference line. This relationship is important when making a drawing. For example, if a view is drawn, such as the front view, dimensions from it can be projected across the reference line to adjacent views (Fig. 6–25). In addition, the location of the other views is determined by their distance from the reference line. For example, in Fig. 6–26, the front of the product is distance X from the frontal plane. It will be this distance in any view in which the frontal plane appears as an edge, the top, right, and left sides and the bottom. The same relationship applies to the other reference planes. A point on a product is the same distance from the reference plane wherever that plane appears in edge view. Since points are the same distance from a reference plane in related views, they can be transferred directly to that view. This is most efficiently done with a divider (Fig. 6–27).

## SELECTING THE VIEWS

The first thing to decide before starting to draw is which views are needed and the best way to position the product on the drawing. The front view is the most important and is selected first. It is the view that shows the most about the product. It is usually the longest view and shows the major shape or profile. It is positioned so that the least number of hidden lines are needed. When the angle plate in Fig. 6–28 was drawn, it was turned so that the open side was visible, reducing the hidden lines.

The object should be drawn in its normal posi-

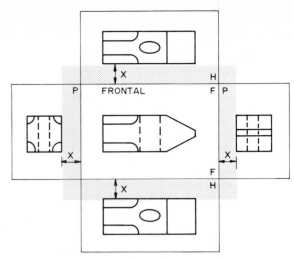

LOCATING AN OBJECT FROM THE FRONTAL PLANE.

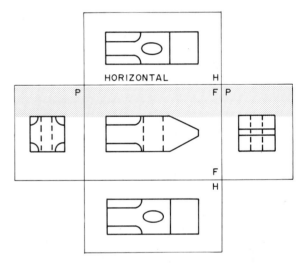

LOCATING AN OBJECT FROM THE HORIZONTAL PLANE.

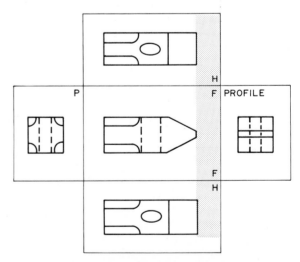

LOCATING AN OBJECT FROM THE PROFILE PLANE.

FIGURE 6–26 The various views are located by their distance from the reference lines.

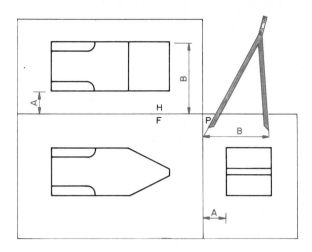

FIGURE 6–27 Points on related views can be transferred by measuring from the reference plane.

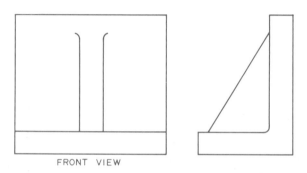

FRONT VIEW

FIGURE 6–28 The front view is the longest view. The product is positioned to have the least number of hidden lines.

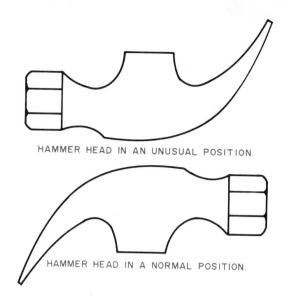

HAMMER HEAD IN AN UNUSUAL POSITION.

HAMMER HEAD IN A NORMAL POSITION.

FIGURE 6–29 Place objects on the drawing so that they appear in a normal position.

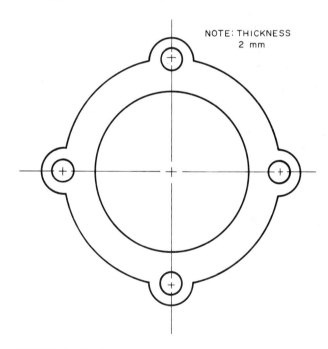

NOTE: THICKNESS 2 mm

FIGURE 6–30 A one-view drawing.

tion. The hammer head in Fig. 6–29 appears as you would expect it.

Most products require three views to describe them fully. The decision as to the number of views involves a close study of the product. Use enough views to describe all the features and no more.

## One-View Drawings

There are many products that can be described with one view and a note. A gasket or other thin, flat product can have a note giving the thickness, thus eliminating the need for all views except the front (Fig. 6–30).

## Two-View Drawings

Products that are symmetrical, conical, or pyramidal in shape can usually be described with two views. The cap in Fig. 6–31 does not need a top, bottom, or left side view. They are identical with the right side view.

## Three-View Drawings

Most products require three-view drawings. The electric pencil sharpener in Fig. 6–32 requires three views to show all the details. The front view is the long view and shows the basic shape. The right side view shows the profile and the unit that holds various-size pencils. The top view shows the point-fineness adjusting wheel and the screw that secures the shell to the base.

While the front, top, and right side views are most frequently used, any of the other regular or-

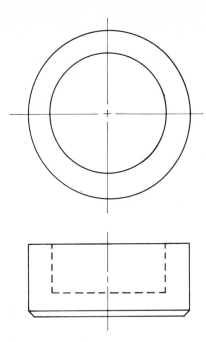

FIGURE 6–31  A two-view drawing.

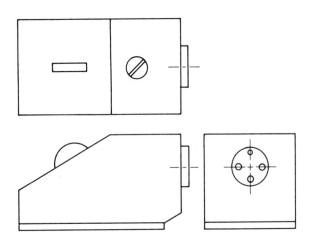

FIGURE 6–32  A three-view drawing requiring front, top, and side views.

FIGURE 6–33  This three-view drawing requires a front and two side views.

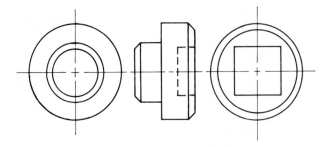

thographic views could be used instead. In Fig. 6–33 is a product requiring right and left side views. No top view is needed because it is the same as the front view.

## More Than Three Views

The designer can use any or all of the six normal orthographic views. The views used must present information that is not clearly shown by the other views. In Fig. 6–34, a bottom view is needed to show the recesses. This view shows they are square. Without this view we would not know their shape.

In addition, a designer might have to use auxiliary views. These are described in detail in Chapter 9.

## Partial Views

Sometimes it is economical to draw only a part of a view. Common partial views include half of a symmetrical object, a portion of one part of an object, or a partial view with a section (Fig. 6–35).

## Alternate Positions of Views

There are times when it is desirable to place views in positions other than the six standard locations. The alternate positions include locating the rear view

FIGURE 6–34  This drawing requires a fourth view, the bottom, to show clearly details located there.

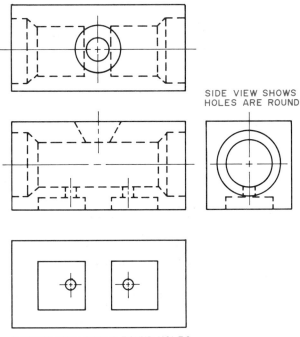

TOP VIEW SHOWS THAT THE CONE SHAPED RECESS IS ROUND.

SIDE VIEW SHOWS HOLES ARE ROUND

BOTTOM VIEW SHOWS ROUND HOLES AND SQUARE RECESSES.

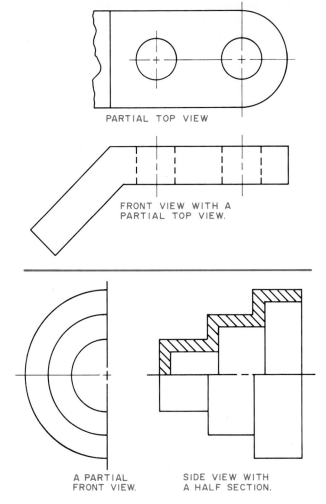

PARTIAL TOP VIEW

FRONT VIEW WITH A
PARTIAL TOP VIEW.

A PARTIAL
FRONT VIEW.

SIDE VIEW WITH
A HALF SECTION.

FIGURE 6–35 Partial views can be used to save drafting time.

off the top view and the right and left side views off the top view (Fig. 6–36).

FIGURE 6–36 Alternate positions for rear and side views.

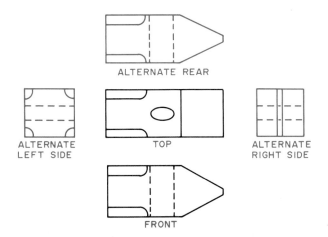

ALTERNATE REAR

ALTERNATE
LEFT SIDE

TOP

ALTERNATE
RIGHT SIDE

FRONT

## LAYING OUT A DRAWING

Once a decision has been made on the views needed to describe a product, the layout for the drawing must be planned. Needed is space for the actual views plus space for dimensions and notes. One way to figure this is to make a freehand sketch of the product and quickly sketch on the dimensions. Since each row of dimensions requires at least 6 mm (¼ in.), the space needed can be determined (Fig. 6–37). Select a piece of drawing paper large enough to handle the drawing plus a border and some clear area on all sides of the drawing. Most companies have standards on the size of drawings they use; in this case, you would have to fit your drawing into one of these sizes.

Arrange the views so that they give the drawing a balanced, uncrowded appearance. The right and left margins are usually kept about equal in size. The top and bottom margins can be equal. Some people prefer the bottom to be slightly larger.

Once the plan is complete, begin by blocking in the views on the drafting paper. Use thin, light construction lines. Cylindrical parts are located by their center lines. Draw these at this point.

After the width, length, and depth are blocked in, begin by drawing the details. After drawing a detail, as a hole, on one view, draw it on all views before going on to another detail. Do not complete one view and then try to complete another view. Carry all views along to completion equally.

After the details are drawn, remove unnecessary lines. If construction lines are very light, they can be left on the drawing. Darken the lines that are to remain and get them to their proper thickness. All lines should be very black. They do vary in thickness but not in blackness.

Following is a summary of this process.

1. Decide on the number of views needed.
2. Select the scale to be used.
3. Calculate the space needed for dimensions and notes.
4. Make a freehand layout and calculate the total space needed.
5. Select a paper large enough to permit the drawing to be made without crowding.
6. Lay out the views as blocks and locate the center lines of cylindrical parts.
7. Begin drawing the details. The following order is suggested.
   a. Center lines
   b. Arcs and circles
   c. Horizontal visible lines
   d. Vertical visible lines
   e. Inclined visible lines
   f. Hidden lines

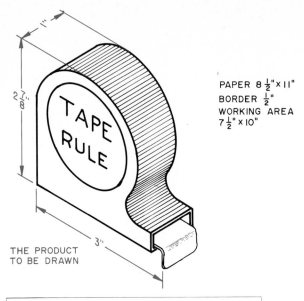

PAPER 8½"×11"
BORDER ½"
WORKING AREA
7½"×10"

THE PRODUCT
TO BE DRAWN

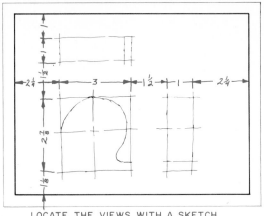

LOCATE THE VIEWS WITH A SKETCH

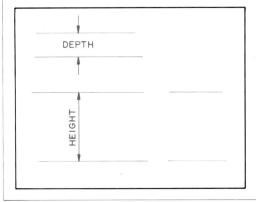

1. LAY OUT THE HEIGHT AND DEPTH.

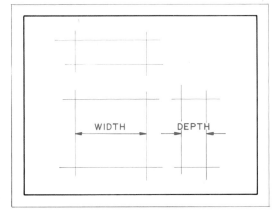

2. LAY OUT THE WIDTH AND DEPTH.

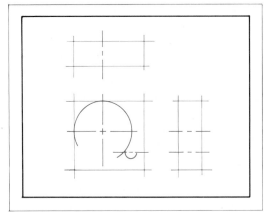

3. LOCATE CENTER LINES AND DRAW CIRCLES
AND ARCS.

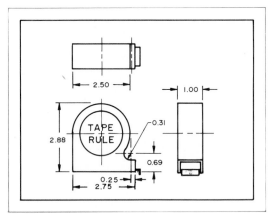

4. COMPLETE THE DETAILS, DARKEN THE LINES,
ADD DIMENSIONS AND NOTES.

FIGURE 6–37 How to plan the layout of a drawing and complete the views.

    g. Section line
    h. Extension and dimension lines
  8. If necessary, draw lines to proper darkness and width.
  9. Letter dimensions and notes.

## PROPER USE OF LINES

Line symbols are described in Chapter 3. The following discussion tells how to use these lines.

    *Center lines* are used to locate the centers of

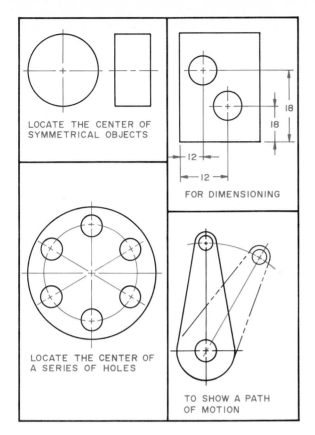

FIGURE 6–38 Typical uses of center lines.

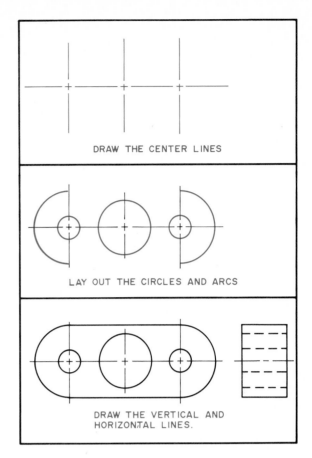

FIGURE 6–39 How to draw a symmetrical object.

symmetrical objects and paths of motion (Fig. 6–38). A symmetrical object is one that is the same on each side of the center line. Center lines are drawn using long and short dashes. The long dash is usually 20 to 35 mm long (¾ to 1½ in.), and the short dash is about 3 mm (⅛ in.) long. The center line must extend 6 mm (¼ in.) beyond the edge of the part.

To draw a symmetrical object, first locate the center lines. Then draw the circles and arcs. Finally, draw the horizontal and vertical lines (Fig. 6–39).

*Phantom lines* are used to show alternate positions for parts that move or a part that is next to something to be drawn but is not actually a part of it (Fig. 6–40).

Phantom lines are a long dash 20 to 35 mm (¾ to 1½ in.) followed by two short dashes each 3 mm (⅛ in.) long. The dashes are about 1 mm (⅛32 in.) apart.

*Hidden lines* are used to indicate details that are hidden from view. The symbol is a series of short dashes about 3 mm (⅛ in.) long spaced about 1 mm (⅛32 in.) apart. There are a number of standard practices for drawing hidden lines (Fig. 6–41).

## Precedence of Lines

Frequently, two lines will fall on top of each other. When this occurs, one must take precedence. Prior-

FIGURE 6–40 Phantom lines are used to show alternate positions.

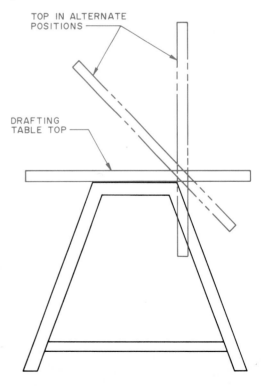

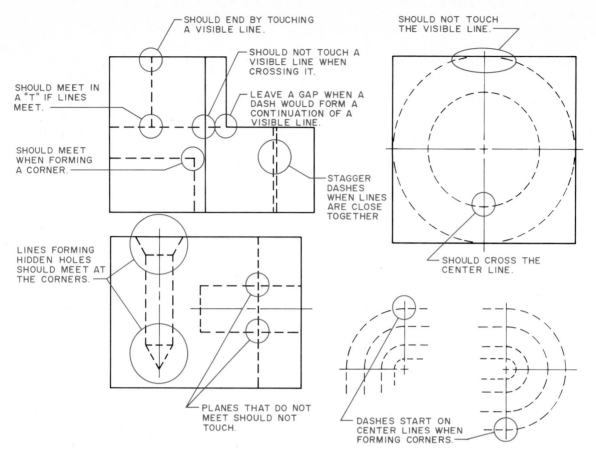

FIGURE 6-41 Standard practices for drawing hidden lines.

ities are according to the following list, in descending order of importance.

1. Visible line.
2. Hidden line.
3. Cutting plane.
4. Center line.

## PLANE AND CURVED SURFACES

When drawing engineering drawings, the objects will contain plane and curved surfaces. As the views are projected onto the planes of projection, these surfaces must be located and drawn. This is done by applying the principles of orthographic projection.

### Inclined Planes

An inclined surface is perpendicular to one plane of projection and at an acute angle (less than 90°) to the other two planes of projection (Fig. 6-42). The plane is shown in its true length on the plane of projection to which it is perpendicular. It is foreshortened (less than true size) in the other planes of projection.

### Oblique Planes

An oblique plane is not perpendicular to or parallel with any of the normal planes of projection. It appears foreshortened on all normal planes (Fig. 6-43).

### Curved Surfaces

The three common types of curved surfaces are cylindrical, conical, and spherical. Both cylindrical and conical surfaces appear as circles in the plane that is perpendicular to their axis. The cylinder appears as a rectangle in the other planes. The cone appears as a triangle in the other planes. The sphere appears as a circle on all planes of projection (Fig. 6-44).

### Projecting Surfaces with Curved Edges

Surfaces with curved edges are projected from one view to the other by locating a series of points along the edge and projecting these points to the views. They are then connected to form the projected edge. The more points used, the more accurate the projection will appear.

A plane with a curved edge is projected in Fig. 6-45.

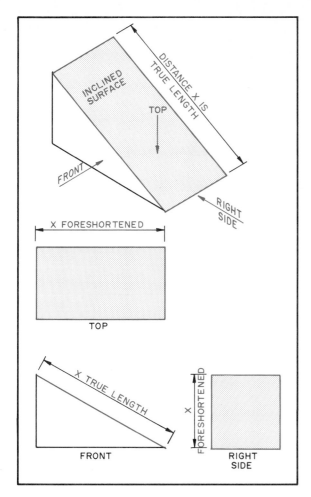

FIGURE 6–42 An inclined plane appears in its true length in the view to which it is perpendicular.

## INTERSECTIONS OF PLANE AND CURVED SURFACES

Even the simplest of products involves the intersection of surfaces of various types. As multiview

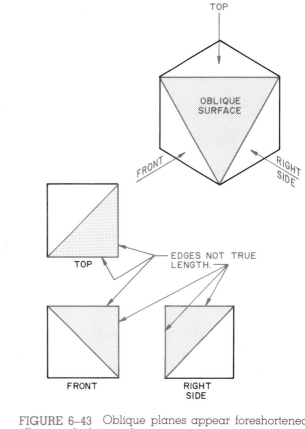

FIGURE 6–43 Oblique planes appear foreshortened in all normal planes of projection.

drawings are made, the intersections between these must be drawn.

*Intersections between a plane and a cylinder* produce plane surfaces. When the intersecting plane is parallel with the axis, a solid cylinder produces a rectangular plane. A hollow cylinder produces two rectangular planes (Fig. 6–46). When the plane is at an angle to the cylinder, an elliptical plane is produced (Fig. 6–47).

*Intersections between a plane and a cone* form four

FIGURE 6–44 Curved surfaces project in various ways.

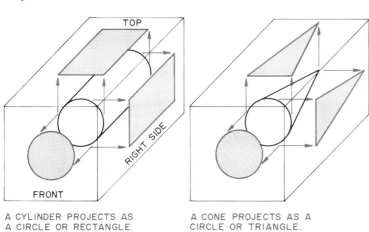

A CYLINDER PROJECTS AS A CIRCLE OR RECTANGLE.

A CONE PROJECTS AS A CIRCLE OR TRIANGLE.

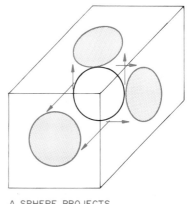

A SPHERE PROJECTS AS A CIRCLE.

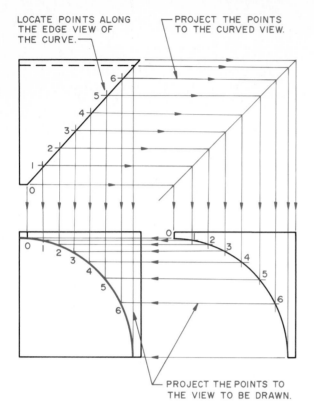

LOCATE POINTS ALONG THE EDGE VIEW OF THE CURVE.

PROJECT THE POINTS TO THE CURVED VIEW.

PROJECT THE POINTS TO THE VIEW TO BE DRAWN.

FIGURE 6–45 How to project a curved surface to normal planes of projection.

FIGURE 6–46 Intersections between planes and cylinders produce rectangular surfaces when the plane is parallel with the axis of the cylinder.

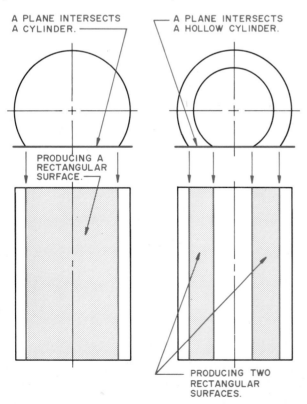

A PLANE INTERSECTS A CYLINDER.

A PLANE INTERSECTS A HOLLOW CYLINDER.

PRODUCING A RECTANGULAR SURFACE.

PRODUCING TWO RECTANGULAR SURFACES.

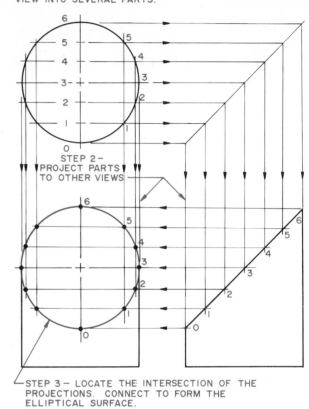

STEP 1 – DIVIDE THE CIRCULAR VIEW INTO SEVERAL PARTS.

STEP 2 – PROJECT PARTS TO OTHER VIEWS

STEP 3 – LOCATE THE INTERSECTION OF THE PROJECTIONS. CONNECT TO FORM THE ELLIPTICAL SURFACE.

FIGURE 6–47 How to project an elliptical surface formed by the intersection of a cylinder and a plane that is at an angle to the axis of the cylinder.

basic surfaces. These are a circle, ellipse, parabola, and hyperbola (Fig. 6–48). The circle is formed when the plane is perpendicular to the axis of the cone. The ellipse is formed when the cone is cut by an inclined plane that makes an angle with the axis greater than the elements. A parabola is formed when the plane is parallel with an element of the cone. A hyperbola is formed when the plane makes an angle with the axis that is smaller than the elements.

*Intersections between plane and curved surfaces* occur frequently on engineering drawings. An actual intersection occurs at a tangent point, so there is no visible line of intersection (Fig. 6–49). However, when two curved surfaces meet, a visible edge is shown. When curved surfaces meet in a continuous slope, no visible edges appear (Fig. 6–50).

*Intersections between cylinders* depend upon the diameter of the cylinders. When the intersecting cylinders are the same size, the intersection projects as a straight line. When one cylinder is very small, the intersection is a straight line. The line of intersection can be approximated by selecting an arbitrary radius and drawing an arc (Fig. 6–51). When diameters are

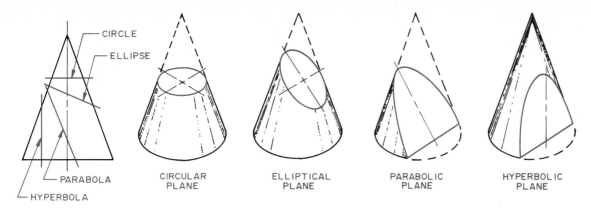

FIGURE 6–48 Intersections between a plane and a cone form circular, elliptical, parabolic, and hyperbolic planes.

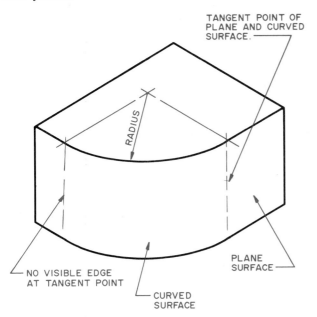

FIGURE 6–49 The intersection between a plane and a curved surface occurs at a tangent point.

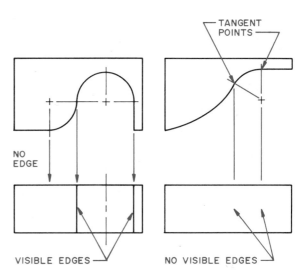

FIGURE 6–50 Intersections between cylinders and other curved surfaces occur at a tangent point.

FIGURE 6–51 How to draw the lines of intersection between cylinders with various diameters.

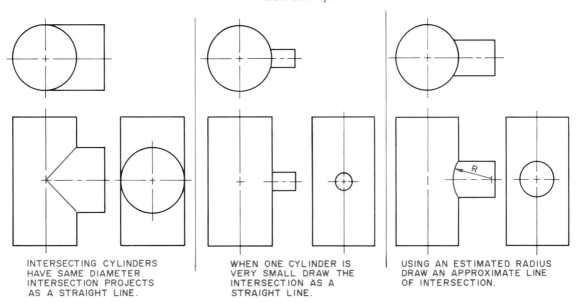

INTERSECTING CYLINDERS HAVE SAME DIAMETER. INTERSECTION PROJECTS AS A STRAIGHT LINE.

WHEN ONE CYLINDER IS VERY SMALL DRAW THE INTERSECTION AS A STRAIGHT LINE.

USING AN ESTIMATED RADIUS DRAW AN APPROXIMATE LINE OF INTERSECTION.

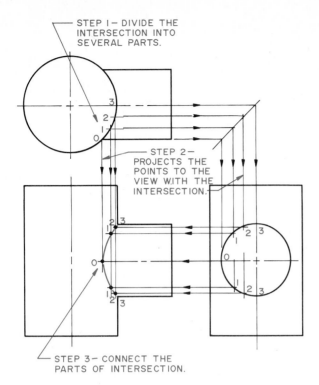

STEP 1 — DIVIDE THE INTERSECTION INTO SEVERAL PARTS.

STEP 2 — PROJECTS THE POINTS TO THE VIEW WITH THE INTERSECTION.

STEP 3 — CONNECT THE PARTS OF INTERSECTION.

FIGURE 6–52 How to plot the intersection between cylinders when their diameters are large.

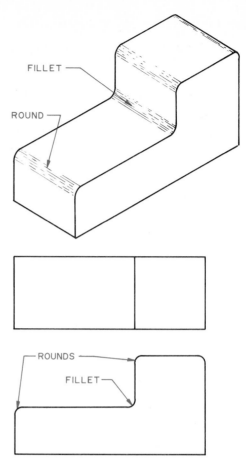

FILLET

ROUND

ROUNDS

FILLET

FIGURE 6–53 Fillets and rounds are external and internal corners used on castings.

FIGURE 6–54 Runouts occur when fillets and rounds meet a plane surface.

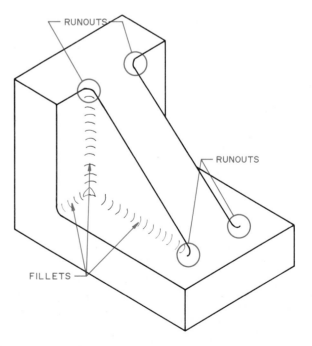

RUNOUTS

RUNOUTS

FILLETS

different and both are large, the intersection is plotted (Fig. 6–52).

## FILLETS AND ROUNDS

External corners on parts made by forging or casting are usually rounded. These corners are called *rounds*. They make it easier to form the part. Internal corners are also usually rounded. They are called *fillets* (Fig. 6–53). They are used to prevent the part from cracking at the corner.

After a part is formed, it is sometimes machined. This removes some of the metal. The rounds are machined away to form sharp corners when needed. Fillets and rounds are usually very small. They often range from 1.5 to 3.0 mm (1/16 to 1/8 in.).

Where fillets and rounds meet with other surfaces, *runouts* are formed. A runout is an extension of the curved surface caused by the fillet or round (Fig. 6–54).

The runout may turn out or in, depending upon the shape of the intersecting parts. Examples are shown in Fig. 6–55.

Small runouts are drawn freehand. Larger runouts are drawn with an irregular curve. The radius of the runout is usually drawn using the radius of the fillet. The arc is drawn about 45° of a circle.

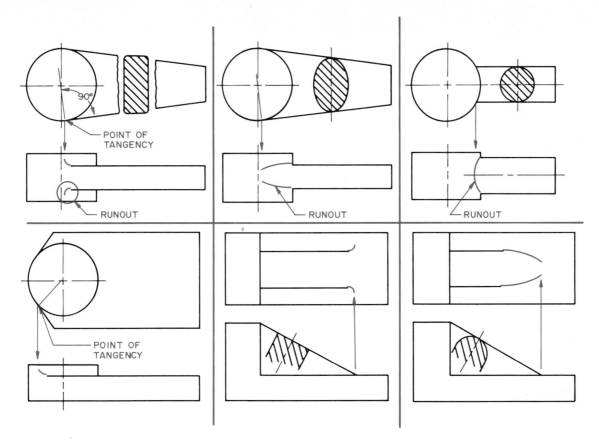

**FIGURE 6–55** Typical runouts formed by the intersection of various shaped surfaces.

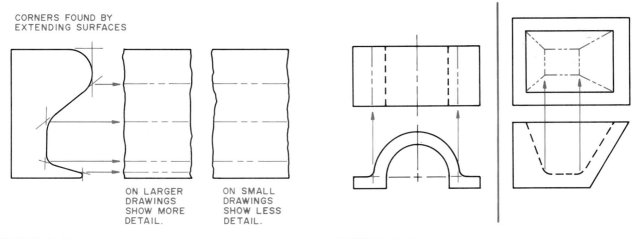

**FIGURE 6–56** The intersection of a round or fillet can be shown with a phantom line if it will clarify the drawing.

**FIGURE 6–57** Typical intersections of curved and plane surfaces indicated with phantom lines.

*Intersections at a round or fillet* are shown when two surfaces meet in order to clarify the drawing even though no visible edge occurs. The corners are found by extending the surfaces until they meet (Fig. 6–56). They are drawn using a phantom line symbol (Fig. 6–57).

## CONVENTIONAL BREAKS

Often it is necessary to break away part of a drawing to show some hidden interior detail. At times a product is too long to fit the paper; if the length is all the same shape, some of it can be broken away.

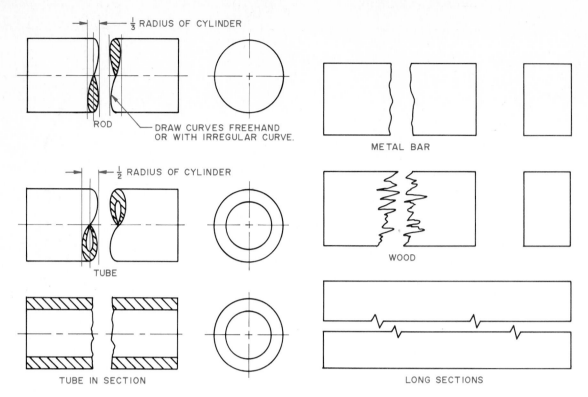

FIGURE 6–58  Symbols used to indicate breaks.

The edges of a break are indicated with a break symbol. Short breaks use a freehand irregular line. It is the same thickness as a visible line. Long breaks use a straight line with a Z symbol inserted every few inches (Fig. 6–58). It is a thin line.

Breaks in cylindrical objects, as a shaft, are made with an S-shaped symbol. The width for solid shafts is about one third the radius of the cylinder. The width for hollow shafts is about one third the radius of the cylinder.

When dimensioning products with breaks, remember to indicate the actual length of the product before the break was made.

## KNURLING

Knurling involves roughing the surface of a round part so that it will not slip when a mating part is forced over it. It is also used on knobs and handles to provide a better grip. The two common types of knurls are straight and diamond. They are drawn as shown in Fig. 6–59.

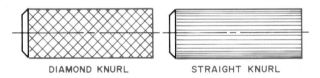

FIGURE 6–59  Types of knurling.

# Problems

The following problems are to be used to study multiview projection. A variety of experiences are presented, including views with missing lines, entire missing views, and decisions about which views are needed and how they should be placed on the paper. The problems range from simple objects to rather complex units.

Group 1.  Missing-Line Problems
The problems in Prob. 6–1 are designed to be sketched or drawn with instruments on graph sheets divided into

4 squares per inch. Sketch the views as shown, then seek out and record the missing lines. One or more lines are missing on each problem. You can get four or six problems on an 8½ × 11 in. page. If you have trouble visualizing the object, make a freehand thumbnail pictorial sketch. This often clarifies the situation.

Group 2.  Missing View Problems
The problems in Probs. 6–2 and Prob. 6–3 are designed to be sketched or drawn with instruments on 8½ × 11 in.

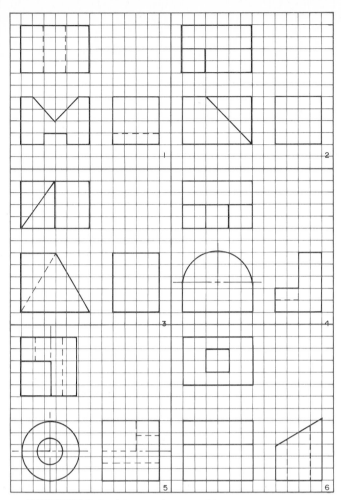

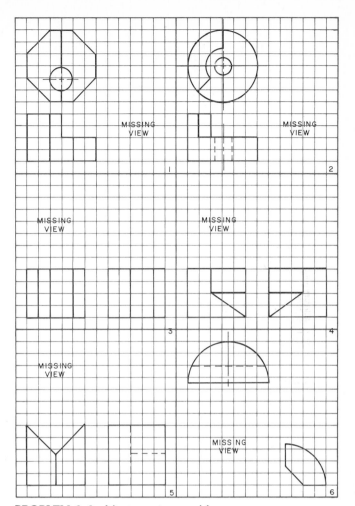

PROBLEM 6–1  Missing line problems.

PROBLEM 6–2  Missing view problems.

PROBLEM 6–3  Missing view problems.

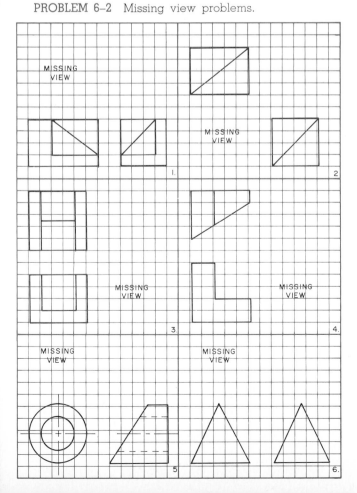

graph sheets divided into 4 squares per inch. Sketch the views shown and add the missing view. Six problems can be placed on one page.

The problems in Probs. 6–4 and Prob. 6–5 are designed to be sketched or drawn with instruments on 8½ × 11 in. graph sheets divided into 4 squares per inch. Sketch the views shown and add the missing view. Omit any dimensions shown. You can fit one problem per page.

Group 3. Working Drawings

The problems in Probs. 6–6 through Prob. 6–35 are designed to be sketched or drawn with instruments on 9 × 12 in. vellum sheets. Use a ½ in. border and title block. Draw the views needed to describe the object completely. You can fit one problem per page. These fit rather tightly on the page. If your instructor will use these later for dimensioning practice draw them on 12 in. × 18 in. vellum leaving room between the views for dimensions.

Group 4. Difficult Problems

The problems in Probs. 6–36 through Prob. 6–43 involve more complex products that require special attention to a multitude of details. Draw the views necessary to describe each product completely. Use auxiliary views, sections, and partial views where necessary. Draw each problem on 12 × 18 in. vellum. If your instructor will what them to be dimensioned, allow room for dimensions between the views.

The Principles of Orthographic Projection  **131**

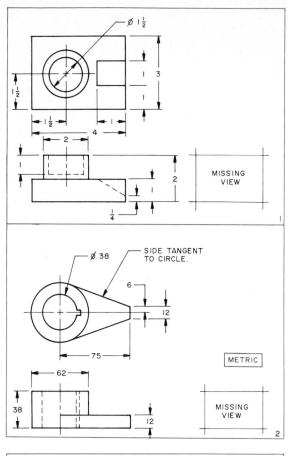

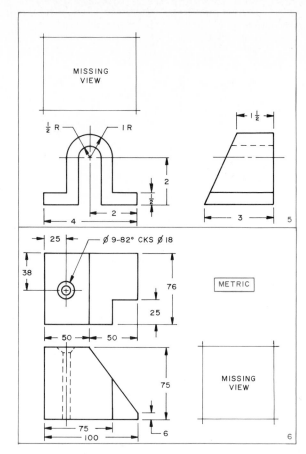

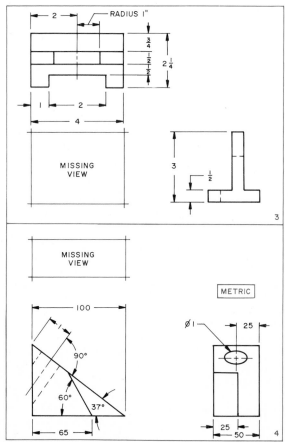

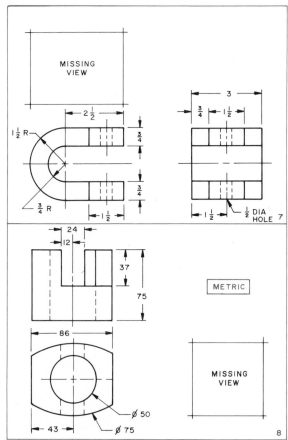

PROBLEM 6–4   Missing view problems.

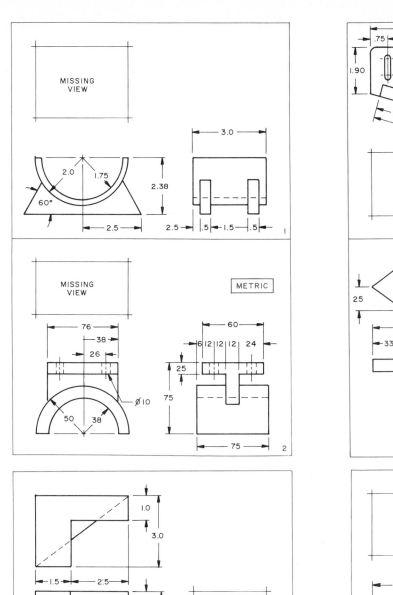

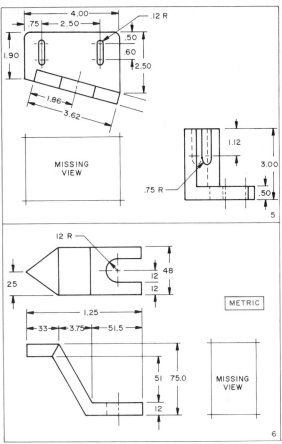

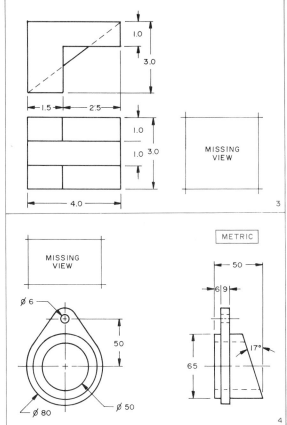

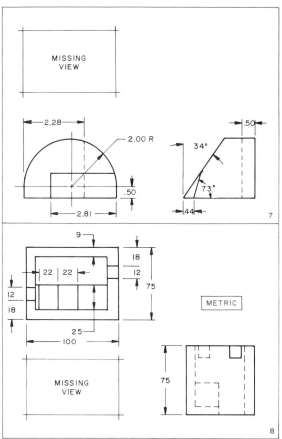

PROBLEM 6–5   Missing view problems.

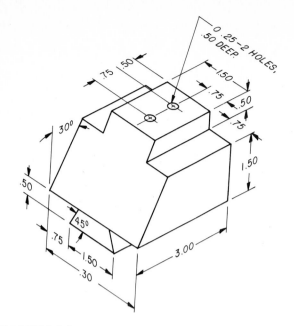

PROBLEM 6–6   Wedge.

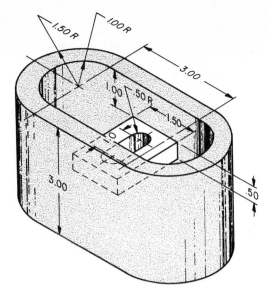

PROBLEM 6–9   Base.

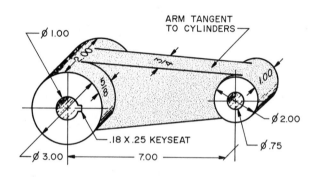

PROBLEM 6–7   Control block.

PROBLEM 6–8   Connecting shaft.

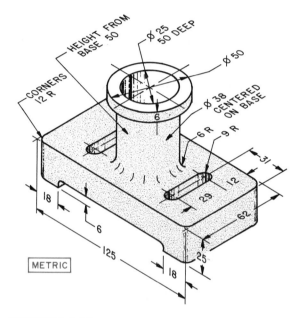

PROBLEM 6–10   Shaft carrier.
PROBLEM 6–11   Swivel connector.

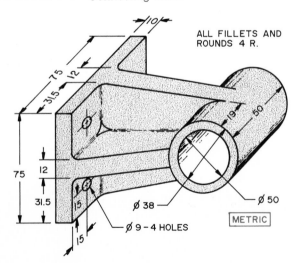

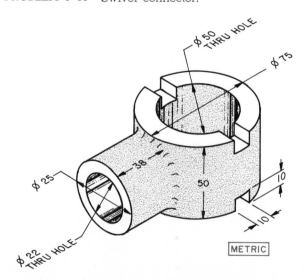

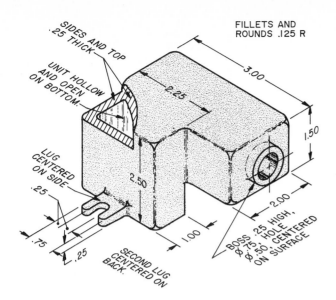

PROBLEM 6–12  Gear case cover.

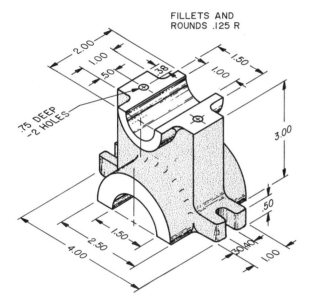

PROBLEM 6–13  Stop pocket.

PROBLEM 6–14  Shaft stabilizer.

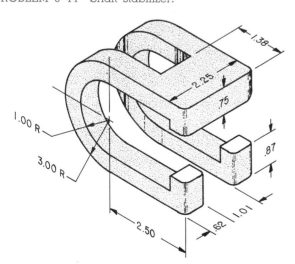

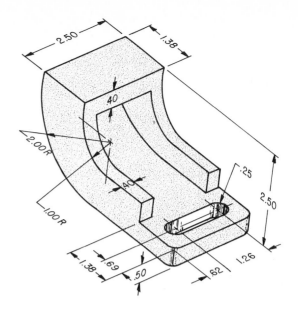

PROBLEM 6–15  Conical base.

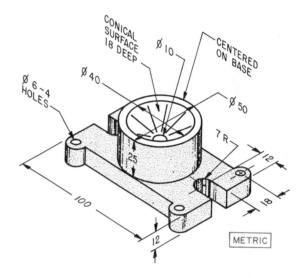

PROBLEM 6–16  Spring clip.

PROBLEM 6–17  Yoke.

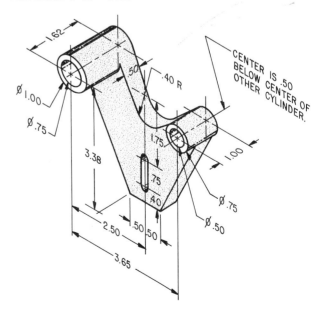

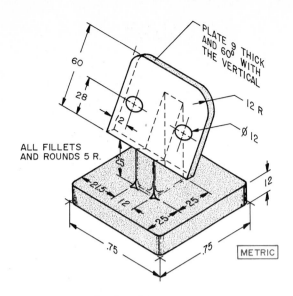

PLATE 9 THICK
AND 60° WITH
THE VERTICAL

60
28
12
12 R
Ø 12

ALL FILLETS
AND ROUNDS 5 R.

25
2.15
12
25
25
25
12
.75
.75

METRIC

PROBLEM 6–18   Angle plate.

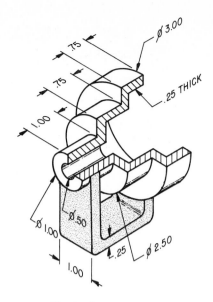

.75
.75
1.00
Ø 3.00
.25 THICK
Ø 1.00
Ø .50
.25
Ø 2.50
1.00

PROBLEM 6–21   Shaft clamp.

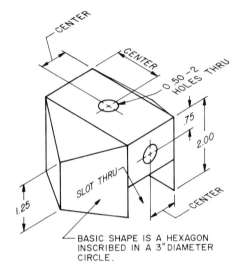

CENTER
CENTER
Ø .50 – 2
HOLES THRU
.75
2.00
SLOT THRU
1.25
CENTER

BASIC SHAPE IS A HEXAGON
INSCRIBED IN A 3" DIAMETER
CIRCLE.

PROBLEM 6–19   Impeller support.

PROBLEM 6–20   Tension fitting.

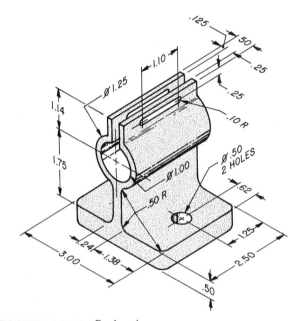

.125
.50
.25
1.10
Ø 1.25
.25
.10 R
1.14
1.75
Ø .50
2 HOLES
Ø 1.00
.62
50 R
.125
.24
1.38
.50
2.50
3.00

PROBLEM 6–22   Brake drum.

PROBLEM 6–23   Friction plate.

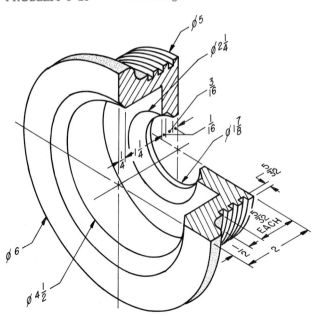

Ø 5
Ø 2¼
3/16
1/16
Ø 1⅞
1¼
¼
1¼
¼
5/32
45°
EACH
½
2
Ø 6
Ø 4½

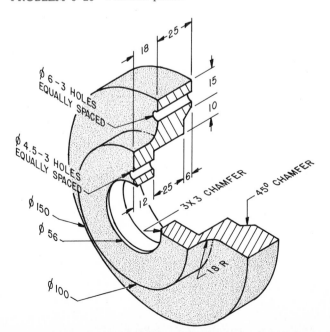

18
25
15
10
Ø 6 – 3 HOLES
EQUALLY SPACED
6
25
Ø 4.5 – 3 HOLES
EQUALLY SPACED
12
3 X 3 CHAMFER
45° CHAMFER
Ø 150
Ø 56
18 R
Ø 100

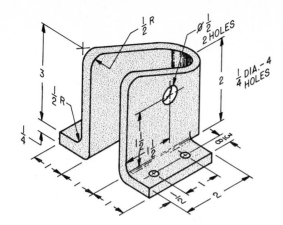

PROBLEM 6–24   Anchor.

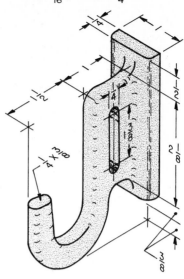

ALL CORNERS ROUNDED
ROUNDS $\frac{1}{16}$ , FILLETS $\frac{1}{4}$

PROBLEM 6–27   Spring housing.

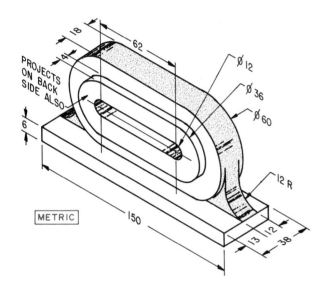

METRIC

PROBLEM 6–25   Hook.

ALL FILLETS AND ROUNDS $\frac{1}{8}$ R.

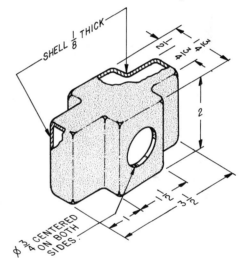

SHELL $\frac{1}{8}$ THICK

$\phi \frac{3}{4}$ CENTERED ON BOTH SIDES.

PROBLEM 6–28   Wedge.

PROBLEM 6–26   Slide guide.

PROBLEM 6–29   Swing arm.

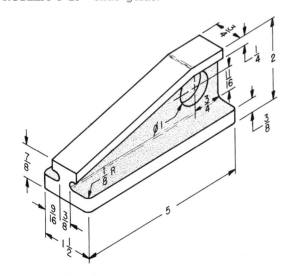

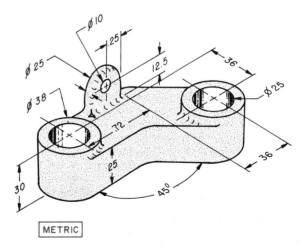

METRIC

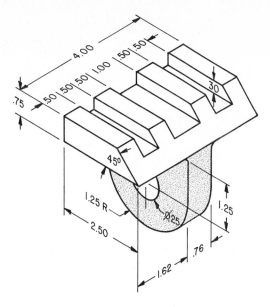

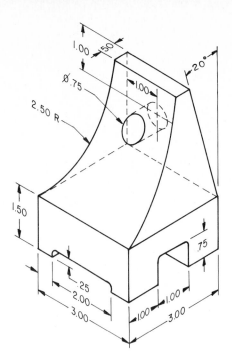

**PROBLEM 6–30** Trolley shoe.

**PROBLEM 6–33** End block.

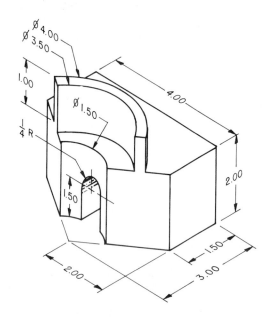

**PROBLEM 6–31** Shaft guide.

**PROBLEM 6–32** Friction plate.

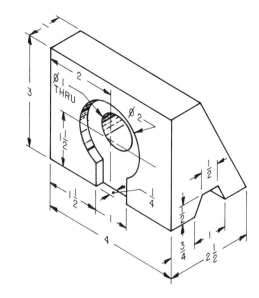

**PROBLEM 6–34** Sliding pivot.

**PROBLEM 6–35** Wedge guide.

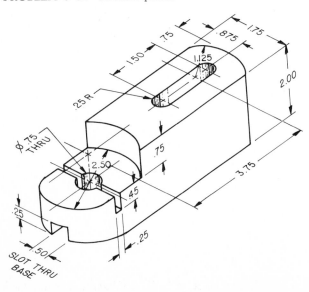

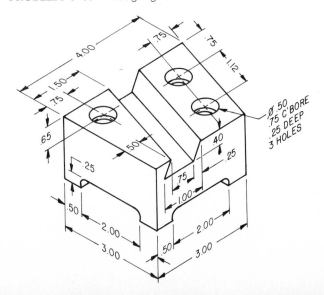

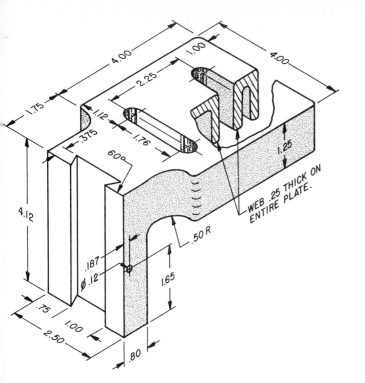

PROBLEM 6–36   Motor mounting bracket.

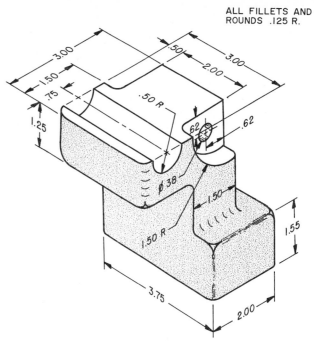

ALL FILLETS AND
ROUNDS .125 R.

PROBLEM 6–38   Drive guide.

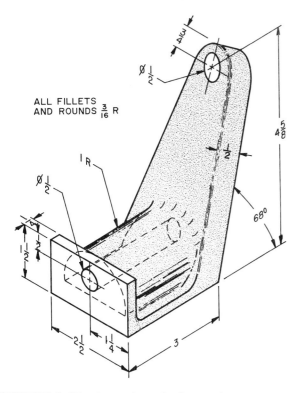

ALL FILLETS
AND ROUNDS $\frac{3}{16}$ R

PROBLEM 6–37   Hydraulic cylinder anchor.

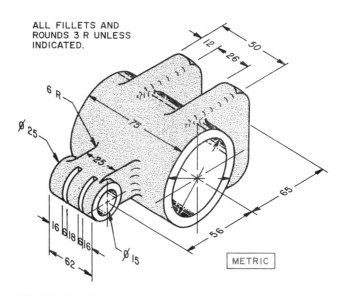

ALL FILLETS AND
ROUNDS 3 R UNLESS
INDICATED.

METRIC

PROBLEM 6–39   Pivot stop.

The Principles of Orthographic Projection   **139**

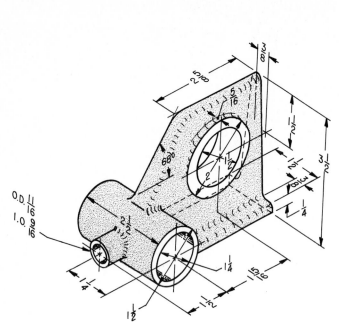

PROBLEM 6–40  Arbor pivot.

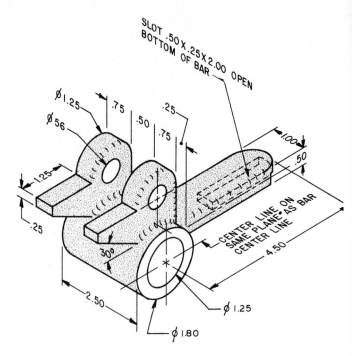

SLOT .50 X .25 X 2.00 OPEN
BOTTOM OF BAR

CENTER LINE ON
SAME PLANE AS BAR
CENTER LINE

PROBLEM 6–42  Plunger bracket.

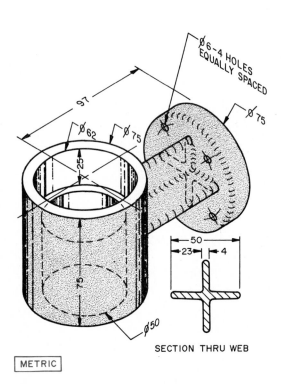

∅6–4 HOLES
EQUALLY SPACED

SECTION THRU WEB

METRIC

PROBLEM 6–41  Control hinge.

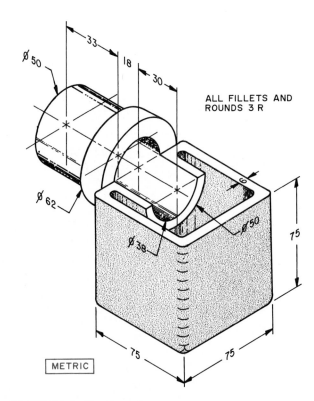

ALL FILLETS AND
ROUNDS 3 R

METRIC

PROBLEM 6–43  Gear box.

# Technical
# Sketching

The ability to make rapid, clear sketches is an important skill for an engineer to possess. No great artistic ability is required. The skill is developed through an understanding of basic sketching techniques and then by practice.

Sketches may be used by an engineer to record an idea quickly. The original idea is preserved. Other possible solutions can be recorded as they are conceived. The original idea can be altered, revised, or changed rather rapidly (Fig. 7–1).

Sketches can be used to show someone how something should be done or designed. The original sketches can be on an envelope, a piece of wrapping paper, or any other available material. After the original concepts are recorded, the engineer refines them by making carefully drawn technical sketches. These follow the suggested guidelines for good sketches and the basic principles of drafting. A good technical sketch is easy to read and in proportion and features proper line symbols and the other characteristics of a good technical drawing. The big difference is that the lines are drawn freehand and are therefore slightly irregular. The sketch should be in proportion but not necessarily to scale. The curves are fairly smooth and the circles rather round. Attention to the quality of lines is also important. They

## TOOLS

The basic tools are soft lead pencils, a soft rubber eraser, and paper. Other implements, such as a ball point or felt tip pen, can be used, but pencils produce the most professional sketch.

Any soft pencil will do. An F or HB drafting pencil is excellent. A regular No. 2 writing pencil is good but will have to be sharpened frequently. A conical point produces the best line. As you sketch, rotate the pencil in your fingers to keep it round (Fig. 7–3). Do not make the point too sharp because it will break. A rounded point is better. If the pencil is too pointed, dull the lead by rubbing it on scrap paper. While any paper will do for sketching, drafting vellum is best. It has a hard surface and erases

FIGURE 7–1 Preliminary ideas (here for a glue dispenser) can be recorded rapidly as freehand sketches.

should be dark and approach the proper line thickness of engineering drawings. A finished technical sketch is neat, clean, and properly executed (Fig. 7–2).

FIGURE 7–2 These sketches exhibit the qualities expected of a good engineering drawing.

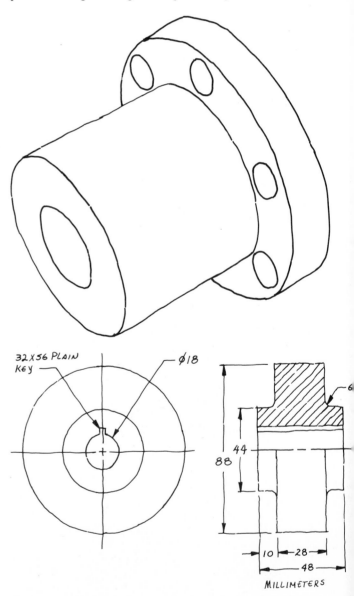

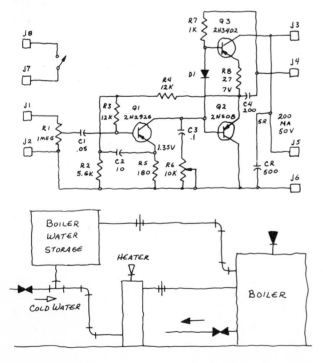

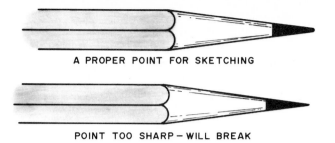

A PROPER POINT FOR SKETCHING

POINT TOO SHARP – WILL BREAK

FIGURE 7–3 Use a slightly rounded point on the pencil.

easily. The sketch can be reproduced on a white-printer if necessary.

The eraser is used sparingly when sketching. Ideally, the sketch is developed quickly without the need to erase. Errors or changes can be corrected easily with an eraser if vellum paper is used.

## SKETCHING TECHNIQUES

The first attempts at recording an idea are usually very rough and hurried. Several variations, such as the sketch in Fig. 7–1, may be made before a design is finalized. The variations are used as the basis for making the finished technical sketch and are then often discarded.

### Holding the Pencil

Hold the pencil in a natural position. The most frequent error is to hold it too close to the point. Most people find it comfortable to hold the pencil 35 to 50 mm from the point (Fig. 7–4). The base of the hand should rest on the paper. Rotate the pencil after each line is drawn.

FIGURE 7–4 When sketching, hold the pencil in a comfortable position, not too close to the point.

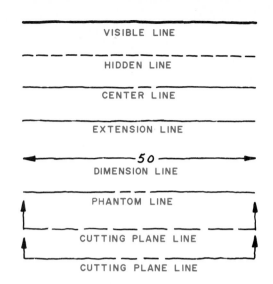

VISIBLE LINE

HIDDEN LINE

CENTER LINE

EXTENSION LINE

50
DIMENSION LINE

PHANTOM LINE

CUTTING PLANE LINE

CUTTING PLANE LINE

FIGURE 7–5 Suggested thickness of lines used on sketches.

### Line Quality

The standard alphabet of lines is used in technical sketching. Attempt to get line thickness near accepted standards. At least make the visible line about twice the width of the hidden lines, dimension lines, and extension lines. This helps the people who will be reading the sketch (Fig. 7–5).

Make the lines dark. While it may not be possible to get them as dark as an instrument drawing, try to approach this quality.

### Sketching Straight Lines

Sketched lines will have some slight irregularities along their length. With practice they can be made quite straight and must run directly to their end point.

Engineers who do a lot of sketching find a pad of graph paper helpful. Some prefer to place a sheet of graph paper under the vellum sheets so that they can see through to it. This gives a guide for straight lines. It also helps establish proportions because the object can be sized by counting squares (Fig. 7–6).

### Sketching Horizontal Lines

The preferred method for sketching horizontal lines is as follows. Locate each end with a point. Place the pencil on the left point and look at the end point at the right. (Left-handed persons may want to reverse these directions.) Move your hand and arm in a single rapid stroke to the end point (Fig. 7–7). Do not look back while making the line. If the line is not the way you want it, return to the first point and redraw it. It is difficult to try to patch up just part of a line. You can pass the pencil between the points without

FIGURE 7–6 Using graph paper when sketching helps keep lines straight and aids in keeping the drawing in proportion.

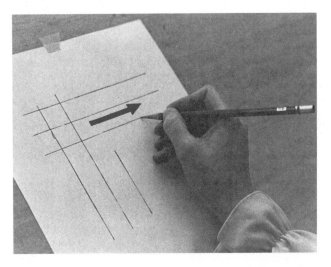

FIGURE 7–7 Sketch horizontal lines from left to right.

touching the paper before drawing the line. This helps establish the arm movement.

Some prefer to sketch using a series of short strokes. The same technique is used, but the line is

FIGURE 7–8 Suggested ways to sketch straight lines.

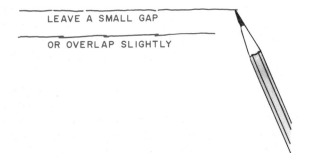

broken. This gives a brief moment to move the hand to keep up with the stroke (Fig. 7–8). Another technique is to overlap the ends of the dashes, producing a ragged but straight line.

If you do not move your arm and elbow with the pencil, you will draw a crowned line. Your elbow serves as a pivot, resulting in a curved line.

Short lines can be drawn by moving only fingers and wrist.

## Sketching Vertical Lines

Vertical lines are usually easiest if drawn down toward your body. The same technique as for horizontal lines is used. Locate the beginning and end points. Draw the line in a single stroke between the two points (Fig. 7–9).

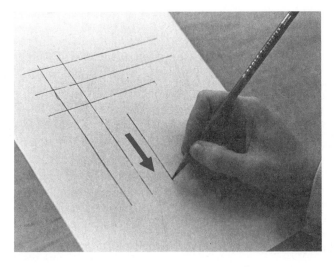

FIGURE 7–9 Sketch vertical lines toward you.

FIGURE 7–10 A finger sliding along the edge of the sketch pad can be used as a guide when sketching straight lines.

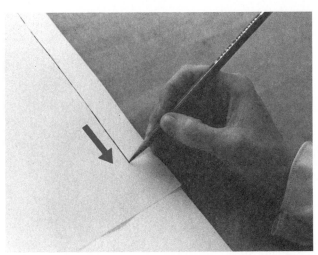

The edge of the note pad or table can be used as a guide. Hold the pencil with the tip touching the paper. Place one or more fingers on the edge of the pad or table. Pull the pencil toward your body (Fig. 7–10). This technique is useful when drawing several parallel lines.

### Sketching Inclined Lines

Inclined lines are sketched as shown in Fig. 7–11. Use the same technique as for horizontal lines. If you have difficulty, you can slant your paper and sketch slanted lines in a horizontal position.

## SKETCHING BASIC GEOMETRIC SHAPES

Manufactured products are made up of a variety of geometric forms, such as cylinders, cones, pyramids, spheres, and prisms of various types (Fig. 7–12). When sketching a product, study it and notice these forms. Then, as you sketch it, develop each form. When assembled, the finished product appears. The engineer can greatly improve sketched drawings by observing these geometric forms (Fig. 7–13).

### Sketching Squares and Rectangles

A square is sketched by laying out the center lines at right angles. Then, mark half the width of the square on one side of the center line. Using a pencil or a piece of paper, transfer this distance to the other parts of the center line. Lightly sketch the horizontal and vertical sides. Complete the square by darkening the lines (Fig. 7–14).

The same technique is used to develop a rectangle. Lay out the length of the long and short

FIGURE 7–11 How to sketch inclined lines.

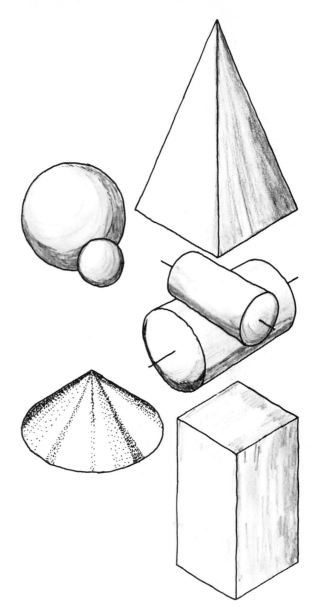

FIGURE 7–12 The basic geometric shapes found in objects to be drawn.

sides. Then, sketch the horizontal and vertical sides (Fig. 7–15).

### Sketching Circles and Arcs

There are two ways to sketch circles. You can sketch the center lines. Then sketch several radial lines through the center. Mark the radius on a piece of paper. Use it to mark the radius on each radial line and the center line. Sketch the circle through the points (Fig. 7–16).

Another way is to sketch a square with sides equal to the diameter of the circle. Sketch center lines and diagonals. Mark the radius on each diagonal. Sketch the circle through the points (Fig. 7–17).

Large-diameter circles can be sketched using a

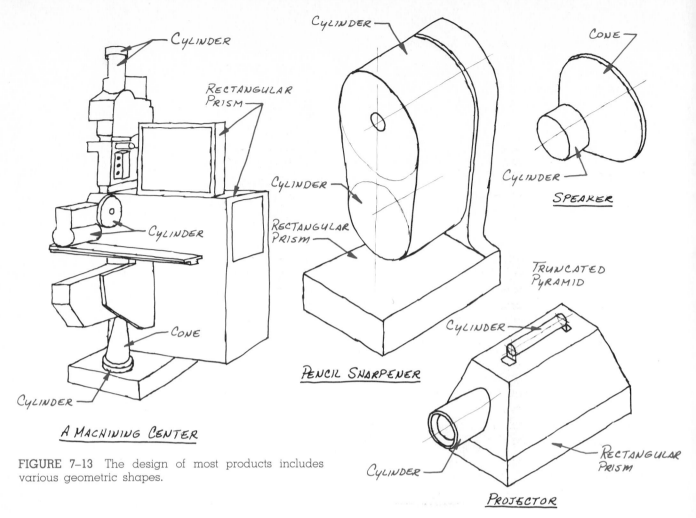

CYLINDER

RECTANGULAR PRISM

CYLINDER

CONE

CYLINDER

A MACHINING CENTER

CYLINDER

CYLINDER

RECTANGULAR PRISM

PENCIL SHARPENER

CONE

CYLINDER

SPEAKER

TRUNCATED PYRAMID

CYLINDER

CYLINDER

RECTANGULAR PRISM

PROJECTOR

**FIGURE 7–13** The design of most products includes various geometric shapes.

**FIGURE 7–14** How to sketch a square.

**FIGURE 7–15** How to sketch a rectangle.

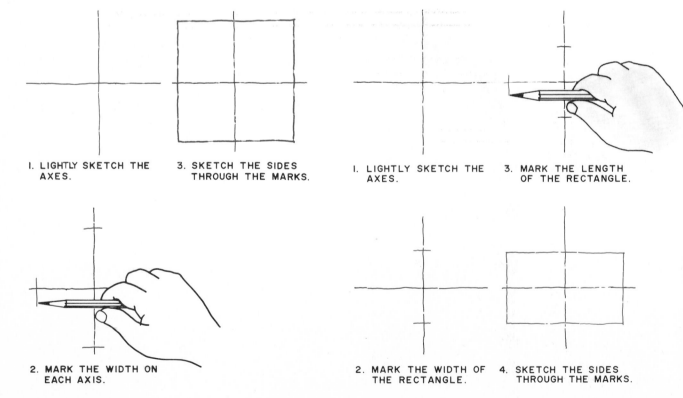

1. LIGHTLY SKETCH THE AXES.

3. SKETCH THE SIDES THROUGH THE MARKS.

2. MARK THE WIDTH ON EACH AXIS.

1. LIGHTLY SKETCH THE AXES.

3. MARK THE LENGTH OF THE RECTANGLE.

2. MARK THE WIDTH OF THE RECTANGLE.

4. SKETCH THE SIDES THROUGH THE MARKS.

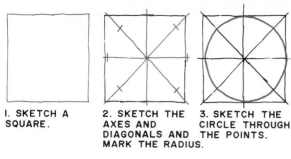

I. SKETCH A SQUARE.

2. SKETCH THE AXES AND DIAGONALS AND MARK THE RADIUS.

3. SKETCH THE CIRCLE THROUGH THE POINTS.

FIGURE 7–16  Sketching a circle using a square.

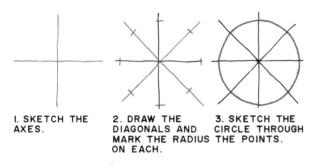

I. SKETCH THE AXES.

2. DRAW THE DIAGONALS AND MARK THE RADIUS ON EACH.

3. SKETCH THE CIRCLE THROUGH THE POINTS

FIGURE 7–17  Sketching a circle using the radius.

strip of paper to locate points on the circumference. First, lay out the radius. Mark a piece of paper with the radius. Holding one mark on the center, rotate the paper and mark points on the circumference. Sketch a line through the points (Fig. 7–18).

FIGURE 7–18  How to lay out and sketch a large-diameter circle.

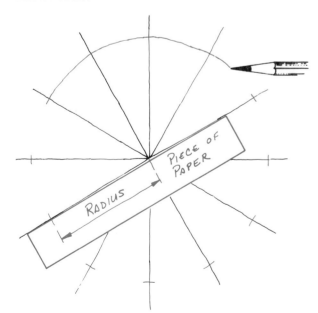

I. SKETCH AXES AND SEVERAL DIAGONALS.
2. MARK RADIUS ON A STRIP OF PAPER. LOCATE RADIUS ON EACH DIAGONAL AND CENTERLINE.
3. SKETCH A CIRCLE THROUGH THE POINTS.

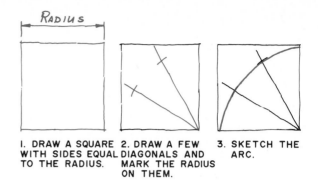

I. DRAW A SQUARE WITH SIDES EQUAL TO THE RADIUS.

2. DRAW A FEW DIAGONALS AND MARK THE RADIUS ON THEM.

3. SKETCH THE ARC.

FIGURE 7–19  Sketch an arc by locating radii.

Arcs are segments of circles. They can be sketched by drawing a square with sides equal to the radius of the arc. Several diagonals can be drawn and the radius marked on them. Sketch the arc through the points (Fig. 7–19).

## Sketching Ellipses

An ellipse has a major and a minor axis (Fig. 7–20). Sketch the center lines and locate the ends of the major and minor axes. Sketch a rectangle through these points. Draw the end and side curves (an ellipse has no straight sections). Connect these curves and darken the line (Fig. 7–21).

## PROPORTION

Although technical sketches are not drawn to exact scale, they must retain the proportions of the object. This means that you must sketch the various parts of the object in the same relative size relationships as exist on the object. For example, if an object measures $3 \times 6$ in., this 1:2 relationship must be retained on the sketch (Fig. 7–22).

To maintain proportion, first sketch a rectangle representing the overall size of the object. Then

FIGURE 7–20  The axes of an ellipse.

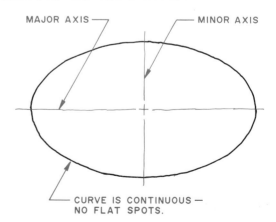

MAJOR AXIS —

— MINOR AXIS

— CURVE IS CONTINUOUS — NO FLAT SPOTS.

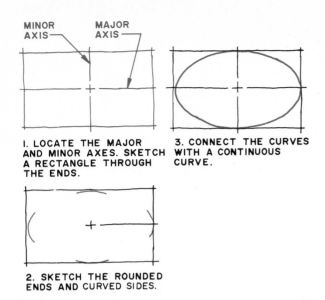

**1. LOCATE THE MAJOR AND MINOR AXES. SKETCH A RECTANGLE THROUGH THE ENDS.**

**3. CONNECT THE CURVES WITH A CONTINUOUS CURVE.**

**2. SKETCH THE ROUNDED ENDS AND CURVED SIDES.**

FIGURE 7–21 How to lay out and sketch an ellipse.

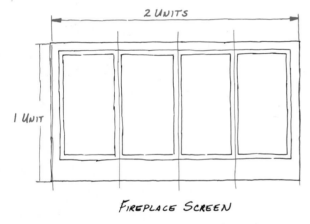

OVERALL PROPORTIONS APPROXIMATELY 1 TO 2. MINOR SUBDIVISIONS INTO 4 EQUAL PARTS.

2 UNITS

1 UNIT

*FIREPLACE SCREEN*

FIGURE 7–22 Sketches must retain the basic proportions of the object.

break it down into the major parts of the object. These divisions can be subdivided into smaller units for sketching details (Fig. 7–23).

Proportions can be determined by actually measuring the object or by using the pencil-and-eye technique. When taking actual measurements and sketching full size, do the sketch on graph paper. The squares are a known size, so the measurements can be located by counting squares (Fig. 7–24). To get proportions by eye, hold a pencil at arm's length from your eye. Line it up with a part of the object. Mark the length on the pencil with your thumb. Record this on paper. Repeat this procedure to obtain other needed measurements (Fig. 7–25). This

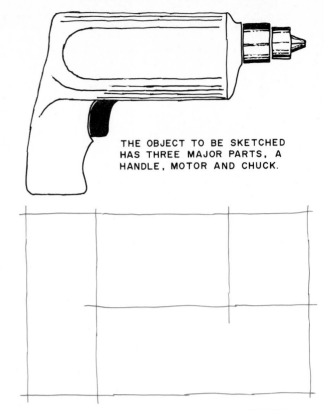

**THE OBJECT TO BE SKETCHED HAS THREE MAJOR PARTS, A HANDLE, MOTOR AND CHUCK.**

**1. SKETCH A RECTANGLE GIVING THE OVERALL LENGTH AND WIDTH. BLOCK IN THE MAJOR ELEMENTS —— THE HANDLE, MOTOR AND CHUCK.**

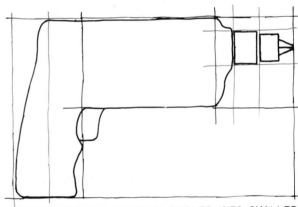

**2. SUBDIVIDE THE MAJOR ELEMENTS INTO SMALLER UNITS TO LOCATE DETAILS. SKETCH THE OBJECT WITHIN THIS GRID.**

FIGURE 7–23 An example of breaking an object down into several parts and keeping these in proportion to one another.

will give a sketch in proportion, though not actual size.

Angles can be drawn by locating the ends and connecting them. In Fig 7–26, the top end is about one third the width of the object. Locate the top end and connect it to the bottom end. It helps if you can learn to estimate angles by degrees. While viewing the object, try to decide if the angle is more or less

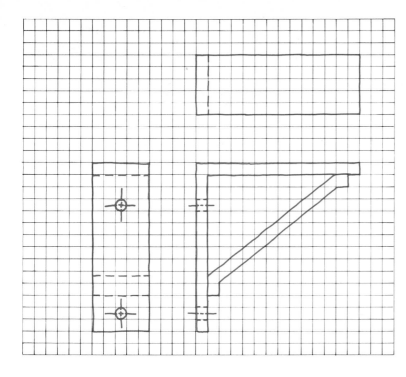

FIGURE 7–24 Proportion can be established when using graph paper by counting the squares. If each square on this sketch represents 5 mm, what is the overall height and length?

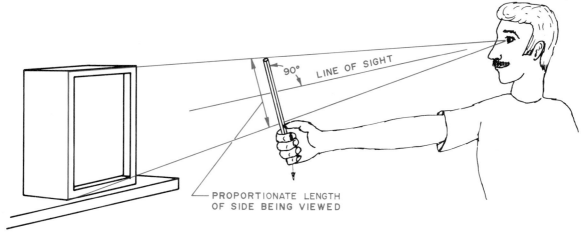

90°  LINE OF SIGHT

PROPORTIONATE LENGTH
OF SIDE BEING VIEWED

FIGURE 7–25 Proportions can be approximated by sighting the object.

FIGURE 7–26 Angles can be sketched by locating the ends of the angles and connecting them or by estimating the angle.

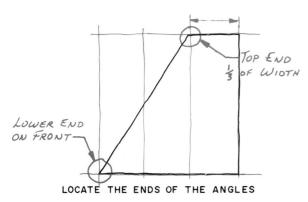

TOP END
⅓ OF WIDTH

LOWER END
ON FRONT

LOCATE THE ENDS OF THE ANGLES

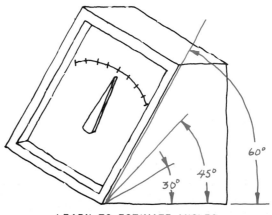

30°  45°  60°

LEARN TO ESTIMATE ANGLES

than 45°. Is it near 30° or 60°? Use this information as you sketch the angle (Fig. 7–26).

The actual size of sketches depends on the object. Some are so large that it is not practical to sketch them full size. Some are so small that it is difficult to get the details if they are sketched actual size. Adjust the size of the sketch to suit the situation.

## ENLARGING AND REDUCING SKETCHES

If it is necessary to enlarge or reduce the size of a sketch, it can be done rapidly by using a grid. Place a grid over the original sketch. The size of the grid depends on the amount of detail. Sketches with a lot of detail will require a smaller grid. To enlarge the drawing, redraw it on graph paper with a larger grid. To reduce it, use a smaller grid. For example, if the original drawing has a ⅛-in. grid and you want to double the size of the drawing, redraw it on a ¼-in. grid.

The drawing is transferred by marking on the enlarged grid the point at which it crosses each grid line on the original. Once the points are located, sketch lines through them (Fig. 7–27).

## ORTHOGRAPHIC SKETCHES

When making orthographic sketches, begin by blocking in the overall sizes of each view. Use very light lines. This establishes the length, width, and height of each view. Then, locate the center lines of cylindrical features. Next, block in the major parts, using the technique described for finding proportion. Circular parts and arcs are sketched first, followed by horizontal and vertical lines. Then, small details are located. If dimensions are needed, they are drawn last. Be certain to leave room for dimensions when the original size is blocked out (Fig. 7–28).

All dimensions and notes must be lettered carefully so that they are easy to read. Poor lettering can lead to errors and confusion.

## PICTORIAL SKETCHES

A great deal of information can be explained with pictorial sketches. They show several sides of a product, and the drawing is rather lifelike. Perspective sketches are more lifelike. Isometric and oblique sketches are the most frequently used. They are easy to make and well suited to sketching techniques.

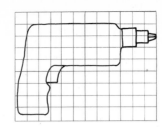

1. DRAW A GRID OVER THE DRAWING TO BE ENLARGED

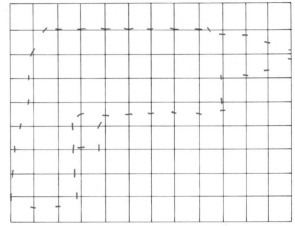

2. DRAW THE ENLARGED GRID. THIS GRID IS TWICE THE SIZE OF THE ORIGINAL. LOCATE EACH POINT WHERE A LINE CROSSES THE GRID.

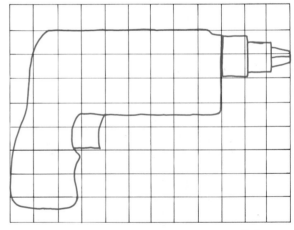

3. CONNECT THE POINTS AND SKETCH IN THE CURVES.

FIGURE 7–27 A grid can be used to enlarge or reduce a drawing.

### Oblique Sketches

To make an oblique sketch, draw an oblique block the size of the overall product. First, sketch the front of the block, then the axes. A 45° axis is most common, but any other angle can be used. Then, add details on the front and side views. Remove unneeded lines and darken in the visible lines (Fig. 7–29).

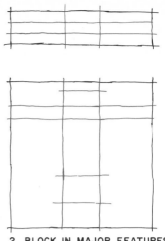

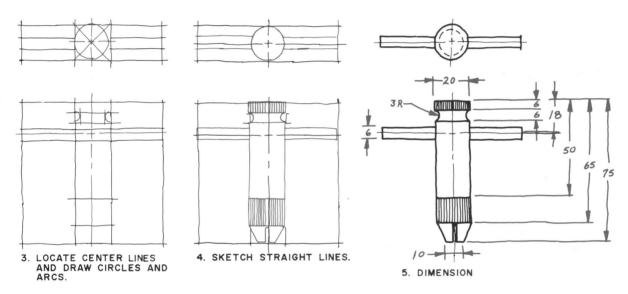

1. BLOCK IN THE AREA NEEDED FOR EACH VIEW.

2. BLOCK IN MAJOR FEATURES TO GET THEM IN PROPER PROPORTION.

3. LOCATE CENTER LINES AND DRAW CIRCLES AND ARCS.

4. SKETCH STRAIGHT LINES.

5. DIMENSION

FIGURE 7–28  Preparing an orthographic sketch.

FIGURE 7–29  Making an oblique sketch.

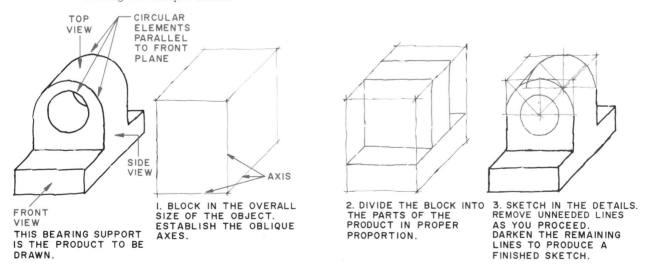

TOP VIEW

CIRCULAR ELEMENTS PARALLEL TO FRONT PLANE

SIDE VIEW

AXIS

FRONT VIEW

THIS BEARING SUPPORT IS THE PRODUCT TO BE DRAWN.

1. BLOCK IN THE OVERALL SIZE OF THE OBJECT. ESTABLISH THE OBLIQUE AXES.

2. DIVIDE THE BLOCK INTO THE PARTS OF THE PRODUCT IN PROPER PROPORTION.

3. SKETCH IN THE DETAILS. REMOVE UNNEEDED LINES AS YOU PROCEED. DARKEN THE REMAINING LINES TO PRODUCE A FINISHED SKETCH.

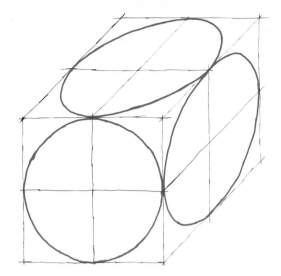

FIGURE 7–30   How to sketch circles on oblique drawings.

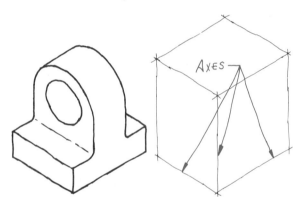

CIRCULAR ELEMENTS
PARALLEL WITH FRONT
PLANE

PROPER USE OF THE
OBLIQUE TYPE
PICTORIAL DRAWING

IMPROPER USE OF THE
OBLIQUE TYPE DRAWING.
IT IS DIFFICULT TO DRAW.

FIGURE 7–31   On oblique sketches, place the major circular elements in the front view.

FIGURE 7–32   The receding axis on oblique drawings is usually full or half scale.

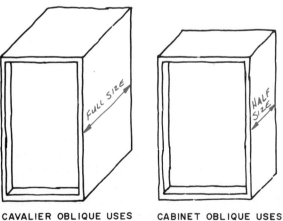

CAVALIER OBLIQUE USES
FULL SIZE SCALE ON
RECEDING AXIS.

CABINET OBLIQUE USES
HALF SIZE SCALE ON
RECEDING AXIS.

Circles that are in the front view or are parallel with it are sketched round. Those in the side view are sketched by first drawing a square whose sides are equal to the diameter of the circle. Draw the center lines. Then, sketch the circle in the square (Fig. 7–30).

Remember, the view containing circular features is placed as the front view because it is drawn to true shape. This is the advantage of using this type of pictorial sketch (Fig. 7–31).

The length of the receding axis can vary. Most commonly it is full size (called *cavalier*) or half size (called *cabinet*). It can have other proportions if the designer so desires (Fig. 7–32).

### Isometric Sketches

To make an isometric sketch, draw an isometric block the size of the overall product. Then, divide the block into the major proportions of the object. Locate the center lines of circular parts and arcs. Sketch these on the drawing. Locate any minor parts

FIGURE 7–33   Making an isometric sketch.

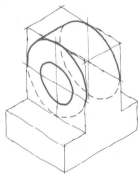

THIS BEARING SUPPORT
IS THE PRODUCT TO BE
DRAWN.

I. BLOCK IN THE OVERALL
SIZE OF THE OBJECT.
ESTABLISH THE ISOMETRIC
AXES.

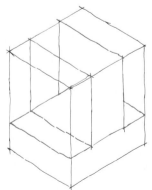

2. DIVIDE THE BLOCK
INTO THE PARTS OF THE
PRODUCT IN PROPER
PROPORTION.

3. SKETCH IN THE DETAILS.
REMOVE UNNEEDED LINES
AS YOU PROCEED. DARKEN
THE REMAINING LINES TO
PRODUCE A FINISHED
SKETCH.

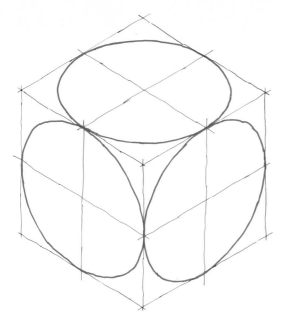

FIGURE 7–34  How to sketch circles in isometric.

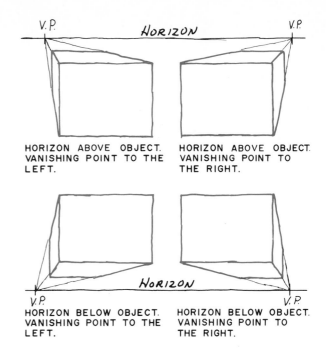

HORIZON ABOVE OBJECT.
VANISHING POINT TO THE
LEFT.

HORIZON ABOVE OBJECT.
VANISHING POINT TO
THE RIGHT.

HORIZON BELOW OBJECT.
VANISHING POINT TO THE
LEFT.

HORIZON BELOW OBJECT.
VANISHING POINT TO
THE RIGHT.

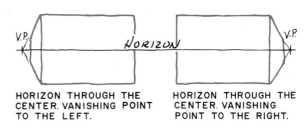

HORIZON THROUGH THE
CENTER. VANISHING POINT
TO THE LEFT.

HORIZON THROUGH THE
CENTER. VANISHING
POINT TO THE RIGHT.

FIGURE 7–35  Common locations for the horizon and vanishing point on one-point-perspective sketches.

yet to be drawn. Darken the visible lines of the finished sketch (Fig. 7–33).

Circles in isometric sketches appear elliptical. To sketch isometric circles, first sketch a square with sides equal to the diameter of the circle. Then sketch the isometric centerlines. Using these as guides, sketch the circle inside the square (Fig. 7–34).

## PERSPECTIVE SKETCHES

Perspective sketches appear more realistic than oblique and isometric sketches, but they are more difficult to make. One- and two-point perspectives are the most common.

To sketch a *one-point perspective*, first decide on the relationship between the horizon, the vanishing point, and the object. The object and the horizon can be placed in many different positions. The object can be above, below, to the right, or to the left of the vanishing point and horizon (Fig. 7–35). Sketch the horizon, the vanishing point, and a block having sides the overall size of the object in the selected relationship. Then, locate the various parts of the object on the block. Remove unneeded lines and complete the perspective drawing (Fig. 7–36).

Two-point perspectives are much like one-point except that they have two vanishing points. The front of the object vanishes to one side, and the side vanishes to the other.

To sketch a *two-point perspective*, locate the horizon and both vanishing points. Their location is estimated so that the finished drawing will appear normal. Experience is needed to make this decision.

Then, block in the overall size of the object in a location properly relating it to the horizon. You have to estimate the lengths of the two sides running toward the vanishing points. They are both less than true length. Finally, remove unneeded lines and finish the perspective sketch (Fig. 7–37).

## SHADING SKETCHES

It is not necessary to shade engineering pictorial sketches, but a few minutes spent doing this can sharpen the drawing. Following are several quick and easy techniques. Exterior-line shading is the easiest. It involves drawing the exterior lines thicker than the other visible lines (Fig. 7–38).

Surface-line shading involves drawing parallel lines on the surface. Assume that a light hits the product from one particular direction. Some surfaces would be in shadow. Those farthest from the light will be darker. Draw the shading lines closer together (Fig. 7–39). Stippling uses the same principle as surface-line shading, except the surface is shaded

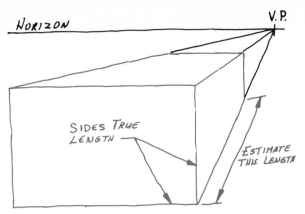

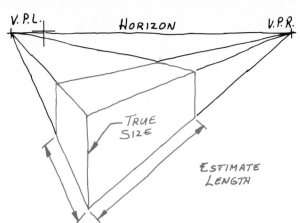

**1. LOCATE THE HORIZON AND VANISHING POINT IN RELATION TO A BLOCK THE SIZE OF THE OVERALL DIMENSIONS OF THE OBJECT. ESTIMATE THE LENGTH OF THE SIDE TOWARD THE VANISHING POINT. IT IS LESS THAN TRUE LENGTH.**

**1. LOCATE A BLOCK WHOSE OVERALL SIZE IS EQUAL TO THE SIZE OF THE OBJECT IN THE DESIRED RELATIONSHIP TO THE HORIZON AND VANISHING POINTS.**

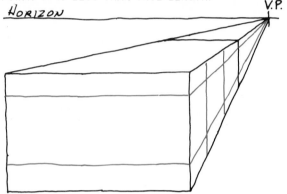

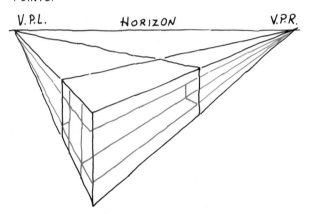

**2. LOCATE THE DETAILS ON THE BLOCK.**

**2. LOCATE THE DETAILS ON THE BLOCK. ESTIMATE THE SIZES OF FEATURES ON THE SIDES TO EACH VANISHING POINT. THEY ARE SLIGHTLY SHORTER THAN TRUE LENGTH.**

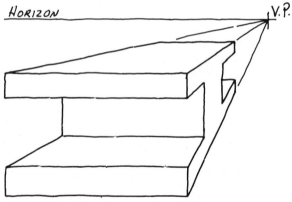

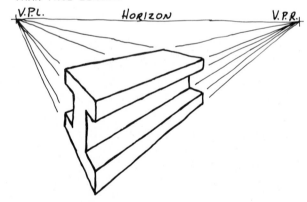

**3. ERASE UNNEEDED LINES AND COMPLETE THE PERSPECTIVE.**

**3. ERASE UNNEEDED LINES AND COMPLETE THE PERSPECTIVE**

FIGURE 7–36   How to sketch a one-point perspective.

FIGURE 7–37   How to sketch a two-point perspective.

using dots. In the dark areas, the dots are closer together (Fig. 7–40).

Smudge shading involves using a soft pencil and rubbing the lead flat on the surface. This produces a wide, thin, gray shading effect. The areas away from the light are made darker (Fig. 7–41).

## TEMPLATES

If a person routinely does a lot of sketching at a desk or drafting table, templates can be used to speed up the process. Hundreds of different types are available, with a wide variety of shapes cut into them. Some are prepared for drawing circles in isometric (Fig. 7–42).

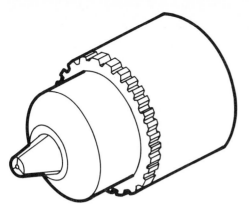

FIGURE 7–38 An example of exterior-line shading.

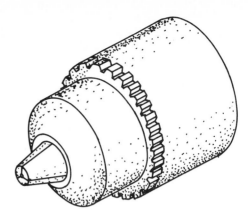

FIGURE 7–40 This pictorial drawing is shaded using stippling.

FIGURE 7–39 Surface-line shading indicates shadows and light areas.

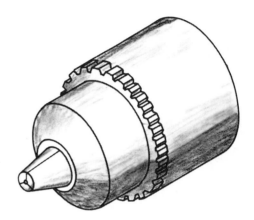

FIGURE 7–41 Smudge shading produces a realistic appearance.

FIGURE 7–42 Templates can speed the pictorial drafting process.

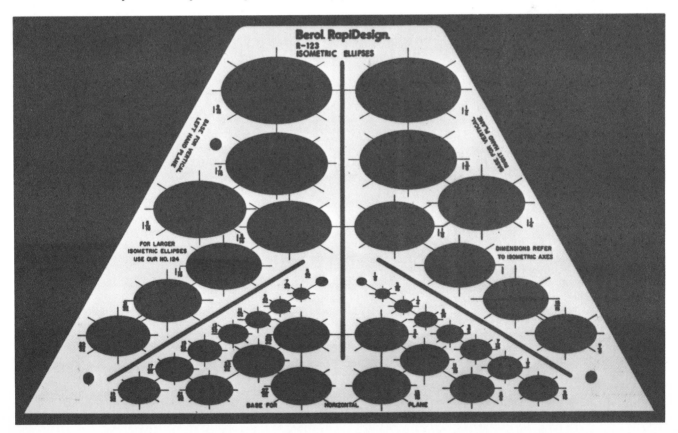

# Problems

The following problems will provide experience in technical sketching plus the opportunity to reinforce the principles of orthographic projection and pictorial drawing. They also provide some design experience.

Prepare your sketches on coordinate paper unless your instructor suggests otherwise. Pictorial sketches should be made on grids designed for this purpose.

1. Select items in the drafting room and prepare technical sketches as directed by the instructor. These can be working drawings or pictorial drawings.

2. Make sketches of the problems at the end of Chapter 6. These range from simple to complex. Copy the problem as shown or sketch the solution to the problem. Dimension only if directed to do so by the instructor.

3. Prepare sketches of proposed packages to be used to store a ceramic cup safely for shipment yet be attractive enough to be used on a retail display. Select any cup you desire.

4. Design a device to hold drafting pencils in a convenient location on the drawing board. It should make them readily available and protect the points from damage. Sketch as many solutions as required by your instructor.

5. Bring to class objects that can serve as models. These can be parts from a machine, an automobile, or any other device. Sketch the necessary views to describe the objects.

6. Find an exterior view of a house or small commercial building. Sketch a front view or pictorial view as directed by your instructor.

7. Visit a machine shop or some type of manufacturing plant. Sketch a variety of jigs and fixtures used in the manufacturing process. Some of these are shown in Probs. 7–1 through 7–8.

8. Visit a construction site. Make a sketch showing a section through the building from the footing to the ridge of the roof. Identify each part with a note.

## SPHERICAL TRUNNION

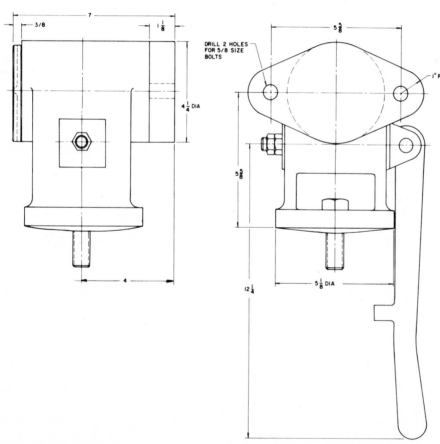

PROBLEM 7–1 Spherical trunnion. (Courtesy of Carr Lane Manufacturing Co.)

# TRUNNION LOCKING

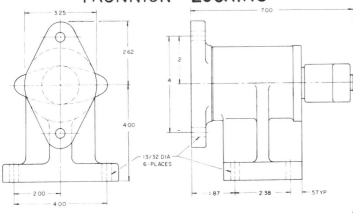

325
262
400
200
400
700
2
4
13/32 DIA
6-PLACES
1.87
2.38
.5 TYP

**PROBLEM 7–2** Trunnion locking. (Courtesy of Carr Lane Manufacturing Co.)

# SAFETY HOIST RING

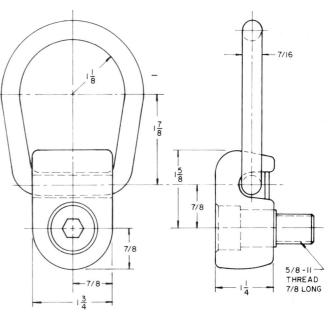

$1\frac{1}{8}$
$1\frac{7}{8}$
$1\frac{5}{8}$
7/8
7/8
7/8
$1\frac{3}{4}$
7/16
$1\frac{1}{4}$
5/8-11 THREAD 7/8 LONG

**PROBLEM 7–3** Safety hoist ring. (Courtesy of Carr Lane Manufacturing Co.)

# CAM CLAMP ASSEMBLY

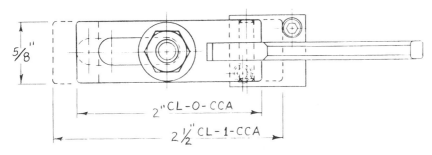

$\frac{5}{8}$"
2"CL-0-CCA
$2\frac{1}{2}$"CL-1-CCA

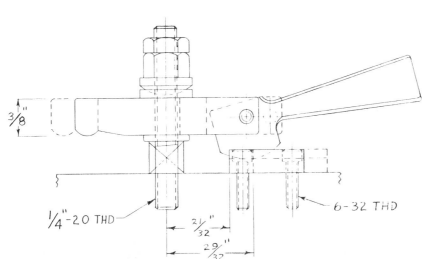

$\frac{3}{8}$"
$\frac{1}{4}$"-20 THD
$2\frac{1}{32}$"
$2\frac{9}{32}$"
6-32 THD

**PROBLEM 7–4** Cam clamp assembly. (Courtesy of Carr Lane Manufacturing Co.)

## STEEL FORGED STRAIGHT CLAMP STRAP

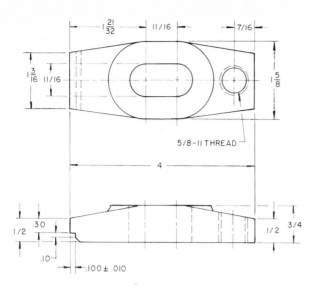

5/8-11 THREAD

PROBLEM 7–5 Steel forged straight clamp strap. (Courtesy of Carr Lane Manufacturing Co.)

## SERRATED ADJUSTABLE CLAMP

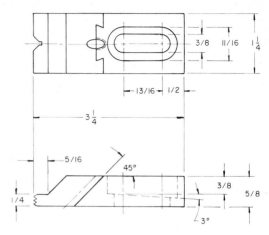

PROBLEM 7–6 Serated adjustable clamp. (Courtesy of Carr Lane Manufacturing Co.)

## ALUMINUM GOOSE NECK CLAMP STRAP

1/2-13 THREAD

PROBLEM 7–7 Aluminum goose neck clamp strap. (Courtesy of Carr Lane Manufacturing Co.)

## SLOTTED HEEL CLAMP ASSEMBLY

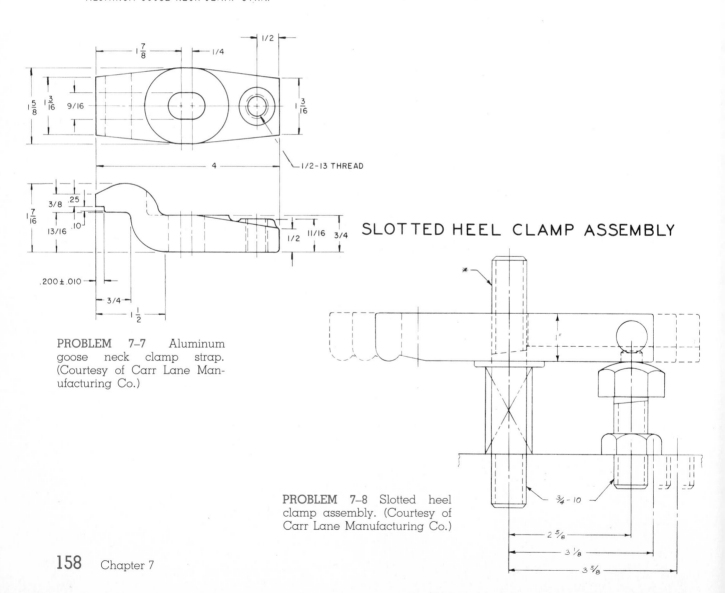

3/4 - 10

PROBLEM 7–8 Slotted heel clamp assembly. (Courtesy of Carr Lane Manufacturing Co.)

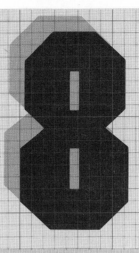

# Descriptive Geometry: Spatial Relationships

Descriptive geometry is a study of points, lines, and planes and their relationships in space. It is the theory upon which engineering drawing is built. A thorough understanding of descriptive geometry is necessary for success in solving engineering problems graphically.

This chapter presents the basic principles involved in spatial relationships. These principles are used in all areas of engineering graphics and are basic to an understanding of the area.

## PRINCIPAL PLANES OF PROJECTION

The six principal planes of projection are explained in detail in Chapter 6. They are reviewed here because they form the basis for all other relationships.

Engineering drawings are based on the principles of orthographic projection. Orthographic projection is the method of representing three-dimensional spatial problems by using views projected perpendicularly on planes of projection. The projection is accomplished by projectors that are parallel. The three principal planes of projection are perpen-

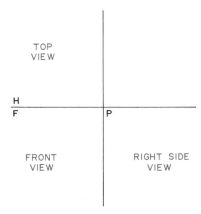

THE PRINCIPAL PLANES UNFOLDED

LOCATION OF VIEWS ON AN
ENGINEERING DRAWING.

dicular to one another (Fig. 8–1). They are called the horizontal, frontal, and profile planes. On drawings they are identified by the letters H, F, and P.

To get the planes from a three-dimensional condition to a single plane (your drawing paper), they unfold as shown in Fig. 8–2. The front view of an object is drawn on the frontal plane. The side view is drawn on the profile plane. The top view is drawn on the horizontal plane (Fig. 8–3).

Sometimes it is not possible to describe an object completely with these three views. The system of orthographic projection permits six principal planes of projection to be used. The six planes are perpendicular to one another. They unfold to a single plane as shown in Fig. 8–4. The top and bottom views are drawn on horizontal planes. The right and left sides are drawn on profile planes. The front and rear views are drawn on frontal planes.

FIGURE 8–2  The principal planes of projection unfold into a plane representing the drafting paper.

FIGURE 8–3  The position of the views in relation to the principal planes.

FIGURE 8–1  The principal planes of projection are perpendicular to one another.

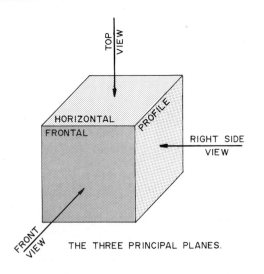

THE THREE PRINCIPAL PLANES.

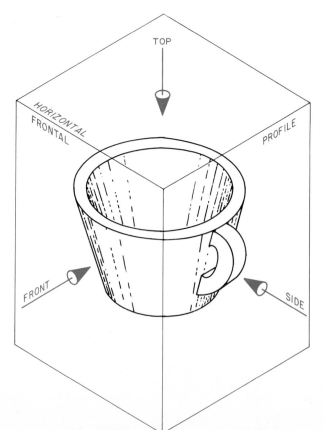

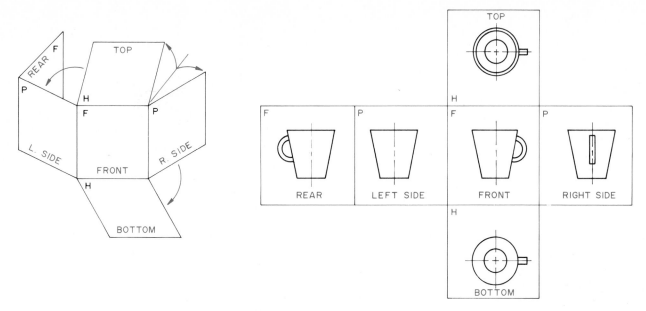

FIGURE 8–4 The six principal planes of projection, unfolded into a plane representing the drafting paper.

## PROJECTION OF A POINT

A point is a geometric element used to establish a specific location. It has no thickness and no dimensions. For purposes of accuracy, it is usually shown on drawings as the intersection of two short dashes.

A point can be projected from its location in space to the principal planes of projection (Fig. 8–5). It must be projected perpendicular to at least two of the planes to establish its position.

In Fig. 8–6, the point is located distance X from the frontal plane, distance Y from the profile plane, and distance Z from the horizontal.

## LINES

A line is a straight distance between two points. It has no width. It may be a principal line or an oblique line.

A *principal line* is a line parallel with the principal planes (Fig. 8–7). It appears as a true-length line on the plane with which it is parallel. It appears as a point on the plane to which it is perpendicular. It can be in any position as long as it is parallel with a principal plane. A *horizontal line* is parallel with the horizontal plane. It appears true length on the horizontal plane (Fig. 8–8).

FIGURE 8–6 The point is located on the principal planes by measuring how far it is from each of them.

FIGURE 8–5 A point projected from its location in space strikes the principal planes at *A*, *B*, and *C*.

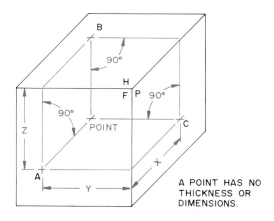

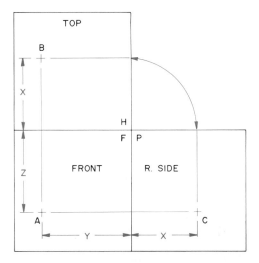

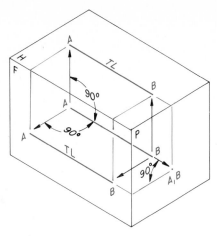

PICTORIAL VIEW OF A PRINCIPAL LINE
AS PROJECTED ON THE PRINCIPAL PLANES.

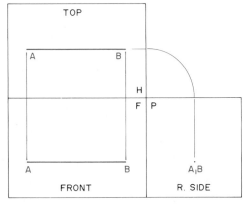

THREE VIEWS OF A PRINCIPAL LINE.

FIGURE 8–7 A principal line is a line that is parallel with two of the principal planes.

A *frontal line* is parallel with the frontal plane. It appears true length on the frontal plane (Fig. 8–9).

A *profile line* is parallel with the profile plane. It appears true length on the profile plane (Fig. 8–10).

An *inclined line* is parallel with one of the principal planes but not parallel or perpendicular to the others. It is true length on the plane to which it is parallel (Fig. 8–11).

An *oblique line* is not parallel to any of the principal planes. It appears foreshortened on all principal planes (Fig. 8–12).

## LOCATING A POINT ON A LINE

If the location of a point on a line is known in one view, it can be found in the other views by projection. Project it perpendicular to the principal planes. Its location is the point where the projection crosses the line (Fig. 8–13). *If a point is known to be on a line in one view, it will show there in all other views.*

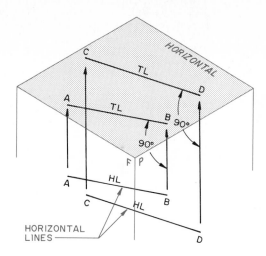

FIGURE 8–8 A horizontal line is parallel with the horizontal plane. It appears true length on the horizontal plane.

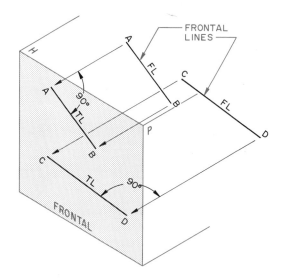

FIGURE 8–9 A frontal line is parallel with the frontal plane. It appears true length on the frontal plane.

FIGURE 8–10 A profile line is parallel with the profile plane. It appears true length on the profile plane.

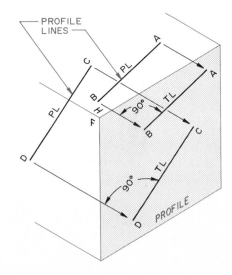

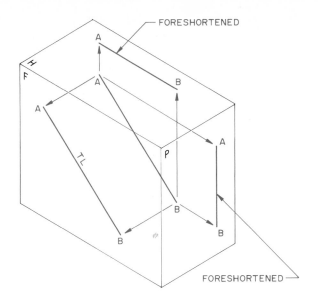

FIGURE 8–11 An inclined line is parallel with only one principal plane.

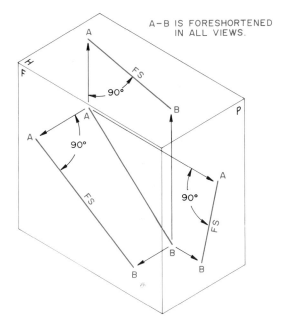

FIGURE 8–12 An oblique line is not parallel with any of the principal planes.

## INTERSECTING AND NONINTERSECTING LINES

Lines may cross in some views, but whether they actually intersect may be uncertain. To check to see if they intersect, project the point of apparent intersection in one view to the other views. If the apparent points of intersection do not project, the lines do not intersect (Fig. 8–14). If the apparent point of intersection does project to the other views, the lines do intersect (Fig. 8–15).

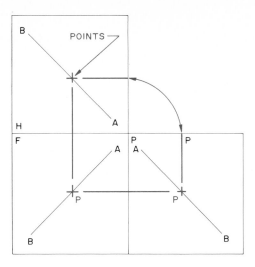

FIGURE 8–13 A point on a line projects to that line in all views.

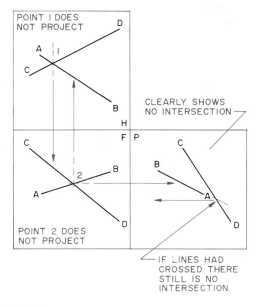

FIGURE 8–14 The apparent points of intersection do not project from view to view; therefore, the lines do not intersect at those points.

FIGURE 8–15 The apparent point of intersection does project to the other views; therefore, these lines do intersect at that point.

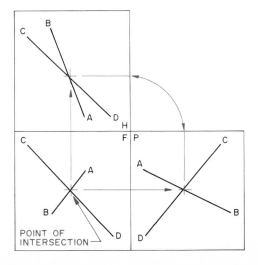

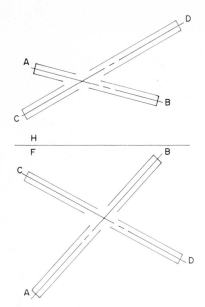

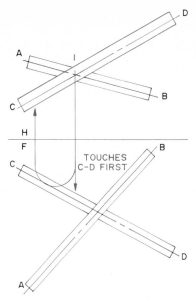

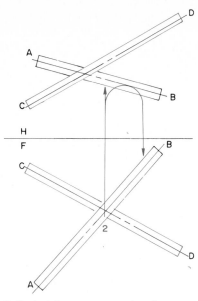

Assignment: Find the visibility of pipes A–B and C–D and complete the drawing.

1. Project one of the apparent points of crossing to the other view. In this step, point 1 in the top view was projected to the front view. It touches C–D first in the front view, so C–D is on top in the top view.

2. Project the apparent point of intersection, 2, in the front view to the top view. It touches A–B first in the top view, so A–B is on top in the front view.

FIGURE 8–16    Ascertaining the visibility of lines.

## VISIBILITY: CROSSING LINES

When it is certain that lines that appear to cross do not intersect, it is then necessary to ascertain which is in front of the other in each view. This is especially important when working with three-dimensional objects such as pipes. Pipes are located by their center line, which is used to check for visibility.

To ascertain visibility, work each view separately. In each view, project the point at which the lines appear to cross to the adjacent view. The line it intersects first in the adjacent view is on top in the view under study (Fig. 8–16).

## VISIBILITY: LINES AND PLANES

Often a line and a plane appear to cross in a view. To complete the view, it is necessary to ascertain which is in front of the other in each view. The procedure is the same as described for visibility of lines. However, each place the line crosses an edge of the plane is checked as a separate solution. Each point of crossing in each view is checked separately. The steps in performing this check are shown in Fig. 8–17.

If, during a check for visibility, one end of a line is found to be on top of a plane and the other

below the plane, you know the line must pierce the plane.

## PIERCING POINT: A LINE AND A PLANE

In some cases a line may appear not to cross a plane but actually pierce it. To complete the views, the piercing point must be found and the visibility of each part of the line ascertained.

If the plane is inclined, the piercing point can be found in the view showing the plane as an edge (Fig. 8–18). The piercing point on the plane is the point where the line crosses the edge view of the plane. This point is projected to the other views.

The steps for finding the piercing point of a line and an oblique plane are shown in Fig. 8–19. Pass an imaginary plane through the line. Project its intersection with the plane to the other view. The point at which it crosses the line is the piercing point. Project this to the other views and ascertain visibility.

## ESTABLISHING PLANES

Theoretically, a plane has no limits. It extends to infinity. For practical purposes, boundaries are set on planes to assist in solving design problems. This es-

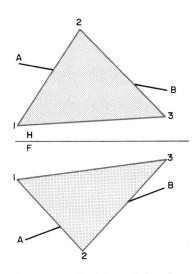

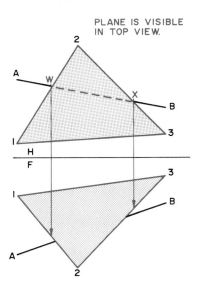

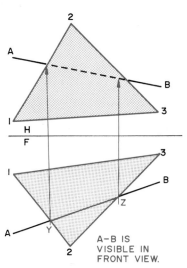

A-B IS
VISIBLE IN
FRONT VIEW.

Assignment: Find the visibility of plane 1–2–3 and line A–B and complete the drawing.

1. In one view project the apparent crossing of line A–B with the edges of plane 1–2–3. In this example, the top view was used. Project points W and X to the front view. Point W hits line 1–2 before A–B, so 1–2 is on top in the top view. Point X hits line 2–3 in the front view before A–B, so 2–3 is on top in the top view. A–B is hidden below the plane in this view.

2. In the front view, project points Y and Z to the top view. Both touch A–B first, so A–B is on top in the front view.

FIGURE 8–17  Ascertaining the visibility of a line and a plane.

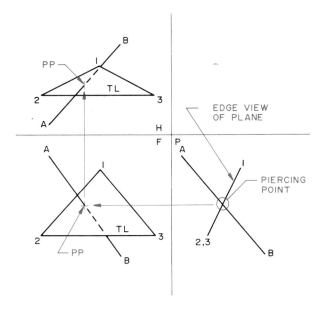

FIGURE 8–18  To find the point where the line pierces an inclined plane, find the plane in edge view.

tablishes the length and width of the plane. A plane has no thickness.

Planes are represented on drawings by several methods. These include three points not in a straight line, two intersecting lines, a point and a line, and two parallel lines (Fig. 8–20). The actual size of the plane is not limited by these methods. They establish that there is a plane and locate it in space.

## TYPES OF PLANES

Planes parallel with the three principal planes of projection (horizontal, frontal, and profile) are identified by the name of the principal plane.

A *horizontal plane* is parallel with the horizontal plane of projection (Fig. 8–21). It is true size in the top view. It appears as an edge on the frontal and profile planes of projection.

A *frontal plane* is parallel with the frontal plane of projection (Fig. 8–22). It is true size on the frontal plane of projection. It appears as an edge on the horizontal and profile planes of projection.

A *profile plane* is parallel with the profile plane of projection (Fig. 8–23). It apears true size on the profile plane of projection. It appears as an edge on the frontal and horizontal planes of projection.

An *inclined plane* is perpendicular to one principal plane of projection and at an acute angle (less than 90°) to the other two principal planes. It appears as an edge and true length on the plane with which it is parallel. It appears foreshortened on the planes to which it is not perpendicular (Fig. 8–24).

Descriptive Geometry: Spatial Relationships  **165**

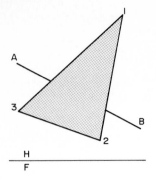
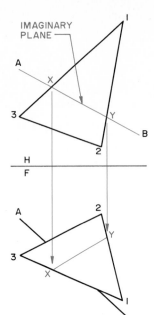
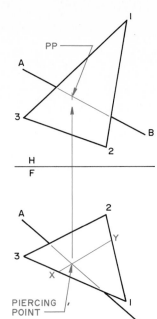
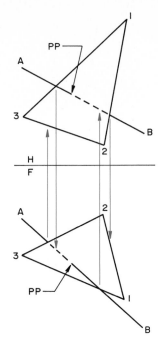

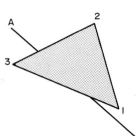
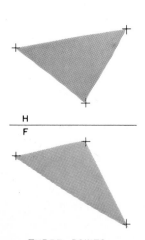

IMAGINARY
PLANE

PP

PP

PIERCING
POINT

Assignment: Find the point where line A–B pierces oblique plane 1–2–3.

1. Pass an imaginary plane through line A–B. In this example, it was done in the top view. Since the line is on the plane, it will pierce plane 1–2–3 at the intersection of the two planes. Project the intersection of the two planes, X and Y, to the front view.

2. Locate the piercing point on the front view. It is where the imaginary plane crosses line A–B. Project the piercing point to the top view.

3. Ascertain visibility for the line and the plane.

FIGURE 8–19 Finding the point where a line pierces an oblique plane.

FIGURE 8–20 Methods for establishing planes.

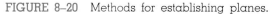
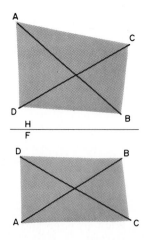
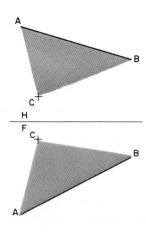
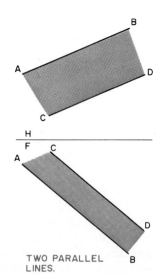

THREE POINTS.

TWO INTERSECTING LINES.

A POINT AND A LINE.

TWO PARALLEL LINES.

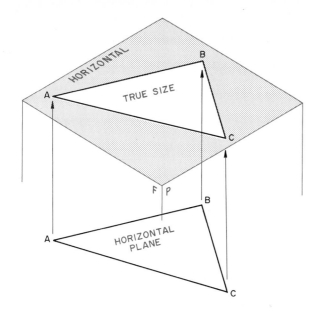

FIGURE 8–21 A horizontal plane is parallel with the horizontal plane of projection.

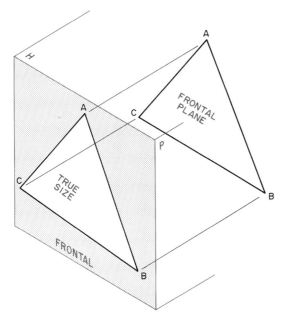

FIGURE 8–22 A frontal plane is parallel with the frontal plane of projection.

An *oblique plane* is not parallel with or perpendicular to any of the principal planes of projection. It appears foreshortened on all principal planes of projection (Fig. 8–25).

## LOCATING A LINE ON A PLANE

When the location of a line on a plane is known in one view, it can be projected to the other views by projecting each end of the line (Fig. 8–26). Connect the end points to form the line.

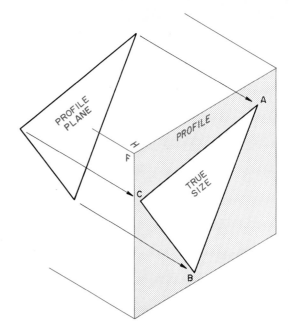

FIGURE 8–23 A profile plane is parallel with the profile plane of projection.

FIGURE 8–24 An inclined plane is perpendicular to one principal plane of projection and at an acute angle to the other two principal planes.

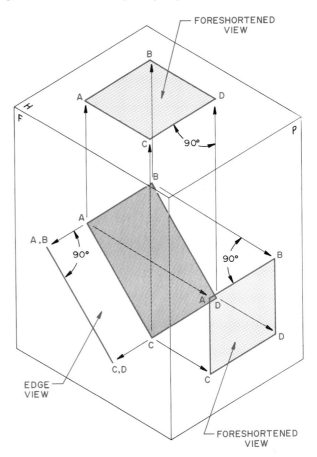

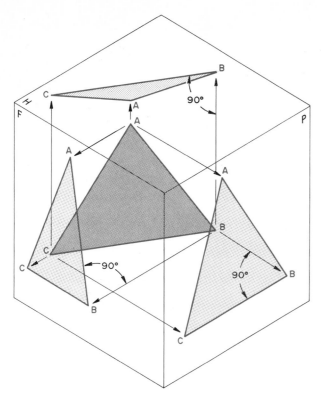

**FIGURE 8-25** An oblique plane is not parallel with or perpendicular to any of the principal planes of projection.

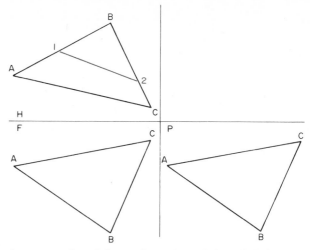

Assignment: Line 1–2 is on the surface of plane A–B–C. Project the line to the other views of the plane.

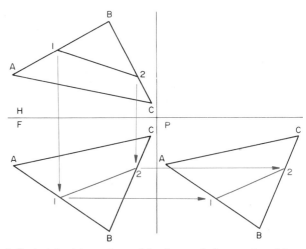

1. Project the intersections of the line and plane at 1 and 2 to the other views. Connect these points to form the line.

**FIGURE 8-26** How to project a line on a plane to the plane in another view.

## LOCATING A POINT ON A PLANE

When the location of a point on a plane is known in one view, it can be projected to the other views by drawing a line through the point to each side of the plane. Locate this line on the other views. Then, project the point to the line in the other views (Fig. 8–27).

## CONSTRUCTING PRINCIPAL LINES ON A PLANE

Some graphic solutions require the use of principal lines. They have the advantage of appearing true length on the plane to which they are parallel. The procedure for locating principal lines on the three principal planes of projection is shown in Figs. 8–28, 8–29, and 8–30. The procedure involves constructing lines parallel with the edge view of the plane upon which the principal lines are needed, then projecting the lines to the plane on the principal plane of projection.

## PARALLELISM: LINES

In some spatial problems, lines may appear parallel. To reach a satisfactory solution it is important to know if they really are parallel. In some cases it is necessary to construct a line parallel with a line on the existing drawing.

Lines can be checked for parallelism by locating them on two of the three principal planes of projection. *If lines are parallel on two principal planes of projection, they are parallel lines* (Fig. 8–31).

The steps for constructing parallel lines are given in Fig. 8–32. Usually, the parallel line must

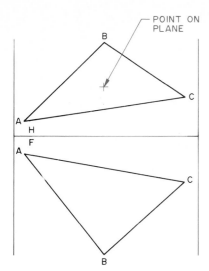

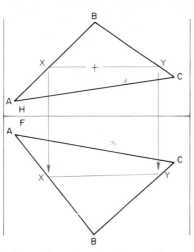

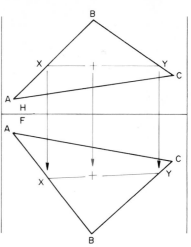

Assignment: Project a point on a plane to other views of the plane.

1. Pass a line through the point that intersects two sides of the plane. Project these intersections, X and Y, to the other views. Form the line by connecting X and Y.

2. Project the point on the line to the line in the other views. This locates the point on the other views of the plane.

FIGURE 8–27 How to project a point on a plane to the plane in another view.

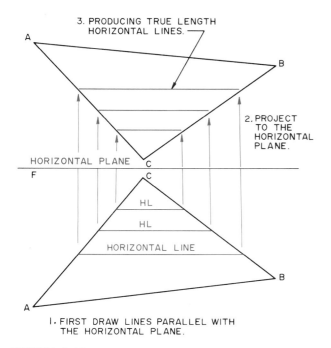

FIGURE 8–28 Projecting a horizontal line on a plane to the plane in another view.

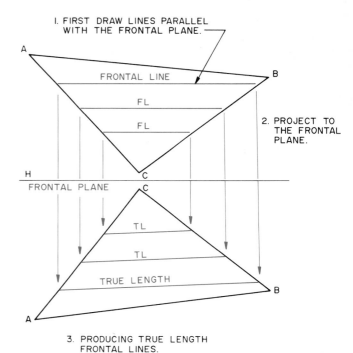

FIGURE 8–29 Projecting a frontal line on a plane to the plane in another view.

pass through a point in space to place it properly in relation to the original line. The procedure involves constructing a line through the point parallel with the given line in each view. When the length is known, the ends of the line in each view can be projected to the other views.

## PARALLELISM: PLANES

*Two planes are parallel if intersecting lines on one plane are parallel to intersecting lines on another plane.* These can be edges of the planes or lines on the surface of

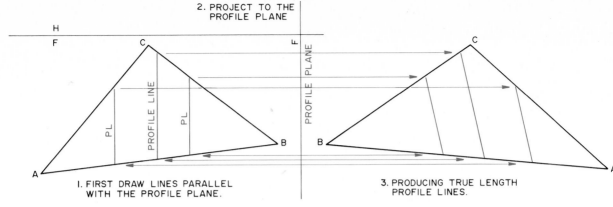

FIGURE 8–30 Projecting a profile line on a plane to the plane in another view.

the planes. Another way to check is to find the planes in edge view. If they are parallel in edge view, they are parallel planes (Fig. 8–33).

The steps for constructing a plane parallel to another through a point in space are shown in Fig. 8–34.

## PARALLELISM: A LINE AND A PLANE

*A line is parallel to a plane if the line is parallel to any line on the plane* (Fig. 8–35). To construct a line parallel with a plane through at point at some predetermined distance from a plane, draw the line through

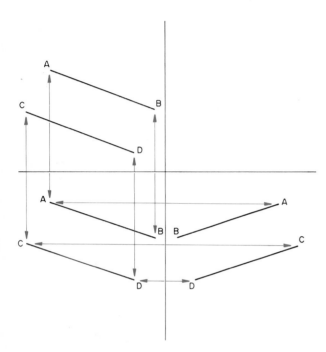

FIGURE 8–31 Lines are parallel if they appear parallel on two principal planes of projection.

FIGURE 8–32 How to construct a line parallel to another through a given point.

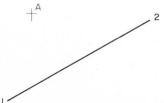

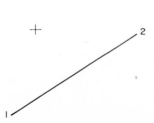

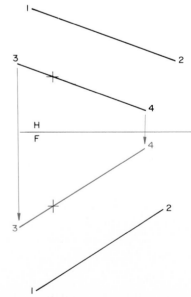

Assignment: Construct a line through point A parallel with line 1–2.

1. Construct a line through point A parallel with line 1–2 in any view. In this example, it is the top view. Establish the desired length of the line, points 3 and 4.

2. Construct a line through point A parallel with line 1–2 in the front view. Then, establish the length. Project the ends from the top view.

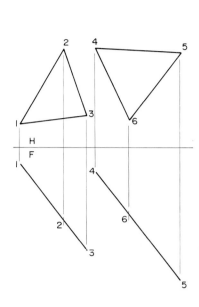

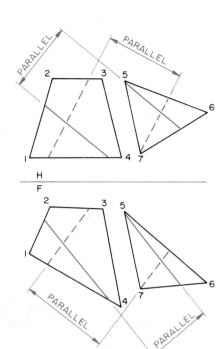

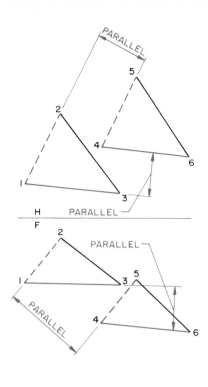

Planes are parallel in edge view.

Intersecting lines on one plane are parallel with intersecting lines on the other plane.

Intersecting edges on one plane are parallel with intersecting edges on the other plane.

FIGURE 8–33 Various ways to tell if two planes are parallel.

FIGURE 8–34 Constructing a plane parallel to a given plane.

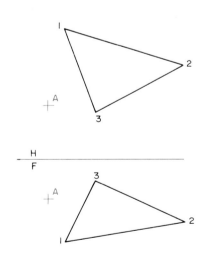

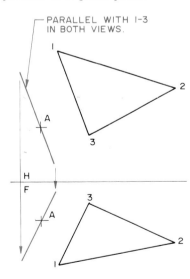

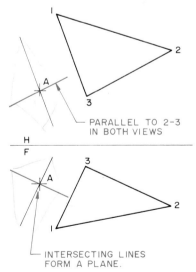

Assignment: Construct a plane through point A that is parallel with plane 1–2–3.

1. Construct a line in the top view parallel with a line on the plane or an edge of the plane through point A. Project it to the front view.

2. Construct a second line in the top view through point A. It must be parallel with another edge or line on the surface that intersects the first line used. Project it to the front view. The two intersecting lines form a plane in space.

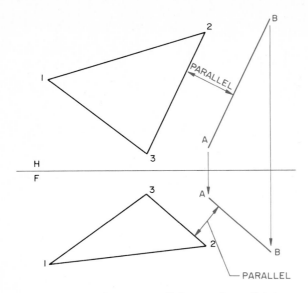

FIGURE 8–35   A line is parallel to a plane if it is parallel to any line on the plane.

the point parallel to a line on the plane in one view. Then, draw the line through the point parallel to the same line on the plane in a second view (Fig. 8–36).

## PERPENDICULARITY: LINES

There are many instances when a designer will need to construct a line perpendicular to a principal or oblique line. *In order to construct a line perpendicular to another, one or both of them must be true length.*

The steps for constructing a line perpendicular

FIGURE 8–36   How to construct a line parallel to a given plane.

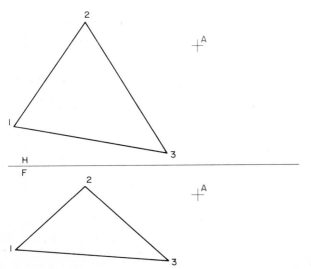

Assignment: Construct a line through point A parallel with plane 1–2–3.

to a principal line are given in Fig. 8–37. The line is parallel with the horizontal plane, so it is true length on the horizontal plane. Draw the line from the desired point perpendicular to the true-length line. Project the intersection to the top view. Connect the known point and the intersection to locate the line in the top view. The same procedure is used to draw lines perpendicular to frontal and profile lines.

The steps for constructing a line perpendicular to an oblique line are shown in Fig. 8–38. Draw the new line so that it appears true length in one view. Draw it perpendicular to the oblique line in the view where the new line is true length.

## PERPENDICULARITY: A LINE TO A PLANE

*A line is known to be perpendicular to a plane if it is perpendicular to two intersecting lines on the plane* (Fig. 8–39). If the line is perpendicular to two intersecting lines on a plane, it is perpendicular to all lines on the plane. The steps for constructing a line perpendicular to a plane are given in Fig. 8–40. On one view, draw a true-length line through the point on the plane where the perpendicular is to touch the plane. Draw a line perpendicular to it. Then, through the intersection of these lines, draw another true-length line in another view. Draw the line perpendicular to it. The line is perpendicular to the plane at the point wanted because it is perpendicular to two true-length lines on the plane passing through the point.

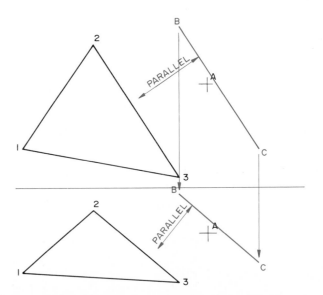

1. Draw a line through A parallel with an edge or a line on the plane. In this example, it was drawn in the top view parallel with 2–3. Then, draw the line in the front view parallel with 2–3. Project the ends of the line if length is important.

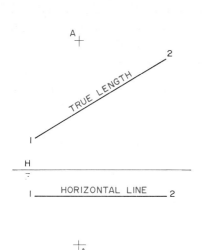

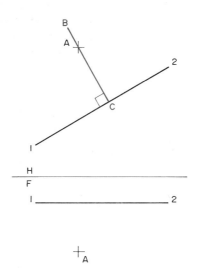

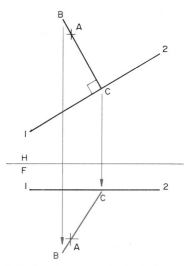

Assignment: Construct a line perpendicular to a horizontal line from a given point.

1. The horizontal line is true length in the top view. Draw a line from point A perpendicular to the true-length line.

2. Project the perpendicular to the front view.

**FIGURE 8–37** How to construct a line perpendicular to a principal line through a given point.

**FIGURE 8–38** How to construct a line perpendicular to an oblique line.

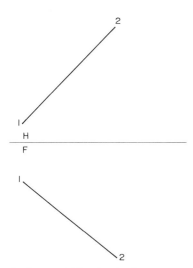

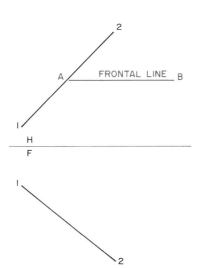

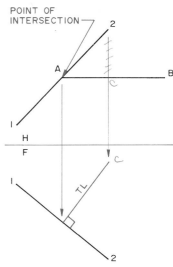

Assignment: Construct a line perpendicular to an oblique line.

1. In any view, construct a principal line that intersects the oblique line. In this example, a frontal line was drawn in the top view.

2. Project the intersection of the two lines to the other view. This was the front view in this example. Draw a line perpendicular to the oblique line, 1–2, at the point of intersection. This line is true length because it was constructed parallel with the frontal plane.

Descriptive Geometry: Spatial Relationships **173**

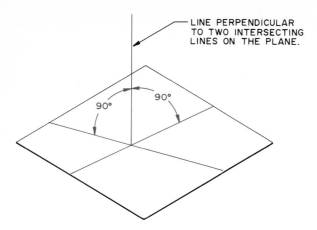

FIGURE 8–39 A line is known to be perpendicular to a plane if it is perpendicular to two interesting lines on the plane.

# PERPENDICULARITY: A PLANE TO AN OBLIQUE LINE

Sometimes a design situation has a line representing part of the product, as a pipe, to which a perpendicular plane must be drawn. The steps for doing this are shown in Fig. 8–41. The principle that a plane can be formed by two intersecting lines is used in this construction.

Draw a principal line through the point that locates the plane in space. Project it to the other view. Here it is true length, so it can be drawn perpendicular to the given line. Next, draw another principal line in another view. Project it to the other view, where it is also true length. Draw it perpendicular to the given line. The two intersecting lines form a plane. The plane is perpendicular to the line because the line is perpendicular to the two intersecting lines.

FIGURE 8–40 How to construct a line perpendicular to a given plane at a given point.

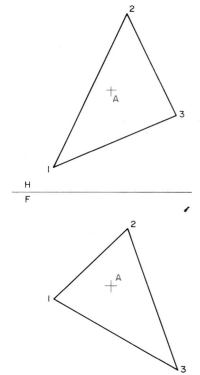

Assignment: Construct a line perpendicular to an oblique plane through a point on the plane.

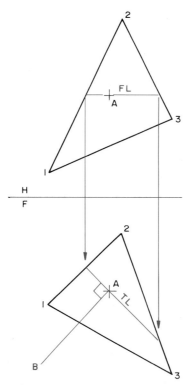

1. Draw a line parallel with one of the principal planes through the joint, A, on the plane. In this example, a frontal line was drawn. Project it to the front view, where it appears true size. Construct a line perpendicular to the true-length line at point A.

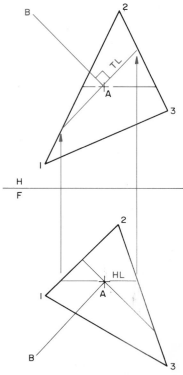

2. In the front view, draw a horizontal line through the point. Project it to the top view, where it appears true size. Construct a line perpendicular to the true-length line at point A in the front view. Since line A–B is now perpendicular to two intersecting lines on the plane, it is known to be perpendicular to the plane.

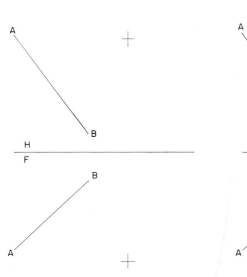

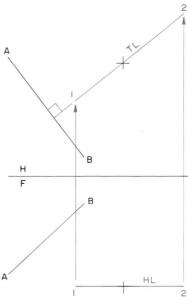

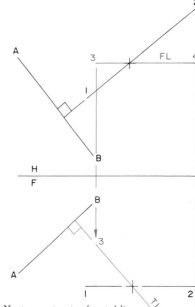

Assignment: Construct a plane through a point in space that is perpendicular to an oblique line.

1. Locate two intersecting lines that are perpendicular to the line when they appear true length. First, construct a principal line through the point. Any view can be used. In this example, a horizontal line was used. Project it to the top view and draw it through the point perpendicular to the line.

2. Next, construct a frontal line through the point. Project it to the front view. Draw it through the point perpendicular to the line. The two intersecting lines, 1–2 and 3–4, form a plane that is perpendicular to the oblique line.

FIGURE 8–41 How to construct a plane through a given point perpendicular to an oblique line.

## PERPENDICULARITY: A PLANE TO A PLANE

*Two planes are perpendicular if a line in one plane is perpendicular to the other plane* (Fig. 8–42). The steps for constructing a plane perpendicular to another plane are shown in Fig. 8–43. Given are a plane and a line locating the second plane. The second plane is to pass through the given line.

Since the given line is known to be in the second plane, it is necessary to draw a second line that intersects the given line to obtain intersecting lines for a plane. This is done by constructing a line that intersects the given line and is also perpendicular to the first plane. The construction is the same as for constructing a line perpendicular to a plane. The intersection between the two lines can occur at any convenient place along the given line.

Because the constructed line is perpendicular to two intersecting lines on the plane, it is perpendicular to the plane. Because the constructed line is also in the second plane, the second plane is perpendicular to the first plane.

FIGURE 8–42 Two planes are perpendicular if a line on one plane is perpendicular to the other plane.

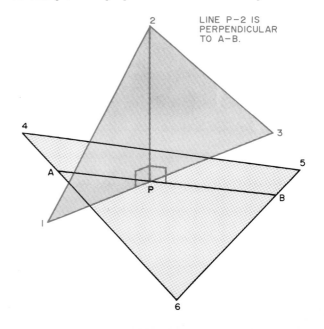

LINE P–2 IS PERPENDICULAR TO A–B.

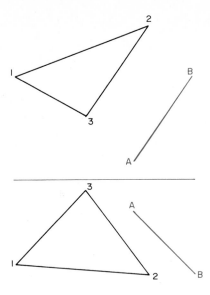
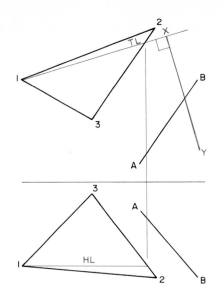
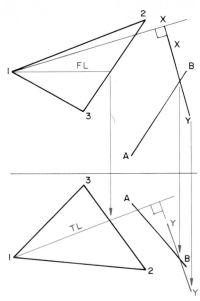

Assignment: Construct a plane through a line in space perpendicular to another plane.

1. Line A–B is known to be on the plane to be constructed. Since two intersecting lines form a plane, this principle will be used. Since the planes are to be perpendicular, the line constructed will be perpendicular to the plane.

Construct a principal line. It can be in any view and cross at any point. In this example, a horizontal line was used. It projects true length in the top view. A line crossing given line A–B was drawn perpendicular to the true-length line.

2. Construct a frontal line that projects true length in the front view. In that view, construct a line perpendicular to the true-length line that crosses line A–B at the point of intersection found in the top view. These two intersecting lines form a plane that is known to be perpendicular to plane 1–2–3.

FIGURE 8–43 How to construct a plane through a line in space perpendicular to a given plane.

# Problems

The following problems are designed to be solved on 8½ × 11 in. coordinate paper divided into four divisions per inch. Locate the given problem on your sheet by counting squares to find the ends of each line. Then, solve the assigned problem on the sheet. Always identify all lines, points, and planes. Leave all construction lines unless directed otherwise by your instructor.

1. Complete the missing views in Probs 8–1, 8–2, 8–3, and 8–4. Identify each true-length line by lettering TL on it and each foreshortened line by lettering FS on it. Then, identify each line as horizontal (H), frontal (F), profile (P), inclined (I), or oblique (O).

2. Add points X, Y, and Z to line AB in Prob. 8–5. Label all points.

3. Draw the front view and right side view of line AB that intersects line CD in Prob. 8–6.

4. Complete the top view and locate point X on line AB in Prob. 8–7.

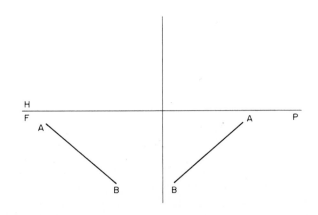

PROBLEM 8–1

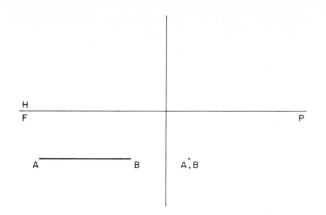

PROBLEM 8–2

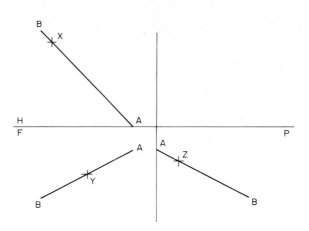

PROBLEM 8–5

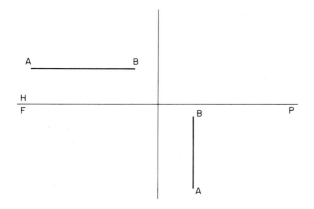

PROBLEM 8–3

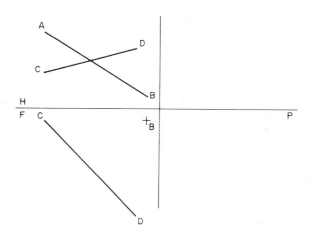

PROBLEM 8–6

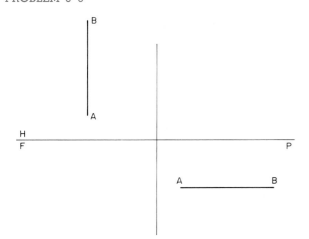

PROBLEM 8–4

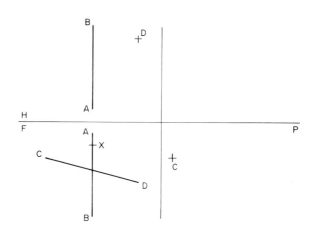

PROBLEM 8–7

5. Complete the top view and locate point X on line AB when line AB and CD intersect as shown in Prob. 8–8.

6. Determine the visibility of the pipes in Prob. 8–9. Complete the drawing showing the pipes with visible and hidden lines.

7. Determine the visibility of the line and plane in each view in Prob. 8–10. Complete each view with the appropriate visible and hidden lines.

8. Find the point where line AB pierces plane CDE in Prob. 8–11. Complete the drawing showing visibility and the piercing point.

9. Find the point where line AB pierces plane CDEF in Prob. 8–12. Complete the drawing showing visibility and the piercing point.

10. Draw the right side view of plane ABC in Prob. 8–13.

11. Draw the top view of plane ABCD in Prob. 8–14.

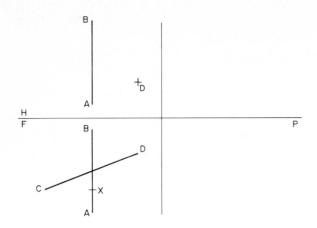

PROBLEM 8-8

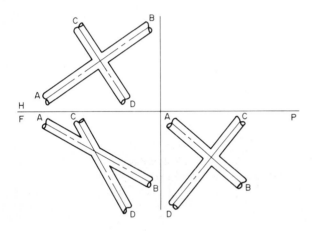

PROBLEM 8-9

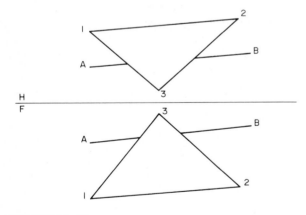

PROBLEM 8-10

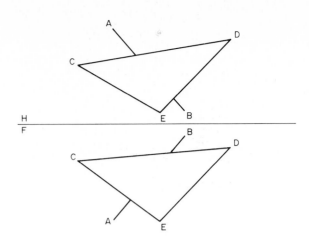

PROBLEM 8-11

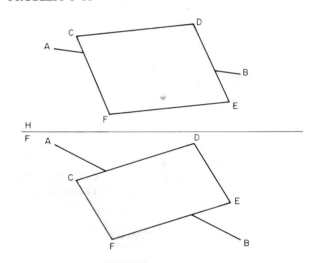

PROBLEM 8-12

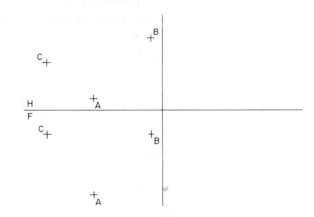

PROBLEM 8-13

12. In Prob. 8-15, locate line AB and point C on all three views of plane DEF.

13. In Prob. 8-16, draw a true-length horizontal, frontal, and profile line on the views given. Identify each true-length line.

14. Lines AB, BC, and CD in Prob. 8-17 are sides of a parallelepiped. Complete the drawing and show all hidden edges.

15. Construct a line of any length through point A that is parallel with line BC in Prob. 8-18 in all three views. Label the ends DE.

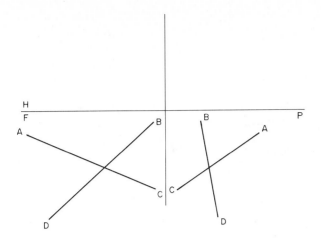

PROBLEM 8–14

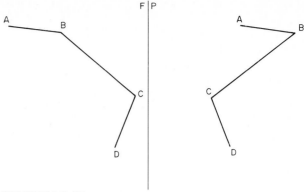

PROBLEM 8–17

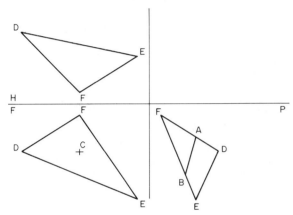

PROBLEM 8–15

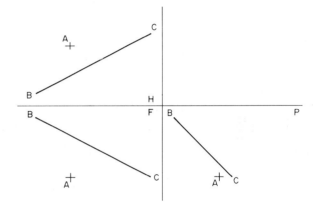

PROBLEM 8–18

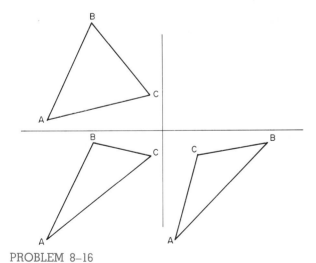

PROBLEM 8–16

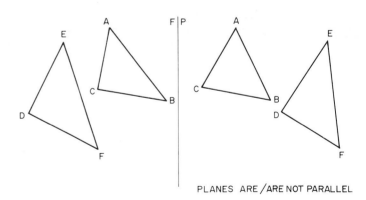

PLANES ARE / ARE NOT PARALLEL

PROBLEM 8–19

16. Ascertain if the two planes in Prob. 8–19 are parallel.

17. Construct a plane through point E parallel with plane ABC in Prob. 8–20.

18. Ascertain if the planes in Prob. 8–21 are parallel by drawing the right side view.

19. Construct lines through points A and B that are parallel with plane DEFG in Prob. 8–22.

20. Construct a line from point A perpendicular to line BC and another from point E perpendicular to line CD in Prob. 8–23.

21. Draw a line perpendicular to plane ABC at point D in all three views in Prob. 8–24. Indicate the 90° angle.

22. In Prob. 8–25, draw a line perpendicular to plane ABC at corner C.

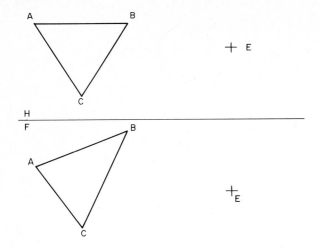

PROBLEM 8–20

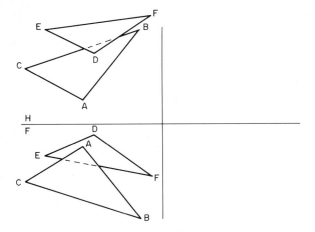

PROBLEM 8–21

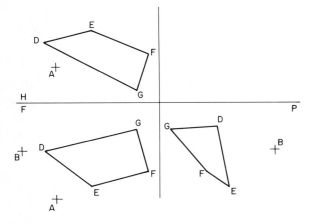

PROBLEM 8–22

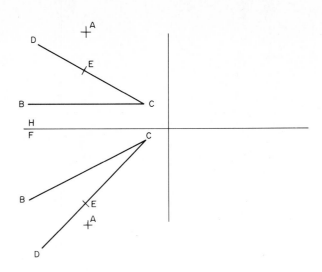

PROBLEM 8–23

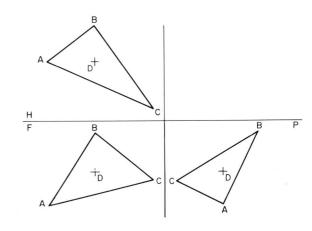

PROBLEM 8–24

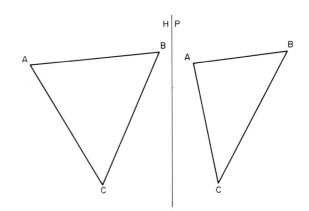

PROBLEM 8–25

23. Construct a plane through point C perpendicular to oblique line AB in Prob. 8–26.

24. Construct a plane through line AB that is perpendicular to plane CDE in Prob. 8–27.

25. Construct a plane through line EF that is perpendicular to plane ABCD in Prob. 8–28.

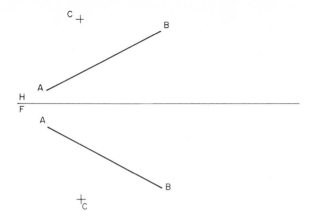

PROBLEM 8–26

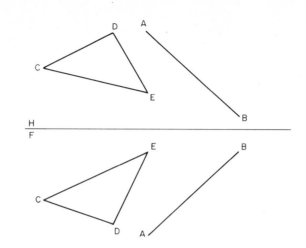

PROBLEM 8–27

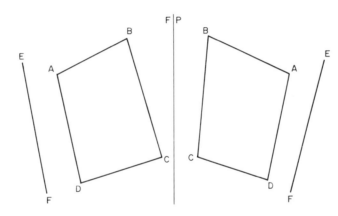

PROBLEM 8–28

# Auxiliary Views

As a product is being designed, surfaces and edges that are inclined or oblique are often involved. Since these do not appear true size in the principal planes of projection, special planes are developed to which they are projected. These are called *auxiliary planes of projection*.

An *inclined* surface is one that is perpendicular to only one of the principal planes of projection and at an acute angle to the other two principal planes (Fig. 9–1). An *oblique* surface is one that is not parallel or perpendicular to any of the principal planes of projection (Fig. 9–2). Inclined and oblique planes do not appear true size when projected to the principal planes of projection.

Figure 9–3 shows a product that contains a variety of surfaces. The engineering designer must be able to find the true sizes of planes easily and quickly. This means that a thorough knowledge of auxiliary views is an important part of an engineer's knowledge. This chapter explains how to use primary and successive auxiliary planes to find the true size of inclined and oblique surfaces and angles.

FIGURE 9–3  This product has a variety of surfaces.

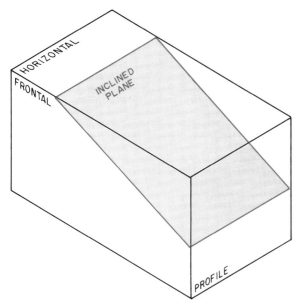

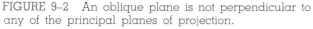

FIGURE 9–1  An inclined plane is perpendicular to only one principal plane of projection. In this example, it is perpendicular to the frontal plane.

FIGURE 9–2  An oblique plane is not perpendicular to any of the principal planes of projection.

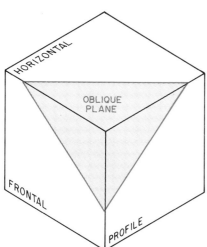

## AUXILIARY VIEWS

An auxiliary view is an orthographic view having a line of sight that is not perpendicular to the principal planes of projection. The line of sight in an auxiliary view is perpendicular to the inclined or oblique surface (Fig. 9–4).

An example of a drawing with an auxiliary view is in Fig. 9–5. The inclined surface is foreshortened in the principal views and true size in the auxiliary view. The inclined plane is shown true length in every view to which its edges are parallel.

Auxiliary views are projected upon an auxiliary plane of projection that is parallel with the surface or line involved (Fig. 9–6).

FIGURE 9–4  To make an auxiliary view, the line of sight must be perpendicular to the inclined or oblique plane.

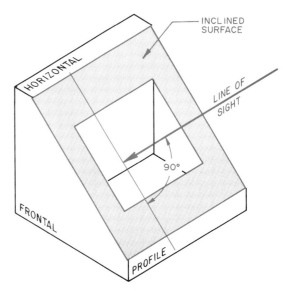

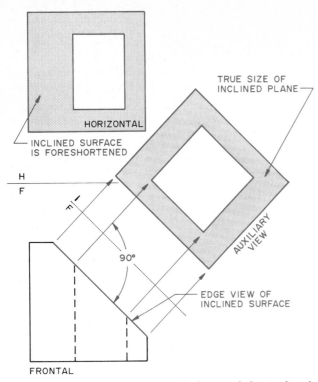

TRUE SIZE OF
INCLINED PLANE

HORIZONTAL

INCLINED SURFACE
IS FORESHORTENED

H
F

F

90°

AUXILIARY
VIEW

EDGE VIEW OF
INCLINED SURFACE

FRONTAL

FIGURE 9–5 The true size and shape of the inclined plane are shown by drawing an auxiliary view.

The basic uses for auxiliary views include the following.

1. Finding the true length of a line.
2. Finding the point projection of a line.
3. Finding the edge view of a plane.
4. Finding the true size and shape of a plane.
5. Finding the true size of an angle.
6. Dimensioning a product.

Auxiliary views are projected following the principles explained for orthographic projection. The lines of sight from the plane in question are perpendicular to the auxiliary plane. The auxiliary view is actually an orthographic view that is drawn upon a plane of projection other than one of the six principal planes.

FIGURE 9–6 The auxiliary view is projected upon an auxiliary plane that is parallel with the inclined surface.

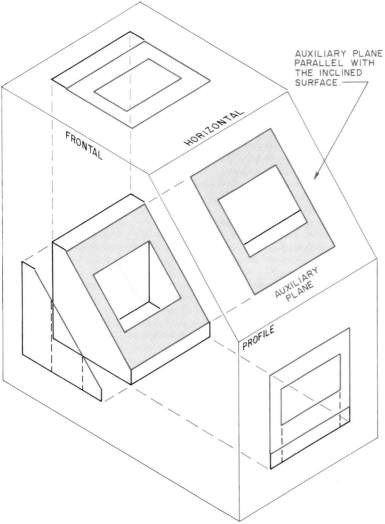

AUXILIARY PLANE
PARALLEL WITH
THE INCLINED
SURFACE.

FRONTAL

HORIZONTAL

AUXILIARY
PLANE

PROFILE

## IDENTIFICATION OF PLANES

It is important to identify carefully each plane upon which a view is to be projected. This helps the engineer to remember in which plane each view is shown.

As shown in the chapter on orthographic projection, the principal planes—horizontal, frontal and profile—are identified by the letters H, F, and P. The primary (or first) auxiliary plane is identified as 1 (Fig. 9–5). A secondary (or second) auxiliary plane is identified with a 2. Successive auxiliary planes are identified with the numbers indicating their relationship with the primary auxiliary. There is no limit to the number of successive auxiliary planes that can be drawn.

## PRIMARY AUXILIARY VIEWS

A primary auxiliary view is an orthographic view that is projected from a principal view. A primary auxiliary view may be projected in any direction from any of the six principal views. Its purpose is to show the true size of a plane, line, or angle that is inclined. The auxiliary view is taken off the regular view in which the inclined surface appears as an edge. This is explained in Figs. 9–7, 9–8, and 9–9.

A primary auxiliary view taken off the top view has the auxiliary plane perpendicular to the horizontal plane. The auxiliary plane appears in edge view in the horizontal plane (Fig. 9–7). The auxiliary view and the front view both show the height true size.

A primary auxiliary view taken off the right or left side view has the auxiliary plane perpendicular to the profile plane. The auxiliary plane appears in edge view in the profile plane. It can be taken off the right or left side (Fig. 9–8). The auxiliary view, the side view, and the top view show the depth true size.

FIGURE 9–8 This inclined plane is perpendicular to the profile plane; therefore, the primary auxiliary view is projected off the side view.

FIGURE 9–7 This inclined plane is perpendicular to the horizontal plane; therefore, the primary auxiliary view is projected off the top view.

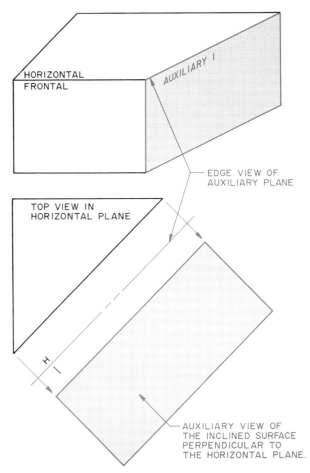

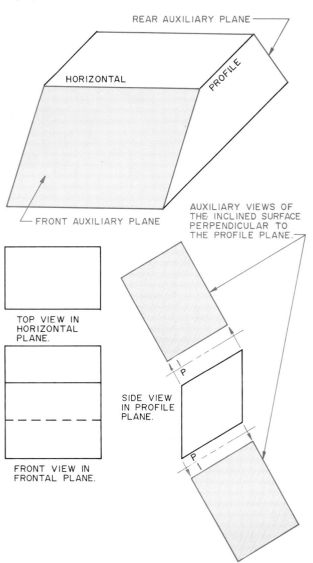

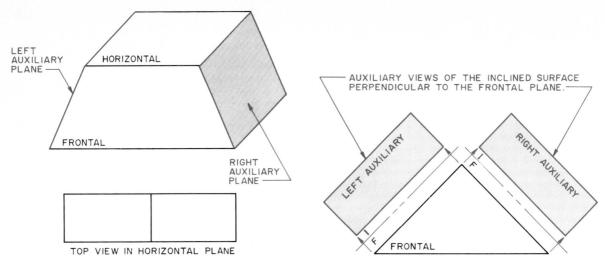

**FIGURE 9–9** This inclined plane is perpendicular to the frontal plane; therefore, the primary auxiliary view is projected off the front view.

**FIGURE 9–10** Drawing a primary auxiliary view of an inclined plane.

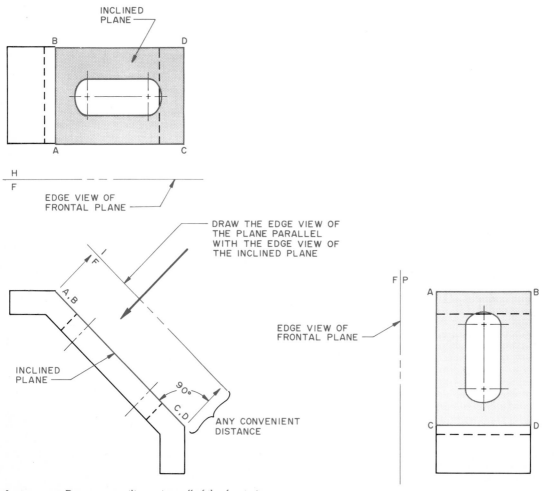

Assignment: Draw an auxiliary view off of the front view.

1. Draw an HF reference plane at any convenient distance between the top and front views. It must be perpendicular to the projectors between the views. Draw a line of sight perpendicular to the edge view of the inclined plane. In this example, it is in the front view.

Draw an edge view of the auxiliary plane parallel with the edge view of the inclined surface. Locate it at any convenient distance from the edge view of the inclined plane.

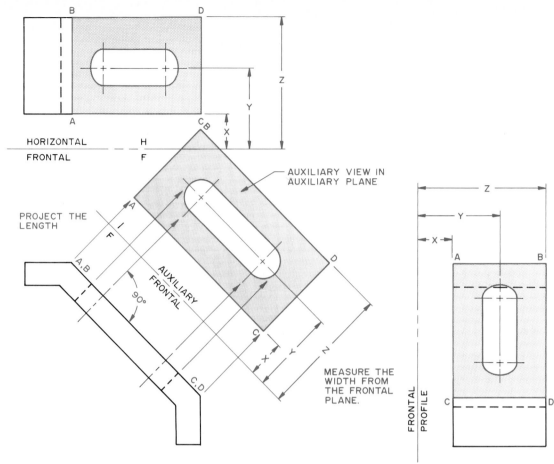

2. Project the inclined surface perpendicular to the auxiliary plane F-1. Locate corners A, B, C, and D by measuring from the edge view of the frontal plane. Connect them to form the auxiliary view of the inclined plane.

FIGURE 9–10 *cont.*

Primary auxiliary views taken off the front or rear views have the auxiliary plane perpendicular to the frontal plane. The auxiliary plane appears in edge view in the frontal plane (Fig. 9–9). The auxiliary view and the top view both show the depth true size.

## Drawing a Primary Auxiliary View

This first example, the drawing of an auxiliary view of a simple inclined plane, demonstrates the principle involved. The regular views of the plane are already drawn. The steps for drawing the auxiliary view are shown in Fig. 9–10.

Rather than laying out the edge view of the principal planes and measuring from there, it is usual practice to locate a reference line on the principal plane and measure from it. Notice in Fig. 9–11 that the reference line was placed on the front surface of the product. Because the inclined plane appears in edge view in the front view, the auxiliary view was taken off the front view.

## Partial Auxiliary Views

Normally, partial auxiliary views are drawn as in Fig. 9–11. The purpose, to show the true size of the surface, was met by showing only the surface. If the remainder of the product were drawn, all other surfaces would be foreshortened (Fig. 9–12). If the surface to be drawn is symmetrical, sometimes only half of it is drawn. This provides enough information to describe the shape. The plane from which measurements are taken is often established through the center of a symmetrical product (Fig. 9–13).

In Fig. 9–14 are examples of how planes can be located when a product is unilateral or bilateral. A *unilateral* product is one that has all projections facing in one direction. A *bilateral* product is one that has projections on two sides.

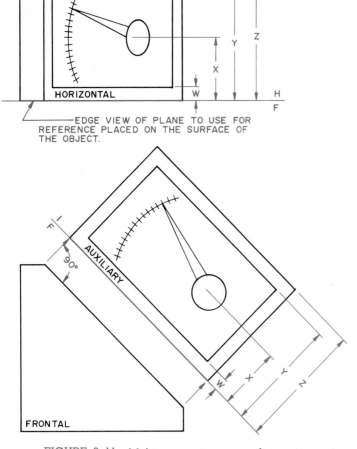

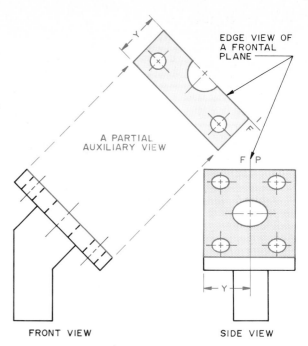

FIGURE 9-13 A partial auxiliary view can be developed around its center line.

FIGURE 9-11 Making a primary auxiliary view using the edge of the object as the frontal plane.

FIGURE 9-12 A complete auxiliary view shows all surfaces, some of which are foreshortened.

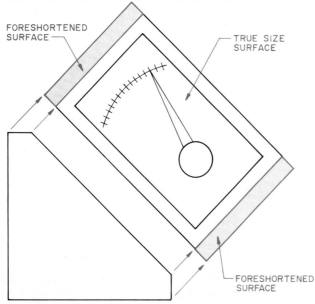

*Partial auxiliary views* may show only a portion of a surface. The view replaces part of a principal view that can also be a partial view. This saves space and time. It removes the need to draw the foreshortened surfaces, which reveal no information and are often difficult to draw (Fig. 9-15).

Because an auxiliary view shows a surface true size, it can be dimensioned. Features that appear in true size or location on normal views are not usually dimensioned on auxiliary views.

## Auxiliary Sections

Auxiliary sections are used to show the true size of some interior feature. The cutting plane is passed through the feature so that it is true size when you look perpendicular to the cutting plane. The cutting plane is actually the edge view of the auxiliary plane. The procedure for drawing the auxiliary section is the same as for other auxiliary views (Fig. 9-16).

## Curved Surfaces

A circular opening appears elliptical if its axis is not perpendicular to the plane to which it is projected. Circular elements perpendicular to inclined planes can be shown true shape by making an auxiliary view of the surface (Fig. 9-17). When an auxiliary view has curved edges, they are plotted by using a series of points. The points are located at any convenient spacing on the curved edge. They are projected to the auxiliary view. They are then connected

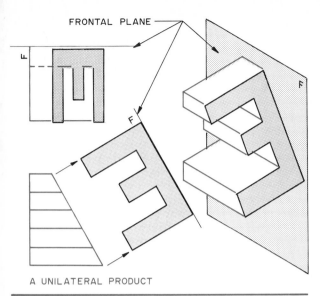

FRONTAL PLANE

A UNILATERAL PRODUCT

FRONTAL PLANE

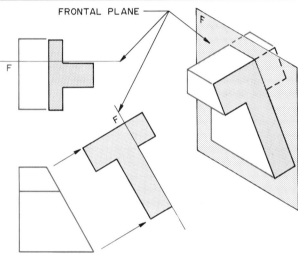

A BILATERAL PRODUCT

FIGURE 9–14 How to locate reference planes when a product is unilateral or bilateral.

with an irregular curve to give the true shape of the curve (Fig. 9–18). The more points used, the more accurate the curve.

## SECONDARY AND SUCCESSIVE AUXILIARY VIEWS

There are certain engineering design problems in which a surface or a relationship between lines or points and lines cannot be described with the principal views or primary auxiliary views. The solution is to use secondary and successive auxiliary views.

FIGURE 9–16 The procedure for drawing an auxiliary section.

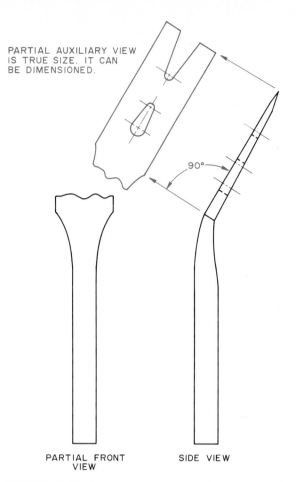

PARTIAL AUXILIARY VIEW IS TRUE SIZE. IT CAN BE DIMENSIONED.

90°

PARTIAL FRONT VIEW          SIDE VIEW

FIGURE 9–15 The working end of this carpenter's ripping chisel can be shown true size by using a partial auxiliary view.

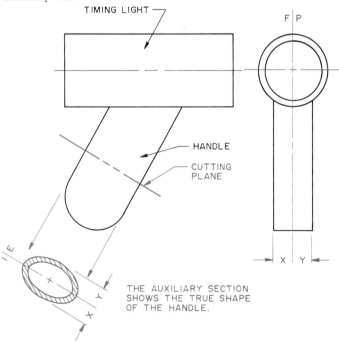

TIMING LIGHT

HANDLE

CUTTING PLANE

THE AUXILIARY SECTION SHOWS THE TRUE SHAPE OF THE HANDLE.

Assignment: Draw an auxiliary section of the handle.

1. Draw a cutting plane perpendicular to the edge of the handle.
2. Project the auxiliary section perpendicular to the cutting plane.
3. Locate the edge view of the frontal plane at any convenient distance from the front view. Since the handle was symmetrical, the frontal plane was placed on its center line.
4. Measure the width of the handle, distances X and Y, from the edge view of the frontal plane.

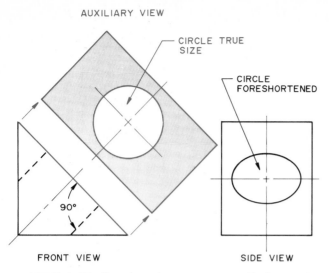

CIRCLE TRUE SIZE

CIRCLE FORESHORTENED

90°

FRONT VIEW

SIDE VIEW

**FIGURE 9–17** Circular elements perpendicular to inclined planes appear true shape in an auxiliary view of the surface.

**FIGURE 9–18** Curved edges are projected to the auxiliary view by projecting points on the edge.

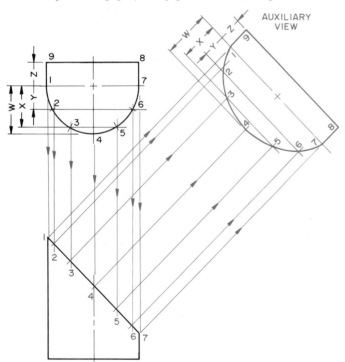

Assignment: Draw the true size of an inclined circular plane.

1. Divide the circular view into several evenly spaced parts.
2. Project these to the edge view of the plane.
3. Establish the auxiliary plane. Project the points onto the auxiliary plane.
4. Locate each point by measuring its distance from the reference plane distances (W, X, Y, and Z in the example).
5. Connect the points to form the curve.

*Primary auxiliary views* are views projected from the frontal, horizontal, or profile views. *Secondary auxiliary views* are projected from primary auxiliary views. *Successive auxiliary views* are projected from secondary auxiliary views or from other successive auxiliary views. There is no limit to the number of successive auxiliary views that can be drawn, as an additional successive auxiliary view can be projected from the previous successive auxiliary view.

The relationship between adjacent auxiliary views is the same as the principal views have with one another. For example, the secondary auxiliary plane is perpendicular to the primary auxiliary plane. The successive auxiliary plane taken off a secondary auxiliary plane is perpendicular to it. Successive auxiliary planes taken off other successive auxiliary planes are perpendicular to each other. *All auxiliary planes of all types are always perpendicular to the preceding plane from which they were projected.*

## SECONDARY AUXILIARY VIEWS

Secondary auxiliary views are used in such design situations as the following.

1. To find the true size of an oblique plane.
2. To find the point view of an oblique line.
3. To find the angle between two planes when their intersection is an oblique line.
4. To find the shortest distance between a point and an oblique line.
5. To find the shortest distance between two skewed lines.
6. To find the shortest level distance between two skewed lines.
7. To draw a line from a point to a line at a given angle to the line.
8. To find the angle between a line and a plane.

### Drawing a Secondary Auxiliary View

The following example shows how to draw a secondary auxiliary view (Fig. 9–19). The object is to find the true size of the oblique plane. The plane is shown in the top and front views. Label each corner of the plane and the principal planes.

The first step is to find the edge view of the plane. This requires a primary auxiliary view. The edge 1–2 is parallel with the horizontal plane, so it is true length in that plane. To find the edge view of a plane, draw an auxiliary plane perpendicular to a true-length line on the plane. Project plane 1–2–3 to the primary auxiliary plane.

The second step is to draw the secondary auxiliary view. Since the plane is in edge view, it can be seen true size by looking perpendicular to it. Draw the secondary auxiliary plane parallel with the edge

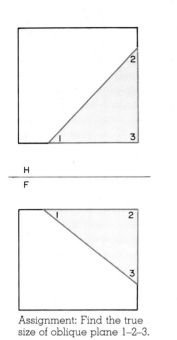

Assignment: Find the true size of oblique plane 1–2–3.

1. Find the plane in edge view in a primary auxiliary view. To do this, use a line of sight parallel with a true-length line on the plane. Edge 1–2 is true length in the top view because it is parallel with the horizontal plane. Draw reference plane H–1 perpendicular to the line of sight to form the primary auxiliary plane.

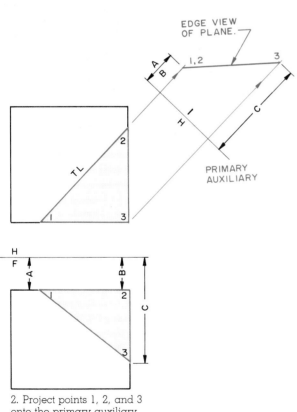

2. Project points 1, 2, and 3 onto the primary auxiliary plane. They are distances A, B, and C from the horizontal plane. This gives an edge view of the plane.

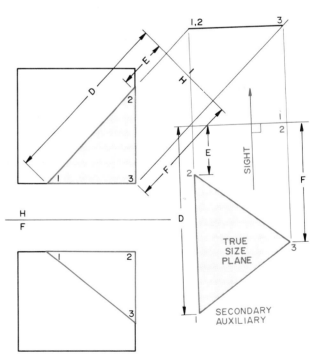

3. The secondary auxiliary is drawn parallel with the edge view of the plane. The line of sight is perpendicular to the edge view. Establish the reference line 1–2 for the secondary auxiliary view. Project the corners of the plane onto 2. They are distances D, E, and F from the primary auxiliary plane, 1. Connect the corners to form the true size of the oblique plane.

FIGURE 9–19  Drawing a secondary auxiliary view.

view of the plane. Identify it as 2. Project it to the secondary auxiliary view.

The corners of the plane are located in the secondary auxiliary view by measuring from the edge of the primary auxiliary plane. This plane appears as an edge in the top view. Measure the distance of each corner from the edge view of the primary auxiliary plane. The measurement must be perpendicular to the edge view of the primary auxiliary plane.

Regardless of the auxiliary view being drawn, the measurements are taken from the second view preceding the auxiliary view being drawn. The only views needed to draw an auxiliary view are the two preceding views.

## SUCCESSIVE AUXILIARY VIEWS

Successive auxiliary views are projected from either a secondary auxiliary view or from another successive auxiliary view. The successive auxiliary plane is always perpendicular to the line of sight used to view the problem. There are an unlimited number of successive views possible. They follow the same principles as primary and secondary auxiliary views.

FIGURE 9–20   Drawing successive auxiliary views.

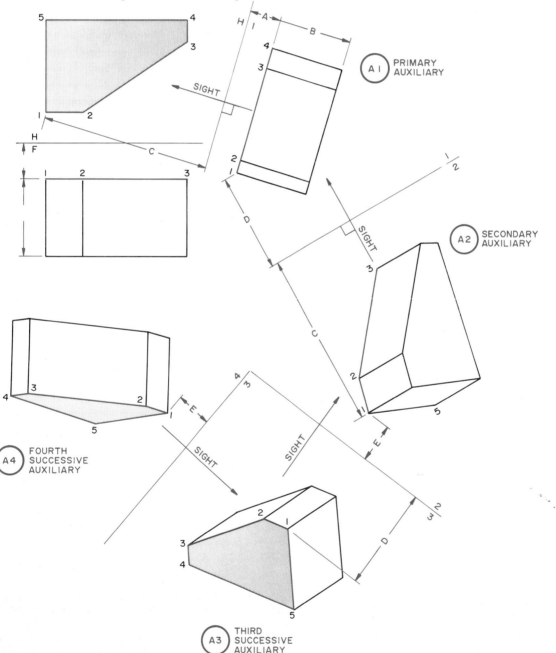

## Drawing Successive Auxiliary Views

A typical problem using successive auxiliary views is shown in Fig. 9–20. After the primary and secondary auxiliary views are drawn, the line of sight for the third auxiliary is decided. The third auxiliary plane is drawn perpendicular to the line of sight. The plane being drawn is projected to the third auxiliary plane with projectors parallel with the line of sight. Each corner is located by securing measurements from the edge view of the secondary auxiliary plane (two views preceding the one being drawn).

Seldom is it necessary to use more than three auxiliary views. However, the same technique is used for taking a fourth, a fifth, or any greater number.

# Problems

Solve the following problems on coordinate paper with four squares per inch or on 8½ × 11 in. vellum, as directed by your instructor. You can determine the sizes of the objects by counting the ¼-in. squares and using them as a scale to measure distances. Most will require one sheet per problem, so your solution can be placed clear of the given problem. Work carefully and accurately so that the auxiliary view shows the true size of the inclined or oblique surface. You can find the size of the products by counting the squares upon which the problems are drawn. Indicate all right angles with the symbol ⌐ and mark true-size surfaces TS.

## Primary Auxiliary Views

1. Draw the top and front views and auxiliary views of the two inclined surfaces in Prob. 9–1.

2. Draw the top and front views and auxiliary views of the two inclined surfaces in Prob. 9–2.

3. Draw two views and auxiliary views of all inclined surfaces in Prob. 9–3.

4. Draw front and side views and an auxiliary view of the entire object in Prob. 9–4. The line of sight should be perpendicular to the inclined surface.

5. Draw top and front views and the indicated auxiliary views in Prob. 9–5.

6. Draw the front and right side views and a partial auxiliary view of the inclined surface in Prob. 9–6.

7. Draw the front and side views and a partial auxiliary view of the inclined surface in Prob. 9–7.

8. Draw the top and front views of the product in Prob. 9–8. Then, take an auxiliary section through the slanted arm to show the true size of its cross section.

9. Draw front and right side views and the two indicated auxiliary views in Prob. 9–9.

10. Draw the top and side views and the indicated auxiliary section in Prob. 9–10.

11. Draw the top and side views and a full auxiliary view as indicated in Prob. 9–11.

12. Find the true size and shape of the groove in the product in Prob. 9–12.

13. Draw top and front views and an auxiliary view of the inclined surface in Prob. 9–13.

14. Draw a full auxiliary view showing the inclined surface in edge view in Prob. 9–14.

15. Draw a full auxiliary view of Prob. 9–15 on which surface A appears true size. Lightly shade the true-size surface.

16. Draw the views necessary to show all surfaces of the object in Prob. 9–16 true size.

17. Draw the front and top views and the true size of the inclined surface in Prob. 9–17.

18. Find the true size of the inclined surface in Prob. 9–18.

19. Draw the front and side views and the indicated auxiliary section in Prob. 9–19.

20. Draw an auxiliary view that shows the true size of the angle between surfaces A and B in Prob. 9–20.

21. Draw a full auxiliary view on which surface B in Prob. 9–21 appears true size.

22. Draw a full section that shows the true shape of the slot in Prob. 9–22.

23. Find the true size of the inclined surface in Prob. 9–23.

24. Find the true size of both of the inclined surfaces in Prob. 9–24.

## Secondary Auxiliary Views

25. Find the true size of the angle between planes A and B in Prob. 9–25. Then, find the true size of plane B in a secondary auxiliary view.

26. Find the true size of the oblique plane that is intersected by the dovetail slot in Prob. 9–26.

27. Find the true size of the angle between surfaces A and B and the true size of surface B in Prob. 9–27.

28. Find the true size of the oblique surface in Prob. 9–28.

## Successive Auxiliary Views

29. Draw three successive auxiliary views of the object in Prob. 9–29. Draw the primary auxiliary view with a line of sight 60° from the left on the front view, the second auxiliary view with a horizontal line of sight from the right, and the third auxiliary view with a vertical line of sight from the bottom of the drawing.

30. Draw three successive auxiliary views of the object in Prob. 9–30. Draw the primary auxiliary view with a line of sight 45° from the upper right on the top view, the secondary auxiliary 45° from the lower right, and the third vertical from the bottom of the drawing.

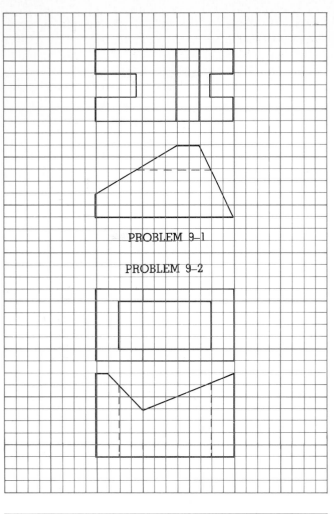

PROBLEM 9–1

PROBLEM 9–2

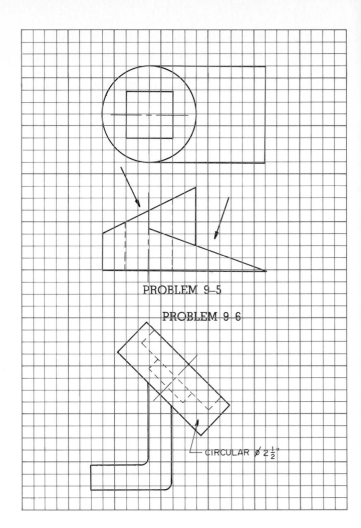

PROBLEM 9–5

PROBLEM 9 6

CIRCULAR ⌀ $2\frac{1}{2}$"

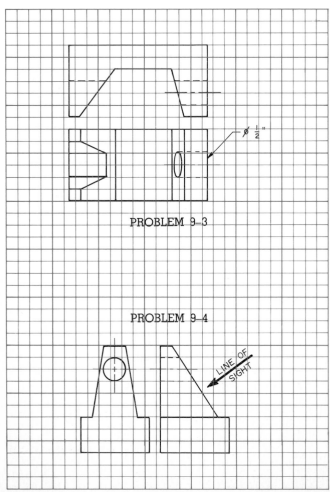

⌀ $\frac{1}{2}$"

PROBLEM 9–3

PROBLEM 9–4

LINE OF SIGHT

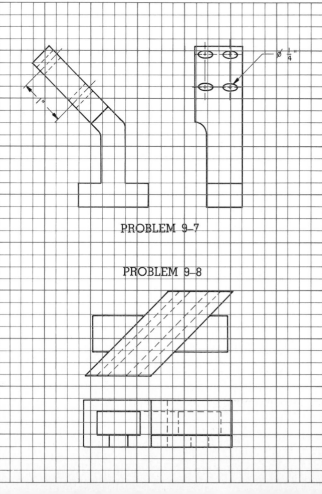

⌀ $\frac{1}{4}$"

PROBLEM 9–7

PROBLEM 9–8

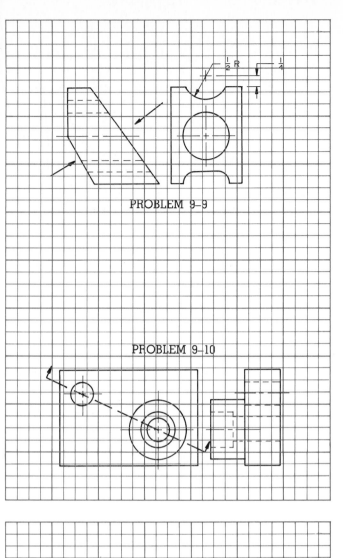

PROBLEM 9-9

PROBLEM 9-10

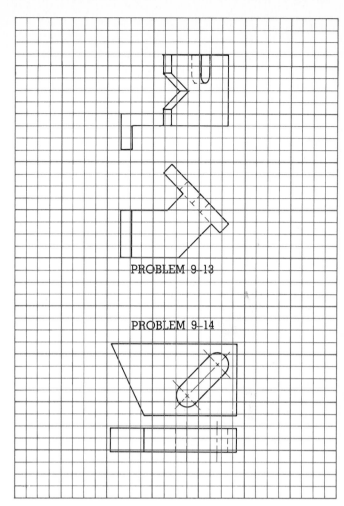

PROBLEM 9-13

PROBLEM 9-14

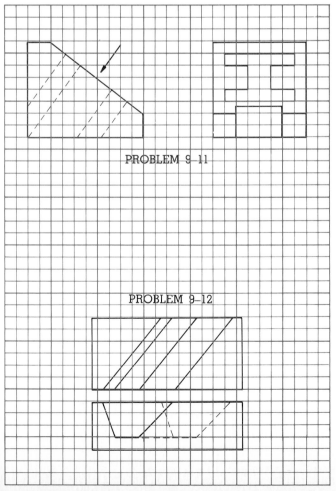

PROBLEM 9-11

PROBLEM 9-12

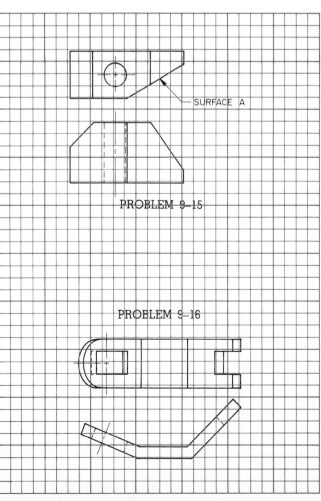

SURFACE A

PROBLEM 9-15

PROBLEM 9-16

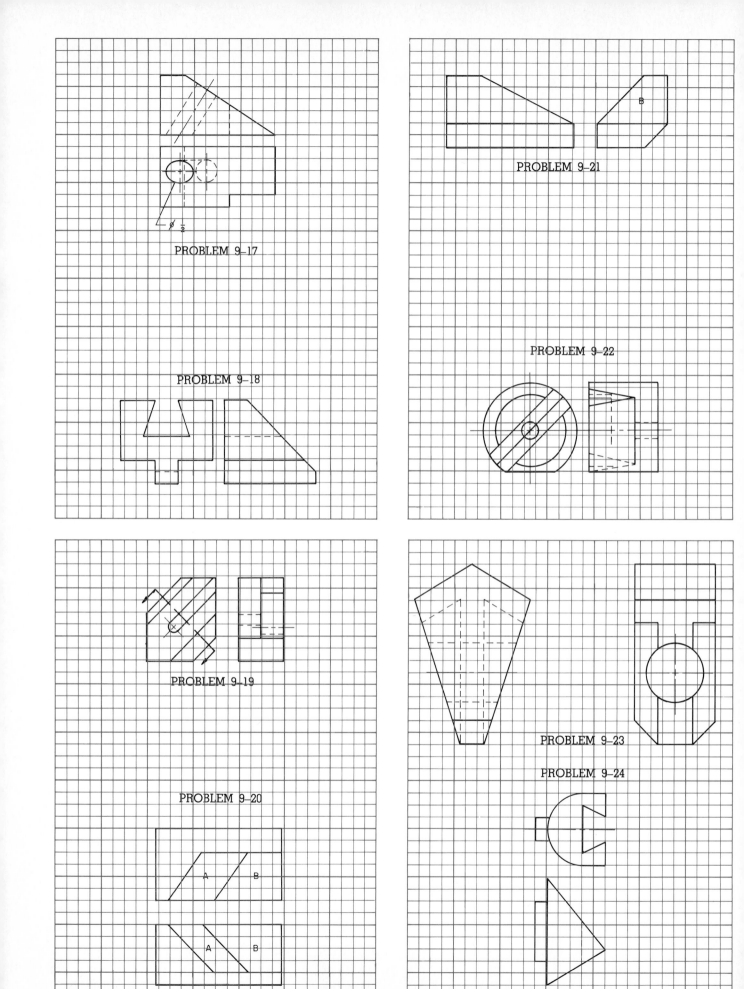

PROBLEM 9-17

PROBLEM 9-18

PROBLEM 9-19

PROBLEM 9-20

PROBLEM 9-21

PROBLEM 9-22

PROBLEM 9-23

PROBLEM 9-24

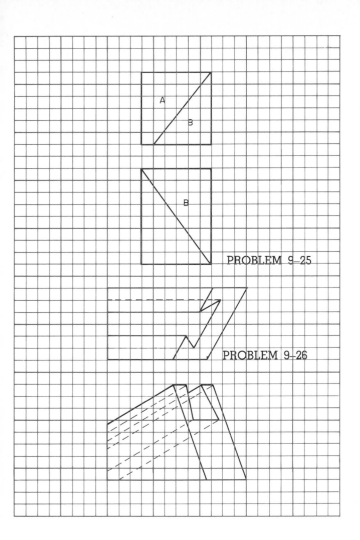

A

B

B

PROBLEM 9-25

PROBLEM 9-26

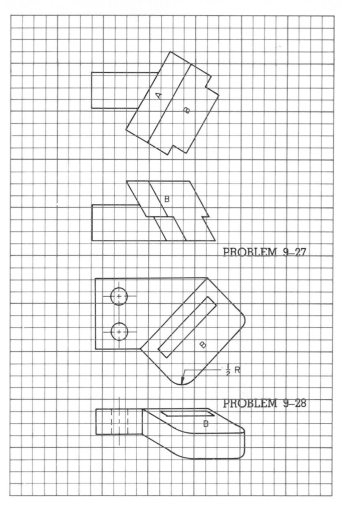

A

B

B

PROBLEM 9-27

B

$\frac{1}{2}$ R

PROBLEM 9-28

B

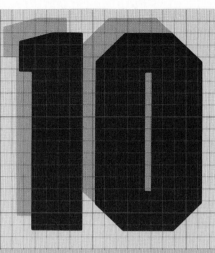

# Descriptive Geometry: Applications

This chapter continues the study of descriptive geometry and shows various procedures for solving problems commonly encountered in engineering design. These solutions involve the use of primary or secondary auxiliary views, explained in Chapter 9, and rely on an understanding of the basic spatial relationship involving lines and planes, explained in Chapter 8.

## MOVING A POINT TO A PRIMARY AUXILIARY PLANE

A point is projected to a principal plane and the primary auxiliary plane following the principles of orthographic projection. In Fig. 10–1, a point is projected to a horizontal and frontal plane. It is also projected to a primary auxiliary plane, 1, which is perpendicular to the horizontal plane. The point is distance X below the horizontal plane. This is shown on the frontal and auxiliary views. If this is difficult to understand, draw the orthographic projection of the point and the planes involved, cut it out, and fold the planes along their lines of intersection. This will help you visualize the projection.

## TRUE LENGTH OF OBLIQUE LINES

Line 1–2 in Fig. 10–2 is an oblique line. To find its true length, an auxiliary view must be constructed parallel to the line. *A line appears true length on any plane to which it is parallel.*

The steps for finding the true length are shown in Fig. 10–3. The auxiliary plane was taken off the front view. A second example was taken off the top view.

To construct the primary auxiliary, draw a line of sight perpendicular to line 1–2. Construct the edge of the auxiliary plane perpendicular to the line of sight. This means it is parallel with line 1–2. Project line 1–2 perpendicular to the auxiliary plane. Locate both ends of the line; they are distances W and X from the frontal plane. The distances are measured from the frontal plane because the auxiliary plane was constructed perpendicular to it. Connect points 1 and 2 to form the true length of the line.

## POINT VIEW OF AN INCLINED LINE

To find the point view of a line, it is necessary to find the line true length and construct a line of sight perpendicular to the end of the line. Inclined lines appear true length in the principal view to which they are parallel.

The steps for finding the point view of an inclined line are given in Fig. 10–4. Because the line is

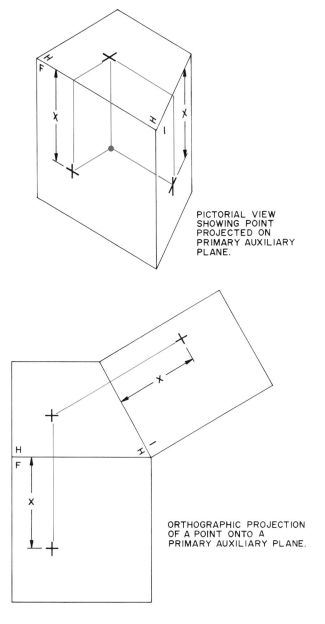

FIGURE 10–1 A point projected to the principal plane and an auxiliary plane following the principles of orthographic projection.

PICTORIAL VIEW SHOWING POINT PROJECTED ON PRIMARY AUXILIARY PLANE.

ORTHOGRAPHIC PROJECTION OF A POINT ONTO A PRIMARY AUXILIARY PLANE.

FIGURE 10–2 An oblique line appears true length when projected upon an auxiliary plane constructed parallel with the line.

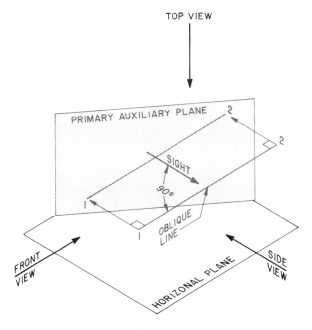

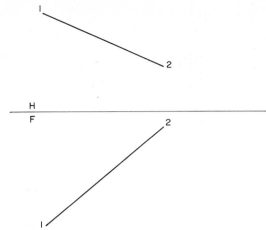

Assignment: Find the true length of oblique line 1–2 by using a primary auxiliary view.

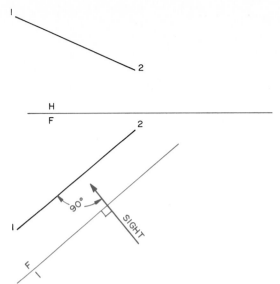

1. Draw a line of sight perpendicular to line 1–2. Draw a reference plane parallel with line 1–2 and perpendicular to the line of sight. Label the reference plane F-1 to identify the planes involved. This auxiliary is being taken off the frontal plane. It could have been taken off the horizontal plane just as easily.

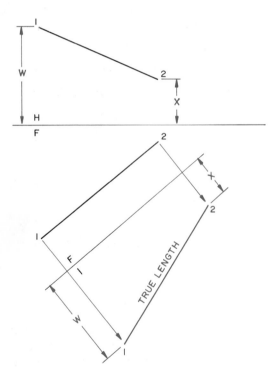

2. Since the primary auxiliary is projected off the front view, the measurements to locate line 1–2 on the auxiliary view are taken from the frontal plane. Project points 1 and 2 perpendicular to the auxiliary plane. Point 1 is distance W and point 2 is distance X from the frontal plane in the top view. Measure these distances perpendicular to the F-1 reference plane. Connect the points to form a true-length line.

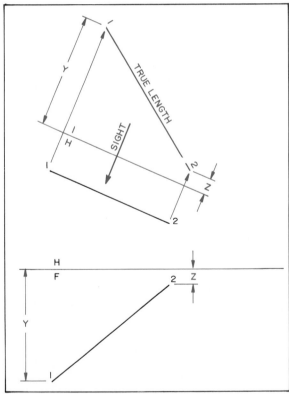

3. This construction shows how to project the auxiliary off the top view. The steps are the same as just described, but the measurements are taken from the horizontal plane.

**FIGURE 10–3** Finding the true length of an oblique line.

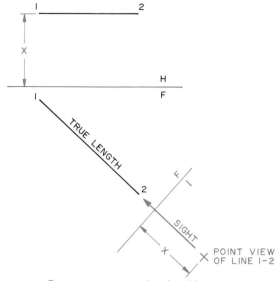

Assignment: Draw a point view of inclined line 1–2.

1. Construct a line of sight perpendicular to the end view of line 1–2 where it appears true length.
2. Construct a reference plane perpendicular to the line of sight.
3. Project line 1–2 perpendicular to the primary auxiliary plane. Measure it from the frontal plane (because the auxiliary was taken off the front view).

FIGURE 10–4 An inclined line can be found in its point view by constructing an auxiliary plane perpendicular to the true length of the line.

parallel with the front view, it is true length in the frontal plane. The primary auxiliary was taken off the frontal plane.

## ANGLE BETWEEN A LINE AND THE PRINCIPAL PLANES

The true size of the angle formed by the intersection of a principal plane and a line is seen *when the plane is in edge view and the line is true length.*

When you draw a primary auxiliary view, the intersecting principal plane appears in edge view. Auxiliary views projected off the front view show the frontal plane as an edge. Auxiliary views projected off the side view show the profile plane as an edge. Auxiliary views projected off the top view show the horizontal plane as an edge.

### Finding the Angle with the Horizontal Plane

The angle between a line and a horizontal plane is shown in Fig. 10–5. Because the line 1–2 is oblique, it does not show the true angle in the front view. To find the angle, take a primary auxiliary off the top view. This will produce a true-length line 1–2 and

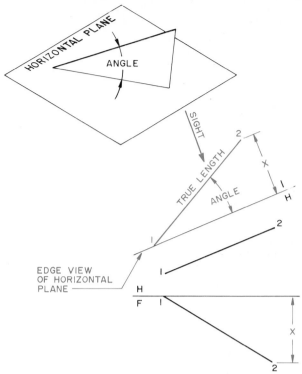

Assignment: Find the true size of the angle between line 1–2 and the horizontal plane.

1. To make it easier to find the angle, draw the horizontal plane through end 1 of the line.
2. Construct a line of sight perpendicular to line 1–2 in the top view.
3. Construct a reference plane perpendicular to the line of sight.
4. Project the ends of line 1–2 to the auxiliary view. Measure their distances from the horizontal plane. End 1 is touching the plane. End 2 is distance X from the plane. Connect the points to form the true-length line. Measure the true size of the angle.

FIGURE 10–5 How to find the true size of an angle between a line and a horizontal plane.

show the horizontal plane as an edge. The angle between them can be measured.

### Finding the Angle with the Frontal Plane

The angle between a line and a frontal plane is shown in Fig. 10–6. Because the line 1–2 is oblique, it does not show true size in the horizontal plane. To find the angle, take a primary auxiliary off the front view. This will produce a true-length line 1–2 and show the frontal plane as an edge. The true angle can be measured.

### Finding the Angle with the Profile Plane

The angle between a line and a profile plane is shown in Fig. 10–7. Because the line 1–2 is oblique, it does not show true size in the front view. To find

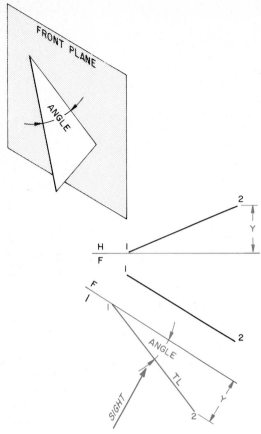

Assignment: Find the true size of the angle between line 1–2 and the frontal plane.

1. Construct a front plane through end 1 of the line.
2. Construct a line of sight perpendicular to line 1–2 in the front view.
3. Construct a reference plane perpendicular to the line of sight.
4. Project the ends of line 1–2 to the auxiliary plane. Measure their distances from the frontal plane. End 1 is touching the plane. End 2 is distance Y from the plane. Connect the ends to form the line. Measure the true size of the angle.

FIGURE 10–6 How to find the true size of an angle between a line and a frontal plane.

the angle, take a primary auxiliary off the side view. This will produce a true-length line 1–2 and show the profile plane as an edge. The true angle can be measured.

## SLOPE

*Slope* is the angle a line makes with the horizontal plane (Fig. 10–8). When a line slopes up from a beginning point, it is identified by a plus (+) sign. When it slopes down, it is given a minus (−) sign. Slope can be expressed in degrees, as a ratio, or as a percent.

When slope is expressed in degrees, decimal degrees, as 30.5°, are commonly used, although

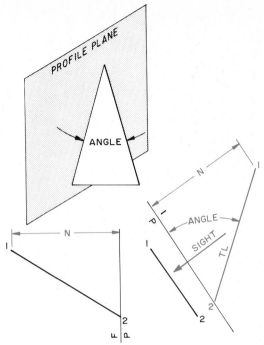

Assignment: Find the true size of the angle between line 1–2 and the profile plane.

1. Construct a profile plane through end 2 of the line.
2. Construct a line of sight perpendicular to line 1–2 in the side view.
3. Construct a reference plane perpendicular to the line of sight.
4. Project the ends of line 1–2 to the auxiliary plane. Measure their distances from the profile plane. End 2 is touching the profile plane. End 1 is distance N from the plane. Connect the ends to form the line. Measure the size of the angle.

FIGURE 10–7 How to find the true size of an angle between a line and a profile plane.

FIGURE 10–8 Slope is the angle a line makes with a horizontal plane.

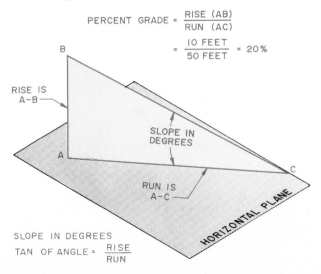

$$\text{PERCENT GRADE} = \frac{\text{RISE (AB)}}{\text{RUN (AC)}}$$

$$= \frac{10 \text{ FEET}}{50 \text{ FEET}} = 20\%$$

SLOPE IN DEGREES

$$\text{TAN OF ANGLE} = \frac{\text{RISE}}{\text{RUN}}$$

FIGURE 10–9 These pipes are used to move water from a lake up a mountain to a storage reservoir above. The pipes, vent tubes, and access tunnels all are built with a grade. (Courtesy of Tennessee Valley Authority.)

minutes and seconds, as 30°30′15″, can be used when appropriate. When the rise and run of a slope are known, the size of the angle can be found by using the tangent of the rise divided by the run. The *rise* is the vertical distance the line rises. The *run* is the horizontal distance the line runs when seen true length.

Slope is expressed as *grade* on highway, railroad, and sewer projects (Fig. 10–9). Grade is expressed as a percent of the rise divided by 100 ft of run. In architectural work, roof slope is expressed as a ratio of rise to run, with the run always a 12-in. unit. The slope of the face of a retaining wall is called batter. *Batter* is a term used to describe a slope that recedes from bottom to top. It is a ratio of the run to the rise, with the rise of a 12-in. unit (Fig. 10–10).

When preparing engineering drawings where slope must be found, the line representing some-

FIGURE 10–10 Slope, grade, batter, and degrees are common ways to specify angles for various engineering applications.

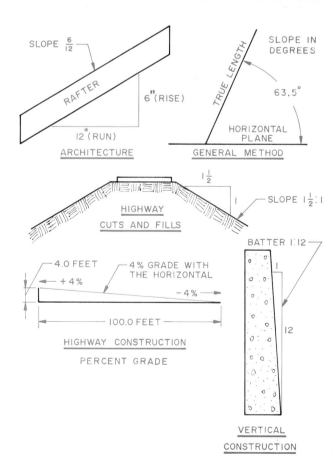

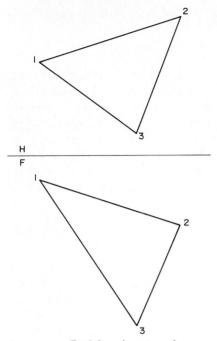

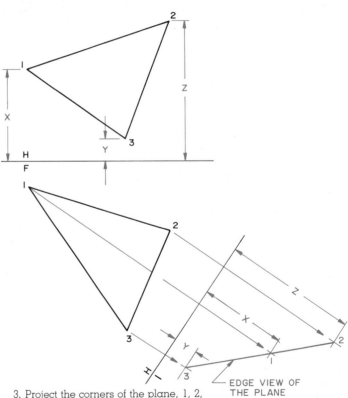

Assignment: Find the edge view of oblique plane 1–2–3.

1. Find a true-length line on the plane. In this solution, a frontal line was drawn in the top view. It is true length in the front view. Any other frontal line would work as well as the one drawn.

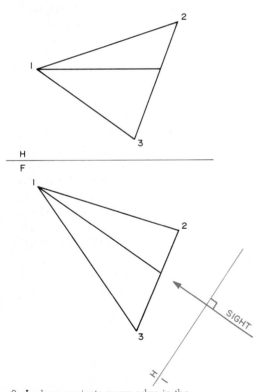

2. A plane projects as an edge in the view where a true-length line on its surface projects as a point. Construct a line of sight parallel with the true-length line. Construct a reference plane perpendicular to the line of sight.

3. Project the corners of the plane, 1, 2, and 3, onto the primary auxiliary plane. Locate each corner by measuring from the frontal plane in the top view. Connect the corners to form the edge view.

FIGURE 10–11 Finding the edge view of an oblique plane.

thing, as the sewage line, must be seen true length. The horizontal plane from which slope is measured must be seen in its edge view. Review the section on how to find the angle between an oblique line and a plane. This procedure is used when solving problems involving slope.

## EDGE VIEW OF AN OBLIQUE PLANE

*A plane will appear as an edge in any view where any line on the plane appears as a point.* The edge view of an oblique plane can be found in any primary auxiliary view by first constructing a true-length line on the plane and then getting the point view of the line.

The steps are shown in Fig. 10–11. The first step is to construct a true-length line on the surface. This is done by drawing a line on one of the views of the plane parallel with a principal plane. In Fig. 10–11, a frontal line was chosen. This is true length when projected to the front view. The primary auxiliary view is taken off the front view. The line of sight is parallel to the true-length line.

## ANGLE BETWEEN TWO PLANES WHEN THE INTERSECTION IS PARALLEL WITH A PRINCIPAL PLANE

The angle between two planes is called the dihedral angle. The need to find this angle true size is a common problem when preparing engineering drawings (Fig. 10–12).

*The true size of the angle between two planes can be found in a view where their line of intersection appears as a point.* To find the point view of a line, it must be found true length. A line of sight parallel with the true-length view will produce a point view in a primary auxiliary view (Fig. 10–13).

FIGURE 10–12 The angle between two intersecting planes is a dihedral angle.

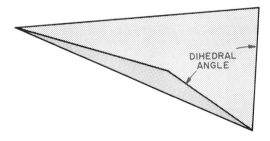

DIHEDRAL ANGLE

## PIERCING POINT OF A LINE ON A PLANE USING A PRIMARY AUXILIARY VIEW

*The point where a line pierces a plane is where the line and the edge view of the plane intersect.* The procedure is to find the edge view of the plane in a primary auxiliary. The point where the line crosses the edge view is the piercing point. Project this point back to the principal views to locate the piercing point on each. Then check the visibility. The steps for finding a piercing point using a primary auxiliary are shown in Fig. 10–14.

## PIERCING POINT OF A LINE ON A PLANE USING PROJECTION

This technique uses a cutting plane drawn perpendicular to the given plane and passed through the given line (Fig. 10–15). *Any line that lies on this cutting plane and intersects the given plane will cross the line of intersection between the two planes.* This locates the point where the line pierces the given plane.

The steps for using the projection method are given in Fig. 10–16. First, pass a cutting plane through the line perpendicular to the given plane. Project it to the top and front views. Where the cutting plane crosses the line is the piercing point.

## LINE PERPENDICULAR TO A PLANE

A line from a point in space perpendicular to a plane is the shortest distance between them.

To find the shortest distance, first find the oblique plane in edge view in a primary auxiliary. Then, construct a line from the given point perpendicular to the edge view of plane. The line is true length in this view. Project the intersection of the line and plane to the principal views. The line will be perpendicular to true-length lines running through the point of intersection of the line and the plane.

*Remember, a line perpendicular to a plane will appear perpendicular in a given view only to those lines on the plane that are true length* (Fig. 10–17).

The steps for drawing a line perpendicular to an oblique plane are shown in Fig. 10–18.

## INTERSECTION OF TWO PLANES USING A PRIMARY AUXILIARY VIEW

The intersection of two planes can be found by finding one of the planes in edge view in a primary auxiliary view. Then, project the points of intersection to the principal views. Connect the points of intersection to form the line of intersection. Finally, ascertain the visibility of the planes.

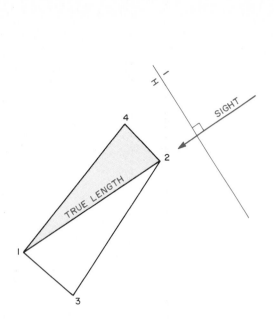

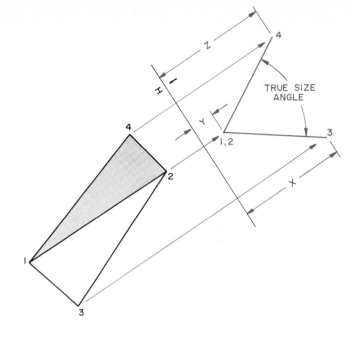

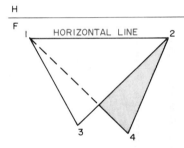

Assignment: Find the true size of the angle between two planes when the intersection of the planes is parallel with a principal plane.

FIGURE 10–13 Finding the true size of an angle between two planes when the intersection is parallel with a principal plane.

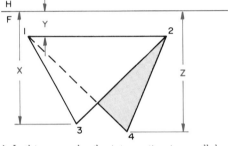

1. In this example, the intersection is parallel with the horizontal plane. It appears true length in the top view.
2. Construct a line of sight parallel with the true-length line.
3. Construct a reference plane perpendicular to the line of sight.
4. Project the corners of the planes onto the primary auxiliary plane.
5. Locate each corner by measuring their distance from the horizontal plane.
6. Connect the corners to form the edge view of the intersecting plane. Measure the angle.

The steps for finding the intersection of two planes using a primary auxiliary view are shown in Fig. 10–19.

## INTERSECTION OF TWO PLANES USING CUTTING PLANES AND PROJECTION

An alternate method of finding the intersection of two planes employs the principle of using projection to find the piercing point of a line on a plane.

To find the intersection of two planes, pass a vertical cutting plane through one edge of one of the planes. The edge, being on the new cutting plane, will pierce the second plane. This gives one end of the line of intersection. Repeat this procedure on a second edge of the same plane. This locates the other end of the line of intersection. Connect the piercing points to form the line of intersection. Finally, ascertain visibility of the edges of each plane.

The steps for finding this intersection are illustrated in Fig. 10–20.

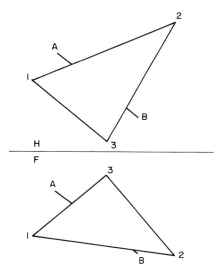

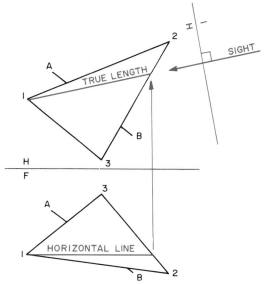

Assignment: Find the piercing point of the line and plane using a primary auxiliary view.

1. Find the plane in edge view. To do this, construct a true-length line on the surface. A horizontal line was used in this example. It is true length in the top view. Construct the line of sight parallel with the true-length line. Construct a reference plane perpendicular to the line of sight.

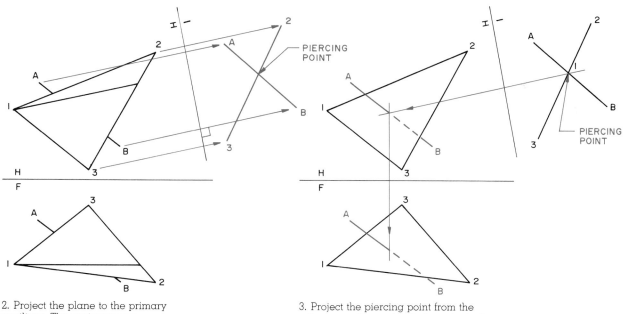

2. Project the plane to the primary auxiliary. The projectors are perpendicular to the H-1 reference plane. Locate each corner from the horizontal plane. This gives the edge view of the plane. Then, project the ends of line A–B to the primary auxiliary view. Draw line A–B. The point at which it crosses the edge view of the plane is the piercing point.

3. Project the piercing point from the primary auxiliary view to the principal planes. The point at which it crosses line A–B is the piercing point in that view. Draw line A–B, checking it for visibility.

**FIGURE 10–14** Finding the piercing point of a line and a plane using a primary auxiliary view.

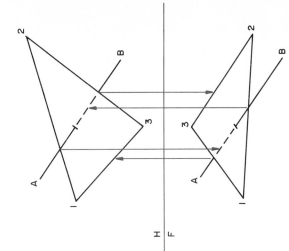

1. Pass an imaginary cutting plane through line A–B. In this example, it was passed in the top view perpendicular to the horizontal plane. Project the line of intersection between the two planes, X–Y, to the front view.

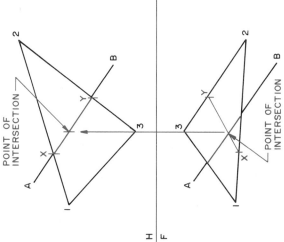

3. Ascertain the visibility of line A–B in each view and draw the line.

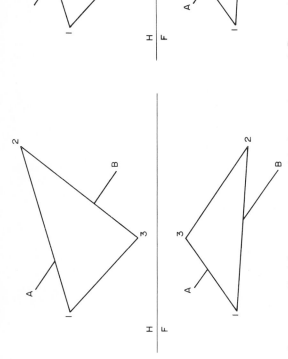

Assignment: Find the piercing point of the line and the plane using projection techniques.

2. The point at which the line of intersection of the two planes crosses line A–B is the piercing point. Project the piercing point to line A–B in the top view.

**FIGURE 10–16** How to find the piercing point of a line and a plane using projection techniques.

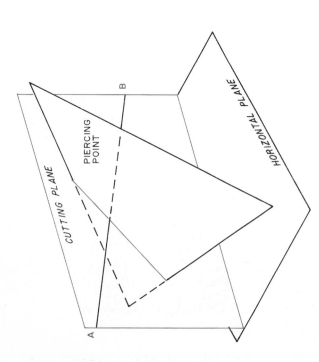

**FIGURE 10–15** The piercing point of a line and a plane can be found by passing a plane through the line and perpendicular to the plane.

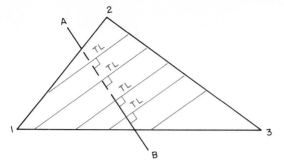

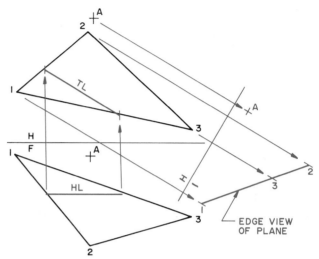

FIGURE 10–17 A line perpendicular to a plane will appear perpendicular to those lines on the plane that appear true length.

FIGURE 10–18 How to draw a line from a point in space perpendicular to a plane.

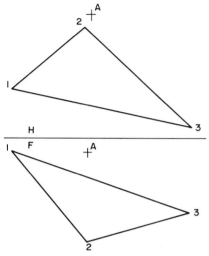

Assignment: Construct a line from point A in space perpendicular to the plane. This is the shortest distance from the point to the plane.

1. Find the edge view of the plane. First, draw a horizontal line in the front view. Project it to the top view, where it appears true length. Project plane 1–2–3 to a primary auxiliary using the true-length line as the line of sight. Project point A to the auxiliary view.

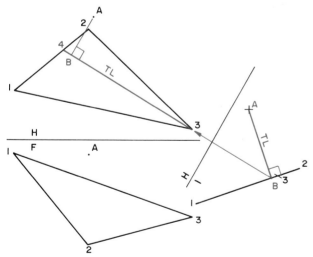

2. Since the plane is in edge view, a line can be drawn from point A perpendicular to it. Since any line parallel with the line of sight is true length on the plane, project the intersection of the line and the plane from the auxiliary view to the top view. Since line 3–4 is true length, a line from point A can be drawn perpendicular to it. This locates the perpendicular line in the top view.

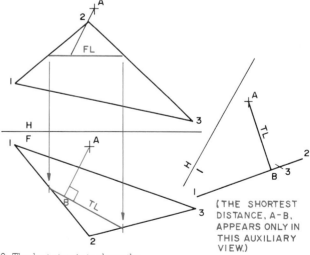

3. The last step is to draw the perpendicular in the front view. To do this, draw a frontal line in the top view. Project it to the front view, where it is true length. Since it is true length in the front view, a perpendicular line can be drawn from point A. This locates the perpendicular line in the front view.

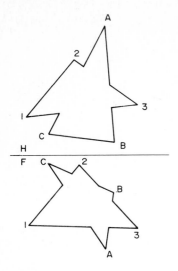

Assignment: Find the line of intersection between the two planes using a primary auxiliary view.

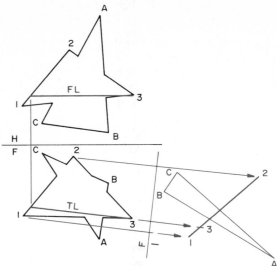

1. Find one of the planes in edge view. In this example, it was decided to find plane 1–2–3 in edge view. The other could have been used. Draw a frontal line that projects true length in the front view. This will enable a primary auxiliary view to be taken off the front view. A horizontal line could have been used if desired. Draw the edge view of plane 1–2–3 and project plane A–B–C to the primary auxiliary view.

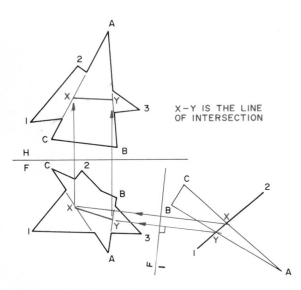

X–Y IS THE LINE OF INTERSECTION

2. Find the line of intersection. This appears in the primary auxiliary. Points X and Y are the points of intersection of edges A–B and A–C. Project these to A–B and A–C in the front and top views. Connect them to form the line of intersection of the two planes.

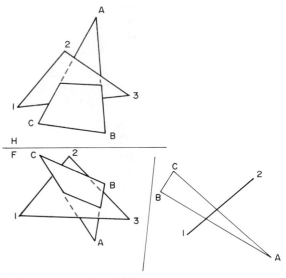

3. Next, ascertain the visibility of the planes. Project the apparent intersections of edges to determine which edge is on top. Draw the planes so that visibility is shown.

**FIGURE 10–19** How to find the line of intersection between two planes using a primary auxiliary view.

## BEARING AND AZIMUTH

A surveying team records the direction of the side of a lot or the center line of a road. These directions are given in terms of compass bearings. The *bearing* is a horizontal direction and appears true size only in a top view. It is independent of the angle a line makes with the horizontal plane (Fig. 10–21).

Bearings are read from the north or the south. When a bearing is between the north and east, say, at 45°, it is described as N45°E. If it is between east and south at 45°, it is S45°E. Likewise, if a bearing is between north and west at 45°, it is N45°W. If it is between west and south, it is S45°W (Fig. 10–22).

An *azimuth* is another means of giving direction in a horizontal plane. It is used to avoid the confusion that sometimes occurs when using compass bearings. An azimuth is measured from the north pole of a compass in a clockwise direction. All measurements are in degrees and in the same direction (Fig. 10–23).

If a line established by a bearing or an azimuth has slope—a pipeline, for example—the direction given refers to the lower end of the line.

When producing maps and geological surveys, north is usually located at the top of the drawing. All measurements are taken from the north-south axis.

## STRIKE AND DIP

*Strike* and *dip* are terms used to describe the position of the strata of the earth. They are used in mining

FIGURE 10–20 How to find the line of intersection between two planes using cutting planes and projection.

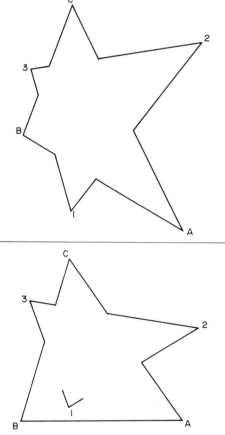

Assignment: Find the line of intersection between the two planes using cutting planes and projection.

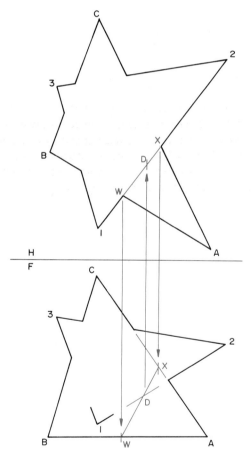

1. Pass a vertical cutting plane through edge 1–2 in the top view. This crosses the other plane at points W and X. Project these intersections to edges AC and AB in the front view. Connect points W and X. Where XY crosses 1–2 in the front view is where 1–2 pierces plane 1–2–3. Project the piercing point, D, to edge 1–2 in the top view. This locates one end of the line of intersection.

Descriptive Geometry: Applications  **211**

FIGURE 10–20 *cont.*

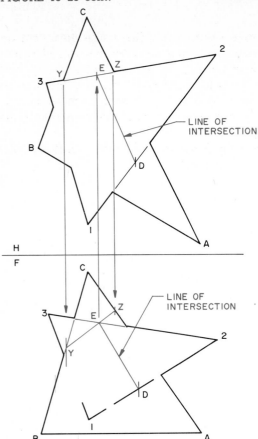

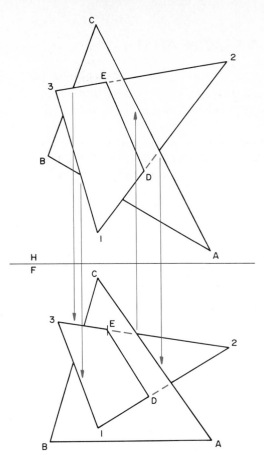

2. Next, find the other end of the line of intersection. Pass a vertical cutting plane through line 2–3 in the top view. This crosses the other plane at points Y and Z. Project these intersections to edges AC and BC in the front view. Connect points Y and Z. Where YZ crosses 2–3 in the front view is where 2–3 pierces plane 1–2–3. Project the piercing point, E, to edge 2–3 in the top view. This locates the other end of the line of intersection. Connect points D and E to form the line of intersection.

3. Finally, ascertain visibility of the planes. Project the apparent intersections of edges to determine which edge is on top. Draw the planes so that visibility is shown.

FIGURE 10–21 Compass bearings are measured from the north-south axis parallel to the horizontal plane.

FIGURE 10–22 Bearings are read from the north or the south.

FIGURE 10–23 The azimuth is measured from the north pole in a clockwise direction.

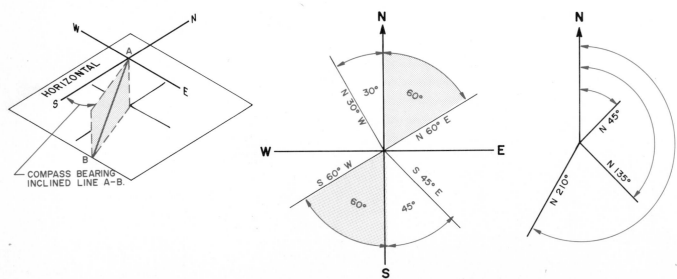

and geology. They are similar to the earlier description of slope and the direction of slope.

A *strike line* is a horizontal line on a plane. *Strike* is the compass bearing of a strike line that appears true length in the top view. It is used to indicate a bearing.

*Dip* is the angle the edge view of a plane makes with the horizontal plane. It also shows the general compass direction. The actual dip angle is found by constructing a primary auxiliary view projected off the top view. The angle between the edge view and the horizontal plane is the dip angle. The general dip direction is found by constructing a line perpendicular to a level line in the plane in the top view. The line should point toward the low side of the plane.

The steps for finding the strike and dip of a plane are given in Fig. 10–24.

FIGURE 10–24 How to find strike and dip.

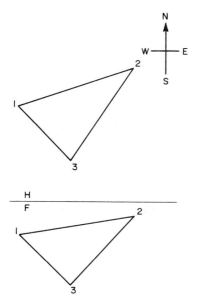

Assignment: Find the strike and dip of an oblique plane.

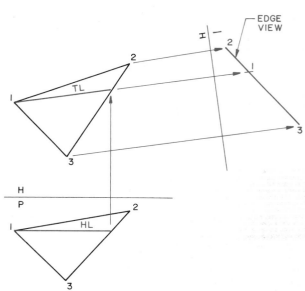

1. First, find the edge view of the plane. Construct a principle line. Use a horizontal line because the edge view of the horizontal plane is needed to find dip. Construct a primary auxiliary view with a line of sight parallel with the true-length line.

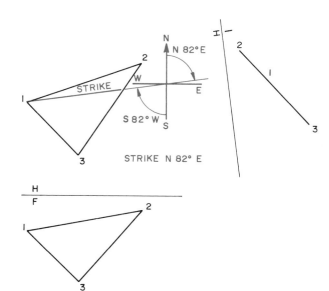

2. Find the strike. It is the compass direction of a true-length horizontal line in the plane. The line is the strike of the plane. It can be either NW or SE.

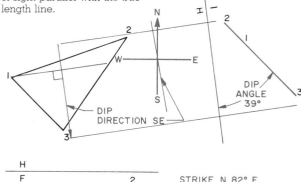
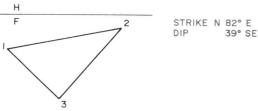

3. Find the dip. Dip is the angle the plane makes with the horizontal plane. Since plane 1–2–3 is in edge view, the dip angle can be measured in the primary auxiliary view.

## SOLUTIONS REQUIRING SECONDARY AND SUCCESSIVE AUXILIARY VIEWS

Most of the following applications require the use of secondary auxiliary views for solution. One requires a third or successive auxiliary. If you do not understand secondary or successive auxiliary views, review the material in Chapter 9.

### Point View of an Oblique Line

Finding the point view of an oblique line is a basic principle that is frequently used in solving spatial problems. The steps for performing this construction are shown in Fig. 10–25.

First, it is necessary to find the true length of the oblique line. This is found in the primary auxiliary. Then, a line of sight parallel to the true-length line is used to construct the second auxiliary. *When you look parallel to a true length line, it appears as a point.*

### Angle Between Oblique Planes

Many designs have planes that meet at a variety of angles. Before a product can be built, the true size of each angle must be known.

To find the true angle between two planes, two steps are required. First, the line of intersection must be found *true length*. This can be found in a primary auxiliary. Then, a *point view* of the line of intersection is found. This can be found in a secondary auxiliary. *The true angle between two planes can be found in the view in which the line of intersection appears as a point and both planes appear in edge view.*

The steps for finding the true angle between intersecting oblique planes are given in Fig. 10–26.

FIGURE 10–25 Finding the point view of an oblique line.

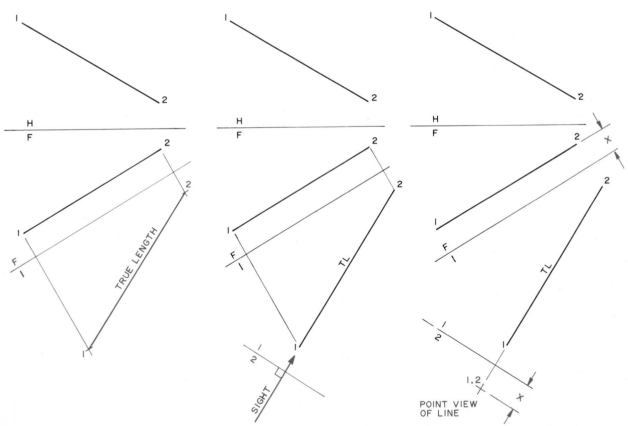

Assignment: Find the point view of the oblique line.
1. First, find the true length of the line in a primary auxiliary. Construct a reference plane parallel with line 1–2. Project line 1–2 to the primary auxiliary.

2. Draw a secondary auxiliary. Begin by constructing a line of sight parallel with the true-length view of line 1–2. Construct a reference plane perpendicular to the line of sight.

3. Project line 1–2 onto the secondary auxiliary view. Measure its distance from the primary auxiliary. This is distance X. Locate the point view of the line.

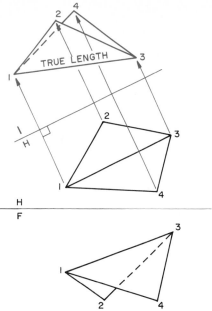

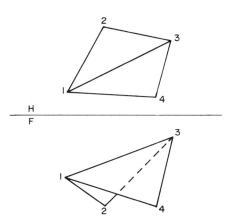

Assignment: Find the true size of the angle between two intersecting oblique planes.

1. First, find the line of intersection true length. Pass a reference plane parallel with the line of intersection. In this example, the top view was used. The front view could have been used instead. Project both planes to the primary auxiliary view.

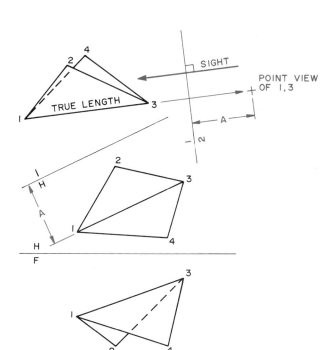

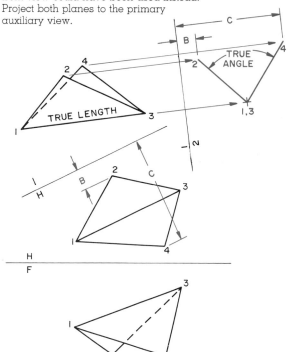

2. Find the point view of the line of intersection, 1–3. Construct a line of sight parallel with the true-length line. Construct a reference plane perpendicular to the line of sight. Project line 1–3 to the secondary auxiliary view. Measure the distance of the point from the primary auxiliary plane. This is distance A.

3. Project the two planes to the secondary auxiliary view. Measure their distance from the primary auxiliary plane, B and C. Measure the true size of the angle.

FIGURE 10–26 The steps for finding the true size of the angle between two intersecting oblique planes.

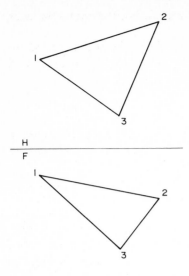

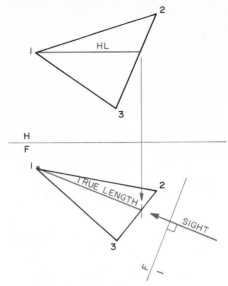

Assignment: Find the true size of the oblique plane.

1. First, construct a true-length line on the plane. Then, construct a line of sight parallel with it. Construct a reference plane perpendicular to the line of sight. This will project the true-length line as a point on the primary auxiliary view.

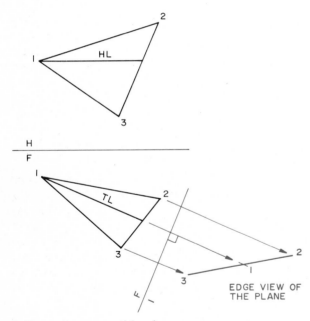

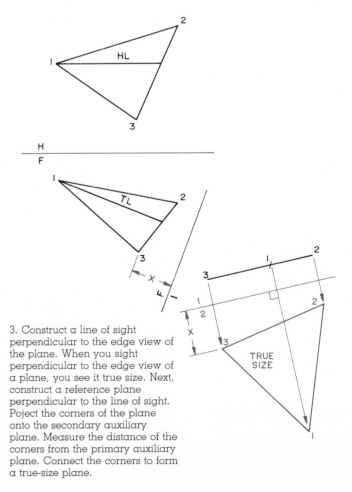

2. Project the corners of the plane to the primary auxiliary view. This gives an edge view of the plane.

3. Construct a line of sight perpendicular to the edge view of the plane. When you sight perpendicular to the edge view of a plane, you see it true size. Next, construct a reference plane perpendicular to the line of sight. Poject the corners of the plane onto the secondary auxiliary plane. Measure the distance of the corners from the primary auxiliary plane. Connect the corners to form a true-size plane.

FIGURE 10–27 How to find the true size of an oblique plane.

## True Size of an Oblique Plane

One of the most frequently occurring constructions is a design on which there are oblique planes. Each of these must be found true size so that it can be dimensioned. This is commonly found by using a secondary auxiliary.

The first step is to find the plane in edge view in a primary auxiliary. Then, construct a secondary auxiliary plane parallel with the edge view. The projection of the plane on the secondary auxiliary is true size.

The steps for performing this construction are given in Fig. 10–27.

## Shortest Distance Between a Point and a Line

Often, it is desired to make a connection, such as from a tank to a passing pipe, using the shortest possible distance. The engineer would prepare a drawing showing the shortest distance from the tank to the pipe (Fig. 10–28).

First, a line is found true length in a primary auxiliary. The point is carried into the auxiliary. Then, the point view of the line is found in a secondary auxiliary. A line can be drawn from the point to the point view of the line. *The perpendicular distance from the point to the line can be seen true length in the view in which the line appears as a point.* The line is then projected back to the primary auxiliary view. Here it will be seen to be perpendicular to the true

length of the original line. It is then projected back to the principal views.

The steps for finding the shortest perpendicular distance from a point in space to a line are shown in Fig. 10–29.

## Shortest Distance Between Parallel Lines

If it is necessary to connect two parallel pipes using the shortest distance possible for the connector, how is this located on a drawing (Fig. 10–30)?

*If you can see both lines in end view, you can see the shortest distance between them.* To do this, first find the lines true length in a primary auxiliary view. Then, construct a secondary auxiliary view with a line of sight parallel with the true-length lines. This will give the point view. Connect the point view to form the shortest distance. The connecting line is drawn perpendicular to the true-length lines and projected back to the principal views.

The steps for performing this construction are illustrated in Fig. 10–31.

## Shortest Distance Between Skew Lines

Skew lines are lines that are neither parallel nor intersecting. Study the skew lines in Fig. 10–32. They appear to intersect. However, when the point of intersection is examined in the top view, it is found that it does not project to the front view. Therefore, these lines do not intersect.

There are many design problems involving skew lines. They could involve large-diameter pipes in a refinery or small-diameter tubing in a tractor hydraulic system. What is the shortest distance between skew lines (Fig. 10–33)?

To find the shortest distance, first find the true length of one skew line in a primary auxiliary. Then, in a secondary auxiliary, find the point view of the true-length line. Project the second line into both auxiliary views. Construct a perpendicular from the point view to the second skew line in the secondary auxiliary. This is the true length of the connecting line. Project the connecting line back to the primary and principal views.

The steps for constructing the shortest distance between two skew lines are shown in Fig. 10–34.

## Shortest Horizontal Distance Between Two Skew Lines

*The shortest horizontal line connecting two skew lines will be true length when the plane that contains one skew line and is parallel to the other appears in edge view (Fig. 10–35).*

The first step is to place one of the skew lines in a plane. One edge of the plane must be parallel with the other skew line. Then, find the edge view

FIGURE 10–28 How do you find the shortest distance between a point and a line?

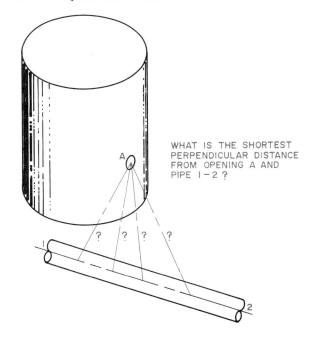

WHAT IS THE SHORTEST PERPENDICULAR DISTANCE FROM OPENING A AND PIPE 1–2 ?

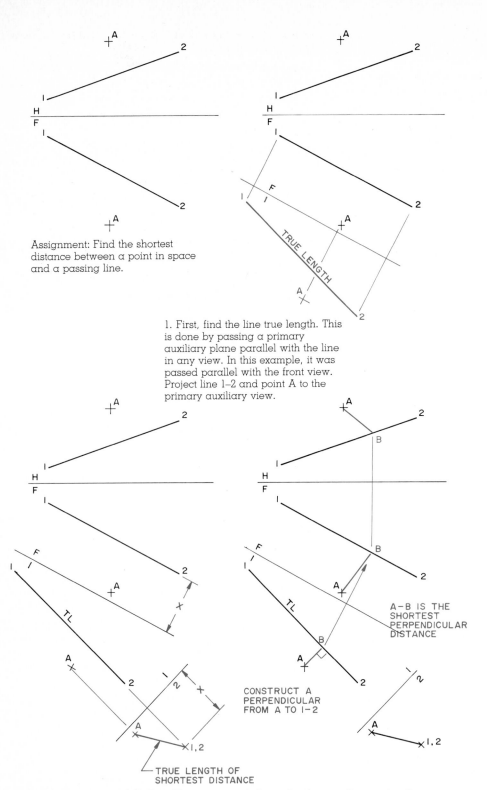

Assignment: Find the shortest distance between a point in space and a passing line.

1. First, find the line true length. This is done by passing a primary auxiliary plane parallel with the line in any view. In this example, it was passed parallel with the front view. Project line 1–2 and point A to the primary auxiliary view.

CONSTRUCT A PERPENDICULAR FROM A TO 1–2

TRUE LENGTH OF SHORTEST DISTANCE

A–B IS THE SHORTEST PERPENDICULAR DISTANCE

2. Find the point view of the line. To do this, construct a line of sight parallel with the true-length view. Construct a secondary auxiliary plane perpendicular with the line of sight. Project line 1–2 and point A to the secondary auxiliary view. Connect point A and the point view of line 1–2. This is the true length of the shortest distance between them.

3. Draw the shortest distance in all views. In the primary auxiliary view, construct a perpendicular from point A to line 1–2. Locate it in any convenient place. This is possible because 1–2 is true length. Whenever two lines are perpendicular and you see one of them true length, they will appear to meet at 90°. Project the shortest line, A–B, to the principal views.

FIGURE 10–29 Finding the shortest distance between a point and a line.

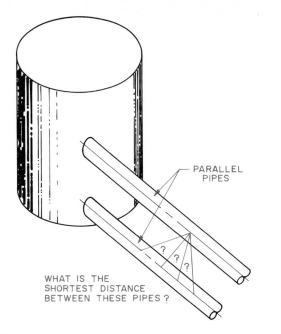

PARALLEL PIPES

WHAT IS THE
SHORTEST DISTANCE
BETWEEN THESE PIPES ?

FIGURE 10–30 How do you find the shortest distance between two parallel lines?

of the plane in a primary auxiliary. The edge view is found by constructing a horizontal line in the front view. This is true length in the top view. The primary auxiliary must be taken off the top view to find the horizontal plane as an edge. The primary auxiliary view shows the edge view of the plane with a skew line.

Next, construct a secondary auxiliary view with the line of sight parallel to the edge view of the horizontal plane in the primary auxiliary. This will place the skew lines in a position where they appear to cross. The point at which they appear to cross is the end view of the shortest horizontal distance. Since this distance is parallel with the horizontal plane, it is true length in the primary auxiliary.

Project the shortest distance, Line AB, to the principal views. The solution can be checked by noticing that line AB is parallel with the horizontal plane in the front view.

The steps for constructing the shortest horizontal distance between two skew lines are given in Fig. 10–36.

FIGURE 10–31 Finding the shortest distance between two parallel lines.

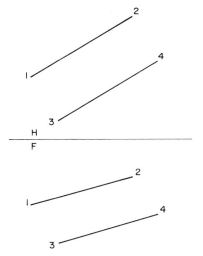

Assignment: Find the shortest
distance between two parallel lines.

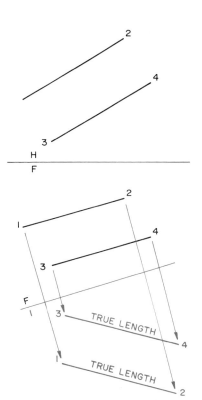

1. First, find the true length of the parallel lines in a primary auxiliary view. The auxiliary view can be constructed parallel with the lines in any principal view. In this example, the front view was chosen.

FIGURE 10–31 *cont.*

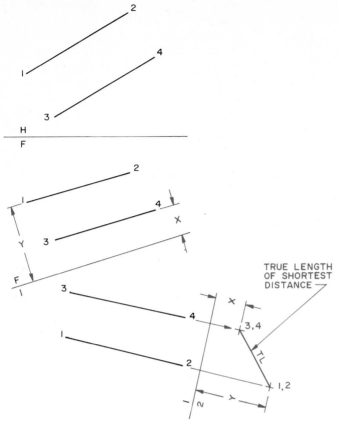

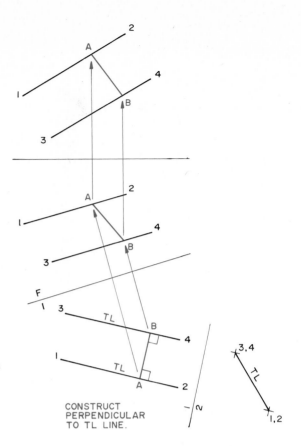

2. Then, find the point view of the parallel lines. To do this, construct a line of sight parallel with the true-length lines. Construct a secondary auxiliary view perpendicular to the line of sight. Project the lines onto the secondary auxiliary. This gives the point view of lines 1–2 and 3–4. Connect the point views to find the true length of the shortest distance between the lines.

3. Draw the shortest distance in all views. Since lines 1–2 and 3–4 are true length in the primary auxiliary, a line can be drawn perpendicular to both. Locate it in any convenient place. This is possible because the lines are true length. Whenever two lines are perpendicular and one of them appears true length, they will appear to meet at 90°. Project the shortest line, A–B, to the principal views.

FIGURE 10–32 Skew lines are not parallel and do not intersect.

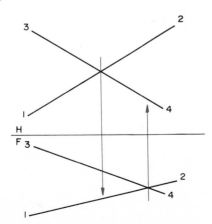

FIGURE 10–33 What is the shortest distance between two skew lines?

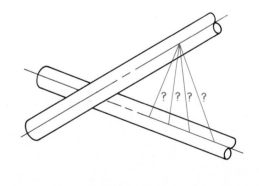

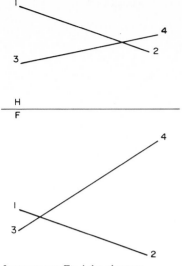

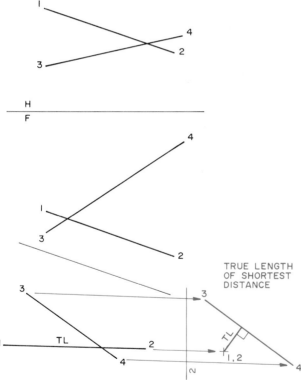

Assignment: Find the shortest distance between two skew lines.

2. Find the point view of the true-length line. When one appears as a point, a perpendicular can be constructed from the point to the other line. The shortest distance between the two lines is a line perpendicular to both.

Construct a line of sight parallel with the true-length line. Construct a secondary auxiliary plane perpendicular to the line of sight. Project both lines to the secondary auxiliary view. Draw the shortest line perpendicular from the point view of line 1–2 to line 3–4. This line is true length.

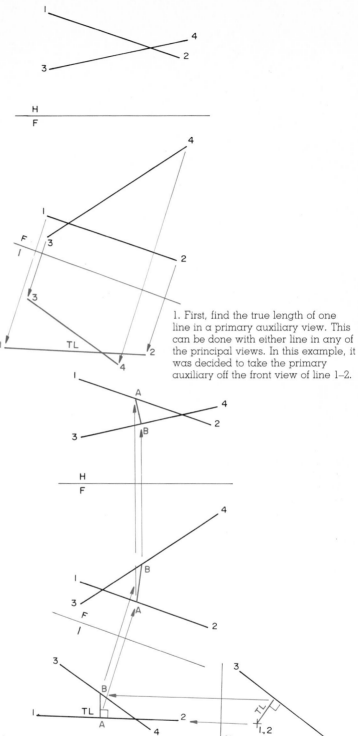

1. First, find the true length of one line in a primary auxiliary view. This can be done with either line in any of the principal views. In this example, it was decided to take the primary auxiliary off the front view of line 1–2.

TRUE LENGTH OF SHORTEST DISTANCE

3. Project the perpendicular line back to the primary view. Since it is true length, it is parallel with reference plane 1-2. It is perpendicular to true-length line 1–2. Project it to the principal views.

**FIGURE 10–34** Finding the shortest distance between two skew lines.

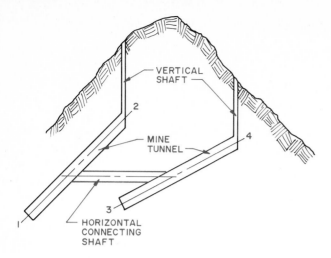

VERTICAL
SHAFT

MINE
TUNNEL

HORIZONTAL
CONNECTING
SHAFT

**FIGURE 10–35** What is the shortest horizontal distance between two skew lines?

## Shortest Line of a Specified Slope Between Two Skew Lines

*The shortest line of a specified slope between two skew lines will be true length when the plane that contains one skew line and is parallel with the other skew line appears in edge view.*

The first step is to place one of the skew lines in a plane. One edge must be parallel with the other skew line. The other edge must be a horizontal line so that it appears true length in the top view. In this way the horizontal plane will appear as an edge in the primary auxiliary. The slope is measured from the horizontal.

**FIGURE 10–36** Finding the shortest horizontal distance between two skew lines.

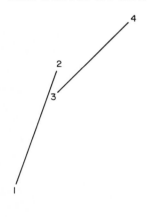

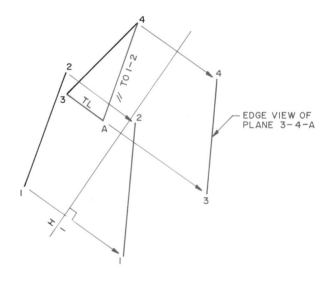

EDGE VIEW OF
PLANE 3–4–A

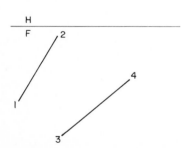

Assignment: Find the shortest horizontal distance between two skew lines.

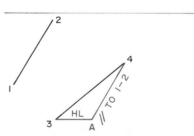

1. Place one of the skew lines in a plane. Line 3–4 was chosen. One edge of the plane is parallel with the other skew line, 1–2. The third edge is a horizontal line. Draw the plane in the front view and project it to the top view.

Find the edge view of the plane. It must be off the horizontal plane. The true-length horizontal line is the line of sight. Construct a primary auxiliary view perpendicular to the line of sight. This shows the plane in edge view and skew line 1–2 parallel with it.

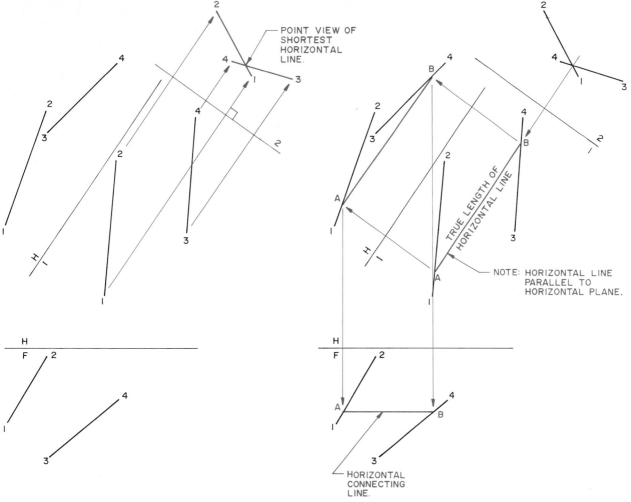

POINT VIEW OF
SHORTEST
HORIZONTAL
LINE.

TRUE LENGTH OF
HORIZONTAL LINE

NOTE: HORIZONTAL LINE
PARALLEL TO
HORIZONTAL PLANE.

HORIZONTAL
CONNECTING
LINE.

2. Since we are looking for a horizontal line, draw a secondary auxiliary view with a line of sight parallel with the horizontal plane. Construct a secondary auxiliary plane perpendicular to this line of sight. Project each line to the secondary auxiliary plane. The point at which the lines appear to cross is the point view of the shortest horizontal line connecting them.

FIGURE 10–36 *cont.*

3. Project the shortest horizontal line to the primary auxiliary view. Since it is parallel with the horizontal plane, it is true length. Project it to the principal views.

Project the plane and skew lines to the primary auxiliary. Here they are parallel. Then, establish the line of sight to set the slope to the connecting line. The slope will be measured from the edge of the horizontal plane in the primary auxiliary view. This establishes the line of sight for the secondary auxiliary. The grade can be drawn in two directions in relation to the H-A1 reference plane. The shortest grade between the two skew lines will be the one drawn in the direction that is most nearly perpendic-

ular to the skew lines. This is ascertained by examining the primary auxiliary view.

Project the skew lines to the secondary auxiliary. Where they appear to cross is the point view of the shortest connecting line at a specified slope. Project this line to the primary auxiliary, where it appears true length. Then project it to the principal views.

The steps for constructing the shortest line of a specified slope between two skew lines are given in Fig. 10–37.

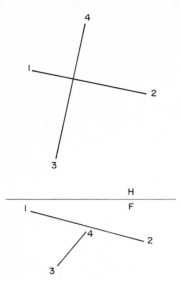

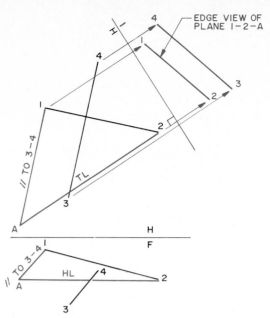

Assignment: Find the shortest line of a specified slope between two skew lines.

1. Place one of the skew lines in a plane. Line 1–2 was chosen. One edge of the plane is parallel with the other skew line, 3–4. The third edge is a horizontal line. Draw the plane in the front view and project it to the top view.

Find the edge view of the plane. It must be off the horizontal plane. The true-length line is the line of sight for the primary auxiliary. Construct a primary auxiliary view perpendicular to the line of sight. This shows the plane in edge view and skew line 3–4 parallel with it.

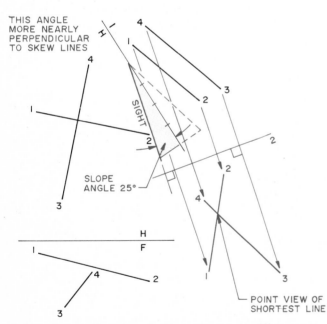

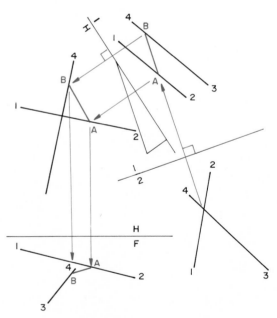

2. Draw the specified slope. In this problem, it was 25°, or a 1:4 slope. The slope can be shown two ways. Use the one that is most nearly perpendicular to the skew lines in the primary auxiliary.

Using the slope line as a line of sight, draw a secondary auxiliary. In this view, the skew lines cross. The point at which they cross is the point view of the shortest line connecting the skew lines at a 25° slope.

3. Project the shortest 25°-sloped line to the primary auxiliary and principal views. Since it is parallel with the line of sight in the primary auxiliary, it appears there true length.

**FIGURE 10–37** The procedure for constructing the shortest line of a specified slope between two skew lines.

## Line Through a Point with a Given Angle to a Line

To design a system having intersecting parts, as two pipes, the designer must show the angle of intersection and the point of intersection. These can be varied by the designer.

To show these on a drawing, first construct a plane that contains the point in space and the line to which the connection must be made. Then, find the edge view of this plane in a primary auxiliary view. This permits finding the true size of the plane in a secondary auxiliary view. In this view the original line 1–2 is true length. Any line drawn from point A to line 1–2 will show the true angle because it is in the same plane. Draw the desired angle and project it to the primary and principal views.

The steps for constructing a given angle from a point in space to a line are shown in Fig. 10–38.

FIGURE 10–38 Constructing a known angle from a point in space to a line.

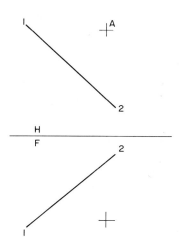

Assignment: Construct a line from a point in space with a given angle to a line.

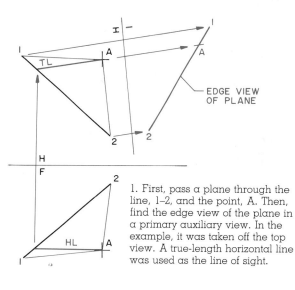

1. First, pass a plane through the line, 1–2, and the point, A. Then, find the edge view of the plane in a primary auxiliary view. In the example, it was taken off the top view. A true-length horizontal line was used as the line of sight.

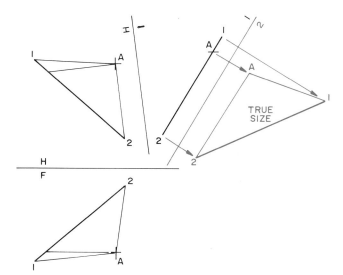

2. Next, find the true size of the plane. Construct a line of sight perpendicular to the edge view of the plane. Construct a second auxiliary plane perpendicular to the line of sight. Project the plane to the secondary auxiliary view. It will appear true size.

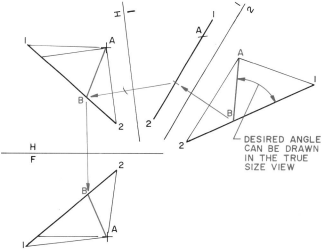

3. Since the true-size plane contains line 1–2 and point A, any angle desired can be drawn true size from A to line 1–2. Project line A–B to the primary and principal views.

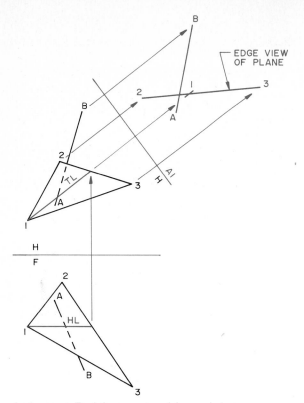

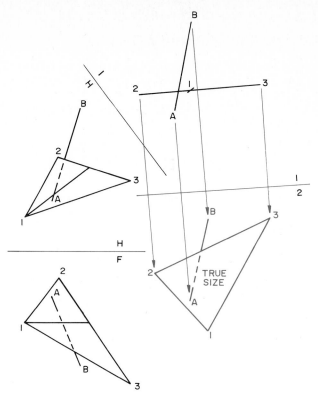

Assignment: Find the true size of the angle between a line and an oblique plane.

1. The first step is to find the edge view of the plane in a primary auxiliary view. It can be taken off any view. In this example, it was taken off the top view by finding a true-length horizontal line of the plane. Project line A–B onto the primary auxiliary view.

2. Then, find the true size of the plane in a secondary auxiliary view. Construct the line of sight perpendicular to the edge view. Construct the secondary auxiliary plane perpendicular to the line of sight. Project line A–B onto the secondary auxiliary view.

3. Next, find line A–B true length. Construct a third auxiliary view having a line of sight perpendicular to line A–B. The third auxiliary is perpendicular to the line of sight. This gives the true length of line A–B. It also shows plane 1–2–3 as an edge. Any view projected from a true-size view of a plane will be an edge view of the plane. Measure the angle between the plane and line A–B.

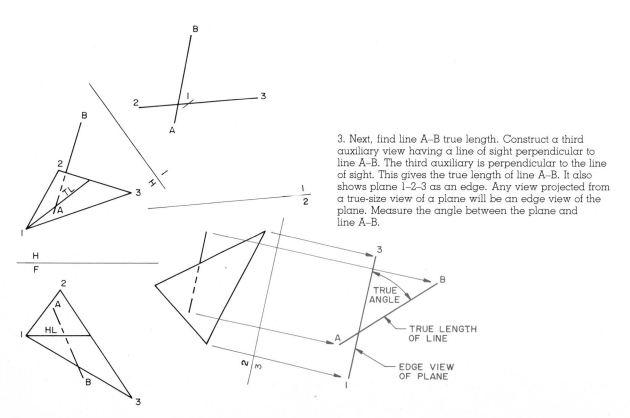

FIGURE 10–39 How to find the true size of the angle between a line and an oblique plane.

## SUCCESSIVE AUXILIARY APPLICATIONS

The following application requires the use of successive auxiliary views. If you do not understand how to construct successive auxiliary views, review Chapter 9.

### Angle Between a Line and an Oblique Plane

It is not unusual when designing a product to have two parts meet at an unusual angle. They have to be manufactured to fit at this angle. Sometimes a brace is needed to reinforce the joint. The true size of the angle must be found.

It requires three auxiliary views to find this angle. The plane is found in edge view in the primary auxiliary. A second auxiliary is taken perpendicular to the edge view. This gives a true-size view of the plane. The third auxiliary is constructed parallel with the line that intersects the plane. Since the plane is in edge view, it will remain in edge view. You now have an edge view of the plane and the true length of the line. Measure the angle between them.

The steps for finding this angle are given in Fig. 10–39.

## Problems

Solve the following problems on coordinate paper with four squares per inch. Locate the features of each problem by counting the ¼-in. squares. Label all reference planes and all points given on the problem with 3-mm (⅛-in.) lettering. Indicate all right angles with the symbol ∟, true-size surfaces with TS, and true-length lines TL. The solutions will fall on the sheet if the problems are located as shown except where otherwise noted.

1. Find the true length of the oblique lines and locate the points on the lines in parts A and B in Prob. 10–1.

2. Find the point view of edges AB and BC in part A in Prob. 10–2. In part B, find the true size of the angle between edge AB and the horizontal, frontal, and profile planes. Find the true size of the angle of edge AC and CB with the horizontal, frontal, and profile planes if so requested by the instructor.

3. Find the grade of the sewer line in part A of Prob. 10–3. Give the results as a percentage, using the scale 1 in. = 100 ft. Show how you found the percentage. In part B, find the slope of the cables supporting the tower.

4. In part A of Prob. 10–4, find the edge view of plane ABC. In part B, find the edge view of surfaces ABCD and EFGH of the object shown.

5. In part A of Prob. 10–5, find the true size of the angle between the oblique planes. In part B, find the true size of the angle between the sides of the pyramid and its base.

6. In part A of Prob. 10–6, find the point at which the line pierces the plane. Locate the piercing point on the plane and ascertain visibility. Solve this using the auxiliary view and the projection methods. In part B, find the points at which the line pierces each plane. Locate the piercing point on each plane and ascertain visibility.

7. In parts A and B of Prob. 10–7, construct a line from point A perpendicular to plane BCD. Project the line to the other views. Label all true-length lines and right angles.

8. Find the intersection of the two planes in parts A and B of Prob. 10–8, using the auxiliary view or cutting planes and projection method. Project the intersection to all views and determine visibility.

9. Find the strike and dip of the plane in parts A and B of Prob. 10–9.

10. Find the angle between the oblique planes in parts A and B of Prob. 10–10. The solution for each part will require a full sheet of paper.

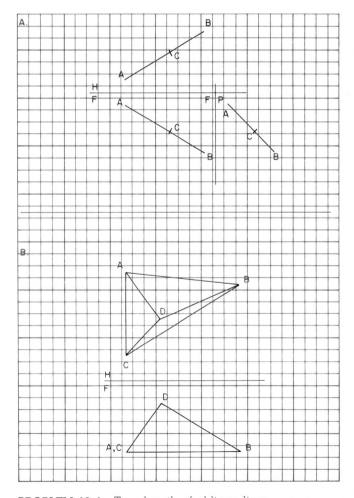

**PROBLEM 10–1** True length of oblique lines.

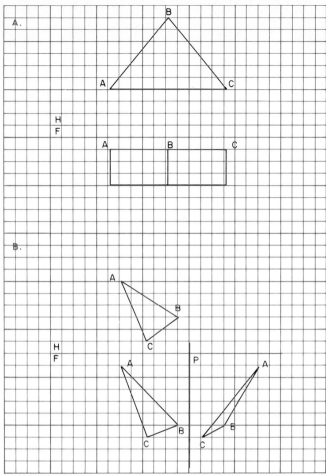

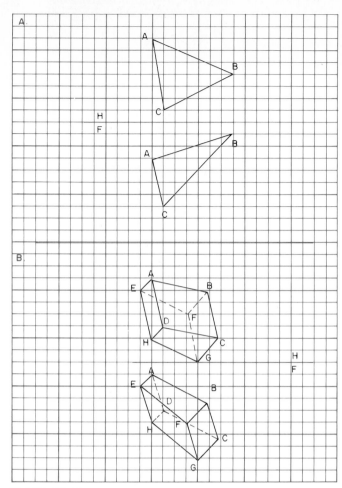

**PROBLEM 10–2**  Point view of a line and true size of angles.

**PROBLEM 10–3**  Grade and scope.

**PROBLEM 10–4**  Edge view of planes.

**PROBLEM 10–5**  True size of angle between oblique planes.

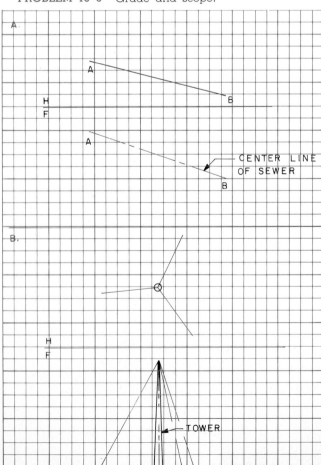

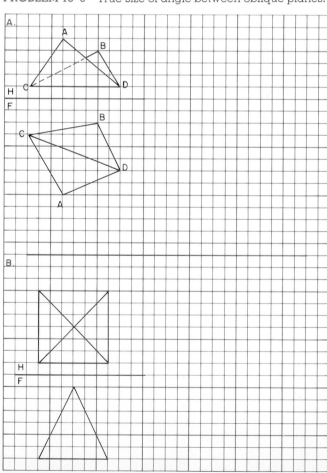

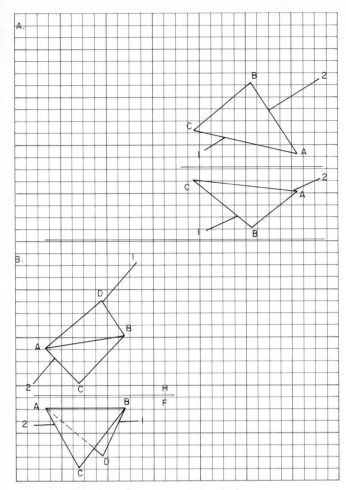

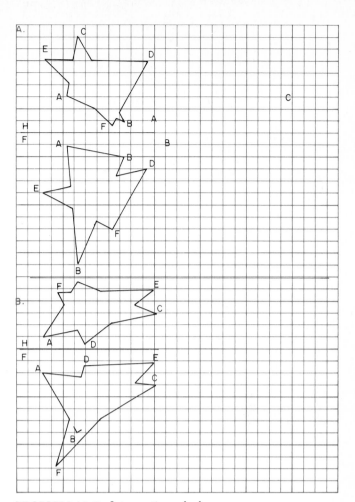

PROBLEM 10–6 Piercing points of planes.

PROBLEM 10–7 Lines perpendicular to planes.

PROBLEM 10–8 Intersection of planes.

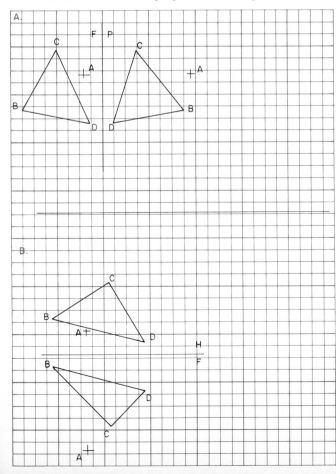

11. In part A of Prob. 10–11, find the true size of the oblique plane. Measure the length of each side in inches and millimeters and record these on the drawing. In part B, find the true size of each of the intersecting planes. Each problem will require a full sheet for the solution.

12. In part A of Prob. 10–12, find the true length of the shortest distance between point A and line BC. Project it perpendicular to line BC and locate it in all views. In part B, find the shortest distance from points C and D to edge EF.

13. In part A of Prob. 10–13, find the shortest distance between the parallel lines. In part B, find the shortest distance between each of the three parallel guidelines.

14. In part A of Prob. 10–14, find the shortest distance between the two skew lines. Locate this distance on all views. In part B, find the shortest horizontal distance between the two skew lines. Locate this distance on all views. Each of these problems will require a full page for the solution.

15. In part A of Prob. 10–15, locate the shortest line that can be drawn at a slope of 30° between the two skew lines. Locate this sloped line in all views. In part B, construct a line from point A to line BC at an angle of 60° to line BC. Project it to all views.

16. In parts A and B of Prob. 10–16, find the true size of the angle between line AB and oblique plane CDE. Place each problem on a separate sheet of paper.

Descriptive Geometry: Applications 229

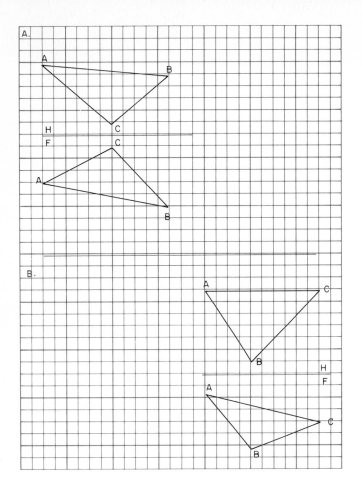

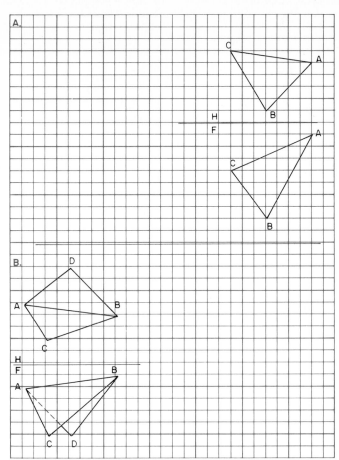

PROBLEM 10–9  Strike and dip.

PROBLEM 10–10  Angle between oblique planes.

PROBLEM 10–11  True size of oblique planes.

PROBLEM 10–12  Shortest distance between a point and a line.

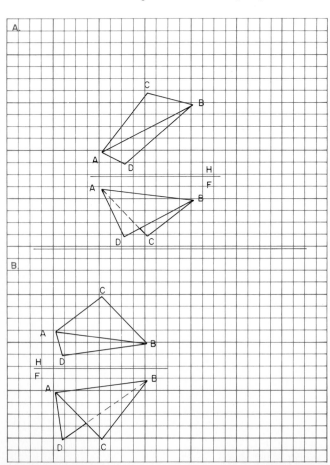

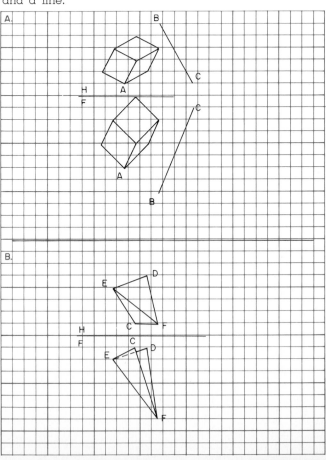

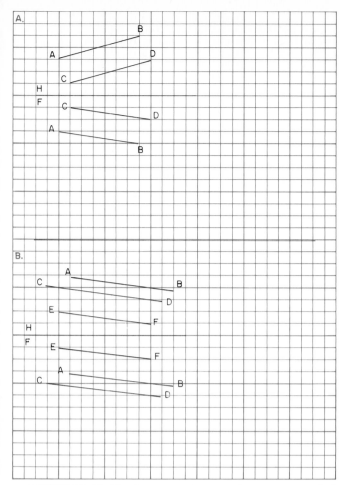

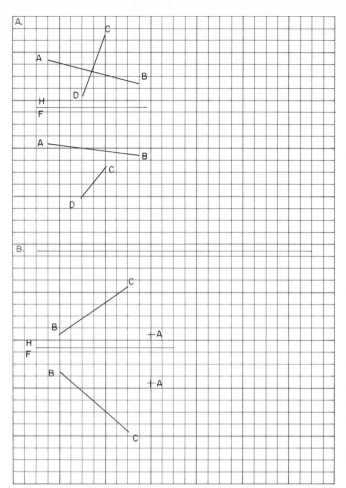

PROBLEM 10–13 Shortest distance between parallel lines.

PROBLEM 10–14 Shortest distance between skewed lines.

PROBLEM 10–15 Constructing sloped lines to known lines.

PROBLEM 10–16 True size of the angle between a line and an oblique plane.

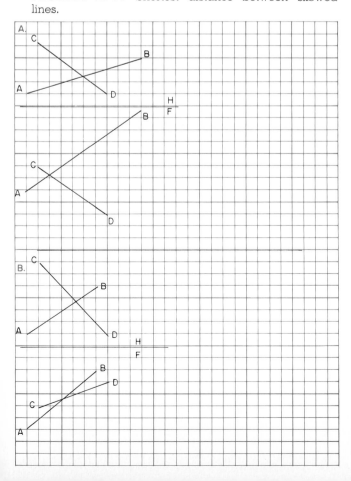

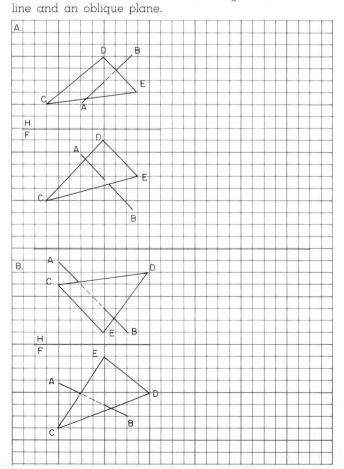

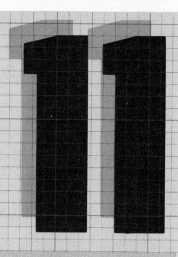

# Revolution

Revolution is a method for solving design problems that could also be solved using auxiliary views. In some cases it might be easier or faster to use revolution than auxiliary views. It is used to provide needed data when checking a design for clearances. It is particularly useful for analysis when a design has a rotating part (Fig. 11–1).

When making an auxiliary view, the view is projected upon a plane constructed parallel with the inclined or oblique edge or surface. The line of sight is moved until it is perpendicular to the line or surface involved. The view is drawn on this special plane, an auxiliary plane (Fig. 11–2).

When solving a problem using revolution, the inclined or oblique line or surface is revolved until it is parallel with a principal plane. The view is projected to the principal plane. The line of sight is the same as that used to draw the principal views (Fig. 11–3). The inclined surface appears true size because it was revolved until it was parallel with a principal plane.

## FUNDAMENTALS OF REVOLUTION

Following are some fundamentals involved in using revolution. It is important that they be understood before using revolution to solve design problems.

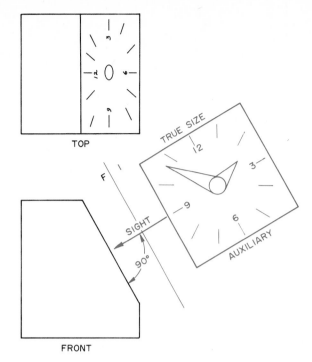

1. A point revolving about an axis follows a circular path. The axis is the center of the circle and appears as a point (Fig. 11-4).

2. When the axis of a circle appears as a line and is parallel with a principal plane, the circular path scribed by the point appears as a line (Fig. 11-4). The length of this line is equal to the diameter of the circle. It is always at right angles to the axis.

FIGURE 11-1 The design of the head of this radial-arm saw required that it rotate about a vertical and a horizontal axis. (Courtesy of DeWalt.)

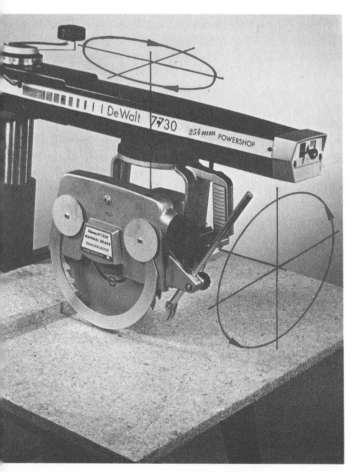

FIGURE 11-2 An auxiliary view has a line of sight perpendicular to the inclined surface. It is drawn on a plane constructed parallel with the inclined surface.

FIGURE 11-3 The inclined face was revolved until it was parallel with the profile plane. It projects true size on that plane. Note that the lower right corner was chosen as the axis for this revolution.

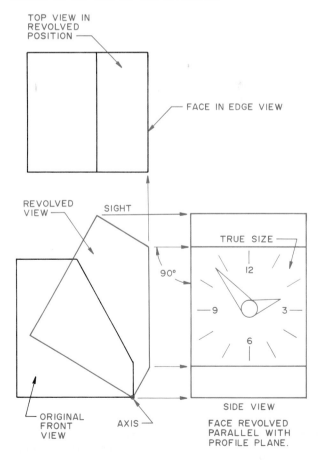

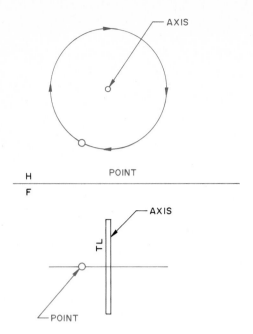

FIGURE 11-4 The path of a point revolving about an axis is a circle when the axis appears as a point. The path is a line when the axis appears as a true-length line.

3. When the axis is inclined to the principal planes of projection, the path formed by a revolving point appears as an ellipse on the principal planes (Fig. 11-5).

4. A line that is parallel with a principal plane and is revolved in that plane remains true length even though it is in a new position (Fig. 11-6).

FIGURE 11-5 The path of a point revolving about an inclined axis appears as an ellipse on the principal planes.

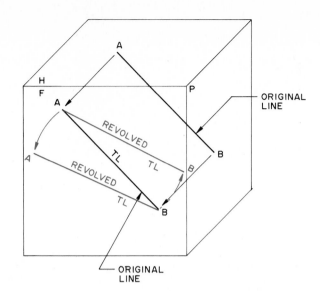

FIGURE 11-6 A line appearing true length on a plane remains true length when it is revolved in the plane.

5. When one end of a line is used as the axis of revolution and the other is revolved, the moving end stays in the original plane (Fig. 11-7). If the end is moved 360°, a cone is formed.

6. The axis of revolution can be placed in any convenient location. It is perpendicular to one of the three principal planes (Fig. 11-8).

FIGURE 11-7 When one end of a line is used as an axis and the other is revolved, the revolved end remains in the original plane.

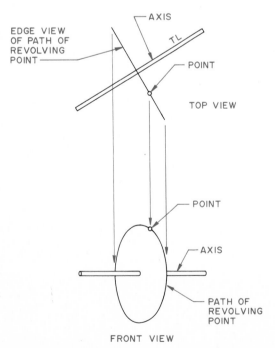

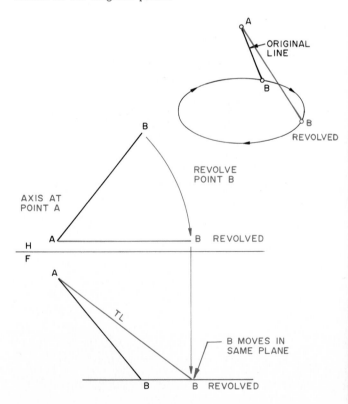

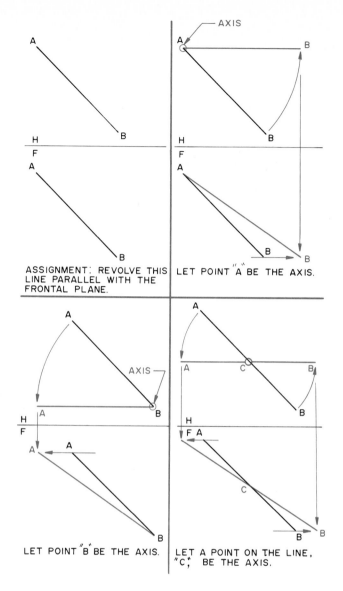

ASSIGNMENT: REVOLVE THIS LINE PARALLEL WITH THE FRONTAL PLANE.

LET POINT "A" BE THE AXIS.

LET POINT "B" BE THE AXIS.

LET A POINT ON THE LINE, "C," BE THE AXIS.

FIGURE 11–8 The axis of revolution can be placed in the most convenient place.

7. A revolution can be made either clockwise or counterclockwise.

8. The dimensions in the view in which the axis of revolution appears as a point do not change (Fig. 11–9).

9. In the view in which the axis shows true length, the dimensions parallel to the axis do not change (Fig. 11–9).

10. Revolution can be used in combination with auxiliary views. For example, after an auxiliary view is drawn, some feature can be revolved in the auxiliary view instead of drawing a secondary auxiliary view.

11. Revolutions, like auxiliary views, can be successive. There is no limit on the number of times an object can be revolved.

## APPLICATIONS OF REVOLUTION

### True Length of a Line

The true length of an oblique line can be found by revolving either end so that the line is parallel with a principal plane. Any line parallel with a principal plane will appear true length on that plane. The steps for finding the true length of an oblique line are given in Fig. 11–10. The revolution could occur in either the horizontal, frontal, or profile plane. In

FIGURE 11–9 In a view where the axis of revolution appears as a point, the dimensions do not change. In a view where the axis is true length, the dimensions parallel to the axis do not change.

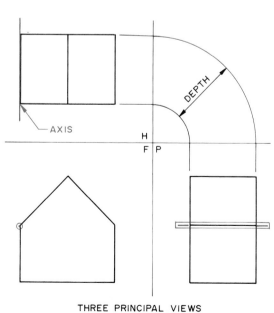

THREE PRINCIPAL VIEWS

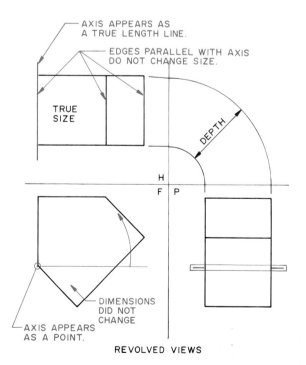

REVOLVED VIEWS

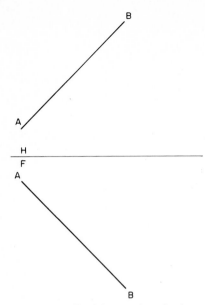

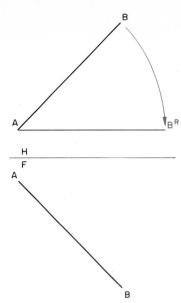

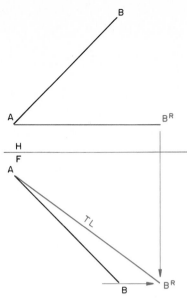

Assignment: Find the true length of an inclined line by revolution.

1. Revolve the line until it is parallel with a principal plane. Here, it was revolved in the top view until it was parallel with the frontal plane.

2. Project the revolved line to the front view. Point B moves parallel to the horizontal plane.

FIGURE 11–10 To find the true length of an oblique line, revolve it until it is parallel with a principal plane.

this example, the line was revolved in the horizontal plane until it was parallel with the frontal plane. When projected to the frontal plane, it appears true length.

This same solution using the horizontal and

FIGURE 11–11 Finding the true length of an oblique line by revolving the line in the horizontal or profile planes.

In this solution, the oblique line was rotated in the front view until it was parallel with the horizontal plane. When projected onto the top view, it appears true length.

Here, the inclined line was rotated in the front view until it was parallel with the profile plane. When projected onto the side view, it appears true length.

profile planes is shown in Fig. 11–11. In part A, the oblique line was rotated in the front view until it was parallel with the horizontal plane. Then it was projected into the top view, where it appears true length. In part B, the inclined line was rotated in the front view until it was parallel with the profile plane. When it was projected into the side view, it appeared true length.

## Angle Between an Oblique Line and a Principal Plane

The angle between an oblique line and a plane will appear true size in any view where the plane appears as an edge and the line is true length. In each principal view, two of the principal planes appear in edge view. To find the true angle, the oblique line must be found true length in a view where the plane desired appears as an edge.

The steps for finding the angle between an oblique line and the principal planes are shown in Fig. 11–12.

## Edge View of an Oblique Plane

The edge view of an oblique plane can be found by revolution. First, construct a true-length line on the plane. Then, revolve the plane until the true-length line is perpendicular to a principal plane. It will project as an edge on the principal plane (Fig. 11–13).

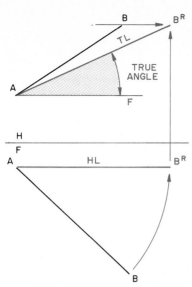

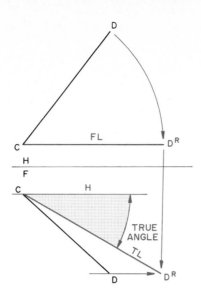

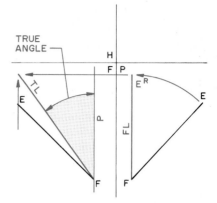

ANGLE WITH FRONTAL PLANE

Assignment: Find the true size of the angle between an oblique line and the principal planes.

1. In this example, the angle with the frontal plane was found by revolving the line in the frontal plane until it was parallel with the horizontal plane. Since in this view the frontal plane appears as an edge and the line is true length, the angle appears true size.

ANGLE WITH HORIZONTAL PLANE

2. In this example, the angle with the horizontal plane was found by revolving the line in the horizontal plane until it was parallel with the frontal plane. Since in this view the frontal plane appears as an edge and the line is true length, the angle appears true size.

FIGURE 11–12 How to find the true size of the angle between an oblique line and the principal planes by revolution.

ANGLE WITH PROFILE PLANE

3. In this example, the angle with the profile plane was found by revolving the line in the profile plane until it was parallel with the frontal plane. Since in this view the profile plane appears as an edge and the line is true length, the angle appears true size.

FIGURE 11–13 Finding the edge view of an oblique plane by revolution.

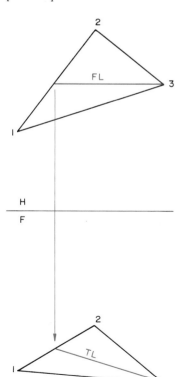

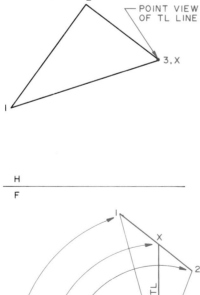

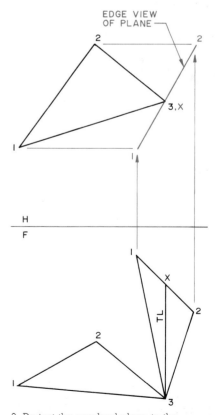

Assignment: Find the edge view of an oblique plane by revolution.

1. First, find a true-length line on the plane. In this example, a frontal line was used.

2. Revolve the plane so that the true-length line appears as a point in a principal view. In this example, it was in the top view.

3. Project the revolved plane to the view in which the true-length line appears as a point. The plane will appear in edge view.

## True Size of an Oblique Plane

The preceding illustrations have shown how to revolve a line. It is just as easy to revolve an entire plane (Fig. 11–14). This example shows how to find the true size of an oblique plane. Two steps are required. First, the plane must be found in edge view. Then, the edge view is revolved until it is parallel with a plane upon which it can be projected.

The plane can be found in edge view by using a primary auxiliary or by revolution. The procedure for finding the edge view was explained earlier. The use of an auxiliary view is shown in this example. The steps for finding the true size of the plane are shown in Fig. 11–15.

## Angle Between Two Planes

Two intersecting planes have their line of intersection in two possible positions: parallel with a principal plane or oblique to the planes.

If the intersection is parallel with a principal plane, it will appear true length on the plane with

FIGURE 11–14 The true shape and size of this tracking and communication antenna could be found by revolving it until it is parallel with a vertical or horizontal plane. (Courtesy of National Aeronautics and Space Administration.)

FIGURE 11–15 How to find the true size of an oblique plane using an auxiliary view and revolution.

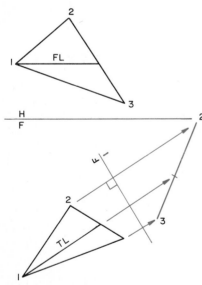

Assignment: Find the true size of the oblique plane using auxiliary views and revolution.

1. Construct a true-length line on the plane. Use it as the line of sight to find the edge view of the plane in a primary auxiliary view.

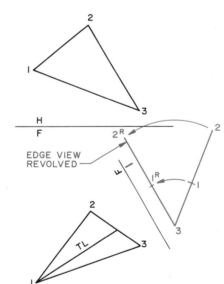

2. Revolve the edge view of the plane until it is parallel with the frontal plane.

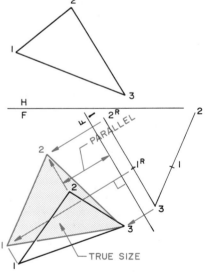

3. Project the revolved plane to the front view. Project the revolved corners of the plane parallel with the edge view of the frontal plane. Connect the corners to form the true size of the plane.

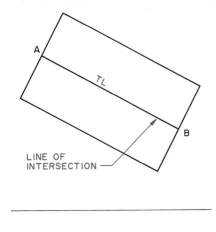

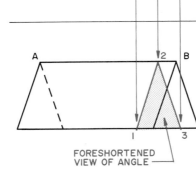

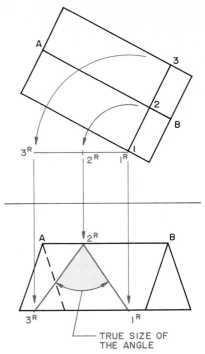

Assignment: Find the true size of the angle between two planes when their line of intersection is parallel with a principal plane.

1. Draw an edge of a slice forming the angle. It is perpendicular to the true-length intersection. In this example, it was in the top view. Project this to the front view.

2. Revolve the edge view of the angle between the intersecting planes until it is parallel with a principal plane. Since it appears in the top view, it was revolved to be parallel with the frontal plane. Project the revolved angle to the front view, where it appears true size.

FIGURE 11–16 How to find the true size of the angle between two planes when their line of intersection is parallel with a principal plane.

which it is parallel. To find the angle between the planes by revolution, draw an edge view of a slice through the planes perpendicular to the true-length intersection. Project this to the other principal view. Then, revolve the edge view drawn so that it is parallel with a principal plane. Project it to that plane to form the true angle between the planes. The steps for finding the true angle are given in Fig. 11–16.

When the intersection between the two planes is an oblique line, it is necessary to find the line of intersection true length in a primary auxiliary or by revolution. Then, construct an edge view of the angle which is perpendicular to the intersection of the planes. Revolve this edge view parallel to a plane and project it to the plane to find the true angle. The steps for finding the true angle between two intersecting planes having an oblique line of intersection are given in Fig. 11–17.

### Revolving a Point About an Oblique Axis

It is frequently necessary to check the clearance needed when some part of a product rotates, such as a handwheel on a machine tool or an arm on a robot (Fig. 11–18). A point on the outermost edge of the machine can be revolved to see if adequate clearance has been provided.

To revolve a point about an oblique axis, first find the axis true length in a primary auxiliary. Then, construct a secondary auxiliary in which the true-length line appears as a point. The point to be revolved about the axis can be rotated in a circular path establishing the clearances needed on all sides of the line. Project the circular path of revolution back to the principal views. It will appear elliptical because the axis is oblique in the principal views. The steps for revolving a point about an oblique axis are shown in Fig. 11–19.

### Extreme Positions on Path of Revolution

Extreme positions on the path of revolution are found in the view in which the path of revolution appears as an edge. The highest and lowest positions are found in an auxiliary view off the top view. Those farthest to the front and to the rear are found in an auxiliary view projected off the front view.

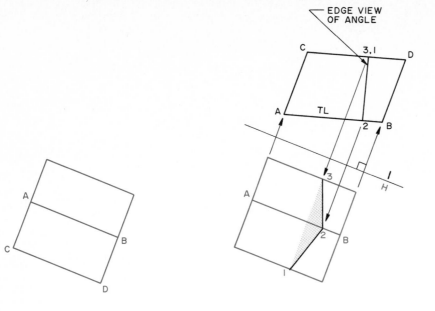

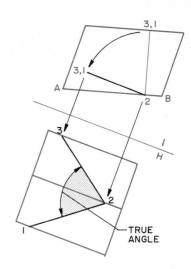

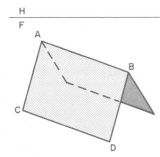

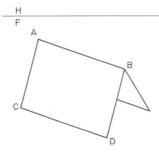

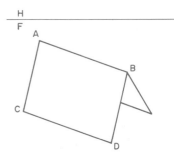

Assignment: Find the true size of the angle between two planes having a line of intersection that is an oblique line.

1. Find the intersection of the two planes true length. In this example, a primary auxiliary view was used. It was constructed off the top view. Then, construct an edge view of the angle perpendicular to the true-length line.

2. Revolve the edge view of the angle until it is parallel with a plane. In this example, it was revolved parallel with the horizontal plane. Project the edge view onto the horizontal plane. Connect each end to form the true size of the angle.

FIGURE 11–17 How to find the true angle between two planes when their line of intersection is oblique.

FIGURE 11–18 The clearance of these handwheels and levers was checked by revolution when the design drawing of the machine was prepared. (Courtesy of Rockwell International, Power Tool Division.)

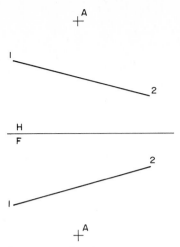

Assignment: Revolve a point, A, in a path that is perpendicular to an oblique line.

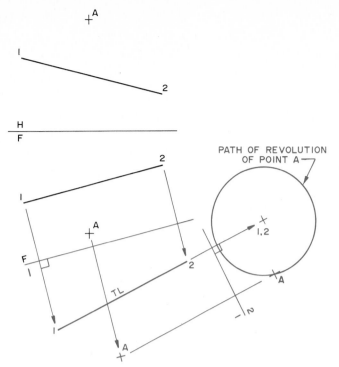

1. First find the line 1–2 which is the axis of revolution, true length in a primary auxiliary view. Find its point view in a secondary auxiliary view. Revolve point A about the point view of line 1–2.

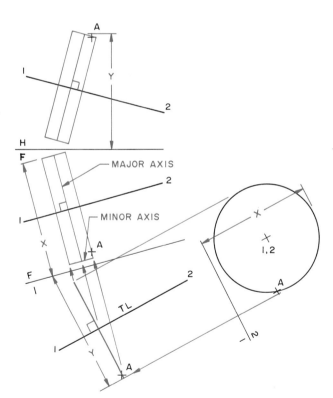

2. Project the diameter of the path of the revolved point back to the primary auxiliary view, where it appears as an edge. Project the major and minor axes to the front and top views. The major axis is perpendicular to line 1–2 because the revolution was made perpendicular to it.

FIGURE 11–19 Revolving a point in a path perpendicular to an oblique line.

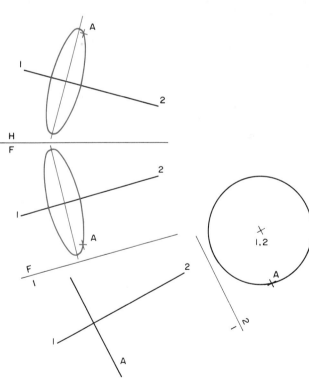

3. Draw the elliptical view of the path of revolution in the front and top views using the axes projected from the primary auxiliary.

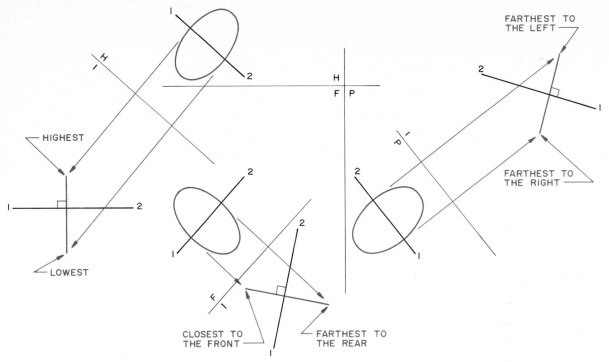

FIGURE 11–20 Locating the extreme positions on a path of revolution.

Those farthest to the right or left are found in an auxiliary projected off the side view (Fig. 11–20).

## Revolving a Line About Another Line

To revolve one line about another, first find one of the lines as a point. In this view, construct a perpendicular from the point view to the other line. Using this as a radius, draw a circle, which is therefore tangent to the other line.

To revolve the line, measure the required number of degrees of rotation from the perpendicular and draw a line to the edge of the circle. Construct the revolved line perpendicular to this at this point.

To find the length of the revolved line, revolve each end until it crosses the revolved line.

The steps for revolving one line about another are shown in Fig. 11–21.

## Angle Between a Line and an Inclined Plane

To find the angle between a line and an inclined plane, it is necessary to find the true size of the plane. In the true-size view, revolve the inclined line until it is parallel with the plane of projection. Project it back to the previous view, where it appears true length. Since the plane is in edge view and the line is true length, the angle between them is true size. The steps for this construction are given in Fig. 11–22.

## Angle Between a Line and an Oblique Plane

The oblique plane must be found true size. This requires the use of a primary and a secondary auxiliary. In the secondary auxiliary, where the plane appears true size, revolve the line parallel to the primary auxiliary plane. Project it to the primary auxiliary, where it will be true length and the true size of the angle can be measured. The steps for this construction are given in Fig. 11–23.

## Revolution of Circles

To revolve a circle, first find it in edge view. Then, revolve the edge view. To assist with projecting the circle to the principal views, divide the circumference into a number of points. Revolve these points with the edge view. Then project them back to the principal plane (Fig. 11–24).

## Revolution of Three-Dimensional Objects

When revolving three-dimensional objects, the axis of revolution can be in any convenient place. The actual location will depend upon the revolved view desired. In Fig. 11–25 are a few selected examples. In example 1, a horizontal axis was placed through the center of the object. It is perpendicular to the frontal plane. The product can be revolved about the axis the number of degrees of revolution desired. *Notice that the view to which the axis is perpendicular does not change shape. The lines in the other views of the views that are parallel with the axis do not change length.*

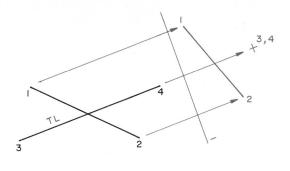

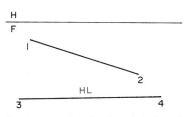

Assignment: Revolve line 1–2 about line 3–4 a total of 120°.

1. First, find line 3–4 in a point view. Since it is a horizontal line, it is true length in the top view. Construct a primary auxiliary using the true-length line as a line of sight.

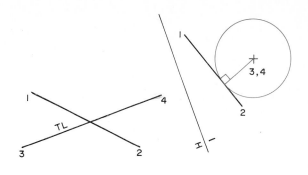

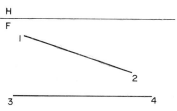

2. Construct a perpendicular from the point view of line 3–4 to line 1–2. Using this distance as a radius, draw a circle. Line 1–2 will always be tangent to this circle as it revolves around line 3–4.

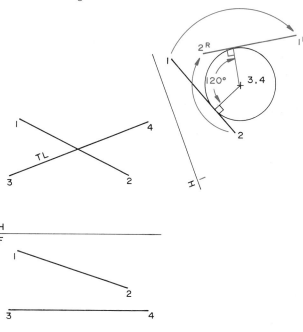

3. Locate the revolved position by drawing a line from 3–4 to the edge of the tangent circle at the required 120°. Construct revolved 1–2 perpendicular to this line. Locate the ends of revolved 1–2 by swinging arcs using as a radius their projection from the top view. The points at which they cross revolved 1–2 locates the ends of the line.

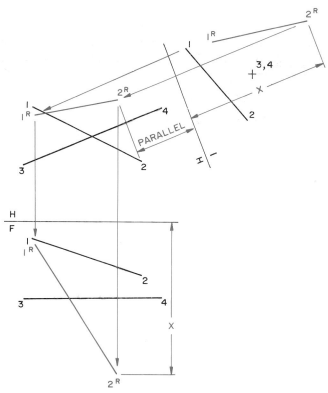

4. Project the revolved line to the principal views. In the top view, the ends of line 1–2 are located by moving them parallel with the horizontal plane.

FIGURE 11–21   Revolving one line about another line.

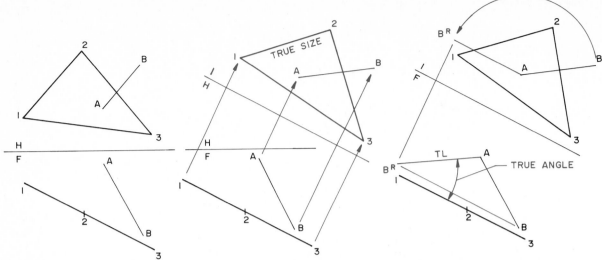

Assignment: Find the true size of the angle between a line and an inclined plane.

1. The plane is in edge view in the front view. Find it true size in a primary auxiliary view. Project the line to the auxiliary view.

2. Revolve line A–B until it is parallel with the frontal plane. Either end could be revolved in either direction. In this example, B was revolved counterclockwise until the revolved line was parallel with the frontal plane. Project revolved point B to the front view. Since the plane is in edge view and the line is true length, the true size of the angle can be measured.

**FIGURE 11–22** Finding the true size of an angle between a line and an inclined plane.

**FIGURE 11–23** Finding the true size of an angle between a line and an oblique plane.

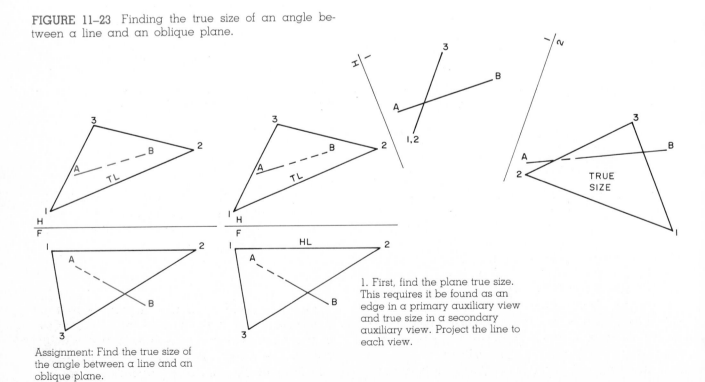

Assignment: Find the true size of the angle between a line and an oblique plane.

1. First, find the plane true size. This requires it be found as an edge in a primary auxiliary view and true size in a secondary auxiliary view. Project the line to each view.

FIGURE 11–23 cont.

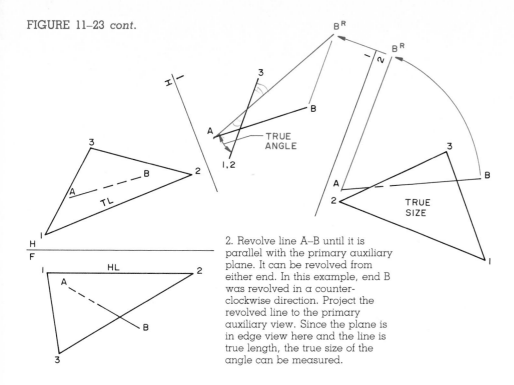

2. Revolve line A–B until it is parallel with the primary auxiliary plane. It can be revolved from either end. In this example, end B was revolved in a counter-clockwise direction. Project the revolved line to the primary auxiliary view. Since the plane is in edge view here and the line is true length, the true size of the angle can be measured.

FIGURE 11–24 A circle can be revolved by first finding it in edge view.

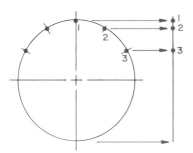

Assignment: Revolve the circle through an angle of 30° from the vertical.

1. First, find the circle in edge view. Then, divide the circle into several parts on the circumference.

2. Revolve the edge view of the circle. Revolve the points along with it.

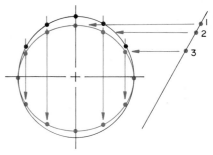

3. Project the points to the revolved view of the circle.

To draw the revolution, first revolve the front view as desired. Since it does not change shape, it is the same size as the principal front view. Then, project the revolved view to the top and right side. Since the width does not change, it can be established by measurement.

In example 2 in Fig. 11–25, an axis is drawn perpendicularly to the horizontal plane. The product is revolved parallel with the horizontal plane. It remains the same shape in the top view. The height remains the same as in the principal views.

In example 3 in Fig. 11–25, an axis is drawn perpendicularly to the profile plane. The product is revolved parallel with the profile plane. It remains the same shape in the side view. The width remains the same as in the principal views.

## Successive Revolutions

The principles presented in this chapter are used to produce successive revolutions. There is no limit to the number of successive revolutions that can be drawn. In the example in Fig. 11–26, the first revolution is made with an axis perpendicular to the

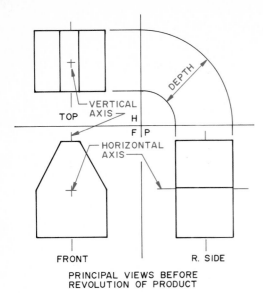

TOP

VERTICAL
AXIS

H
F | P

HORIZONTAL
AXIS

DEPTH

FRONT

R. SIDE

**PRINCIPAL VIEWS BEFORE
REVOLUTION OF PRODUCT**

Assignment: Given are the principal views. Construct revolutions with axes perpendicular to the frontal, horizontal, and profile planes. Revolve the product 30°.

EXAMPLE 1

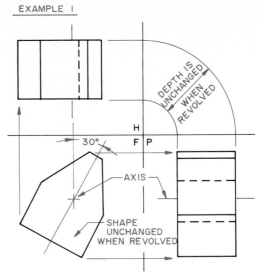

H
F | P

30°

DEPTH IS UNCHANGED WHEN REVOLVED

AXIS

SHAPE
UNCHANGED
WHEN REVOLVED

**REVOLVED 30° FROM THE VERTICAL
WITH AXIS PERPENDICULAR TO THE
FRONT VIEW.**

1. In this example, the axis is perpendicular to the frontal plane. The shape of the view to which the axis is perpendicular is unchanged by the revolution. In this example, the shape of the front view is unchanged. The depth is also unchanged because it is parallel with the axis.

EXAMPLE 2

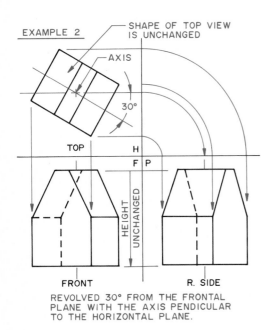

SHAPE OF TOP VIEW
IS UNCHANGED

AXIS

30°

TOP

H
F | P

HEIGHT UNCHANGED

FRONT

R. SIDE

**REVOLVED 30° FROM THE FRONTAL
PLANE WITH THE AXIS PENDICULAR
TO THE HORIZONTAL PLANE.**

2. In this example, the axis is perpendicular to the horizontal plane. The shape of the top view is unchanged. The height in the front and profile views is unchanged because it is parallel with the vertical axis.

FIGURE 11-25 When revolving three-dimensional objects, the view to which the axis is perpendicular does not change shape. The axis can be placed in any convenient location.

EXAMPLE 3

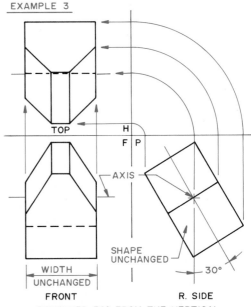

TOP

H
F | P

AXIS

SHAPE
UNCHANGED

30°

WIDTH
UNCHANGED

FRONT

R. SIDE

**REVOLVED 30° FROM THE VERTICAL
WITH THE AXIS PERPENDICULAR TO
PROFILE PLANE.**

3. In this example, the axis is perpendicular to the profile plane. The shape of the side view is unchanged. The width in the front and top views is unchanged because it is parallel with the vertical axis.

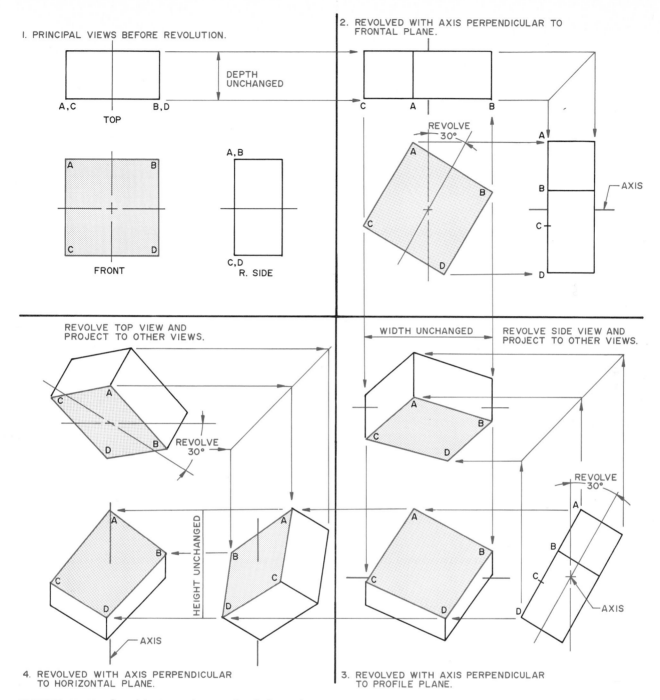

FIGURE 11-26  An object can be revolved through an
unlimited number of successive revolutions.

frontal plane and in the center of the prism. It was
revolved 30° from the vertical. The depth of the revolved views remained the same as it appeared in
the principal views. The second revolution was
made with the axis perpendicular to the profile
plane, and the prism in the first revolution was revolved 30° with the vertical again. The width of the
second revolved position was the same as it appeared in the first revolution. The second revolved

position was revolved a third time with the axis perpendicular to the horizontal plane. The prism was
revolved 30° from a frontal plane through its center.
The height of the third revolved prism was the same
as it appeared in the second revolved position because the height was parallel with the axis in the
third revolution. This procedure could continue in-
definitely.

# Problems

The revolution problems that follow are designed to be solved using standard coordinate paper with a grid of ¼-in. squares. Locate the problems on the paper by counting the squares in the given problems. Label all reference lines and points on the problems.

1. In Prob. 11-1, find the true length of each line by revolution. In part A, find the true length in the horizontal plane and its angle with the frontal plane; in part B, find the true length in the frontal plane and its angle with the horizontal plane; in part C, find the true length in the frontal plane and its angle with the profile plane; and in part D, find its true length in the profile plane and its angle with the frontal plane.

2. In parts A and B of Prob. 11-2, find the edge view of the plane by revolution.

3. In part A of Prob. 11-3, find the true size of the oblique plane by revolution. In part B, find the true size by using a primary auxiliary and then revolving the plane.

4. In parts A and B of Prob. 11-4, find the true angle between the intersecting planes that have their line of intersection parallel with a principal plane.

5. In part A of Prob. 11-5, find the true angle between the intersecting oblique planes by revolution. In part B, use a primary auxiliary and then revolution to find the true size of the angle.

6. In part A of Prob. 11-6, locate in all views the path of revolution of a point that is ½ in. from the given line. In part B, show the path of rotation of point A around the given line.

7. In part A of Prob. 11-7, find the points of the path of revolution that are farthest to the rear, closest to the front, and farthest to the right and left of the axis. In part B, find the highest and lowest points on the path of revolution and those farthest to the front and rear of the line.

8. In part A of Prob. 11-8, rotate line 1-2 90° counterclockwise about line AB. Project it back to all views. In part B, revolve line 1-2 45° clockwise around line AB. Notice that in this problem both lines are oblique.

9. In part A of Prob. 11-9, find the true size of the angle between the line and the inclined plane. In part B, find the angle between the line and the oblique plane.

10. In part A of Prob. 11-10, revolve the circle 45° to the frontal plane. In part B, revolve it 60° to the profile plane.

11. In part A of Prob. 11-11, revolve the object 30° from the frontal plane, keeping the axis perpendicular to the horizontal plane. Locate the axis at corner B. Draw the top, front, and right side views in the revolved position. Show only the visible edges. In part B, revolve the letter H 30° from the vertical, with the axis perpendicular to the profile plane. Locate the axis at point A. Draw the top, front, and right side views in the revolved position. Show only the visible edges.

12. In part A of Prob. 11-12, find the true size of the oblique surface, ABC. Revolve the entire object each time and show all hidden edges. In part B, revolve the object three times. To get the solution on the paper, tape two sheets of 8½ × 11 in. coordinate paper together along the short edge with the border on that edge cut away, or use an 11 × 18 in. sheet. The solution requires 15 in.

*First revolution.* Locate the axis at edge AB perpendicular to the profile plane. Rotate it 30° from the vertical, with edge CD toward the frontal plane. Draw three views and show all hidden edges.

*Second revolution.* Revolve the rotated top view in the first revolution so that it makes an angle of 30° with the frontal plane. The axis is perpendicular to the horizontal plane. Locate the axis at point B. Revolve the object so that corner C is toward the frontal plane. Project the height from the front view of the first revolution to the second revolution. Draw three views and show hidden lines.

*Third revolution.* Revolve the front view of the second revolution with an axis perpendicular to the frontal plane until edge DC is at an angle of 30° with the horizontal plane. Revolve it so that corner D is toward the horizontal plane. Locate the axis at corner A. Project the depth from the second revolution to the third revolution. Draw three views and show all hidden edges.

13. In part A of Prob. 11-13, show the true size of the inclined surface of the cylinder. In part B, find the true angle between the block and the cable. Also find the piercing point.

14. In part A of Prob. 11-14, find the true length of each guy wire. Use the scale ¼ in. = 10 ft. Record the length on the true-length view. In part B, find the true length of the guy wires holding the sign and the true angle between them. Also find the true angle of the wires with the wall. Record these by the true size on the drawing. Use the scale ¼ in. = 1 ft.

15. In part A of Prob. 11-15, find the true size of the inclined surface of the guide plate. In part B, find the true size of every surface of the wedge.

16. In part A of Prob. 11-16, find the true length of each oblique corner of the pylon and the true angles between these corners. In part B, find the true shape of the triangular arms of the support bracket.

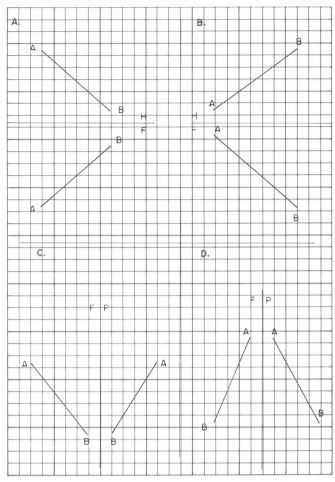

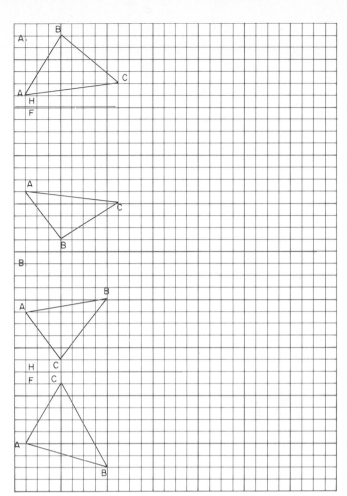

**PROBLEM 11-1**  True length of lines and angles.
**PROBLEM 11-2**  Edge view of an oblique plane.

**PROBLEM 11-3**  True size of oblique plane.
**PROBLEM 11-4**  True size of angles.

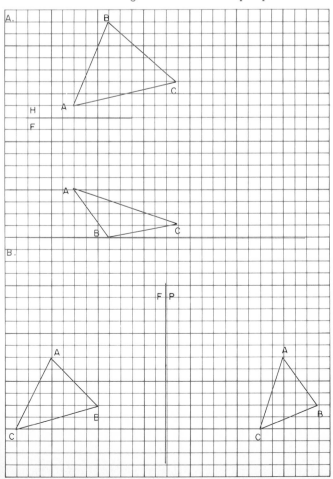

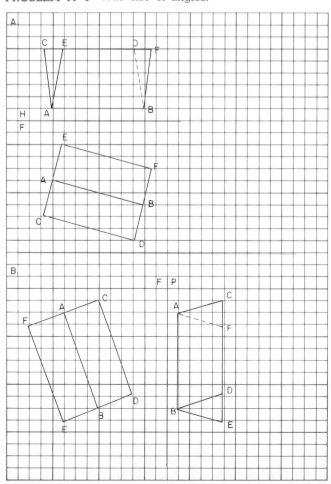

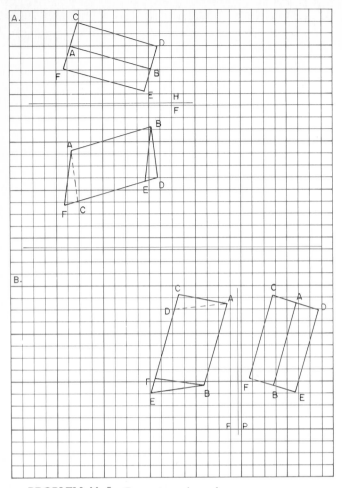

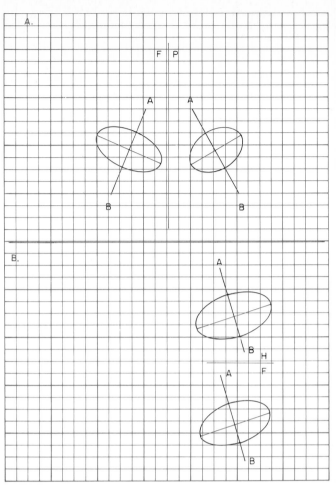

PROBLEM 11–5  True size of angles.

PROBLEM 11–7  Extreme position of path of revolution.

PROBLEM 11–6  Revolution of a point about a line.

PROBLEM 11–8  Revolving one line around another.

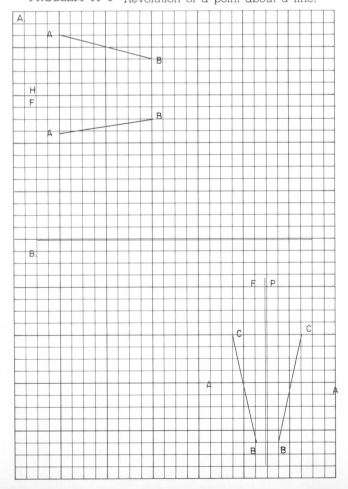

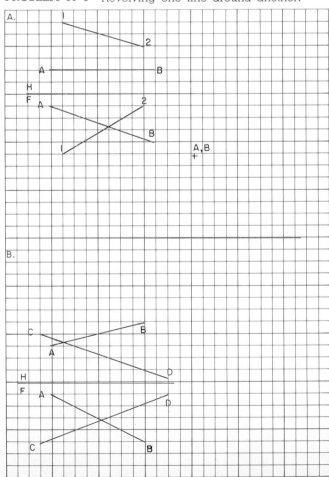

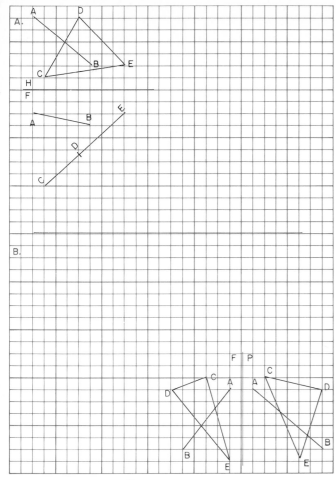

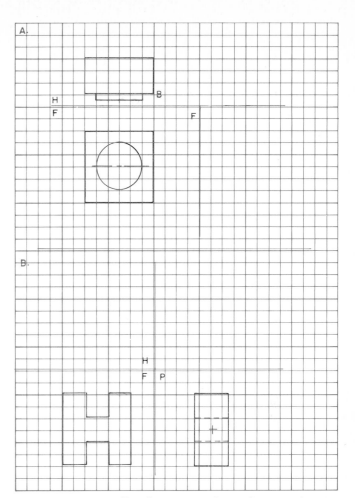

PROBLEM 11–9 True size of an angle between a plane and a line.

PROBLEM 11–10 Revolving circles.

PROBLEM 11–11 Revolving complete objects when the axis is perpendicular to a normal plane.

PROBLEM 11–12 Successive revolutions.

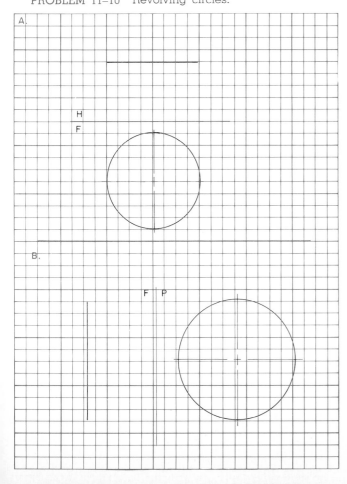

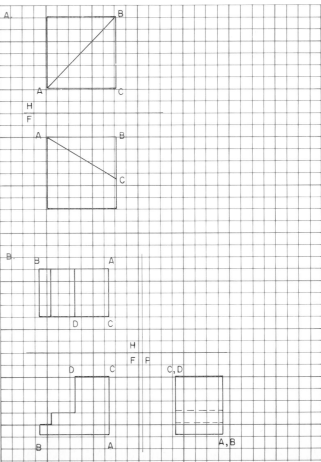

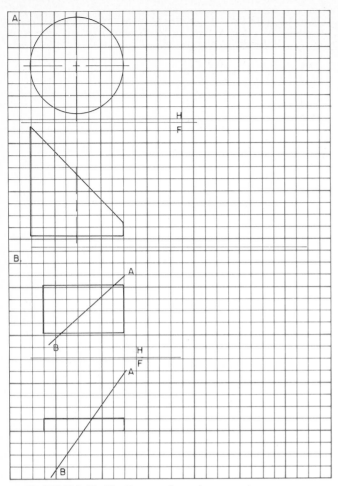

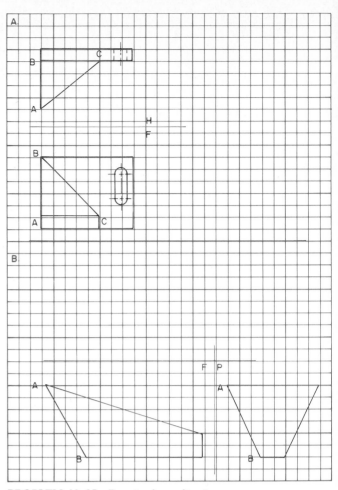

**PROBLEM 11–13** Practical applications.
**PROBLEM 11–14** Practical applications.

**PROBLEM 11–15** Practical applications.
**PROBLEM 11–16** Practical applications.

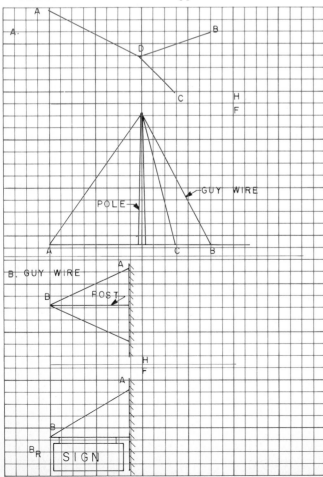

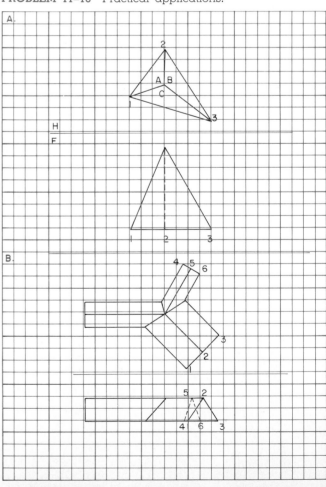

# Pictorial Presentation of Engineering Designs

As an engineer works on a design project, some of the first original sketches are often in pictorial form. This records an idea or possible solution quickly and in a form that is easy to understand.

A major value of pictorial drawings is that they show objects three-dimensionally. People are accustomed to seeing actual objects and can visualize the drawing. Engineering drawings, with all their dimensions and multiple views, are much more difficult to read. An untrained worker can view a pictorial drawing and understand it (Fig. 12–1).

Pictorial drawings have many industrial uses. They are often included on engineering drawings to clarify a detail. Sometimes, a section is shown pictorially. Maintenance manuals and parts manuals rely heavily on pictorial drawings. Information such as where and how to lubricate is given. Often, an assembly is drawn in an exploded view (Fig. 12–2). This shows the relationships of the parts and their order of assembly. Shaded pictorial drawings are used extensively in sales literature. Architectural perspectives show the exteriors of buildings, the areas surrounding the building, and the interiors of rooms. Most pictorial drawings are made by a special type of drafter, the technical illustrator. These individuals have artistic abilities and special prepa-

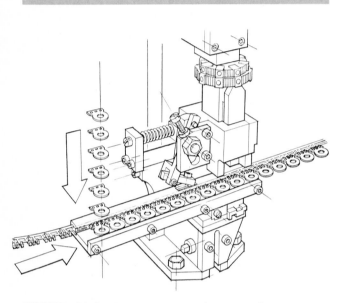

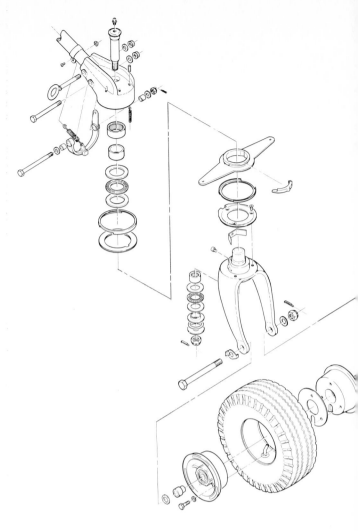

FIGURE 12–1 A perspective drawing illustrating a manufacturing proposal. It was developed with the Perspective Sciences perspective system. (Courtesy of Perspective Sciences.)

FIGURE 12–2 An isometric projection of an aircraft tailwheel and lock assembly. This is an exploded pictorial commonly used in aircraft maintenance manuals. (Courtesy of Cessna Aircraft Company.)

ration. The engineer and engineering drafter generally produce line pictorial drawings.

## TYPES OF PICTORIAL PROJECTIONS

There are three kinds of *pictorial projections:* axonometric, which includes isometric, dimetric, and trimetric; oblique, which includes cabinet and cavalier; and perspective (Fig. 12–3).

*Axonometric projection* is the projection of an object onto a picture plane with the object positioned so that it appears three-dimensional. The object is turned and inclined to the projection plane. The projectors are parallel and are perpendicular to the picture plane. This is the same projection technique as is used for orthographic projection. The term *picture plane* is used to describe the plane upon which pic-

torials are drawn. In actuality, it is your drawing paper.

*Oblique projection* is the projection of an object onto a picture plane. The object is positioned with the front view parallel with the projection plane. In this way it appears true size. The receding surfaces are drawn on an angle to give a pictorial appearance. The projectors are at an angle to the picture plane. The image produced is the least realistic of the three types.

*Perspective projection* produces the most realistic pictorial. The projectors converge at a point (the viewer's eye) and therefore make a variety of angles with the picture plane. The sides of the object converge toward a distant point (the horizon). This produces the illusion of the object's getting smaller the farther it is from the picture plane. This is the way objects really appear to the viewer.

These three types are projections from known

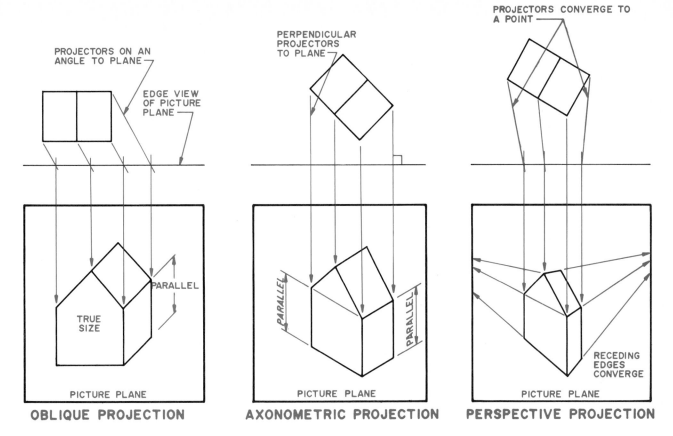

FIGURE 12–3 The three basic types of pictorial projections.

views (orthographic views). Pictorials made this way are called *projections.* Simplified ways of producing pictorials that appear about the same are in wide use in industry. These are called *pictorial drawings.*

## AXONOMETRIC PROJECTION

Axonometric projection follows the principles of orthographic projection. The object is projected perpendicularly to the projection plane (called a picture plane) with parallel projectors. The object is revolved before projection to produce a three-dimensional view.

The three types of axonometric projection are isometric, dimetric, and trimetric. In each, the object to be drawn is inclined to the picture plane. The lengths of the edges and the sizes of the angles will vary as the position of the object is changed.

Isometric, dimetric, and trimetric projections are shown in Fig. 12–4. The isometric is inclined so that the angles between the three axes are equal and the sides are foreshortened equally. The dimetric has two angles equal and two sides foreshortened equally. The trimetric has no angles equal, and each side is foreshortened a different amount.

To produce an isometric projection, the top view is revolved 45° and the side view is revolved 35°16' from the vertical. To produce a dimetric projection, the top view is revolved 45° and the side view any angle from the vertical except 35°16'. To produce a trimetric projection, the top view is revolved any angle except 45°. The side view is revolved any angle desired (Fig. 12–5).

### Isometric Projection

The word *isometric* means "having equality of measure." This means that the object is placed with its major edges at equal angles with the picture plane. Because the edges are at equal angles to the picture plane, they are foreshortened equally.

An isometric projection can be found most easily by revolution (Fig. 12–6). The object is placed with the vertical axis toward the picture plane and the receding sides revolved until the diagonal of the object is perpendicular to the vertical plane. This places the front edges at an angle of 35°16' to the picture plane. The corners of the revolved object are then projected to the front view to form the isometric projection.

A projection of this type will meet the picture plane on an angle of 35°16'. The angles between the axes will be 120°. The edges of the object will be foreshortened to 82% of actual size.

ISOMETRIC

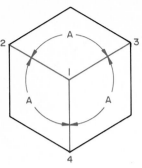

THREE EQUAL ANGLES
SIDE 1-2, 1-3, 1-4
EQUAL.

DIMETRIC

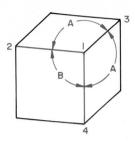

TWO EQUAL ANGLES
SIDE 1-2, 1-4 EQUAL.

TRIMETRIC

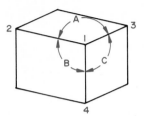

NO EQUAL ANGLES
NO EQUAL SIDES.

FIGURE 12–4   The three types of axonometric projections.

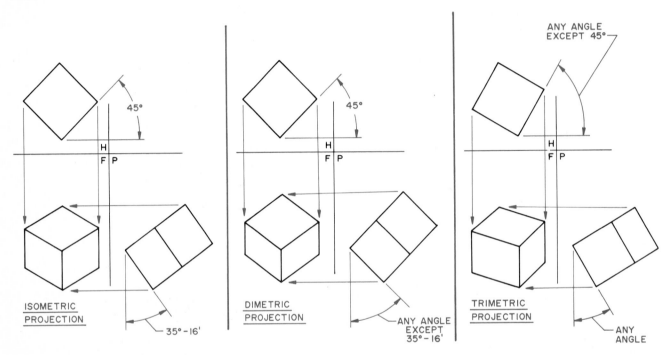

FIGURE 12–5   Axonometric projection follows the principles of orthographic projection.

The distances on an isometric projection parallel with the axis are measured with an isometric scale. It is 82% of a full-size scale. An isometric scale can be laid out as shown in Fig. 12–7.

## Isometric Drawing

The major difference between an isometric projection and an isometric drawing is the scale used to measure distances along the isometric axes. An *isometric projection* has equal axes, 120°, and uses an isometric scale for linear measuring. The scale is 82% of full size. An *isometric drawing* also has equal axes, 120°, but uses full scale for linear measurements (Fig. 12–8). It appears similar to an isometric projec-

tion but is not a true projection. It is slightly larger but for most purposes is just as useful. It has the advantage of using the same scale used to produce the original engineering drawing.

## Making an Isometric Drawing

An isometric drawing is built around isometric axes. The position of the axes used depends on how you want to view the object (Fig. 12–9). Position A shows the top, while B shows the bottom. Positions C and D are alternate positions.

The steps for laying out an isometric drawing are given in Fig. 12–10. After laying out the axes, it is helpful to draw very lightly an isometric block us-

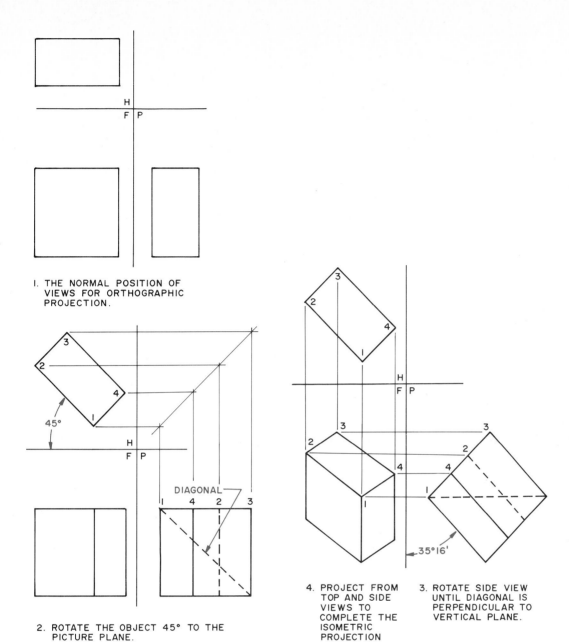

1. THE NORMAL POSITION OF VIEWS FOR ORTHOGRAPHIC PROJECTION.

45°

DIAGONAL

2. ROTATE THE OBJECT 45° TO THE PICTURE PLANE.

35°16'

4. PROJECT FROM TOP AND SIDE VIEWS TO COMPLETE THE ISOMETRIC PROJECTION

3. ROTATE SIDE VIEW UNTIL DIAGONAL IS PERPENDICULAR TO VERTICAL PLANE.

FIGURE 12–6  An isometric projection is developed by projecting from rotated orthographic views.

FIGURE 12–7  The technique for developing a scale for isometric projections.

FIGURE 12–8  Comparison of an isometric projection and an isometric drawing.

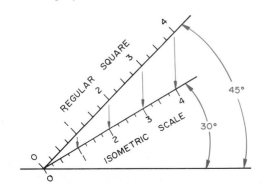

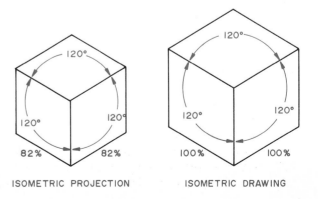

ISOMETRIC PROJECTION

ISOMETRIC DRAWING

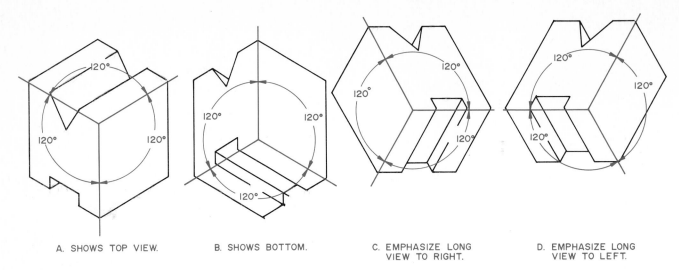

A. SHOWS TOP VIEW.　　B. SHOWS BOTTOM.　　C. EMPHASIZE LONG VIEW TO RIGHT.　　D. EMPHASIZE LONG VIEW TO LEFT.

FIGURE 12-9　The isometric axes can be positioned to emphasize specific parts of an object.

FIGURE 12-10　Laying out and completing an isometric drawing using an isometric block.

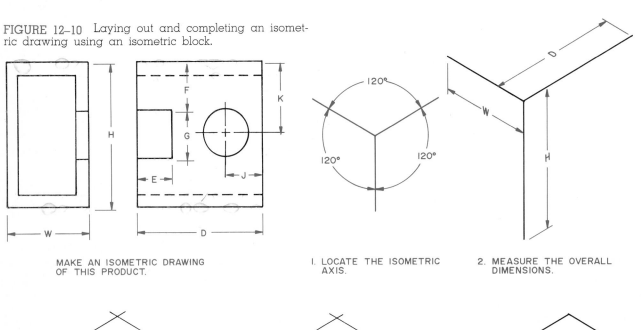

MAKE AN ISOMETRIC DRAWING OF THIS PRODUCT.

1. LOCATE THE ISOMETRIC AXIS.

2. MEASURE THE OVERALL DIMENSIONS.

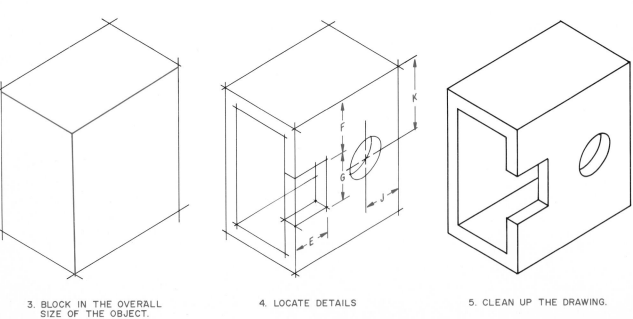

3. BLOCK IN THE OVERALL SIZE OF THE OBJECT.

4. LOCATE DETAILS

5. CLEAN UP THE DRAWING.

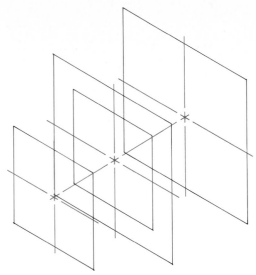

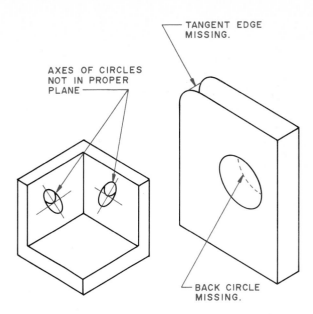

Assignment: Lay out a number of circular faces in parallel planes.

1. Locate the center line of each cylindrical face. If the four-center ellipse is to be constructed, draw isometric squares equal to the diameter of each circular face.

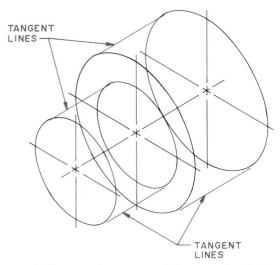

TANGENT LINES

TANGENT LINES

2. Construct the isometric circles on each axis. Draw the tangent lines connecting the circles.

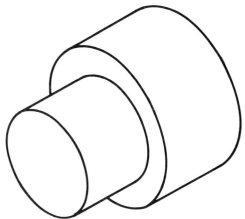

3. Remove center lines and unneeded lines to complete the isometric drawing.

FIGURE 12–11 Cylindrical objects can be laid out using the center lines.

TANGENT EDGE MISSING.

AXES OF CIRCLES NOT IN PROPER PLANE

BACK CIRCLE MISSING.

FIGURE 12–12 Common errors made when drawing circles and arcs in isometric.

ing the overall dimensions of the object. When laying out measurements, you must always measure parallel to an axis. Lines parallel to the axes are called *isometric lines*. *All measurements must be made along isometric lines.* After the basic block is drawn, begin to locate details. Notches are located by measuring parallel with the axes. Circular features, such as holes, are located by their center lines.

If the object consists of a number of circular faces lying in the same plane or in parallel planes, it is easier to use only the center lines for the layout (Fig. 12–11). First, locate the center lines, measuring distances along them. If the circles are to be drawn using the four-center method, isometric squares are drawn; if a template is used, only the center lines are needed. Finish by drawing the circles. Connect them with tangent lines parallel with the axis.

Three common errors made when drawing circles and arcs are omitting the tangent line, forgetting the back side of a hole, and getting the axis of the circle in the wrong plane (Fig. 12–12).

## Hidden Lines in Isometric

Hidden lines are generally not used on isometric drawings. Any pictorial drawing, because of its relationship to the picture plane, would have many hidden edges. The addition of hidden lines would confuse the drawing and reduce its effectiveness. They are used only when necessary to clarify some feature on a drawing.

## Angles in Isometric

A line on an angle is not parallel with the isometric axes. It is called a *nonisometric* line. Angles cannot be

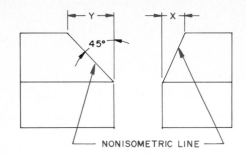

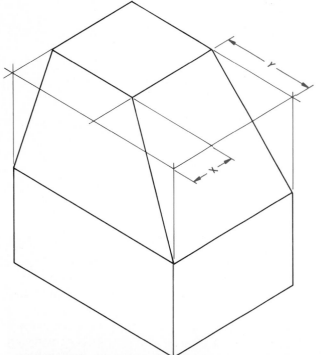

NONISOMETRIC LINE

FIGURE 12–13 To draw an angle (a nonisometric line), locate the ends of the line forming the angle.

laid out true size directly on the drawing. Because the object is inclined to the picture plane, the angles are not true size. There are special isometric protractors available that can be used for this purpose.

To draw a nonisometric line, locate each end by measuring on lines parallel with the axes (Fig. 12–13). Connect the ends to form the line.

## Circles and Curves in Isometric

Circles on isometric drawings appear elliptical. Arcs appear as parts of an ellipse.

Circles in isometric are made with isometric circle templates (Fig. 12–14). They are available in a wide range of hole diameters and are the fastest and most accurate way to draw circles. The template is located by lining it up with the center lines of the circle. Center lines are always used to locate circles, though they are sometimes removed from the finished drawing. They are used if it is necessary to indicate symmetry and for dimensioning. The center lines parallel the isometric axes (Fig. 12–15). The position of an isometric circle depends upon the plane on which it appears.

Isometric circles can be constructed by using an approximate four-center ellipse. This is based on the

FIGURE 12–14 An isometric circle template. (Courtesy of Berol USA/RapiDesign.)

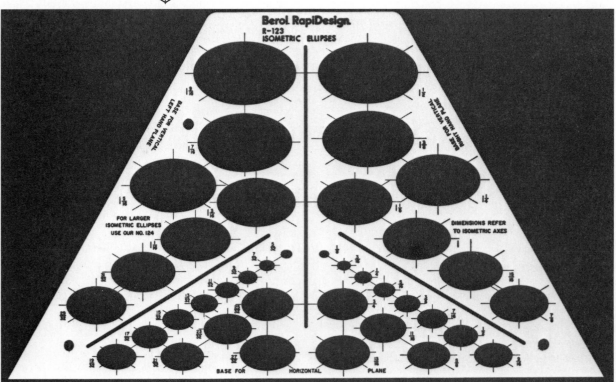

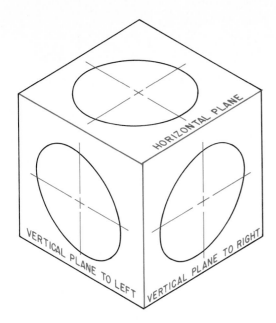

FIGURE 12–15 The center lines of isometric circles parallel the isometric axes.

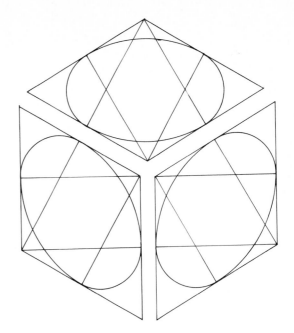

FIGURE 12–17 Drawing isometric circles in each isometric plane.

fact that a circle inscribed in a square will touch the square at its midpoints. The steps are shown in Fig. 12–16. Draw the square surrounding the circle in the desired isometric plane. Then, find the midpoint of each side. Connect the midpoints with perpendicular lines. They meet at the corners opposite them. This locates two of the centers. Where the perpendiculars cross are the other two centers. Draw arcs from these centers tangent to the sides of the square.

The construction for the four-center isometric circle in several planes is shown in Fig. 12–17.

This same technique can be used to draw arcs. It is not necessary to draw the entire isometric square. Lay out the parts needed to draw the arc. Remember, the radius is located perpendicular to the midpoint of the opposite side (Fig. 12–18).

Irregular curves in isometric are drawn by plotting points along the curve and locating these points using coordinates parallel with the isometric axes. The more points used, the more accurately the curve

FIGURE 12–16 Isometric circles can be constructed using the approximate four-center method.

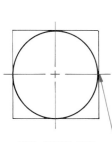

MID–POINT AND TANGENT POINT—

Assignment: Draw the above circle in isometric.

1. A circle inscribed in a square touches the square at the midpoint of each side.

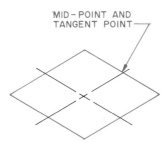

MID–POINT AND TANGENT POINT—

2. Draw an isometric square in the plane desired. The length of the sides equals the diameter of the circle. The center lines cross at the midpoint of each side.

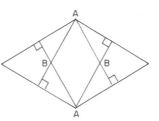

3. Draw perpendicular lines at each midpoint. They meet at the corners opposite each side. This forms two of the needed centers, A. The point at which these lines cross locates the other two centers, B.

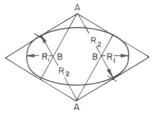

4. From center A, the radius is across the square to the midpoint of the opposite side. Draw the side arcs from one midpoint to the other. The midpoint is the point of tangency.

From B, set the radius to the midpoint and draw the ends of the circle.

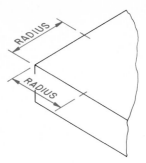
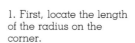
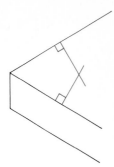
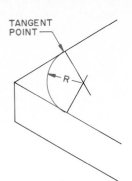
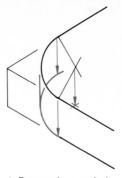

Assignment: Draw in isometric a corner with a known radius.

1. First, locate the length of the radius on the corner.

2. Erect perpendiculars to the points located by the radius.

TANGENT POINT

3. Draw the arc from the point of intersection. The perpendicular lines also locate the tangent point.

4. Project the needed points to the lower level. The distance is equal to the thickness of the part. The projection is parallel to an isometric axis.

FIGURE 12–18 Arcs in isometric can be drawn by locating the needed centers.

can be drawn (Fig. 12–19). The curve is formed by connecting the points with an irregular curve.

## Sections in Isometric

Sections are used to show interior detail. They help to clarify details that may be hidden in orthographic drawings. There are many types of sections. These are explained in Chapter 13.

To draw an isometric section, it is usually easier to block in the entire object. Then, staying parallel with the isometric axis, remove the parts needed to produce the section. A typical isometric half section is shown in Fig. 12–20.

## Dimensioning Isometric and Other Pictorial Drawings

The rules for dimensioning a pictorial drawing are the same as for multiview engineering drawings. Following are some general directions (Fig. 12–21 and 12–22).

1. Dimension lines and extension lines are drawn parallel with the axes.
2. If using the aligned system, letter dimensions parallel with the axes.
3. If using the unidirectional system, letter the dimensions vertically so that they read from the bottom of the drawing.

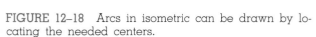
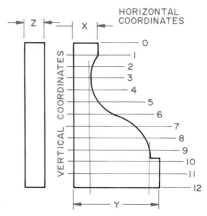
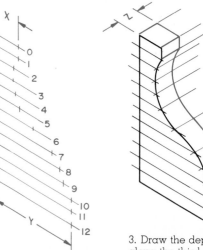
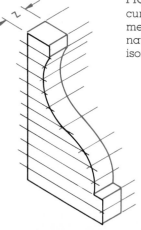

Assignment: Make an isometric drawing of the molding above, which contains irregular curves.

1. Divide the object into vertical and horizontal coordinates.

2. Draw the needed isometric axes. Then, locate the vertical and horizontal coordinates along the isometric coordinate lines.

3. Draw the depth along the third isometric axis. Project the coordinate lines parallel with the third axis. Measure the depth of the object, 2, on these lines. Connect the points located with an irregular curve.

FIGURE 12–19 Irregular curves can be drawn in isometric by projecting coordinates drawn parallel with the isometric axes.

262 Chapter 12

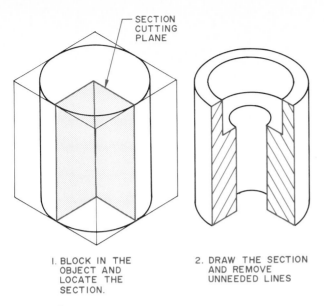

1. BLOCK IN THE OBJECT AND LOCATE THE SECTION.

2. DRAW THE SECTION AND REMOVE UNNEEDED LINES

FIGURE 12–20 An isometric section is drawn with the cutting planes parallel with the isometric axes.

4. Dimensions on sloped lines are lettered parallel with the dimension line.
5. Arrowheads are drawn so that they lie in the same plane as the extension line.
6. Notes are lettered parallel with the bottom edge of the paper.
7. Dimensions are lettered in the same plane as the feature they are describing.

FIGURE 12–21 Dimension lines on isometric drawings are drawn parallel with the isometric axes. The arrowheads, numbers, and letters are also drawn in isometric.

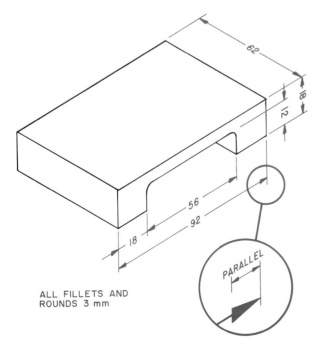

ALL FILLETS AND ROUNDS 3 mm

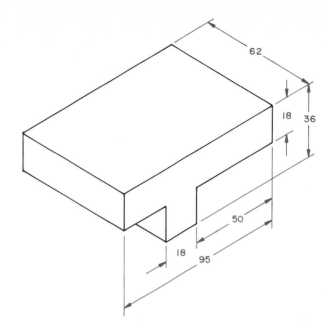

FIGURE 12–22 Dimensions on isometric drawings can be lettered parallel with the horizontal plane.

## Dimetric Projection

A dimetric projection is an axonometric projection of an object that has been positioned so that the angles of two of the axes are equal and greater than 90° and the third angle is different. Two of the axes are foreshortened equally, while the third is foreshortened by a different amount. This permits an object to be turned so that one face may receive additional emphasis (Fig. 12–23). A dimetric projection is projected in the same manner as an isometric projection (Fig. 12–6). The side view is revolved at any angle except 35°16′, which would produce an isometric.

FIGURE 12–23 In a dimetric projection, two of the angles of the axes are equal and greater than 90°, and the third is of a different size.

EACH PLANE SHOWS EQUALLY

FRONT PLANE GIVEN INCREASED EMPHASIS

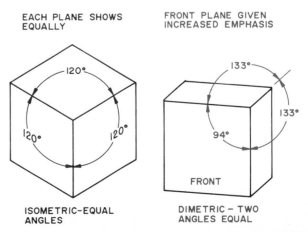

ISOMETRIC-EQUAL ANGLES

DIMETRIC – TWO ANGLES EQUAL

Pictorial Presentation of Engineering Designs 263

## Dimetric Drawing

Dimetric drawings are made using angles and scales that are approximately the same as those developed by true projection but are easier to draw. The production of a dimetric drawing includes choosing a suitable set of angles and scales for the foreshortened axes. If there are circular elements, the proper sizes of ellipses must be determined.

The selection of angles is made by deciding in what position the object is best shown. Some commonly used angles are given in Fig. 12–24. After the angles are selected, the scales for drawing the foreshortened axes can be found in the table.

After the angles and scale are developed, ellipses of the proper size must be selected for drawing the circular elements in each plane. They can be obtained from tables listing sizes calculated for various angles or from templates designed for producing dimetric drawings (Fig. 12–24).

When locating an ellipse, the major axis is perpendicular to the extension of the axis that is perpendicular to the surface upon which the ellipse is drawn (Fig. 12–25). Axis 1–2 is perpendicular to plane A. Extend axis 1–2 and draw a perpendicular to it. This is the direction of the major axis of the

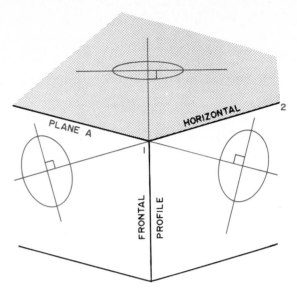

FIGURE 12–25 The major axis of a dimetric circle ellipse is found by constructing a perpendicular to the extention of the axis that is perpendicular to the surface upon which the ellipse is to be drawn.

ellipse. The other two planes are handled in the same way.

A typical dimetric drawing appears in Fig. 12–26.

FIGURE 12–24 Some commonly used axis angles, axis proportions, and circle ellipse guides for making dimetric drawings.

FIGURE 12–26 A typical dimetric drawing showing the angles of the axis, axis proportions, and circle ellipse template angles.

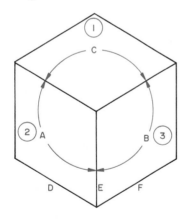

| AXIS ANGLES | | | AXIS SCALES | | | ELLIPSE GUIDES BY PLANES | | |
|---|---|---|---|---|---|---|---|---|
| A | B | C | D | E | F | 1 | 2 | 3 |
| 100 | 130 | 130 | 1 | 1 | ½ | 20° | 65° | 20° |
| 130 | 130 | 100 | 1 | ½ | 1 | 60° | 20° | 20° |
| 94 | 133 | 133 | 1 | ½ | 1 | 15° | 70° | 15° |
| 150 | 105 | 105 | ¾ | ¾ | 1 | 45° | 15° | 45° |
| 105 | 150 | 105 | ¾ | 1 | ¾ | 15° | 45° | 45° |

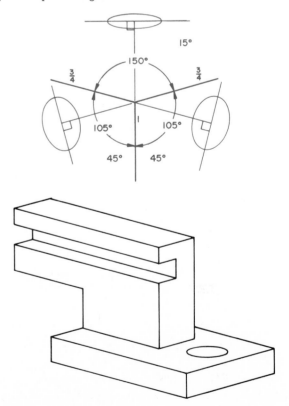

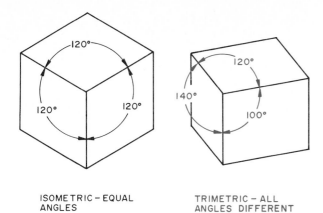

ISOMETRIC – EQUAL
ANGLES

TRIMETRIC – ALL
ANGLES DIFFERENT

FIGURE 12–27 In a trimetric drawing, all angles and axis scales are different.

## Trimetric Projection

A trimetric projection is an axonometric projection of an object that has been positioned so that all three axes make different angles with the picture plane. Since the axes are all on different angles, each axis is foreshortened by a different amount (Fig. 12–27).

Trimetric projections are constructed in the same manner as an isometric projection (Fig. 12–6). The top view is revolved at any angle desired. The side view is then revolved as needed to show the planes as desired.

FIGURE 12–28 Some commonly used axis angles, axis proportions, and circle ellipse guides for trimetric drawings.

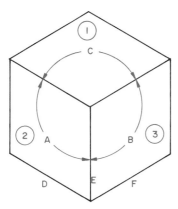

## Trimetric Drawing

Trimetric drawings are made using angles and scales that are approximately the same as those developed by true projection. These are altered slightly so that the drawings are easier to make. The production of a trimetric drawing includes choosing a suitable set of angles and scales for the foreshortened axes. If there are circular elements, the proper sizes of ellipses must be determined.

The selection of angles is made by deciding in what position the object is best shown. Some commonly used angles are given in Fig. 12–28. After the angles are selected, the scales for drawing the foreshortened axes can be found in the tables. This scale is shown as a proportion of true size.

After the angles and scales are developed, ellipses of the proper size must be selected for the circular elements in each plane. They can be found in the table in Fig. 12–28.

Review Fig. 12–25 for how to locate the ellipse on each plane.

A typical trimetric drawing appears in Fig. 12–29.

FIGURE 12–29 A typical trimetric drawing showing the angles of the axis, axis proportions, and circle ellipse template angles.

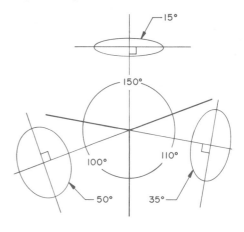

| AXIS ANGLES | | | AXIS SCALES (PERCENT OF FULL SIZE) | | | ELLIPSE GUIDES BY PLANES | | |
|---|---|---|---|---|---|---|---|---|
| A | B | C | D | E | F | 1 | 2 | 3 |
| 100 | 110 | 150 | 84 | 96 | 60 | 15° | 50° | 35° |
| 110 | 120 | 130 | 84 | 88 | 72 | 25° | 45° | 35° |
| 120 | 135 | 105 | 92 | 65 | 86 | 50° | 30° | 25° |
| 140 | 100 | 120 | 56 | 88 | 94 | 25° | 20° | 55° |

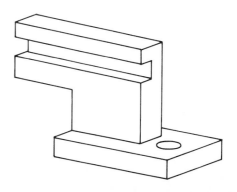

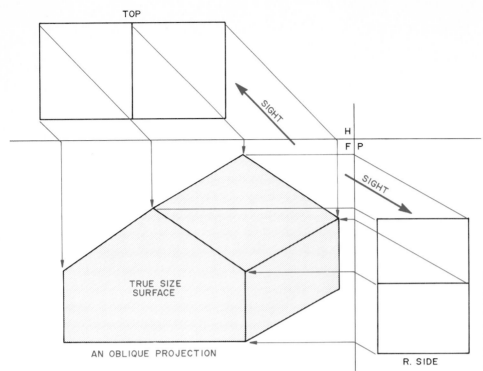

TOP

SIGHT

H
F P

SIGHT

TRUE SIZE SURFACE

AN OBLIQUE PROJECTION

R. SIDE

**FIGURE 12–30** An oblique projection is a pictorial view drawn using parallel projectors that are oblique to the picture plane. As the angle of the line of sight changes, the length of the elements on the receding axis changes.

## OBLIQUE PROJECTION

Oblique projection is a method of projecting an object to produce a pictorial view using parallel projectors drawn oblique to the picture plane (Fig. 12–30). The line of sight is oblique to the projection planes in the top and side views. Parallel projectors are drawn from the points at which the oblique projectors strike the projection planes. This locates points on the object forming the oblique projection. One face is parallel with the picture plane and appears true size and shape on it. The receding sides are foreshortened.

The amount of foreshortening depends on the angle of the line of sight. When the angle is 45°, the receding axis is the same scale as the horizontal and vertical axes. This is called a *cavalier oblique*. When the line of sight is 63°26′, the receding axis is half the horizontal and vertical axes. This is called a *cabinet oblique* (Fig. 12–31). All other lines of sight produce oblique drawings that are referred to as *general oblique*.

True oblique projection is seldom used. Simplified construction techniques that produce *oblique drawings* are more usual.

### Oblique Drawings

Oblique drawings produce a pictorial view very much like that produced by oblique projection. The techniques used are much simpler, and the pictorial produced is just as descriptive as an oblique projection.

The most important face of an object is placed parallel with the picture plane. It appears true size. For this reason, circular features are often placed in this position. They appear as true circles and can be drawn easily.

The receding axis can be drawn on any angle. Those most frequently used are 30° and 45°. The depth is measured along the receding axis. It can be full size or reduced, as desired (Fig. 12–32).

Circles appearing parallel with the picture plane are true size and can be drawn with a compass or template. Those in the other two planes appear as ellipses and are drawn using the four-center method (Fig. 12–33). A perpendicular is drawn to the center of each side of a rhombus having sides equal to the diameter of the circle. This locates four points used to draw the elliptical view of the circle (Fig. 12–34). The center lines may be left on the drawing if they aid clarity; otherwise, they are removed.

Oblique drawings are uniquely suited for showing objects that are primarily circular. Because the circular elements parallel with the picture plane appear true size, they are easy to draw. Each circular element is drawn by locating its center on an axis, which is a receding line (Fig. 12–35). Pictorial sections can also be drawn to show interior details (Fig. 12–36).

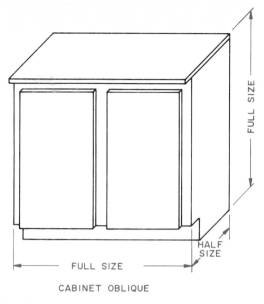

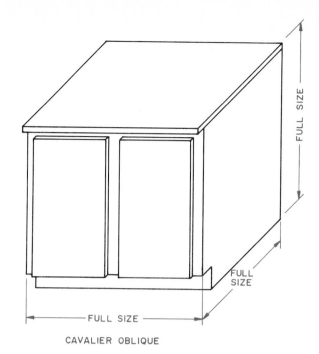

FIGURE 12–31  Cabinet and cavalier oblique drawings.

CABINET OBLIQUE

CAVALIER OBLIQUE

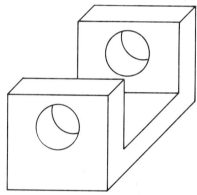

Assignment: Produce an oblique drawing of an object.

1. Lay out an oblique block the overall size of the object to be drawn.

2. Locate details on the block outline. The circles were located by their center lines.

3. Remove unneeded lines and darken the drawing.

FIGURE 12–32  An oblique drawing can have the receding axis on any convenient angle, and it can be drawn true size or to scale.

FIGURE 12–33  Circles on the planes forming an oblique drawing.

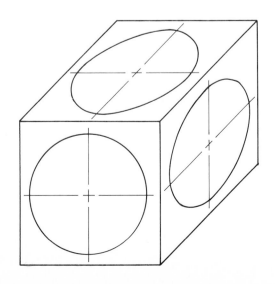

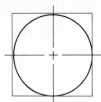
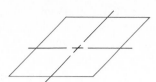

Assignment: Draw a circle in a receding plane on an oblique drawing.

1. Draw the center lines of the circle and a rhombus that circumscribes the circle.

**FIGURE 12–34** Circles on oblique drawings are laid out using the four-center ellipse method.

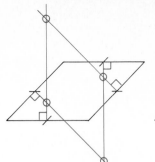
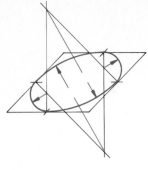

2. Draw perpendiculars to the midpoints of the sides of the rhombus. The points at which they intersect are the points for drawing the oblique circle.

3. Using the four points as centers, draw the oblique circle. The perpendicular lines are the tangent points.

**FIGURE 12–35** Circular elements are drawn by locating each along a center line.

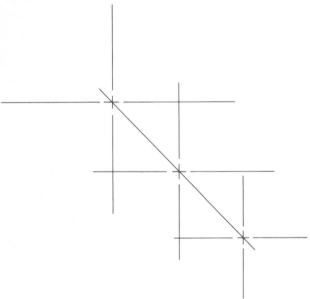

Assignment: Layout an object in oblique that has parallel circular surfaces.

1. Establish the location of the center line of each circular element.

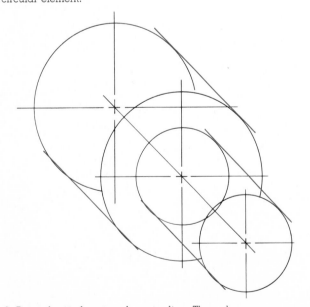
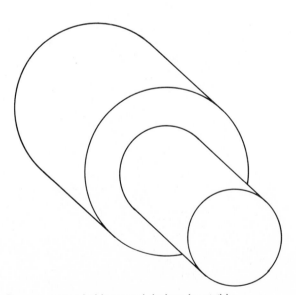

2. Draw the circles at each center line. Then, draw the receding sides tangent to the circles.

3. Remove unneeded lines and darken the visible lines.

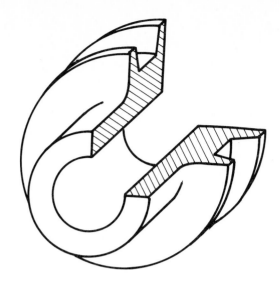

FIGURE 12-36  An oblique section drawing.

## PERSPECTIVE PROJECTION

The most realistic form of pictorial drawing is the perspective. This is because it allows for the reduction in size of an object as it recedes from the viewer. This is the image we are accustomed to seeing as we view a building or other object (Fig. 12-37).

Perspective drawings are used extensively in architectural design to portray the finished structure

FIGURE 12-37  A perspective is the most realistic of the various types of pictorial drawing. (Courtesy of Home Planners, Inc.)

in a realistic manner. The industrial designer also uses perspectives to illustrate proposed product designs and to illustrate advertising copy. Engineers will occasionally use perspective drawings and should understand the principles involved. They should also realize the possibilities presented by moving the elements of a perspective, permitting the product to be viewed from various positions (Fig. 12-38).

In addition to the drafting techniques shown in this chapter, computer drafting devices can produce perspectives rapidly and can show the product in an infinite variety of positions (Fig. 12-39). If an engineer understands the principles of perspective presented in this chapter, the transition to computer procedures will be enhanced. Additional information on computer drafting is given in Chapter 23.

### Principles of Perspectives

A perspective has four main elements: (1) the point from which the product is viewed, called the station point; (2) the product itself; (3) the plane upon which the product is projected, called the plane of projection or picture plane; and (4) the projectors, called visual rays, which run from the station point to the product. The picture plane is placed between the station point and the product being observed (Fig. 12-40). The points where the visual rays pierce the picture plane locate points on the perspective.

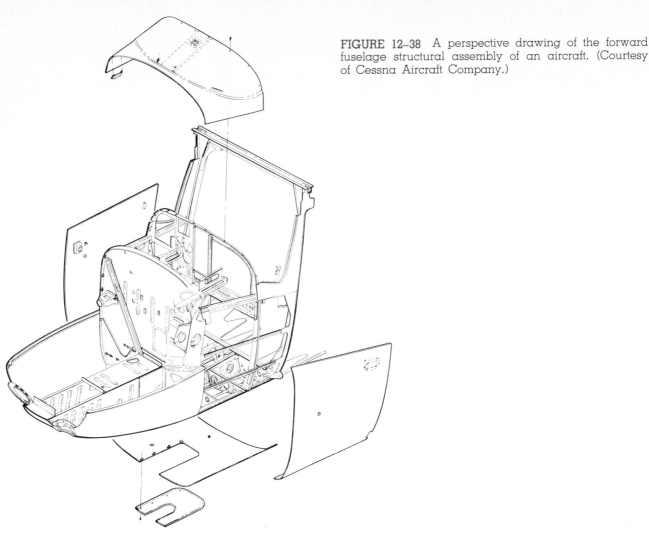

FIGURE 12–39  Perspectives and other pictorial drawings can be drawn with computer drafting systems. This is being drawn on a digital plotter. (Courtesy of Houston Instrument Division, Bausch and Lomb.)

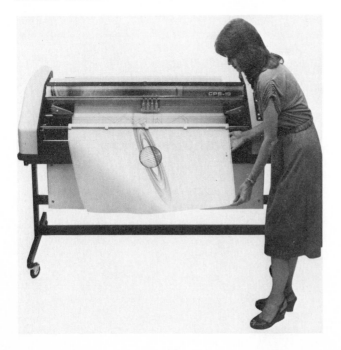

The points where the visual rays pierce the picture plane are perspectives of the original points. The *center of vision* is a point located directly in front of the viewer in a horizontal plane. The center of vision is located at the center of the area on the product that is of greatest interest to the viewer.

In Fig. 12–41, the perspective of the object is shown as it pierces the picture plane. If the object were farther from the picture plane, it would appear smaller. If it were on the other side of the picture plane, it would appear larger than true size. If the object touches the picture plane, it would appear true size. The perspective view is shown in Fig. 12–42.

A two-point perspective is shown in Fig. 12–43. The visual rays from the object converge at the *station point*, which is the location of the person viewing the object. This is noted on drawings as *SP*. The points at which the visual rays pierce the picture plane locate points on the perspective. The object rests on the *ground plane*. It intersects the picture plane at a *ground line*. This is usually noted on drawings as *GL*. The *horizon* is a line in the distance at

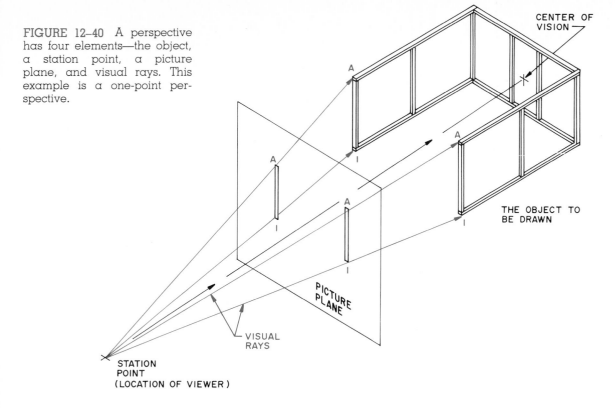

FIGURE 12–40 A perspective has four elements—the object, a station point, a picture plane, and visual rays. This example is a one-point perspective.

FIGURE 12–41 In a perspective, the farther an object is behind the picture plane, the smaller it appears on the drawing. This example shows projections for a one-point perspective.

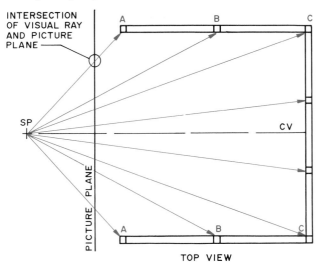

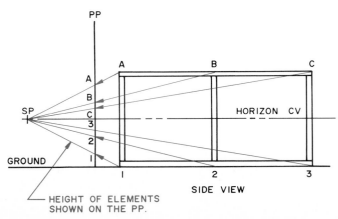

which the receding edges of the object appear to meet. Theoretically the horizon is at infinity. It is the same height above the ground line as the height of the viewer's eye. The *vanishing points* are where the receding edges meet the horizon. The actual perspective drawing appears on the picture plane. This is represented by the drawing paper.

The object can be placed on, above, or below the horizon. The choice depends upon where it will show the object to best advantage (Fig. 12–44). The

FIGURE 12–42 The perspective of the frame wall in Fig. 12–40. This example is a one-point perspective.

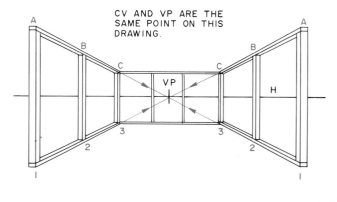

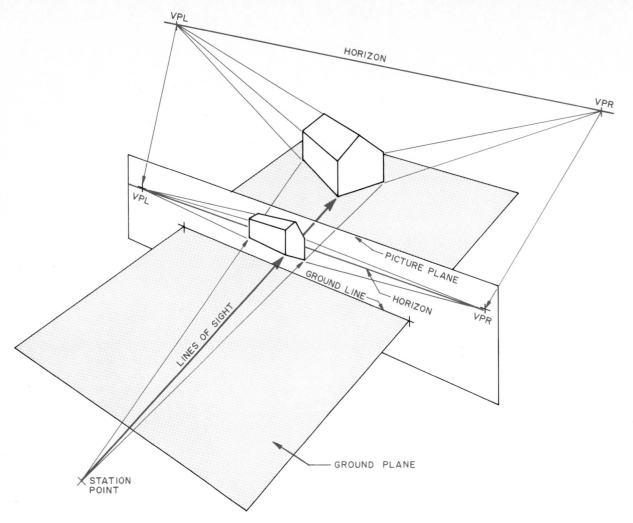

FIGURE 12-43 Terms used to describe the parts of a perspective layout. This is a two-point perspective.

FIGURE 12-44 The object is positioned so as to produce the best illustration of the product.

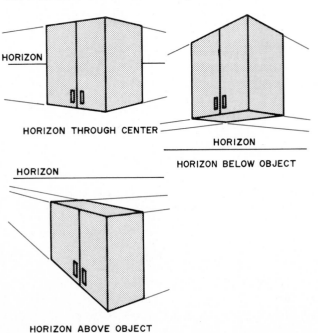

station point can also be moved. As it is moved, the view of the object will vary. It can be moved to the right or left of center (Fig. 12-45).

## Types of Perspectives

Perspective drawings may be classified into three types, according to the number of vanishing points required. If the object is placed so that the face or front view is parallel with the picture plane and the drawing has one vanishing point, it is a *one-point* or *parallel perspective* (Fig. 12-46). One-point perspectives are used extensively for room interiors (Fig. 12-47).

If the object is placed so that its faces are at an angle to the picture plane and the drawing has two vanishing points, it is a *two-point* or *angular perspective*. All planes of a two-point perspective recede to a right or left vanishing point. Vertical lines, while foreshortened because of the perspective, remain vertical and parallel with the picture plane. This type of perspective is widely used for architectural drawings and engineering illustrations (Fig. 12-48). The

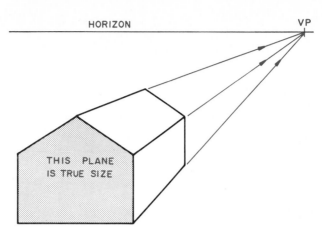

**FIGURE 12–46** The elements in a one-point perspective flow to a single vanishing point.

house in Fig. 12–37 is drawn in two-point perspective.

If the object is placed so that no series of parallel edges is parallel to the picture plane and the drawing has three vanishing points, it is a *three-point perspective* (Fig. 12–49). All principal lines vanish to one of these three vanishing points. Horizontal lines

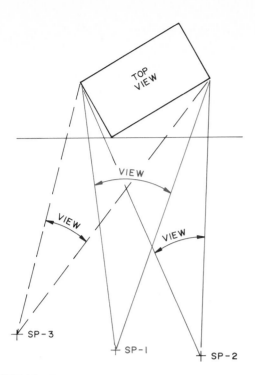

**FIGURE 12–45** The position of the station point determines which surfaces receive the greatest emphasis.

**FIGURE 12–47** A one-point perspective focusing attention on the house entrance. (Courtesy of Home Planners, Inc.)

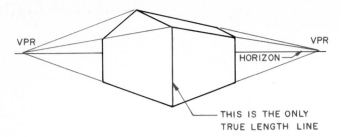

FIGURE 12-48 The elements in a two-point perspective flow to right and left vanishing points.

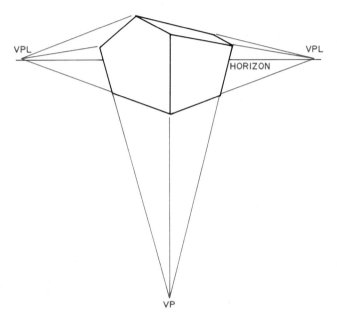

FIGURE 12-49 A three-point perspective flows to three vanishing points.

vanish right or left to a horizon. Vertical lines vanish to a third vanishing point above or below the horizon. This type of perspective is used when drawing large structures that are viewed from high above or far below. It has some use for large buildings and other architectural products but has limited use in general engineering.

## Location of the Picture Plane

The picture plane is placed between the object and the station point. When drawing a parallel perspective, the plane is generally passed through the face of the object, which then appears true size and shape (Fig. 12-50). In angular perspective, a corner of the object is usually located touching the plane. This gives a true-length line that is useful in locating other elements of the perspective.

## Location of the Station Point

The location of the station point is important because it influences the appearance of the finished

perspective drawing. The location chosen depends upon the surfaces of the object that are to be stressed. For most perspectives, the station point is located off center so that some of the end of the product can be seen. Generally, it is located above or below the center of the object. This permits some of the top or bottom to be seen.

The distance between the station point and the picture plane is also important. If it is too close, the object will appear distorted. Generally, this distance should be equal to at least twice the maximum dimension (width, height, or depth) of the object.

The angle of viewing must also be considered. This is the angle in degrees between the visual rays to the extreme sides of the object. A wide angle of view does not provide a natural perspective. If the angle of vision is kept to 30° or less, the perspective will be more pleasing.

## Locating the Object

The object to be drawn as an angular perspective should be at an angle to the picture plane. It should be placed so that the faces to be seen do not make the same angle with the plane. Angles commonly used are 15°, 30°, and 45°.

When drawing a parallel perspective, the object is placed parallel with the picture plane and usually touching it (Fig. 12-50).

## Drawing a One-Point Perspective

The one-point perspective has one face parallel with the plane of projection. The receding edges converge toward a single vanishing point on the horizon. To draw a one-point perspective, draw the top view of the object touching the picture plane (Fig. 12-50). This face will be true size on the picture plane. Then, locate the station point. Its location is important because it can be used to emphasize one face. It is normally to one side so that the front and side of the object are seen. The station point should be away from the picture plane a distance of at least two or three times the width of the object.

Next, locate the ground line on the area where the perspective is to be drawn. Draw the front view on the picture plane by projecting from the top view.

Locate the horizon. Its position is determined by the desired view of the object (Fig. 12-51). Notice that as the horizon is moved above center, more of the top of the object is visible. As it moves below center, the top is not visible and only the front and side can be seen. When drawing a one-point perspective of something, as a room interior, locate the horizon so that the surfaces that should be emphasized are shown.

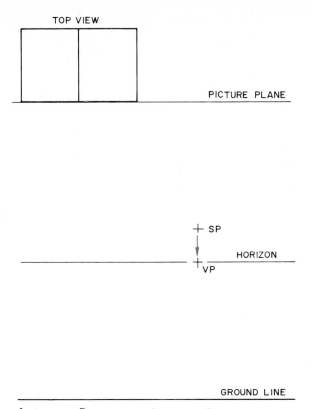

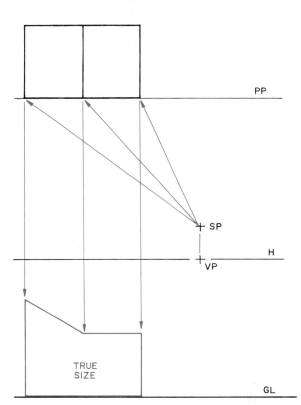

Assignment: Draw a one-point perspective.

1. First, locate the top view of the object parallel with and touching the picture plane. Locate the station point as desired. In the area where the pictorial drawing is to appear, locate the ground line and horizon. Then, project the station point perpendicular to the horizon to find the vanishing point.

2. Project the front view to the picture-plane area. It is true length and height.

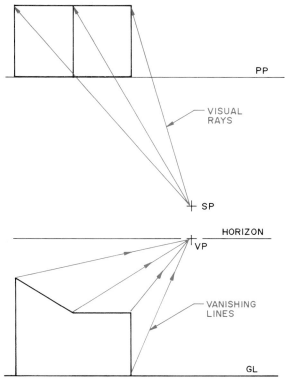

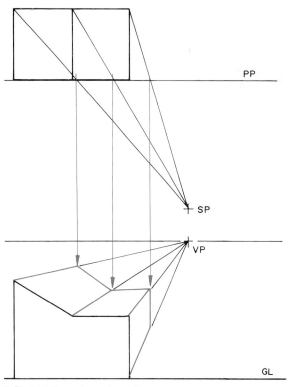

3. Draw visual rays from the station point to the other corners of the object. Draw vanishing lines from each corner of the object to the vanishing point.

FIGURE 12–50  Drawing a one-point perspective.

4. Project the points where the visual rays cross the picture plane in the top view to the pictorial view. Where they intersect, the lines leading to the vanishing point locate points on the object. Connect the points to form the one-point perspective view.

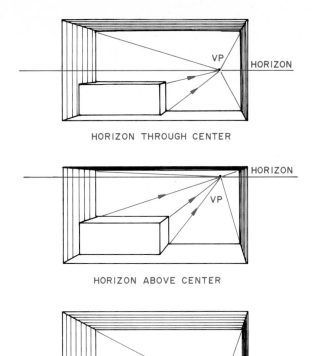

HORIZON THROUGH CENTER

HORIZON ABOVE CENTER

HORIZON BELOW CENTER

FIGURE 12–51 The location of the horizon influences the appearance of the one-point perspective.

Locate the vanishing point. It is found by drawing a perpendicular from the station point to the picture plane.

Draw visual rays from the corners of the front view to the vanishing point. Then, draw visual rays to each corner of the top view of the object from the station point. The points at which they cross the picture plane are projected to the pictorial view. The point at which they intersect with the vanishing line is a corner on the pictorial. As corners are located, connect them to complete the pictorial.

## Drawing a Two-Point Perspective

The steps for drawing a two-point perspective are shown in Fig. 12–52. First, draw the edge view of the picture plane. Then, draw the top view of the object at an angle to it. Let the corner touch the picture plane. This corner is the only true-length line on the perspective.

Next, locate the station point so that the entire object is within the 30° visual angle. If it is not, move the station point away from the object.

Draw the ground line and the horizon. The distance between them equals the distance of the viewer's eye from the ground. Then, find the vanishing points. To do this, draw lines parallel with the sides

of the top view from the station point to the picture plane. Where they meet the picture plane, project them perpendicularly to the horizon. This locates the vanishing points on the horizon.

Project the corner touching the picture plane in the top view to the pictorial view. Measure its true height. Draw vanishing lines to each vanishing point from this line. Then, from the station point, draw visual rays to each corner of the object in the top view. Project the point where they cross the picture plane to the pictorial view. The points at which these projectors cross vanishing lines are corners of the object. Locate all corners in this manner to complete the two-point perspective.

## Multiview Perspective

As shown in the preceding section, a perspective is formed on the picture plane by visual rays from the observer through the picture plane to the object. Applying this basic principle, a perspective can be drawn using orthographic projection. It is not used extensively because, while it is easy to understand, it involves the use of many projection lines, which tend to complicate the drawing.

To draw a perspective using orthographic projection, draw the top and side views, placing the top view on the desired angle and projecting it to form the side view. Locate the picture plane and station point in each view. The location chosen for the station point depends on the desired view of the object. From each station point, draw visual rays to the object in that view. The points at which these pierce the picture plane project to the perspective view. The point at which the projections of a point from the top and side views cross locates a point on the perspective. For example, point B in the top view and side view is projected from the edge view of the picture plane to the perspective drawing (Fig. 12–53).

## Circles and Curves in Perspective

When a circle is parallel with the picture plane, it appears on the perspective as a circle. When a curve is parallel with the picture plane, it appears true size and shape. If the circle or curve is inclined to the picture plane, its perspective will resemble an ellipse. The circle in perspective is constructed by dividing it into several parts—usually, 30° and 60° angles are sufficient. Locate these on the top view of the object. Project each point to the perspective. The point at which the projectors cross is a point on the circle in perspective (Fig. 12–54). The same technique is used to draw curves in perspective (Fig. 12–55).

When a circle is parallel with the plane of projection and touches the plane, it is true size and can

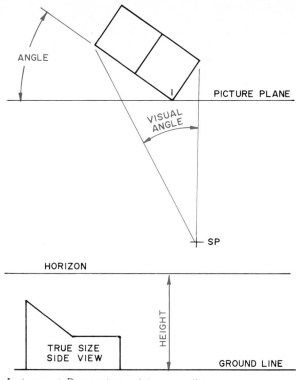

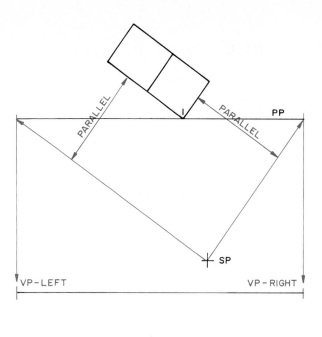

Assignment: Draw a two-point perspective.

1. First, locate the top view of the object at an angle to the picture plane. Then, locate the station point. Check to see that the visual ray is 30° or less. Locate the ground line and horizon. The higher the horizon is above the ground line, the more the top of the object will be seen.

2. Find the vanishing points. From the station point, draw lines parallel with the two sides of the object. The points at which they cross the picture plane are projected perpendicular to the horizon. This locates the two vanishing points.

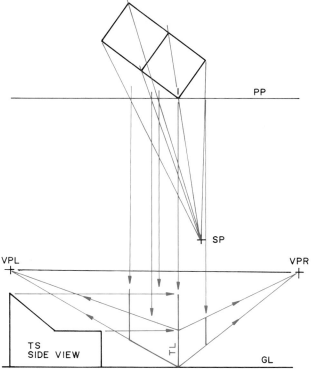

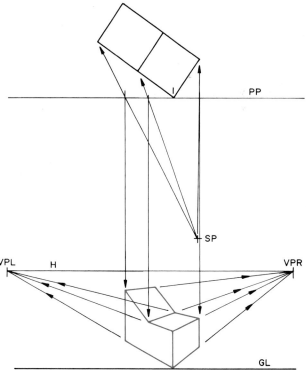

3. Project corner 1 to the pictorial view. It is true length because it is touching the picture plane. Draw vanishing lines from it to the vanishing points. The true height can be projected from the side view if it is drawn as shown.

Next, draw visual rays to the corners of the top view. The points at which they cross the picture plane are projected to the pictorial view. The points at which they meet the vanishing line are points on the object.

4. Project all the corners of the object to the pictorial view. Connect the corners to complete the two-point perspective.

FIGURE 12–52  Drawing a two-point perspective.

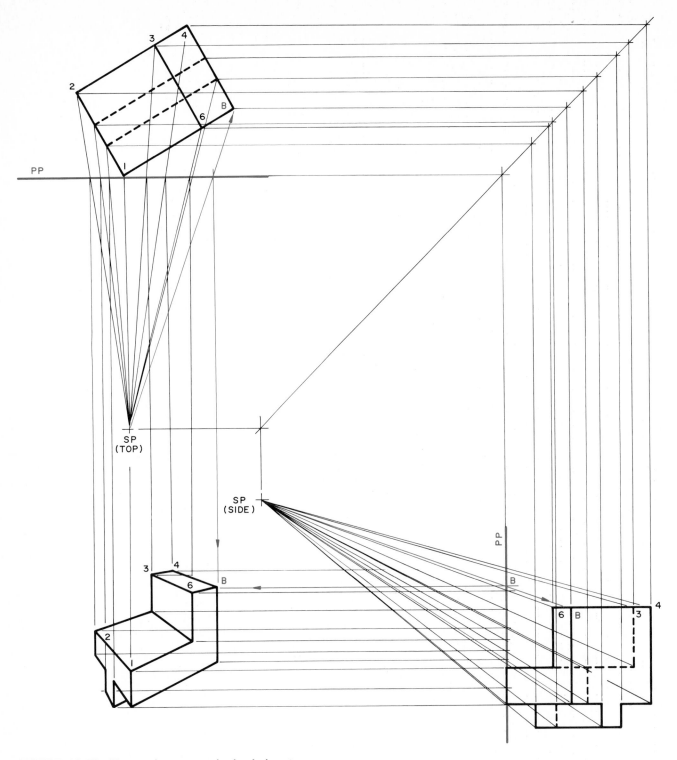

FIGURE 12–53 The multiview method of drawing a perspective.

be drawn with a compass. When it is behind the plane, it will still appear as a circle, but the diameter must be found (Fig. 12–56). The diameter is found by projecting it to the picture plane and drawing the circle with a compass.

## Vanishing Points of Inclined Lines

The perspective of an inclined line can be found by locating on the perspective the end points of the line and connecting them. This is usually accomplished

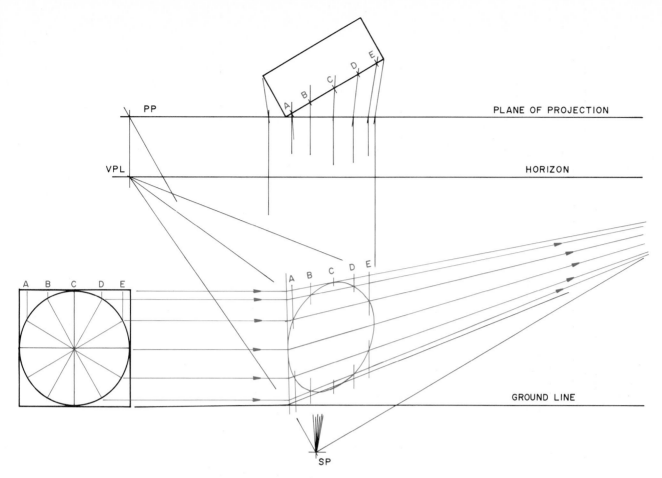

FIGURE 12-54 Drawing a circle in perspective.

FIGURE 12-55 Drawing a curve in perspective.

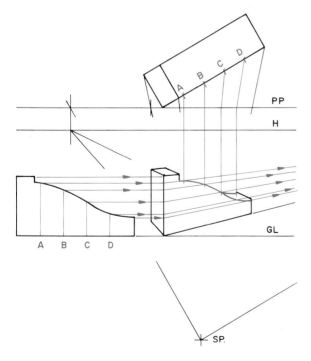

FIGURE 12-56 Sizing circles parallel with the picture plane in a perspective.

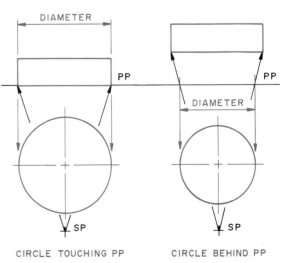

CIRCLE TOUCHING PP                    CIRCLE BEHIND PP

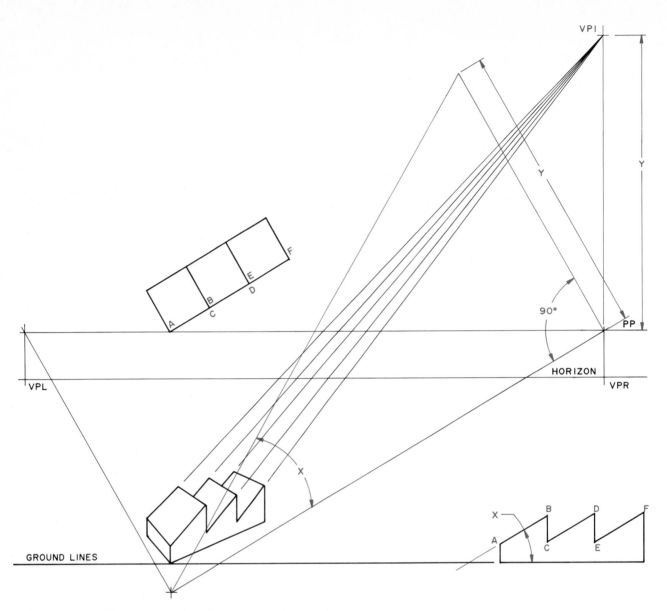

FIGURE 12–57 Finding the vanishing points for inclined lines.

by finding the intersection of two horizontal lines at
the end of the inclined line.

If a product has several inclined lines vanishing
to the same vanishing point, it is more accurate and
often faster to find a vanishing point for the inclined
line itself.

The vanishing point for an inclined line can be
established by finding the *piercing point* in the picture
plane of a line through the *station point* parallel with
the inclined line. The principle involved is that vertical planes vanish in vertical lines and a line in a
vertical plane will vanish at a point on the vanishing
line of the plane.

The construction is shown in Fig. 12–57. Lines

AB, CD, and EF are inclined lines. They lie in the
same plane as other planes that vanish to the right.
Therefore, the vanishing point of the inclined line
(VPI) will lie on a vertical line through the right vanishing point (VPR). The VPI is the point at which a
line drawn through the station point (SP) and parallel to the inclined lines pierces the plane of projection.

The distance of the VPI from the VPR is found
by constructing a right triangle with the 90° angle at
the VPR. The angle from the SP (X) is equal to the
slope angle of the inclined line. The short leg of the
triangle produced (Y) is the distance of the VPI
above the VPR. This locates the VPI.

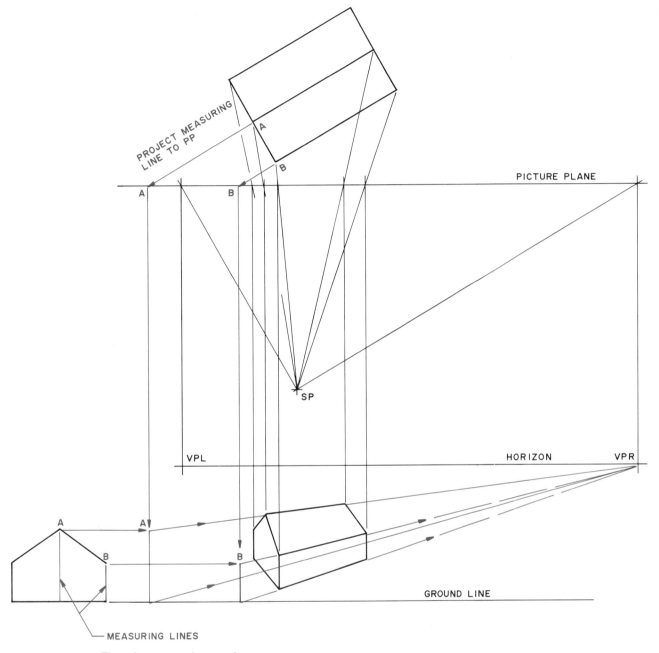

FIGURE 12–58 The element to be used as a measuring line is projected to the picture plane.

## Measuring Lines

When the vertical front corner of an object is placed touching the plane of projection, it appears true length on the perspective and serves as a measuring line. When the vertical edge lies behind the plane, the heights are found by using a measuring line. A *measuring line* is a true-length line from which true-height measurements can be taken. To develop a measuring line, project the vertical edge on the angle of the perspective until it intersects the picture plane. Project this to the perspective view; it is true length at that location (Fig. 12–58). The height can be projected to the proper vanishing point. The sections of these lines needed on the perspective are found by the usual method of sighting the object from the station point.

## Measuring Points

Measuring points are used on a perspective drawing when there are a series of measurements, such as a

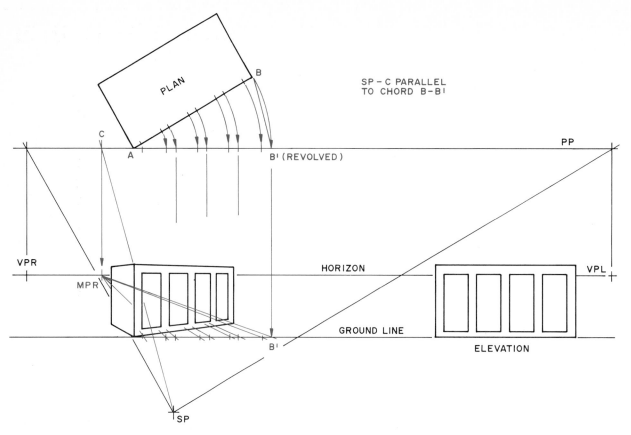

FIGURE 12–59 When using measuring points, the edge of the object is rotated parallel with the picture plane.

row of columns. This avoids the confusion of the many projection lines required when using the procedure presented earlier.

*Measuring points* are special vanishing points used to establish distances along lines in the perspective. To find them, first rotate the face of the object until it is parallel with the picture plane. This permits direct measurements to be projected to the horizontal ground line. Then, find the measuring point by drawing a line from the SP parallel to cordal distance BB' created by the revolution. At the point at which this pierces the picture plane, drop a perpendicular to the horizon. This locates the measuring point for the vertical divisions. Project these divisions to the measuring point. The points at which they cross the horizontal base locates the vertical elements (Fig. 12–59).

## Perspective Drawing Aids

One system uses a series of charts. The charts permit the selection of 116 positions with the object

tilted from 0° to 85° and rotated 0° to 85° (Fig. 12–60). The illustrator selects the chart desired and positions it on the drawing board. Using special perspective guides and the Perspective Science Instrument, the perspective can be drawn (Fig. 12–61).

Another aid to drawing perspectives is the perspective grid. A variety of grids gives a choice as to the position and surfaces emphasized. The grids are usually printed on heavy paper (Fig. 12–62). The drafter places the drawing sheet over the grid, and the lines show through (Fig. 12–63). For making quick freehand perspective sketches, perspective graph paper is available.

Another perspective drawing device is shown in Fig. 12–64. It includes a drawing surface with a variety of scales and a special T-square. The head of the T-square slides in a curved groove in the left side of the board. The top edge of the T-square is used to project and draw lines. Vertical lines are drawn by placing the head of the T-square on the top of the board.

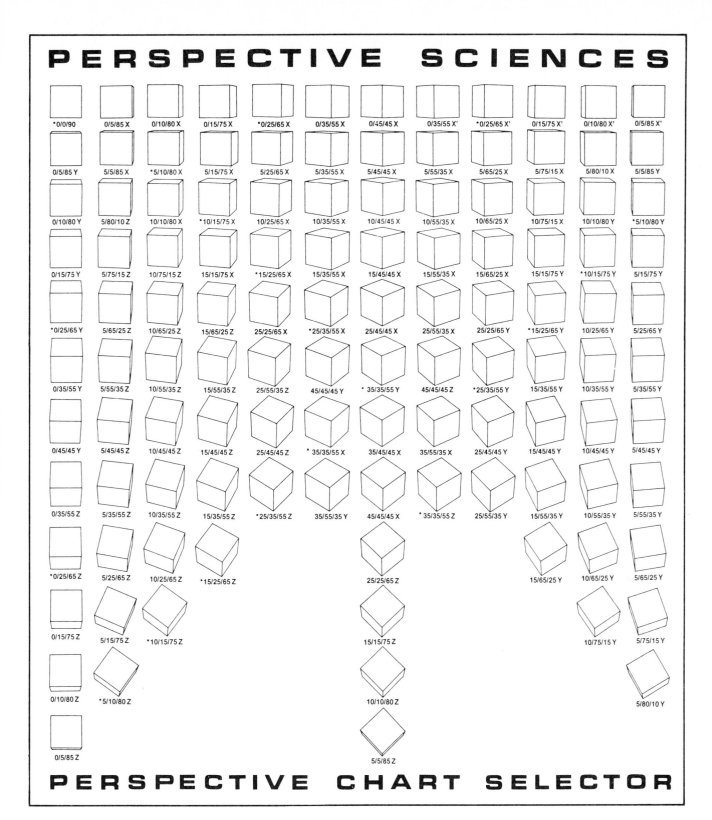

FIGURE 12–60 The Perspective Sciences system permits the illustrator to select from 116 positions for a perspective. This illustration shows the view from the top of the object. By turning the page upside down, 116 worm's-eye views are available. (Courtesy of Perspective Sciences.)

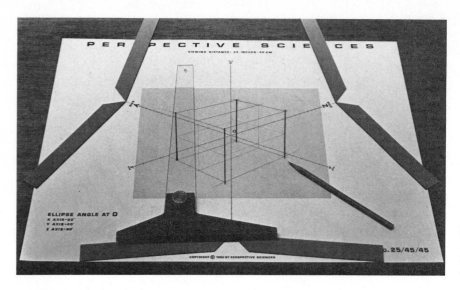

FIGURE 12–61 The components of a perspective drawing system. (Courtesy of Perspective Sciences.)

FIGURE 12–62 Perspective grids. (Courtesy of Graphi-Craft.)

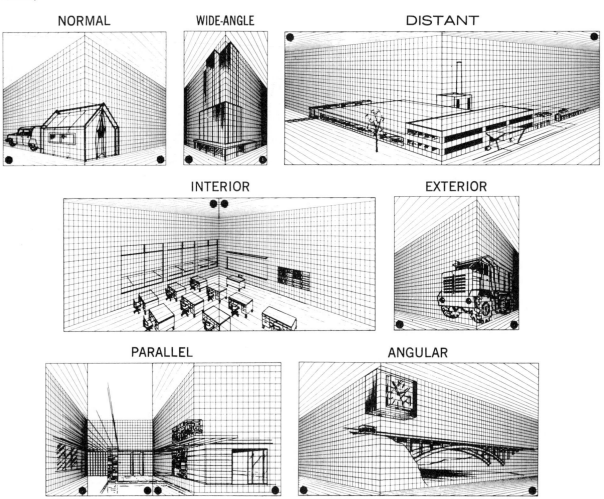

NORMAL

WIDE-ANGLE

DISTANT

INTERIOR

EXTERIOR

PARALLEL

ANGULAR

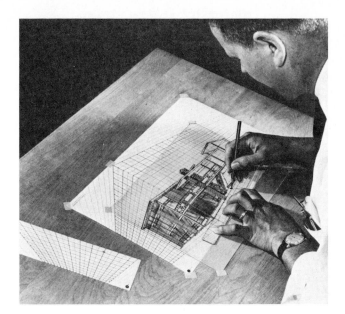

FIGURE 12–63 The illustrator places the grid below the drawing paper and uses it as a guide for drawing the perspective. (Courtesy of GraphiCraft.)

FIGURE 12–64 The Klok Perspector permits the drawing of perspectives with a minimum of training. (Courtesy of the Utley Company.)

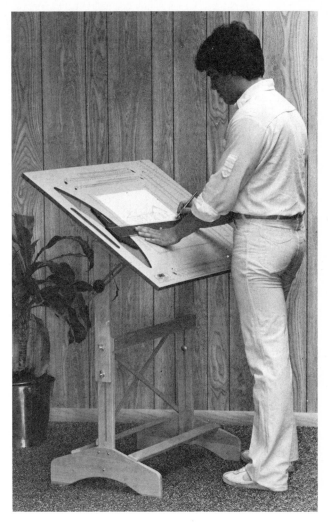

# Problems

Some problems are printed over a grid of ¼-in. squares. The size of the various parts can be found by counting the squares. Estimate those sizes falling between squares as well as all fillets and rounds. Plan your work so that the solutions are balanced on the page.

Solutions are to be on 8½ × 11 in. coordinate paper with ¼-in. squares or plain paper having a ½-in. (12-mm) border on all sides and a ½-in. (12-mm) title strip laid out as follows:

1. Make freehand or instrument isometric drawings of the objects in Prob. 12–1. Select axes that you think will most clearly show the object. With careful planning you can get two or three per sheet.

2. Make freehand or instrument isometric drawings of the objects in Prob. 12–2. Select axes that you think will most clearly show the object. With careful planning you can get two or three per sheet.

3. On the top of a sheet, develop a scale for laying out isometric projections in inches. Below this draw Prob. 12–3 as an isometric projection using the scale developed. Draw the object twice—once with the top visible and again with the bottom visible. These will fit on one sheet.

4. Draw the object in Prob. 12–3 as an isometric projection using revolution. Position it so that the top is seen. Check the sizes on this drawing with those in Prob. 12–3. They should be exactly the same, 82% smaller than the original object.

5. Draw Prob. 12–4 as an isometric projection viewing it from the top and a second drawing with the axes positioned to emphasize the long view to the right. Draw them twice the size given. Both drawings will fit on one sheet.

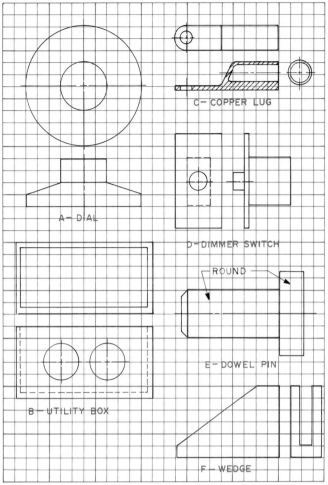

PROBLEM 12–1

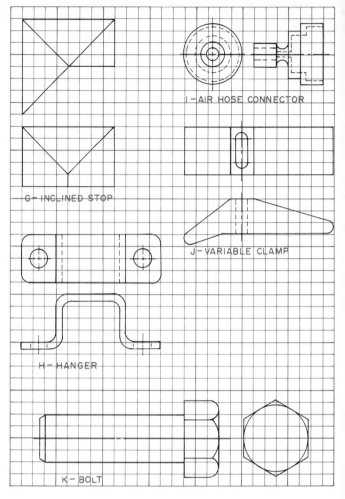

PROBLEM 12–2

6. Draw the object used in Prob. 12–4 as an isometric projection using the method of revolution. Position it so the top is seen.

7. Make freehand pictorial sketches of the products in the photos in Prob. 12–5. Estimate the sizes and keep them in proper proportion.

8. Using Probs. 12–6 through 12–37, make isometric, dimetric, trimetric, or oblique drawings as assigned by your instructor. Place on 8½ × 11 in. or 297 × 420 mm paper with a ½-in. (12-mm) border on all sides and a ½-in. (12-mm) title strip as used on earlier problems. Dimension if required by your instructor. Select axes that show the object in the clearest position.

9. Draw Probs. 12–7, 12–13, 12–21, and 12–27 as isometric half sections.

10. Draw Probs. 12–13, 12–15, 12–18, 12–25, 12–27, and 12–28 as one-point perspectives. Locate the vanishing point where it most clearly shows the object. Draw the product twice the given size if your instructor approves.

11. Use Prob. 12–38 to produce a one-point perspective.

12. Draw Probs. 12–39 through 12–41 as two-point perspectives. Locate the station point and vanishing point where it shows the object to best advantage. It is possible that vanishing points may fall off the sheet. If they do tape a piece of paper at the location so the points can be noted. The top view can also be placed on a separate paper and taped to the drawing board above the perspective drawing sheet. Use 11 × 17 in. or 297 × 420 mm paper or a larger size. Put on a ½-in. or 12-mm border and title strip.

13. There are many problems in Chapter 6, Principles of Orthographic Projection, that can be used for pictorial drafting experience. Complete those assigned by the instructor.

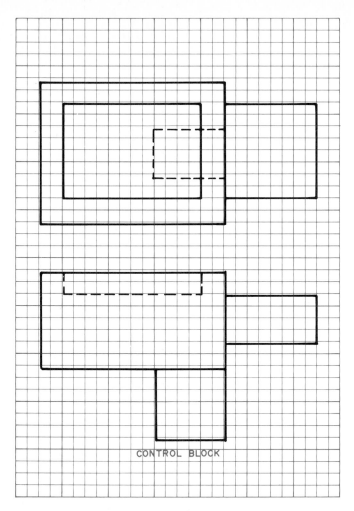

CONTROL BLOCK

PROBLEM 12-3

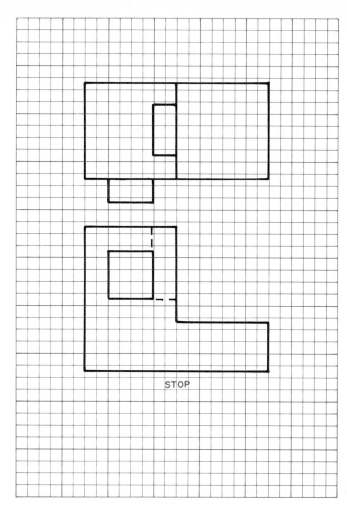

STOP

PROBLEM 12-4

PROBLEM 12-5

(Courtesy Tektronix)

(Courtesy Kroy, Inc.)

(Courtesy Caterpillar Tractor Co.)

(Courtesy General Electric)

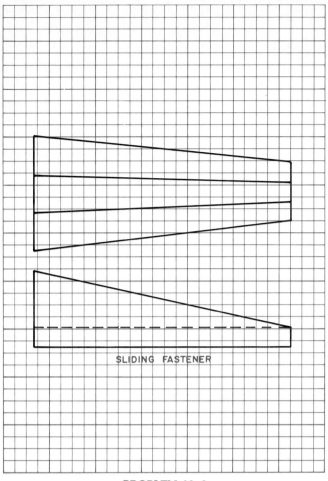

SLIDING FASTENER

PROBLEM 12–6

PROBLEM 12–7

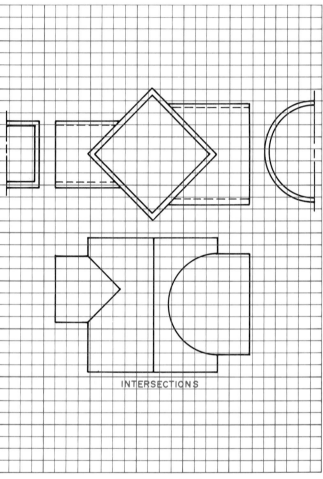

INTERSECTIONS

PROBLEM 12–8

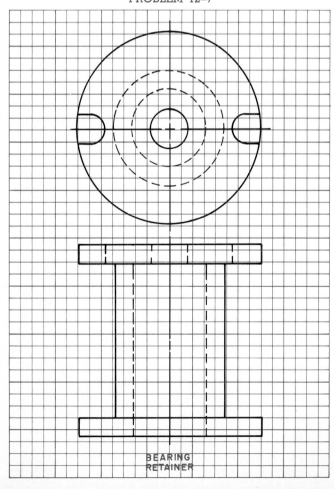

BEARING
RETAINER

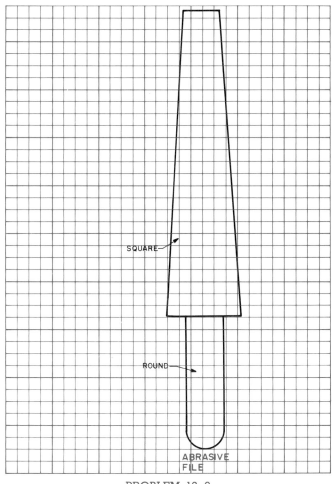

PROBLEM 12–9

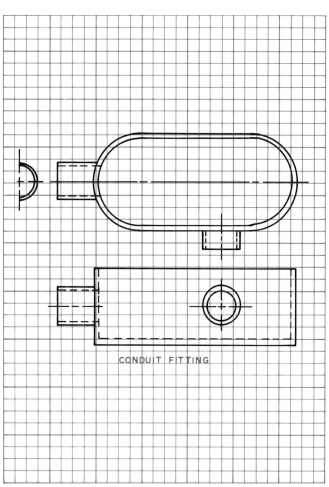

PROBLEM 12–10

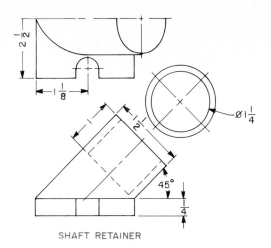

SHAFT RETAINER

PROBLEM 12-12

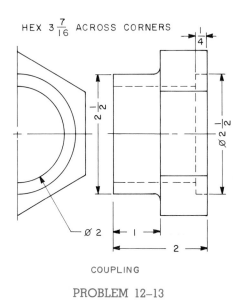

COUPLING

PROBLEM 12-13

PROBLEM 12-14

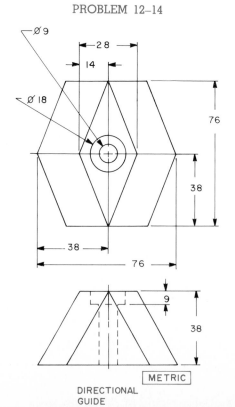

DIRECTIONAL
GUIDE

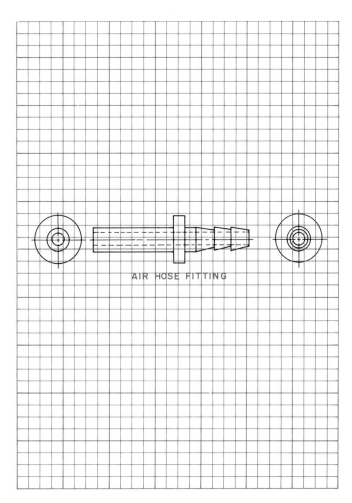

AIR HOSE FITTING

PROBLEM 12-11

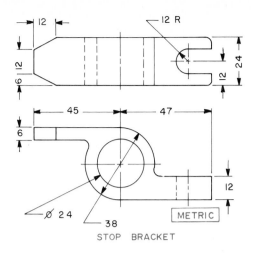

STOP BRACKET

PROBLEM 12–15

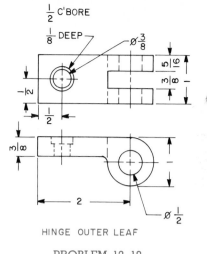

HINGE OUTER LEAF

PROBLEM 12–18

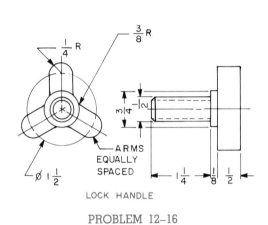

LOCK HANDLE

PROBLEM 12–16

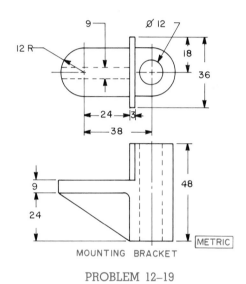

MOUNTING BRACKET

PROBLEM 12–19

PROBLEM 12–20

PROBLEM 12–17

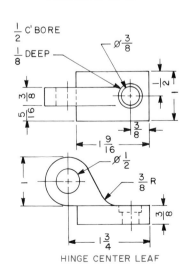

HINGE CENTER LEAF

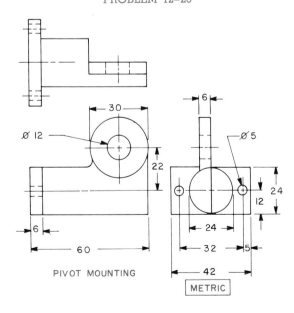

PIVOT MOUNTING

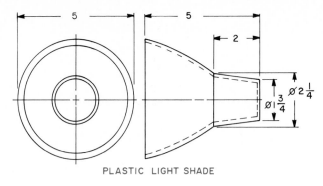

PLASTIC LIGHT SHADE

PROBLEM 12-21

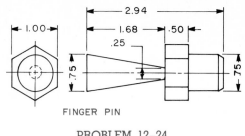

FINGER PIN

PROBLEM 12-24

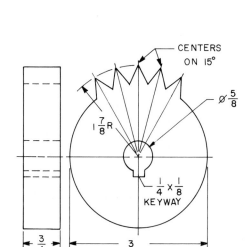

CONTROL GEAR

PROBLEM 12-22

PROBLEM 12-23

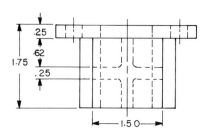

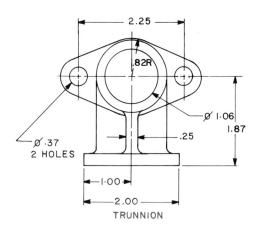

TRUNNION

PROBLEM 12-25

PROBLEM 12-26

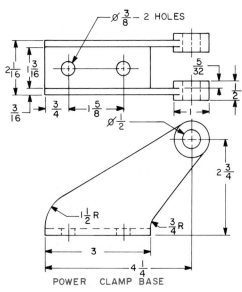

POWER CLAMP BASE

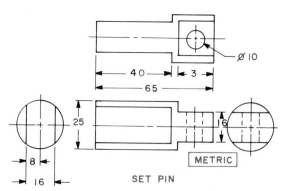

SET PIN

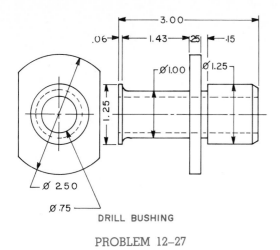

DRILL BUSHING

PROBLEM 12–27

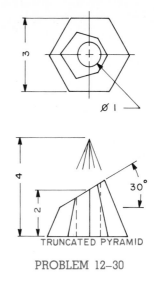

TRUNCATED PYRAMID

PROBLEM 12–30

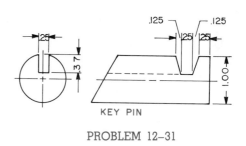

ROCKER ARM

METRIC

PROBLEM 12–28

KEY PIN

PROBLEM 12–31

PROBLEM 12–29

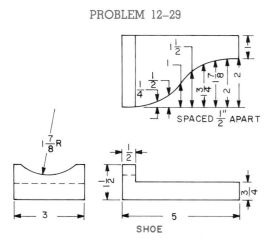

SHOE

PROBLEM 12–32

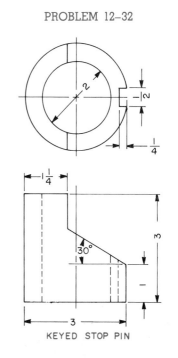

KEYED STOP PIN

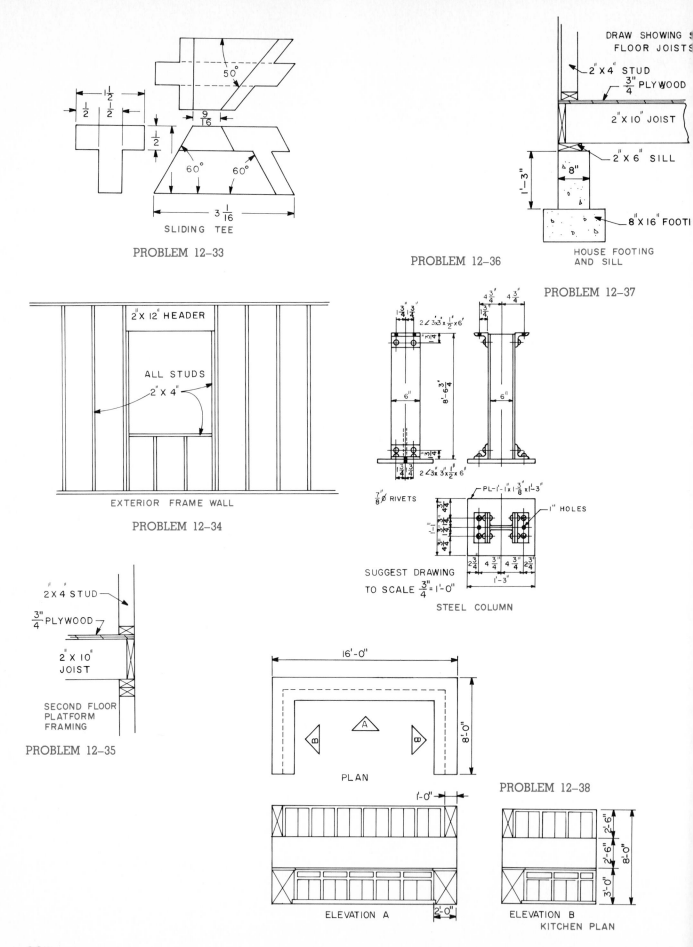

**PROBLEM 12–33**

SLIDING TEE

50°

1½

½

2½

9/16

½

60°   60°

3 1/16

**PROBLEM 12–36**

DRAW SHOWING S
FLOOR JOISTS

2" X 4" STUD

3/4" PLYWOOD

2" X 10" JOIST

2" X 6" SILL

1'-3"

8"

8" X 16" FOOTI

HOUSE FOOTING
AND SILL

**PROBLEM 12–34**

2" X 12" HEADER

ALL STUDS
2" X 4"

EXTERIOR FRAME WALL

**PROBLEM 12–37**

2 ∠ 3x3x 1½ x6'

6"

8'-6 3/4"

6"

2 ∠3x 3½ x 6"

7/8" ∅ RIVETS

PL-1'-1 x 1 3/8 x 1'-3"

1" HOLES

2¾  4¾  4¾  2¾

1'-3"

SUGGEST DRAWING
TO SCALE 3/4" = 1'-0"

STEEL COLUMN

**PROBLEM 12–35**

2" X 4" STUD

3/4" PLYWOOD

2" X 10"
JOIST

SECOND FLOOR
PLATFORM
FRAMING

**PROBLEM 12–38**

16'-0"

A

B        B

8'-0"

PLAN

1'-0"

2'-6"

2'-6"

2'-6"

3'-0"

8'-0"

ELEVATION A        2'-0"

ELEVATION B
KITCHEN PLAN

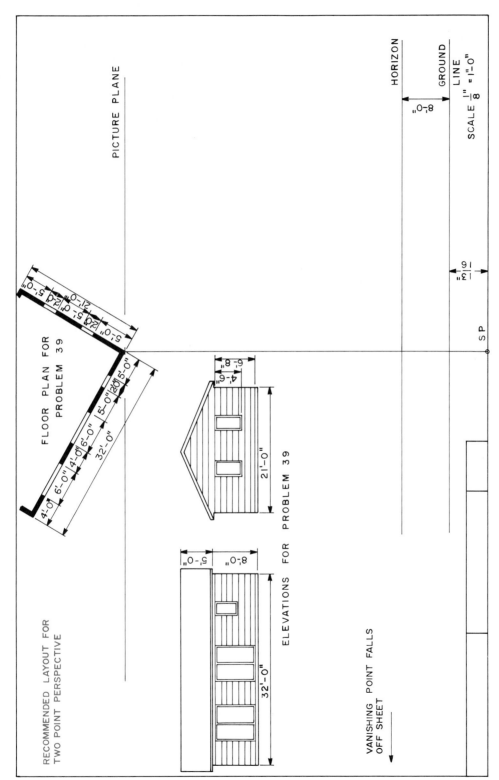

RECOMMENDED LAYOUT FOR
TWO POINT PERSPECTIVE

PICTURE PLANE

FLOOR PLAN FOR
PROBLEM 39

ELEVATIONS FOR PROBLEM 39

VANISHING POINT FALLS
OFF SHEET

HORIZON

GROUND
LINE
SCALE $\frac{1}{8}$" = 1'-0"

8'-0"

SP

PROBLEM 12-39

295

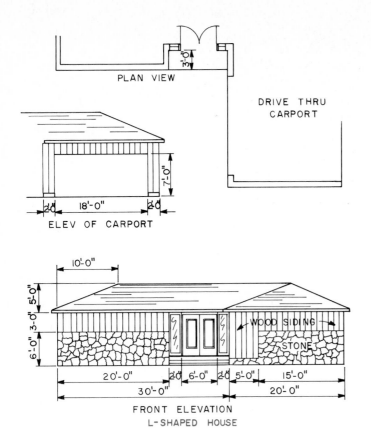

PLAN VIEW

DRIVE THRU
CARPORT

ELEV OF CARPORT

7'-0"

2'0" 18'-0" 2'0"

10'-0"

5'-0"

3'-0"

6'-0"

WOOD SIDING

STONE

20'-0" 2'0" 6'-0" 2'0" 5'-0" 15'-0"

30'-0" 20'-0"

FRONT ELEVATION
L-SHAPED HOUSE

PROBLEM 12-40

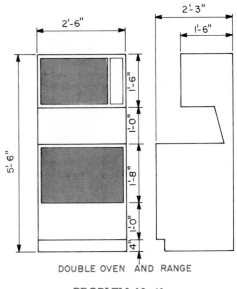

2'-6"

2'-3"

1'-6"

1'-6"

1'-0"

5'-6"

1'-8"

1'-0"

4"

DOUBLE OVEN AND RANGE

PROBLEM 12-41

# Sectional Views on Engineering Drawings

An orthographic projection of a product will show the exterior surfaces. Interior details are hidden and are represented with the hidden-line symbol. If the hidden features are complex, these hidden lines do not show them clearly. In such cases, a sectional view is drawn.

A *sectional view* is made by passing an imaginary cutting plane through the product. One part is removed, exposing the interior details to view (Fig. 13–1). The location of the imaginary cut is shown on a drawing with a cutting-plane line.

## CUTTING-PLANE LINES

The edge view of the cutting plane is shown on a drawing with a cutting-plane line (Fig. 13–2). Two symbols are in use. One is a series of dashes, and the other is a repeating sequence of a long dash followed by two short dashes (Fig. 13–3).

The cutting plane is drawn as a thick line on the view in which it appears as an edge. The sectional view is projected orthographically from this view. Often, the cutting plane line has arrows on each end, drawn at right angles to the line. The arrows indicate the direction of the line of sight used to draw the sectional view.

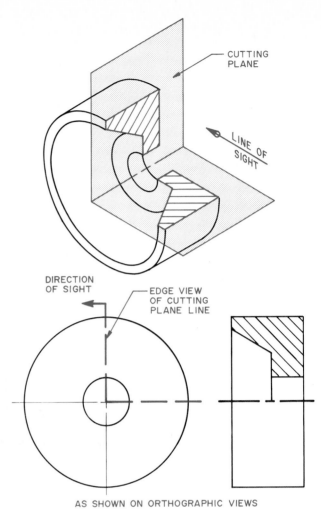

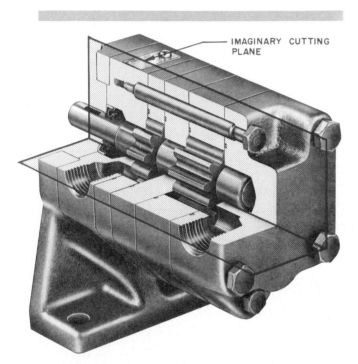

FIGURE 13–1 A sectional view is formed by passing an imaginary cutting plane through an object. Viking Pump Division, Houdaille Industries, Inc.

DIRECTION OF SIGHT

EDGE VIEW OF CUTTING PLANE LINE

AS SHOWN ON ORTHOGRAPHIC VIEWS

FIGURE 13–2 The cutting plane is shown wherever it appears in edge view.

Often, a section is identified by a capital letter beside each arrowhead (Fig. 13–4). This lettering is especially important when several sections have been taken. If it is obvious where each section has been taken, however, the letters can be omitted.

When the cutting-plane line coincides with a center line, the cutting-plane line takes precedence unless the location of the section is clear without it.

## SECTION LINING

The surface cut by the cutting plane and exposed in the sectional view is indicated by drawing section lining on it. Section lining is drawn as a thin line.

Section-lining symbols may be used to indicate the type of material in the sectioned part. Some of

FIGURE 13–3 Standard cutting-plane symbols.

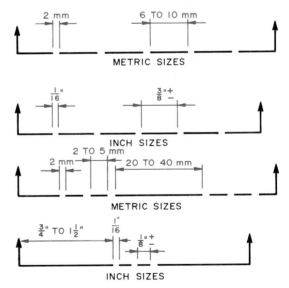

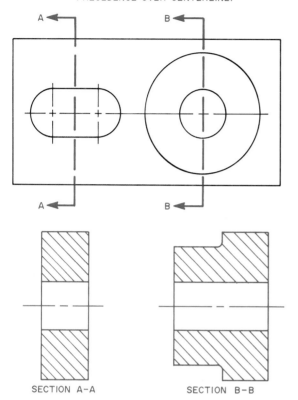

CUTTING PLANE SYMBOL TAKES
PRECEDENCE OVER CENTERLINE.

SECTION A-A          SECTION B-B

FIGURE 13-4 A cutting plane takes precedence over a center line when it is decided that a cutting plane is needed.

these are indicated in Fig. 13–5. A complete listing is available in the American National Standards Institute publication "Line Conventions and Lettering," Y14.2M-1979. These symbols can only show the general type of material, such as plastic. They do not identify the specific material. Specific information about materials is given on the drawing in a note (Fig. 13–6).

If material symbols are not going to be used, it is standard drafting practice to use the general symbol for all materials. This is the same symbol as is used for cast iron (Fig. 13–7).

The section lines are normally drawn on a 45° angle. They must never parallel the edge of a surface (Fig. 13–8). Any other angle can be used when necessary. If a product has several parts, the section lining must be on a different angle on each part (Fig. 13–9). Adjacent parts should have section lining at

FIGURE 13–5 A few of the material symbols used on surfaces cut when making a section view. (From "Line Conventions and Lettering," ANSI Y14.2M-1979 with permission of the publisher, the American Society of Mechanical Engineers.)

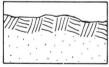

CAST OR MALLEABLE IRON AND GENERAL USE FOR ALL MATERIALS.

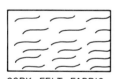

CORK, FELT, FABRIC LEATHER AND FIBER.

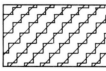

MARBLE, SLATE, GLASS, PORCELAIN, ETC.

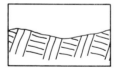

STEEL

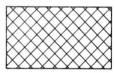

SOUND INSULATION.

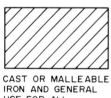

EARTH

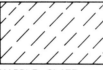

BRONZE, BRASS, COPPER AND COMPOSITIONS.

THERMAL INSULATION.

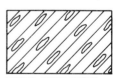

ROCK

WHITE METAL, ZINC, LEAD, BABBITT AND ALLOYS.

TITANIUM AND REFRACTORY MATERIAL

SAND

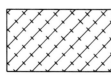

MAGNESIUM, ALUMINUM AND ALUMINUM ALLOYS.

ELECTRIC WINDINGS, ELECTROMAGNETS, RESISTANCE.

WATER AND OTHER LIQUIDS.

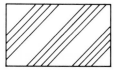

RUBBER, PLASTIC AND ELECTRICAL INSULATION.

CONCRETE

ACROSS GRAIN WOOD.

WITH GRAIN WOOD.

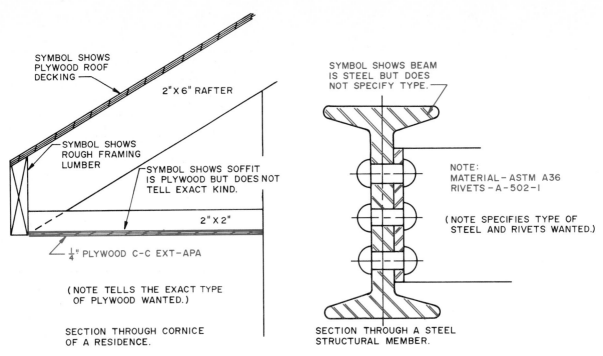

SYMBOL SHOWS
PLYWOOD ROOF
DECKING

2" X 6" RAFTER

SYMBOL SHOWS
ROUGH FRAMING
LUMBER

SYMBOL SHOWS SOFFIT
IS PLYWOOD BUT DOES NOT
TELL EXACT KIND.

2" X 2"

¼" PLYWOOD C-C EXT-APA

(NOTE TELLS THE EXACT TYPE
OF PLYWOOD WANTED.)

SECTION THROUGH CORNICE
OF A RESIDENCE.

SYMBOL SHOWS BEAM
IS STEEL BUT DOES
NOT SPECIFY TYPE.

NOTE:
MATERIAL – ASTM A36
RIVETS – A-502-1

(NOTE SPECIFIES TYPE OF
STEEL AND RIVETS WANTED.)

SECTION THROUGH A STEEL
STRUCTURAL MEMBER.

FIGURE 13–6 Specific information about materials can be given in a note.

nearly opposite angles. This helps clarify the drawing. When the part being section-lined appears in two or more separated pieces, the section lining on them is identical.

On most drawings, the spacing between section lines is about 2.5 mm (³⁄₃₂ in.). They are spaced farther apart on larger areas. Spacing normally ranges from 1.5 mm (¹⁄₁₆ in.) to 3.0 mm (⅛ in.)(Fig. 13–10). The spacing between section lines is drawn by eye. Do not take time to measure the location of each line. Thin parts, such as gaskets, are sectioned solid black. If several thin pieces are touching, a small space is left between the solid sections (Fig. 13–11). Very large areas can be section-lined by using outline sectioning. The symbol is placed around the outside edges of the area (Fig. 13–12).

FIGURE 13–7 General and specific symbols can be used to indicate materials in section.

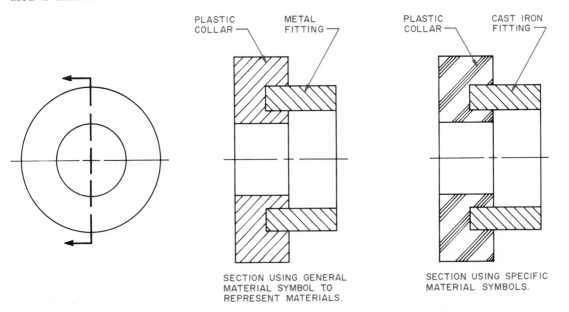

PLASTIC COLLAR    METAL FITTING

PLASTIC COLLAR    CAST IRON FITTING

SECTION USING GENERAL
MATERIAL SYMBOL TO
REPRESENT MATERIALS.

SECTION USING SPECIFIC
MATERIAL SYMBOLS.

YES    NO    NO

FIGURE 13–8  Section lines should be drawn at an angle to the sides of the area sectioned.

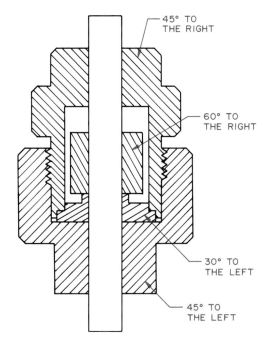

45° TO THE RIGHT

60° TO THE RIGHT

30° TO THE LEFT

45° TO THE LEFT

FIGURE 13–9  The angle of the section lines is different on each individual part of a product.

FIGURE 13–10  The spacing between section lines varies with the area.

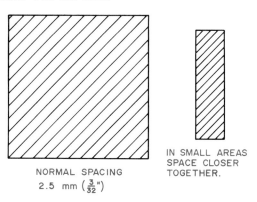

NORMAL SPACING
2.5 mm ($\frac{3}{32}$")

IN SMALL AREAS SPACE CLOSER TOGETHER.

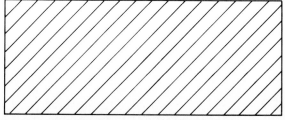

IN LARGE AREAS SPACE FARTHER APART.

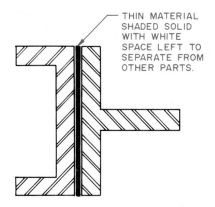

THIN MATERIAL SHADED SOLID WITH WHITE SPACE LEFT TO SEPARATE FROM OTHER PARTS.

FIGURE 13–11  Thin materials are sectioned solid.

FIGURE 13–12  Large areas use outline sectioning.

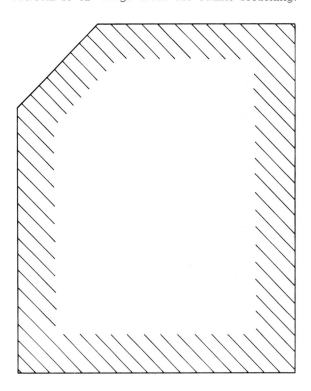

## CONVENTIONAL BREAKS

In some types of sections, it is necessary to break away only a part of an object. These breaks are shown using conventional break symbols (Fig. 13–13). Cylindrical objects use a symbol that clearly shows they are round. It also shows if they are solid or tubular. The width of the symbol for solid objects is one third the radius of the cylinder. Tubular breaks are half the radius. The size of the small inside loop is estimated and can be drawn freehand. The larger curves can be drawn with an irregular curve.

Breaks in long surfaces are made with a freehand Z-shaped mark inserted at intervals. These are drawn as thin lines. Short breaks in rectangular ob-

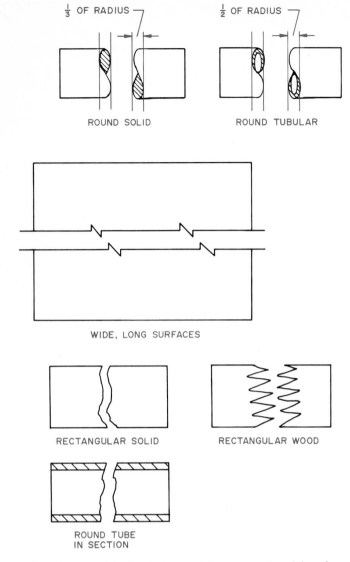

FIGURE 13–13 Symbols used for conventional breaks.

## LINES IN SECTIONAL VIEWS

All visible edges and surfaces that can be seen behind the cut surface forming the section should be drawn. Since the section is drawn to clarify interior details, *hidden lines are omitted*. The surfaces that are hidden are shown on some other view. The hidden lines would only confuse the section and reduce its effectiveness (Fig. 13–15). Hidden lines are used only if necessary to clarify something connected with the sectional view.

Center lines are placed on sectional views. The edges of the cut surfaces are drawn as visible lines.

## PARTS NOT SECTIONED

In an attempt to produce a sectional view that is as clear as possible, it is not necessary to section everything cut by the cutting plane. Items that are not section-lined even though they are cut include fasteners (bolts, washers, nuts, screws, rivets, etc.) Other parts not section lined include shafts, keys, ribs, spokes, webs, set screws, pins, and ball and roller bearings. Since they have no hidden interior features, a section would not add to the clarity of the drawing. Examples of sections through these are given in Fig. 13–16.

## FULL SECTIONS

A *full section* is one in which the cutting plane cuts the object in two parts. One part is removed, exposing the internal features of the other part (Fig. 13–17). The cutting plane is not shown if it is obvious where the cutting plane was passed. A full section is used when a product is not symmetrical (Fig. 13–18).

## HALF SECTIONS

A *half section* is one in which the cutting plane cuts halfway through the product. It is usually used when the product is symmetrical (Fig. 13–19). The portion cut by the cutting plane is removed, exposing the interior detail (Fig. 13–20). The cutting-plane line is not needed when it coincides with a center line.

The half that is not sectioned shows the exterior surfaces. Hidden lines are omitted from both halves of a half section unless they are needed for dimensioning. The two halves are separated by a center line because the cutting-plane line is imaginary. If it were drawn solid, it would represent a visible edge, which does not correctly describe the situation. The cutting-plane line can be omitted if the location of the section is obvious.

jects other than wood are made with an irregular freehand line. This is also used for breaks in round objects in section. It is the same thickness as a visible line. Short breaks in wood are made with a sharply pointed jagged line. These are made freehand.

## INTERSECTIONS IN SECTIONING

The standard practices for showing sections through circular and rectangular holes are in Fig. 13–14. When the circular or rectangular hole is small, it is shown with no regard for the curved elements. If the rectangular hole is large, the edges can be projected to the section. The width of a large circular hole is projected, and the curves drawn with an irregular curve. If the large hole has the same diameter as the hole it intersects, the intersection is a straight line.

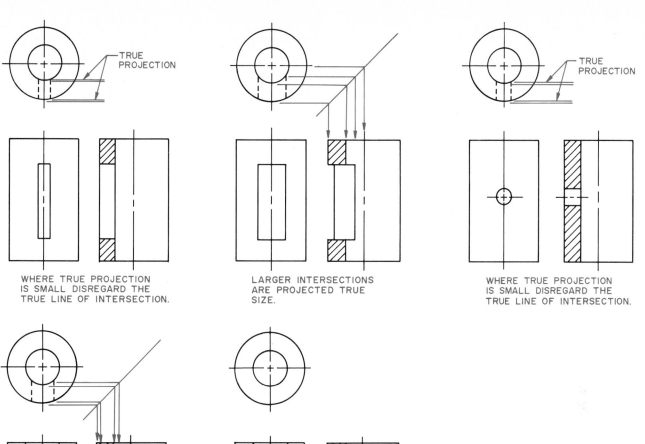

WHERE TRUE PROJECTION
IS SMALL DISREGARD THE
TRUE LINE OF INTERSECTION.

LARGER INTERSECTIONS
ARE PROJECTED TRUE
SIZE.

WHERE TRUE PROJECTION
IS SMALL DISREGARD THE
TRUE LINE OF INTERSECTION.

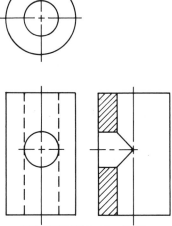

LARGER INTERSECTIONS
ARE PROJECTED TRUE
SIZE.

INTERSECTIONS OF HOLES
THE SAME SIZE APPEAR
AS STRAIGHT LINES.

FIGURE 13–14 Standard practices for showing sections
through circular and rectangular holes.

FIGURE 13–15 Hidden lines
are seldom used on views
having sections.

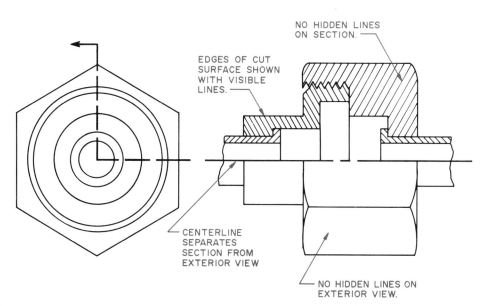

NO HIDDEN LINES
ON SECTION.

EDGES OF CUT
SURFACE SHOWN
WITH VISIBLE
LINES.

CENTERLINE
SEPARATES
SECTION FROM
EXTERIOR VIEW

NO HIDDEN LINES ON
EXTERIOR VIEW.

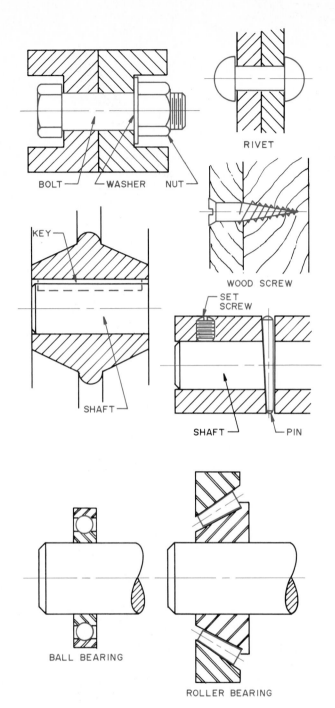

BOLT — WASHER — NUT

RIVET

WOOD SCREW

KEY

SHAFT

SET SCREW

SHAFT — PIN

BALL BEARING

ROLLER BEARING

FIGURE 13–16  Some items are not drawn in section even though the cutting plane passes through them.

FIGURE 13–17  A full section is used when a product is cut fully in two parts.

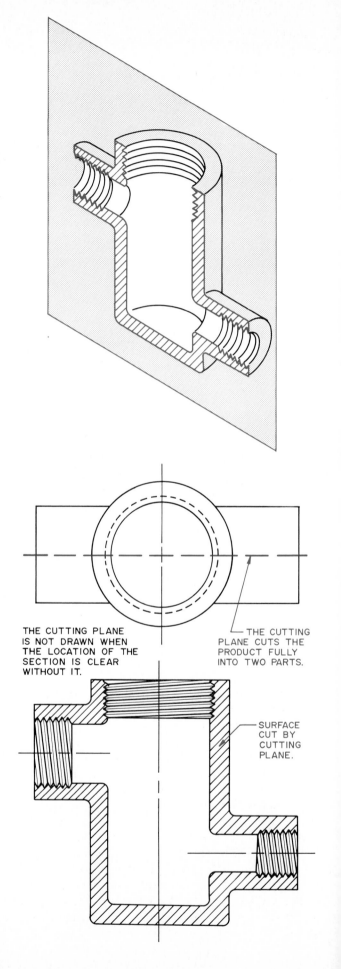

THE CUTTING PLANE IS NOT DRAWN WHEN THE LOCATION OF THE SECTION IS CLEAR WITHOUT IT.

THE CUTTING PLANE CUTS THE PRODUCT FULLY INTO TWO PARTS.

SURFACE CUT BY CUTTING PLANE.

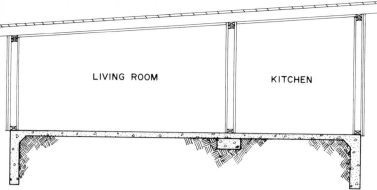

LIVING ROOM

KITCHEN

FIGURE 13–18 Full sections are especially useful when a product is not symmetrical.

FIGURE 13–19 This product is symmetrical and can be described by cutting a section halfway through it. (Courtesy of Andantex, USA, Inc.)

## WEBS, SPOKES, AND RIBS IN SECTION

A web, spoke, and rib are shown in Fig. 13–21. They require special attention when drawing a sectional view. When sectioning a circular object, any portion not solid around the axis, such as a spoke or a rib, is not section-lined. This avoids giving the wrong appearance; if they were section-lined, they could be assumed to be continuous.

When a cutting plane passes parallel with the face of a rib, it appears as in Fig. 13–22. If it passes perpendicular to the rib, the rib is section-lined (Fig. 13–23). An alternate way of showing a rib in section is in Fig. 13–22. The outline of the rib is shown with a dashed line. Every other section line is continued over the rib area. This has the advantage of calling attention to the rib, which might otherwise be overlooked.

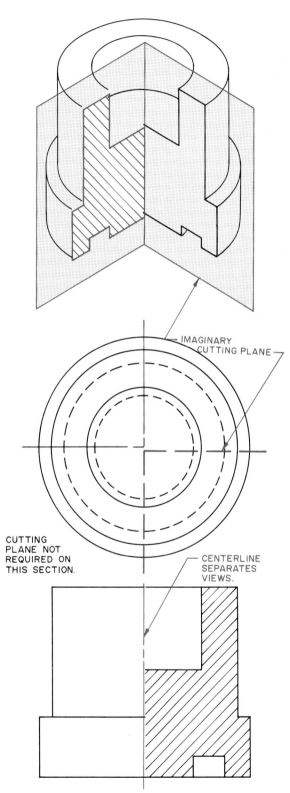

IMAGINARY CUTTING PLANE

CUTTING PLANE NOT REQUIRED ON THIS SECTION.

CENTERLINE SEPARATES VIEWS.

FIGURE 13–20 A half section is used when a product is cut halfway through to expose interior details. The remaining half shows exterior details.

305

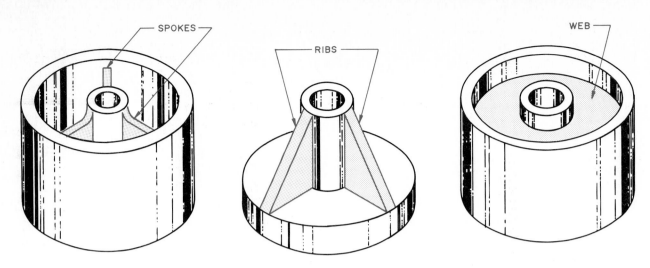

FIGURE 13–21 Spokes, webs, and ribs are commonly used in machine design.

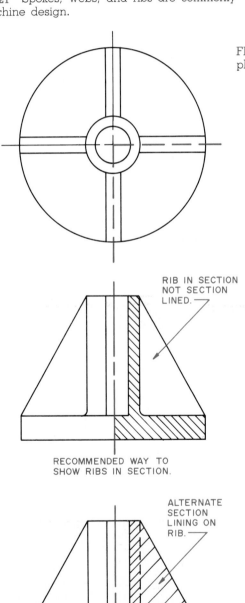

RIB IN SECTION NOT SECTION LINED.

RECOMMENDED WAY TO SHOW RIBS IN SECTION.

ALTERNATE SECTION LINING ON RIB.

ALTERNATE METHOD FOR SHOWING RIBS IN SECTION.

FIGURE 13–22 A rib is not section-lined if the cutting plane passes through it parallel with its face.

FIGURE 13–23 A rib is section-lined when the cutting plane is perpendicular to it.

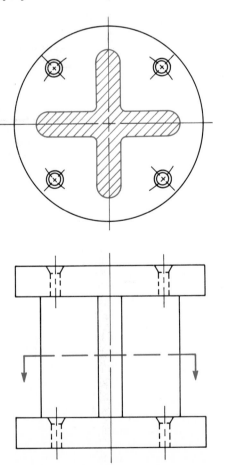

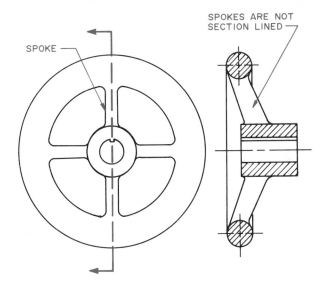

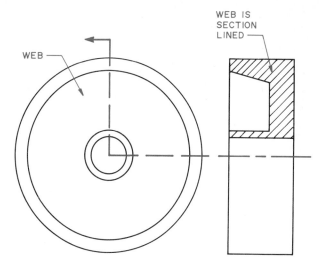

FIGURE 13–24 Spokes are not section-lined when the cutting plane passes parallel through them.

FIGURE 13–25 Since webs are solid, they are section-lined when the cutting plane passes through them.

When a cutting plane passes parallel through a spoke, it is not section-lined (Fig. 13–24).

When a cutting plane passes parallel through a web, it is section-lined because the web is continuous, and the section portrays this fact (Fig. 13–25).

## ALIGNED SECTIONS

Some features of a product appear more normal to the viewer if they are sectioned in a revolved position. These are called *aligned sections*. Examples of

these are spokes, ribs, lugs, holes, and slots. A revolved spoke is shown in Fig. 13–26. If projected orthographically to the section, the lower spoke would appear foreshortened and be confusing. Revolved so that it is parallel with the sectional view, however, it appears true size. It is standard practice to revolve such features. A revolved rib is shown in Fig. 13–27. Notice that in the revolved position it appears true size. A revolved slot is shown in Fig. 13–28. Again, it is true size in the revolved position. Revolved lugs and holes are shown in Figs. 13–29 and 13–30.

FIGURE 13–26 Features that would appear foreshortened in a section drawing are revolved until they appear true size.

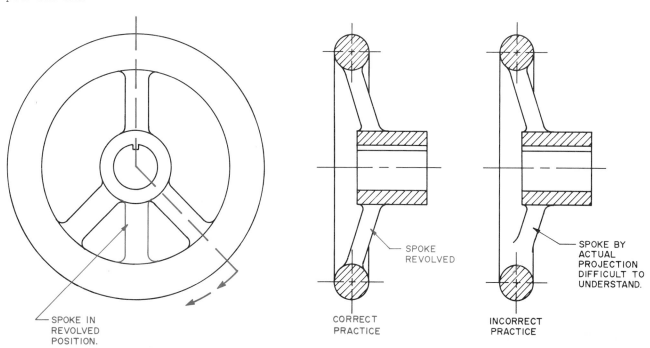

SPOKE IN REVOLVED POSITION.

SPOKE REVOLVED

CORRECT PRACTICE

SPOKE BY ACTUAL PROJECTION DIFFICULT TO UNDERSTAND.

INCORRECT PRACTICE

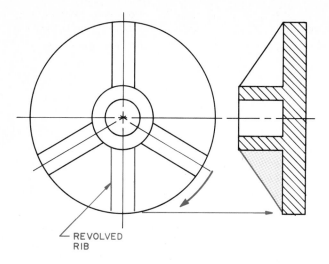

FIGURE 13–27 This rib has been revolved so that it appears true size in the section view.

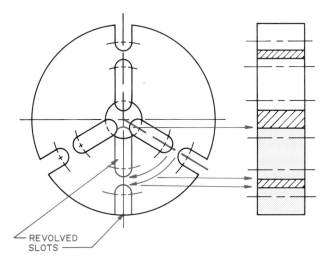

FIGURE 13–28 A slot should be revolved when a section is passed through it.

FIGURE 13–29 Lugs are commonly revolved when section views are drawn.

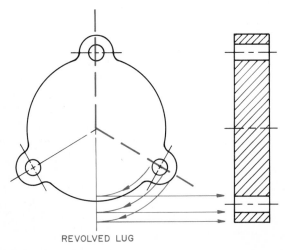

REVOLVED LUG

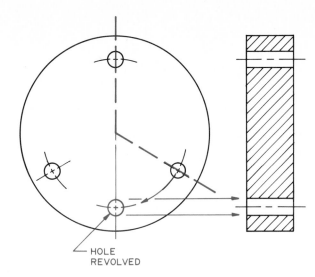

FIGURE 13–30 This hole has a cutting plane through it and must be revolved in the section view.

## OFFSET SECTIONS

An offset section is generally a full section. The unique feature is that the cutting plane changes direction in order to pass through several features that are not in a single plane. The cutting plane is offset as needed to cut these features. The offset is always made at a 90° angle. Offsets are not shown on the section view. This section gives a better picture of the structure of the part than a regular full section (Fig. 13–31).

## REVOLVED SECTIONS

A revolved section is used to show the cross section of a product at a particular location. The location of the section is indicated on the part with a center line. Then, either the cross section at this point is drawn on top of the orthographic view, or a section is broken away (Fig. 13–32).

## REMOVED SECTIONS

A removed section is the same as a revolved section except that the sectional view is drawn not on the orthographic view but near the view. The location of the section is marked on the orthographic view with arrows or a cutting-plane line. Removed sections are especially useful when several sections are to be made and if revolved sections would crowd the orthographic view. The sections can be identified as shown in Fig. 13–33. They are arranged in alphabetical order from left to right. If the sections are small, they can be drawn to a larger scale than the orthographic view. Be certain to note the larger scale by the sectional views.

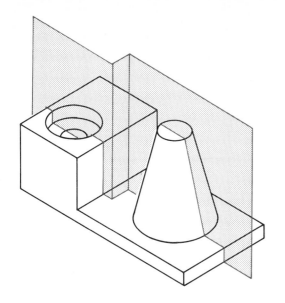

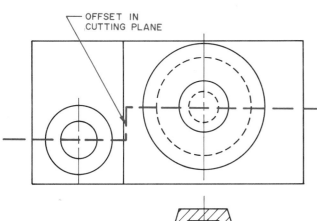

OFFSET IN
CUTTING PLANE

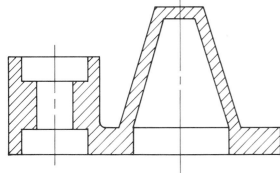

FIGURE 13–31 An offset section has a cutting plane that passes through several features not all in the same plane.

FIGURE 13–32 Revolved sections are drawn on top of the orthographic view.

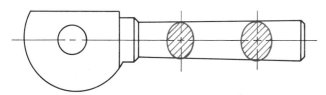

It is also possible to draw removed views using this same technique. If some feature cannot be seen in any of the normal views, a special removed view can be drawn (Fig. 13–34).

## BROKEN-OUT SECTIONS

When some small interior detail needs to be shown, a broken-out section can be used. This can often eliminate the need to draw a larger full or half section. The cutting plane is assumed to pass through the desired feature. It is not drawn on the drawing. A freehand short-break line is drawn to separate the section from the rest of the view. It is section-lined the same as other sections (Fig. 13–35). A broken-out section drawn on a photograph of a product is shown in Fig. 13–36.

## AUXILIARY SECTIONS

An auxiliary section can be used to clarify a feature that will not appear true size on sections made parallel with the principal planes of projection. The cutting plane is located on an orthographic view parallel with the feature to be shown. The section is drawn with a line of sight perpendicular to the cutting plane (Fig. 13–37). Review Chapter 9 if you do not understand how to project to an auxiliary plane.

## PHANTOM SECTIONS

A phantom section is drawn on top of a regular orthographic view. The interior features are outlined with dashed lines. The section lining is also made up of short dashes (Fig. 13–38).

## ASSEMBLIES IN SECTION

As a product is being designed, it is important that the various parts fit together in such a way that the product will operate satisfactorily. The engineering designer makes extensive use of assembly drawings for this purpose. An assembly drawing shows the product put together in its final form. It usually involves sections to check interior fits. Additional information on the types of assembly drawings appears in Chapter 18.

When making an assembly, the product could be shown as a full or half section. If it is symmetrical, a half section is often used. A center line separates the two halves of the drawing.

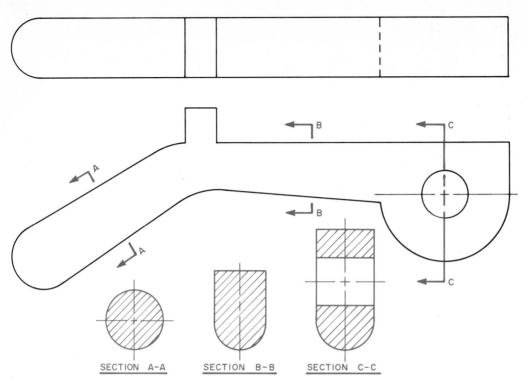

FIGURE 13–33 Removed sections are drawn off but near the orthographic view.

Of importance in an assembly section is the proper use of section lining. Each part must have the section lining on a different angle. Adjacent parts must be section-lined at opposite angles, when pos-sible. The general symbol can be used for all parts; however, the use of material symbols helps separate the various parts. A typical general assembly draw-ing is shown in Fig. 13–39.

FIGURE 13–34 Removed views can be used to show exterior details.

FIGURE 13–35 A broken-out section shows partial in-terior details.

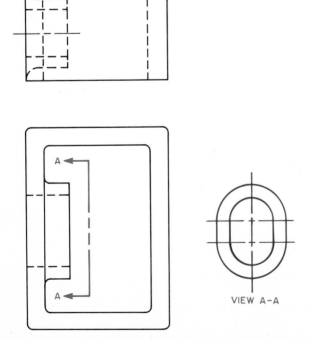

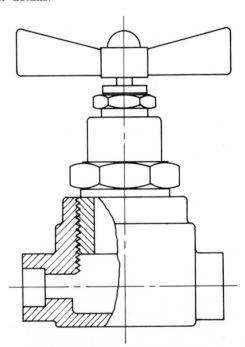

FIGURE 13–36 This broken-out section was drawn by altering a photograph so that interior and exterior details can be seen. (Courtesy of AMP Incorporated.)

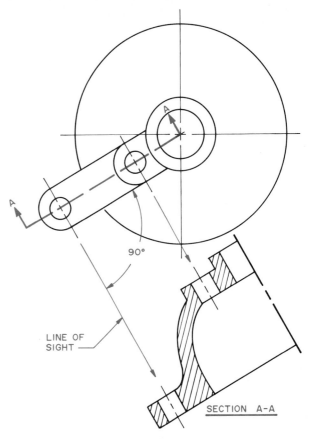

SECTION A-A

FIGURE 13–37 Auxiliary sections are made by projecting the section view perpendicularly from the orthographic view.

FIGURE 13–38 A phantom section is drawn on top of the orthographic view.

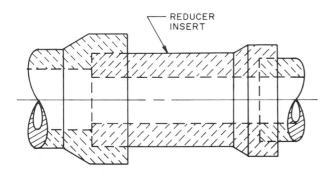

REDUCER INSERT

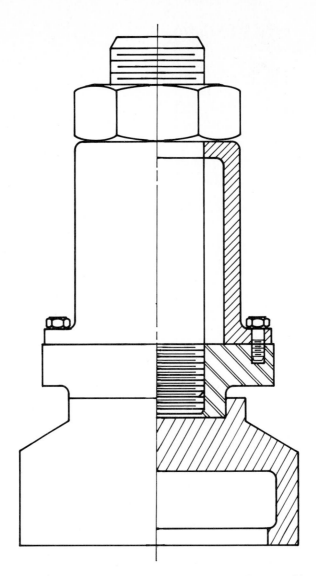

**FIGURE 13–39** One of several types of assembly drawings.

# Problems

The following problems provide a variety of experiences in sectioning engineering drawings. They can be solved with freehand or instrument drawings as assigned by the instructor.

1. On 8½ × 11 in. coordinate paper with ¼-in. squares or 297 × 420 mm coordinate paper with 5-mm squares, sketch the views shown in Probs. 13–1 through 13–4 and draw the indicated sections. Estimate the size of fillets

and rounds. Other sizes can be found by counting the squares. Be certain to show the cutting plane.

2. Using Probs. 13–5 through 13–22, make sections as directed. Place on 8½ × 11 in. or 297 × 420 mm paper with a ½-in. (12-mm) border on all sides and a ½-in. (12-mm) title strip as used in earlier problems. Select a scale that will enable the drawing to fit on the paper.

3. Using the problems in Chapters 6, 12, and 14, make sections as assigned by the instructor.

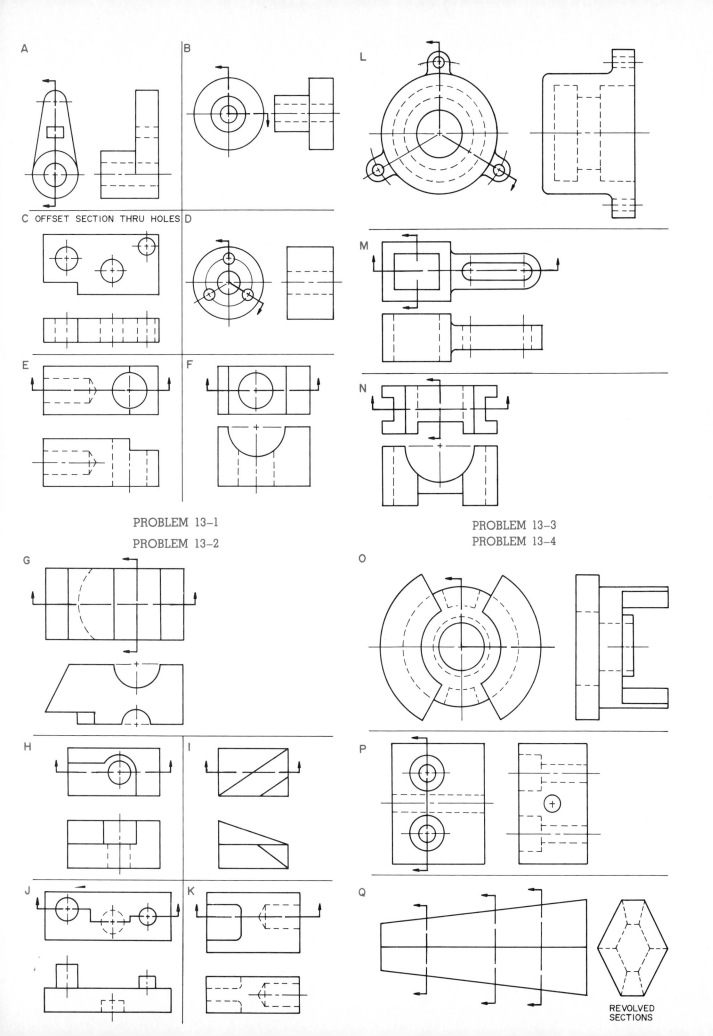

A

B

L

C OFFSET SECTION THRU HOLES

D

M

E

F

N

PROBLEM 13-1

PROBLEM 13-2

PROBLEM 13-3

PROBLEM 13-4

G

O

H

I

P

J

K

Q

REVOLVED
SECTIONS

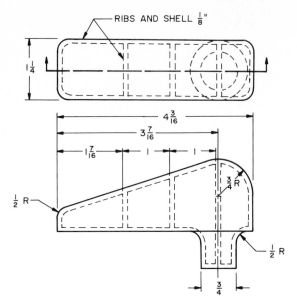

RIBS AND SHELL ⅛"

$1\frac{1}{4}$

$4\frac{3}{16}$

$3\frac{7}{16}$

$1\frac{7}{16}$   1   1

$\frac{1}{2}$ R

$\frac{3}{4}$ R

$\frac{1}{2}$ R

$\frac{3}{4}$

PLASTIC ROUTER HANDLE.

PROBLEM 13–5

PROBLEM 13–6

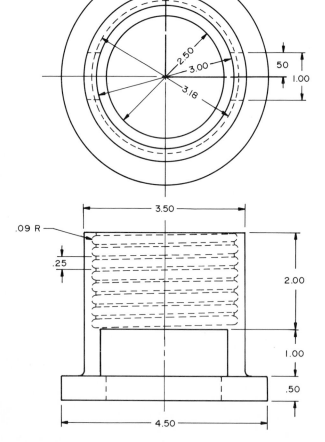

2.50
3.00
3.18

50
1.00

3.50

.09 R

.25

2.00

1.00

.50

4.50

ROUTER BASE

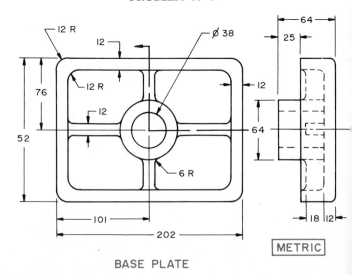

Ø 1.50
Ø .50
Ø .25

1.00

.48
.13
.12
1.25

⅛ DIA.
RIVET

2.00

CIRCLE CUTTER

PROBLEM 13–7

PROBLEM 13–8

12 R
12
12 R
Ø 38
12
12
76
52
6 R
101
202

64
25
12
64
18  12

METRIC

BASE PLATE

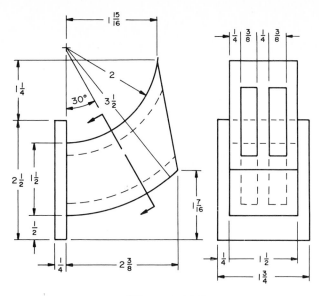

WALL MOUNT                    PROBLEM 13–9

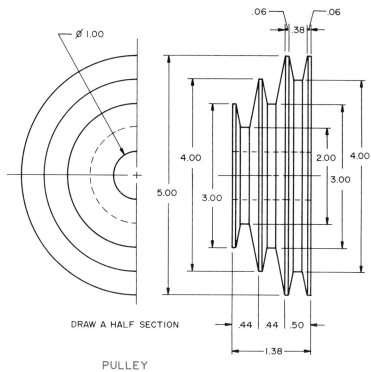

DRAW A HALF SECTION

PULLEY                        PROBLEM 13–10

PROBLEM 13–11

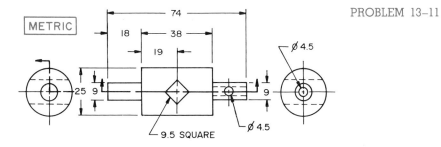

CIRCLE CUTTER BIT TOOL

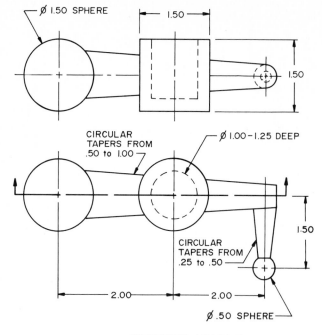

MORTISER HANDLE.

PROBLEM 13-12

PROBLEM 13-13

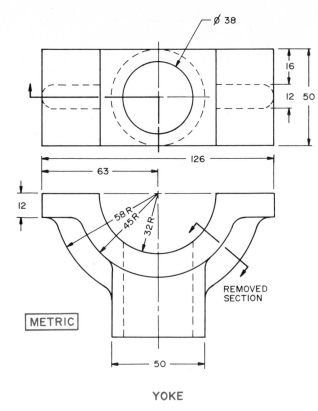

METRIC

YOKE

PROBLEM 13-14

PROBLEM 13-15

DRAW THE
REMOVED
SECTIONS
INDICATED

ROTATING SHOE

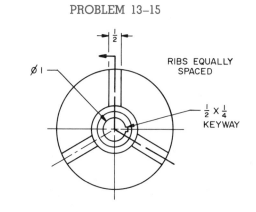

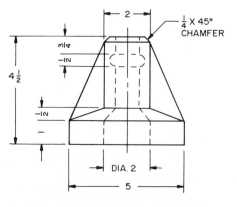

BEARING CAP

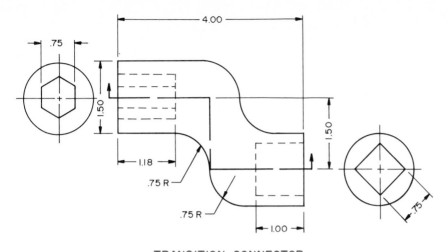

TRANSITION CONNECTOR

PROBLEM 13–16

PROBLEM 13–17

DRAW A HALF SECTION

PROBLEM 13–18

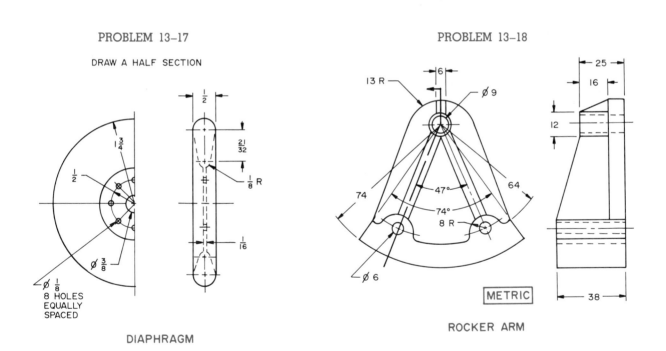

DIAPHRAGM

METRIC

ROCKER ARM

DRAW A FULL SECTION

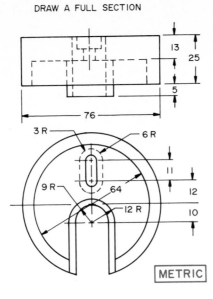

SPACER BLOCK

PROBLEM 13-19

METRIC

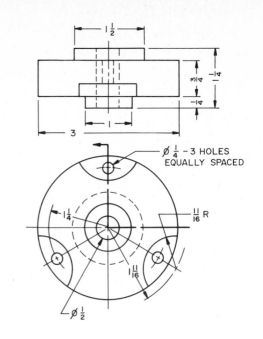

Ø ¼ - 3 HOLES
EQUALLY SPACED

CAP PLATE

PROBLEM 13-21

PROBLEM 13-20

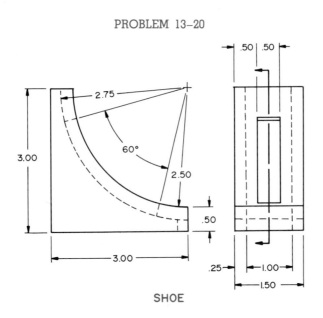

SHOE

PROBLEM 13-22

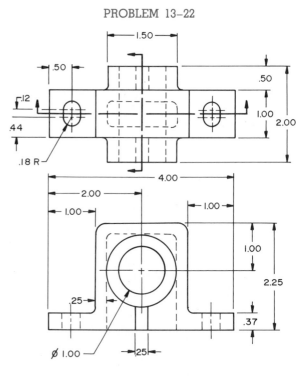

BEARING HOUSING

# 14

# Dimensioning

An essential part of the product design process is recording the sizes of the various components and the precise location of holes and mating parts. The engineering drawings show the shape of the object and each part of it. They show both external and internal details using special drawings such as sections and auxiliary views. However, these are of limited value unless they are carefully and accurately dimensioned.

Dimensions are used to specify the size of the various features of a part, the location of holes and other features, and the relationship between the parts of a product. They are related to the manufacturing process to be used to make the part. For example, if a part is to be made from a casting, the patternmaker will use the drawings to make the original pattern. It must be larger than the desired casting because metal shrinks when it cools. Then, the machinist uses the drawing to machine the casting to its finished size. The cooled casting must be larger than the finished product so that there is some metal that can be removed during machining. Both casting shrinkage and machining allowances must be considered by the engineer as the engineering drawings are dimensioned.

Since engineering drawings made in one location may be used in a factory in another location,

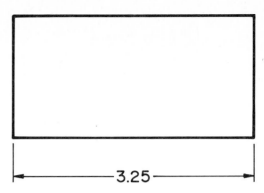

**FIGURE 14–1** Dimension lines are thin lines and are therefore less prominent than visible lines on the object.

dimensioning practices must be standardized. To facilitate a common interpretation of engineering drawings for products designed in the United States and subject to worldwide manufacturing and distribution, the American National Standards Institute participates in meetings with representatives from other countries to produce and keep current dimensioning and tolerancing standards for use in the United States. This standard is titled "Dimensioning and Tolerancing," ANSI Y14.5-1973. This standard does not apply to architectural and civil engineering drawing practices. It is copyrighted by the American Society of Mechanical Engineers. Close liaison is maintained with the British Standards Institute and the Canadian Standards Association. Their national standards on dimensioning and tolerancing are specified as BS308 and CSA B78.2.

## BASIC FACTORS IN DIMENSIONING

Dimensions are recorded as linear distances, angles, and notes. Following are the basic factors that must be considered as a drawing is dimensioned.

1. *Lines, symbols, and abbreviations.* Essential to success is the use of the proper lines of the proper weight and at the proper spacing. There must be a sharp contrast between the lines representing the product and those recording dimensions (Fig. 14–1). The use of the proper symbols and abbreviations is an important part of the completed drawing.

2. *Selection of dimensions.* The choice of the dimensions is critical. Those used must enable the product to function as designed. A second consideration is the manufacturing process involved. When possible, the product is dimensioned to assist those who will make it. Function usually receives first consideration, followed by the consideration of manufacturing processes.

3. *Placement of dimensions.* The placement of dimensions involves following standard drafting practices plus a logical analysis of the solution. Dimensions

must be arranged in a logical and orderly fashion. They must be located where those using the drawings are accustomed to looking for them. Basic to successful placement of dimensions is a drawing that is absolutely clear and easy to read.

4. *Standard dimensioning practices.* Standard dimensioning practices are established for dimensioning features such as chamfers, angles, holes, fasteners, geometric shapes, tapers, round-end shapes, surface finishes, and a variety of notes. They must be observed so that an orderly and easily read drawing is produced.

5. *Tolerances.* The amount of variation in the size of parts of a product varies according to the precision needed for it to function properly. Parts are designed with tolerances as large as possible that still permit the product to operate as designed. Close tolerances increase the cost of manufacturing the product. The actual size of each part is controlled by its dimensions. Standard methods for indicating tolerances are followed.

6. *Manufacturing methods.* The dimensioning of each part is related to the method to be used to manufacture it. Typical methods include casting, stamping, forging, and machining. The dimensions and notes on a drawing give information needed by those performing the various manufacturing operations.

## LINES, SYMBOLS, AND ABBREVIATIONS

The sizes of the various features of a part are shown with dimensions using extension and dimension lines or with notes that often are connected to the feature with a leader (Fig. 14–2).

A dimension gives the actual size or distance between two points, lines, or planes, or some combination of these. The numerical value is shown as a number. The *dimension line* shows direction and extent, and the *extension line* relates the dimension to the object. *Arrowheads* are used on the ends of dimension lines to show the points at which the numerical value terminates.

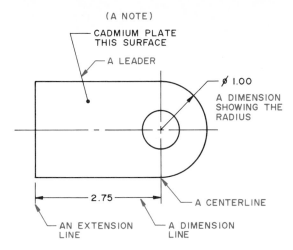

(A NOTE)

CADMIUM PLATE
THIS SURFACE

A LEADER

∅ 1.00

A DIMENSION
SHOWING THE
RADIUS

2.75

A CENTERLINE

AN EXTENSION
LINE

A DIMENSION
LINE

FIGURE 14–2 The size and location of the features of a part are given with notes and leaders or with dimensions using dimension and extension lines.

## Notes

Notes are worded statements used to provide information that cannot be shown with the dimensions and views. They are very brief and must be so clearly written that only one interpretation is possible. Notes are always lettered horizontally in capital Gothic letters. Whenever space permits, they are placed around the outside perimeter of the drawing, avoiding the areas between views. Sufficient space should be kept between notes so that they are not confusing. Do not letter notes in a crowded area or permit them to cross a part of the drawing. Place them so that leaders, if used, can be kept short.

Two types of notes are used: specific and general.

*Specific notes* apply to a particular feature and are connected to it with a leader. For example, a chamfer can be described as "1.5 × 45° CHAMFER," or a knurled surface could be indicated as "96 DP DIAMOND KNURL, RAISED." The leader starts at midheight on the lettering and can be directed from the first or last word of the note. Approved abbreviations, such as DIA, CIR, and FIN, are used freely in notes to reduce their length.

*General notes* apply to the product in total. They are not connected to the drawing with a leader but are usually lettered above or to the left of the title block in the lower right-hand corner of the drawing. Sometimes, they are placed in a central location in an uncrowded area of the drawing. Typical examples include "FINISH ALL OVER," "ALL DIM IN mm," and "ALL DRAFT ANGLES 7° UNLESS OTHERWISE SPECIFIED."

Under most conditions, it is advisable to letter notes after the drawing has been dimensioned. This enables all dimensions to be put in the best possible positions. Sometimes, the need for a note is eliminated by an idea developed while dimensioning the drawing.

## Dimension Lines

Dimension lines show the direction and extent of the numerical size. They are thin lines and terminate with an arrowhead at each end. Although they are thin, they must be drawn solid black. The dimension line nearest the object is placed at least 10 mm (0.40 in.) from the object. Dimension lines parallel to it are at least 6 mm (0.24 in.) apart (Fig. 14–3). The dimension line is broken so that the numerical value can be inserted. Dimension lines should not cross each other. If this is unavoidable, they should not be broken. Dimension lines should be aligned and grouped to produce a uniform appearance (Fig. 14–4). They are drawn parallel with the direction of measurement. When several dimension lines are parallel, the numbers should be staggered to aid in reading the drawing (Fig. 14–5).

A center line, extension line, or line that is part of the outline of the object should never be used as a dimension line. A dimension line is never used as an extension line (Fig. 14–6).

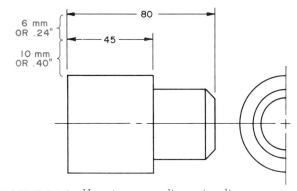

6 mm
OR .24"

10 mm
OR .40"

80

45

FIGURE 14–3 How to space dimension lines.

FIGURE 14–4 Dimensions that are consecutive must be aligned.

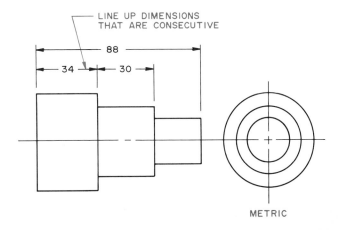

LINE UP DIMENSIONS
THAT ARE CONSECUTIVE

88

34

30

METRIC

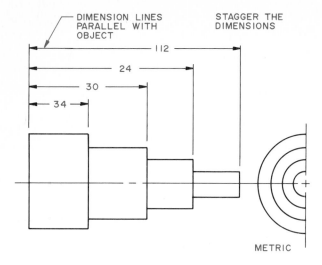

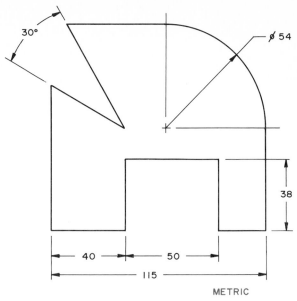

FIGURE 14–5 Stagger parallel dimensions to avoid crowding.

When dimensioning angles, the dimension line is an arc drawn with its center at the apex of the angle. It terminates at extension lines from each side (Fig. 14–6).

## Extension Lines

Extension lines are used to extend a surface or a point beyond the outline of the object so that a numerical size can be related to that surface. Each starts with a 3-mm (0.13-in.) gap from the visible outline and extends 1.5 mm (0.06 in.) past the last dimension line. Extension lines are drawn perpendicular to the dimension lines. In special cases, they may be drawn at an oblique angle if it helps clarify their location. When oblique extension lines are used, the dimension lines are still drawn in the direction in which they apply (Fig. 14–7).

Extension lines are thin, solid black lines of the same width as dimension lines.

It is best if extension lines do not cross each other. They must not cross dimension lines. If it is necessary for an extension line to cross another extension line, dimension line, or visible line, it is not broken (Fig. 14–8). If an extension line crosses an arrowhead or is close to one, the extension line can be broken (Fig. 14–9). If a point is located by an extension line, the line must pass through the point (Fig. 14–10).

Center lines serve as extension lines when dimensioning cylindrical objects (Fig. 14–11).

## Leaders

A leader is used to connect a dimension, note, or symbol to a specific place on a drawing. It is a thin, straight, solid black line of the same width as a dimension line. It is never drawn as a curved line. A

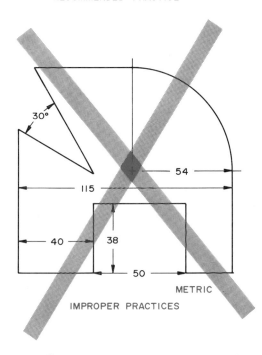

FIGURE 14–6 Center lines, extention lines, and visible lines are never used as dimension lines.

leader always terminates in an arrowhead, except when it is referring to a surface by ending within the outline of the surface. In this case it terminates with a dot (Fig. 14–12).

A leader always starts with a short horizontal line beginning at the midheight of the first letter of the first word or the last letter of the last word of a note. The second part of the leader is on an angle to the point of reference. An angle of 45° to 60° is the most acceptable. Avoid small angles between the leader and the object. A leader should never be hor-

322    Chapter 14

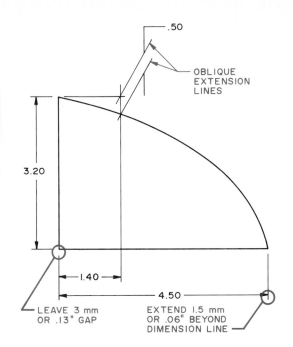

FIGURE 14-7 Extension lines are used to refer the dimension to the product.

izontal or vertical. Adjacent leaders are drawn parallel (Fig. 14-13).

In a situation where too many leaders would confuse a drawing or long multiple angle leaders would have to drawn, alike features can be identified with letter symbols (Fig. 14-14).

FIGURE 14-8 When extension lines cross each other or a visible line, they are not broken.

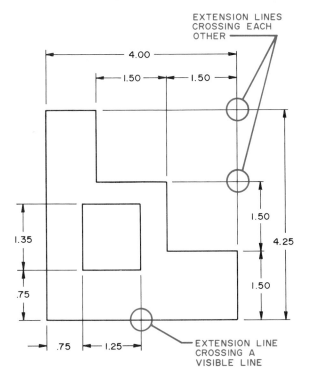

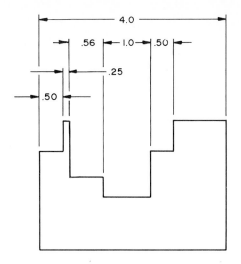

FIGURE 14-9 Extension lines can be broken to make room for a dimension.

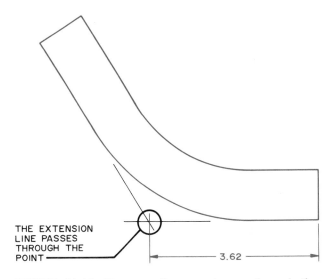

FIGURE 14-10 Extension lines must pass through the point they are locating.

FIGURE 14-11 Center lines serve as extension lines when dimensioning cylindrical objects.

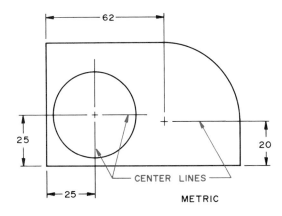

METRIC

Dimensioning **323**

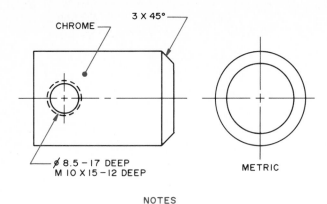

FIGURE 14–12   A leader is used to connect a dimension, note, or symbol to the drawing.

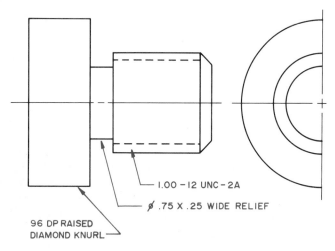

FIGURE 14–13   A leader leaves the note with a horizontal line at the midheight of the first or last word.

FIGURE 14–14   Letter symbols can be used to identify alike features.

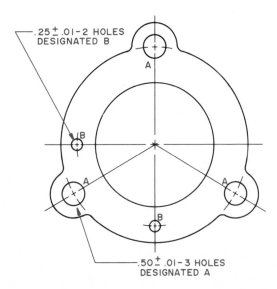

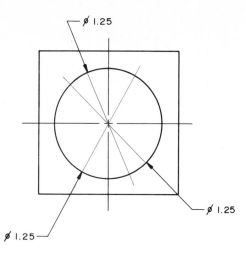

FIGURE 14–15.   All leaders directed to a cylindrical form must point to the center.

When a leader is directed to a circle or an arc, it should be directed toward the center (Fig. 14–15).

Leaders should not cross other leaders or dimension lines. They can cross object lines and are unbroken.

Long leaders should be avoided. They should not be drawn parallel to adjacent dimension lines, extension lines, or section lines.

## Arrowheads

Arrowheads are used to indicate a sense of direction and the termination of a dimension line or leader. They are drawn freehand. They may be drawn open or solid. An open type can be drawn with one or two strokes (Fig. 14–16). A solid arrowhead can be made with a duller pencil and closed with a short stroke (Fig. 14–17). The solid type is usually a little narrower than the open type. The sides are straight rather than curved like the open type. For general

FIGURE 14–16   An open arrowhead can be drawn with two freehand strokes.

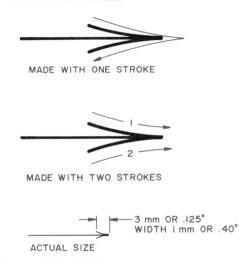

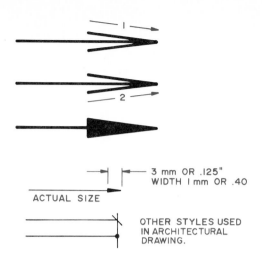

FIGURE 14–17 A solid arrowhead is drawn with three strokes.

| SIZE OF DRAWING | HEIGHT OF LETTERING (INCHES) |
|---|---|
| A, B | .125 |
| C | .150 |
| D, E | .170 |
| Drawing titles | .250 |

FIGURE 14–18 Recommended lettering heights for drawings dimensioned in inches.

| SIZE OF DRAWING | HEIGHT OF LETTERING (MILLIMETERS) |
|---|---|
| A4, A3 | 3.5 |
| A2 | 4.0 |
| A1, A0 | 4.5 |
| Drawing titles | 7.0 |

FIGURE 14–19 Recommended lettering heights for drawings dimensioned in millimeters.

work, arrowheads are 3 mm (0.125 in.) long. On larger drawings, they can be 5 mm (0.19 in.) long. They are about one third as wide as they are long. In architectural work, the dot and diagonal are often used.

It is necessary to practice forming acceptable arrowheads. Improperly made arrowheads spoil the overall appearance of a drawing.

## Letters and Numerals

Vertical and inclined lettering is used on engineering drawings. Generally, capital Gothic letters are used. Certain symbols, as *mm* for millimeter, require the use of lowercase letters. Clearly formed lettering is

essential to a satisfactory drawing. If a note is not legible or a number cannot be read, expensive errors may result.

On engineering drawings using inch dimensions, letter sizes range from 0.125 to 0.170 in., depending on the size of the drawing (Fig. 14–18). Architectural drawings use the same sizes. On metric drawings, lettering heights vary from 3.5 mm to 7.0 mm (Fig. 14–19). These sizes are acceptable for drawings to be microfilmed.

FIGURE 14–20 How to place a dimension on engineering and architectural drawings.

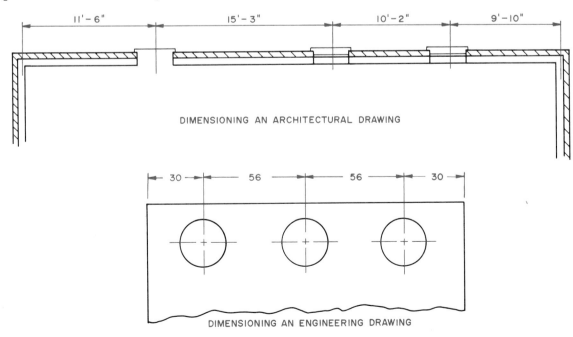

DIMENSIONING AN ARCHITECTURAL DRAWING

DIMENSIONING AN ENGINEERING DRAWING

When dimensioning engineering drawings, break the dimension line for the numerals. The dimension line comes at the midheight of the letters. On architectural and structural drawings, dimensions are placed above the continuous dimension line. They should not touch the dimension line (Fig. 14–20).

Common fractions are lettered with the fraction bar parallel with the lettering guide lines. The height of the complete fraction is equal to the height of two normal numerals (Fig. 14–21). The numerals in decimals are lettered at full letter height. When placed one above the other, as in a tolerance, they are spaced 0.06 in. apart (Fig. 14–22).

When indicating feet and inches, always show both, even if the inch value is 0. This makes it clear that the inches measurement was not inadvertently omitted. Feet are indicated with a single prime (') and inches with a double prime (") (Fig. 14–23). Most industries require that all dimensions up to and including 72 in. be given in inches. Those above that are given in feet and inches. There are exceptions, such as the aerospace industry, where inches (and millimeters) are used for all distances. Metric drawings are dimensioned in millimeters. The symbol for millimeters is two lowercase *m*s (mm) (Fig. 14–23).

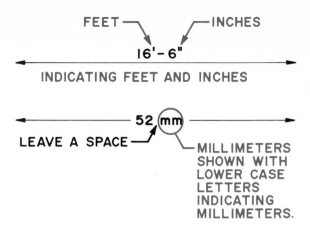

FIGURE 14–23 Use the proper symbols to indicate a dimension in feet and inches or in millimeters.

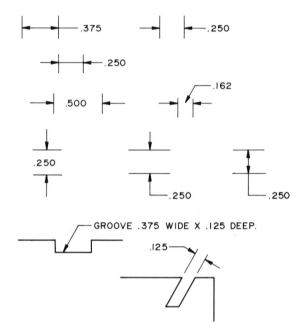

FIGURE 14–24 Suggested placement of dimensions when space is crowded.

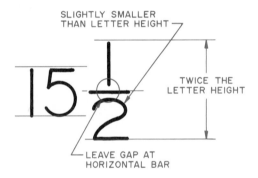

FIGURE 14–21 Common fractions are twice the height of the dimension.

FIGURE 14–22 Decimals are lettered at the same height as the dimension.

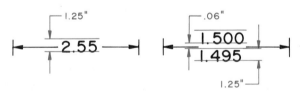

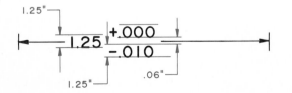

When all dimensions are in inches, no inch symbols are used. The note "ALL DIMENSIONS IN INCHES UNLESS OTHERWISE NOTED" is placed on the drawing. If all are in millimeters, the note "ALL DIMENSIONS IN MILLIMETERS UNLESS OTHERWISE NOTED" is placed on the drawing.

Never crowd dimensions or notes so that legibility is sacrificed. Figure 14–24 shows ways to handle dimensions in crowded spaces.

## Abbreviations

Abbreviations are used on engineering drawings to save time and space. Many, such as DIA, THD, DEG, and R, are frequently used and easily recognized. Others, such as OXD and SFXD, are not frequently encountered. Their use should be carefully considered because they may confuse persons who must use the drawing. A list of approved abbreviations appears in Appendix A.

## Decimal Points

Decimal points are used with both inch and metric dimensions. They must be absolutely clear and sharp. They occupy one letter space and are placed in line with the bottom edge of letters and digits.

## Dashes

Dashes are used to separate yet tie together selected standard notes, such as the thread note "0.375-16UNC-A." They are one letter space long, are at the midheight of the letters and digits, and are drawn parallel with the baseline of the letters.

## Finish Marks

Finish marks are used to indicate those surfaces of metal products that require some type of machining or finishing operation involving the removal of metal. They are not used on parts made from sheet stock or rolled stock that have surfaces that are not machined. Finish marks are not used on surfaces formed by drilling, reaming, or counterboring. When tolerances are given, the finish mark is not needed because the tolerances establish the required accuracy of the surface.

The recommended finish mark is a V. The point touches the edge view of the surface to be finished. The V is placed off the part on the same side as the dimensions. It is drawn about 3 mm (0.12 in.) high (Fig. 14–25). This mark only indicates that the surface is to be machined. It does not give any information about the quality of the finish.

Finish marks are placed on the edge view of the surface, wherever it appears, on all views, including hidden surfaces represented by dashed lines. If an object is finished on all surfaces, the finish marks can be omitted and the general note "FINISH ALL OVER" or the abbreviation "FAO" is used. Some parts require that specific data relating to the degree of smoothness of the finish be given. This is discussed later in this chapter.

## Symbols

Symbols are used to represent complex features that are difficult to show by drawing literally. They speed up the drafting process and simplify the drawing. For example, items such as a diode or an electric light are represented on drawings with a simply drawn symbol. The meaning is clear without having to draw a picture of the specific diode or light (Fig. 14–26).

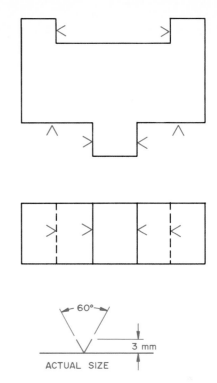

FIGURE 14–25 The finish symbol is placed on the edge view of the surface to be finished in every view in which it appears as an edge.

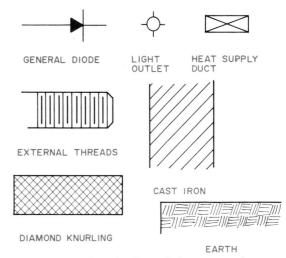

FIGURE 14–26 Standard symbols are used to represent complex items.

There are hundreds of standardized symbols. These are recorded in standards publications of the American National Standards Institute. Many are used and identified as they occur in the various chapters of the text.

## SELECTION OF DIMENSIONS

As a part is drawn, decisions must be made on the dimensions needed to describe it accurately. Most products can be broken down into the basic geomet-

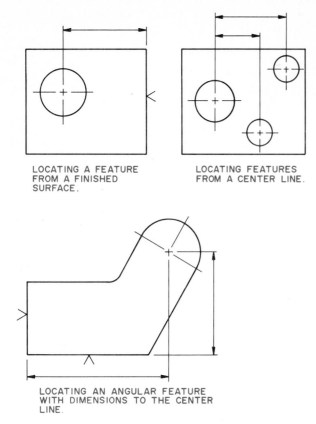

LOCATING A FEATURE
FROM A FINISHED
SURFACE.

LOCATING FEATURES
FROM A CENTER LINE.

LOCATING AN ANGULAR FEATURE
WITH DIMENSIONS TO THE CENTER
LINE.

FIGURE 14–27  Preferred ways to locate features.

giving the location as an angle in degrees (Fig. 14–27).

## Size and Location Dimensions

Dimensioning includes recording the *sizes* of positive and negative features and their *locations*. Size dimensions give the dimensions of the geometric shapes. Location dimensions show the position of these shapes in relation to each other (Fig. 14–28).

## Dimensioning Geometric Shapes

For purposes of dimensioning, an object is broken down into a number of basic geometric shapes, such as a prism, cylinder, and cone. Some parts will be only a part of a geometric shape but still recognizable. The shape may be positive or negative. A void, such as a hole, is a negative shape (Fig. 14–29). A positive shape is a solid.

Geometric shapes are described by their height, width, and depth (Fig. 14–30). Notice that a cone and a cylinder have two dimensions because the diameter is both the width and the depth. A sphere has only one dimension.

FIGURE 14–28  The dimensioning system includes size and location dimensions.

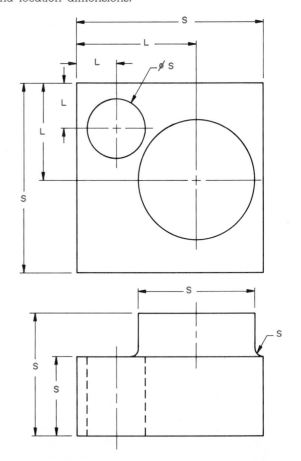

ric shapes and dimensioned following recommended practices. Frequently other factors must be considered, and these influence the selection of dimensions. For example, the processes to be used to make the part will dictate which dimensions are needed. The part may have to mate with or relate to another part, and its dimensions must provide the specific information needed for this to occur.

Anyone who is preparing engineering drawings must understand the manufacturing process to be used to make the product. This includes a wide range of activities such as extruding, sand casting, permanent mold casting, machining, welding, forging, hot and cold forming, and standard assembly procedures. Usually, the dimensions derived from a breakdown of the part into basic geometric shapes will provide the people in manufacturing with the size information they need.

Location dimension selection is also carefully related to the way the part will be manufactured. For example, a surface may be specified to be machined. All location distances might then be measured from this surface because it is a finished surface. Seldom are location dimensions taken from an unfinished surface. Center lines often serve as the dimension from which other locations and size dimensions are taken. It is preferable to locate offset features by measuring from center lines or surfaces rather than

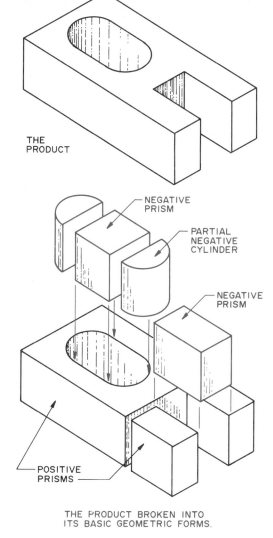

THE PRODUCT

NEGATIVE PRISM

PARTIAL NEGATIVE CYLINDER

NEGATIVE PRISM

POSITIVE PRISMS

THE PRODUCT BROKEN INTO ITS BASIC GEOMETRIC FORMS.

FIGURE 14–29 To assist in dimensioning, a product can be broken into basic geometric shapes.

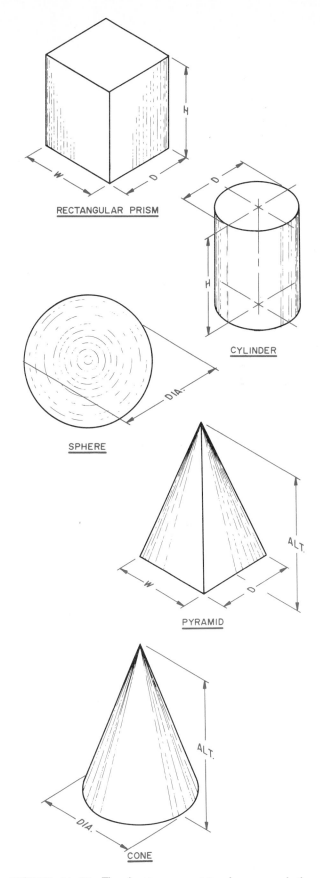

RECTANGULAR PRISM

CYLINDER

SPHERE

PYRAMID

CONE

FIGURE 14–30 The basic geometric shapes and the dimensions needed to describe them.

The prism is one of the most frequently encountered shapes. The rectangular prism requires three dimensions—height, width, and depth. The views can be dimensioned as shown in Fig. 14–31. The height and width are shown on the front view. When the front and top views are given, the width is shown between the views because it relates to both. The depth is shown on the top view. When the front and side views are shown, the width can be shown on the top or bottom of the front view. The depth is shown on the side view and is lined up with the width dimension. The height is shown between the two views because it relates to both.

Regular hexagonal and octagonal prisms are dimensioned by giving their height in the front view and the across-the-corners or across-the-flats dimension in the top view (Fig. 14–32).

Triangular prisms are dimensioned by giving

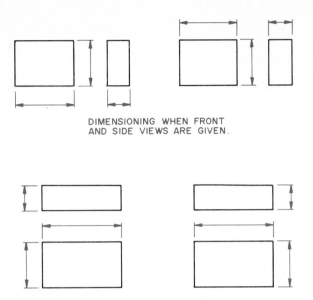

DIMENSIONING WHEN FRONT
AND SIDE VIEWS ARE GIVEN.

DIMENSIONING WHEN FRONT
AND TOP VIEWS ARE GIVEN.

FIGURE 14-31 Recommended ways to dimension a prism.

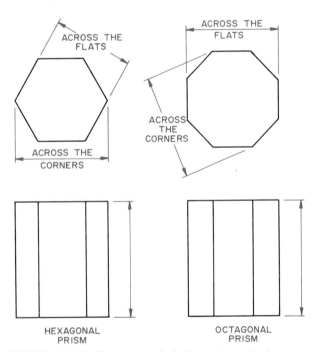

ACROSS THE FLATS

ACROSS THE FLATS

ACROSS THE CORNERS

ACROSS THE CORNERS

HEXAGONAL PRISM

OCTAGONAL PRISM

FIGURE 14-32 Recommended dimensions for hexagonal and octagonal prisms.

the height and width and the location of the apex in the front view. The depth is in the top or side view (Fig. 14-33).

A pyramid is dimensioned by giving the width, depth, and location of the vertex in the top view. The altitude is given on the front view. If it is a frustum, the location of the top surface is given in the front view (Fig. 14-34).

Right cylinders are also commonly found on industrial products. Shafts, holes, and fasteners are

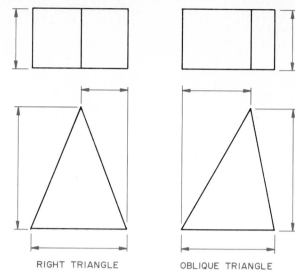

RIGHT TRIANGLE

OBLIQUE TRIANGLE

FIGURE 14-33 Recommended dimensions for triangular prisms.

FIGURE 14-34 Recommended dimensions for pyramids.

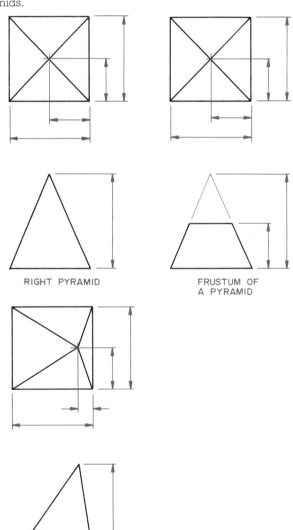

RIGHT PYRAMID

FRUSTUM OF A PYRAMID

OBLIQUE PYRAMID

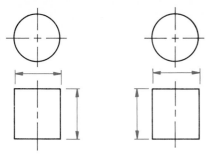

CYLINDERS IN VERTICAL POSITION

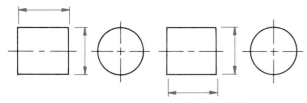

CYLINDERS IN HORIZONTAL POSITION

FIGURE 14–35 Recommended dimensions for positive cylinders.

usually cylinders. A cylinder is dimensioned by giving its diameter and length. The diameter and length of positive cylinders are shown on the rectangular view. In other words, the diameter of a positive cylinder is shown on the edge view of the cylinder. Acceptable locations for dimensions are shown in Fig. 14–35. The diameter is placed between the views because it relates to both. When it is obvious that the dimension is a diameter, no symbol is used. If it is not clear, the abbreviation DIA or the symbol for diameter is used. The diameter of negative cylinders (such as holes) is given on the circular view (Fig. 14–36).

A cylinder is never dimensioned by giving the radius. Tools used to measure cylindrical objects are designed to measure the diameter. The worker would have to double the radius to find the diameter. The possibility of mathematical errors makes this an undesirable practice.

Right cones are dimensioned by giving the di-

FIGURE 14–36 The diameter symbol is not used when it is obvious that the dimension is a diameter. Negative cylinders are dimensioned to the circular view.

FIGURE 14–37 Recommended dimensions for cones.

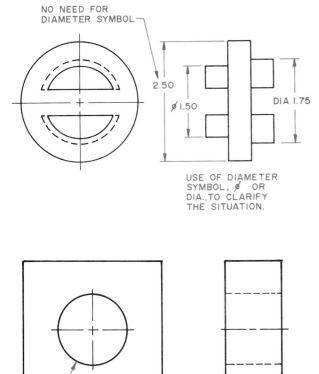

NO NEED FOR DIAMETER SYMBOL

2.50

⌀1.50

DIA 1.75

USE OF DIAMETER SYMBOL, ⌀ OR DIA., TO CLARIFY THE SITUATION.

.75±.02

NEGATIVE CYLINDERS ARE DIMENSIONED TO THE CIRCULAR VIEW

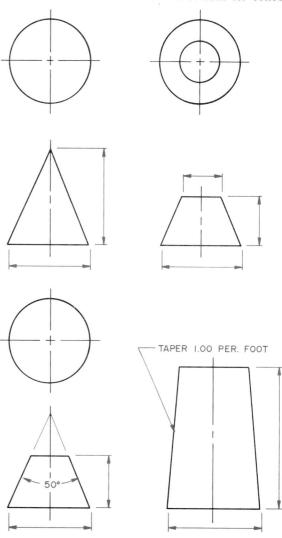

TAPER 1.00 PER. FOOT

50°

ameter of the base and the altitude. When they are frustums, they require the diameter of both ends and the altitude. Sometimes, it is clearer to dimension a cone with degrees or by giving the amount of taper (Fig. 14–37).

Spheres require that the diameter be given. If the dimension is marked "SPHERE," only one view is required. When a spherical end is dimensioned, the spherical radius is given (Fig. 14–38).

A torus can be dimensioned in either of two ways, depending on how it is to be manufactured. If it is formed by bending a rod around a mandrel,

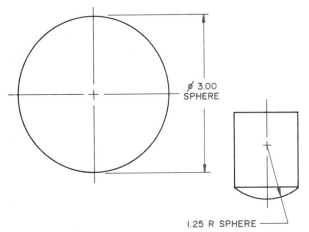

FIGURE 14–38  Recommended dimensions for spheres.

FIGURE 14–39  Two ways to dimension a torus.

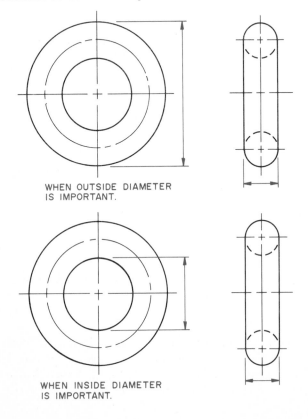

WHEN OUTSIDE DIAMETER IS IMPORTANT.

WHEN INSIDE DIAMETER IS IMPORTANT.

the diameter of the inside is needed. If it is formed to fit inside a cylinder, the outside diameter is required (Fig. 14–39).

## Mating Parts

When two parts are required to join, their size and location dimensions are involved. The sizes of each feature must be selected so that the desired fit is accomplished. The location dimensions regulate where this occurs on the part so that the product functions as expected. Usually, mating parts require dimensions specified with more accuracy than other features, such as unfinished surfaces. Figure 14–40 shows dimensions pertaining to a mating part. They are specified in three decimals, while other, less critical features are in two decimals. Part A could be machined from rectangular stock. The tongue fits into the groove in part B and is dimensioned to three decimal places. The other surfaces do not contact other parts of the product, so they do not need this degree of accuracy. The groove in part B forms the mating part and must be dimensioned so that it can receive the tongue and provide the minimum clearances desired. Additional information on this dimensioning situation, called tolerancing, is given in Chapter 15.

## Excess Dimensions

When making engineering drawings, great care should be taken to avoid excess dimensions. An excess dimension is any dimension that is not needed to manufacture the part. If a feature is dimensioned on one view, do not repeat the dimension on another view. If the drawing is revised, it is possible that duplicate dimensions might be overlooked. The drawing would then show the feature in two different sizes. In some cases, sizes or locations are intended to be measured from a particular surface. If all the dimensions related to the feature are given, this information is lost. One dimension in a series is always omitted (Fig. 14–41).

Circular features, such as holes, if they are all the same size, can be dimensioned with a single note. When the corners have the same radius, only one need be dimensioned (Fig. 14–42). Often, the radius of fillets and rounds is given with a note such as "ALL FILLETS AND ROUNDS 3 mm R." Any information given on notes is not repeated on the drawing. A summary of common dimensioning errors is in Fig. 14–43.

In architectural and structural drawings, it is common practice to put all dimensions in a series and repeat dimensions on various parts of the drawing (Fig. 14–44). Since these projects do not involve interchangeable parts, no difficulty is caused. Ac-

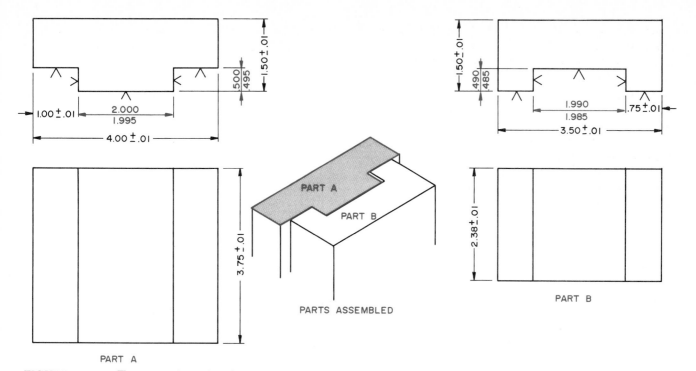

PART A

PART B

FIGURE 14-40 The procedure for dimensioning mating parts.

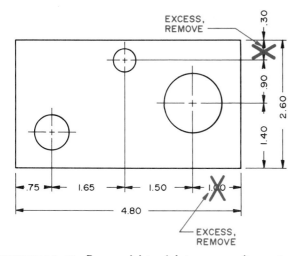

FIGURE 14-41 Be careful to delete excess dimensions.

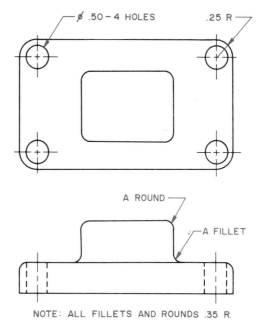

NOTE: ALL FILLETS AND ROUNDS .35 R.

FIGURE 14-42 Dimensioning holes and round corners.

tually, the extra dimensions are a help in checking the correctness of the overall dimensions and other parts of the structure.

## Reference Dimensions

There are occasions when it may be considered desirable to place an excess dimension on a drawing so that it can be used as a reference for checking. The preferred way to indicate that it is a reference dimension is to enclose it in parentheses. An acceptable alternative is to place the abbreviation REF immediately following the dimension or directly under it. This indicates that the dimension is without tolerance and is not to be used in the manufacture of the part (Fig. 14-45).

## Basic Dimensions

When a dimension on a drawing is specified as "BASIC," it is a theoretical value used to describe the exact size, shape, or position of a feature. It is a reference from which predetermined variations are

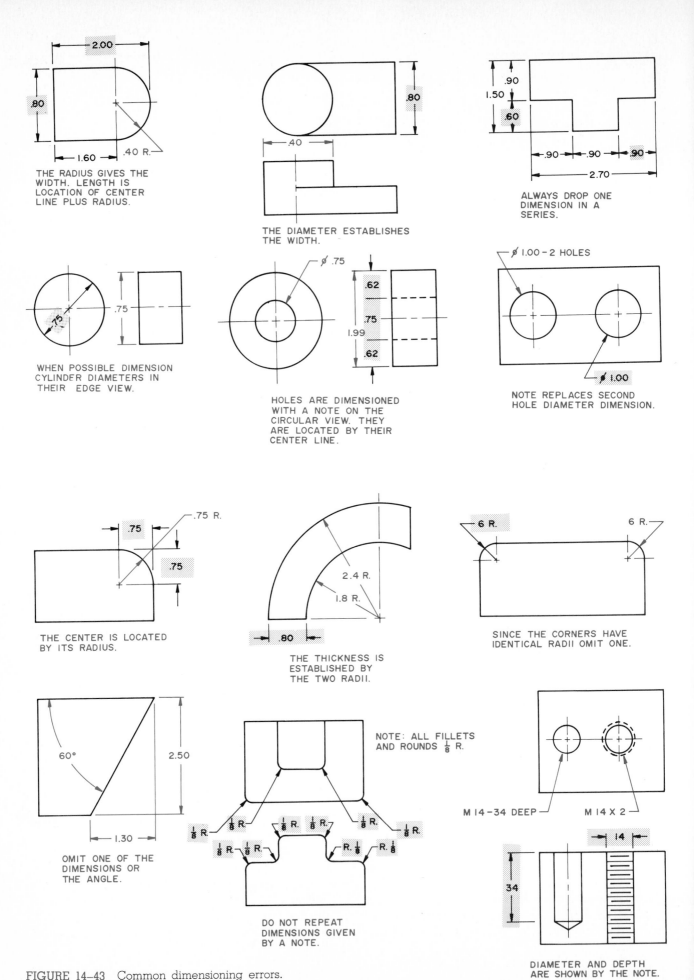

2.00

.80

1.60

.40 R.

THE RADIUS GIVES THE
WIDTH. LENGTH IS
LOCATION OF CENTER
LINE PLUS RADIUS.

.80

.40

THE DIAMETER ESTABLISHES
THE WIDTH.

.90

1.50

.60

.90  .90  .90

2.70

ALWAYS DROP ONE
DIMENSION IN A
SERIES.

.75

.75

WHEN POSSIBLE DIMENSION
CYLINDER DIAMETERS IN
THEIR EDGE VIEW.

∅ .75

.62

.75

.62

1.99

HOLES ARE DIMENSIONED
WITH A NOTE ON THE
CIRCULAR VIEW. THEY
ARE LOCATED BY THEIR
CENTER LINE.

∅ 1.00 - 2 HOLES

∅ 1.00

NOTE REPLACES SECOND
HOLE DIAMETER DIMENSION.

.75

.75 R.

.75

THE CENTER IS LOCATED
BY ITS RADIUS.

2.4 R.

1.8 R.

.80

THE THICKNESS IS
ESTABLISHED BY
THE TWO RADII.

6 R.

6 R.

SINCE THE CORNERS HAVE
IDENTICAL RADII OMIT ONE.

60°

2.50

1.30

OMIT ONE OF THE
DIMENSIONS OR
THE ANGLE.

NOTE: ALL FILLETS
AND ROUNDS $\frac{1}{8}$ R.

$\frac{1}{8}$ R.  $\frac{1}{8}$ R.  $\frac{1}{8}$ R.  $\frac{1}{8}$ R.

$\frac{1}{8}$ R.  $\frac{1}{8}$ R.  R. $\frac{1}{8}$  R. $\frac{1}{8}$

DO NOT REPEAT
DIMENSIONS GIVEN
BY A NOTE.

M 14-34 DEEP

M 14 X 2

14

34

DIAMETER AND DEPTH
ARE SHOWN BY THE NOTE.

FIGURE 14-43  Common dimensioning errors.

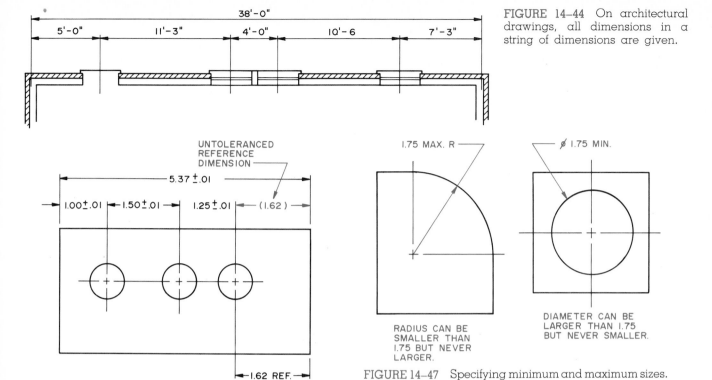

FIGURE 14-44 On architectural drawings, all dimensions in a string of dimensions are given.

FIGURE 14-44 On architectural drawings, all dimensions in a string of dimensions are given.

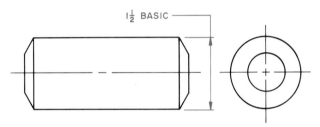

FIGURE 14-45 Two ways to indicate reference dimensions.

FIGURE 14-47 Specifying minimum and maximum sizes.

FIGURE 14-46 The size from which variations are taken is identified as the basic size.

means the radius can be smaller than this but never larger. If a hole is marked "2.00" DEEP MIN," it means the hole can be deeper but no shallower (Fig. 14-47). This practice is frequently used when dimensioning chamfers, radii, lengths of threads, depths of holes, and counterbores.

## Dimensions Not to Scale

When it is necessary to indicate a dimension on a feature that has not been drawn to the scale indicated, *underline the dimension with a straight line.* (A wavy line was used for years, but the straight line is now the accepted form in international practice.)

established. For example, a design solution may require a 1½-in.-diameter shaft; this is the basic size, with no tolerances applied (Fig. 14-46).

## Nonmandatory Dimensions

It is acceptable to identify as nonmandatory any processing dimensions that provide for finish allowance, shrinkage allowance, and other such requirements as long as the final dimensions are given on the drawing or on a higher assembly drawing. These nonmandatory processing dimensions are identified by a note such as "NONMANDATORY (MANU-FACTURING DATA)."

## Maximum and Minimum Sizes

Maximum and minimum sizes establish a design limit beyond which the size cannot be permitted to vary. For example, a radius marked "0.1875 MAX R"

## PLACEMENT OF DIMENSIONS

When a part has been designed and the sizes and locations of its features have been decided, they must be positioned properly on the drawing. This involves deciding which view is to receive each dimension and where it will be placed on that view. Some of these decisions are made by observing the principles in the preceding section of this chapter. For example, the basic geometric forms have commonly accepted locations for dimensions.

The following recommendations for placing dimensions are accepted as standard. People who use the drawings are accustomed to looking for dimensions in the places recommended. To violate these

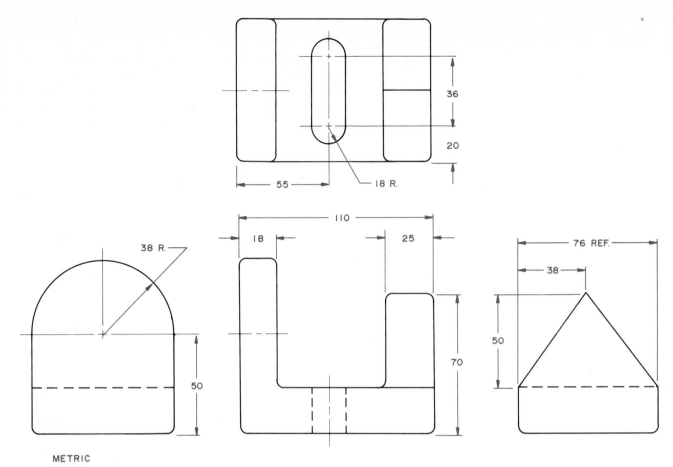

METRIC

**FIGURE 14–48** Dimensions are placed where the various features are shown in profile. The front view normally has the largest number of dimensions.

standards would make the drawing more difficult to use. An important consideration is clarity. The placement of dimensions must contribute to the clarity of the drawing. This involves judgment and experience.

The front view shows the most important contour and places the object in its natural position. Generally, it will receive the most dimensions because dimensions are placed where the characteristics of the feature occur. Examples of this are shown in Fig. 14–48. Here, the basic profile is shown in the front view. The overall length and height are dimensioned. They are placed between the views because they relate to both views. The profile of the left side is shown in the left side view. The radius of the curved feature plus the location of the center are dimensioned here because they are in profile. The depth is shown on the front view. The profile of the right end is in the right side view. The altitude and length of the triangular solid are dimensioned here because they are in profile and this is the accepted location for these dimensions. The depth is shown on the front view. The slot in the base appears in

profile in the top view and is therefore dimensioned there. Its location dimensions relate to the profile view and are therefore also shown on the top view. This example shows the use of the principle of dimensioning where the profile appears plus observing the procedures for dimensioning geometric solids. Review the material in the preceding section of this chapter to see the proper placement of dimensions on geometric shapes.

Whenever possible, dimensions are placed between views. Even though they are connected to only one view by extension lines, they can still be easily related to the other view.

### Aligned and Unidirectional Dimensioning Systems

The aligned dimensioning system has the numerals positioned so that they are perpendicular to the dimension line. They are arranged so that they are read from either the *bottom* or the *right side* of the drawing. When a surface is inclined, the dimension line is drawn parallel to the surface (Fig. 14–49).

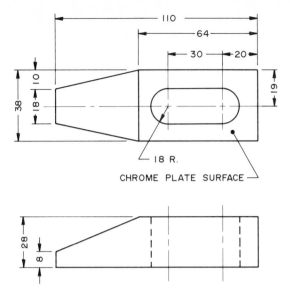

THE ALIGNED DIMENSIONING SYSTEM.

FIGURE 14–49 Dimensions in the aligned dimensioning system are read from the bottom or the right side.

Avoid placing dimensions in the upper left quadrant. Here they appear upside down and are difficult to read (Fig. 14–50).

FIGURE 14–50 Proper quadrant placement of numerals for the aligned and unidirectional dimensioning systems.

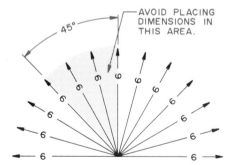

PLACEMENT WHEN USING THE ALIGNED SYSTEM.

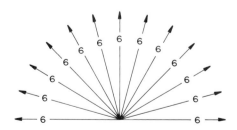

PLACEMENT WHEN USING THE UNIDIRECTIONAL SYSTEM.

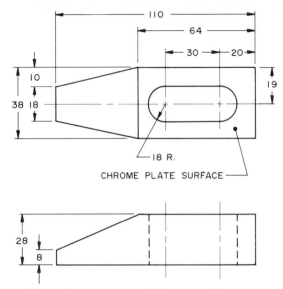

THE UNIDIRECTIONAL DIMENSIONING SYSTEM.

FIGURE 14–51 All dimensions in the unidirectional dimensioning system are read from the bottom of the drawing.

The unidirectional dimensioning system positions all numerals so that they are read from the *bottom* of the drawing. In some cases, they are perpendicular with the dimension line and are parallel with it in other locations. Inclined surfaces are dimensioned by keeping the dimension parallel with the horizontal (Fig. 14–51). Since the dimensions are parallel with the horizontal, they may be placed in any quadrant (Fig. 14–50).

In both systems, notes are lettered horizontally so that they can be read from the bottom of the drawing.

## Order of Dimensioning

To prevent mistakes and to avoid frequent erasing of dimensions that must be moved to make room for other dimensions, a planned, orderly procedure for placing dimensions is necessary. The exact procedure may vary with the desires of the drafter, but a standard practice must be developed. Remember, decisions related to where dimensions are to be placed must be made before the engineering drawing is begun so that the necessary space can be provided between views. Here is a suggested order for placing dimensions (Fig. 14–52).

1. Draw the extension lines lightly and extend any center lines to be used for dimensioning.
2. Locate and draw lightly the dimension lines and leaders.
3. Examine the layout to see that all features to be dimensioned are in place.
4. Draw guide lines and letter the dimensions.

Dimensioning **337**

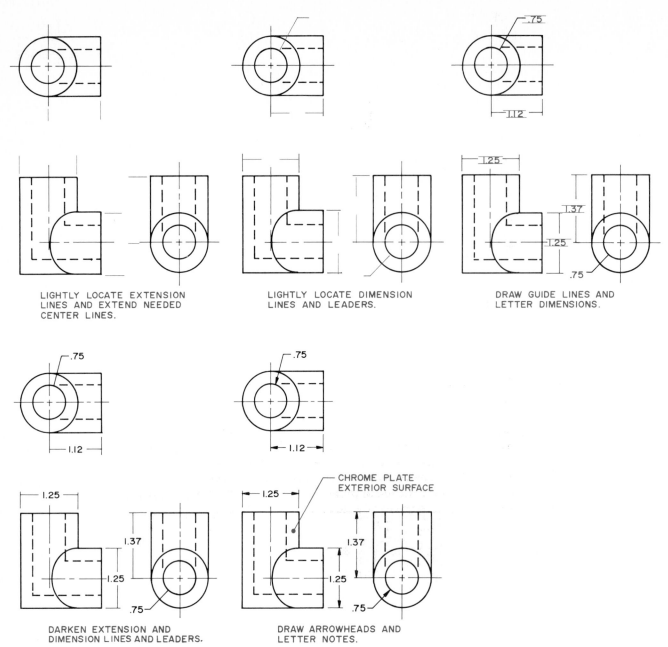

FIGURE 14-52 A suggested order for placing dimensions on a drawing.

5. Darken the extension and dimension lines and leaders and adjust their length as needed.

6. Draw arrowheads.

7. Letter any notes.

## Placement on Sectional Views

Dimensions on sectional views should be placed off the view. Placement follows the procedures for dimensioning any other view. If it is necessary to place a dimension on a cut surface, omit the section lines in the area to contain the dimension. Inside diameters are lettered on the view. The dimension line extends beyond the center line so that it is not confused with a radius. The diameter numeral is followed by DIA or the symbol for diameter (Fig. 14-53).

Occasionally, it is necessary to place other dimensions on the view. To do so, visible edges can be used as extension lines. The decision to do this is based on the need for clarity. If there is no alternative, hidden edges can be drawn on the unsectioned side of a half section and dimensions extended to them.

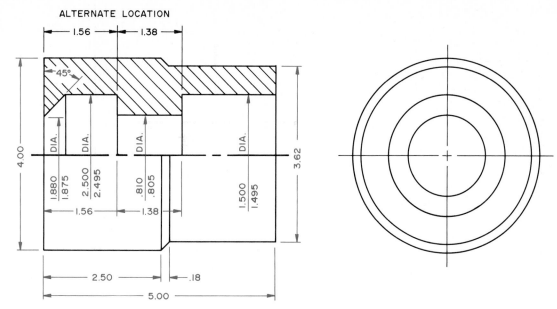

FIGURE 14–53 Procedure for dimensioning a sectional view.

## Placement on Auxiliary Views

Dimensions on auxiliary views follow the same principles as those for any other view. Because the auxiliary view is made specifically to show a feature true size, any dimensions that relate to this feature are shown on the auxiliary view. The dimensions are

placed parallel with the edges of the auxiliary view (Fig. 14–54).

## Placement on Pictorial Views

A pictorial view contains only one view of a part. All dimensions must be related to a position on this one

FIGURE 14–54 Auxiliary views are dimensioned just like other views. The dimensions parallel the edges of the view.

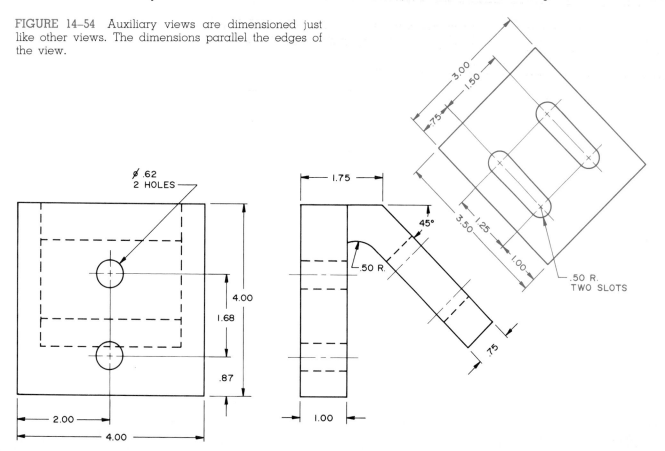

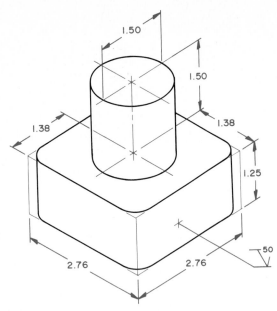

FIGURE 14–55 A pictorial drawing dimensioned using the unidirectional system.

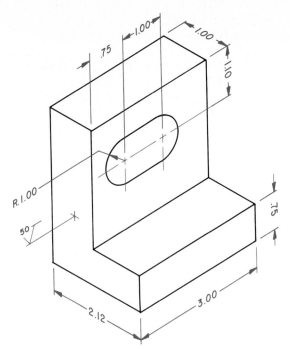

FIGURE 14–56 This drawing is dimensioned using the pictorial plane system. The leaders, dimensions, and arrowheads are parallel with the plane being dimensioned.

view. In general, the basic practices for placing dimensions on orthographic drawings are followed. When the unidirectional system is used, the dimension and extension lines must lie in the same plane as the feature being dimensioned. The numerals and letters are placed so that they are vertical. Leaders and notes are placed so that they lie in or are parallel with the picture plane and are lettered horizontally. Arrowheads are made as on conventional drawings (Fig. 14–55).

When the pictorial plane system is used, the dimensions should lie in one of the pictorial planes. This plane is usually parallel with the plane of the surface being dimensioned. On occasion, it may be necessary to place it perpendicular to the surface being dimensioned. Arrowheads are made in the pictorial plane (Fig. 14–56).

If a finish mark is needed, it is placed perpendicular to the surface. The symbol is drawn so that it is parallel with one of the pictorial axes.

## Summary of Placement Practices

Following is a summary of recommended practices pertaining to the placement of dimensions on an engineering drawing.

1. Place dimensions outside the view. Only when this is not possible or when it would improve the clarity of the situation should dimensions be placed on a view.

2. Dimensions are related to the view where the profile of the feature is most clearly shown. Keep dimensions off the section-lined area on sectional views. If this is not possible, omit the section lining in the area to contain the dimension.

3. Place dimensions between the views. They can be related to both views this way.

4. Place dimensions so that they relate to only one view. Extension lines never extend from one view to another.

5. Place dimensions on the view that shows the distance true length. On inclined and oblique surfaces, this may be an auxiliary view.

6. Place the first dimension 10 mm (0.40 in.) from the outline of the view. Other rows of dimensions paralleling this are placed at least 6 mm (0.24 in.) apart.

7. Place dimensions in the center of the dimension line unless it becomes necessary to stagger the dimensions.

8. Dimensions that fall in strings should have their dimension lines form a continuous row.

9. Place the dimension so that the dimension line is at the midheight of the figures.

10. Place longer or overall dimensions outside shorter dimensions.

11. Place dimensions so that dimension lines do not cross each other or extension lines will not cross dimension lines.

12. Place dimensions so that an extension line or leader does not have to cross a dimension.

13. Do not crowd dimensions; crowding makes them difficult to read.

14. When placing dimensions, utilize center lines to locate features.

15. Do not place a dimension on a center line.

16. Do not use a visible edge of a view as a dimension line.

17. Do not use a visible edge of a view as an extension line unless there is no other choice.

18. Place notes in clear, open areas of the drawing. Letter them horizontally regardless of the dimensioning system being used.

19. Place dimensions so that it is not necessary to dimension to a hidden line. There will be occasions when there is no choice but to dimension to a hidden line, but if this is extensive, make a section view and dimension it.

## DIMENSIONING STANDARD FEATURES

Some features of a part have unique methods of dimensioning. The following material presents the accepted methods for dimensioning these features.

### Diameters

In a one-view drawing in which it is not clear that the feature being dimensioned is a cylinder, place the abbreviation DIA or the diameter symbol after the numeral. If the diameter is dimensioned in the circular view, the abbreviation or symbol is omitted. If the diameters of several concentric cylindrical features are required, they are indicated in the rectangular view in which the cylinders appear as an edge whenever possible (Fig. 14–57). If it is necessary to dimension them or other circular features on the cir-

FIGURE 14–57 The diameters of cylindrical objects are dimensioned in the view in which the cylinder appears as an edge.

CYLINDER DIMENSIONED TO EDGE VIEW.

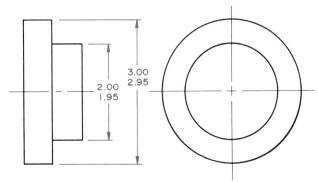

DIAMETER SYMBOL OR ABBREVIATION NOT REQUIRED.

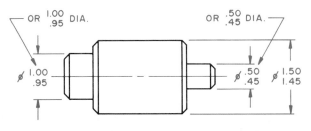

DIAMETER SYMBOL OR ABBREVIATION REQUIRED.

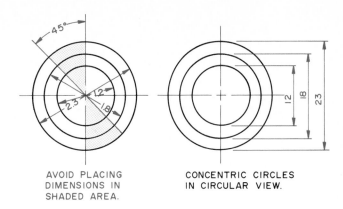

AVOID PLACING DIMENSIONS IN SHADED AREA.

CONCENTRIC CIRCLES IN CIRCULAR VIEW.

DIAMETER SYMBOL OR ABBREVIATION NOT NEEDED.

FIGURE 14–58 Avoid placing dimensions in the shaded area because they are difficult to read.

cular view, the dimensions may be placed on or off the object. When using the aligned dimensioning system, avoid placing the dimensions in the shaded area shown in Fig. 14–58 because they are hard to read in this location.

### Radii

An arc is dimensioned by giving the radius. The dimension is connected to the arc with a leader that has an arrowhead touching the arc. If the location of the center is important and there is sufficient room, the dimension is placed on a dimension line running from the center to the radius with an arrow touching the radius. The numerical value of the radius is always followed by the abbreviation R (Fig. 14–59).

When the center of the radius is located with location dimensions, it is shown with a cross. Extension and dimension lines then locate the center in relation to some part of the object (Fig. 14–60). If the location of the center is not important, the arcs can be located with tangent lines. The extension lines run from the points of tangency between the arcs. The center of the radius is not shown (Fig. 14–61).

If the center of the radius is outside the drawing or extends into another view, the radius dimen-

FIGURE 14–59 How to dimension arcs.

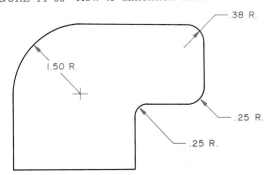

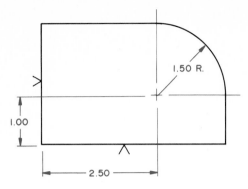

FIGURE 14–60 The center of a radius can be located in relation to some part of the product.

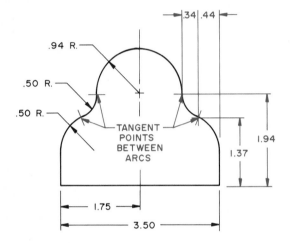

FIGURE 14–61 These arcs are located by dimensioning their tangent points.

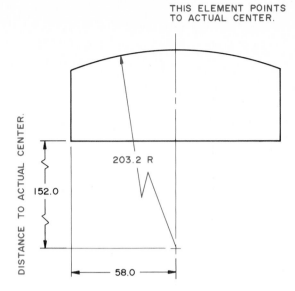

FIGURE 14–62 The radius line for a center that is outside the drawing is shortened using the break symbol.

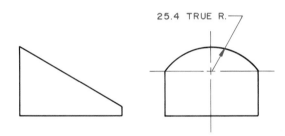

FIGURE 14–63 When an arc is dimensioned in a view in which it is not true size, the dimension shown is identified as a "true radius."

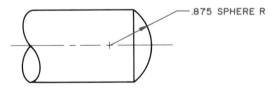

FIGURE 14–64 The radius of a spherical surface is identified as a "spherical radius."

sion line is shortened. The part of the dimension line touching the curve points toward the actual center. The dimension lines that carry the dimensions to locate the center are also shortened, using a break-line symbol (Fig. 14–62).

If a radius must be dimensioned in a view where it does not show true size, "TRUE R" is written after the dimension (Fig. 14–63).

When a part has several radii of the same size, the radius is given in a note rather than individually. A typical note would read "ALL RADII 0.25"; if they are fillets and rounds, it would read "ALL FILLETS AND ROUNDS 0.125 R."

Spherical surfaces that are dimensioned by radii have the abbreviation SPHER R, following the dimension (Fig. 14–64).

## Chords, Arcs, and Angles

A chord is the straight-line distance between two points on a curve. An arc is a distance along the actual curve. When dimensioning these, it is necessary to place the term CHORD or ARC after the dimension so that its intent is clear. Angles are dimen-

sioned in degrees. Their extension lines are radial, extending toward the apex of the angle. When circular features are dimensioned by their angle, the dimensions are placed as in Fig. 14–65. Other ways to dimension angles are shown in Fig. 14–66. Angles are usually dimensioned by giving the size of the angle in degrees, minutes, and seconds. When necessary, dimensions can be given in decimal parts of a degree. Angles can be toleranced as shown in Fig. 14–66. When the number of minutes or seconds is 0, this is shown as part of the dimension, although "zero seconds" may be omitted if it occurs in all angle measurements.

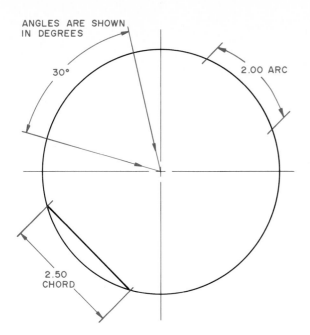

FIGURE 14-65 Chords, arcs, and angles are used when dimensioning curves.

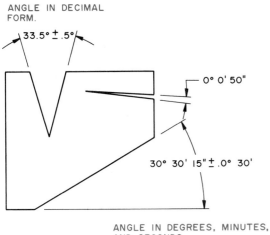

FIGURE 14-66 Other techniques used to dimension angles.

The sides of the angle are extended by extension lines to receive the dimension.

## Batter, Grade, Slope, and Bevel

Batter and slope are used on civil engineering drawings. A *batter* is a deviation from the vertical, such as the face of a retaining wall. A *slope* is a deviation from the horizontal. Both are expressed as a ratio, with one side of the ratio equal to 1 (Fig. 14–67). *Grade,* as used in highway construction, is the same as slope except that the inclination from the horizontal is expressed in a percentage of rise per 100 feet of run. A 15% grade has a rise of 15 feet for every 100 feet of run (Fig. 14–68).

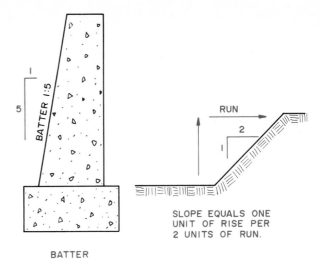

FIGURE 14-67 Batter and slope are used on civil engineering drawings.

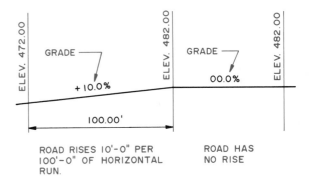

FIGURE 14-68 Grade is used on highway drawings.

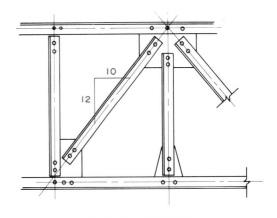

BEVEL OF A MEMBER OF A STEEL TRUSS.

FIGURE 14-69 The angles of structural steel members are called bevels.

The angle a structural member makes with the horizontal is expressed as a *bevel*. A *bevel* is any angle between two members except 90°. Bevels are expressed as a ratio of inches of rise to 12 inches of run (Fig. 14–69).

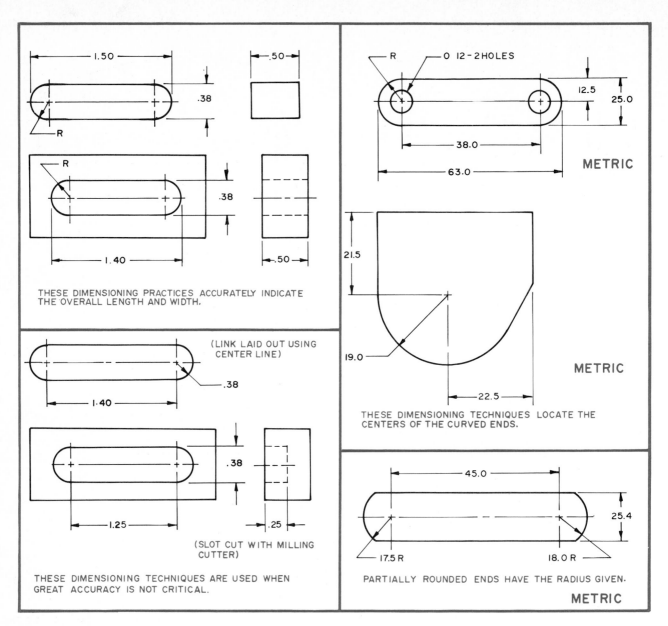

FIGURE 14–70 Dimensioning parts with rounded ends.

## Parts with Rounded Ends

The method used to dimension parts with rounded ends depends on the manufacturing process or the degree of accuracy necessary. When the overall length must be accurate, the dimension is given over the curved ends and the radius is indicated but not given a specific value. A part with partially rounded ends requires that the radius be given. When the location of the centers of the rounded ends is important, the center-to-center dimensions are given, or the center can be located from surfaces on the product (Fig. 14–70).

When precise accuracy is not required, the rounded ends can be dimensioned to aid the people

who are making the part. The parts are laid out in the same manner as they would be in the shop. The link in Fig. 14–70 would be laid out by locating the centers of the curved ends along the center line. The part with the slot would have it milled out. The machinist would like to know the distance from center to center because this is the length of travel of the milling cutter. The diameter of the cutter establishes the width of the slot and the curved end.

Often, a hole is located in a rounded corner. The hole and the radius of the corner share the same center. When the location of one is more important than the other, each is toleranced separately (Fig. 14–71).

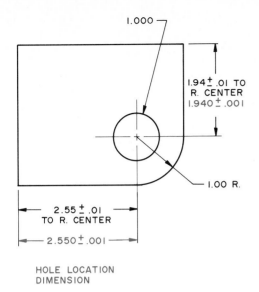

FIGURE 14–71 The location of the center of a hole and a rounded corner can each be toleranced separately when necessary for accuracy.

## Rounded Corners

Rounded corners are dimensioned by locating the edges of the part. The rounded corner is drawn tangent to these edges (Fig. 14–72). If the location of the corner radius must be toleranced, it is dimensioned as in Fig. 14–71.

## Curved Outlines Composed of Arcs

A curved outline of several arcs is dimensioned by locating the center of each arc and giving the radius. If a radius can be located using its point of tangency, the center is not located (Fig. 14–73).

## Irregular Curves

Irregular curves are dimensioned by dimensioning points on the curve from a datum (Fig. 14–74) or locating vertical or horizontal coordinates from a datum (Fig. 14–75). A datum is a reference line, surface, or point from which distances are measured.

FIGURE 14–72 Rounded corners are dimensioned by locating the edges of the part.

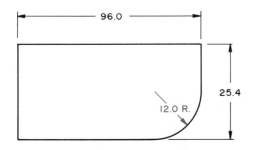

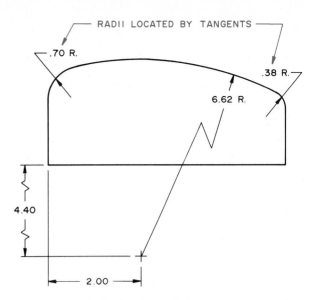

FIGURE 14–73 Radii located by tangents need not have their centers located.

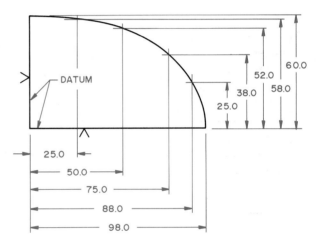

FIGURE 14–74 Irregular curves are dimensioned by locating points on the curve from the datum.

FIGURE 14–75 Coordinates can be used to locate points forming an irregular curve.

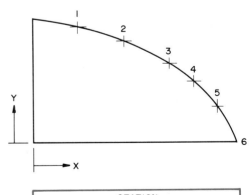

| STATION | | | | | |
|---|---|---|---|---|---|
| | 1 | 2 | 3 | 4 | 5 |
| X | 1.00 | 2.00 | 3.00 | 3.50 | 4.00 |
| Y | 2.44 | 2.12 | 1.68 | 1.31 | .75 |

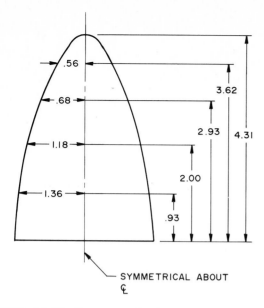

FIGURE 14–76 Symmetrical objects can be dimensioned using points on the curve and recording them on one side of the center line.

## Symmetrical Outlines

Symmetrical outlines are dimensioned about the center line. Selected points on the curve are located by a dimension from a datum. The distance from the center line to the outline is recorded for each point. The note "SYMM ABOUT CL" indicates that the other curved side has the same dimensions (Fig. 14–76).

## Round Holes

Round holes are dimensioned by giving their diameter. They are usually dimensioned to the circular view. The process used to form the hole, such as drilling, is not specified. Round holes may be dimensioned in section if this adds to the clarity of the drawing. When this is done, the dimension is followed by the abbreviation DIA or the diameter symbol. The depth of the hole follows the dimension. If it is a through hole, the dimension is followed by the word THRU. On section views, a hole can be dimensioned from the longitudinal view of the hole (Fig. 14–77).

## Slotted Holes

A slotted hole of regular shape is dimensioned by giving its length and width. The radius of the end is indicated but not dimensioned. The slot is located by the end of the slot or to the center line (Fig. 14–78).

## Countersunk Holes

A countersunk hole is specified by the diameter of the hole, the diameter of the countersink, and the

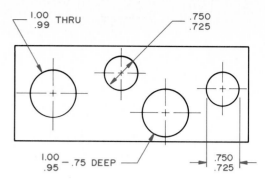

DIMENSIONING CIRCULAR VIEWS OF HOLES.

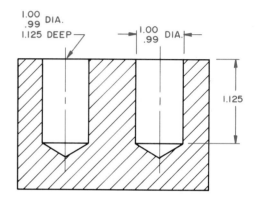

DIMENSIONING ON SECTIONAL VIEWS.

FIGURE 14–77  Techniques for dimensioning round holes.

FIGURE 14–78  Slotted holes are dimensioned with a note or by projecting off the circular views.

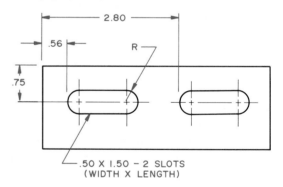

SLOTTED HOLES DIMENSIONED BY GIVING WIDTH AND LENGTH IN A NOTE.

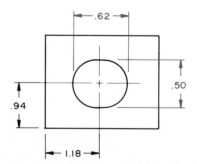

SLOTTED HOLE LOCATED BY CENTER LINES. WIDTH AND LENGTH PROJECTED OFF CIRCULAR VIEWS.

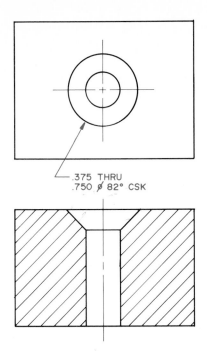

FIGURE 14-79 Countersunk holes are specified with a note off the circular view.

included angle of the countersink. This is shown with a note to the circular view (Fig. 14-79).

## Counterdrilled Holes

A counterdrilled hole is specified by the diameter of the hole, the diameter of the counterdrill, the included angle of the counterdrill, and its depth (Fig. 14-80). The note is to the circular view.

## Counterbored Holes

A counterbored hole is specified by the diameter of the hole, the diameter and depth of the counterbore, and its corner radius. If the thickness of the material remaining below the counterbore is important, it is given on a section view rather than as the depth of the counterbore (Fig. 14-81). The note is connected to the circular view.

## Spotfaces

Spotfaces are specified by the diameter of the flat, circular area and the depth or the desired remaining thickness below it. It is shown by a note if not drawn on the drawing. Frequently, a spotface is used to clean up a surface around a hole. In this case, no depth is shown (Fig. 14-82).

## Chamfers

Chamfers are specified by their angle and length. The length is measured parallel with the axis of the part. A note can be used to specify chamfers on a

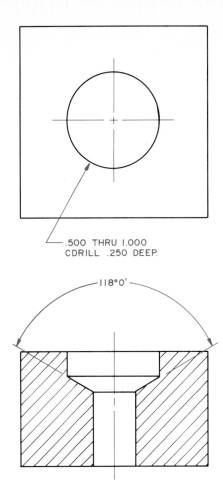

FIGURE 14-80 Counterdrilled holes are specified with a note off the circular view.

FIGURE 14-81 Counterbored holes are specified with a note off the circular view.

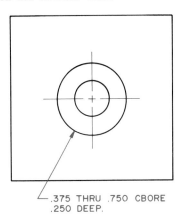

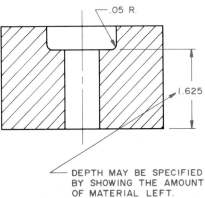

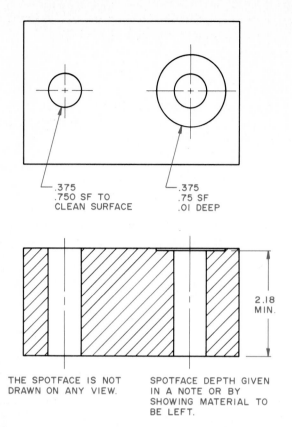

.375
.750 SF TO
CLEAN SURFACE

.375
.75 SF
.01 DEEP

2.18
MIN.

THE SPOTFACE IS NOT
DRAWN ON ANY VIEW.

SPOTFACE DEPTH GIVEN
IN A NOTE OR BY
SHOWING MATERIAL TO
BE LEFT.

FIGURE 14–82  Spotfaces are specified with a note off the circular view.

45° angle. In this case, the longitudinal and radial dimension are the same (Fig. 14–83). When the edge of a round hole is chamfered on 45°, the diameter and angle of the countersink can be given in a note

FIGURE 14–83  A chamfer may be specified by angle and length. Only 45° chamfers may be specified with a note.

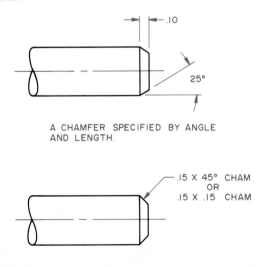

.10

25°

A CHAMFER SPECIFIED BY ANGLE
AND LENGTH.

.15 X 45° CHAM
OR
.15 X .15  CHAM

DIMENSIONING A 45° CHAMFER.

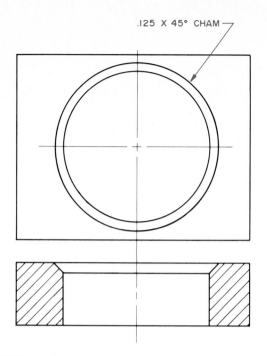

.125 X 45° CHAM

FIGURE 14–84  When the edge of a round hole is chamfered on an angle of 45°, the angle and diameter are given.

(Fig. 14–84). If it is necessary to control the diameter of the chamfer, the diameter and angle are given as shown in Fig. 14–85.

FIGURE 14–85  When the diameter of a chamfer in a round hole must be controlled, the diameter and angle must be given.

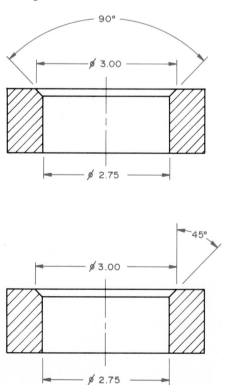

90°

∅ 3.00

∅ 2.75

45°

∅ 3.00

∅ 2.75

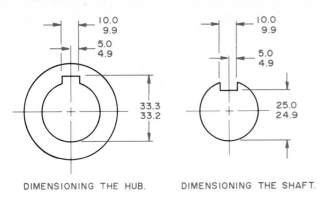

DIMENSIONING THE HUB.    DIMENSIONING THE SHAFT.

FIGURE 14–86   How to specify keyseats in hubs and shafts.

## Keyseats

Keyseats are specified by giving their width, depth, location, and length if they are not cut through the part. The depth of the keyseat in a shaft is dimensioned from the side opposite the bottom. In the mating part, it is dimensioned from the opposite side over the top of the keyseat (Fig. 14–86).

## Knurling

Knurling is specified by type, pitch, and the diameter of the part before and after knurling. If the finished diameter is not important, the after-knurling

FIGURE 14–87   Knurling is specified with a note.

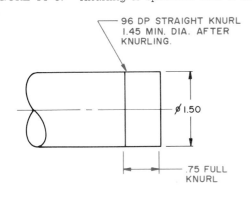

A KNURL FOR A PRESS FIT.

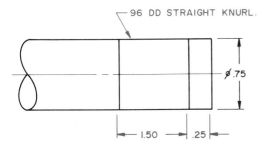

A KNURL FOR DECORATION OR GRIPPING.

dimension is omitted. The location of the knurl on the part is also dimensioned (Fig. 14–87).

## Conical Tapers

A taper is a conical surface. Conical tapers are specified using some combination of the following dimensions and tolerances.

1. The diameter at one end of the taper.
2. The length of the taper.
3. The diameter at a cross-sectional plane. This is a basic dimension from which tolerances are taken.
4. A location dimension locating the position of the cross-sectional plane.
5. The rate of taper.
6. The included angle.

FIGURE 14–88   Ways to dimension conical tapers.

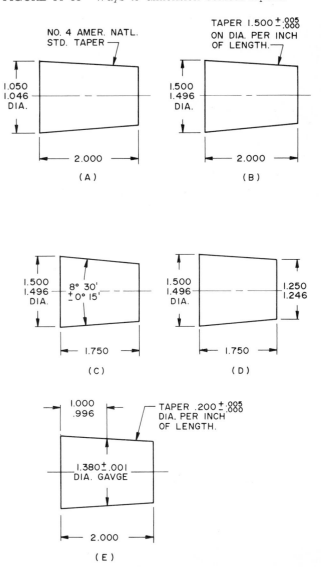

Conical tapers are dimensioned in several ways, depending on the method to be used to manufacture them (Fig. 14–88). One way is to specify a standard taper plus the diameter at one end and the length (A). Standard tapers are commonly used on spindles, tool shanks, and pins. They are shown in the appendix. They are given in "inches per foot." This is the amount in inches the diameter varies per linear foot of length.

Another method is to specify the amount of taper per inch of length plus the diameter at one end and the length (B). The "taper per inch" is the amount in inches the diameter varies in one inch of length.

Tapers may be specified by giving the diameter of the large end, the included angle, and the length (C) or the diameter at both ends plus the length (D). A more accurate method is to specify the diameter at a gauge line. This is a cross-sectional plane at which the basic size is specified. The taper can be specified in inches per inch of length (E). The diameters at the ends are controlled by the tolerances on the angle.

## Repetitive Dimensions

When several identical features, such as a series of holes, are spaced equally, a notation can be used to indicate the spacing rather than dimensioning each individually. If there are tolerances involved, these should be specified. Typical notations are shown in Fig. 14–89.

## Coordinate Dimensions for Symmetrical Features

Various symmetrical features can be located by giving the distances or directions of the features from a center line or a datum. Typical coordinate dimensions are shown in Fig. 14–90. If coordinate dimensions without dimension lines are used, they originate from a datum plane which is the zero point. The horizontal and vertical distances from the datum are shown with extensions of the center lines. No dimension lines or arrowheads are required (Fig. 14–91). The holes are identified by a size symbol, usually a letter. The sizes are shown in a table beside the drawing.

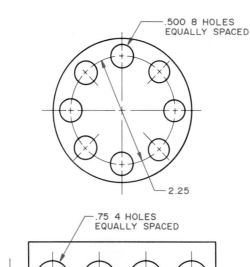

FIGURE 14–89 Equally spaced features can be dimensioned with a note.

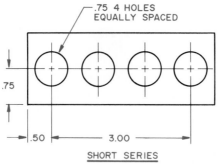

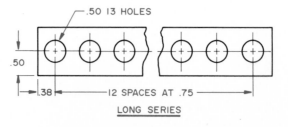

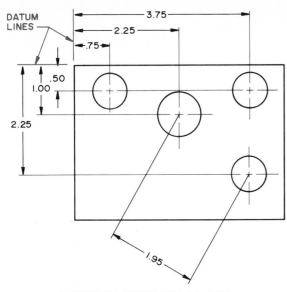

COORDINATE DIMENSIONS LOCATED FROM A DATUM.

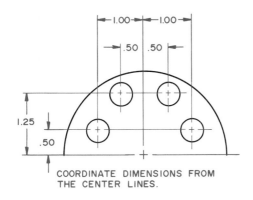

COORDINATE DIMENSIONS FROM THE CENTER LINES.

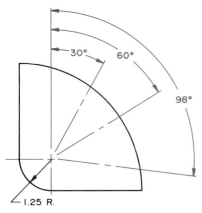

POLAR COORDINATE DIMENSIONS FROM A CENTER LINE.

**FIGURE 14–90** Coordinate dimensions are used to dimension symmetrical features.

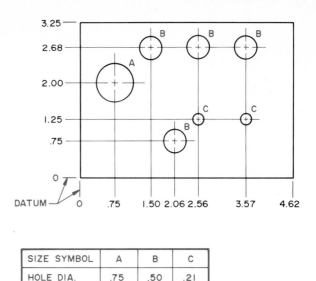

| SIZE SYMBOL | A | B | C |
|---|---|---|---|
| HOLE DIA. | .75 | .50 | .21 |

**FIGURE 14–91** Coordinate dimensions can be taken from the datum.

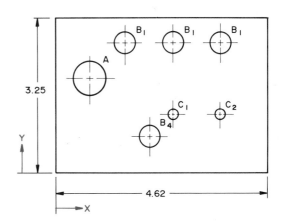

| NO. REQ'D | | 1 | 4 | 2 |
|---|---|---|---|---|
| HOLE DIA. | | .710 | .438 | .250 |
| POSITION | | HOLE | SYMBOL | |
| X → | Y ↑ | A | B | C |
| .75 | 2.00 | A₁ | | |
| 1.50 | 2.68 | | B₁ | |
| 2.56 | 2.68 | | B₂ | |
| 3.57 | 2.68 | | B₃ | |
| 2.56 | 1.25 | | | C₁ |
| 3.57 | 1.25 | | | C₂ |

**FIGURE 14–92** Tabular dimensioning locates features with dimensions from the X and Y axes.

## Tabular Dimensioning

Tabular dimensioning is another system for placing dimensions. It uses the principles of rectangular coordinates described in the preceding paragraphs. The dimensions locating the center of each feature are taken from mutually perpendicular datum planes. The horizontal distance is the X axis. The vertical distance is the Y axis. The coordinates for each feature are given in a table by recording their X and Y values. This is particularly useful when there are a large number of features to locate (Fig. 14–92).

# Problems

The following problems provide a variety of experiences in dimensioning. When laying out the problem, be certain to allow sufficient room between views for dimensions. In many cases, the examples need more space than shown. Those with inch and metric scales can be dimensioned using either. They can be dimensioned using the aligned or unidirectional system. Your instructor will indicate which of these you are to use.

1. Sketch the objects in Probs. 14–1 through 14–4 on 8½ × 11 in. coordinate paper with ¼-in. squares. Then, sketch the dimensions with extension lines and dimension lines. You may leave off the actual dimensions because the important part of this assignment is the selection and location of the dimensions rather than the actual size. Space is tight, so work carefully. A little preplanning on scrap paper before dimensioning the final drawing would be advisable.

2. Draw full size and dimension Probs. 14–5 through 14–10. Printed by each problem is a scale in inches and millimeters. Use it to measure the printed drawing to determine the sizes. Record the actual dimensions in inches or millimeters as assigned by your instructor. If it helps to dimension an object, draw a special view, such as a section, auxiliary, or partial view. Before starting to work, sketch out the object to actual size and locate the dimensions so that you can plan to fit it acceptably on the paper. Problems can be drawn on 8½ × 11 in. or 297 × 420 mm coordinate or plain paper, as assigned.

3. Dimension the problems you solved in Chapter 6 or solve any of the problems assigned and dimension them.

4. Bring products to the classroom. Dismantle them and produce dimensioned drawings of each part. If assigned, redraw a product manufactured in inches as a metric product.

PROBLEM 14–1                    PROBLEM 14–2

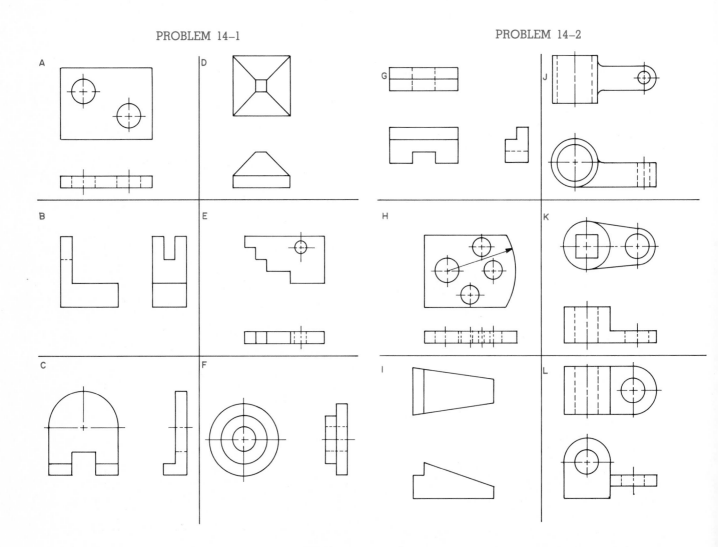

M

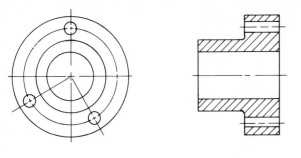

N

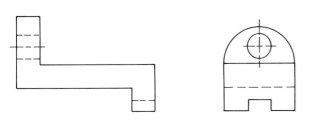

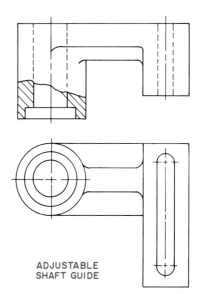

ADJUSTABLE
SHAFT GUIDE

PROBLEM 14-5

PROBLEM 14-6

O

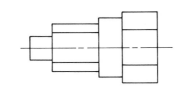

PROBLEM 14-3

PROBLEM 14-4

P

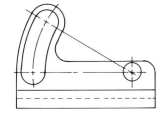

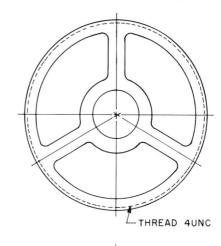

THREAD 4UNC

Q

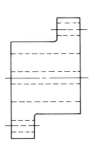

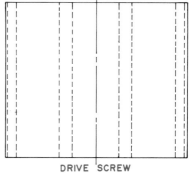

DRIVE SCREW

R

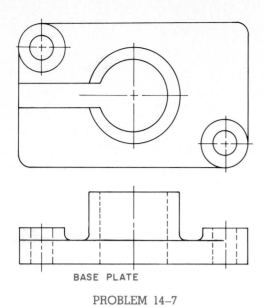

BASE PLATE

PROBLEM 14-7

PROBLEM 14-8

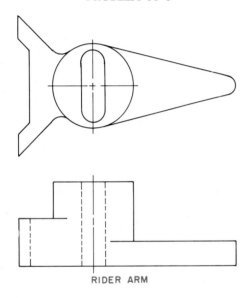

RIDER ARM

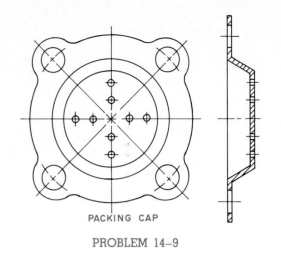

PACKING CAP

PROBLEM 14-9

PROBLEM 14-10

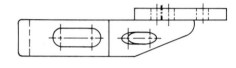

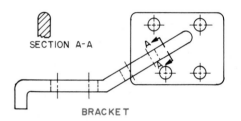

SECTION A-A

BRACKET

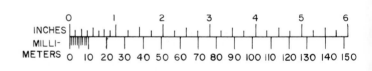

| INCHES | 0 | 1 | 2 | 3 | 4 | 5 | 6 |
|---|---|---|---|---|---|---|---|
| MILLI-METERS | 0 10 20 30 | 40 50 60 70 80 | 90 100 110 | 120 130 | 140 150 | | |

# Tolerancing and Surface Quality

The sizes of parts shown on a drawing are directly related to where and how the part fits into the product for which it is intended. If a part must fit against or into another part, both parts must be dimensioned to a size that permits them to be assembled. Most products are made of parts that are built in different factories. When these parts are assembled at a central location, any one must fit into the assembly. In other words, the parts must be interchangeable. To ensure that parts are interchangeable, their sizes are *toleranced* on the engineering drawings. *Tolerancing* means that each dimension is allowed to vary in size within a specified range. This assures that the parts will fit when assembled and reduces the cost of producing them. The size of the tolerance depends on the degree of precision needed for the surface being dimensioned.

Each dimension on each part is toleranced. Dimensions that are not critical to the final assembly have a general tolerance. The *general tolerance* is given on the drawing with a note or in supplementary documents. Such a note might read, "TOLERANCE ± 0.01 UNLESS OTHERWISE SPECIFIED." Often, the general tolerance is shown in the title block. Parts that must join, called mating parts, usually have closer tolerances. These are applied directly to these dimensions (Fig. 15–1). Dimensions

ries: form tolerances and positional tolerances. Standards for tolerancing are available in publication ANSI Y14.5-1973, "Dimensioning and Tolerancing," published by the American Society of Mechanical Engineers, United Engineering Center, 345 East 47th Street, New York, N.Y. 10017.

## TERMINOLOGY

The terms used in connection with tolerancing are standardized, permitting a common language to be used on drawings, in specifications, and in discussions of tolerancing.

*Actual size*   is the measured size of a finished part.

*Nominal size*   is an approximate size used for general identification. For example, a shaft could be described as having a nominal size of 1 in., while in reality tolerances might limit it to something smaller (Fig. 15–2).

labeled BSC (basic), REF (reference), MAX (maximum), and MIN (minimum) do not have tolerances because they are absolute dimensions upon which tolerances are based.

Tolerancing dimensions fall into two catego-

FIGURE 15–1   General tolerances apply to all parts of the product except those with specific, individual tolerances.

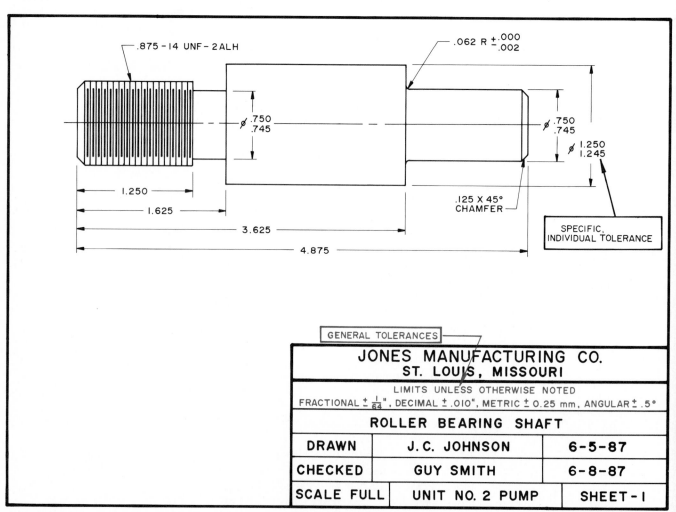

356   Chapter 15

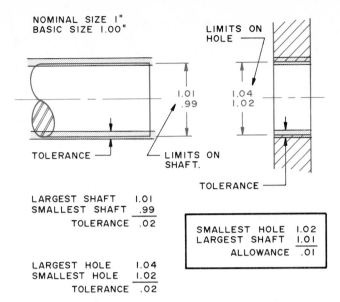

NOMINAL SIZE 1"
BASIC SIZE 1.00"

LIMITS ON HOLE

1.01
.99 — LIMITS ON SHAFT.

TOLERANCE

1.04
1.02

TOLERANCE

| LARGEST SHAFT | 1.01 |
| SMALLEST SHAFT | .99 |
| TOLERANCE | .02 |

| SMALLEST HOLE | 1.02 |
| LARGEST SHAFT | 1.01 |
| ALLOWANCE | .01 |

| LARGEST HOLE | 1.04 |
| SMALLEST HOLE | 1.02 |
| TOLERANCE | .02 |

FIGURE 15–2 The application of limits to the basic size produces the tolerances.

**Basic size** is the exact, theoretical size from which limits are derived by the application of tolerances and allowances (Fig. 15–2).

**Design size** is the size from which the size limits are taken by the application of tolerances. When there is no allowance, the design size and the basic size are the same.

**Limits of size** are the maximum and minimum sizes (Fig. 15–2).

**Tolerance** is the difference between the limits of size for each dimension (Fig. 15–2).

**Allowance** is the tightest fit between two mating parts, such as the largest shaft and the smallest hole (Fig. 15–2). It is the minimum clearance (positive allowance) or the maximum interference (negative allowance) between mating parts.

**Single limits** are dimensions that are designated either MIN (minimum) or MAX (maximum) but not both. For example, the minimum length of a thread could be given, but not the maximum, because it is not important to the product (Fig. 15–3). Depths of holes, corner radii, and chamfers are often given single limits. They are used where the intent of the dimension is clear and the unspecified limit can ap-

proach zero or infinity without creating conditions that are detrimental to the design.

**Fit** is a general term used to specify the range of tightness that will result from the application of a particular combination of allowances and tolerances applied to mating parts. The three general types of fits are clearance, interference, and transition.

**Clearance fit** is a fit that provides for clearance between two assembled mating parts. Even when the parts are at their tightest condition, they will still have clearance (Fig. 15–4).

**Interference fit** is a fit where the limits specified always result in an interference between the mating parts when they are assembled (Fig. 15–5).

**Transition fit** is a fit that can result in two assembled mating parts' having either a clearance fit or an interference fit (Fig. 15–6).

**Unilateral tolerances** are tolerances that vary in only one direction from the design size (Fig. 15–7).

**Bilateral tolerances** are tolerances in which variations larger and smaller than the design size are permitted (Fig. 15–7).

The **basic hole system** is a system of fits in which the design size of the hole is the same as the basic

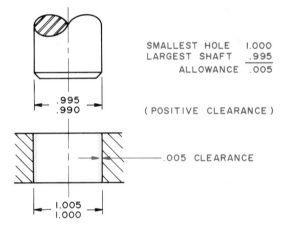

| SMALLEST HOLE | 1.000 |
| LARGEST SHAFT | .995 |
| ALLOWANCE | .005 |

( POSITIVE CLEARANCE )

.995
.990

.005 CLEARANCE

1.005
1.000

FIGURE 15–4 A clearance fit allows positive clearance between mating parts.

FIGURE 15–5 An interference fit always provides negative allowance (interference) between mating parts.

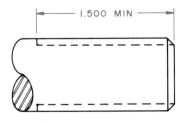

FIGURE 15–3 A single-limit dimension is designated as maximum or minimum.

1.500 MIN

DIMENSION MEANS THE THREAD LENGTH MUST BE AT LEAST 1.500 IN. LONG.

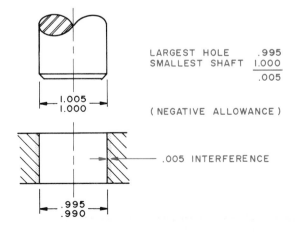

| LARGEST HOLE | .995 |
| SMALLEST SHAFT | 1.000 |
| | .005 |

( NEGATIVE ALLOWANCE )

1.005
1.000

.005 INTERFERENCE

.995
.990

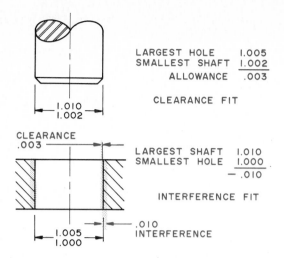

LARGEST HOLE 1.005
SMALLEST SHAFT 1.002
ALLOWANCE .003

CLEARANCE FIT

1.010
1.002

CLEARANCE .003

LARGEST SHAFT 1.010
SMALLEST HOLE 1.000
− .010

INTERFERENCE FIT

.010
INTERFERENCE

1.005
1.000

FIGURE 15–6  A transition fit can result in a positive (clearance) or negative (interference) fit.

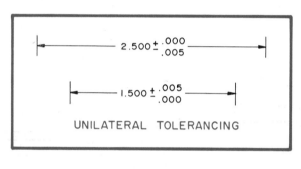

$2.500 {}^{+.000}_{-.005}$

$1.500 {}^{+.005}_{-.000}$

UNILATERAL TOLERANCING

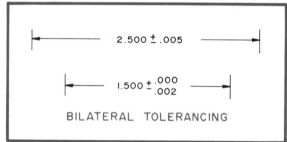

$2.500 \pm .005$

$1.500 {}^{+.000}_{-.002}$

BILATERAL TOLERANCING

FIGURE 15–7  Tolerances may be bilateral or unilateral.

size and the allowance is applied to the shaft (Fig. 15–8).

The **basic shaft system** is a system of fits in which the design size of the shaft is the same as the basic size and the allowance is applied to the hole (Fig. 15–9).

## TOLERANCE EXPRESSION

Tolerance limits (the plus-or-minus tolerance expression) and the related dimension must always be given in the same form and to the same number of decimal points (Fig. 15–10).

General tolerances, usually given in the title block, apply to all dimensions not carrying a specific

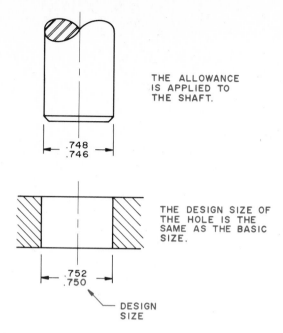

THE ALLOWANCE IS APPLIED TO THE SHAFT.

.748
.746

THE DESIGN SIZE OF THE HOLE IS THE SAME AS THE BASIC SIZE.

.752
.750

DESIGN SIZE

FIGURE 15–8  The basic hole system of tolerances uses the design size of the hole as the basic size.

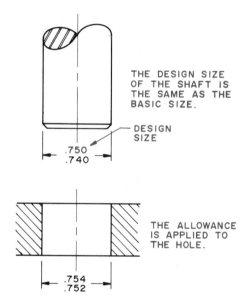

THE DESIGN SIZE OF THE SHAFT IS THE SAME AS THE BASIC SIZE.

DESIGN SIZE

.750
.740

THE ALLOWANCE IS APPLIED TO THE HOLE.

.754
.752

FIGURE 15–9  The basic shaft system of tolerances uses the design size of the shaft as the basic size.

tolerance. They may be expressed as a common fraction (such as ± 1/64 in.), a decimal fraction (± 0.01 in.), or in millimeters (± 0.25 mm). Angular tolerances are given in degrees (± 1/2°).

## TOLERANCING METHODS

Tolerances applied directly to the dimension can be expressed as limit dimensions or plus-or-minus dimensions. When using *limit dimensioning*, the high

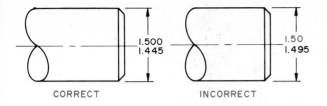

CORRECT                    INCORRECT

BOTH LIMITS SHOULD HAVE THE SAME
NUMBER OF DECIMAL PLACES.

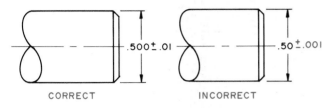

CORRECT                    INCORRECT

LIMITS SHOULD HAVE THE SAME NUMBER
OF DECIMAL PLACES AS THE DIMENSION.

FIGURE 15–10 Tolerance limits and the related design size should have the same number of decimal places.

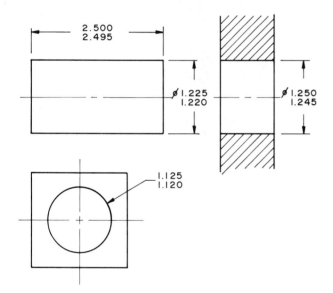

FIGURE 15–11 When applying limit dimensions, the larger limit is placed above the smaller limit or to the right of the lower limit when they are in line.

limit (maximum size) is placed above the low limit (minimum size). When expressed in a single line, the low limit precedes the high limit, and they are separated by a dash (Fig. 15–11).

*Plus-or-minus* tolerances are expressed by giving the dimension first, followed by a plus-or-minus symbol and the tolerance (Fig. 15–7).

## SELECTIVE ASSEMBLY

Selective assembly is used when the fit desired between mating parts is so close that the cost to produce interchangeable parts is prohibitive. The toler-ances used are as small as practicable. The parts are produced and carefully measured to establish their actual size. Those that are placed together in an assembly are selected from the sizes available to produce the allowance desired.

## TOLERANCE SELECTION

The selection of tolerances involves many factors. The engineer must understand the manufacturing processes to be used and the limitations these impose on accuracy. Parts can be cast to certain tolerances and machined to closer tolerances. There is no use specifying that a part be machined to tolerances that are impossible to attain with a particular process. The cost of producing parts with various tolerances must be considered. Tolerances should be no closer than needed for the product to function properly.

Experience in designing and specifying tolerances is essential to the final decision. For example, how much allowance is required to maintain the desired oil film on a rotating part? Other factors to consider are the temperatures at which the product will function, the speed of rotation, material used, humidity, and pressures expected. Sometimes, experimental models are built and thoroughly tested. Tables of recommended fits for general work, such as the ANSI table of cylindrical fits, are available (see Appendices G and H).

Some general guidelines for selecting tolerances for parts to be finished by basic machine tools or formed by casting or forging are shown in the tables on pg. 360.

## APPLICATION OF TOLERANCES

The engineer must ascertain the minimum and maximum clearance or interference that can exist between mating parts and still permit them to function satisfactorily. The difference between the maximum and minimum limits will be the tolerance. For example, assume a 1-in. shaft in a journal must have at least 0.003 in. clearance to allow for lubrication and expansion due to heat. It can have a maximum clearance of 0.010 in. The tolerance is:

| Maximum clearance | 0.010 in. |
| Minimum clearance | 0.003 in. |
| Tolerance | 0.007 in. |

If half of this is on the shaft and half on the journal, each could have a tolerance of 0.0035 in. However, a four-decimal tolerance is excessive. Also, if turned on a lath, the shaft can only be made with a tolerance of 0.003 in. The journal may be reamed and can therefore have a tolerance of 0.001 in. Since the hole

| MACHINING | INCHES | MILLIMETERS |
|---|---|---|
| Grinding | | |
|   Cylindrical and Surface | | |
|   Grinding | ± 0.0005 | 0.013 |
| Milling | | |
|   Milling a single surface | ± 0.002 to 0.003 | ± 0.05 to 0.08 |
|   Milling two surfaces | | |
|     Most important surface | ± 0.002 | ± 0.05 |
|     Other surfaces | ± 0.005 | ± 0.13 |
| Planing and Shaping | | 0.13 to 0.25 |
|   Used to finish larger parts | 0.005 to 0.010 | |
| Sand Casting | | |
|   Surfaces that will not be machined | | |
|     Small castings | ± 1/32 | 0.8 |
|     Medium-size castings | ± 1/16 | 1.6 |
| Die Casting and Plastic Molding | | |
|   Small and medium-size | ± 1/64 | 0.4 |
|   Large | ± 1/32 | 0.8 |
|   Some alloys can be cast | ± 0.001 | 0.02 |

| DRILLING | TOLERANCE IN INCHES | METRIC DRILLS | TOLERANCE IN MILLIMETERS |
|---|---|---|---|
| Drills No. 60 to No. 30 | + 0.002  − 0.000 | 0.500 to  3.000 mm | + 0.05  − 0.000 |
|   No. 29 to No. 1 | + 0.004  − 0.000 | 3.000 to  6.000 mm | + 0.10  − 0.000 |
|   1/4 to 1/2 in. | + 0.005  − 0.000 | 6.000 to 12.000 mm | + 0.13  − 0.000 |
|   1/2 to 3/4 in. | + 0.008  − 0.000 | 12.000 to 18.000 mm | + 0.20  − 0.000 |
|   3/4 to 1 in. | + 0.010  − 0.000 | 18.000 to 25.000 mm | + 0.25  − 0.000 |
|   1 to 2 in. | + 0.015  − 0.000 | 25.000 to 50.000 mm | + 0.40  − 0.000 |
| Reaming | | | |
|   Diameter up to 1/2 in. | + 0.0005 − 0.000 | Up to 12.000 mm | + 0.013 − 0.000 |
|   1/2 to 1 in. | + 0.001  − 0.000 | 12.000 to 25.000 mm | + 0.025 − 0.000 |
|   1 in. and larger | + 0.0015 − 0.000 | 25.000 mm and larger | + 0.040 − 0.000 |
| Finish Turning on a Lathe | Total Tolerance | | Total Tolerance |
|   DIA 1/4 to 1/2 in. | 0.002 | DIA  6.00 to 12.00 mm | + 0.05 |
|   1/2 to 1 in. | 0.003 | 12.00 to 25.00 mm | 0.08 |
|   1 to 2 in. | 0.005 | 25.00 to 50.00 mm | 0.13 |
|   Over 2 in. | 0.007 | Over 50.00 mm | 0.18 |
| Broaching | | | |
|   Diameters up to 1 in. | ± 0.001 | Up to 25.000 mm | ± 0.025 |
|   1 to 2 in. | ± 0.002 | 25.000 to 50.000 mm | ± 0.050 |
|   2 to 4 in. | ± 0.003 | 50.000 to 100.000 mm | ± 0.080 |
| Drop Forgings | | | |
|   Less than 1 lb | ± 1/32 | Less than 0.453 kg | 0.800 |
|   1 to 10 lb | ± 1/16 | 0.453 to 4.530 kg | 1.600 |
|   10 to 60 lb | ± 1/8 | 4.530 to 27.200 kg | 3.100 |

in the journal can be reamed to a closer tolerance than the shaft can be turned, it could be chosen as the design size. The hole could have a tolerance of 0.002 in. and the shaft 0.005 in. The application of this reasoning is shown in Fig. 15–12.

## ANSI PREFERRED LIMITS AND FITS FOR CYLINDRICAL PARTS

Limits and fits for general work are recorded in American National Standards Institute publication ANSI B4.1-1967 (R1979), "Preferred Limits and Fits for Cylindrical Parts." These tables appear in Appendices G and H. The limits are given in thousandths of an inch. The limits for the hole and shaft are applied algebraically to the basic size to obtain the limits. The basic hole system is used.

The standard classes of fits are as follows.

  RC  Running or sliding fit
  LC  Locational clearance fit
  LT  Locational transitional fit
  LN  Locational interference fit
  FN  Force or shrink fit

These letter symbols are used with numbers to represent the class of fit. A designation FN4 means

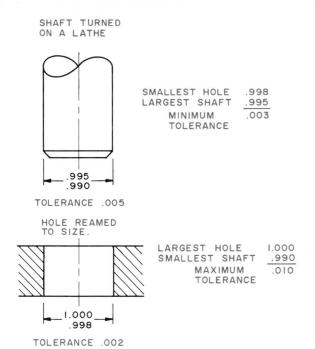

SHAFT TURNED
ON A LATHE

| SMALLEST HOLE | .998 |
| LARGEST SHAFT | .995 |
| MINIMUM TOLERANCE | .003 |

.995
.990

TOLERANCE .005

HOLE REAMED
TO SIZE.

| LARGEST HOLE | 1.000 |
| SMALLEST SHAFT | .990 |
| MAXIMUM TOLERANCE | .010 |

1.000
.998

TOLERANCE .002

FIGURE 15–12 The selection of limits sometimes depends on the process to be used to produce the part. In this example, the hole can be reamed to a closer tolerance; therefore, it is dimensioned with a smaller tolerance.

a class 4 force fit. The symbols and numbers do not appear on the drawing. The limits obtained from the tables are used to size the piece on the drawing.

# DESCRIPTION OF FITS

The classes of fits are arranged in three general groups: running and sliding fits, locational fits, and force fits.

## Running and Sliding Fits (RC)

Running and sliding fits are intended to provide a running performance with suitable lubrication allowance in all sizes. The clearances of the first two classes, used chiefly as slide fits, increase more slowly with diameter than other classes.

*Close sliding fits (RC1)* are intended for accurate location of parts that must be assembled without perceptible play (noticeable movement).

*Sliding fits (RC2)* are intended for accurate location but with greater maximum clearance than RC1 parts that are made to fit, move, and turn easily but are not intended to run freely. In larger sizes, this fit may stick fast due to a temperature change.

*Precision running fits (RC3)* are about the closest fits that can be expected to run freely. They are intended for precision work at slow speeds and light journal pressures. The RC3 fit is not suitable when

appreciable temperature differences are likely to occur.

*Close running fits (RC4)* are intended chiefly for running fits on accurate machinery with moderate surface speeds and journal pressures. They are used where accurate location and minimum play are desired.

*Medium running fits (RC5 and RC6)* are intended for higher running speeds and/or heavy journal pressures.

*Free running fits (RC7)* are used where accuracy is not essential or where large temperature variations are likely to occur.

*Loose running fits (RC8 and RC9)* are used when materials such as cold-rolled shafting and tubing made to commercial standards are involved.

## Locational Fits (LC, LT, and LN)

Locational fits are intended to determine only the location of mating parts. They may provide rigid or accurate location, as with interference fits, or provide some freedom of location, as with clearance fits. Locational fits are divided into three groups: clearance fits (LC), transitional fits (LT), and interference fits (LN).

*Locational clearance fits (LC)* are intended for parts which are normally stationary. These fits can be freely assembled or dissassembled. They range from snug fits through medium-clearance fits to loose fastener fits when freedom of assembly is of prime importance.

*Locational transitional fits (LT)* are a compromise between clearance and interference fits. They are used when accuracy of location is important but a small amount of either clearance or interference is permissible.

*Locational interference fits (LN)* are used where accuracy of location is of prime importance. They are also used for parts requiring rigidity and alignment with no requirements for bore pressure. These fits are not intended for parts designed to transmit frictional loads from one part to another. By virtue of tightness of fit, these latter conditions are covered by force fits.

## Force Fits and Shrink Fits (FN)

Force and shrink fits are a special type of interference fit. These are characterized by maintenance of constant bore pressure throughout the range of sizes. The interference varies almost directly with the diameter. The difference between these values is small to maintain the resulting pressures within reasonable limits.

*Light drive fits (FN1)* are those that require light assembly pressures and produce more or less permanent assemblies. They are suitable for thin sections, long fits, or use in cast-iron external members.

*Medium drive fits (FN2)* are suitable for ordinary steel parts or for shrink fits on light sections. They are about the tightest fits that can be used with high-grade cast-iron external members.

*Heavy drive fits (FN3)* are suitable for heavier steel parts or for shorter fits in medium sections.

*Force fits (FN4 and FN5)* are suitable for parts that can be highly stressed or for shrink fits when the heavy pressing forces required are impracticable.

## DIMENSIONING CYLINDRICAL PARTS USING ANSI FITS

Following are several examples to show how to use the ANSI tables. Assume a 1.5-in. diameter shaft is to run at moderate speeds and journal pressures. An RC4 fit would be chosen. The table in Appendix H shows the following:

$$\begin{array}{r} \text{Hole} & + 1.0 \\ & - 0.0 \\ \text{Shaft} & - 1.0 \\ & - 2.0 \end{array}$$

Since these values are in thousandths of an inch and the table uses the basic hole system, the following limits are found:

$$\begin{array}{rl} \text{Hole} & 1.501 \\ & 1.500 \\ \text{Shaft} & 1.499 \\ & 1.498 \end{array}$$

The allowance between the largest shaft and the smallest hole is 0.001 in. The allowance between the smallest shaft and the largest hole is 0.003 in. These are shown in the table of fits under the heading "Limits of Clearance."

Another example could be shown using a force fit. If a 1.0-in. steel shaft is to be joined to a bearing with a medium drive fit, the FN2 table is used. The following limits are found:

$$\begin{array}{r} \text{Hole} & + 0.8 \\ & - 0.0 \\ \text{Shaft} & + 1.9 \\ & + 1.4 \end{array}$$

These data lead to the following limits:

$$\begin{array}{rl} \text{Hole} & 1.0008 \\ & 1.0000 \\ \text{Shaft} & 1.0019 \\ & 1.0014 \end{array}$$

The interference between the largest shaft and the smallest hole is 0.0019 in. The interference between the largest hole and the smallest shaft is 0.0006 in. These are shown in the table under the heading "Limits of Interference."

## ISO SYSTEM OF LIMITS AND FITS

The ISO metric system of limits and fits is a series of standard tolerances for shafts and holes. The terms *shaft* and *hole* refer to the space containing or contained by two parallel faces of any part, such as the thickness of a key, the width of a slot, or the diameter of a pin.

### Tolerance Symbols

The nominal statement of an ISO fit includes the basic size, the tolerance position, a letter, and the grade number. The grade number establishes the magnitude of the tolerance zone or the amount of variation allowed for internal and external dimensions of the part. The actual toleranced size is therefore defined by giving the basic size of the part, followed by the symbol composed of a letter and a number (Fig. 15–13).

When a fit between mating parts is indicated, the basic size is common to both. It is followed by the hole symbol and the shaft symbol (Fig. 15–14).

### Preferred Basic Sizes and Fits

When mating parts are designed, preferred basic sizes should be used whenever possible (Fig. 15–15). The preferred tolerance zones are those agreed on as first choice when designing a product. The preferred fits are based on the preferred tolerance zones. Detailed information on these can be found in "Preferred Metric Limits and Fits," ANSI B4.2-1978, published by the American Society of Mechanical

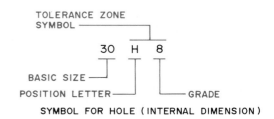

FIGURE 15–13 Examples of hole and shaft tolerance indications.

FIGURE 15–14 The symbol used to indicate the basic size and the fit between the hole and the shaft.

Preferred Metric Basic Sizes (millimeters)

| FIRST CHOICE | SECOND CHOICE | FIRST CHOICE | SECOND CHOICE | FIRST CHOICE | SECOND CHOICE |
|---|---|---|---|---|---|
| 1 | | 10 | | 100 | |
| | 1.1 | | 11 | | 110 |
| 1.2 | | 12 | | 120 | |
| | 1.4 | | 14 | | 140 |
| 1.6 | | 16 | | 160 | |
| | 1.8 | | 18 | | 180 |
| 2 | | 20 | | 200 | |
| | 2.2 | | 22 | | 220 |
| 2.5 | | 25 | | 250 | |
| | 2.8 | | 28 | | 280 |
| 3 | | 30 | | 300 | |
| | 3.5 | | 35 | | 350 |
| 4 | | 40 | | 400 | |
| | 4.5 | | 45 | | 450 |
| 5 | | 50 | | 500 | |
| | 5.5 | | 55 | | 550 |
| 6 | | 60 | | 600 | |
| | 7 | | 70 | | 700 |
| 8 | | 80 | | 800 | |
| | 9 | | 90 | | 900 |
| | | | | 1000 | |

**FIGURE 15–15** The preferred metric design sizes. (Reproduced from ANSI B4.2-1978 with the permission of the publisher, The American Society of Mechanical Engineers.)

Engineers. These preferred fits are used in the tolerancing information that follows.

When determining the fits between mating parts, such as a shaft and a hole, the *hole-basis* or *shaft-basis* systems can be used. In the hole-basis system, the design size of the hole is the basic size and the tolerances are applied to the shaft. The fundamental deviation for the hole-basis system is specified as *H*, as shown in Appendix G. Since negative areas, such as holes or keyways, are generally made with standard machine tools, such as drills and milling cutters, the hole-basis system is usually preferred in the design of mating parts. In the shaft-basis system the design size of the shaft is the basic size and the tolerances are applied to the hole. The fundamental deviation for the shaft-basis system is specified as *h*, as is shown in Appendix G. If a common shaft has to mate with several holes, the shaft-bases system would be preferred. The shaft diameter is the basic size and the size of the mating holes is adjusted to secure the desired fit.

There are 10 preferred basic hole fits and 10 basic shaft fits (Fig. 15–16). These range from a loose clearance fit through transition fits to interference fits.

### Dimensioning Parts Using ISO Metric Fits

Tables giving the limits for basic hole and basic shaft fits appear in Appendix G. Following is an example showing how to read these tables. Assume an engineer has designed a part with a diameter of 10 mm and a fit of H8/f7. The symbol would read 10 H8/f7. Assuming the hole-basis system was used, the limits would be found by entering the table at 10 mm on the left (see the abbreviated table in Fig. 15–17). Move to the right until the fit column H8-f7 is found. Here it shows that the maximum hole is 10.022 mm, the minimum hole is 10.000 mm, the maximum shaft is 9.987 mm, and the minimum shaft is 9.972 mm. The MMC (maximum material condition) is 0.050 and the LMC (least material condition) is 0.013. These limits are placed on a drawing using normal dimensioning techniques (Fig. 15–18).

## TOLERANCE ACCUMULATION

Tolerance accumulation results from three different methods of dimensioning.

When *chain dimensioning* is used, the maximum variation between any two features is equal to the sum of the tolerances on all intermediate distances. In Fig. 15–19, hole B has a cumulative tolerance of 0.002 in. from surface X, while hole C has a tolerance of 0.006 in. Surfaces E and F also have cumulative tolerances of 0.04 and 0.06 in., respectively, from surface X. The cumulative tolerance between two features, surfaces D and F, is 0.04 in.

When using *datum dimensioning*, the maximum variation between the datum and the point being dimensioned is equal to the tolerance applied to that

| | ISO SYMBOL | | DESCRIPTION |
|---|---|---|---|
| | Hole Basis | Shaft Basis | |
| | H11/c11 | C11/h11 | *Loose running* fit for wide commercial tolerances or allowances on external members. |
| | H9/d9 | D9/h9 | *Free running* fit not for use where accuracy is essential, but good for large temperature variations, high running speeds, or heavy journal pressures. |
| | H8/f7 | F8/h7 | *Close running* fit for running on accurate machines and for accurate location at moderate speeds and journal pressures. |
| | H7/g6 | G7/h6 | *Sliding* fit not intended to run freely but to move and turn freely and locate accurately. |
| | H7/h6 | H7/h6 | *Locational clearance* fit provides snug fit for locating stationary parts but can be freely assembled and disassembled. |
| | H7/k6 | K7/h6 | *Locational transition* fit for accurate location, a compromise between clearance and interference. |
| | H7/n6 | N7/h6 | *Locational transition* fit for more accurate location where greater interference is permissible. |
| | H7/p6 | P7/h6 | *Locational interference* fit for parts requiring rigidity and alignment with prime accuracy of location but without special bore pressure requirements. |
| | H7/s6 | S7/h6 | *Medium drive* fit for ordinary steel parts or shrink fits on light sections; the tightest fit usable with cast iron. |
| | H7/u6 | U7/h6 | *Force* fit suitable for parts that can be highly stressed or for shrink fits where the heavy pressing forces required are impractical. |

*Left margin labels (top to bottom):* Interference Fits, Transition Fits, Clearance Fits
*Right margin labels (top to bottom):* More Clearance, More Interference

FIGURE 15–16 Descriptions of preferred metric fits for mating parts. (Reproduced from ANSI B4.2-1978 with the permission of the publisher, The American Society of Mechanical Engineers.)

FIGURE 15–17 A partial table of preferred hole-basis clearance fits. A complete table is in Appendix G.

| BASIC SIZE | | LOOSE RUNNING | | | FREE RUNNING | | | CLOSE RUNNING | | |
|---|---|---|---|---|---|---|---|---|---|---|
| | | Hole H11 | Shaft c11 | Fit | Hole H9 | Shaft d9 | Fit | Hole H8 | Shaft f7 | Fit |
| | | | | | | | | 1.014 | 0.994 | 0.030 |
| | | | | | | | | 1.000 | 0.984 | 0.006 |
| | | | | | | | | 1.214 | 1.194 | 0.030 |
| | | | | | | | | 1.200 | 1.184 | 0.006 |
| | | | | | | | | 1.614 | 1.594 | 0.030 |
| | | | | | | | | 1.600 | 1.584 | 0.006 |
| | | | | | | | | 2.014 | 1.994 | 0.030 |
| | | | | | | | | 2.000 | 1.984 | 0.006 |
| | | | | | | | | 2.514 | 2.494 | 0.030 |
| | | | | | | | | 2.500 | 2.484 | 0.006 |
| | | | | | | | | 3.014 | 2.994 | 0.030 |
| | | | | | | | | 3.000 | 2.984 | 0.006 |
| | | | | | | | | 4.018 | 3.990 | 0.040 |
| | | | | | | | | 4.000 | 3.978 | 0.010 |
| | | | | | | | | 5.018 | 4.990 | 0.040 |
| | | | | | | | | 5.000 | 4.978 | 0.010 |
| 6 | MAX | 6.075 | 5.930 | 0.220 | 6.030 | 5.970 | 0.090 | 6.018 | 5.990 | 0.040 |
| | MIN | 6.000 | 5.855 | 0.070 | 6.000 | 5.940 | 0.030 | 6.000 | 5.978 | 0.010 |
| 8 | MAX | 8.090 | 7.920 | 0.260 | 8.036 | 7.960 | 0.112 | 8.022 | 7.987 | 0.050 |
| | MIN | 8.000 | 7.830 | 0.080 | 8.000 | 7.924 | 0.040 | 8.000 | 7.972 | 0.013 |
| 10 | MAX | 10.090 | 9.920 | 0.260 | 10.036 | 9.960 | 0.112 | 10.022 | 9.987 | 0.050 |
| | MIN | 10.000 | 9.830 | 0.080 | 10.000 | 9.924 | 0.040 | 10.000 | 9.972 | 0.013 |

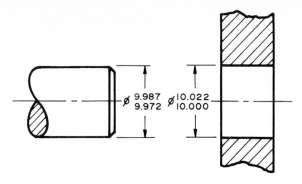

FIGURE 15–18 Applying tolerances when the preferred hole-basis clearance-fit system is used.

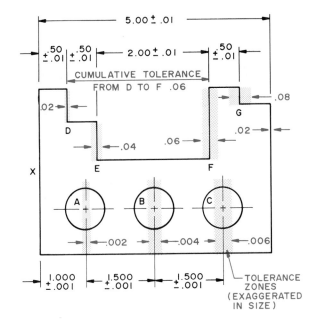

FIGURE 15–19 When using chain dimensioning, the total tolerance is the sum of the tolerances of all the features in the distance involved.

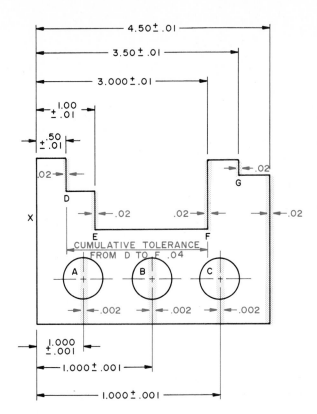

FIGURE 15–20 When using datum dimensioning, the maximum variation between the datum and the point being dimensioned is equal to the tolerance applied to that one dimension.

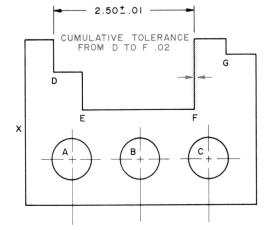

FIGURE 15–21 Direct dimensioning produces the least accumulation of tolerances.

one dimension (Fig. 15–20). The maximum variation between any two features, such as two holes, is equal to the sum of the tolerances on the two dimensions *from the datum to the feature.* For example, in Fig. 15–20, the relationship between surfaces D and F has a tolerance of 0.04 in. because each surface has a tolerance of ± 0.01 in.

In *direct dimensioning,* the maximum variation between any two features is controlled by the tolerance on the dimension between the features (Fig. 15–21). This produces a tolerance of 0.02 in. from surfaces D and F. This method produces the least variation between the two features.

## TOLERANCING BETWEEN CENTERS

A common feature in product design is pins or shafts that must mate with holes in another part. The tolerance used on the pins and the holes influ-

ences the between-centers tolerance. An example is given in Fig. 15–22. The pins and holes have a tolerance of 0.002 in. If the largest pin and the smallest hole are assembled, the allowance would be 0.002 in. on each. As the tolerance on the pin and the hole

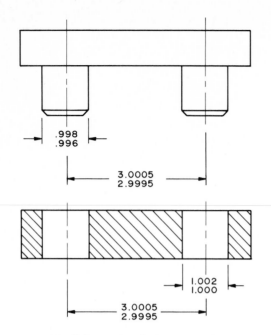

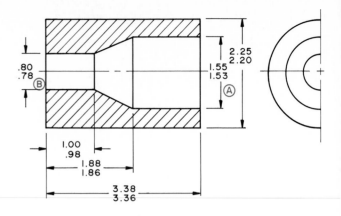

FIGURE 15–22 Tolerancing between centers.

DIMENSIONS A AND B TO BE CONCENTRIC
WITHIN .0005 FULL INDICATOR.

FIGURE 15–24 A tolerance of concentricity applied
with a general note. Each surface involved is identified with a reference letter.

becomes smaller, the tolerance on the center-to-
center dimension would have to be smaller to permit
interchangeability of parts.

## TOLERANCE OF CONCENTRICITY

If a part has several concentric features, such as cylinders or cones, and the relationship of each to the
axis is important, it is necessary to give the acceptable deviation from concentricity. This is indicated
with a general note or a note connected by leaders
to the surfaces involved (Fig. 15–23). When a general note is used, the surfaces involved are identified
with reference letters (Fig. 15–24).

## TOLERANCING ANGULAR DIMENSIONS

Angular surfaces can be defined by using a linear
dimension plus an angle in degrees or completely
with linear dimensions. When the linear dimension
plus an angle is used, the surface being located lies
between the planes represented by the area of the
tolerance zone. The ends of the surface are located
with toleranced linear dimensions. The angle is toleranced in degrees (Fig. 15–25). One disadvantage
to this method is that the tolerance zone gets wider
the farther it gets from the apex of the angle.

The basic-angle method overcomes the widen-
ing tolerance zone by producing a zone with parallel
boundaries (Fig. 15–26). The angle does not have a
tolerance. The angular variation permitted is defined
by the tolerances on the linear dimensions and the
untoleranced basic angle. This produces a tolerance
zone with parallel boundaries at the basic angle. The
actual angle produced when the surface is formed
cannot fall outside the tolerance zone.

FIGURE 15–23 A tolerance of concentricity applied
with a note connected by leaders to the surfaces involved.

## TOLERANCING CONICAL TAPERS

A conical taper can be specified by one of several
combinations of dimensions and tolerances, including the following.

1. The diameter at one end of the tapered feature (Fig.
   15–27).
2. The length of the tapered feature (Fig. 15–28).
3. The diameter at a selected cross-sectional plane,
   which may be within or outside the tapered feature.
   The position of this plane is shown with a basic dimension (Fig. 15–29).

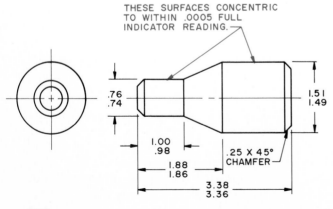

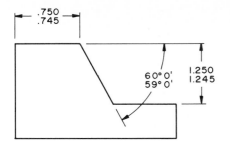

THIS ON THE DRAWING

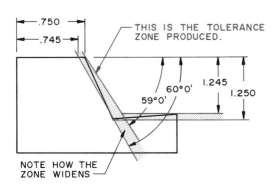

THIS IS THE TOLERANCE ZONE PRODUCED.

NOTE HOW THE ZONE WIDENS

THE ANGULAR SURFACE CAN LIE ANYWHERE WITHIN THE TOLERANCE ZONE AND ITS ANGLE WITH THE HORIZONTAL CANNOT BE GREATER THAN 60°0' OR LESS THAN 59°0'.

PRODUCES THIS TOLERANCE ZONE

**FIGURE 15–25** Tolerancing an angular surface using toleranced linear dimensions and toleranced angles in degrees. (Reproduced from ANSI Y14.5-1973 with the permission of the publisher, The American Society of Mechanical Engineers.)

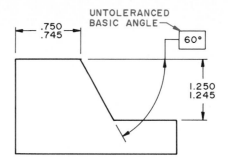

UNTOLERANCED BASIC ANGLE

THIS ON THE DRAWING

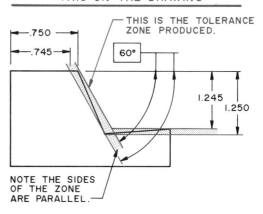

THIS IS THE TOLERANCE ZONE PRODUCED.

NOTE THE SIDES OF THE ZONE ARE PARALLEL.

THE ANGULAR SURFACE CAN LIE ANYWHERE WITHIN THE TOLERANCE ZONE THAT HAS PARALLEL BOUNDARIES INCLINED AT THE BASIC ANGLE.

PRODUCES THE TOLERANCE ZONE

**FIGURE 15–26** Tolerancing an angular surface using only toleranced linear dimensions. This produces a zone with parallel boundaries. (Reproduced from ANSI Y14.5-1973 with the permission of the publisher, The American Society of Mechanical Engineers.)

**FIGURE 15–27** Using a toleranced diameter at one end of the tapered feature plus the *basic taper* desired. Any variation must fall within the tolerance zone created by the limits on the diameter. (Reproduced from ANSI Y14.5-1973 with the permission of the publisher, The American Society of Mechanical Engineers.)

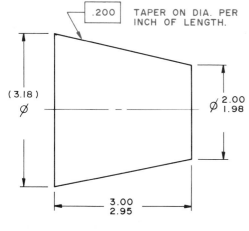

.200 TAPER ON DIA. PER INCH OF LENGTH.

THIS ON THE DRAWING

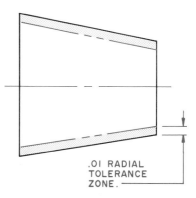

.01 RADIAL TOLERANCE ZONE.

PRODUCES THIS RADIAL TOLERANCE ZONE.

Tolerancing and Surface Quality **367**

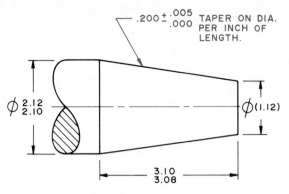

FIGURE 15–28 Conical tapers can be specified by giving the length of the tapered feature and the taper per inch of length on the taper. (Reproduced from ANSI Y14.5-1973 with the permission of the publisher, The American Society of Mechanical Engineers.)

SHAFT TOLERANCE

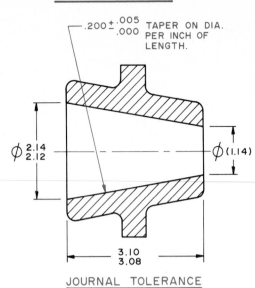

JOURNAL TOLERANCE

FIGURE 15–29 Specifying a *basic taper* and a *basic diameter*. The basic diameter controls the size of the tapered section and its longitudinal position in relation to another surface. (Reproduced from ANSI Y14.5-1973 with the permission of the publisher, The American Society of Mechanical Engineers.)

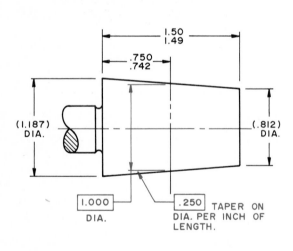

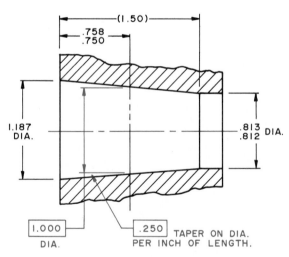

THESE SHAFT AND JOURNAL TOLERANCES ON THE DRAWING

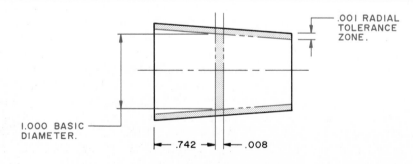

PRODUCES THESE TOLERANCE ZONES

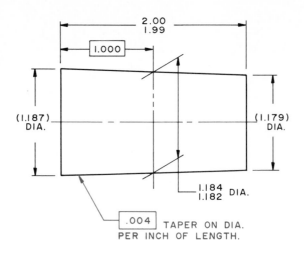

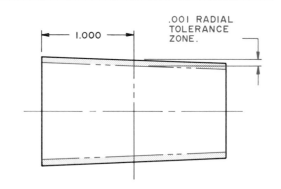

FIGURE 15–30 Specifying a *basic taper* and a *basic length*. The basic length locates the basic location, and a tolerance is applied to the diameter at that location. (Reproduced from ANSI Y14.5-1973 with the permission of the publisher, The American Society of Mechanical Engineers.)

4. A dimension locating a cross-sectional plane at which a basic diameter is specified (Fig. 15–29).

5. The rate of taper (Fig. 15–30).

6. The included angle.

## TOLERANCING FLAT TAPERS

The methods used to dimension conical tapers are adapted for use on flat tapers. An example is shown in Fig. 15–31.

## TOLERANCING RADII

The toleranced radius with an unlocated center is defined by arcs tangent to adjacent surfaces. The arc on the part must fall between these two arcs and meet the adjacent surfaces without reversals (Fig. 15–32).

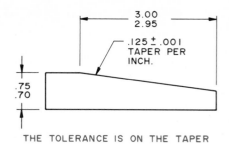

FIGURE 15–31 Specifying a flat taper with a note giving a tolerance on the taper.

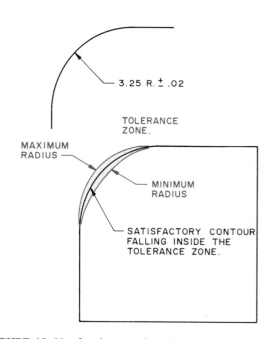

FIGURE 15–32 A toleranced radius with an unlocated center is defined by arcs tangent to the adjacent surfaces. (Reproduced from ANSI Y14.5-1973 with the permission of the publisher, The American Society of Mechanical Engineers.)

## GEOMETRIC TOLERANCING

Geometric tolerancing is used in conjunction with size dimensions to increase the interchangeability of manufactured parts. Geometric tolerancing is used to specify the position and form of features.

The geometric characteristic of the part or its location is specified by standardized symbols and by the use of notes (Fig. 15–33). Complete details concerning the application of these symbols and their many uses are explained in detail in publication ANSI Y14.15-1973, "Dimensioning and Tolerancing," published by the American Society of Mechanical Engineers. The material in this section was developed from this publication.

| Characteristics | American ANSI Y14.5 | British BS 308 | Canadian CSA B78.2 | International ISO R1101 |
|---|---|---|---|---|
| Straightness | — | Same | Same | Same |
| Flatness | ▱ | Same | Same | Same |
| Roundness (Circularity) | ◯ | Same | Same | Same |
| Cylindricity | ⌭ | Same | Same | Same |
| Profile of a Line | ⌒ | Same | Same | Same |
| Profile of a Surface | ⌓ | Same | Same | Same |
| Parallelism | // | Same | Same | Same |
| Perpendicularity (Squareness) | ⊥ | Same | Same | Same |
| Angularity | ∠ | Same | Same | Same |
| Position | ⊕ | Same | Same | Same |
| Concentricity (Coaxiality) | ◎ | Same | Same | Same |
| Symmetry | ≡ | Same | Same | Same |
| Maximum Material Condition | Ⓜ | Same | Same | Same |
| Diameter | ⌀ | Same | Same | Same |
| Circular Runout | ↗ | Same | Same | Same |
| Total Runout | ↗↗ ☆ | None | None | None |
| Datum Identification | -A- | A or ⌐ | -A- or ⌐ | A or ⌐ |
| Reference Dimension | (5.000) | (127) | (5.000) | (127) |
| Basic Dimension | 5.000 | 127 | 5.000 | 127 |
| Regardless of Feature Size | Ⓢ | None | None | None |
| Projected Tolerance Zone | Ⓟ | None | Ⓟ | None |
| Datum Target | (A/1) | (A/1) | (A/1) | None |
| Part Symmetry | None | ⊣⊢·⊣⊢ | ⊣⊢·⊣⊢ | ⊣⊢·⊣⊢ |
| Shape of the tolerance zone | Zone is total width. ⌀ specified where zone is circular or cylindrical. | Zone is a total width in direction of leader arrow. ⌀ specified where zone is circular or cylindrical. | Zone shape evident from chacteristic being controlled. | Zone is a total width in direction of leader arrow. ⌀ specified where zone is circular or cylindrical. |
| Sequence within the feature control symbol | ⊕ A B C ⌀.02 Ⓜ or ⊕ ⌀.02 Ⓜ A B C | ⊕ ⌀ 0.5 Ⓜ A B C | ⊕ .02 Ⓜ A B C | ⊕ ⌀ 0.5 Ⓜ A B C |

☆ "TOTAL" specified under the feature control symbol.

FIGURE 15–33 Geometric characteristic symbols. (Reproduced from ANSI Y14.5-1973 with the permission of the publisher, The American Society of Mechanical Engineers.)

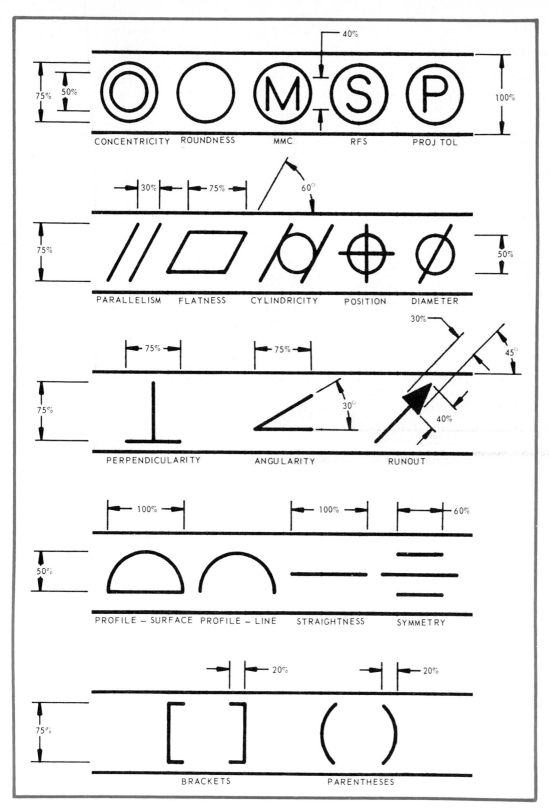

FIGURE 15–34 Recommended form and proportions for geometric characteristic symbols. (Reproduced from ANSI Y14.5-1973 with the permission of the publisher, The American Society of Mechanical Engineers.)

## Symbols and Symbol Construction

The symbols should be carefully lettered so that they can be easily read. They should meet legibility and reproduction standards for letter heights as specified for engineering drawings. The lettering should be a minimum of 3.8 mm or 0.15 in. high.

When a situation arises in which the precise geometric requirement cannot be given by symbols, a note can be used. This can supplement a symbol or explain the requirement.

*Geometric characteristic symbols* The symbols used to indicate the geometric characteristics are shown in Fig. 15-33, which gives a comparison of those used in several countries and the ISO standards. The recommended forms and proportions for geometric characteristic symbols are shown in Fig. 15-34. While most symbols will be drawn with the aid of templates, this figure shows the recommended sizes. The preferred proportions are based on modules of 50% and 75% of the basic frame height. For example, the diameter of the symbol representing position is 75% of the size of the frame. The frame is normally 8 mm or 0.3 in. high (Fig. 15-35).

*Datum identifying symbols* A datum is the theoretically exact line, surface, or point from which features are located. The datum identifying symbol has a frame with a datum reference letter inside. The letter is preceded and followed by a dash (Fig. 15-36).

All letters of the alphabet except I, O, and Q are used as datum reference letters. Each of the first 23 datum features requiring identification is assigned a different reference letter. If there are more datum references, doubled letters (AA, BB, etc.) are used.

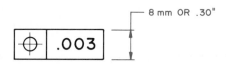

FIGURE 15-35 Recommended frame height. (Reproduced from ANSI Y14.5-1973 with the permission of the publisher, The American Society of Mechanical Engineers.)

FIGURE 15-36 The datum identifying symbol. (Reproduced from ANSI Y14.5-1973 with the permission of the publisher, The American Society of Mechanical Engineers.)

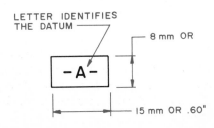

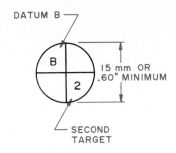

FIGURE 15-37 The datum target symbol. (Reproduced from ANSI Y14.5-1973 with the permission of the publisher, The American Society of Mechanical Engineers.)

*Datum target symbol* Datum targets define lines, points, or areas on a part and can be used in defining a datum surface. The datum target symbol is divided into four quadrants. The reference letter placed in the upper left quadrant identifies its associated datum feature. The numeral in the lower right quadrant identifies the specific target (Fig. 15-37). The datum symbol is placed on a drawing when circumstances are such that failure to do so could result in the selection of the wrong feature.

The method for designating datum targets is illustrated in Fig. 15-38. A datum may be represented

FIGURE 15-38 Ways to designate datum targets.

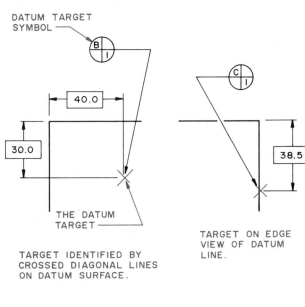

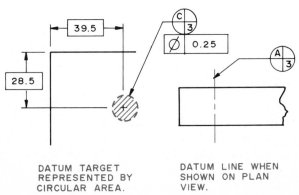

372

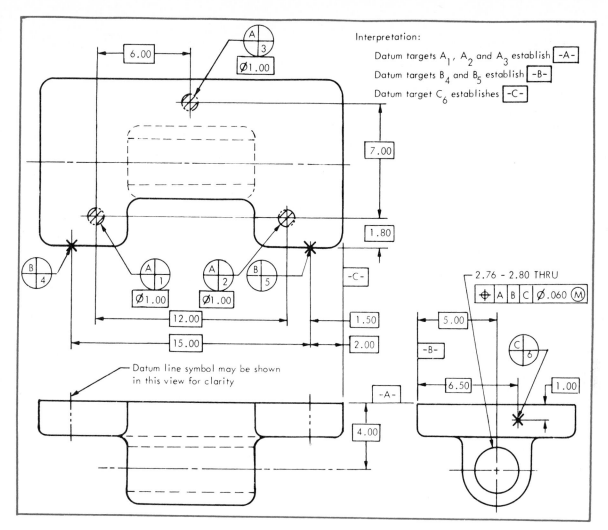

FIGURE 15–39 Datum point, line, and area targets are used to define a datum surface. (Reproduced from ANSI Y14.5-1973 with the permission of the publisher, The American Society of Mechanical Engineers.)

by one, two, or three targets. Each target is dimensioned on engineering drawings using basic or toleranced dimensions. Dimensions locating datum targets in a single set should be dimensionally related to one another. The application of datum targets is shown in Fig. 15–39.

*Basic dimension symbol* A basic dimension is identified by enclosing it in a frame (Fig. 15–40). A basic dimension is the numerical value used to describe the theoretically exact size, shape, or location of a feature or datum target.

*Diameter symbol* The symbol used to specify a diameter is shown in Fig. 15–41. It is placed before the tolerance in the feature control symbol. It is also used instead of the abbreviation DIA on engineering drawings.

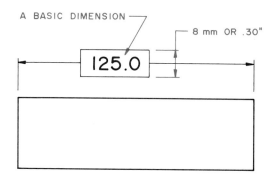

FIGURE 15–40 A basic dimension is identified by enclosing it in a frame.

*MMF and RFS symbols* The symbols used to specify "maximum material condition" and "regardless of feature size" are shown in Fig. 15–41. The abbreviations MMS and RFS are used in notes, or the words may be spelled out.

| TERM | ABBREVIATION | SYMBOL |
|---|---|---|
| MAXIMUM MATERIAL CONDITION | MMC | Ⓜ |
| REGARDLESS OF FEATURE SIZE | RFS | Ⓢ |
| DIAMETER | DIA. | ⌀ |
| PROJECTED TOLERANCE ZONE | TOL ZONE PROJ. | P |
| REFERENCE | REF | ( 1.75 ) |
| BASIC | BSC | 2.55 |

FIGURE 15–41  Other symbols used in geometric tolerancing. (Reproduced from ANSI Y14.5-1973 with the permission of the publisher, The American Society of Mechanical Engineers.)

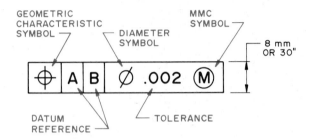

FIGURE 15–42  A feature control symbol is used to specify position or form tolerances. (Reproduced from ANSI Y14.5-1973 with the permission of the publisher, The American Society of Mechanical Engineers.)

*Maximum material condition* is the condition wherein a size feature contains the maximum amount of material within the stated size limits. For example, the smallest hole diameter and largest shaft diameter produce a maximum material condition.

*Regardless of feature size* indicates that a form or positioned tolerance applies (1) at any increment of size of the feature within its size tolerance or (2) at the actual size of a datum feature. If no modifiers are shown in the feature control symbol, the tolerance is assumed to be at RFS, except in the case of position tolerance, which is understood to be at MMC.

*Projected tolerance zone symbol*  The symbol used to designate a projected tolerance zone is shown in Fig. 15–41. A projected tolerance zone exists where a positional or perpendicularity tolerance zone is established for a part that will project beyond the surface, such as a stud or bolt.

*Reference dimension symbol*  A reference value is designated by enclosing it in parentheses (Fig. 15–41).

Combined Symbols

The individual symbols, tolerances, and datum reference letters are combined to express a tolerance symbolically.

*Feature control symbol*  The *feature control symbol* is used to specify the tolerance of position or form. It consists of a frame containing the geometric characteristic followed by the allowable tolerance. These elements are separated by a vertical line (Fig. 15–42). The first symbol on the left specifies the geometric characteristic being tolerated. Next, one or more datums are usually required, with tolerancing of location, such as position, concentricity, and symmetry; tolerances of relationship, such as angularity, perpendicularity, and parallelism; and runout. Datum references are sometimes required with profile tolerances of a line or a surface. All of these are explained later in the chapter. The third specification in the feature control symbol is the size of the toler-

ance zone. This may be a diameter or a linear measurement. When applicable, the diameter symbol is placed before the tolerance. When applicable, modifiers are used to indicate "maximum material condition" or "regardless of feature size."

*Feature control symbol incorporating datum references*  When a tolerance of form or position is related to a datum, the relationship is stated in the feature control symbol by placing the datum reference letter after either the geometric characteristic symbol or the tolerance. Each letter or symbol is separated by a vertical line. The datum reference entry includes the MMS or RFS symbol when applicable. The length of the frame is increased as needed. Two accepted methods for referencing datums are shown in Fig. 15–43. The international method places the datum references on the right end of the symbol. Do not intermix the two approaches on a drawing.

Datum reference letters are recorded in the desired order of preference from left to right. They need not be in alphabetical order. When a single datum reference is established by multiple datum features, the datum reference letters are separated by a dash between the letters (Fig. 15–44).

When more than one tolerance of a given geometric characteristic applies to the same feature, a composite feature control symbol is used (Fig. 15–45). A single geometric characteristic symbol is followed by each tolerance requirement, one above the other.

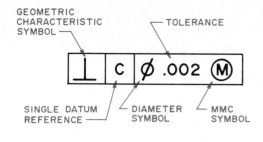

GEOMETRIC CHARACTERISTIC SYMBOL — — TOLERANCE

SINGLE DATUM REFERENCE — — DIAMETER SYMBOL — MMC SYMBOL

FIGURE 15-43 Feature control symbols incorporating datum references. (Reproduced from ANSI Y14.5-1973 with the permission of the publisher, The American Society of Mechanical Engineers.)

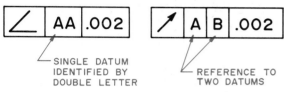

SINGLE DATUM IDENTIFIED BY DOUBLE LETTER — REFERENCE TO TWO DATUMS

## USUAL SEQUENCE OF DATA

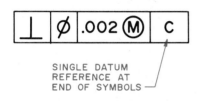

SINGLE DATUM REFERENCE AT END OF SYMBOLS

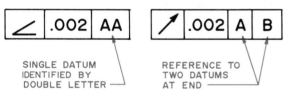

SINGLE DATUM IDENTIFIED BY DOUBLE LETTER — REFERENCE TO TWO DATUMS AT END

## INTERNATIONAL SEQUENCE OF DATA

FIGURE 15-44 The recommended order of precedence of datum references. (Reproduced from ANSI Y14.5-1973 with the permission of the publisher, The American Society of Mechanical Engineers.)

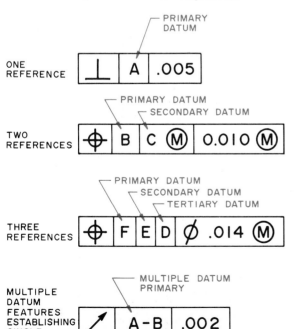

ONE REFERENCE — PRIMARY DATUM

TWO REFERENCES — PRIMARY DATUM — SECONDARY DATUM

THREE REFERENCES — PRIMARY DATUM — SECONDARY DATUM — TERTIARY DATUM

MULTIPLE DATUM FEATURES ESTABLISHING SINGLE DATUM REFERENCE — MULTIPLE DATUM PRIMARY

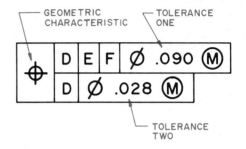

GEOMETRIC CHARACTERISTIC — TOLERANCE ONE

TOLERANCE TWO

FIGURE 15-45 A composite feature control symbol is used when more than one tolerance is applied to give geometric characteristics. (Reproduced from ANSI Y14.5-1973 with the permission of the publisher, The American Society of Mechanical Engineers.)

**Combined feature control and datum identifying symbol** When a feature is controlled by a positional or form tolerance and serves as a datum feature, the feature control and datum identifying symbols are combined (Fig. 15-46). The length of the frame for the datum identifying symbol should be at least 0.60 in. and may be the same length as the feature control frame.

The datums in the feature control symbol are not considered part of the datum identifying symbol.

**Combined feature control and projected tolerance zone symbol** When a positional or perpendicularity tolerance is specified as a projected tolerance zone, the frame containing the projected height and symbol is placed below the feature control symbol (Fig. 15-47).

FIGURE 15-46 A combined feature control and datum identifying symbol is used when a feature is controlled by a positional or form tolerance and serves as a datum feature. (Reproduced from ANSI Y14.5-1973 with the permission of the publisher, The American Society of Mechanical Engineers.)

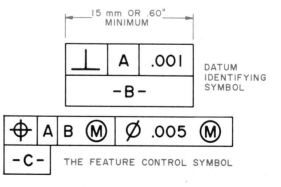

15 mm OR .60" MINIMUM

DATUM IDENTIFYING SYMBOL

THE FEATURE CONTROL SYMBOL

Tolerancing and Surface Quality  **375**

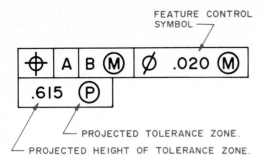

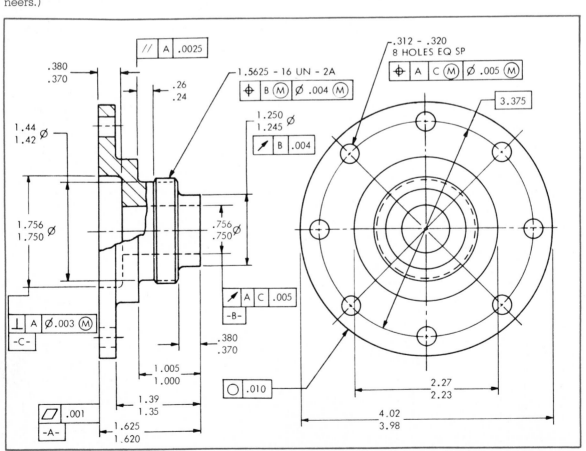

**FIGURE 15–47** The feature control symbol used when positional or perpendicularity tolerances are specified as a projected tolerance zone. (Reproduced from ANSI Y14.5-1973 with the permission of the publisher, The American Society of Mechanical Engineers.)

### Feature Control Symbol Placement

The feature control symbol is placed on a drawing by one of the following methods (Fig. 15–48).

1. By adding the symbol to a note or dimension that specifies the size of the feature.
2. By running a leader from the symbol to the feature.

**FIGURE 15–48** Methods for placing geometric characteristic symbols on engineering drawings. (Reproduced from ANSI Y14.5-1973 with the permission of the publisher, The American Society of Mechanical Engineers.)

3. By attaching a side, end, or corner of the symbol frame to the feature with an extension line.
4. By attaching a side, corner, or end of the symbol frame to a dimension line pertaining to the feature.

## TOLERANCES OF LOCATION

Tolerances of location include positional, concentricity, and symmetry tolerances used to control the following.

1. Center-to-center distances between features such as holes, slots, bosses, and tabs.
2. Location of features from datum features such as planes and cylindrical surfaces.
3. Coaxiality (the existence of a common axis) between features.
4. Features with center distances equally arranged about a datum axis or plane.

### Position Tolerance

A *positional tolerance* specifies a zone within which an axis or the center plane of a feature is permitted to

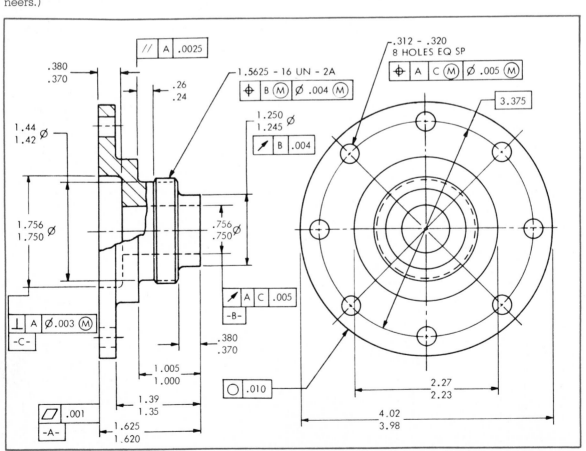

vary from the true (theoretically exact) position. Basic dimensions establish the true position from specified datum features and between interrelated features. The positional tolerances are part of the feature control symbol. Positional tolerancing is identified by its symbol, which becomes part of the feature control symbol. It can also be specified by

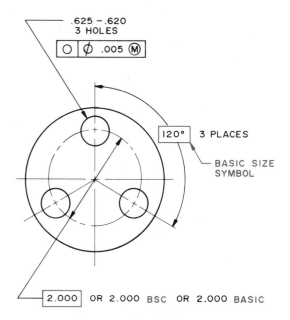

FIGURE 15–49 How to identify basic dimensions on an engineering drawing. (Reproduced from ANSI Y14.5-1973 with the permission of the publisher, The American Society of Mechanical Engineers.)

FIGURE 15–50 Basic dimensions can be indicated by a general note. (Reproduced from ANSI Y14.5-1973 with the permission of the publisher, The American Society of Mechanical Engineers.)

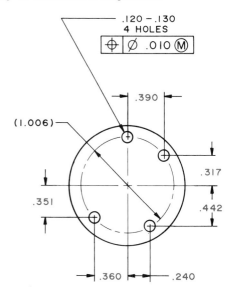

UNTOLERANCED DIM. LOCATING TP ARE BASIC.

placing the abbreviation BSC or spelling out the word BASIC after the dimension (Fig. 15–49). Basic dimensions may be indicated by placing the following note on the drawing: "UNTOLERANCED DIM LOCATING TP ARE BASIC." This means that the basic dimensions locating a true position are not identified on the drawing but are understood to be basic (Fig. 15–50).

Geometric positional tolerances establish a circular zone (Fig. 15–51). The true position of the hole is established by basic dimensions. The *circular tolerance zone* around the true position has a diameter equal to the amount of tolerance. In Fig. 15–51, the total tolerance is 1.00 mm. Any feature center located within this zone lies within the acceptable variation range from the true position.

Geometric positional tolerancing establishes a *cylindrical tolerance zone*. It applies to the true position at the surface of a feature and to the full depth of the feature being located (Fig. 15–52).

Positional tolerancing methods can also be used for noncircular features such as a slot or a tab. When positioning noncircular features, the positional tolerance is usually applied to the surfaces related to the center plane of the feature. The center plane is located in true position. The positional tolerance is the total width of the tolerance zone within which the center plane of the feature must lie. The center plane of the tolerance zone is at the true position. An example of positional tolerancing of *tabs* is in Fig. 15–53. Positional tolerancing of *slots* is shown in Fig. 15–54. The tolerance value in the feature con-

FIGURE 15–51 Geometric positional tolerancing establishes a circular tolerance zone around the true position.

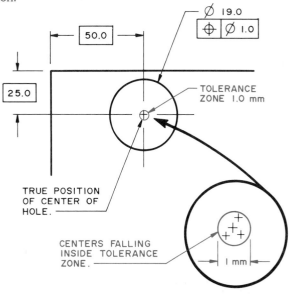

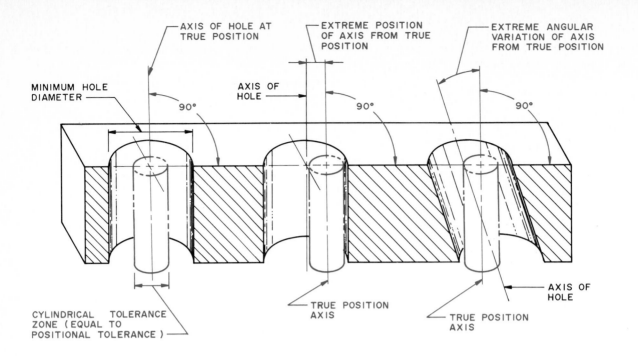

AXIS OF HOLE AT TRUE POSITION

EXTREME POSITION OF AXIS FROM TRUE POSITION

EXTREME ANGULAR VARIATION OF AXIS FROM TRUE POSITION

MINIMUM HOLE DIAMETER

AXIS OF HOLE

90°

90°

90°

CYLINDRICAL TOLERANCE ZONE (EQUAL TO POSITIONAL TOLERANCE)

TRUE POSITION AXIS

TRUE POSITION AXIS

AXIS OF HOLE

**THE AXIS OF THE HOLE COINCIDES WITH THE TRUE POSITION AXIS.**

**THE AXIS OF THE HOLE IS SHOWN LOCATED AT AN EXTREME POSITION FROM THE TRUE POSITION BUT STILL WITHIN THE TOLERANCE ZONE.**

**THE AXIS OF THE HOLE IS INCLINED TO AN EXTREME POSITION BUT IS STILL WITHIN THE TOLERANCE ZONE.**

**FIGURE 15–52** Geometric positional tolerances establish cylindrical tolerance zones. (Reproduced from ANSI Y14.5-1973 with the permission of the publisher, The American Society of Mechanical Engineers.)

**FIGURE 15–53** Positional tolerancing of tabs. (Reproduced from ANSI Y14.5-1973 with the permission of the publisher, The American Society of Mechanical Engineers.)

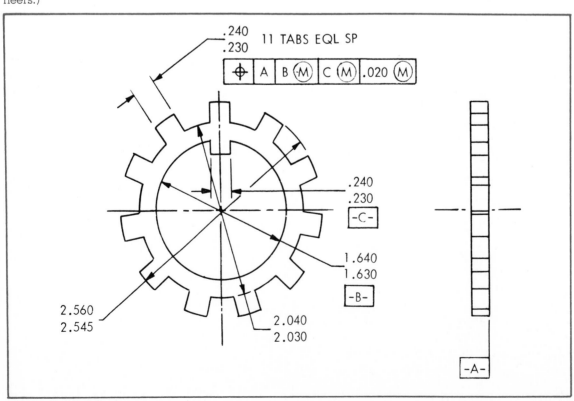

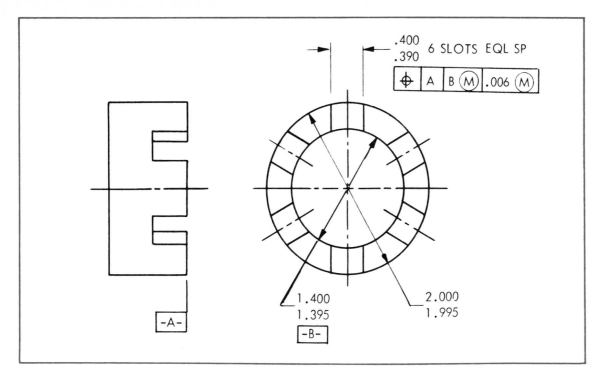

FIGURE 15–54 Positional tolerancing of slots. (Reproduced from ANSI Y14.5-1973 with the permission of the publisher, The American Society of Mechanical Engineers.)

trol symbol represents a distance between parallel planes. It is a linear distance, so the diameter symbol is not used. The slot at MMC must have its axis or center plane within the tolerance zone. This tolerance zone is the distance between two planes located at equal distances from the true position (Fig. 15–55).

### Concentricity Tolerance

Cylindrical features that have a common axis are concentric. The axis of a feature must be within a cylindrical zone whose diameter is equal to the concentricity tolerance and whose axis coincides with the datum axis. A concentricity feature symbol is shown in Fig. 15–56. If it is difficult to establish the axis due to irregularities in the form (as out of round), it is recommended that the surface tolerance be controlled by positional or runout tolerances. Runout tolerances are explained later in this chapter.

### Symmetry Tolerance

Symmetry is the condition whereby a feature is the same on both sides of the center of a datum feature. Symmetry drawing callout and its interpretation are shown in Fig. 15–57. When a feature is symmetrical about a datum and the tolerance is stated at MMC, a positional tolerance is used (Fig. 15–58).

## TOLERANCES OF FORM

Form tolerances control straightness, flatness, roundness, cylindricity, profile of a surface or line, angularity, parallelism, and perpendicularity. Surface texture and tolerances of size and location control form to a certain degree.

FIGURE 15–55 The tolerance zone for the center plane of the slot at maximum material condition.

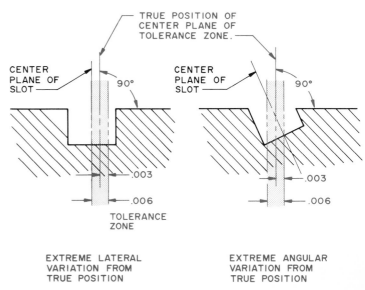

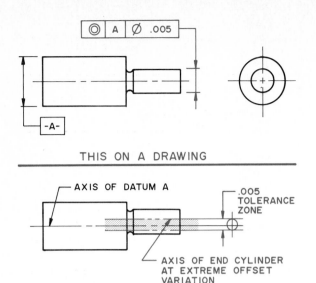

THIS ON A DRAWING

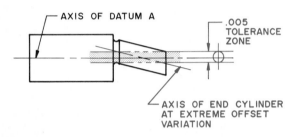

PRODUCES THIS OFFSET TOLERANCE ZONE

AXIS OF DATUM A

.005 TOLERANCE ZONE

AXIS OF END CYLINDER AT EXTREME OFFSET VARIATION

PRODUCES THIS ANGULAR VARIATION TOLERANCE

**FIGURE 15–56** How to specify concentricity, datum, and related features regardless of feature size. (Reproduced from ANSI Y14.5-1973 with the permission of the publisher, The American Society of Mechanical Engineers.)

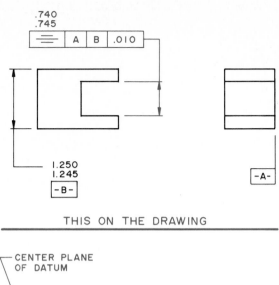

THIS ON THE DRAWING

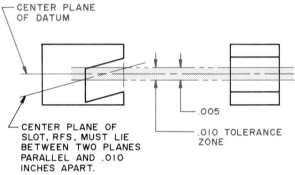

CENTER PLANE OF SLOT, RFS, MUST LIE BETWEEN TWO PLANES PARALLEL AND .010 INCHES APART.

.005

.010 TOLERANCE ZONE

PRODUCES THIS TOLERANCE ZONE

**FIGURE 15–57** How to specify symmetry, datum, and related features regardless of feature size. (Reproduced from ANSI Y14.5-1973 with the permission of the publisher, The American Society of Mechanical Engineers.)

Form tolerances are specified for features critical to product function and interchangeability where tolerances of size and location do not provide the desired control.

A form tolerance specifies a *tolerance zone* within which the feature, its axis, or the center plane must be contained. If the form tolerance represents the diameter of a cylindrical zone, it is preceded by a diameter symbol. In all other conditions, no symbol is used, and the form tolerance value represents a linear distance between two geometric boundaries.

## Straightness Tolerance

Straightness is a condition whereby an element of a surface or an axis is a straight line. The straightness tolerance specifies a tolerance zone within which an axis or all points of the element must lie. The straightness tolerance is applied to the view where the elements to be controlled appear as a straight line. In Fig. 15–59, all the circular elements of the

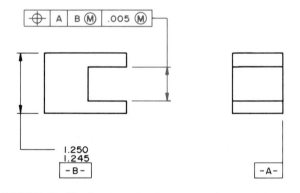

**FIGURE 15–58** Positional tolerancing for symmetry at maximum material condition. (Reproduced from ANSI Y14.5-1973 with the permission of the publisher, The American Society of Mechanical Engineers.)

surface must lie within the specified tolerance zone. The feature symbol is directed to the feature surface or to an extension line rather than to the dimension.

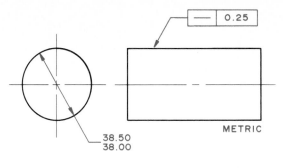

THIS ON THE DRAWING

PRODUCES THE FOLLOWING

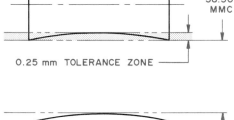

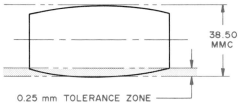

FIGURE 15–59 The application of straightness tolerance. (Reproduced from ANSI Y14.5-1973 with the permission of the publisher, The American Society of Mechanical Engineers.)

## Flatness Tolerance

Flatness is a condition of a surface that has all its elements in one plane. A flatness tolerance specifies a tolerance zone that is defined by two parallel planes within which the surface must lie. The symbol is attached by a leader to the surface or to an extension line of the surface. The symbol is placed in the view in which the surface elements to be controlled are represented by a line (Fig. 15–60).

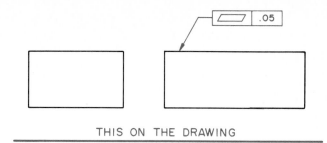

THIS ON THE DRAWING

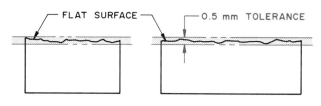

PRODUCES THIS TOLERANCE ZONE

FIGURE 15–60 The application of flatness tolerance. (Reproduced from ANSI Y14.5-1973 with the permission of the publisher, The American Society of Mechanical Engineers.)

FIGURE 15–61 Specifying roundness for a cylinder. (Reproduced from ANSI Y14.5-1973 with the permission of the publisher, The American Society of Mechanical Engineers.)

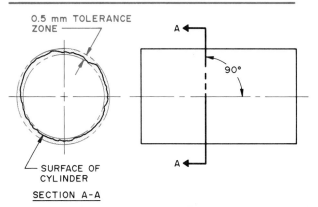

THIS ON THE DRAWING

PRODUCES THIS TOLERANCE ZONE

## Roundness Tolerance

Roundness is a condition of circularity. Roundness is a condition of a surface of revolution in either of the following situations:

1. On a cylinder or cone, when all points of the surface intersected by any plane perpendicular to a common axis are equidistant from that axis.
2. On a sphere, when all points of the surface intersected by any plane passing through a common center are equidistant from that center.

A roundness tolerance specifies a tolerance zone defined by two concentric circles within which each circular element of the surface must lie. The specification and interpretation of roundness of a cylinder are illustrated in Fig. 15–61, of a cone in Fig. 15–62, and of a sphere in Fig. 15–63.

## Cylindricity

Cylindricity is a condition of a surface of revolution on which all points of the surface are equidistant from a common axis. It controls both the roundness

FIGURE 15–62 Specifying roundness for a cone. (Reproduced from ANSI Y14.5-1973 with the permission of the publisher, The American Society of Mechanical Engineers.)

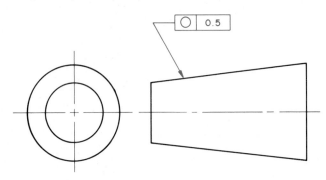

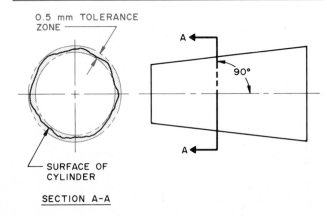

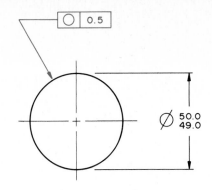

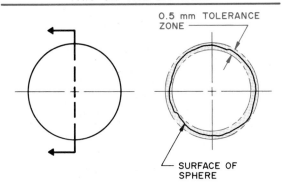

FIGURE 15–63 Specifying roundness for a sphere. (Reproduced from ANSI Y14.5-1973 with the permission of the publisher, The American Society of Mechanical Engineers.)

and the straightness of a cylinder. The tolerance zone is made up of two concentric cylinders around the axis. All specified diameters must be within these cylinders (Fig. 15–64).

FIGURE 15–64 Specifying cylindricity. (Reproduced from ANSI Y14.5-1973 with the permission of the publisher, The American Society of Mechanical Engineers.)

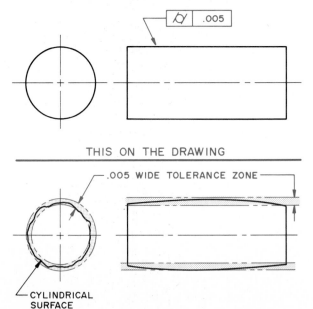

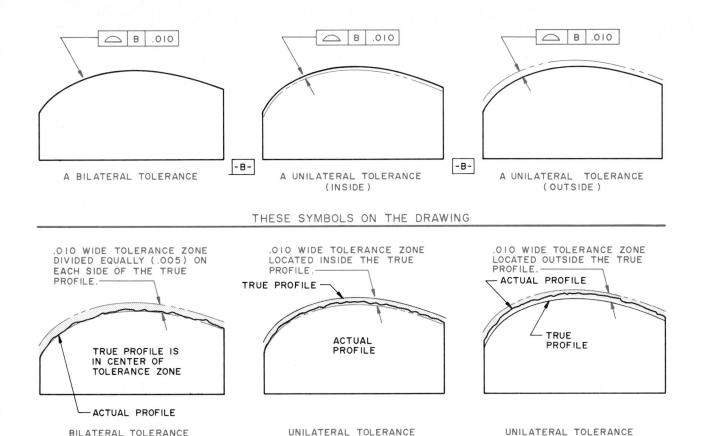

THESE SYMBOLS ON THE DRAWING

PRODUCE THESE TOLERANCES

FIGURE 15–65 Specifying a profile tolerance. (Reproduced from ANSI Y14.5-1973 with the permission of the publisher, The American Society of Mechanical Engineers.)

## Profile Tolerance

A profile is the two-dimensional outline of an object. It is formed on a specified plane. Profiles are formed by projecting a three-dimensional figure onto a plane or by taking sections through it. The elements of a profile may be straight lines, arcs, and/or curved lines. The tolerance can be applied bilaterally or unilaterally around the basic contour definition. When the tolerance is to be applied unilaterally, the direction of application is shown with a phantom line. When no direction is indicated, it is assumed that the tolerance is bilateral (Fig. 15–65). All points on the profile must lie within the tolerance zone indicated.

## FORM TOLERANCES FOR RELATED FEATURES

Form tolerances applicable to related features include angularity, parallelism, perpendicularity, and, in some cases, profile. They control the attitude of features to one another.

## Angularity Tolerance

Angularity is the condition of a surface or axis at a specified angle other than 90° from a datum plane or axis. Angularity tolerances are specified for planes and axes. A tolerance zone is established at a basic angle within which a surface element or an axis must lie (Figs. 15–66 and 15–67).

## Perpendicularity Tolerance

Perpendicularity is the condition of a surface, median plane, or axis that is at a right angle to a datum plane or axis. Perpendicularity tolerances are specified for planes and axes. The tolerance zone is established at a basic 90°. The surface or axes specified must lie within the tolerance zone (Figs. 15–68 and 15–69).

## Parallelism Tolerance

Parallelism is the condition of a surface or axis that is equidistant at all points from a datum plane or axis. A parallelism tolerance specifies a tolerance

Tolerancing and Surface Quality　383

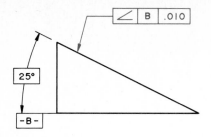

THIS ON THE DRAWING

THE SURFACE MUST LIE WITHIN
THE SPECIFIED TOLERANCE OF
SIZE AND BETWEEN THE
PARALLEL PLANES.

.010 TOLERANCE
ZONE

25°

PRODUCES THIS TOLERANCE ZONE

**FIGURE 15–66** Specifying angularity for a plane surface. (Reproduced from ANSI Y14.5-1973 with the permission of the publisher, The American Society of Mechanical Engineers.)

**FIGURE 15–67** Specifying angularity for an axis. (Reproduced from ANSI Y14.5-1973 with the permission of the publisher, The American Society of Mechanical Engineers.)

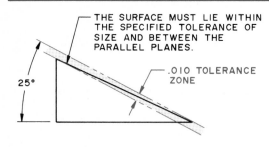

THIS ON THE DRAWING

THE AXIS MUST LIE WITHIN THE SPECIFIED
TOLERANCE OF LOCATION AND BETWEEN
THE PARALLEL PLANES.

.010 WIDE
TOLERANCE
ZONE

60°

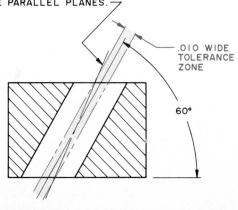

PRODUCES THIS TOLERANCE ZONE

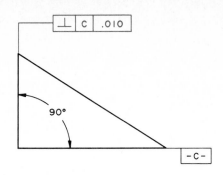

THIS ON THE DRAWING

THE ACTUAL SURFACE MUST
LIE WITHIN THE PARALLEL
PLANES.

.010 WIDE TOLERANCE ZONE

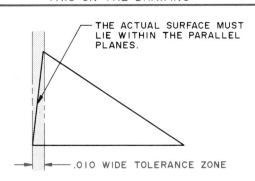

PRODUCES THIS TOLERANCE ZONE

**FIGURE 15–68** Specifying perpendicularity for a plane surface. (Reproduced from ANSI Y14.5-1973 with the permission of the publisher, The American Society of Mechanical Engineers.)

**FIGURE 15–69** Specifying perpendicularity for an axis. (Reproduced from ANSI Y14.5-1973 with the permission of the publisher, The American Society of Mechanical Engineers.)

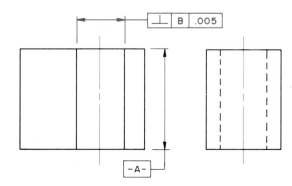

THIS ON THE DRAWING

.005 WIDE
TOLERANCE
ZONE

THE ACTUAL AXIS MUST
LIE WITHIN THE SPECIFIED
TOLERANCE OF LOCATION
AND BETWEEN THE
PARALLEL PLANES.

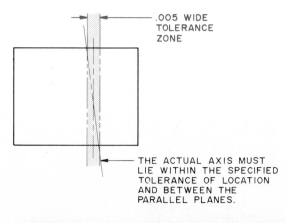

PRODUCES THIS TOLERANCE ZONE

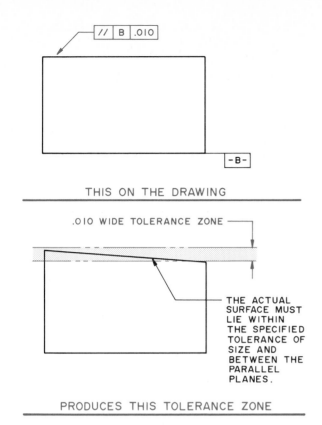

FIGURE 15–70 Specifying parallelism for a plane surface. (Reproduced from ANSI Y14.5-1973 with the permission of the publisher, The American Society of Mechanical Engineers.)

zone defined by two planes or lines parallel to a datum or axis within which the surface or axis being toleranced must lie (Fig. 15–70). It could also specify a cylindrical tolerance zone parallel to a datum axis within which the axis being toleranced must be (Fig. 15–71).

## RUNOUT TOLERANCE

Runout is a composite tolerance used to control the functional relationship of one or more features of a part to a datum axis. Features controlled by runout tolerances include surfaces constructed around a datum axis and those constructed at right angles to a datum axis (Fig. 15–72). Each feature must be within its runout tolerance when rotated about the datum axis. The tolerance specified is the total tolerance. This is measured by a dial indicator while the part is rotating. The total tolerance is the full indicator movement (FIM) registered as the part rotates.

There are two types of runout control, circular and total.

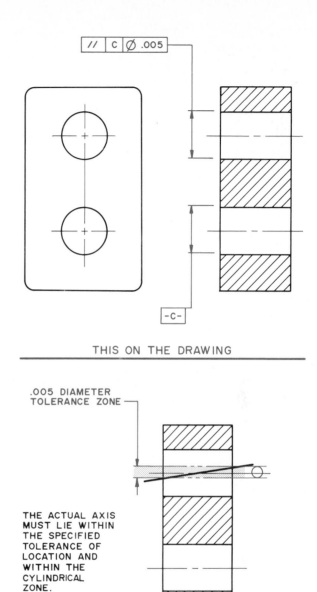

FIGURE 15–71 Specifying parallelism for an axis. (Reproduced from ANSI Y14.5-1973 with the permission of the publisher, The American Society of Mechanical Engineers.)

### Circular Runout Tolerance

Circular runout provides composite control of individual circular elements of a surface at right angles to or around the datum axis. When applied to surfaces constructed around the datum axis, circular runout controls the cumulative variation of roundness and concentricity. When applied to surfaces constructed at right angles to the datum axis, circu-

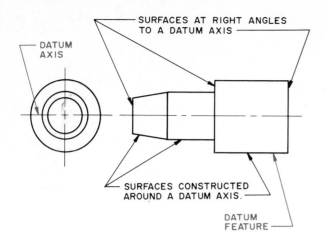

FIGURE 15–72 Features applicable to runout tolerancing.

lar runout controls surface elements of perpendicularity (Fig. 15–73).

## Total Runout Tolerance

Total runout provides control of all surface elements. The FIM is registered as the part revolves about its axis as the indicator is moved along the surface of the part. When applied to surfaces constructed around a datum axis, total runout controls cumulative variations of roundness, straightness, concentricity, angularity, and profile of a surface. When applied to surfaces constructed at right angles to the datum axis, total runout controls perpendicularity and flatness (Fig. 15–74).

FIGURE 15–73 Specifying circular runout relative to a datum diameter. (Reproduced from ANSI Y14.5-1973 with the permission of the publisher, The American Society of Mechanical Engineers.)

FIGURE 15–74 Specifying total runout relative to a datum diameter. (Reproduced from ANSI Y14.5-1973 with the permission of the publisher, The American Society of Mechanical Engineers.)

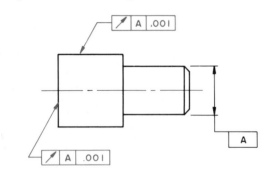

THIS ON THE DRAWING

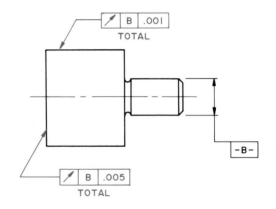

THIS ON THE DRAWING

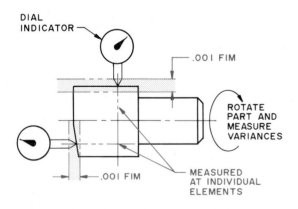

THE SURFACES, AT ANY ELEMENT, MUST BE WITHIN THE RUNOUT TOLERANCE ZONE WHEN THE PART IS ROTATED 360° ABOUT THE DATUM AXIS.

PRODUCES THESE RUNOUT TOLERANCE ZONES

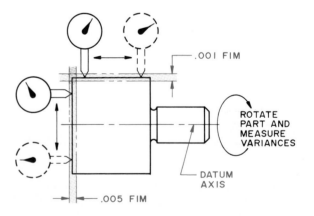

THE TOTAL SURFACE MUST BE WITHIN THE RUNOUT TOLERANCE ZONE WHEN THE PART IS ROTATED 360° ABOUT THE DATUM AXIS.

PRODUCES THESE RUNOUT TOLERANCE ZONES

# CONTROL OF SURFACE QUALITY

Designating finish on surfaces of products is an important part of the dimensioning process. The type of finish required varies with the purpose the surface will serve. The finish on the walls of cylinders in the engine of an automobile will be of a higher quality than that on the wrenches used to repair the vehicle. A system of symbols and notes is used to convey the specifications of surface quality. The system is described in publications ANSI Y14.36-1978, "Surface Texture Symbols," and ANSI B46.1-1978, "Surface Texture," published by the American Society of Mechanical Engineers. The standards establish the method to designate controls for surface texture but do not specify the means by which the surface texture is to be produced or measured. The units are expressed in SI metric and are to be regarded as standard. Approximate nonmetric equivalents are given for reference.

When no surface texture control is specified, the surface produced by normal manufacturing processes is satisfactory if it produces a part within the limits of size as specified by tolerances on the drawing. It is poor practice to produce a drawing with no surface texture control indication. Some maximum value should always be given even, if only by a general note.

## Definitions

The terms used to describe surface texture are shown in Fig. 15–75. These are defined as follows.

*Surface texture* describes repetitive or random deviations from the normal surface that form the pattern of the surface. These deviations include roughness, waviness, lay, and flaws.

*Roughness* is the *height* of the finer surface irregularities produced during the manufacturing process.

The average roughness height is measured in micrometers (millionths of a meter) or microinches (millionths of an inch).

*Roughness width* is the distance between the roughness peaks. This is measured in decimal fractions of a millimeter or an inch.

*Waviness* describes irregularities that are spaced farther apart than roughness peaks. It is the result of variations in the machine operation, such as vibration or warping. Roughness is superimposed on a wavy surface. Waviness, height, and width are measured in decimal fractions of a millimeter or an inch. Waviness is not currently in ISO standards but is used in the United States.

*Lay* is the direction of the predominant surface pattern produced during manufacture. It depends on the method of production. The symbols used to designate lay are shown in Fig. 15–76.

*Flaws* are irregularities that may occur at only one place on the surface. They are accidental defects such as cracks, scratches, holes, checks, and ridges. The effect of flaws is not included in the roughness height measurement.

*Contact area* is the surface area on one part that will make contact with its mating surface.

## Surface Texture Symbols

The surface texture symbol is used to designate the control of surface irregularities. It is used to specify roughness. A horizontal extension is used when needed to specify values other than roughness (Fig. 15–77). The basic surface texture symbol is used to specify that a surface is to be finished. When the material is to be removed by machining, a horizontal bar is drawn closing the V opening. The amount of material to be removed is specified in millimeters or inches to the left of the short V leg. When a surface is to be produced without machining, a circle is drawn in the V. When values other than roughness (such as waviness) are given, a horizontal bar is drawn above the long side of the V.

FIGURE 15–75 The terms used to describe surface texture.

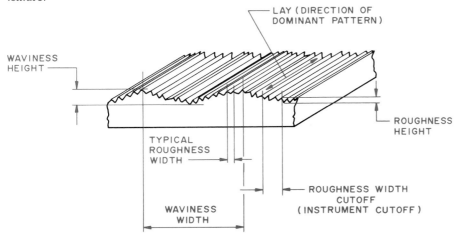

| Lay Symbol | Meaning | Example Showing Direction of Tool Marks |
|---|---|---|
| — | Lay approximately parallel to the line representing the surface to which the symbol is applied. | 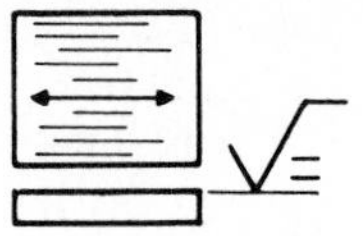 |
| ⊥ | Lay approximately perpendicular to the line representing the surface to which the symbol is applied. | 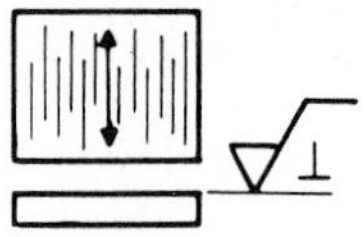 |
| X | Lay angular in both directions to line representing the surface to which the symbol is applied. | 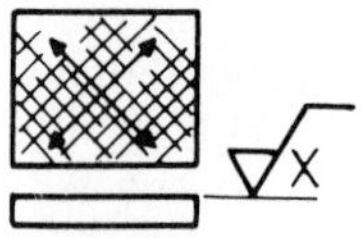 |
| M | Lay multidirectional. |  |
| C | Lay approximately circular relative to the center of the surface to which the symbol is applied. | 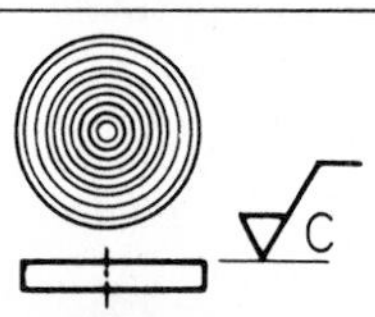 |
| R | Lay approximately radial relative to the center of the surface to which the symbol is applied. | 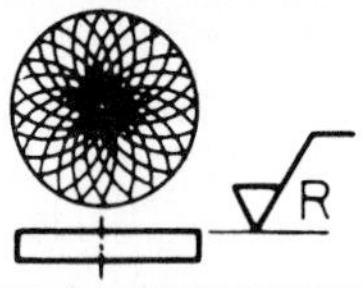 |
| P[3] | Lay particulate, non-directional, or protuberant. | 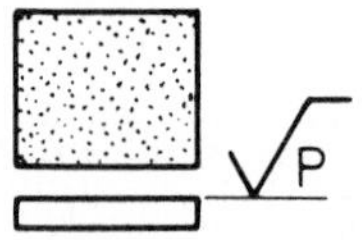 |

[3]The "P" symbol is not currently shown in ISO Standards. American National Standards Commitee B46 (Surface Texture) has proposed its inclusion in ISO 1302—"Methods of indicating surface texture on drawings."

FIGURE 15–76 Symbols used to identify lay. (Reproduced from ANSI Y14.36-1978 with the permission of the publisher, The American Society of Mechanical Engineers.)

FIGURE 15–77 Surface texture symbols. (Reproduced from ANSI Y14, 36–1978 with the permission of the publisher, The American Society of Mechanical Engineers.)

| SYMBOL | MEANING |
|---|---|
|  | **Basic Surface Texture Symbol.** Surface may be produced by any method except when the bar or circle is specified. |
|  | **Material Removal by Machining Is Required.** The horizontal bar indicates that material removal by machining is required to produce the surface and that material must be provided for that purpose. |
| 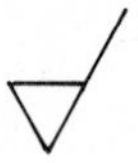 3.0 | **Material Removal Allowance.** The number indicates the amount of stock to be removed by machining in millimeters (or inches). Tolerances may be added to the basic value shown or in a general note. |
|  | **Material Removal Prohibited.** The circle in the v indicates that the surface must be produced by processes such as casting, forging, hot finishing, cold finishing, die casting, powder metallurgy, or injection molding without subsequent removal of material. |
| 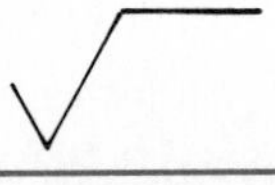 | **Surface Texture Symbol.** To be used when any surface characteristics are specified above the horizontal line or to the right of the symbol. Surface may be produced by any method except when the bar or circle is specified. |

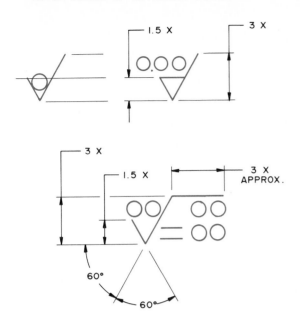

NOTE: HEIGHT OF LETTERING EQUALS X

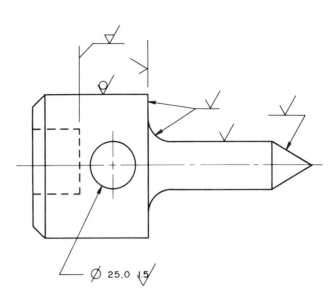

**FIGURE 15–78** The construction of the surface texture symbol. (Reproduced from ANSI Y14.36-1978 with the permission of the publisher, The American Society of Mechanical Engineers.)

**FIGURE 15–79** Application of surface texture symbols. (Reproduced from ANSI Y14.36-1978 with the permission of the publisher, The American Society of Mechanical Engineers.)

**FIGURE 15–80** The application of surface texture values to the symbol. (Reproduced from ANSI Y14.36-1978 with the permission of the publisher, The American Society of Mechanical Engineers.)

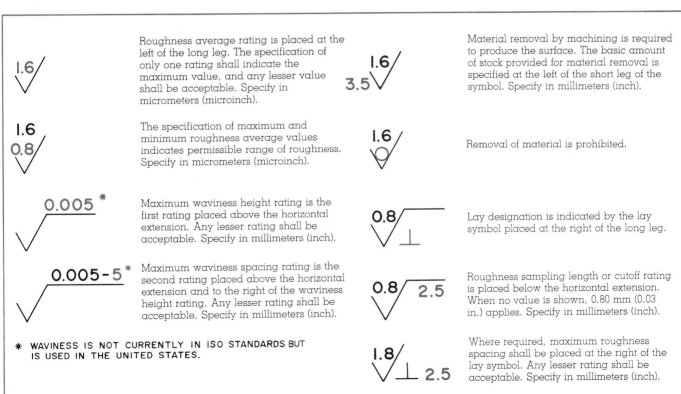

Roughness average rating is placed at the left of the long leg. The specification of only one rating shall indicate the maximum value, and any lesser value shall be acceptable. Specify in micrometers (microinch).

The specification of maximum and minimum roughness average values indicates permissible range of roughness. Specify in micrometers (microinch).

Maximum waviness height rating is the first rating placed above the horizontal extension. Any lesser rating shall be acceptable. Specify in millimeters (inch).

Maximum waviness spacing rating is the second rating placed above the horizontal extension and to the right of the waviness height rating. Any lesser rating shall be acceptable. Specify in millimeters (inch).

**\* WAVINESS IS NOT CURRENTLY IN ISO STANDARDS BUT IS USED IN THE UNITED STATES.**

Material removal by machining is required to produce the surface. The basic amount of stock provided for material removal is specified at the left of the short leg of the symbol. Specify in millimeters (inch).

Removal of material is prohibited.

Lay designation is indicated by the lay symbol placed at the right of the long leg.

Roughness sampling length or cutoff rating is placed below the horizontal extension. When no value is shown, 0.80 mm (0.03 in.) applies. Specify in millimeters (inch).

Where required, maximum roughness spacing shall be placed at the right of the lay symbol. Any lesser rating shall be acceptable. Specify in millimeters (inch).

The proportions for drawing the surface texture symbol are illustrated in Fig. 15–78. The size is based on the lettering height used on the drawing.

The point of the symbol should touch a line representing the surface, an extension line of the surface, a leader directed to the surface, or an extension of such a leader. The long leg and extension must appear to the right as the drawing is read. A note can be used to specify surface control for surfaces without values attached directly to them (Fig. 15–79).

## Application of Surface Texture Values to Symbol

The examples in Fig. 15–80 show how to designate roughness, waviness, and lay by inserting the values in the proper location on the symbol. Roughness and waviness measurements apply in the direction that gives the maximum reading, usually across the lay. Surface texture values apply to the entire surface unless otherwise specified. When parts are to be coated or plated, the drawing should indicate if the surface texture values apply before or after the coating or plating. A summary is given in Fig. 15–81.

Preferred specified roughness average values are shown in Fig. 15–82. Standard values for cutoff or roughness sampling length are shown in Fig. 15–83. The preferred series of maximum waviness height values is shown in Fig. 15–84. When specifying surface roughness or waviness, use the recommended values; do not choose values in between.

## Production Methods and Surface Roughness

The degree of surface roughness depends on the manufacturing process used. Figure 15–85 shows some typical processes and the range of surface roughness produced. Surface roughness will vary,

even for the same process. The skill of the operator, the quality of the tools, and the accuracy of the machine all influence the final outcome.

| MICROMETERS—μm | | MICROINCHES—μin. | |
|---|---|---|---|
| μm | μin. | μm | μin. |
| 0.012 | 0.5 | 1.25 | 50 |
| 0.025* | 1* | 1.60* | 63* |
| 0.050* | 2* | 2.0 | 80 |
| 0.075 | 3 | 2.5 | 100 |
| 0.10* | 4* | 3.2* | 125* |
| 0.125 | 5 | 4.0 | 160 |
| 0.15 | 6 | 5.0 | 200 |
| 0.20* | 8* | 6.3* | 250* |
| 0.25 | 10 | 8.0 | 320 |
| 0.32 | 13 | 10.0 | 400 |
| 0.40* | 16* | 12.5* | 500* |
| 0.50 | 20 | 15 | 600 |
| 0.63 | 25 | 20 | 800 |
| 0.80* | 32* | 25* | 1000* |
| 1.00 | 40 | | |

*Recommended

FIGURE 15–82 Preferred roughness average values. (Reproduced from ANSI Y14.36-1978 with the permission of the publisher, The American Society of Mechanical Engineers.)

| MILLIMETERS—mm | | INCHES—in. | |
|---|---|---|---|
| mm | in. | mm | in. |
| 0.08 | 0.003 | 2.5 | 0.1 |
| 0.25 | 0.010 | 8.0 | 0.3 |
| 0.80 | 0.030 | 25.0 | 1.0 |

FIGURE 15–83 Standard roughness sampling-length (cutoff) values. (Reproduced from ANSI Y14.36-1978 with the permission of the publisher, The American Society of Mechanical Engineers.)

FIGURE 15–84 Preferred series of maximum waviness height values. (Reproduced from ANSI Y14.36-1978 with the permission of the publisher, The American Society of Mechanical Engineers.)

| mm | in. | mm | in. | mm | in. |
|---|---|---|---|---|---|
| 0.0005 | 0.00002 | 0.008 | 0.0003 | 0.12 | 0.005 |
| 0.0008 | 0.00003 | 0.012 | 0.0005 | 0.20 | 0.008 |
| 0.0012 | 0.00005 | 0.020 | 0.0008 | 0.25 | 0.010 |
| 0.0020 | 0.00008 | 0.025 | 0.001 | 0.38 | 0.015 |
| 0.0025 | 0.0001 | 0.05 | 0.002 | 0.50 | 0.020 |
| 0.005 | 0.0002 | 0.08 | 0.003 | 0.80 | 0.030 |

FIGURE 15–81 A summary of data application to the surface texture symbol.

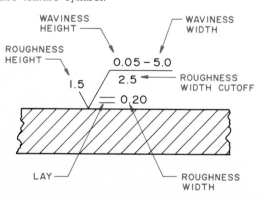

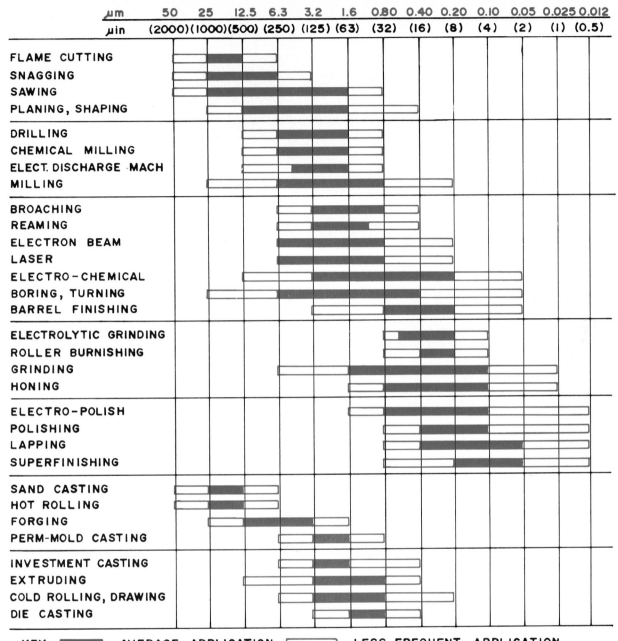

SURFACE ROUGHNESS AVERAGE OBTAINABLE BY COMMON PRODUCTION METHODS

ROUGHNESS HEIGHT RATING MICROMETRES, μm (MICROINCHES, μin) AA

KEY ▬ AVERAGE APPLICATION ▭ LESS FREQUENT APPLICATION

THE RANGES SHOWN ABOVE ARE TYPICAL OF THE PROCESSES LISTED.
HIGHER OR LOWER VALUES MAY BE OBTAINED UNDER SPECIAL CONDITIONS.

FIGURE 15–85 Typical machining processes and the range of surface roughness they produce. (Reproduced from ANSI B46.1-1978 with the permission of the publisher, The American Society of Mechanical Engineers.)

# Mechanical Fasteners: bolts, keys, nuts, screws, springs, and washers

Industrial products could not be assembled or function without some form of mechanical fastener, the use of threads, springs, or keys. These devices hold the product together, permit power to be transmitted, enable the product to be disassembled for repair and maintenance, apply tension or friction when needed, and permit parts to be adjusted in relation to one another. They are manufactured in thousands of types and sizes from a variety of materials. The designer attempts to use these mass-produced fasteners whenever possible because they cost less than fasteners designed especially for a single operation. Any time a fastener has to be custom-made, the cost soars.

The devices discussed in this chapter include some of those most frequently used, and many others that are not discussed are available. The manufacturing of fasteners is a large industry in itself. The designer should keep on hand for ready reference catalogs from these companies that give the specifications of all the fasteners they manufacture. Some companies specialize in segments of industry, such as the manufacturing of fasteners for the aircraft industry.

While not all fasteners utilize threads, this is the single most frequently used fastening system. A designer must be familiar with the standard metric and inch thread systems because they are used daily in mechanical design.

## SCREW THREADS

The threads used on screws, bolts, and nuts are standardized. The standards for threads and various fastening devices are established by the International Standardization Organization (ISO), the American National Standards Institute (ANSI), and the Industrial Fasteners Institute (IFI). The ISO is a worldwide body involved with the establishment of engineering and product standards. The ANSI works with engineering organizations in the United States to develop and publish engineering standards. It represents the United States on ISO committees. The IFI develops standards for mechanical fasteners.

Accepted national standards are published by these organizations. For example, ANSI publishes the standard "Screw Thread Representation, Engineering Drawing, and Related Documentation Practice," ANSI Y14.6-1978.

## THREAD TERMINOLOGY

The engineering designer will frequently have to refer to threads and parts of a thread. A knowledge of thread terminology is therefore important (Fig. 16–1).

*Screw thread:*  A helical ridge having a uniform section that can be on the outside of a cylindrical shaft or inside a cylindrical hole.

*External thread:*  A thread on the surface of a cylindrical shaft.

*Internal thread:*  A thread on the inside of a cylindrical hole.

*Straight thread:*  A thread formed on a cylinder.

*Taper thread:*  A thread formed on a cone.

*Right-hand thread:*  A thread that, when viewed parallel with its axis, winds clockwise and in a receding direction (Fig. 16–2). Threads are always considered right-hand unless otherwise specified.

*Left-hand thread:*  A thread that, when viewed parallel with its axis, winds counterclockwise and in a receding direction (Fig. 16–2).

*Major diameter:*  The largest diameter of an external or internal thread.

*Minor diameter:*  The smallest diameter on an external or internal thread.

*Pitch:*  The distance from a point on one thread to the same point on the next thread. It is measured parallel to the thread axis.

*Pitch diameter:*  The diameter of an imaginary cylinder passing through the threads where the width of the thread and the width of the space between threads is the same.

*Lead:*  The distance a screw thread will advance when turned one revolution.

*Angle of thread:*  The angle between the sides of the thread.

*Crest:*  The edge or surface occurring where the sides of a thread meet to form the major diameter.

*Root:*  The edge or surface occurring where the sides of a thread meet to form the minor diameter.

FIGURE 16–1   Terms used to identify parts of a thread.

AN INTERNAL THREAD

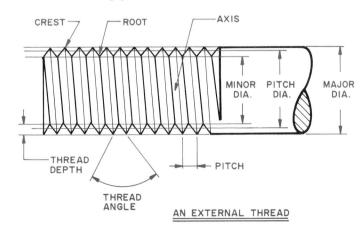

AN EXTERNAL THREAD

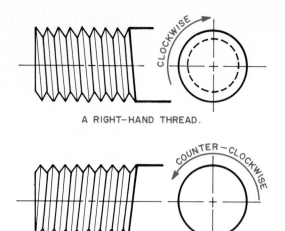

FIGURE 16–2  Right-hand and left-hand threads.

*Side:*  The surface connecting the crest and the root.

*Axis:*  The longitudinal center line through a screw.

*Depth of thread:*  The distance between the crest and the root, measured perpendicular to the axis.

*Thread form:*  The shape of a thread in cross section.

*Thread series:*  The thread series indicates the basic thread form and its nominal size and thread pitch combinations. These include the American National Unified Thread Series and the metric thread series. These are shown in Appendices E and F.

*Single thread:*  A thread formed by one helix around a cylinder (Fig. 16–3).

FIGURE 16–3  Single and multiple threads.

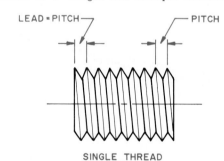

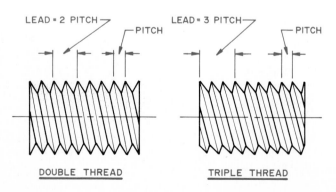

*Multiple threads:*  A thread with the same thread form on two or more helices running side by side on the shaft. The lead is equal to more than one pitch. For example, a double thread has a lead equal to two pitch distances (Fig. 16–3). A multiple thread permits a more rapid advance of a part than a single thread. A double thread will advance twice as far in one revolution as a single thread with the same thread form. A triple thread will advance three times as far.

## SCREW THREAD FORMS

A variety of screw thread forms have been developed to serve different purposes. Some are best for transmitting power, while others are better for making fine adjustments. The basic thread forms are shown in Fig. 16–4. Shown is the pitch of each thread form and the relationship between the pitch and the other dimensions producing the thread form.

*V-type threads* are used extensively on mechanical fasteners and devices used to fine-adjust parts of a product. The inch thread form used in the United States is the American National Unified Thread Series. The crest may be rounded or flat, while the root is rounded.

The *metric thread* is also a V type and has been standardized. It is widely used in the United States and will eventually replace the Unified American Standard thread.

The *Acme thread* is generally preferred for power transmission. It is a stronger form, is easier to manufacture, and permits the use of a split nut for engaging it to a device to which it transmits power. Acme screw threads are of two types, centralizing and general purpose. *Acme centralizing threads* have a limited clearance at major diameters of external and internal threads. The bearing at the major diameter maintains the alignment of the thread axis and prevents wedging on the flanks of the thread. *Acme general-purpose threads* have clearances on all diameters, permitting free movement. They are used in assemblies where both internal and external members are supported to prevent movement.

Acme centralizing threads have five classes of fits, and general-purpose threads have four classes. The preferred series is listed in "Acme Screw Threads," ANSI B1.5-1977, and appears in Appendix F.

The *stub Acme thread* form is used when a shallow thread form is needed in thin stock. The depth of the thread is equal to 0.433 of the pitch. Stub Acme screw threads have one class, class 2G. It is a general-purpose thread using two threads with modified thread depths. The preferred series is

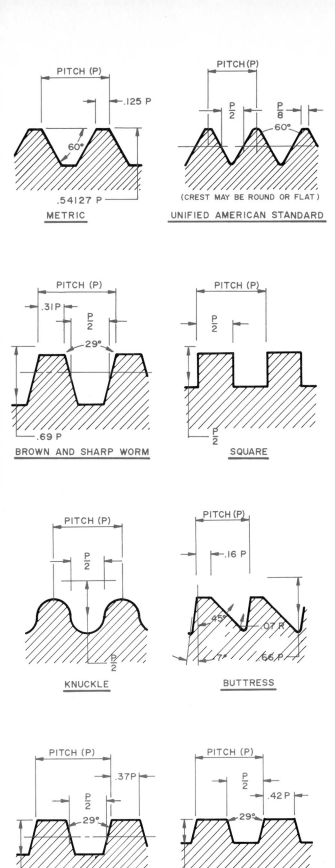

FIGURE 16–4 The basic screw thread forms and their proportions.

listed in "Stub Acme Screw Threads," ANSI B1.8-1977, and appears in Appendix F.

The *square thread* form is used to transmit power. The thread transmits power almost parallel to its axis. Generally, the square thread is manufactured with a 5° taper on the sides. It is not a standardized thread. Because it is difficult to make and to engage to the device needing power, it is not widely used.

The *buttress thread* is used to transmit power in one direction. It is an efficient and strong thread. It is made with a 7° slope on the pressure flank of the tooth.

Buttress screw threads usually require special design considerations due to the varied ways they are used. No diameter pitch series is recommended; however, preferred diameter and pitch series are reported in "Buttress Screw Threads," ANSI B1.9-1973, which appears in Appendix F.

There are two classes of buttress threads. Class 2 is the standard grade, and class 3 is the precision grade.

The *worm thread* form is similar to the Acme but is deeper. It is cut on shafts that carry power to gears having teeth cut to match the worm thread.

The *knuckle thread* form is used where threads are molded or rolled in sheet metal. It is not suitable for mechanical fasteners or power transmission. Common uses include electric light bulbs and metal lids for glass jars.

## THE THREAD SERIES

The thread series specifies the basic thread form and its nominal size (outside diameter) and pitch combinations.

The various thread sizes are designated by abbreviations (Fig. 16–5). The symbols are used on drawings to specify threads and in text material when discussing them.

### Metric Thread Series

Metric thread sizes are standardized by the ISO. The United States adopted 25 of the recommended diameter-pitch combinations; therefore, the metric threads used in the United States are compatible with the international ISO metric thread. This metric thread is identified as the M profile and all data are given in millimeters. These diameter-pitch combinations are shown in Appendix E.

### Unified Thread Series

The Unified thread is designated in inches. Details are listed in "Unified Screw Threads," ANSI B1.1-1974. The Unified Thread Series is classified into

| DESIGNATION USED ON THREAD NOTE | THREAD SERIES |
|---|---|
| ACME-C | Acme threads, centralizing |
| ACME-G | Acme threads, general-purpose |
| BUTT | Buttress thread, pull-type |
| PUSH-BUTT | Buttress threads, push-type |
| M | ISO metric threads, 6 mm and larger |
| S | ISO metric threads, up to and including 5 mm |
| Stub Acme | Stub Acme threads |
| UN, UNJ, UNR | The three Unified thread series |
| UNM | Unified miniature thread series |

FIGURE 16–5 The abbreviations used to designate the various thread series. (Reproduced from ANSI Y14.6-1978 with the permission of the publisher, The American Society of Mechanical Engineers.)

three major groups, designated as UN, UNJ, and UNR (Fig. 16–6). The UN series has four types, UN (constant pitch), UNC (coarse), UNF (fine), and UNEF (extra fine). The UNJ series has a mandatory limit on the radius root on the external thread. It has five types, UNJ (constant pitch), UNJC (coarse), UNJF (fine), UNJEF (extra fine), and UNJS (special). The UNR series also has mandatory limits specified for the radius root on the external thread. It has the same five types as shown for the UNJ thread. The Unified Thread Series is detailed in Appendix F.

The *Unified National Coarse* (UNC) thread is used for general-purpose inch fasteners. It has fewer threads per inch than the UNF and UNEF series. This permits easy assembly and disassembly. The *Unified National Fine* (UNF) thread is a fine thread that will not vibrate loose as easily as the UNC thread and is stronger because there are more threads per inch. The *Unified National Extra Fine* (UNEF) has more threads per inch than the UNF and is therefore a finer thread. It is used where parts are under severe vibration or under great stress.

The *constant-pitch* Unified threads have the same number of threads per linear inch regardless of the diameter. For example, the 8UN thread has 8 threads per inch for all diameters. The pitches avail-

able include 4, 6, 8, 12, 16, 20, 28, and 32. These are shown in Appendix F.

## THREAD CLASSES

The class of a thread is the standard allowance or the amount of tolerance and allowance of the pitch diameters. For example, the threads in a nut generally have a slightly larger diameter than those on the screw so that the parts can join together. A large allowance gives a loose fit between the internal and external threads. A small allowance permits a close fit. The class of fit to use depends on how the fastener will be used. A fastener to be subjected to vibration should have a small allowance on the threads so that they fit tightly and are less likely to vibrate loose.

### Metric Classes of Thread Fit

There are two commonly used classes of fit for metric threads. One is for general-purpose applications and contains tolerance classes 6H/6g. 6H is the class of fit for the internal thread, and 6g is the class of fit for the external thread. The other class is for precision threads where close tolerances are required. It

FIGURE 16–6 The designations used for the Unified Thread Series. (Reproduced from ANSI Y14.6-1978 with the permission of the publisher, The American Society of Mechanical Engineers.)

| BASIC THREAD SERIES | EXTERNAL THREAD ROOT | CONSTANT PITCH | COARSE | FINE | EXTRA FINE | SPECIAL DIAMETERS, PITCHES, OR LENGTHS OF ENGAGEMENT |
|---|---|---|---|---|---|---|
| UN | With optional radius root on external thread. | UN | UNC | UNF | UNEF | UNS |
| UNJ | With 0.15011 p to 0.18042 p mandatory radius root on external thread | UNJ | UNJC | UNJF | UNJEF | UNJS |
| UNR | With 0.10825 p to 0.14434 p radius root on external thread | UNR | UNRC | UNRF | UNREF | UNRS |

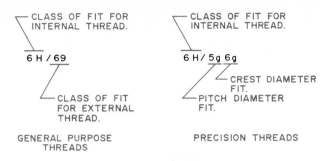

FIGURE 16–7  How to specify the class of fit for metric threads.

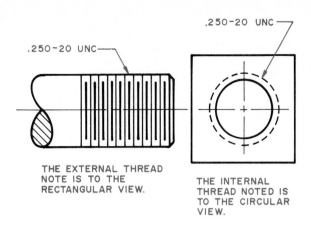

FIGURE 16–8  How to dimension internal and external threads.

is the tolerance class 6H/5g6g. 6H is the tolerance on the internal threads, 5g is the pitch diameter tolerance, and 6g is the crest diameter tolerance (Fig. 16–7).These tolerance classes can be found in Appendix G. They are among the ISO limits of tolerance used for sizing holes and shafts. Additional information is given in Chapter 15.

## Unified Thread Classes of Fit

The Unified thread series has three classes of fit. External threads have classes 1A, 2A, and 3A. Internal threads have classes 1B, 2B, and 3B. The letter A indicates an external thread, while B indicates an internal thread.

Classes 1A and 1B have a large allowance and provide a loose fit. They are used when a part must be easily and quickly assembled or disassembled. They are not used where vibration will occur. Classes 2A and 2B are used for general-purpose fasteners. Most inch-type bolts and nuts use this class of thread. Classes 3A and 3B have the smallest allowance and are used where a precision fit is needed.

## SPECIFYING THREADS

Information about a thread is given on the drawing with a note. The symbol used on the drawing to indicate a thread does not give specific data. The note identifies the thread series, diameter, pitch, and class of fit. All threads are assumed to be right-hand single threads. Left-hand threads are indicated by adding LH at the end of the note. A multiple thread is indicated by lettering DOUBLE or TRIPLE at the end of the note.

External thread data are usually given in the rectangular view of the threaded part. Internal thread data are given on the circular view. The note is connected to the part with a leader (Fig. 16–8). Sometimes the diameter of the drill used to form the hole to be threaded is given as well as the length of the thread and the depth of the hole. Drill sizes are listed in Appendix D.

## Specifying Metric Threads

The metric thread note begins with a capital M, which indicates that the threads are ISO metric. This is followed by the diameter and pitch in millimeters (for example, M6 × 1). This is the only information needed for the first class of threads, the general-purpose class. If it is not clear on the drawing whether the thread is external or internal, the note is followed by EXT for external and INT for internal. The length of engagement between internal and external threads can be given following the tolerance class designation. If the threaded part is to be coated or plated, the note should indicate whether the thread class-of-fit designation applies before or after coating or plating (for example, "M14 × 2—6H/5g6g BEFORE PLATING" or "M14 × 2—6H/5g6g AFTER PLATING"). Typical metric thread notes are shown in Fig. 16–9.

## Specifying Unified Threads

The standard thread designation is shown in Fig. 16–10. The first number specifies the nominal thread diameter in inches. This is followed by the number of threads per inch (or the pitch and lead), the thread series, the number and letter of the thread class, and any qualifying information (such as left-hand thread). If it is not clear whether it is an external or internal thread, the class is followed by A (external) or B (internal). The thread length, the hole size, and the chamfer or countersink may be part of the note. The diameter is in decimal inches unless a fractional size is specifically required. The decimal size must be to three decimal places and not more than four. If the fourth place is 0, it is omitted. The method of forming the thread is not specified on the drawing.

It is a good practice to indicate on drawings the name and number of the thread standard used when specifying threads. It is recommended that the controlling organization and the thread standard be

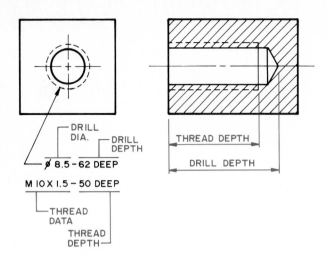

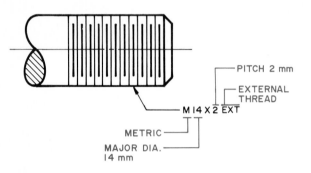

**FIGURE 16–9** Typical metric thread notes.

specified as part of the note or as a general note on the drawing as follows:

> 0.500—13 UNC—2A (thread data)
> ANSI B1.1-1974 (controlling organization)
>
> 0.250—24 UNS—3A (thread data)
> MIL-S-8879 (controlling organization)

**FIGURE 16–10** The symbol for specifying threads in the Unified Series. (Reproduced from ANSI Y14.6-1978 with the permission of the publisher, The American Society of Mechanical Engineers.)

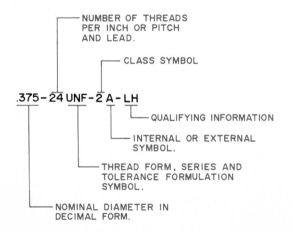

A. .250—20 UNC—2A
   A standard coarse, external thread, class 2.
B. 1½—12 UNF—2B
   A standard fine, internal thread, class 2.
C. .187–.190 DIA BEFORE THD
   .216—28 UNF—2B
   An internal thread, hole limits set before threading, fine, thread, class 2.
D. .500—20 UNF—SE2A
   PD .4435–0.4475
   MINOR DIA .4038 MAX
   LE 1.00
   Special length of engagement required, pitch diameter (PD), maximum minor diameter, and length of engagement (LE) given.
E. .625—18 UNF—2A MOD
   MAJOR DIA .614–0.620 MOD
   A modified thread, diameter reduced to modified limits designated.
F. .500—12 UNS—2A
   MAJOR DIA .4886–0.5000
   PD 0.4419–.4459
   MINOR DIA .3978 MAX
   A special thread with required special limits of size on major, pitch, and minor diameters.
G. .80 UNM (.0315 in.)
   A unified miniature screw thread in millimeters with inches in parentheses.
H. .500—20 UNJF—3A
   MIL-S-8879
   A thread with a controlled radius root following military standard 8879.

**FIGURE 16–11** Typical thread notes for Unified threads.

Figure 16–11 gives examples of thread notes for various situations involving Unified threads. Example A is a note for a standard coarse thread that requires no modification. Example B is a Unified thread with the nominal diameter given as a fraction. Example C shows a note for an internal hole. The limits on the hole diameter before threading are specified, followed by the thread information.

Example D shows how to designate threads having a special length of engagement (LE). This is necessary when the length of engagement needed exceeds that for which the standard pitch diameter tolerances are applicable. The letters SE (special engagement) are placed before the class symbol to indicate the increased length of engagement and tolerance. It is necessary to specify pitch diameter limits of size and length of engagement.

Example E shows how to indicate a modification of the limits of size of a thread. The established thread designation is followed by the modified diameter limits. The symbol MOD indicates that these are modified threads.

Example F shows the specification of special threads (UNS). The basic thread designation is followed by the required limits of size.

Example G illustrates the specification of Unified miniature screw threads (UNM). The designation includes the nominal size in hundredths of a millimeter, followed by the thread series symbol and possibly the decimal inch equivalent in parentheses.

Example H shows the specification of a controlled-radius root thread (UNJ). It is specified by the nominal diameter, number of threads per inch, thread form, series symbol, thread class symbol, and the controlling standard.

## Specifying Acme Threads

Acme threads are designated by giving, in order, the nominal diameter, the number of threads per inch, the thread form symbol, and the thread class symbol (Fig. 16–12). The thread form and class symbols identify the controlling thread standard and establish the details of the thread design, dimensions, and tolerances that are not specified on the drawing. The hole size for internal threads and the chamfer or countersink may be part of the note or dimensioned on the drawing.

Figure 16–13 gives some examples of the use of Acme thread designations. Example A shows a typical thread note. Example B shows that if the note is used in the text or not connected to a drawing with a leader, the symbol EXT is used to indicate an exterior thread and INT an interior thread. Example C

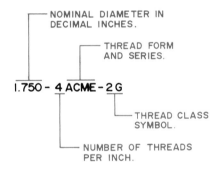

FIGURE 16–12 The symbol used to specify Acme threads. (Reproduced from ANSI Y14.6-1978 with the permission of the publisher, The American Society of Mechanical Engineers.)

FIGURE 16–13 Typical Acme thread notes.

A. .750—6 ACME—2G
A typical Acme thread note.
B. .500—16 ACME—3G EXT
A note indicating this is an exterior thread.
C. .500—16 ACME—3G—SPL
MAJOR DIA .4950–.5000
PD .4586–.4641
MINOR DIA .4193–.4275
LE 1.75
A special Acme thread note indicating the diameter limits and the length of engagement.

A. .500—10 STUB ACME
A typical thread note.
B. .625—8 STUB ACME—LH
A stub Acme thread modified to show it is a left-hand thread.

FIGURE 16–14 Specifying stub Acme threads.

shows a special diameter-pitch combination Acme thread. The limits for the major, minor, and pitch diameters are indicated; the length of engagement is also shown.

## Specifying Stub Acme Threads

Stub Acme threads are specified in the same manner as Acme threads. They have only one class, so they do not have a thread class symbol in the note. They can be modified as described for other threads (Fig. 16–14).

## Specifying Buttress Threads

Buttress threads are specified by giving the nominal diameter, threads per inch, thread form symbol, and class of fit. Modifications, such as left-hand threads, can follow the class designation (Fig. 16–15).

Buttress threads are standard when the opposite flank angles are 7° and 45°, the basic height is 0.6P, the tolerances and allowances are as specified in ANSI B1.9-1973, and the length of engagement is 10P or less When the thread is a pull thread, the designation BUTT is used. When it is a push thread, the designation PUSH-BUTT is used. Typical buttress thread notes appear in Fig. 16–16. The pitch can be given instead of the threads per inch.

## Specifying Taper Threads

External taper threads are specified by including in the note the nominal diameter, number of threads per inch, thread form, and thread pipe series symbol

FIGURE 16–15 A note for a buttress thread. (Reproduced from ANSI Y14.6-1978 with the permission of the publisher, The American Society of Mechanical Engineers.)

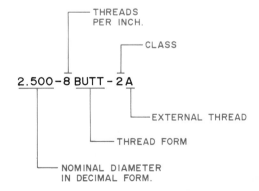

FIGURE 16-16  Typical buttress thread notes.

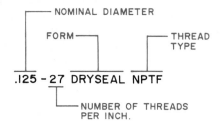

FIGURE 16-17  Specifications for an external taper thread. (Reproduced from ANSI Y14.6-1978 with the permission of the publisher, The American Society of Mechanical Engineers.)

FIGURE 16-18  Typical taper thread notes.

(Fig. 16–17). Interior taper thread notes may include the hole diameter and depth, the chamfer, the countersink, and the length of the minimum full or effective thread (Fig. 16–18).

The American National Standard dryseal taper pipe thread series is designated as NPTF. This includes external and internal taper pipe threads for pipe joints in practically every type of service. Aeronautic taper pipe thread requirements are specified in Military Specification MIL-P-7105 and are designated ANPT.

## DRAWING THREADS

Since threads are in the form of a helix, they are very difficult to draw as they really appear. There is seldom a need for a true projection; consequently, symbolic representations of threads are generally used. The three representations used are detailed, schematic, and simplified.

### Detailed Representation

*Detailed representation* produces a drawing of a thread that is very similar to its actual appearance. However, straight lines are used to show the crest and

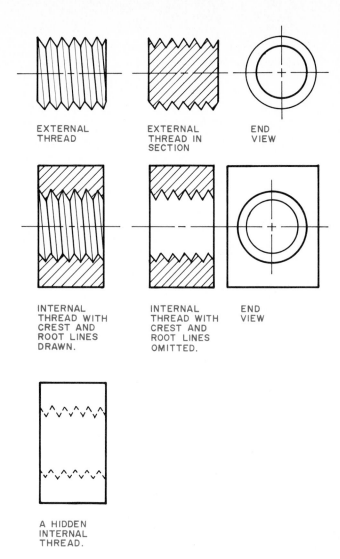

FIGURE 16-19  The detailed representation method of drawing internal threads.

root lines rather than the actual helical curves. The sides are drawn to a sharp V, even though they will be flat or rounded when actually cut. This representation is usually used for threads 1 in. and larger in diameter. Internal threads may be shown with hidden lines or in section. If in section, the crest and root lines are not usually drawn (Fig. 16–19). The detailed representation symbol is used on both detail and assembly drawings.

The procedure for drawing a detailed representation of Unified and metric threads is given in Fig. 16–20. The pitch is found by dividing 1 in. by the number of threads per inch. Since this is only a graphic representation, the actual number of threads per inch need not be used. For example, if 9 threads per inch is specified, the drawing could use 8 because it is easier to lay out and draw. If the number of threads per inch is large, say, 16, they would be very small and hard to draw. They could be drawn as 10 or 12 to make them larger and easier to draw.

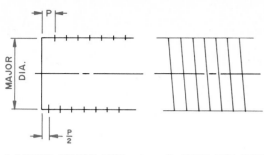

1. LOCATE CENTER LINE
AND MAJOR DIAMETER.
MARK PITCH DISTANCES.
THIS IS A RIGHT-HAND
THREAD.

2. DRAW THE CREST
LINES.

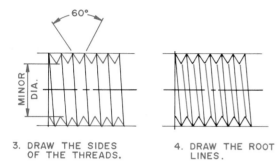

3. DRAW THE SIDES
OF THE THREADS.

4. DRAW THE ROOT
LINES.

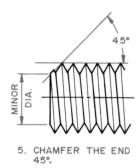

5. CHAMFER THE END
45°.

FIGURE 16-20 Drawing unified and metric threads using the detailed representation method.

The crest and root lines can be sloped to represent right- or left-hand threads.

Threaded parts usually have a chamfer on the end. This makes it easier to screw the shaft into its mating part. The chamfer is usually drawn at 45°, beginning at the minor diameter.

The procedure for drawing detailed representation of Acme threads is given in Fig. 16-21. The thread depth is half the pitch. The thread width at the pitch diameter is half the pitch. The root lines are not parallel with the crest lines in the completed drawing.

## Schematic Representation

Schematic thread representation of external and internal threads is shown in Fig. 16-22. This symbol is used on diameters of less than 1 in. The crest line

is drawn thin, while the root line is drawn thick. These lines are perpendicular to the axis of the thread. The crest lines are spaced according to the number of threads per inch. To simplify the drawing, the actual number can be changed to make the layout easier. For example, 7 threads to the inch can be drawn as 8 because it is easier to lay out 8 threads. The length of the root line is equal to the minor diameter. The end of the thread is chamfered on a 45° angle (Fig. 16-23). Since the symbol does not indicate left- or right-hand threads, this information is given in the thread note. Schematic representation is not used for hidden internal threads or external threads in section.

## Simplified Representation

Simplified thread representation of external and internal threads is illustrated in Fig. 16-24. It is recommended for straight and tapered V-form threads, Acme, stub Acme, buttress, and other thread forms, except where detailed representations are required. It is the easiest and fastest method available. It is used for diameters of less than 1 in. When drawing internal threads in section, visible lines are used to indicate the minor diameter and hidden lines to indicate the major diameter. When drawing external threads, the reverse is true. Detailed information about the thread is given in a note.

## THREADED FASTENERS

The commonly used threaded fasteners include the bolt, cap screw, machine screw, set screw, and stud (Fig. 16-25). A *bolt* has a head on one end, usually hexagonal, with the other end threaded. A *nut* is designed to screw on the threaded end. A *cap screw* is much like a bolt; however, it is designed to screw into a threaded hole in a product rather than use a nut. A *machine screw* is like a cap screw but smaller. It is often used with a nut. A *set screw* is used to lock one part, such as a pulley, to another, such as a shaft. A *stud* has threads on both ends. One end screws into a threaded hole in a product. A nut is screwed on the other end.

### Bolts and Nuts

Bolts and nuts are available with hexagon and square heads. They are classified as finished, semifinished, and unfinished (Fig. 16-26). The finished and semifinished heads have a washer face machined on the bottom of the head. The finished head is machined to meet standard tolerances. None of the surfaces on the unfinished is machined. Standard bolts are available in light, regular, and heavy series. The thickness and dimensions of the head

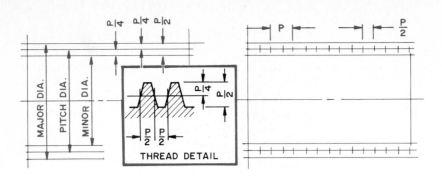

1. LAY OUT MAJOR, MINOR AND PITCH DIAMETERS.

2. LOCATE PITCH AND HALF PITCH DISTANCES ON THE PITCH DIAMETER.

3. MEASURE THE ANGLE OF THE THREAD PROFILE AND DRAW THROUGH THE HALF PITCH DISTANCES ON THE PITCH DIAMETER.

4. DRAW THE CREST LINES.

5. DRAW THE ROOT LINES.

6. REMOVE CONSTRUCTION LINES.

FIGURE 16–21 Drawing Acme threads using the detailed representation method.

FIGURE 16–22 Schematic representation of internal and external threads. These are used on diameters under 1 in.

FIGURE 16–23 Drawing a schematic representation of screw threads.

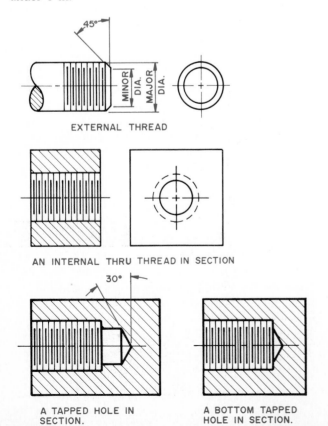

EXTERNAL THREAD

AN INTERNAL THRU THREAD IN SECTION

A TAPPED HOLE IN SECTION.

A BOTTOM TAPPED HOLE IN SECTION.

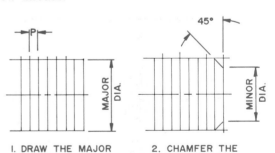

1. DRAW THE MAJOR DIAMETER AND LOCATE THE CREST LINES.

2. CHAMFER THE FINISHED END AND LOCATE THE ENDS OF THE ROOT LINES.

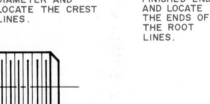

3. DRAW THE ROOT LINES.

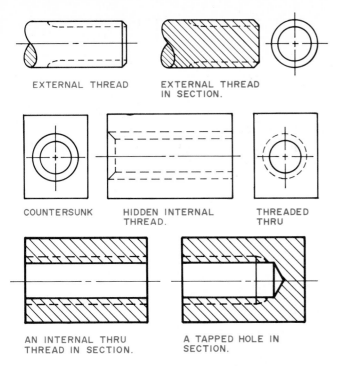

EXTERNAL THREAD     EXTERNAL THREAD IN SECTION.

COUNTERSUNK     HIDDEN INTERNAL THREAD.     THREADED THRU

AN INTERNAL THRU THREAD IN SECTION.     A TAPPED HOLE IN SECTION.

FIGURE 16–24 The simplified representation method of drawing screw threads. It is used on threads under 1 in. in diameter.

FIGURE 16–25 Typical threaded fasteners.

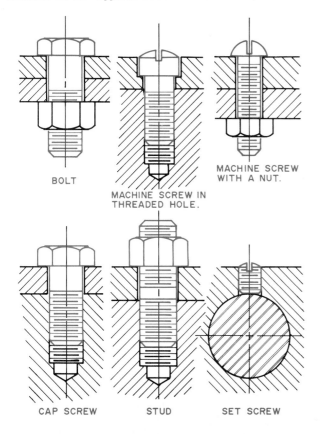

BOLT

MACHINE SCREW IN THREADED HOLE.

MACHINE SCREW WITH A NUT.

CAP SCREW     STUD     SET SCREW

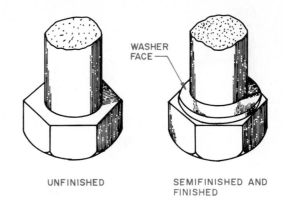

WASHER FACE

UNFINISHED     SEMIFINISHED AND FINISHED

FIGURE 16–26 Bolts and nuts are made in three classifications.

vary. The heavy series has the thickest heads and is used for the heaviest applications. Standard nuts are available in square, hexagon, jam, castle, and hex-slotted types (Fig. 16–27). A jam nut is used to lock another nut in place. It has the same basic specifications as a standard nut but is thinner. Castle and hex-slotted nuts are prepared to receive a cotter pin. The pin goes through the slot in the nut and a hole in the shaft. This keeps the nut from unscrewing (Fig. 16–28). Standard bolt and nut sizes are given in Appendices E and F.

Bolts are specified on a drawing in a note. The note shows the diameter, number of threads per inch, type of thread, class of fit, length, finish, and type of head. If the bolt is of the regular series, nothing is stated in the note. If it is light or heavy, this is made part of the note. If it is unfinished, nothing is stated. If it is semifinished or finished, this is made part of the note. If it is made of steel, nothing is said in the note. If it is of another material, such as brass, this is made part of the note.

FIGURE 16–27 Typical American Standard nuts.

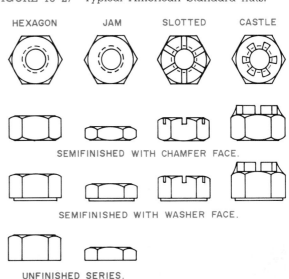

HEXAGON     JAM     SLOTTED     CASTLE

SEMIFINISHED WITH CHAMFER FACE.

SEMIFINISHED WITH WASHER FACE.

UNFINISHED SERIES.

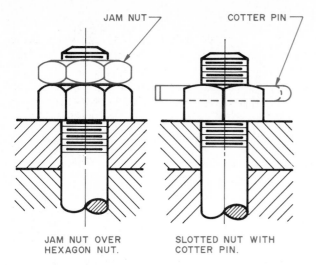

JAM NUT ─┐                COTTER PIN ─┐

JAM NUT OVER            SLOTTED NUT WITH
HEXAGON NUT.            COTTER PIN.

FIGURE 16–28 Jam nuts and cotter pins are used to keep nuts from vibrating loose.

Following are examples of notes for bolts and nuts in customary sizes. A typical note reads, "¾—16UNF—2A × 3 BRASS HEAVY SEMIFIN HEX BOLT." This means the diameter is ¾ in., 16 threads per inch, Unified National Fine Thread, class 2A fit, 3 in. long, made of brass, heavy type, semifinished hexagon bolt.

A note for a customary nut could read, "¾—16UNF—2B HEX NUT." This means the nut is for a ¾-in.-diameter bolt, 16 threads per inch, Unified National Fine Thread, class 2B fit, steel material, unfinished hexagon nut.

Following are examples of notes for bolts and nuts in metric sizes. A typical note reads, "M14 × 2 × 25 BRASS SEMIFIN HEX BOLT AND NUT." This means the bolt is metric, with a diameter of 14 mm, a pitch of 2 mm, and a length of 25 mm, made of brass, and is a semifinished hex-head bolt. A nut would be specified "M14 × 2 × 25."

The steps for drawing hexagon- and square-head bolts are shown in Figs. 16–29 and 16–30. They are usually drawn with the across-the-corners distance parallel with the plane of projection. Nuts are drawn in the same manner as the heads of the bolts (Fig. 16–27).

Another bolt finding increasing use is the spline-head. It has a strong head and is easily driven (Fig. 16–31). Sizes are given in Appendix F.

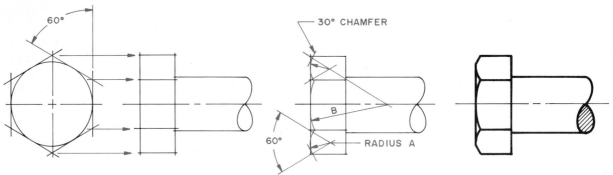

1. Draw the top view of the bolt head. Project to the front view.

2. Draw the thickness of the head. Locate the width by projection from the top view.

3. Locate the radius, A, and B as shown. Draw a 30° chamfer on each end.

4. Remove construction lines and darken all lines.

FIGURE 16–29 Drawing a hexagon bolt.

FIGURE 16–30 Drawing a square-head bolt.

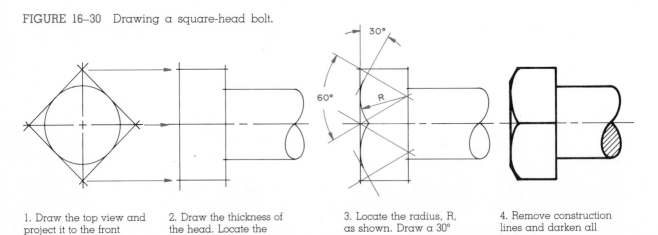

1. Draw the top view and project it to the front view.

2. Draw the thickness of the head. Locate the width by projection.

3. Locate the radius, R, as shown. Draw a 30° chamfer on each end.

4. Remove construction lines and darken all lines.

FIGURE 16–31 A 12-spline flange bolt.

## Cap Screws

The common head types for cap screws include round, flat, hexagon, fillister, spline socket, and hexagon socket. The shorter cap screws are threaded to the head. Longer lengths are threaded partway to the head (Fig. 16–32).

Customary-type cap screws are specified by giving the diameter, number of threads, thread type, class of fit, length, head type, and type of screw. An example would be "⅜—24UNF—2A × 1 HEX HD CAP SCR."

Metric cap screws are specified by giving the metric symbol, diameter, thread pitch, length, type of head, and type of screw. An example would be "M8 × 1.25 × 30 HEX HD CAP SCR." Standard sizes are shown in Appendix E.

## Machine Screws

Common head types for machine screws include flat, round, pan, oval, and fillister (Fig. 16–33). They are usually used with a nut to join thinner materials but can be used in threaded holes the same as cap screws. They are made in sizes smaller than cap screws. Inch-type machine screws are specified by giving the diameter, number of threads, type of threads, class of fit, length, head type, and type of fastener. An example would be "¼—28UNF—2A × ½ FLAT HD MACH SCR."

Metric machine screws are specified by giving the metric symbol, screw diameter, thread pitch, length, type of head, and type of screw. An example would be "M4—0.7 × 10 FIL HD MACH SCR." Standard sizes are shown in Appendices E and F.

## Studs

National standards for studs have not been established. Those in use have been standardized for particular industries. Sometimes a stud will have coarser threads on one end than the other. The coarse end is screwed into the product, and a standard nut is screwed onto the other end. Inch-type studs are dimensioned by giving the diameter, threads per inch, type of thread, class of fit, length of the stud, and thread length on each end. Metric studs are specified by giving the diameter, thread pitch, stud length, and total thread length. If studs are of any material other than steel, this must be indicated. If they have a finish, as some type of plating, this is also noted (Fig. 16–34). One standard is given in Appendix F.

## Set Screws

Set screws are made with square, slotted, hexagon socket, and spline socket heads. They are available

FIGURE 16–32 American Standard cap screws.

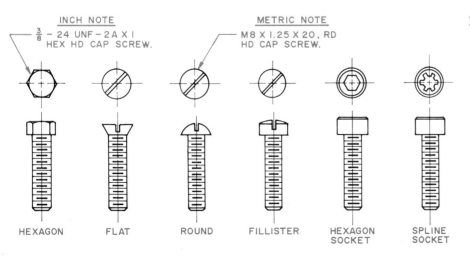

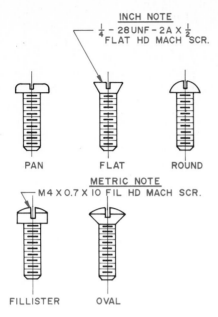

INCH NOTE
¼ – 28 UNF – 2A X ½
FLAT HD MACH SCR.

PAN     FLAT     ROUND

METRIC NOTE
M4 X 0.7 X 10 FIL HD MACH SCR.

FILLISTER     OVAL

FIGURE 16–33 American Standard machine screws.

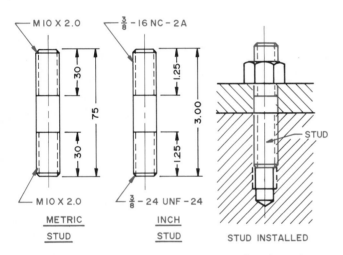

M10 X 2.0    ⅜ – 16 NC – 2A

30
75
30

1.25
3.00
1.25

STUD

M10 X 2.0    ⅜ – 24 UNF – 24

METRIC
STUD

INCH
STUD

STUD INSTALLED

FIGURE 16–34 Dimensioning metric and inch studs.

with a variety of points (Fig. 16–35). A type of heavy-duty set screw has a square head that enables it to be tightened with a wrench. If parts are under heavy load, keys are used instead of set screws.

Inch-type set screws are specified on a drawing by giving the diameter, number of threads, thread type, class of fit, length, type of head, type of point, and type of screw. An example would be "⅜–16UNC–2A × ⅝ SLOTTED FLAT PT SET SCR." A metric set screw is specified by giving the metric symbol, diameter, pitch of threads, length, type of head, type of point, and name of screw. An example would be "M10 × 1.5 × 10 SLOTTED FL PT SET SCREW." Standard sizes are given in Appendix F.

## KEYS AND KEYSEATS

A *key* is a metal part that fits into a keyseat cut into two parts that are to be held together, such as a pulley and a shaft. A *keyseat* is a rectangular groove machined into each part (Fig. 16–36). The common types of keys include the square, rectangular, plain taper, Gib-head, and Woodruff (Fig. 16–37). Square keys are preferred on shafts 6½ in. in diameter or smaller. Rectangular keys are used with shafts over 6½ in. in diameter. *Plain taper* keys are made in square and rectangular shapes. They may be of two types. One tapers the distance the key is in contact with the part. The other tapers the entire length of the key. *Gib-head* keys are tapered square or rectangular keys that have a Gib head on one end. The head helps to remove the key from the keyseat. *Woodruff* keys are almost half a cylinder. The keyseat in the shaft is half round. In the hub it is rectangular.

The size of key to use depends on the shaft diameter. Standard key sizes and recommended

FIGURE 16–35 American Standard set-screw heads and joints.

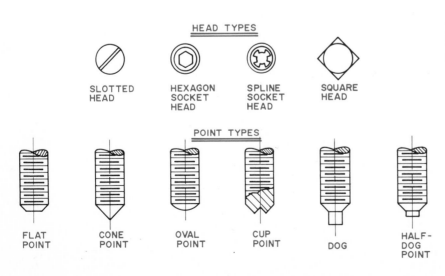

HEAD TYPES

SLOTTED HEAD    HEXAGON SOCKET HEAD    SPLINE SOCKET HEAD    SQUARE HEAD

POINT TYPES

FLAT POINT    CONE POINT    OVAL POINT    CUP POINT    DOG    HALF-DOG POINT

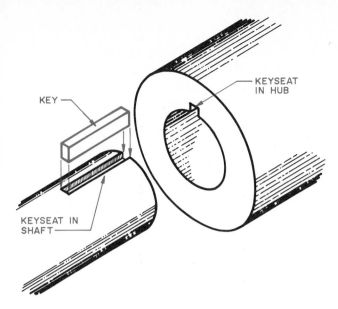

FIGURE 16–36 A key fits into the keyseats to prevent the two parts from slipping under load.

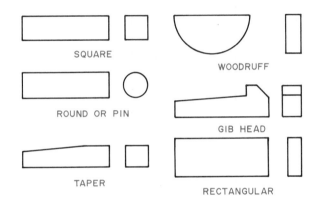

FIGURE 16–37 Common types of keys.

shaft diameter sizes for each are given in Appendices E and F.

Keys are specified by size and type. Square, rectangular, plain taper, and Gib-head keys are specified by giving the body size, length, and type, such as "¼ × ½ × 1½ FLAT KEY" or, in metric units, "6 × 6 × 10 SQ KEY." Woodruff keys are specified by number, such as "NO. 204 WOOD-RUFF KEY."

Keyseat depths for inch square, rectangular, plain taper, and Gib-head keys are in general half the thickness of the key. Metric depths are given in Appendix E. For the greater accuracy needed in manufacturing, formulas for calculating control of keyseats can be found in publication ANSI B17.1-1967 (R1973), "Keys and Keyseats."

## PINS

Pins are used to hold parts, such as collars, to a shaft. They pass through the part and the shaft (Fig. 16–38). They are round and may be straight or ta-

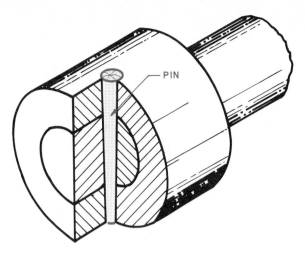

FIGURE 16–38 Pins are used to hold parts together and prevent slipping under load.

pered. They are used when light pressures are encountered. Common types include taper, straight, dowel, clevis, and cotter pins (Fig. 16–39).

*Taper pins* fit into a tapered hole. They are held in place by friction. They are removed by punching them out. Some suggested taper pin sizes for use with shafts of various diameters are in Appendix F. The pin diameters are identified by numbers.

*Straight pins* have either chamfered or straight ends. They are less costly to install because they fit in a cylindrical hole.

FIGURE 16–39 Standard types of pins.

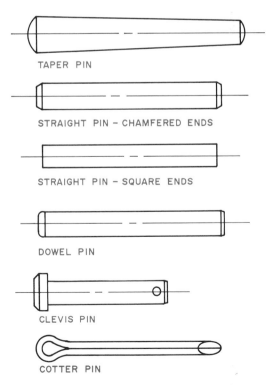

TAPER PIN

STRAIGHT PIN – CHAMFERED ENDS

STRAIGHT PIN – SQUARE ENDS

DOWEL PIN

CLEVIS PIN

COTTER PIN

*Dowel pins* have one end chamfered and the other round. The chamfered end is slightly smaller in diameter. They are used to hold parts in position or alignment.

*Clevis pins* are straight pins with a head. They are held in place with a cotter pin.

*Cotter pins* are used to keep parts from accidentally coming apart. They are placed through a hole in the part, and the ends are spread. They are available with mitered, beveled, and chisel points.

Standard sizes are given in Appendices E and F.

## WASHERS

Washers are placed between the heads and nuts of mechanical fasteners and the material being joined. They provide a broader base to press against the surface than the head or nut provides. The common types of washers are plain, lock, and tooth lock (Fig. 16–40).

*Plain washers* are flat and have standard hole diameters, outside diameters, and thicknesses. They are specified on a drawing by giving the inside and outside diameters and the thickness. A typical note for customary washers would read, "0.562 × 1.375 × 0.109 TYPE A PLAIN WASHER." A metric note would read, "19.0 × 34.0 × 4.5 PLAIN WASHER."

*Lock washers* are split and bent into a helical shape. When the nut is tightened, they are forced flat. The upward pressure they exert tends to keep the nut from accidentally coming loose. They are available in regular, extra-duty, and high-collar series. Customary sizes are specified by giving the nominal size (hole size) and the series, such as "0.500 REGULAR HELICAL SPRING LOCK WASHER." A metric note would read, "10 REGULAR HELICAL SPRING LOCK WASHER."

FIGURE 16–40 Standard washers.

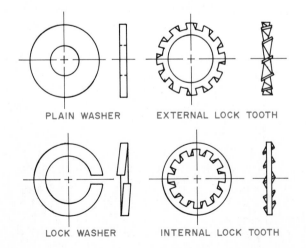

PLAIN WASHER      EXTERNAL LOCK TOOTH

LOCK WASHER      INTERNAL LOCK TOOTH

*Tooth lock washers* are made of hardened steel. They have teeth that are twisted at an angle to the face of the washer. When a nut is tightened on them, the teeth grip the nut, preventing accidental loosening. They are available in three types—internal, external, and internal-external. They are specified by giving the nominal size (hole size), type of teeth, and type of fastener. For example, a customary note might read, "0.375 INTERNAL TOOTH, TYPE A." A metric note would read, "10 INTERNAL TOOTH LOCK WASHER."

Standard washer sizes are given in Appendices E and F.

## SPRINGS

Springs are devices that can store energy when deflected from their at-rest position and push or pull upon the unit that is causing the deflection. There are two basic types of springs, helical and flat. The three types of helical springs are extension, compression, and torsion. *Extension springs* apply force when they are pulled. In their natural condition, the coils are almost touching (Fig. 16–41). Extension springs may have any of several ends. Typically, some form of loop or hook is used. The loop or hook may be on the center or side of the spring. *Compression springs* apply force when they are compressed. In their natural condition, the coils are spaced apart to allow for compression. The ends of compression springs may be squared, plain end-ground, or squared end-ground (Fig. 16–42). *Torsion springs* apply energy when one end of the coil is moved in a circular path perpendicular to the axis of the spring (Fig. 16–43).

### Technical Terms Relating to Helical Springs

*Free length* is the length of the spring when it is in its natural, unloaded condition. The full length of a compression spring includes the entire spring. The free length of an extension spring is measured inside the hooks.

*Loaded length* is the length of a spring when under a known load.

*Solid length* is the length of a compression spring when it is compressed to the point that the coils touch.

*Outside diameter* is the overall diameter of the outside of the coil.

*Inside diameter* is the inside diameter of the coil.

*Wire size* is the diameter of the wire used to make the spring.

A *coil* is one turn of the wire through 360°.

*Left-hand* and *right-hand coils* refer to the direction of the coils. A left-hand coil is formed counterclockwise; a right-hand coil is formed clockwise.

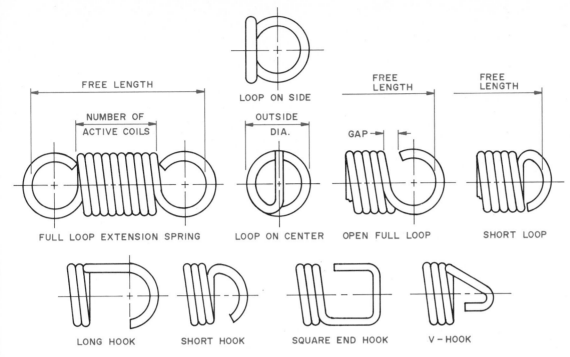

FIGURE 16–41  Extension springs and examples of a few of the hooks and loops in use.

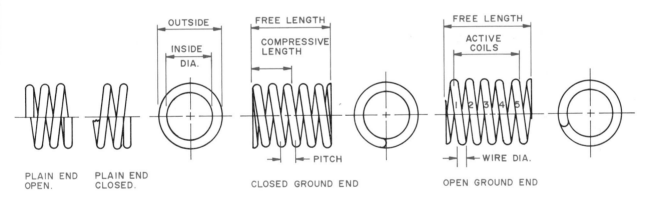

FIGURE 16–42  Compression springs.

*Pitch* is the horizontal distance from a point on one coil to the same point on the next coil.

*Active coils* refers to the total number of coils in a spring, less those rendered inactive because of the design or application of the spring. On compression springs, the coil on each end is not considered an active coil.

## How to Draw Helical Springs

Helical springs can be drawn with a schematic or detailed representation.

The *schematic representation* is the easiest and saves drafting time. The coils are indicated by straight lines. The distance from one point to the next represents the pitch or one coil (Fig. 16–44).

On compression springs, one coil on each end is closed or squared off and is therefore lost. An eight-coil spring has only six active coils.

All coils are active in extension springs. The number of points indicates the number of active coils. All coils on a torsion spring are active, and it is drawn showing the ends extended beyond the coil.

*Detailed representation* produces a drawing that appears much like the actual spring. However, the sides of the wire are drawn with straight lines rather than the actual helical curve. The steps for drawing helical springs using detailed representation are shown in Fig. 16–45.

Generally, it is too time-consuming to draw all the coils on a spring on a detailed drawing. A few are drawn on each end, and phantom lines are used to represent those omitted (Fig. 16–46). Springs can also be drawn in section, as shown in Fig. 16–47. The front half of the coil is removed, revealing the

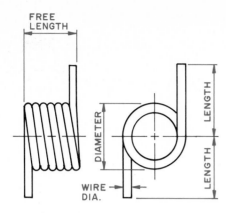

6 COILS WIND RIGHT HAND WITH NO INITIAL TENSION.

LOAD TO BE 10⁺ LB. INCH TORQUE WITH 180° DEFLECTION.

MUST WIND UP 270° WITHOUT PERMANENT SET.

MATERIAL : OIL TEMPERED STEEL WIRE.

HEAT TREAT TO RELIEVE STRESSES.

FINISH — NONE.

FIGURE 16–43  Drawing of a typical torsion spring.

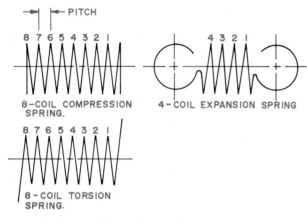

8-COIL COMPRESSION SPRING.

4-COIL EXPANSION SPRING

8-COIL TORSION SPRING.

FIGURE 16–44  The schematic representation method of drawing springs.

rear half of the coil. On detail drawings, compression springs may also be represented by a rectangle with diagonals, and the specifications of the spring are given (Fig. 16–48).

## Specifying Helical Springs

The following information is needed to describe fully a compression spring.

1. Outside diameter if spring is to fit inside a tube.
2. Inside diameter if spring is to fit over a shaft.
3. Wire size.
4. Material.
5. Number of coils.

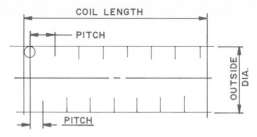

1. Locate the center line of the coil.
2. Draw the coil length.
3. Draw the inside or outside diameter.
4. Draw a wire diameter.
5. Locate and draw lightly the pitch distance for each coil.

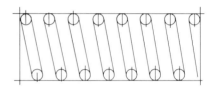

6. Lightly draw the circles for the wire diameter.
7. Connect the wire-diameter circles with straight lines. This is a right-hand spring. These are the coils on the front of the spring.

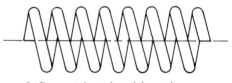

8. Connect the sides of the coils on the back of the spring.
9. Draw the ends as needed.
10. Remove construction lines.

FIGURE 16–45  Drawing a detailed representation of a coil spring.

6. Right-hand or left-hand orientation.
7. Style of ends.
8. The load at a specified deflected length.

FIGURE 16–46  Phantom lines can be used to represent coils in the area where they are not drawn.

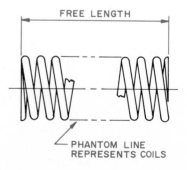

FREE LENGTH

PHANTOM LINE REPRESENTS COILS

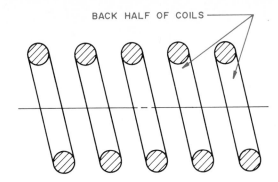

FIGURE 16–47   A coil spring in section.

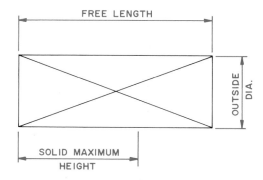

MATERIAL: .150 HARD DRAWN SPRING
STEEL 10 COILS 8 ACTIVE
CLOSED ENDS GROUND.

FIGURE 16–48   Simplified representation of a compression spring.

9. The load rate desired over a specified compression distance.

10. The minimum compressed height when in use.

The following information is needed to describe fully an extension spring.

1. Free length overall, over the coil, or inside the hooks.
2. Outside diameter or inside diameter.
3. Wire size.
4. Material.
5. Number of coils.
6. Right-hand or left-hand orientation.
7. Style of ends.
8. Load at inside hooks.
9. Load rate in pounds per inch of deflection.
10. Maximum extended length.

The following information is needed to describe fully a torsion spring.

1. Free length.
2. Outside diameter or inside diameter.
3. Wire size.
4. Material.

5. Number of coils.
6. Right-hand or left-hand orientation.
7. Torque in pounds at a specified number of degrees of deflection.
8. Maximum deflection in degrees from the unloaded position.
9. Style of ends.

## Flat Springs

*Flat springs* are made from spring steel, plastic, and other materials. They may be a single flat piece which when subjected to a force will react by opposing the force (Fig. 16–49). *Belleville springs* are cup-shaped to store energy when deflected (Fig. 16–50). *Leaf springs* are semielliptic and usually made of several pieces. They straighten as a load is applied (Fig. 16–51).

Flat springs are not standardized; they are designed especially for each use. To describe a flat spring, the following information is needed.

1. A dimensioned detail drawing.
2. Material to be used.
3. Heat treatment desired.
4. Finish.

A typical drawing of a flat spring appears in Fig. 16–49.

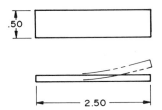

SPRING .150 THICK
MATERIAL – C.R. SPRING STEEL
SAE 1080 HEAT TREAT ROCKWELL
C 40 – 50
LOAD – 10 LBS ± 1 LB. AT .50

FIGURE 16–49   Detail of a typical flat spring.

FIGURE 16–50   Drawing of a typical Belleville spring.

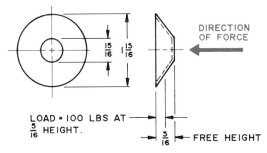

LOAD = 100 LBS AT
$\frac{5}{16}$ HEIGHT.

MATERIAL – C.R. SPRING STEEL SAE 1370
HEAT TREAT ROCKWELL C 38 – 43
THICKNESS .062

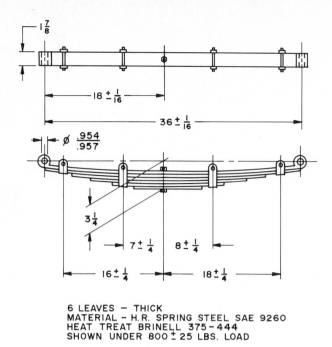

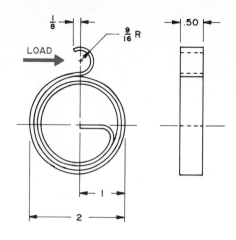

FIGURE 16–51   Detail drawing of a leaf spring.

6 LEAVES — THICK
MATERIAL — H.R. SPRING STEEL SAE 9260
HEAT TREAT BRINELL 375–444
SHOWN UNDER 800 ± 25 LBS. LOAD

2 – COILS
MATERIAL — C.R. SPRING STEEL SAE 1080
HEAT TREAT ROCKWELL C 40–50
THICKNESS .062
LOAD 15 LBS ± 2 LBS AT $1\frac{3}{4}$

FIGURE 16–52   Detail drawing of a flat coil spring.

## Power Springs

Power springs are made from flat material wound
into a coil. Winding causes a power spring to store
up energy. It can deliver power by torque through a
central shaft to which it is connected or through a
drum in which it is secured. The involute develops
power when it is compressed. It transmits power
parallel with its central axis (Figs. 16–52 and 16–53).

## SCREWS

Tapping screws and wood screws are used extensively in manufactured products. *Tapping screws* are
designed to cut their own threads into metal or plastic parts as they are screwed tight. They pass
through a clearance hole in one part and into a
smaller-diameter hole in the other part, where they
cut threads in the side of the hole (Fig. 16–54). Tapping screws are available with a variety of slotted,
recessed, and hexagon heads and a variety of points.
Type AB tapping screws are used for joining thin
metal to wood. Type B join thin metal, nonferrous
castings, and plastics. Types BF, BT, D, and F have
blunt points and cut threads in aluminum, sheet
steel, cast iron, brass, and plastic. Types D, F, and
T have pitch combinations and thread profile forms
as found on machine screws.

Tapping screws are dimensioned by giving the
nominal screw size and pitch, length, fastener name,
head type, and finish. For example, a metric screw
note would read, "M8 × 2.12 × 30 TYPE AB SLOT-
TED PAN HEAD TAPPING SCREW NICKEL FIN."
Design sizes are given in Appendix E.

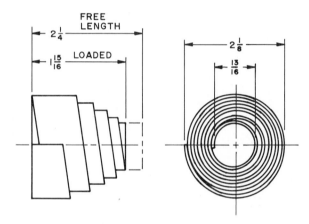

5 COMPLETE COILS
ENDS CLOSED AND GROUND
MATERIAL — C.R. SPRING STEEL SAE 1080
THICKNESS .062
WIDTH $1\frac{1}{16}$
HEAT TREAT ROCKWELL C 40–50
LOAD 150 LBS. ± 10 LBS. AT $1\frac{15}{16}$

FIGURE 16–53   Detail drawing of a volute spring.

*Wood screws* are used to secure two pieces of
wood together or some other material to wood. The
screw threads into the wood. The top material has a
hole larger than the screw diameter. The second part
has a hole about half the diameter of the screw. The
wood screw cuts threads in the side of the hole (Fig.
16–55).

Wood screws are made with round, flat, and
oval heads, which may be slotted or recessed. They
are made from brass, aluminum, and steel.

Wood screws are dimensioned by giving the
diameter of the body (specified by a wire-diameter
number), length, type of head, type of screw, and
material. A typical note is "No. 8 × 1½ FLAT HEAD
WOOD SCREW BRASS." Design sizes are given in
Appendix F.

| Nominal Screw Size | D—Clearance Hole Diameter, Basic | | |
| --- | --- | --- | --- |
| | Close Clearance | Normal Clearance (Preferred) | Loose Clearance |
| 2.2 | 2.4 | 2.6 | 2.8 |
| 2.9 | 3.1 | 3.3 | 3.6 |
| 3.5 | 3.8 | 4.0 | 4.4 |
| 4.2 | 4.5 | 4.8 | 5.2 |
| 4.8 | 5.2 | 5.4 | 5.8 |
| 5.5 | 5.8 | 6.2 | 6.5 |
| 6.3 | 6.7 | 7.1 | 7.5 |
| 8.0 | 8.5 | 9.0 | 9.5 |
| 9.5 | 10.0 | 10.5 | 11.5 |

All dimensions are in millimetres.

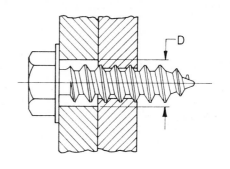

**Clearance holes for metric tapping screws.**

FIGURE 16–54 Metric tapping screw sizes.

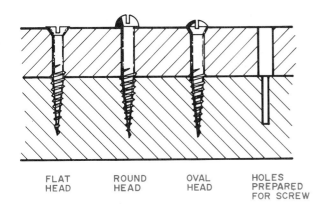

FLAT HEAD    ROUND HEAD    OVAL HEAD    HOLES PREPARED FOR SCREW

FIGURE 16–55 The basic types of wood screws.

# Problems

As the class prepares drawings for design projects, fasteners and threads will become a major part. The information in this chapter is to be used as a reference for selecting, sizing, and drawing these things. This will provide a realistic experience with threads and fasteners. Following are some exercises that provide some drafting experience.

1. In Prob. 16–1, draw the indicated external and internal threads full size. Do a little preplanning so that the views fit on the paper.

2. In Prob. 16–2, draw the indicated external and internal threads full size. Do a little preplanning so that the views fit on the paper.

3. In Prob. 16–3, draw the indicated internal and external threads. Then, show a 1-in. hole threaded in elevation and in section under the conditions specified in the problem. Do a little preplanning so that the views fit on the paper.

4. In Prob. 16–4, draw the indicated threaded fasteners full size.

5. In Prob. 16–5, draw the indicated springs full size.

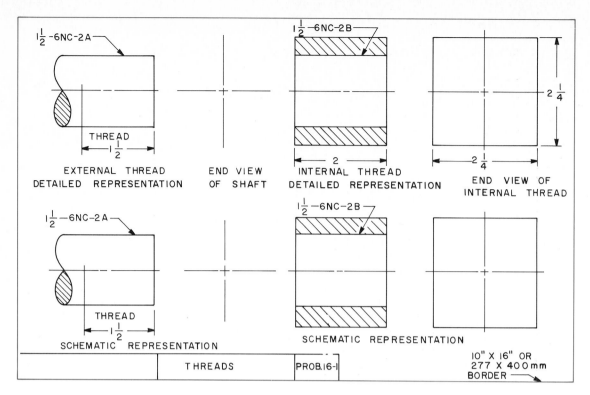

PROBLEM 16-1

PROBLEM 16-2

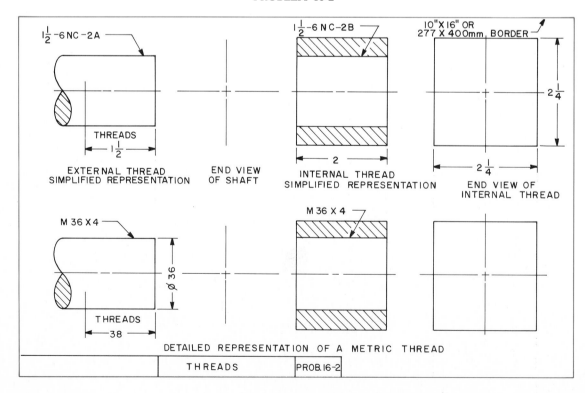

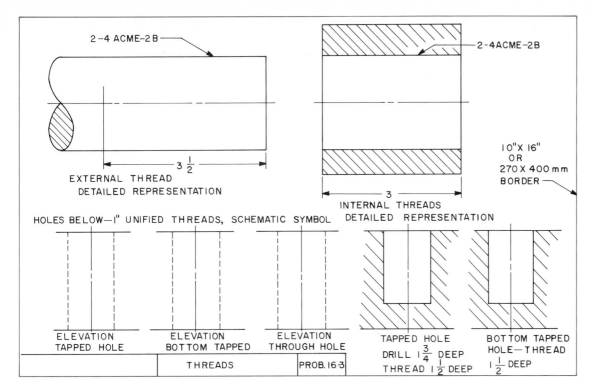

PROBLEM 16-3

PROBLEM 16-4

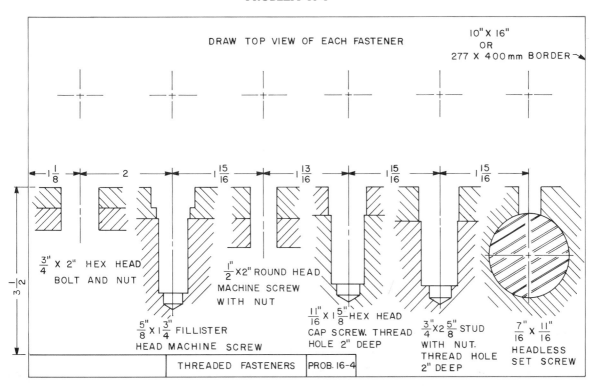

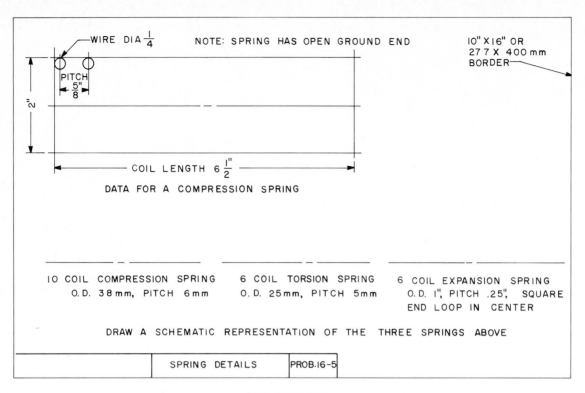

WIRE DIA $\frac{1}{4}$  NOTE: SPRING HAS OPEN GROUND END  10" X 16" OR
27.7 X 400 mm
BORDER

PITCH $\frac{5}{8}$"

2"

COIL LENGTH $6\frac{1}{2}$"

DATA FOR A COMPRESSION SPRING

10 COIL COMPRESSION SPRING
O.D. 38 mm, PITCH 6 mm

6 COIL TORSION SPRING
O.D. 25 mm, PITCH 5 mm

6 COIL EXPANSION SPRING
O.D. 1", PITCH .25", SQUARE
END LOOP IN CENTER

DRAW A SCHEMATIC REPRESENTATION OF THE THREE SPRINGS ABOVE

| | SPRING DETAILS | PROB.16-5 |

PROBLEM 16-5

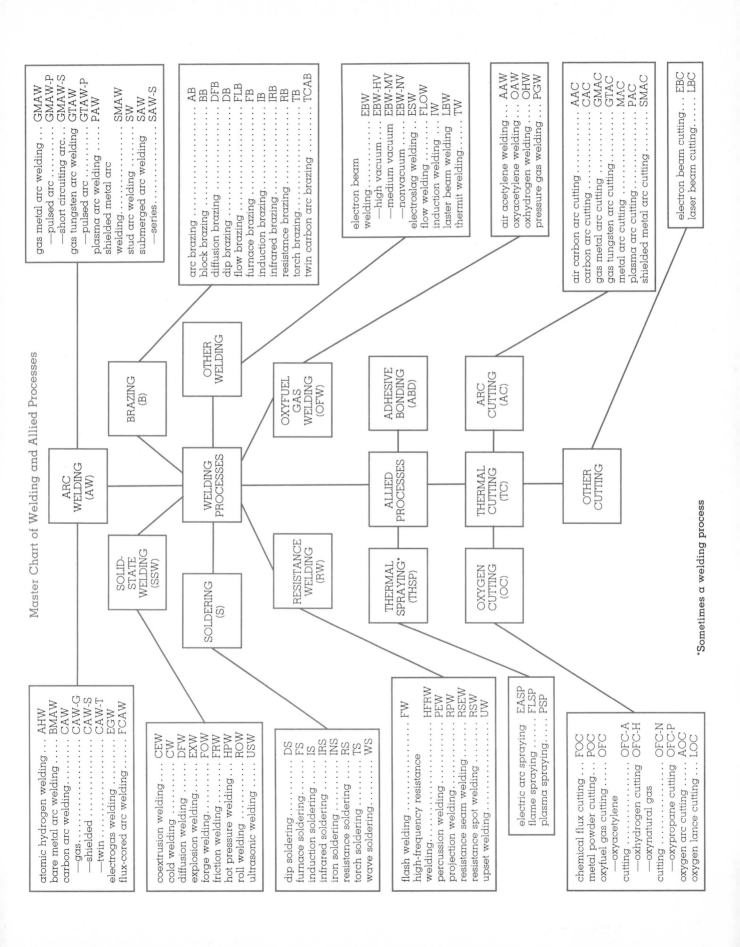

Master Chart of Welding and Allied Processes

# Permanent Fasteners: Welding and Riveting

Welding is a major means of permanently fastening parts together and is used extensively in mass-produced products. It generally produces parts that cost less and weigh less than those made by casting or forging. The rapid development of robots that can perform repetitive welding processes in mass-produced products has lowered the cost and increased the consistency of the quality of welds (Fig. 17–1). The two basic welding processes are fusion and resistance welding.

## FUSION WELDING

Fusion welding involves heating the materials to be welded to a molten state and joining them with another molten metal. The most commonly used types of fusion welding are oxyacetylene, shielded-arc, gas shielded-arc, gas metal-arc, buried-arc $CO_2$, and submerged-arc.

*Oxyacetylene welding*, often called gas welding, involves directing a gas flame over the parts to be welded and inserting a welding rod, which melts and intermingles with the melted metal. The most common gas used is acetylene, which is mixed with oxygen (Fig. 17–2).

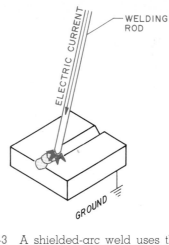

FIGURE 17–3 A shielded-arc weld uses the heat from an electric arc to melt the metal to produce fusion.

*Shielded-arc welding,* often called simply arc welding, involves producing an electric arc between the material to be welded and a coated metallic electrode. The electrode is kept slightly above the work, producing an arc that melts the metal and mixes the melted electrode with it, fusing the metal parts together. The coating produces a gas that shields the arc from the atmosphere (Fig. 17–3).

*Gas shielded-arc welding* involves two major processes, tungsten-arc and metal-arc welding. *Tungsten-arc welding,* often called Tig, involves heating the parts to be welded with a nonconsumable tungsten electrode. During the welding process, the electrode produces an electric arc in a gas-filled atmosphere, which prevents oxidation of the electrode, the melted base metal, and the heated metal around the weld area. Fusion occurs by joining the melted base metals only because the electrode does not melt (Fig. 17–4). A filler metal can be used if necessary.

*Gas metal-arc welding,* often called Mig or metal-inert gas welding, involves striking an electric arc

FIGURE 17–1 Robots can rapidly and accurately weld products on a production line. (Courtesy of Unimation Inc.)

FIGURE 17–2 Oxyacetylene welding involves joining two pieces of metal by fusing them together with a molten metal heated by a gas flame.

FIGURE 17–4 Gas tungsten-arc welding (Tig) uses a nonconsumable electrode in a gas atmosphere.

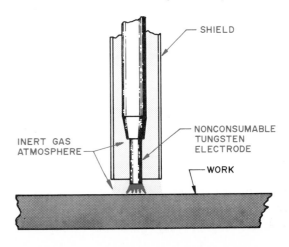

Permanent Fasteners: Welding and Riveting **419**

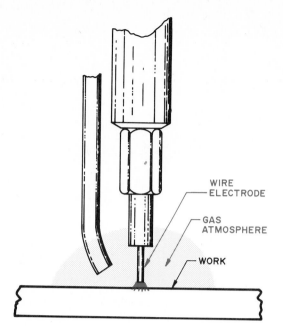

FIGURE 17–5 Gas metal-arc welding (Mig) uses a consumable wire electrode in a gas atmosphere.

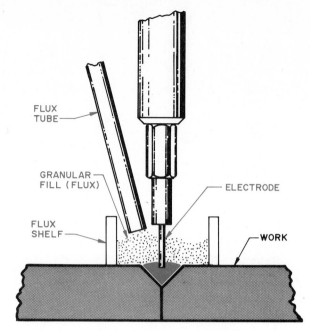

FIGURE 17–7 Submerged-arc welding shields the molten metal with a consumable granular material.

between the work and an electrode, producing heat that fuses the metal work piece and the melted electrode in a shielded-gas atmosphere, which protects the melted material from atmospheric contamination. The electrode, in the form of a wire, is fed at controlled rates from a reel (Fig. 17–5).

*Buried-arc CO$_2$ welding* involves placing an electrode level with or below the surface of the work and almost touching it. Mig welding equipment is used. The arc produced is shielded with pure carbon dioxide gas (Fig. 17–6).

*Submerged-arc welding* involves producing an electric arc that melts and fuses the metal. The arc is submerged under a granular material, called flux, which protects the arc and the material being welded from the atmosphere. The granular material around the arc fuses and covers the molten metal.

FIGURE 17–6 Buried-arc CO$_2$ welding shields the arc with carbon dioxide gas.

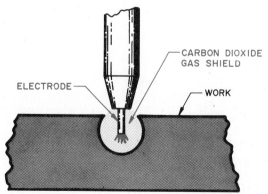

After the weld has cooled, it is easily brushed away (Fig. 17–7).

## RESISTANCE WELDING

Resistance welding is done by passing an electric current through the metal to be joined at the point where a weld is desired. The types of resistance welding include spot, seam, projection, flash butt, upset butt, and high-frequency.

A *spot weld* is formed by pressing electrodes against opposite sides of the parts and passing an electric current through them. The pressure plus the heat from the electric current fuses the sheets together (Fig. 17–8).

A *seam weld* is like a spot weld except the weld is usually continuous over the entire length. Sometimes spots are spaced so that they do not touch (Fig. 17–9).

A *projection weld* has one part with a dimple stamped into it. This is placed in contact with the member to be joined to it. An electric current is passed through the projection, fusing the two parts together (Fig. 17–10).

A *flash butt weld* is made by passing an electric current through two members that are spaced so that they do not quite touch. An electric arc jumps the gap, melting the edges. The members are then forced together under pressure (Fig. 17–11). An *upset butt weld* is made by passing an electric current through two members that are in firm contact with

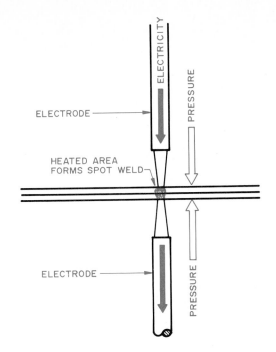

FIGURE 17–8  A spot weld is formed by the heat from the flow of electric current plus mechanical pressure from the electrodes.

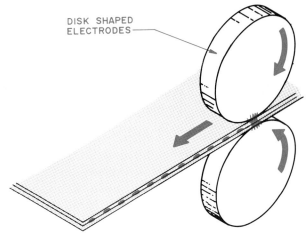

FIGURE 17–9  A seam weld is a continuous weld made of overlapping or uniformly spaced spot welds.

FIGURE 17–10  A projection weld is made by passing an electric current through a dimple stamped in the metal and applying pressure on it.

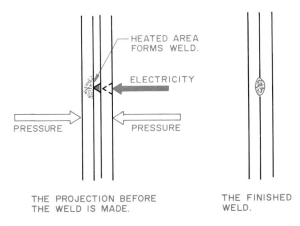

THE PROJECTION BEFORE THE WELD IS MADE.

THE FINISHED WELD.

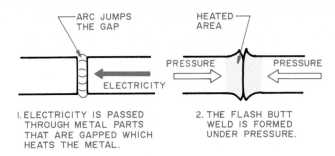

1. ELECTRICITY IS PASSED THROUGH METAL PARTS THAT ARE GAPPED WHICH HEATS THE METAL.

2. THE FLASH BUTT WELD IS FORMED UNDER PRESSURE.

FIGURE 17–11  A flash butt weld.

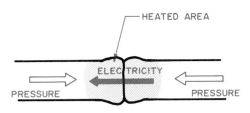

ELECTRICITY IS PASSED THROUGH TOUCHING PARTS AND PRESSURE IS APPLIED.

FIGURE 17–12  An upset butt weld.

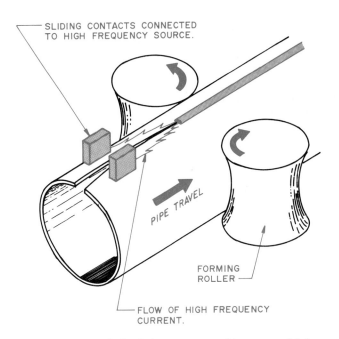

FIGURE 17–13  A high-frequency weld passes high-frequency current through a part to heat the edges to form the weld.

each other. The current heats the two parts, which fuse together under pressure (Fig. 17–12).

*High-frequency resistance welding* uses high-frequency current to produce heat for bonding edges of metal. A major use is in the fabricating of pipe (Fig. 17–13). The current flows from one contact along the surface of the weld to the root of the metal. It returns along the surface of the other edge of the joint to the second contact. The two pieces are heated by this flow of electricity. The molten edges are brought together and forge-welded under pressure.

## SPECIAL WELDING PROCESSES

Special welding processes include electron beam, laser, ultrasonic, plasma, and inertia.

*Electron beam welding* is basically a fusion welding process. A high-power beam of electrons is focused on the material to be joined. When it strikes the material, the kinetic energy of the high-velocity electrons changes to thermal energy, which heats and fuses the metal (Fig. 17–14).

*Laser welding* involves focusing a highly concentrated beam to a spot the diameter of a human hair. This concentrated beam produces heat in the material, causing it to fuse. Because the area of the weld is so small, a weld can be made close to other materials, such as glass, with no damage caused by excess heat. It can also be used to join dissimilar materials, such as copper, nickel, stainless steel, and aluminum (Fig. 17–15).

*Ultrasonic welding* involves applying vibratory energy to the surfaces to be joined. This process is used to join flat surfaces that are as smooth and flat as it is possible to get them. The vibratory action enables the metal atoms in one piece to unite with at-

FIGURE 17–14 Electron beam welding focuses a high-power beam of electrons on the parts to be joined.

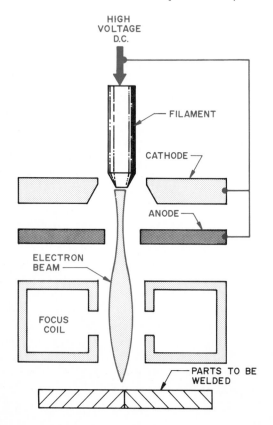

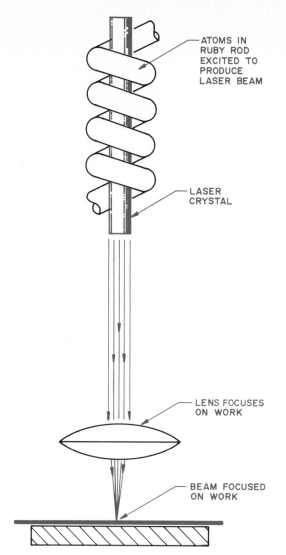

FIGURE 17–15 Laser welding focuses a concentrated beam the diameter of a human hair on the spot to be welded.

oms in the other, thus forming a bond. This process does not involve the application of heat or any type of filler material. It is especially useful for joining electrical and electronic components because of the absence of heat, fumes, and filler material.

*Plasma welding* involves a central core of plasma gas at a high temperature surrounded by a layer of cool shielding gas. The heat for fusion is produced by an arc that heats the plasma gas until it becomes ionized. The end of the electrode is in a nozzle in the central core and regulates the jet of heated plasma gas, which exits at a high velocity and intense heat. This superheated arc can make butt welds in steel up to ½ in. in thickness in one pass with no filler rod (Fig. 17–16).

*Inertia welding*, sometimes called friction welding, involves the use of stored kinetic energy to produce heat for fusion. The members to be welded are aligned. One is held stationary, while the other is

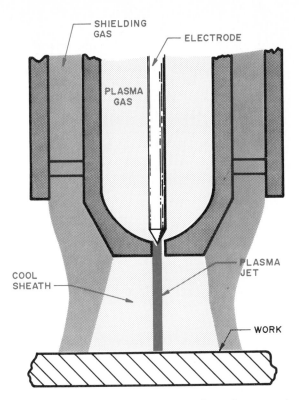

FIGURE 17–16 Plasma welding involves the use of a jet of high-temperature, high-velocity plasma gas.

placed in a rotating spindle. The spindle is rotated at the required speed, disconnected from the source of power, and brought into contact with the stationary piece at a predetermined thrust load. The kinetic energy in the rotating part produces heat due to friction, fusing the two parts together (Fig. 17–17).

## WELDED JOINTS

The basic joints used in welding are shown in Fig. 17–18. They are joined by one of the many types of welding processes.

The basic types of welds are shown in Fig. 17–19. The *back weld* is on the surface of two parts butted together. This is often used on thin material. Thicker material will require a *square weld,* where the pieces are spaced apart and the weld penetrates to the back side. The *fillet weld* requires no special preparation. *Bevel, U, V,* and *J welds* require that the joining pieces be ground to provide space for the molten welding rod. *Plug* and *slot welds* are made through holes cut in the top member to be welded.

### Welding Symbols

The type of weld needed for the design is shown on engineering drawings with standard welding symbols (Fig. 17–20). These were developed by the American Welding Society and are available in pub-

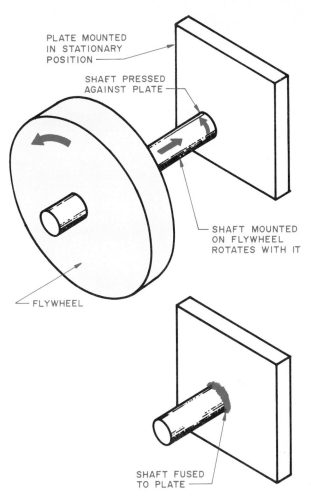

FIGURE 17–17 Inertia welding uses the heat produced by the friction between parts to fuse them together.

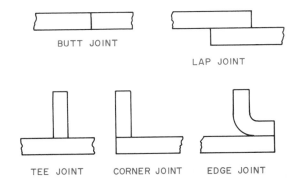

lication ANSI/AWS A2.4-1979, "Symbols for Welding and Nondestructive Testing." See Appendix L. The basic welding symbol is shown in Fig. 17–21. It is an assembly of elements necessary to describe and dimension a weld. The eight possible elements are (1) a reference line, (2) an arrow, (3) basic weld symbols, (4) dimensions, (5) supplementary symbols, (6) finish symbols, (7) tail data, and (8) specification

FIGURE 17–18 The basic joints used in welding.

BUTT JOINT

LAP JOINT

TEE JOINT          CORNER JOINT          EDGE JOINT

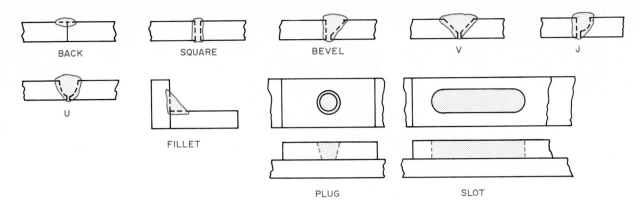

FIGURE 17–19 The basic types of welds.

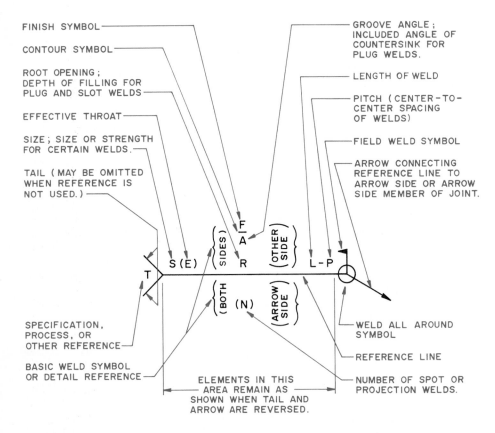

FINISH SYMBOL

CONTOUR SYMBOL

ROOT OPENING;
DEPTH OF FILLING FOR
PLUG AND SLOT WELDS

EFFECTIVE THROAT

SIZE; SIZE OR STRENGTH
FOR CERTAIN WELDS.

TAIL ( MAY BE OMITTED
WHEN REFERENCE IS
NOT USED.)

GROOVE ANGLE;
INCLUDED ANGLE OF
COUNTERSINK FOR
PLUG WELDS.

LENGTH OF WELD

PITCH (CENTER-TO-
CENTER SPACING
OF WELDS)

FIELD WELD SYMBOL

ARROW CONNECTING
REFERENCE LINE TO
ARROW SIDE OR ARROW
SIDE MEMBER OF JOINT.

F
A

R

S (E)   (SIDES)   (OTHER SIDE)   L - P

T

(BOTH)   (N)   (ARROW SIDE)

WELD ALL AROUND
SYMBOL

SPECIFICATION,
PROCESS, OR
OTHER REFERENCE

BASIC WELD SYMBOL
OR DETAIL REFERENCE

ELEMENTS IN THIS
AREA REMAIN AS
SHOWN WHEN TAIL AND
ARROW ARE REVERSED.

REFERENCE LINE

NUMBER OF SPOT OR
PROJECTION WELDS.

FIGURE 17–20 The location of welding symbol elements on a standard welding symbol. This symbol is used on the drawing to specify the weld characteristics. (Reproduced from AWS A2.4-79, "Symbols for Welding and Nondestructive Testing," by permission of the American Welding Society.)

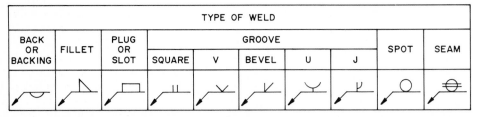

| TYPE OF WELD | | | | | | | | | |
|---|---|---|---|---|---|---|---|---|---|
| BACK OR BACKING | FILLET | PLUG OR SLOT | GROOVE | | | | | SPOT | SEAM |
| | | | SQUARE | V | BEVEL | U | J | | |
| | | | | | | | | | |

| SUPPLEMENTARY SYMBOLS | | | | |
|---|---|---|---|---|
| WELD ALL AROUND | FIELD WELD | CONTOUR | | |
| | | FLUSH | CONVEX | CONCAVE |
| | | | | |

FIGURE 17–21 Some standard welding symbols. (Reproduced from AWS A2.4-79, "Symbols for Welding and Nondestructive Testing," by permission of the American Welding Society.)

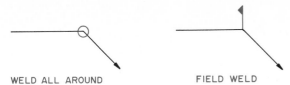

WELD ALL AROUND          FIELD WELD

**FIGURE 17–22** The use of the field weld and weld-all-around symbols.

**FIGURE 17–23** Designation of welding and allied processes by letters. (Reproduced from AWS A2.4-79, "Symbols for Welding and Nondestructive Testing," by permission of the American Welding Society.)

| LETTER DESIGNATION | WELDING AND ALLIED PROCESSES | LETTER DESIGNATION | WELDING AND ALLIED PROCESSES |
|---|---|---|---|
| AAC | air carbon arc cutting | HPW | hot pressure welding |
| AAW | air acetylene welding | IB | induction brazing |
| ABD | adhesive bonding | INS | iron soldering |
| AB | arc brazing | IRB | infrared brazing |
| AC | arc cutting | IRS | infrared soldering |
| AHW | atomic hydrogen welding | IS | induction soldering |
| AOC | oxygen arc cutting | IW | induction welding |
| AW | arc welding | LBC | laser beam cutting |
| B | brazing | LBW | laser beam welding |
| BB | block brazing | LOC | oxygen lance cutting |
| BMAW | bare metal arc welding | MAC | metal arc cutting |
| CAC | carbon arc cutting | OAW | oxyacetylene welding |
| CAW | carbon arc welding | OC | oxygen cutting |
| CAW-G | gas carbon arc welding | OFC | oxyfuel gas cutting |
| CAW-S | shielded carbon arc welding | OFC-A | oxyacetylene cutting |
| CAW-T | twin carbon arc welding | OFC-H | oxyhydrogen cutting |
| CEW | coextrusion welding | OFC-N | oxynatural gas cutting |
| CW | cold welding | OFC-P | oxypropane cutting |
| DB | dip brazing | OFW | oxyfuel gas welding |
| DFB | diffusion brazing | OHW | oxyhydrogen welding |
| DFW | diffusion welding | PAC | plasma arc cutting |
| DS | dip soldering | PAW | plasma arc welding |
| EASP | electric arc spraying | PEW | percussion welding |
| EBC | electron beam cutting | PGW | pressure gas welding |
| EGW | electrogas welding | POC | metal powder cutting |
| EBW | electron beam welding | PSP | plasma spraying |
| EBW-HV | electron beam welding—high vacuum | RB | resistance brazing |
| EBW-MV | electron beam welding—medium vacuum | RPW | projection welding |
| EBW-NV | electron beam welding—nonvacuum | RS | resistance soldering |
| ESW | electroslag welding | RSEW | resistance seam welding |
| EXW | explosion welding | RSW | resistance spot welding |
| FB | furnace brazing | ROW | roll welding |
| FCAW | flux-cored arc welding | RW | resistance welding |
| FLB | flow brazing | S | soldering |
| FLOW | flow welding | SAW | submerged arc welding |
| FLSP | flame spraying | SAW-S | series submerged arc welding |
| FOC | chemical flux cutting | SMAC | shielded metal arc cutting |
| FOW | forge welding | SMAW | shielded metal arc welding |
| FRW | friction welding | SSW | solid-state welding |
| FS | furnace soldering | SW | stud arc welding |
| FW | flash welding | TB | torch brazing |
| GMAC | gas metal arc cutting | TC | thermal cutting |
| GMAW | gas metal arc welding | TCAB | twin carbon arc brazing |
| GMAW-P | gas metal arc welding—pulsed arc | THSP | thermal spraying |
| GMAW-S | gas metal arc welding—short circuiting arc | TS | torch soldering |
| | | TW | thermit welding |
| GTAC | gas tungsten arc cutting | USW | ultrasonic welding |
| GTAW | gas tungsten arc welding | UW | upset welding |
| GTAW-P | gas tungsten arc welding—pulsed arc | WS | wave soldering |
| HFRW | high-frequency resistance welding | | |

process or other related reference. Not all eight elements are needed every time a weld is specified.

*Field weld symbol* The field weld symbol indicates that the joint involved is to be welded in the field rather than in a shop where the device is being built. Often, a product is too large to ship completely assembled. It is built in several sections, and these are welded together on the job (Fig. 17–22).

*Weld-all-around symbol* The weld-all-around symbol means the indicated joint is welded on all sides (Fig. 17–22).

*Symbol tail* If a weld has special specifications, the abbreviation for these is placed in the tail. Some commonly used abbreviations are given in Fig. 17–23. If no specifications are needed, the tail is left off the symbol.

*Location of the weld symbol* When the symbol indicating the type of weld is placed above the reference line, it means that the weld is to be made on the side opposite the arrow. If the weld symbol is below the reference line, the weld is to be made on the same side as the arrow. When on both sides, the weld is to be made on both sides of the joint (Fig. 17–24). Spot, seam, flash, and upset weld symbols have no arrow significance.

*Location of the arrow* When the material to be welded has a groove, the welding symbol arrow points toward the side to have the groove (Fig. 17–25).

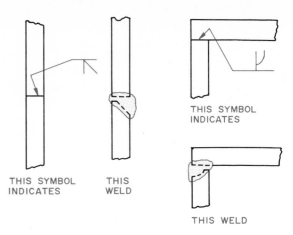

FIGURE 17–25 The arrow points toward the surface that will have the groove.

Notice that the vertical side of the weld symbol always faces to the left regardless of the location of the symbol. The weld symbol should have a definite break to indicate clearly which side is to have the groove.

*Dimensions on welds* Inch, millimeter, degree, pounds per square inch, and newton indications can be used on welding symbols if needed for clarity. If the welding data are covered in a note, such as "UNLESS OTHERWISE INDICATED, ALL FILLET WELDS ARE 6 mm," this information need not be repeated on the weld symbols. All welds are assumed to be continous unless otherwise indicated.

FIGURE 17–24 The position of the weld symbol indicates the location of the weld.

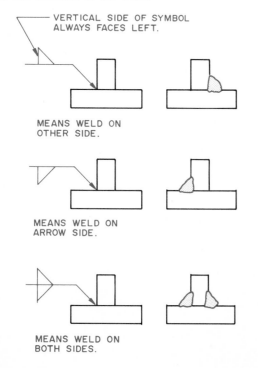

MEANS WELD ON OTHER SIDE.

MEANS WELD ON ARROW SIDE.

MEANS WELD ON BOTH SIDES.

FIGURE 17–26 Dimensions for spot and seam welds are placed on the drawing; other types of welds are dimensioned on the weld symbol.

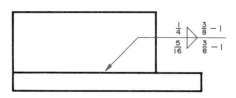

WELD DIMENSIONS ARE PLACED ON THE WELDING SYMBOL.

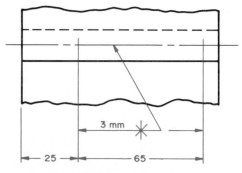

WELD DIMENSIONS FOR SPOT AND SEAM WELDS MAY BE PLACED ON THE DRAWING.

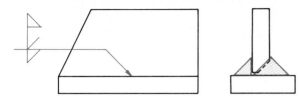

FIGURE 17-27 A weld symbol is shown for each weld at a joint.

The finish of welds, other than cleaning, is shown by contour and finish symbols. When dimensioning spot and seam welds, the dimensions may be placed only on the drawing. All others are placed on the welding symbol reference line (Fig. 17-26). If a joint has more than one weld, a symbol is shown for each weld (Fig. 17-27).

## Fillet Welds

The application of the basic welding symbol to a variety of fillet welds is shown in Fig. 17-28. The size of the fillet weld is shown to the left of the symbol and on the same side of the line. If welds on both sides are the same size, one or both may be dimensioned. If the weld has legs of unequal size, they are enclosed in parentheses. A dimension is given to show which leg is vertical. The length of fillet welds is shown to the right of the weld symbol. If nothing is shown, it is assumed to be continuous for the length of the part. If a series of short welds is needed, their pitch is given to the right of the weld length. It is assumed that the first weld increment begins at the end of the dimensioned length. When it is desired to show the extent of a fillet weld on a drawing, crosshatching can be used and the lengths dimensioned (Fig. 17-29). If it is necessary to show

FIGURE 17-28 The application of the basic welding symbol to a variety of fillet welds.

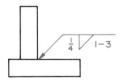

FILLET WELD BOTH LEGS ¼", WELDS 1" LONG, SPACED 3" ON CENTER, ARROW SIDE.

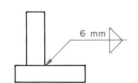

FILLET WELD, BOTH LEGS 6 MILLIMETERS, CONTINUOUS BOTH SIDES.

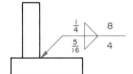

FILLET WELD, OTHER SIDE LEGS ¼" — WELD 8" LONG, ARROW SIDE LEGS 5/16" LONG — WELD 4" LONG, CONTINUOUS.

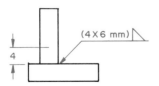

FILLET WELD, 4 MILLIMETER VERTICAL LEG, 6 MILLIMETER HORIZONTAL LEG, CONTINUOUS, OTHER SIDE.

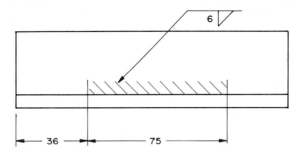

FIGURE 17-29 The extent of a fillet weld can be shown graphically by using crosshatching to represent the weld location.

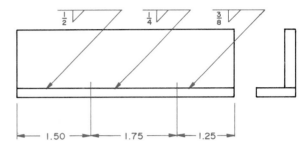

FIGURE 17-30 Specific lengths of fillet welds may be shown by symbols in conjunction with location dimensions on the drawing.

the length of continuous welds that have a change in specification, the specific lengths can be dimensioned (Fig. 17-30).

The surface contour of fillet welds is indicated with surface contour symbols. When a fillet weld is to have a flat, inclined face without any mechanical finish, the flush contour symbol is used. If the flat face is to be produced by mechanical means, the flush contour symbol is used and the standard finish symbol is placed above it. Fillet welds to be mechanically finished to a convex contour take the convex contour symbol plus the standard finish symbol (Fig. 17-31).

FIGURE 17-31 Finish symbols are used on fillet welds.

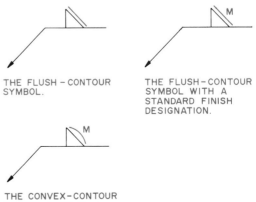

THE FLUSH-CONTOUR SYMBOL.

THE FLUSH-CONTOUR SYMBOL WITH A STANDARD FINISH DESIGNATION.

THE CONVEX-CONTOUR SYMBOL WITH A STANDARD FINISH DESIGNATION.

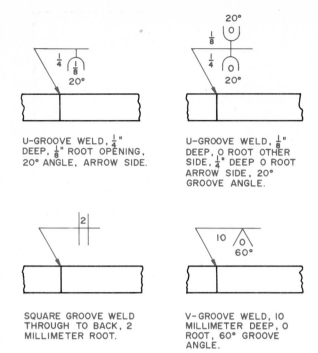

U-GROOVE WELD, $\frac{1}{4}$"
DEEP, $\frac{1}{8}$" ROOT OPENING,
20° ANGLE, ARROW SIDE.

U-GROOVE WELD, $\frac{1}{8}$"
DEEP, 0 ROOT OTHER
SIDE, $\frac{1}{4}$" DEEP 0 ROOT
ARROW SIDE, 20°
GROOVE ANGLE.

SQUARE GROOVE WELD
THROUGH TO BACK, 2
MILLIMETER ROOT.

V-GROOVE WELD, 10
MILLIMETER DEEP, 0
ROOT, 60° GROOVE
ANGLE.

FIGURE 17–32   Typical groove weld symbols and their applications.

## Groove Welds

The application of the basic groove welding symbol to a variety of groove welds is shown in Fig. 17–32. The size of the groove weld that extends only partly through the member being welded is shown to the left of the weld symbol. If the groove is the same on both sides, one or both sizes are given. If the grooves are different on each side, both dimensions must be given. If a groove extends completely

FIGURE 17–33   Finish symbols are used on groove welds.

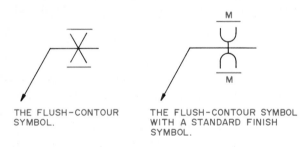

THE FLUSH-CONTOUR
SYMBOL.

THE FLUSH-CONTOUR SYMBOL
WITH A STANDARD FINISH
SYMBOL.

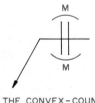

THE CONVEX-COUNTOUR
SYMBOL WITH A STANDARD
FINISH SYMBOL.

through the members being joined, the size need not be given. Groove angles are given outside the opening of the weld symbol. The root opening is given inside the weld symbol. A root opening is the space between the two members being welded.

If all grooves have the same angle or are the same size, a general note can be given, such as "ALL V-GROOVE WELDS 60° GROOVE ANGLE UNLESS OTHERWISE SPECIFIED" or "ALL V-GROOVE WELDS 4 mm DEEP."

The surface contour of groove welds is shown above or below the welding symbol. If the weld is to be ground flush without specifying the method, the flush contour symbol is used. If the weld is to be ground flush by mechanical means, the flush contour symbol and the standard finish symbol are used. If the groove weld is to be finished to a convex contour, the convex contour symbol and the standard finish symbol are used (Fig. 17–33).

### Back or Backing Welds

Single back welds are used to back up single groove welds. The single bead symbol is placed on the side of the reference line opposite the groove weld symbol. The dimensions are not shown on the weld symbol. If they are necessary, they are shown on the drawing (Fig. 17–34).

The surface contour of backing welds is shown as explained for groove welds. The symbols used are shown in Fig. 17–35. If a surface is to be built up (thickened) by welding adjoining rows of backing welds, this is indicated with a dual bead weld symbol. Since a back weld does not form a joint, the arrow has no other-side significance. The symbol is drawn on the bottom side of the reference line, and the arrow points to the surface on which the weld is to be made. The size of the surface built up by welding is shown by giving the minimum height of the weld deposit to the left of the weld symbol. When an entire surface is to be built up, no dimension other than the height of deposit is given. When a specific area is to be built up, the location and size of the area are dimensioned on the drawing (Fig. 17–36).

FIGURE 17–34   Typical back weld symbol applications.

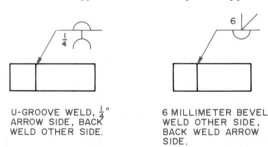

U-GROOVE WELD, $\frac{1}{4}$"
ARROW SIDE, BACK
WELD OTHER SIDE.

6 MILLIMETER BEVEL
WELD OTHER SIDE,
BACK WELD ARROW
SIDE.

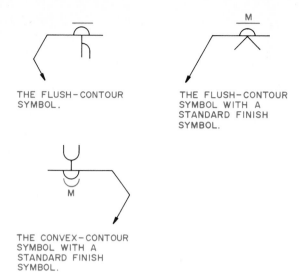

FIGURE 17–35 Finish symbols as used on back weld symbols.

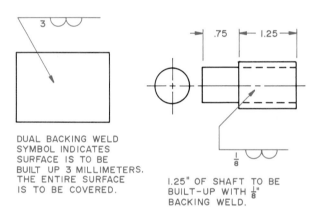

FIGURE 17–36 Using the back weld symbol to indicate built-up surfaces.

## Plug Welds

A plug weld is made through a circular opening in one piece that allows the weld to join with the metal below. If the weld symbol is on top of the reference line, the plug is on the side opposite the arrow. If the weld symbol is below the reference line, the plug is on the arrow side (Fig. 17–37). The diameter of a plug weld is shown to the left of the weld symbol and on the same side of the reference line. The angle of the countersink of a plug weld is shown directly above or below the weld symbol. If no depth of filling of the plug is given, it is assumed to fill it completely. When the depth is less than a full plug, it is given in inches or millimeters inside the weld symbol. The pitch (center-to-center distance) of a plug weld is given to the right of the weld symbol. Plug welds that are to be welded flush with no specified method of finishing are indicated with the flush contour symbol. If they are to be finished by mechani-

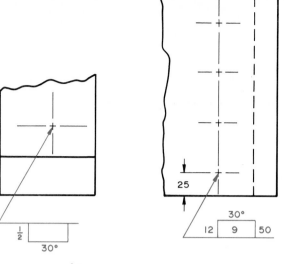

FIGURE 17–37 Typical applications of plug weld symbols.

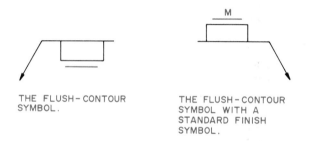

FIGURE 17–38 Finish symbols as used on plug weld symbols.

cal means, the flush contour symbol is used and the standard finish symbol is placed above it (Fig. 17–38).

## Slot Welds

A slot weld is made through a slot cut in one piece that allows the weld to join with the metal below. If the weld symbol is on top of the reference line, the slot is on the side opposite the arrow. If the weld symbol is below the reference line, the slot is on the arrow side (Fig. 17–39). The depth of filling of slot welds is assumed to be full if no information is given. When the depth is less than full, it is given inside the weld symbol. The length, width spacing and included angle of countersink, orientation, and location of slot welds are not shown on the symbol. They are shown on the drawing or by a detail with a reference to it on the welding symbol (Fig. 17–40). Slot welds that are to be welded flush with no spec-

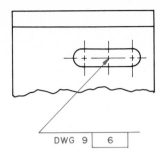

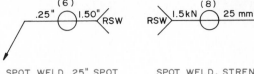

SPOT WELD .25" SPOT
DIAMETER, SPACED 1.50"
CENTER–TO–CENTER, 6
WELDS.

SPOT WELD, STRENGTH
1.5 KILONEWTONS, SPACED
25 MILLIMETERS, CENTER–
TO–CENTER, 8 WELDS.

SLOT WELD, DETAILS ON
DRAWING NO. 9, FILL SLOT
6 MILLIMETERS.

FIGURE 17–39  A typical slot weld symbol application.

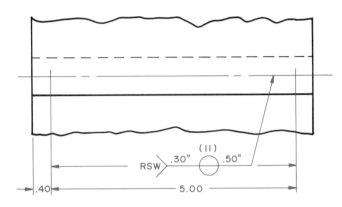

SPOT WELD, .30" DIAMETER, 11 WELDS, .50"
APART IN A 5.00" DISTANCE.

FIGURE 17–41  Typical spot weld applications.

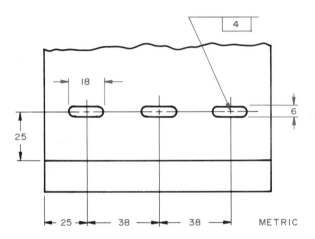

FIGURE 17–40  Slot welds sized and located on the drawing.

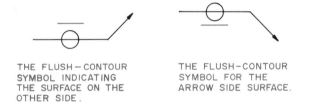

THE FLUSH–CONTOUR
SYMBOL INDICATING
THE SURFACE ON THE
OTHER SIDE.

THE FLUSH–CONTOUR
SYMBOL FOR THE
ARROW SIDE SURFACE.

FIGURE 17–42  Finish symbols as used on a spot weld symbol.

ified method of finishing are indicated with the flush contour symbol. If they are to be finished by mechanical means, the flush contour symbol is used and the standard finish symbol is placed above it. This is indicated the same as for plug welds.

## Spot Welds

Spot weld symbols have no arrow-side or other-side significance. The symbols are drawn centered on the reference line (Fig. 17–41). The dimensions can be above or below the line. The size of the spot weld is given as a diameter of the weld and is expressed as a decimal part of an inch or in millimeters. The inch mark (″) or millimeter abbreviation (mm) is shown. This specification is located to the left of the weld symbol. If the *strength* of the weld is given instead of diameter, it represents the minimum acceptable shear strength in pounds per square inch (psi) or newtons (N) and is placed to the left of the weld symbol. The *pitch* (center-to-center spacing) of spot welds is shown to the right of the weld symbol, expressed in inches on customary drawings and in

millimeters on metric drawings. If the spot weld symbols are placed on the drawing, the pitch can be dimensioned on the drawing. When a specific *number* of spot welds is wanted, the number is shown in parentheses above or below the weld symbol. When spot welding is limited to a certain length on a part, this length is dimensioned on the drawing.

When the exposed surface of a spot-welded joint is to be flush, this is indicated by adding the flush contour symbol to the weld symbol. If the surface concerned is on the same side as the arrow, the flush contour symbol is placed below the weld symbol. It is placed above the weld symbol if it relates to the other side (Fig. 17–42).

## Seam Welds

Seam welds have no arrow-side or other-side significance. The symbols are drawn centered on the reference line (Fig. 17–43). The size of the seam weld

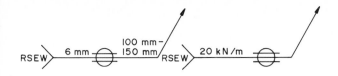

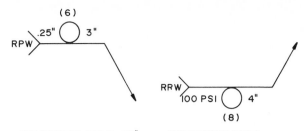

SEAM WELD, 6 MILLIMETERS WIDE, 100 mm LONG, SPACED 150 mm ON CENTER.

SEAM WELD, STRENGTH 20 KILONEWTONS PER METER OF LENGTH.

PROJECTION WELD, 25" DIA., 3" ON CENTER, 6 WELDS, PROJECTION ON OTHER SIDE.

PROJECTION WELD STRENGTH 100 PSI, 4" ON CENTER, 8 WELDS, PROJECTION ON ARROW SIDE.

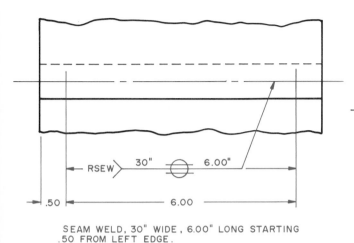

SEAM WELD, 30" WIDE, 6.00" LONG STARTING .50 FROM LEFT EDGE.

FIGURE 17–43 Typical seam weld symbol applications.

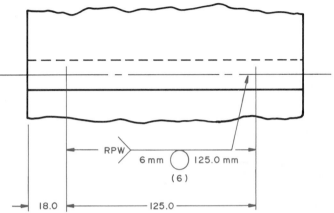

PROJECTION WELD, 6 MILLIMETER DIAMETER, 6 WELDS IN 125 MILLIMETER LENGTH, PROJECTIONS ARROW SIDE.

FIGURE 17–44 Typical projection weld symbol applications.

is the width expressed as a decimal part of an inch or in millimeters. The inch mark or millimeter abbreviation is shown. The specification is located to the left of the weld symbol. If the *strength* of seam welds is given instead of the width, it represents the minimum acceptable shear strength in pounds per square inch or newtons and is shown to the left of the weld symbol. The length of a seam weld is given in inches or millimeters to the right of the weld symbol. When a seam weld extends the full length of a part, no length dimension is needed. If it is less than full length, it is dimensioned on the drawing.

Seam welding can be done intermittently, and the pitch of each seam is shown to the right of the length dimension. The pitch is the center-to-center distance of each seam. The dimensioning of flush seam welds is the same as for spot welds.

## Projection Welds

If the weld symbol is below the reference line, the projection is in the arrow-side piece. If the weld symbol is above the reference line, the projection is in the other-side piece (Fig. 17–44). The *size* of the projection weld is the diameter of the weld given in decimal parts of an inch or in millimeters to the left of the weld symbol. The inch or millimeter abbreviation is used. The size is located on the same side of the reference line as is the weld symbol. If *strength*

is given instead of diameter, it is the minimum acceptable shear strength in pounds per square inch per weld or in newtons and is shown to the left of the weld symbol. The *pitch* is shown to the right of the weld symbol and is in inches or millimeters.

If a projection weld does not extend the full length of a part, its length is dimensioned on the drawing. When the number of projections must be known, they are indicated above or below the symbol in parentheses. A flush surface is indicated with the flush contour symbol as for spot welds.

## Flash and Upset Butt Welds

Flash and upset butt weld symbols have no arrow-side or other-side significance. The weld symbols are centered on the reference line (Fig. 17–45). No dimensions are necessary on the welding symbol. The process to be used is indicated in the tail of the symbol. The flash butt weld symbol is FW, and the uspet butt weld symbol is UW. Welds that are to be made flush by mechanical means have the flush contour symbol above the weld symbol and the standard finish symbol above this. If the surface is to be convex, the convex contour symbol and finish symbol are used.

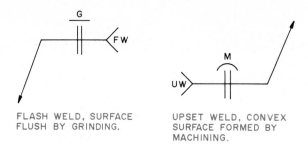

FLASH WELD, SURFACE
FLUSH BY GRINDING.

UPSET WELD, CONVEX
SURFACE FORMED BY
MACHINING.

FIGURE 17–45 Flash and upset welds are centered on the reference line.

## RIVETS

Rivets are permanent fasteners used to join sheet metal parts and steel plate. They have a round body and a head on one end. They are placed through holes in the material, and a head is then formed on the other end (Fig. 17–46). Some rivets are heated to help form the head, while others are cold-formed. The holes in metal parts may be formed by punching, punching and reaming, or drilling. The size of the hole is important because as the second head is formed, the body of the rivet is enlarged by the forming pressure. The hole is made larger than the body of the rivet to handle this expansion. The rivet should be long enough to provide material to fill the hole completely and form the second head. The length of the rivet depends on the grip (thickness of material to be joined). This and other information on standard types and sizes of rivets is available in publications from the American National Standards Institute, in the "Manual of Steel Construction," and in *Structural Steel Detailing,* published by the American Institute of Steel Construction.

FIGURE 17–46 Rivets are used to join permanently two or more pieces of material.

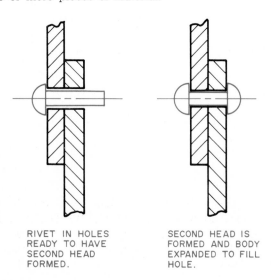

RIVET IN HOLES
READY TO HAVE
SECOND HEAD
FORMED.

SECOND HEAD IS
FORMED AND BODY
EXPANDED TO FILL
HOLE.

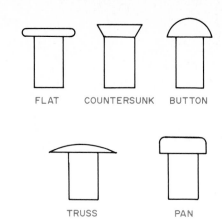

FLAT    COUNTERSUNK    BUTTON

TRUSS    PAN

FIGURE 17–47 ANSI small-rivet head types.

Small rivets are available with flat, countersunk, button, pan, and truss heads (Fig. 17–47). Body diameters range from $\frac{1}{16}$ to $\frac{7}{16}$ in. in $\frac{1}{32}$-in. increments. Design sizes are given in Appendix F. They are used for sheet metal and light structural parts.

Large rivets are available with button, high-button, cone, flat-top countersunk, round-top countersunk, and pan heads (Fig. 17–48). Body diameters range from $\frac{1}{2}$ to $1\frac{3}{4}$ in. in $\frac{1}{8}$-in. increments. Large rivets are used in structural steel work in construction, boilers, and tanks. Design sizes are given in Appendix F.

In addition to these standardized rivet sizes, many specialty rivets are manufactured. Some are made for a special purpose or a limited industry, such as aircraft rivets. The designer must have manufacturers' catalogs available for ready reference when deciding on specific rivet specifications.

FIGURE 17–48 Head types for large-diameter rivets.

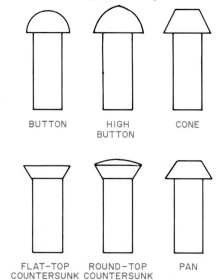

BUTTON    HIGH
BUTTON    CONE

FLAT-TOP    ROUND-TOP    PAN
COUNTERSUNK  COUNTERSUNK

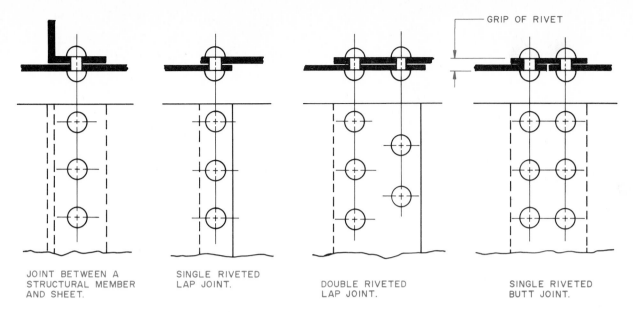

FIGURE 17–49  Typical riveted joints.

## Riveted Joints

Whenever two pieces of material overlap or butt, the possibility of joining them with a rivet exists. Typical joints joined with rivets are shown in Fig. 17–49.

## Indicating Rivets on Drawings

Standard symbols are used to represent rivets on drawings (Fig. 17–50). They are divided into two types—field rivets and shop rivets. The field rivet symbol indicates rivets to be put in place on the job. Shop rivet symbols represent rivets to be put in place as the product is manufactured.

The location of rivets is given by dimensioning to the center lines of the rivet holes. The size of the rivet is specified by a note giving the body diameter, length, and head type. A metric rivet would be indicated as "∅6 × 12 PAN HEAD RIVET," a customary rivet as "∅⅜ × 1½ ROUND HEAD RIVET" (Fig. 17–51).

## KNURLING

*Knurling* is a term used to describe the process that forms a uniformly roughened surface on a cylindrical part. It is used to roughen the surface of a shaft so that when a part, such as a knob, is forced onto the shaft, the knob will turn the shaft without slipping. This forms a permanent union between the parts. Surfaces are also knurled to help provide a better grip. The surface of the knob could be knurled. The common types of knurls are straight and diamond (Fig. 17–52). Knurls used for a pressed fit are specified by a note giving the diameter before knurling, minimum diameter after knurling, pitch,

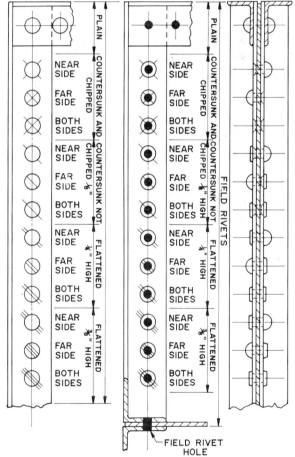

FIGURE 17–50  Standard rivet symbols used on drawings.

length of knurled area, and type of knurl. Those providing a rough surface for gripping require only pitch of knurl, type of knurl, and length of knurl (Fig. 17–53). The pitch of knurl is the distance between grooves.

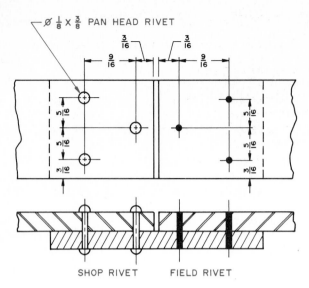

∅ ⅛ X ⅜ PAN HEAD RIVET

SHOP RIVET    FIELD RIVET

FIGURE 17–51 How to locate rivets on a shop drawing.

FIGURE 17–52 The two types of knurling commonly used to join parts together permanently.

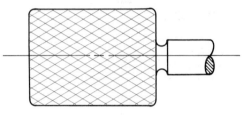

DIAMOND KNURL

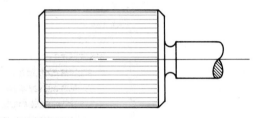

STRAIGHT KNURL

FIGURE 17–53 How to draw and dimension knurls.

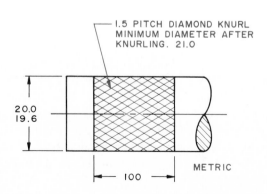

1/16 PITCH STRAIGHT KNURL

∅ .80

1.50

A KNURL TO BE USED FOR GRIPPING OR DECORATION.

1.5 PITCH DIAMOND KNURL
MINIMUM DIAMETER AFTER
KNURLING. 21.0

20.0
19.6

100    METRIC

434    Chapter 17

# Problems

The following problems can in most cases be drawn full size. Careful planning will enable several problems to be placed on an A or A4 sheet.

1. Study the welding symbols in Prob. 17–1 and write a full explanation of what each means.

2. Draw the items in Prob. 17–2 and indicate the welding information described below it.

3. Study the welding symbols in Prob. 17–3 and write a full explanation of what each means.

4. Draw the items in Prob. 17–4 and indicate the welding information described below it.

5. Study the welding symbols in Prob. 17–5 and write a full explanation of what each means.

6. Draw the items in Prob. 17–6 and indicate the welding information described below it.

7. Study the welding symbols in Prob. 17–7 and write a full explanation of what each means.

8. Draw the items in Prob. 17–8 and indicate the welding information described below it.

9. Study the welding symbols in Prob. 17–9 and write a full explanation of what each means.

10. Draw the items in Prob. 17–10 and indicate the welding information described below it.

11. In Probs. 17–11 through 17–17, draw whatever views are necessary to describe the product and add the welding symbols. More than one symbol could be correctly

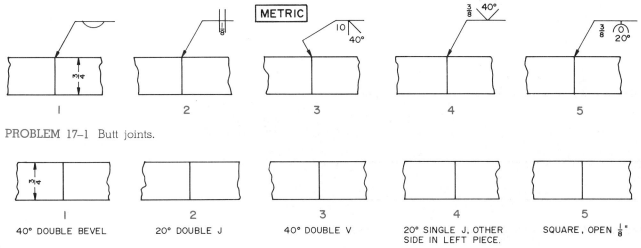

PROBLEM 17–1  Butt joints.

| 1 | 2 | 3 | 4 | 5 |
|---|---|---|---|---|
| 40° DOUBLE BEVEL | 20° DOUBLE J | 40° DOUBLE V | 20° SINGLE J, OTHER SIDE IN LEFT PIECE. | SQUARE, OPEN $\frac{1}{8}$" |

PROBLEM 17–2  Butt joints.

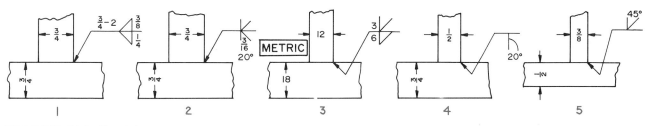

PROBLEM 17–3  Tee joints.

PROBLEM 17–4  Tee joints.

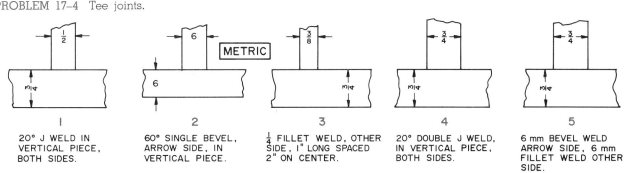

| 1 | 2 | 3 | 4 | 5 |
|---|---|---|---|---|
| 20° J WELD IN VERTICAL PIECE, BOTH SIDES. | 60° SINGLE BEVEL, ARROW SIDE, IN VERTICAL PIECE. | $\frac{1}{4}$ FILLET WELD, OTHER SIDE, 1" LONG SPACED 2" ON CENTER. | 20° DOUBLE J WELD, IN VERTICAL PIECE, BOTH SIDES. | 6 mm BEVEL WELD ARROW SIDE, 6 mm FILLET WELD OTHER SIDE. |

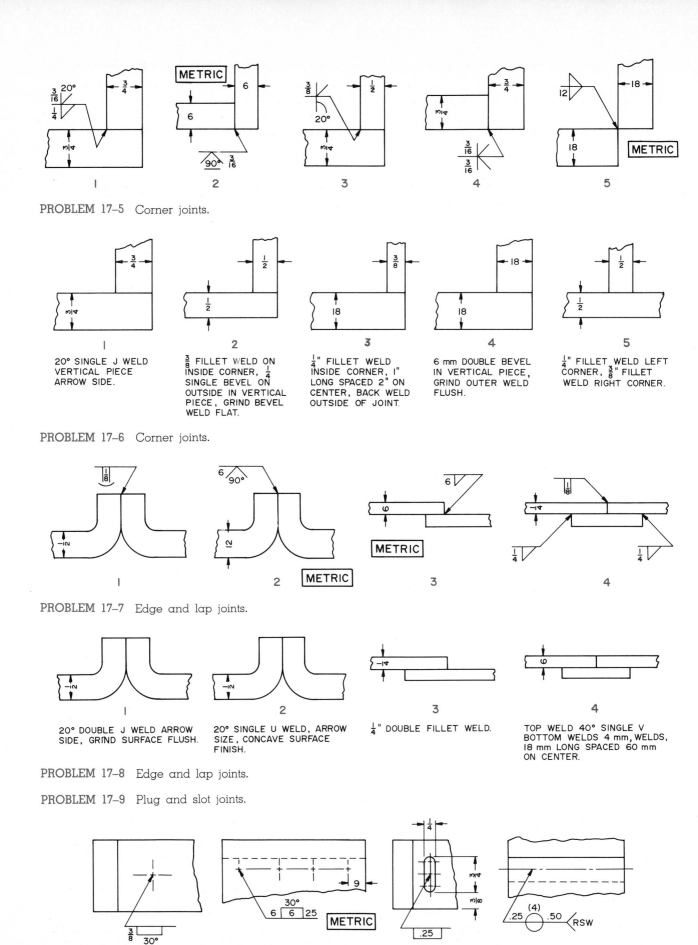

PROBLEM 17–5  Corner joints.

**1**
20° SINGLE J WELD VERTICAL PIECE ARROW SIDE.

**2**
3/8 FILLET WELD ON INSIDE CORNER, 1/4 SINGLE BEVEL ON OUTSIDE IN VERTICAL PIECE, GRIND BEVEL WELD FLAT.

**3**
1/4" FILLET WELD INSIDE CORNER, 1" LONG SPACED 2" ON CENTER, BACK WELD OUTSIDE OF JOINT.

**4**
6 mm DOUBLE BEVEL IN VERTICAL PIECE, GRIND OUTER WELD FLUSH.

**5**
1/4" FILLET WELD LEFT CORNER, 3/8" FILLET WELD RIGHT CORNER.

PROBLEM 17–6  Corner joints.

PROBLEM 17–7  Edge and lap joints.

**1**
20° DOUBLE J WELD ARROW SIDE, GRIND SURFACE FLUSH.

**2**
20° SINGLE U WELD, ARROW SIZE, CONCAVE SURFACE FINISH.

**3**
1/4" DOUBLE FILLET WELD.

**4**
TOP WELD 40° SINGLE V BOTTOM WELDS 4 mm, WELDS, 18 mm LONG SPACED 60 mm ON CENTER.

PROBLEM 17–8  Edge and lap joints.

PROBLEM 17–9  Plug and slot joints.

436

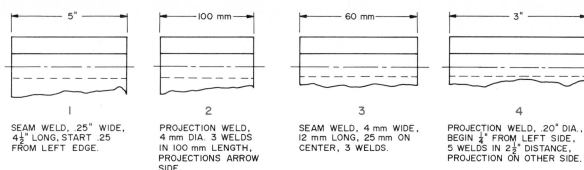

**1**

SEAM WELD, .25" WIDE,
4½" LONG, START .25
FROM LEFT EDGE.

**2**

PROJECTION WELD,
4 mm DIA. 3 WELDS
IN 100 mm LENGTH,
PROJECTIONS ARROW
SIDE.

**3**

SEAM WELD, 4 mm WIDE,
12 mm LONG, 25 mm ON
CENTER, 3 WELDS.

**4**

PROJECTION WELD, .20" DIA.,
BEGIN ¼" FROM LEFT SIDE,
5 WELDS IN 2½" DISTANCE,
PROJECTION ON OTHER SIDE.

PROBLEM 17–10  Seam and projection welds.

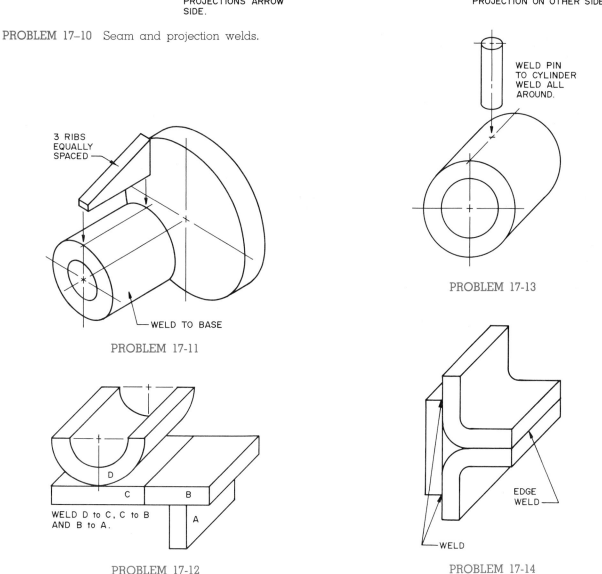

3 RIBS
EQUALLY
SPACED

WELD TO BASE

PROBLEM 17-11

WELD PIN
TO CYLINDER
WELD ALL
AROUND.

PROBLEM 17-13

WELD D to C, C to B
AND B to A.

PROBLEM 17-12

EDGE
WELD

WELD

PROBLEM 17-14

used for each joint, so all solutions will not necessarily be the same. Draw twice as large as shown. Measure the printed drawing with your dividers to determine sizes.

12. Following is a list of problems from Chapter 6. Change them so they are made from several parts to be welded together: Prob. 6–7, 6–8, 6–10, 6–13, 6–15, 6–17, 6–18, 6–39, 6–41, and 6–42.

13. Draw the top view and section of the riveted joint shown in Prob. 17–18. Use 5⁄16-in. button-head rivets that are installed in the shop. The pitch is 2 in., and there are three rivets in each row. They are set in ½ in. from each end of the top plate.

14. Draw the top view and section of the riveted joint shown in Prob. 17–19. Use 3⁄8-in. cone-head rivets. Those in the left joint are shop-riveted. Those in the right joint are field-riveted. There are two rivets in each row. They are set in ¾ in. from the edge of the plate and have a pitch of 1 in.

15. Draw the top view and section of the riveted joint in Prob. 17–20. There is one row of three rivets. They are set in 12 mm from the end of the plate and have a pitch of 18 mm. The rivet is 6 mm in diameter with a flat-top countersunk head. Install the rivets so that the head is flush with the plate.

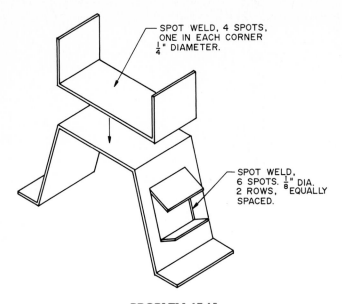

SPOT WELD, 4 SPOTS, ONE IN EACH CORNER ¼" DIAMETER.

SPOT WELD, 6 SPOTS. ⅛" DIA. 2 ROWS, EQUALLY SPACED.

PROBLEM 17-15

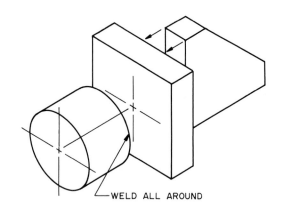

WELD ALL AROUND

PROBLEM 17-16

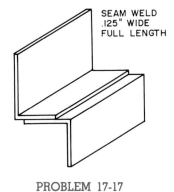

SEAM WELD .125" WIDE FULL LENGTH

PROBLEM 17-17

SLOT WELD, 6 mm FROM EACH SIDE, 6 mm WIDE, FILL SLOT 4 mm

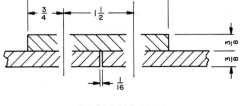

PROBLEM 17-18

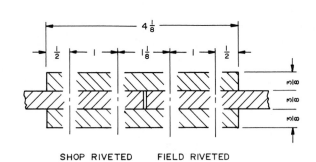

PROBLEM 17-19

PROBLEM 17-20

SHOP RIVETED    FIELD RIVETED

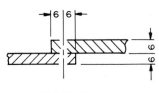

PROBLEM 17-21

# Preparation of Production Drawings

## DESIGN DRAWINGS

The design process, described in detail in Chapter 2, includes the production of freehand and instrument-made design drawings. These, plus any notes and calculations developed, are used to produce an accurate assembly drawing of the final design. This drawing, called a *design assembly*, is drawn very accurately, and full size whenever possible. It shows the exact shape and location of each part of the product. No attempt is made to show all the minute details of each part or to dimension them. The only dimensions shown are those that may be needed to show the relationship between critical parts, such as distances between center lines of parts or a size that is important. Notes needed to set forth specifications, such as material, heat treatment, finish, or clearances, may be lettered on this drawing or attached to it as a separate report.

Drafters use the design assembly and accompanying design specifications when making detail drawings of each part. They observe the engineering and drafting standards of their employer and keep the designs in line with them. The final drawings contain everything needed by manufacturing to make each part.

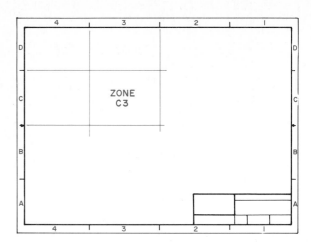

FIGURE 18–1 Sections of a drawing are identified by zones.

The detail drawings are also used by the tool designers when they design the tooling, jigs, and fixtures to be used in the manufacturing process. Drawings of these are produced so that the jigs and fixtures can be made.

## PRODUCTION DRAWING FEATURES

The following features of a production drawing are presented in a general context. They will vary somewhat from industry to industry and company to company. Engineers and drafters must become familiar with the procedures used in their company.

### Monodetail and Multidetail Drawing

When a device is small and has only a few parts, these can all be completely detailed on a single sheet, called a *multidetail drawing*. Larger devices with many parts may require several multidetail sheets, with an assembly on a separate drawing. Some industries prefer to place one detail per sheet. This is a *monodetail*. The detail is complete and does not rely on other drawings for interpretation.

### Zoning

A drawing can have its borders ruled with equally spaced lines to form a grid or series of zones across the drawing. The zones are identified by letters and numbers starting at the lower right-hand corner and moving up and to the left. These create imaginary rectangular zones across the drawing and are used to facilitate the location of changes, sections, and other items. For example, a change may have been made in zone C3 (Fig. 18–1). The zone indication is recorded on the schedule of revisions.

### Revising Drawings

Frequently, it is necessary to change a drawing because of a design improvement, a customer request,

or some other reason. It is important that an accurate record of all changes be kept. The record should show the nature of the change, who made it, the date, and the location (zone).

Changes are made by erasing the original, by adding additional information to the original, or by redrawing the detail. It is possible to make a duplicate drawing photographically and make the changes on it, thus preserving the original. The original could be put on microfilm before it is corrected so that a permanent record of the original design is available. It is necessary to issue new prints of a drawing each time a change is made.

If it is necessary to redraw a drawing, the old original is stamped "OBSOLETE," and in the title block, under "SUPERSEDED BY," is written the number of the new drawing. The original is filed in the obsolete file as a permanent record. The new drawing shows the number of the old drawing under the title "SUPERSEDES."

When changes are made, they are recorded in the change-record block on the drawing. The zone is lettered beside the change. If the zone system is not used, a number or letter enclosed in a circle is placed on the drawing next to the change. This number or letter is recorded after the change in the change-record block (Fig. 18–2). A brief description of the change is lettered in the change-record block. Zoning is used on drawings that are C size and larger.

### Checking Drawings

It is absolutely essential that drawings be complete, clear, and accurate, conform to standards, and ensure that the product is operational. Responsibility for this is shared by the design engineer, drafters, checkers, and others responsible for reviewing them before they are released to production. The drafter has the initial responsibility for quality, even though

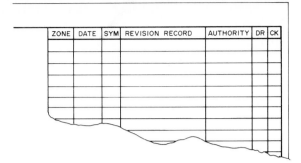

FIGURE 18-2 A typical change-record block.

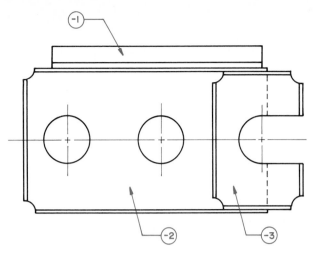

CONTROL CABLE STRUT ASSEMBLY
DRAWING NO. 6422112

FIGURE 18-3 Dash numbers are used to identify parts of an assembly.

a checker will review the drawings. The drafter's signature on a drawing indicates that it is correct. The checker not only reviews for errors but also looks for missing information.

*Checking* is the act of examining and verifying the accuracy, completeness, and efficiency of drawings. Every new and revised drawing should be checked. The checker should be familiar with the manufacturing practices of the industry and drafting standards. A checker may suggest a better way to design the part or product and is expected to mention this to a supervisor.Since the checker is usually the last person to examine a drawing before it is released, the final quality depends on the checker's work and is verified by the checker's signature in the title block.

It is advisable for a checker to work on a print rather than the original drawing. This avoids marking the original and provides a permanent record, from which a final check can be made after corrections are completed. A plan for making corrections should be followed. For example, military drafting standards require a yellow check by each correct dimension, note, or callout or a yellow line drawn through the applicable data. Incorrect data are encircled or otherwise marked in red, and the correct information is lettered next to the point in question. A checker's suggestions, general comments on features of design, simplification possibilities, or notes are made in blue or black.

## Numbering Drawings and Parts

Every drawing is assigned a number. This is vital to information filing and retrieval. There are many numbering systems in use. They have a logical relationship to the industry and the product.

Numbers should be assigned to drawings by a central staff member so that numbers are in a sequence and none are skipped or duplicated. Sometimes, a block of numbers will be set aside for a particular project. Records should show for each number assigned the drawing title, name of the drafter, name or code number of the project, date of drawing number release, and date of drawing release. Drawing numbers may include arabic numerals, uppercase letters, and dashes. The letters *I, O, Q, S, X,* and *Z* are not used, nor is any form of punctuation.

*Part numbers* are assigned to identify a specific item or part. A *dash number* is a numerical identification limited to three digits. It is added to a drawing number to identify individual parts or assemblies.

Following is one number system in use in the aircraft industry. The company uses a seven-digit number to identify each drawing. The *first two digits* indicate the model number assigned to the aircraft being designed. The *third digit* indicates a major assembly, such as a wing. The *fourth digit* indicates a minor assembly, such as a flap on a wing. The last three digits are assigned by the drafting supervisor to specific detail drawings. Following the seven-digit number is a dash number. This is used when a drawing has more than one piece shown. Each piece is numbered so that it can be identified. The dash number is placed inside a 12-mm (½-in.) circle on the drawing and connected to the part with a leader or placed on the drawing as a note. In Fig. 18-3, the number of the first part is 6422112-1, the second part 6422112-2, and the third part 642212-3.

## Titles of Drawing

A drawing title should be brief and descriptive. It should describe the item and distinguish between drawings of similar items. Be certain that the title is consistent with the title of the drawing of its next assembly.

A drawing title should consist of a noun or noun phrase followed by primary modifiers or a qualifying word to distinguish it from similar parts. If a noun is insufficient, the title should be expanded by adding secondary modifiers. The following examples will show this development.

Drawing Title Development

| NOUN | PRIMARY MODIFIER | QUALIFYING WORD | SECONDARY MODIFIER |
|------|------------------|-----------------|--------------------|
| FILTER | OIL | | |
| BRAKE | | PAD | |
| INSERT | | SCREW | 6 mm BRASS |

When an assembly or installation drawing is titled, "ASSEMBLY" or "INSTALLATION" appears as the last word, as in "BRAKE ASSEMBLY" or "WIRING INSTALLATION." The *primary modifier* indicates characteristics such as type, grade, or variety. The *secondary modifier* is used to distinguish between similar items, such as "SCREW, STEEL." The primary modifier is separated from the noun by a dash, while the secondary modifier is separated from the primary with a comma, as "LEAF SPRING—UNDERSLUNG, REAR."

## PRODUCTION DRAWINGS

Many kinds of drawings are used in the design, manufacture, and assembly of a product. They fall in general into two classes, detail drawings and assembly drawings. Together they provide all the information needed to produce a finished product.

## DETAIL DRAWINGS

A *detail drawing* is a drawing that gives all the information needed to make a single part. It describes its shape, gives the dimensions, indicates finish and material, sets tolerances, and gives any other information needed by manufacturing to make the part.

Detailing practices vary with industry and company policies. In mechanical work, it is recommended that the detail drawing for each part be on a separate sheet, called a monodetail drawing (Fig. 18–4). If policy permits placing several parts on one sheet, some companies prefer that they be located in approximately the same relative position as they appear on the design assembly drawing. Parts made of the same material are often grouped together. For example, parts to be forged could be grouped on one

FIGURE 18–4 A typical monodetail drawing.

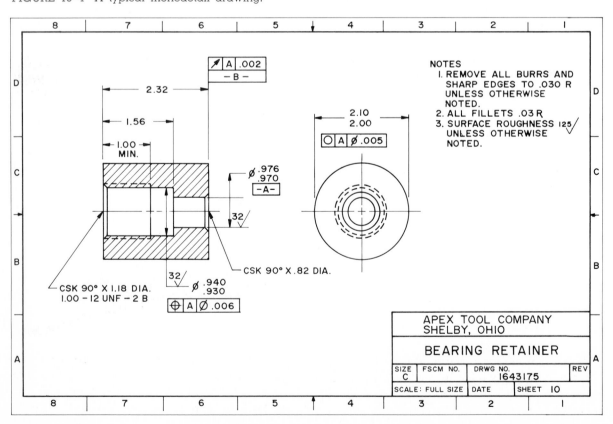

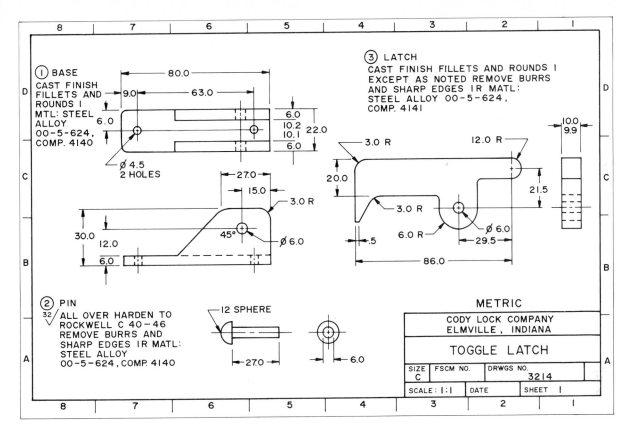

FIGURE 18-5  A typical multidetail drawing.

sheet, while those to be machined from stock material are detailed on another sheet. Each part must have a subtitle, giving its name, part number, and other necessary descriptive information. These are called multidetail drawings (Fig. 18–5).

Some companies use a single-drawing system, while others use a multiple-drawing system. The *single-drawing system* uses one drawing for all phases of production. The *multiple-drawing system* uses separate drawings for the various phases of manufacturing, such as a casting drawing for the foundry, a machining drawing for the machine shop, and a pattern drawing for use by the patternmakers. Regardless of the system used, a detail drawing must be complete in every way for the purpose for which it is intended. It must not be dependent on another drawing for any information.

Some of the commonly used detail drawings are pattern, machining, casting, forging, and stamping.

## PATTERN DETAIL DRAWINGS

A *pattern detail drawing* gives the information needed by a patternmaker to build a pattern. A *pattern* is a full-size duplicate of a part that is to be cast in metal. Patterns are made from wood, metal, or plastics. A pattern has to take into account draft, shrinkage, and machining. *Draft* is a slight tapering on the casting that helps in removing it from the mold. The draft angle is specified on the drawing in a note. Since cast metal shrinks as it cools, the pattern must be made larger than the desired finished size to allow for *shrinkage*. Shrinkage allowances depend on the metal to be used. For example, a part to be made from cast iron requires a smaller shrinkage allowance than one made from copper. The amount of shrinkage is not shown on the pattern drawing, but the material to be cast is shown. The patternmaker knows how much to enlarge the pattern by noting the material to be used (Fig. 18–6).

*Machining* involves the removal of metal from the surface of the part after it is cast. Extra material must be allowed wherever a part is to be machined so that after machining the part is the desired size. The surfaces to be machined are not indicated on the pattern drawing because the allowance is included in the dimensions. The patternmaker makes the pattern the size shown, knowing that this will allow extra material for machining where it is needed Fig. 18–7).

Pattern detail drawings show the holes or cavities that are to be formed by cores. A *core* is a hard, baked-sand, three-dimensional form that is inserted in the mold to form a cavity on the interior of the

| MATERIAL | INCHES PER FOOT | MILLIMETERS PER mm |
|---|---|---|
| Cast iron, malleable iron | ⅛ | 0.0104 |
| Steel | ¼ | 0.0208 |
| Brass, copper, aluminum, bronze | ³⁄₁₆ | 0.0156 |

FIGURE 18–6 Shrinkage allowances for commonly used metals.

| PATTERN SIZE | SURFACE FINISH ALLOWANCE | |
|---|---|---|
| | Inches | Millimeters |
| up to 8 in. (200 mm) | 0.07 | 2.0 |
| 8 in. (200 mm) to 16 in. (400 mm) | 0.10 | 2.5 |
| 16 in. (400 mm) to 24 in. (600 mm) | 0.14 | 3.6 |
| 24 in. (600 mm) to 32 in. (800 mm) | 0.18 | 4.6 |
| over 32 in. (800 mm) | 0.24 | 6.0 |

FIGURE 18–7 Finish allowance for casting bronze, aluminum, and cast iron.

casting. Round holes can also be cored; those with small diameters, however, are drilled after the casting is made and are not shown on the pattern detail drawing.

It is difficult to cast sharp corners, and they tend to fracture if they are cast. For this reason, internal and external corners are rounded. The radii of these rounds and fillets are shown on the pattern detail drawing.

The tolerance recommended for casting patterns depends on the size of the casting and the material to be used. Examples for two metals are shown in Fig. 18–8.

Figure 18–9 shows a casting after it has been *machined* to its finish size. The *pattern drawing* for this product is dimensioned to show the desired *finish size of the casting* before it is machined. The draft angle, tolerance, and material are given in notes. The patternmaker uses these as the pattern is built so

FIGURE 18–8 Casting tolerances for cast iron.

| CASTING SIZE | TOLERANCES ± | |
|---|---|---|
| | Inches | Millimeters |
| up to 8 in. (200 mm) | 0.03 | 0.8 |
| 8 in. (200 mm) to 16 in. (400 mm) | 0.05 | 1.3 |
| 16 in. (400 mm) to 24 in. (600 mm) | 0.07 | 1.8 |
| 24 in. (600 mm) to 32 in. (800 mm) | 0.09 | 2.3 |
| over 32 in. (800 mm) | 0.12 | 3.0 |

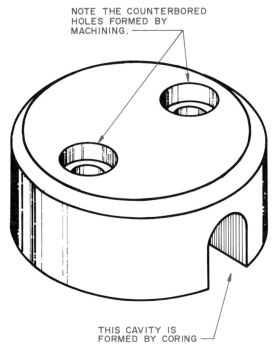

NOTE THE COUNTERBORED HOLES FORMED BY MACHINING.

THIS CAVITY IS FORMED BY CORING

FIGURE 18–9 A finished product after it has been cast and machined to the desired dimensions.

that the finished casting is the desired size (Fig. 18–10).

When a part to be cast is simple in design, the pattern and machining information are combined on one drawing. This is called a casting detail and is discussed later in this chapter.

## MACHINING DETAIL DRAWINGS

A *machining detail drawing* shows only the details needed by the machinist to perform the machining operations. These include the finish size, location and size of holes, surface finish, and tolerances. The parts of the casting that are not machined need not be dimensioned (Fig. 18–11). When dimensioning a machining drawing, it is important to relate size and location dimensions to a controlling surface or feature. In this way, they are related to one another, thus controlling the finish size with greater accuracy. Since cast holes and surfaces cannot be accurately formed, finished dimensions for machining cannot be located from them.

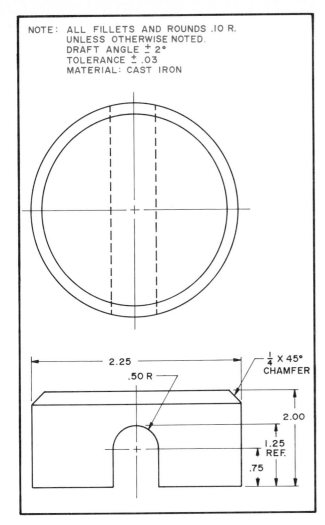

**FIGURE 18–10** A pattern detail drawing with the machining allowance added to the dimensions. Shrinkage is added by the patternmaker.

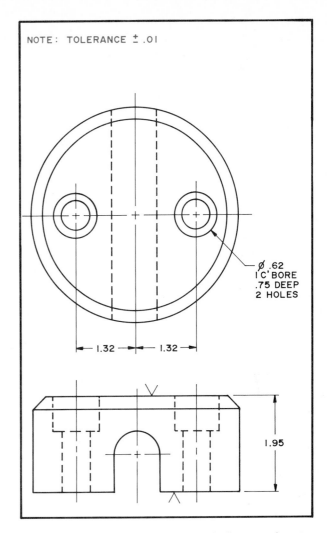

**FIGURE 18–11** A machining detail drawing locating and dimensioning the parts of the casting that require machining.

## CASTING DETAIL DRAWINGS

Simple cast parts can have the pattern and machining information on one drawing. These are called *casting detail drawings*. The drawing contains the overall sizes of the rough casting and the precision finish sizes needed by the machinist. The notes include draft information as well as finish tolerances (Fig. 18–12).

## FORGING DETAIL DRAWINGS

A *forging detail drawing* shows the information needed to produce a part by the forging process. *Forging* is a process in which a metal blank is heated and then formed to shape in a die with a power-driven hammer. A *die* is a metal form with a cavity the shape of the part to be formed. The heated metal is placed between the dies, which are driven together, forcing the heated metal to fill the cavity in the dies (Fig. 18–13).

The corners of forgings are made rounded because it is difficult to forge sharp corners. When shown in elevation, they are drawn rounded. When shown on the plan view, they are located where the corner would be if it were sharp (Fig. 18–14). A forging has draft so that it can be removed from the die after it is formed. *Draft* is a slope given to the exterior surfaces. Standard draft for most forgings is 7°. This is shown on the drawing in a note.

Forgings have a *parting line,* which is the line where the two dies meet. It is shown on a drawing using the center line symbol and labeling it "PL" (Fig. 18–14). A forged part is shown in Fig. 18–15.

### Forging Tolerances

The tolerances used on forgings include thickness, shrinkage, die-wear, mismatch, and machining tol-

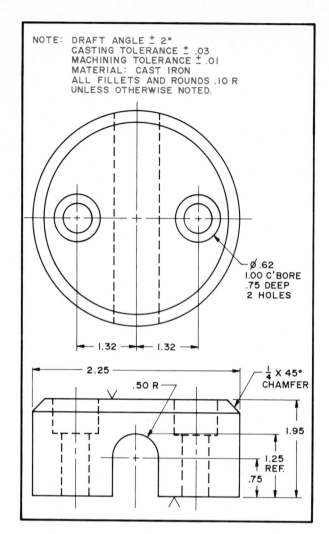

FIGURE 18–12 A casting detail drawing containing pattern and machining information.

erances. *Thickness tolerances* relate to the overall height of the part. This size can vary depending on how close the dies come when fully closed. The thickness tolerances depend on the weight of the

FIGURE 18–13 A forging is formed by hammering a heated metal blank between dies containing a cavity the shape of the part.

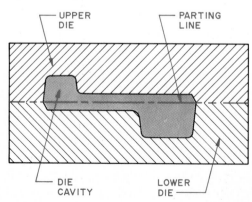

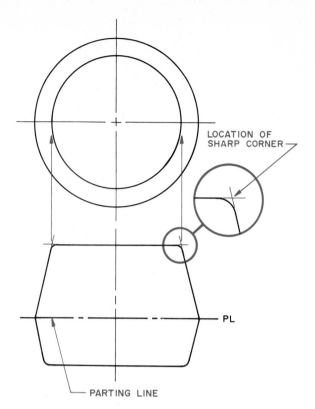

FIGURE 18–14 The rounded corners on a forging are projected and drawn from the location of the sharp corner.

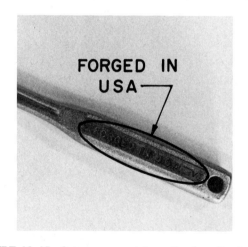

FIGURE 18–15 It is apparent from the handle that this socket handle was forged.

casting (Fig. 18–16). *Shrinkage tolerances* are used to allow for shrinkage in the part that occurs when it cools. Typical tolerances are shown in Fig. 18–17. They are figured in the direction parallel with the parting line. *Die-wear tolerances* provide an allowance for the wear on dies that become larger as they are used. These are figured parallel with the parting line (Fig. 18–17). Mismatch occurs when the dies do not line up properly when they close over a piece of heated metal. *Mismatch tolerances* are shown in Fig.

**Thickness Tolerances for Forgings (inches)**

| MAXIMUM NET WEIGHT (pounds) | COMMERCIAL − | COMMERCIAL + |
|---|---|---|
| 0.2 | .008 | .024 |
| 0.4 | .009 | .027 |
| 0.6 | .010 | .030 |
| 0.8 | .011 | .033 |
| 1 | .012 | .036 |
| 2 | .015 | .045 |
| 3 | .017 | .051 |
| 4 | .018 | .054 |
| 5 | .019 | .057 |
| 10 | .022 | .066 |
| 20 | .026 | .078 |
| 30 | .030 | .090 |
| 40 | .034 | .102 |
| 50 | .038 | .114 |
| 60 | .042 | .126 |
| 70 | .046 | .138 |
| 80 | .050 | .150 |
| 90 | .054 | .162 |
| 100 | .058 | .174 |

**Thickness Tolerances for Forgings (millimeters)\***

| MAXIMUM NET WEIGHT (kilograms) | COMMERCIAL − | COMMERCIAL + |
|---|---|---|
| 0.09 | 0.20 | 0.61 |
| 0.18 | 0.23 | 0.69 |
| 0.27 | 0.25 | 0.76 |
| 0.36 | 0.28 | 0.84 |
| 0.45 | 0.30 | 0.91 |
| 0.90 | 0.38 | 1.14 |
| 1.36 | 0.43 | 1.30 |
| 1.81 | 0.46 | 1.37 |
| 2.27 | 0.48 | 1.45 |
| 4.54 | 0.56 | 1.68 |
| 9.07 | 0.66 | 1.98 |
| 13.61 | 0.76 | 2.29 |
| 18.14 | 0.86 | 2.59 |
| 22.68 | 0.97 | 2.90 |
| 27.22 | 1.07 | 3.20 |
| 31.76 | 1.17 | 3.50 |
| 36.29 | 1.27 | 3.81 |
| 40.82 | 1.37 | 4.11 |
| 45.40 | 1.47 | 4.42 |

\*Soft conversion in millimeters of accompanying inch table.

FIGURE 18–16 Forging tolerances depend on the weight of the part. (Reproduced from ANSI Y14.9-1958 with permission of the publishers, The American Society of Mechanical Engineers.)

**Shrinkage and Die Wear Tolerances for Forgings (inches)**

| | SHRINKAGE | | DIE WEAR |
|---|---|---|---|
| Maximum Length or Width (inches) | Commercial + or − | Maximum Net Weight (pounds) | Commercial + or − |
| 1 | .003 | 1 | .032 |
| 2 | .006 | 3 | .035 |
| 3 | .009 | 5 | .038 |
| 4 | .012 | 7 | .041 |
| 5 | .015 | 9 | .044 |
| 6 | .018 | 11 | .047 |
| Each additional inch add | .003 | Each additional 2 lb add | .003 |

FIGURE 18–17 Tolerances for forgings for shrinkage and die wear. (Reproduced from ANSI Y14.9-1958 with permission of the publishers, The American Society of Mechanical Engineers.)

**Shrinkage and Die Wear Tolerances for Forgings (Millimeters)\***

| | SHRINKAGE | | DIE WEAR |
|---|---|---|---|
| Maximum Length or Width (millimeters) | Commercial + or − | Maximum Net Weight (kilograms) | Commercial + or − |
| 25.4 | 0.08 | 0.45 | 0.81 |
| 50.8 | 0.15 | 1.36 | 0.89 |
| 76.2 | 0.23 | 2.27 | 0.97 |
| 101.6 | 0.30 | 3.18 | 1.04 |
| 127.0 | 0.38 | 4.08 | 1.12 |
| 152.4 | 0.46 | 4.99 | 1.19 |
| Each additional millimeter add | 0.003 | Each additional kilogram add | 0.04 |

\*Soft conversion in millimeters of accompanying inch table.

**Mismatch Tolerances for Forgings**
(inches)

| MAXIMUM NET WEIGHT (pounds) | COMMERCIAL |
|---|---|
| 1 | .015 |
| 7 | .018 |
| 13 | .021 |
| 19 | .024 |
| Each additional 6 lb add | .003 |

**Mismatch Tolerances for Forgings**
(millimeters)*

| MAXIMUM NET WEIGHT (kilograms) | COMMERCIAL |
|---|---|
| 0.45 | 0.38 |
| 3.18 | 0.46 |
| 5.90 | 0.53 |
| 8.62 | 0.61 |
| Each additional 2.70 kg add | 0.08 |

*Soft conversion of accompanying inch table.

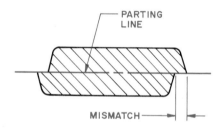

FIGURE 18–18 Tolerances for mismatch on forgings. (Reproduced from ANSI Y14.9-1958 with permission of the publishers, The American Society of Mechanical Engineers.)

18–18. Mismatch, shrinkage, and die-wear tolerances are added to produce a single total tolerance (Fig. 18–19). To this is added an allowance for machining if the forging is to be machined to finish size. Standard machining tolerances for forgings are shown in Fig. 18–20.

The *length* of the actual forging is the sum of the desired finished length plus shrinkage, die wear, and mismatch, plus a machining allowance if needed on each end of the part. The *height* of the actual forging is the desired finished length plus thickness tolerance and a machining allowance if needed on each side of the part.

### Radii on Forgings

The sizes of the radii of fillets and rounds are shown in Fig. 18–21. It is important to get the correct size

because small radii increase die wear and produce seams on the forging.

### Forging Drawings

Forging drawings may be made as a single drawing or combined with a machining drawing. When combined with a machining drawing, the lines of the forging are shown with a dashed line and the machined surfaces with visible lines. The difference between the dashed and visible lines is the tolerance (Fig. 18–22).

If the forging information is given on a single drawing, the outline is shown with visible lines. Important machined surfaces are shown with phantom lines that can be used as references for dimensioning. The forged outline is dimensioned by referencing it to the machined surface. Whenever possible, the forging drawing and the machining drawing are made side by side, with the forging drawing on the left (Fig. 18–23).

Forging drawings are dimensioned using the datum system. This helps to avoid the accumulation of tolerances. The datum should be some important feature from which all dimensions are taken. All dimensions parallel with the parting line should refer to the size at the bottom of the die cavity because this is the narrowest part. The draft makes the die cavity widest at the parting line. All dimensions at right angles to the parting line should refer to the parting line, using it as a datum. Whenever possible, make forging drawings full size. Sectional views are frequently used to describe complex shapes.

Forging drawings specify draft angles, parting lines, fillets, rounds, tolerances, allowances for machining, material, heat treatment, and part identification name and numbers.

## STAMPING DETAIL DRAWINGS

Parts produced by the stamping process are formed by pressing sheet metal between various dies. The dies cut the metal and bend it to the shapes desired (Fig. 18–24).

### Stamping Processes

Some of the processes involved include bending and forming, drawing, coining, blanking, punching, and trimming (Fig. 18–25). *Bending* and *forming* involve folding, twisting, or bending the sheet metal to shape. The thickness of the metal usually is unchanged. *Drawing* involves stretching the metal over a form that has the shape of the finished part. This stretching thins the metal in the section drawn. *Coining* causes metal to flow under great pressure. The area coined is thinned, and the metal flows to

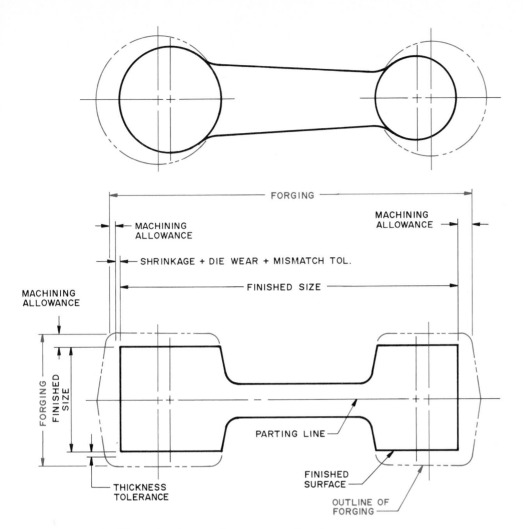

FIGURE 18-19 How shrinkage, die-wear, and mismatch tolerances are applied to the finish size. (Reproduced from ANSI Y14.9-1958 with permission of the publishers, The American Society of Mechanical Engineers.)

**Machining Allowance (inches)**

| WEIGHT OF PART | UNDER 1 LB | 1 TO 10 LB | 11 TO 40 LB | 41 TO 100 LB | 101 TO 200 LB |
|---|---|---|---|---|---|
| Compact parts, gears, discs | 1/32 | 3/64 | 1/16 | 3/32 | 1/8 |
| Thin, extended parts | 1/16 | 1/16 | 3/32 | 1/8 | 5/32 |
| Long parts, shafts | 1/16 | 1/16 | 3/32 | 1/8 | 3/16 |

FIGURE 18-20 Machining allowances for forgings.

**Machining Allowance (millimeters)***

| WEIGHT OF PART | UNDER 0.45 KG | 0.45 KG TO 4.5 KG | 4.95 KG TO 18.0 KG | 18.45 KG to 45.00 KG | 45.45 KG TO 90.00 KG |
|---|---|---|---|---|---|
| Compact parts, gears, discs | 0.8 | 1.2 | 1.6 | 2.4 | 3.2 |
| Thin, extended parts | 1.6 | 1.6 | 2.4 | 3.2 | 4.0 |
| Long parts, shafts | 1.6 | 1.6 | 2.4 | 3.2 | 4.8 |

*Soft conversion of accompanying inch table.

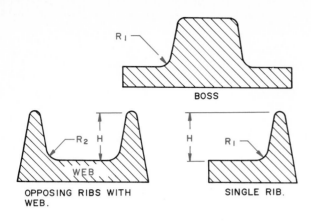

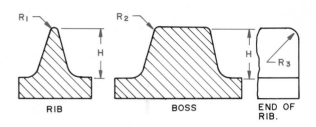

Minimum Fillet Radii

| INCHES | | | | | | | |
|---|---|---|---|---|---|---|---|
| $H$ | $1/4$ | $1/2$ | $1$ | $2$ | $3$ | $4$ | $5$ | $6$ |
| $R_1$ | $1/8$ | $1/8$ | $1/4$ | $1/2$ | $3/4$ | $1$ | $1\,1/4$ | $1\,1/2$ |
| $R_2$ | $1/8$ | $1/8$ | $3/8$ | $3/8$ | $1$ | $1\,3/8$ | $1\,3/4$ | $2$ |

| MILLIMETERS | | | | | | | |
|---|---|---|---|---|---|---|---|
| $H$ | $6$ | $12$ | $25$ | $50$ | $75$ | $100$ | $125$ | $150$ |
| $R_1$ | $3$ | $3$ | $6$ | $12$ | $19$ | $25$ | $31$ | $37$ |
| $R_2$ | $3$ | $3$ | $9.5$ | $16$ | $25$ | $34.5$ | $44$ | $50$ |

Minimum Corner Radii

| INCHES | | | | | | | |
|---|---|---|---|---|---|---|---|
| $H$ | $1/4$ | $1/2$ | $1$ | $2$ | $3$ | $4$ | $5$ | $6$ |
| $R_1$ | $1/16$ | $1/16$ | $1/8$ | $3/16$ | $1/4$ | $3/16$ | $3/8$ | $7/16$ |
| $R_2$ | $1/16$ | $1/16$ | $1/8$ | $1/4$ | $5/16$ | $7/16$ | $1/2$ | $5/8$ |
| $R_3$ | $3/16$ | $3/16$ | $3/8$ | $1/2$ | $3/4$ | $1$ | $1\,1/8$ | $1\,1/4$ |

| MILLIMETERS | | | | | | | |
|---|---|---|---|---|---|---|---|
| $H$ | $6$ | $12$ | $25$ | $50$ | $75$ | $100$ | $125$ | $150$ |
| $R_1$ | $1.5$ | $1.5$ | $3$ | $4.7$ | $6$ | $8$ | $9.5$ | $11$ |
| $R_2$ | $1.5$ | $1.5$ | $3$ | $6$ | $8$ | $11$ | $12$ | $15.8$ |
| $R_3$ | $4.7$ | $4.7$ | $9.5$ | $12$ | $19$ | $25$ | $28$ | $31$ |

FIGURE 18–21  Recommended sizes for radii for fillets and rounds on forgings. (Reproduced from ANSI Y14.9-1958 with permission of the publishers, The American Society of Mechanical Engineers.)

areas not coined. *Blanking* involves cutting the sheet metal blank to the size and shape desired. *Punching* is the process of internal blanking. It might cut a hole or another opening in the metal. *Trimming* is the process used to cut off excess material using dies.

Holes are formed in stampings by punching, extruding, or piercing. *Punched holes* are made by a metal punch the shape of the hole. It shears a clean hole of the finished size. *Extruded holes* are punched using a punch that first forms a hole smaller than desired and is tapered to form a hole with a flange. *Pierced holes* are made with a punch having a sharp point. It pierces the metal, leaving a flange with a ragged edge (Fig. 18–26).

Notches have either a sharp V or a rounded vertex (Fig. 18–27). The V notch tends to crack under stress. High-stress parts use the rounded vertex.

The drafter must also figure bend allowance for many types of stampings. When metal is formed,

the length is shortened due to the bend. To have proper length after forming, bend allowance must be added to the desired size.

Bend allowance is found by using a bend allowance chart (see Appendix I). This chart gives the amount of material needed to make a straight bend in flat metal through an angle of 1°. The left column gives this radius of the bend. Across the top of the chart are various thicknesses of sheet metal. For example, a bend of 30° in metal .040 in. thick with a 1-in. radius requires that the pattern be made .53220 in. longer. The bend allowance for 1° was .01774. The total bend is .01774 × 30° = .53220 (Fig. 18–28). The thickness and type of material influence the bend radius used when forming sheet metal.

There are many ways to form corners on stampings. A few of these are shown in Fig. 18–29. On some jobs, forming is facilitated by using corner relief holes. These must be planned as the stamping drawing is made. One general type of corner relief

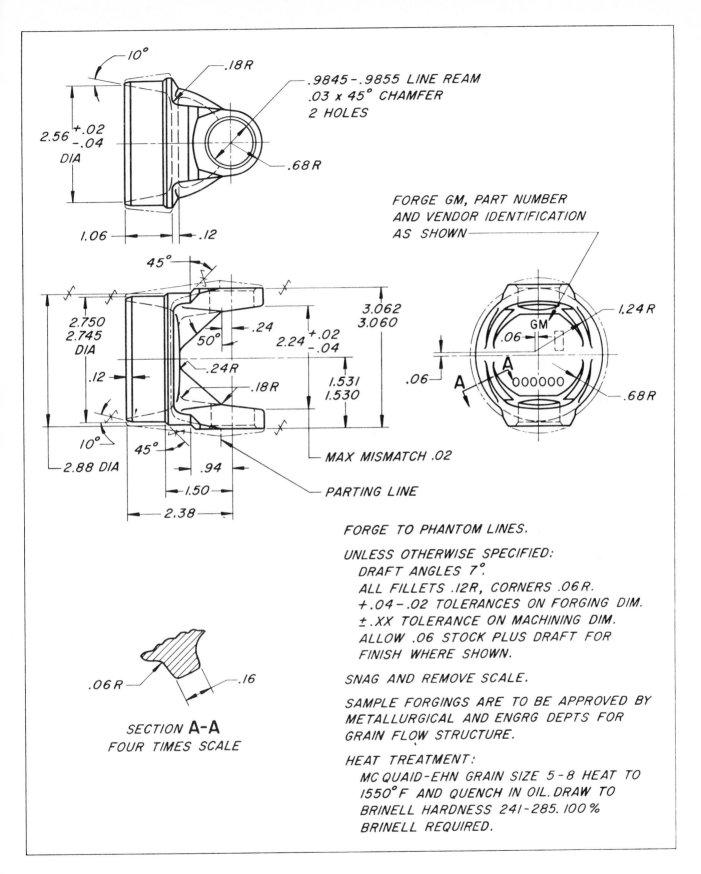

.18R

.9845-.9855 LINE REAM
.03 x 45° CHAMFER
2 HOLES

10°

2.56 +.02 -.04 DIA

.68R

1.06     .12

FORGE GM, PART NUMBER
AND VENDOR IDENTIFICATION
AS SHOWN

45°

2.750
2.745
DIA

50°     .24

.12

.24R

.18R

10°     45°

2.88 DIA

.94

1.50

2.38

3.062
3.060

2.24 +.02 -.04

1.531
1.530

.06

MAX MISMATCH .02

PARTING LINE

1.24R

GM

.06

A     A

000000

.68R

FORGE TO PHANTOM LINES.

UNLESS OTHERWISE SPECIFIED:
   DRAFT ANGLES 7°.
   ALL FILLETS .12R, CORNERS .06R.
   +.04 -.02 TOLERANCES ON FORGING DIM.
   ±.XX TOLERANCE ON MACHINING DIM.
   ALLOW .06 STOCK PLUS DRAFT FOR
   FINISH WHERE SHOWN.

SNAG AND REMOVE SCALE.

SAMPLE FORGINGS ARE TO BE APPROVED BY
METALLURGICAL AND ENGRG DEPTS FOR
GRAIN FLOW STRUCTURE.

HEAT TREATMENT:
   MC QUAID-EHN GRAIN SIZE 5-8 HEAT TO
   1550°F AND QUENCH IN OIL. DRAW TO
   BRINELL HARDNESS 241-285. 100%
   BRINELL REQUIRED.

.06R     .16

SECTION A-A
FOUR TIMES SCALE

FIGURE 18–22   A composite forging drawing showing
the forging and machining information. (Courtesy of
General Motors Corporation.)

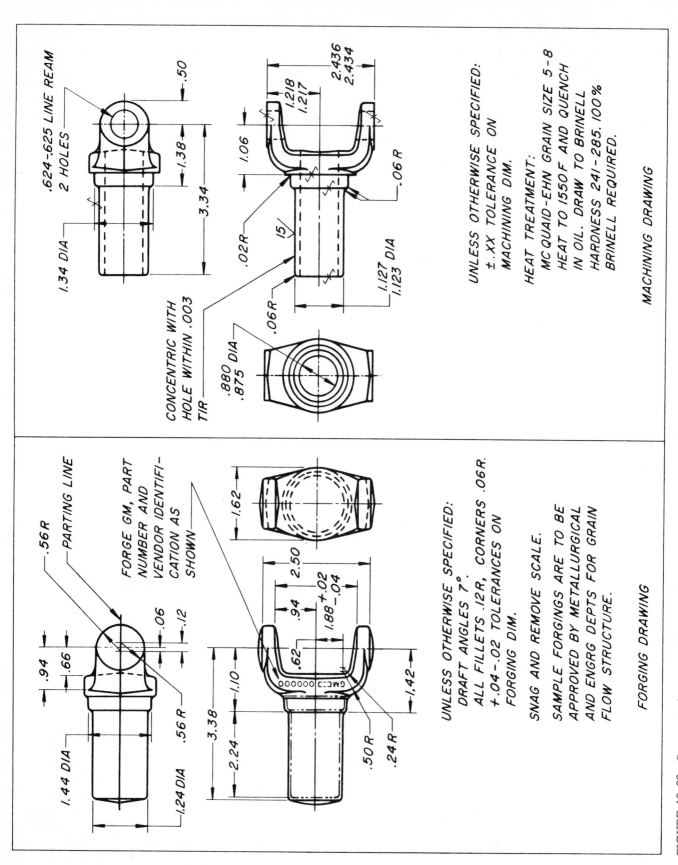

FIGURE 18-23 Separate drawings showing forging and machining information. (Courtesy of General Motors Corporation.)

FIGURE 18–24 This hot-air-heater diffuser was formed by stamping it from a single metal sheet.

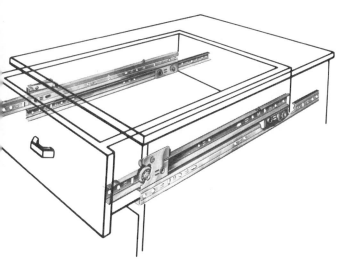

FIGURE 18–25 These metal drawer runners have holes of various shapes produced by punching before the parts are formed to shape. (Courtesy of Knape and Vogt Manufacturing Co.)

FIGURE 18–26 Holes are formed in stamped parts by punching, extruding, or piercing.

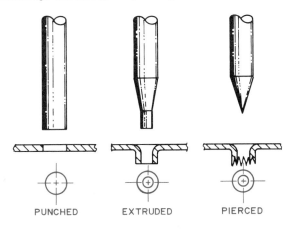

PUNCHED          EXTRUDED          PIERCED

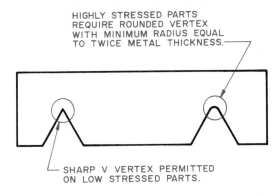

HIGHLY STRESSED PARTS REQUIRE ROUNDED VERTEX WITH MINIMUM RADIUS EQUAL TO TWICE METAL THICKNESS.

SHARP V VERTEX PERMITTED ON LOW STRESSED PARTS.

FIGURE 18–27 Notches in stampings use either a sharp V or a rounded vertex.

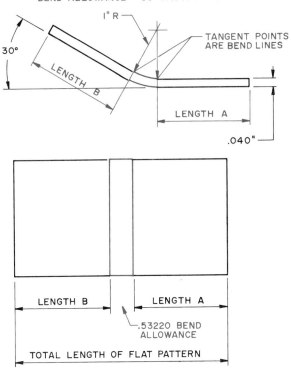

BEND ALLOWANCE = 30° X .01774 = .53220

30°

1" R

TANGENT POINTS ARE BEND LINES

LENGTH B

LENGTH A

.040"

LENGTH B          LENGTH A

.53220 BEND ALLOWANCE

TOTAL LENGTH OF FLAT PATTERN

FIGURE 18–28 How to figure bend allowance for sheet metal parts.

hole is shown in Fig. 18–30. The steps for drawing a general corner relief hole are shown in Fig. 18–31.

## Stamping Drawings

A stamping drawing shows the part in its final, assembled condition. It must be carefully dimensioned so that a pattern can be drawn using the stamping drawing as the basis for size and location dimensions. Dimensions are placed in the same manner as in other drawings. Hole locations and other critical dimensions should be located from one another, not from the outside edge of the stamping (Fig. 18–32).

The patterns developed from stamping drawings follow the procedures described in Chapter 19.

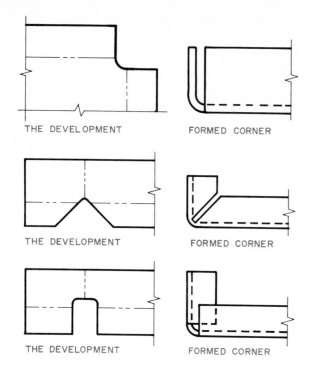

FIGURE 18–29 Some frequently used ways to prepare stamping to form the corners.

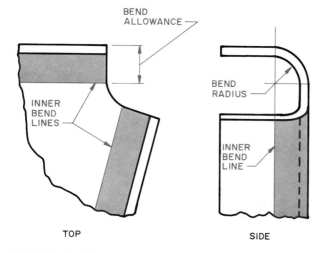

FIGURE 18–30 A typical general relief hole at the corner of a sheet metal part.

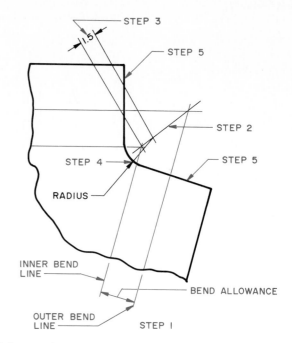

1. Lay out the pattern and locate the inner and outer bend line.
2. Draw a line connecting the points of intersection of the inner and outer bend lines. The center of the relief hole is on this line.
3. Measure 1.5 mm (.06 in.) along the center line from the intersection of the inner bend line. This is the center of the relief hole.
4. Draw the relief hole by swinging the arc until it reaches the tangent point, which is the inner bend line.
5. Draw a perpendicular from this tangent point to the outer edge of the pattern.

FIGURE 18–31 Drawing a flat pattern for a general relief hole at a corner.

## ASSEMBLY DRAWINGS

After the design assembly is made, the drafter uses it plus specifications and calculations to develop detail drawings of each part of a product. Once this is done, various assembly drawings may be made to help describe the product and facilitate its manufacture, assembly, and use.

The common types of assembly drawings include a check assembly, general assembly, working drawing assembly, installation assembly, mainte-

nance assembly, pictorial assembly, and catalog assembly.

### Procedures for Drawing Assembly Drawings

An assembly drawing shows the various parts of a product put together to form a finished unit. It is made from the information given on the detail drawings; sometimes, the details are traced onto the assembly drawing.

The drafter must understand how the product is supposed to function and how the various parts relate to one another. A decision must be made as to the best views to draw to show the assembly. Preliminary work can be done by making a few freehand sketches of the basic parts of the assembled device before making the final assembly drawing. The drafter will call upon a wide range of knowledge and techniques, such as drawing sections, fasteners, and material symbols.

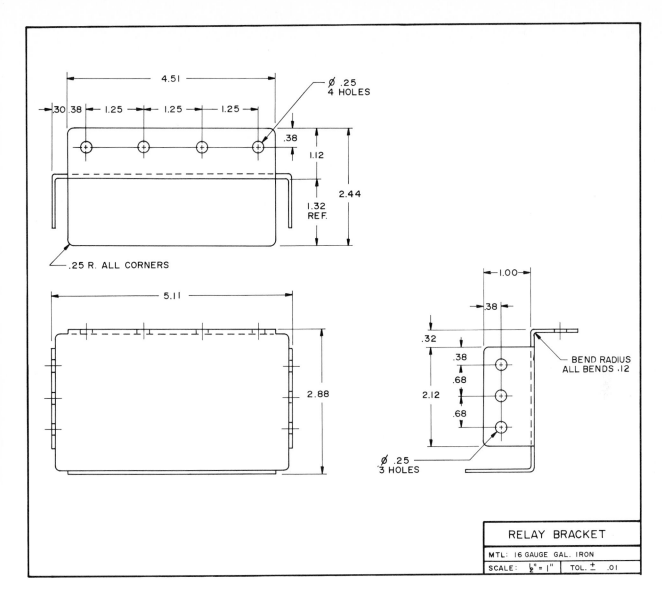

FIGURE 18–32  A typical stamping drawing.

Most assembly drawings follow the principles used to make multiview drawings. After the views have been established, the main features are drawn in outline form. Often, a center line or a series of center lines establishes the location of the various parts. Then, the details of each part are drawn. Do not completely draw one view and then start on the next view. Build all views together to completion. For example, if a bearing is drawn on the front view, draw it immediately on the other views. Then repeat this for all the other parts (Fig. 18–33).

## Check Assembly Drawings

After a product has been designed and each part detailed, it is necessary to check the detail drawings for accuracy. This is often done by *inspection*. A drafting checker goes over each drawing, checks it for completeness, and ascertains if it will fit with adjoining parts. This involves calculations to check tolerances and other dimensions. An aid to this validation process is a check assembly. A *check assembly* shows the various parts drawn very accurately in their assembled position. Each part is not completely detailed. What is shown are those portions needed to check to see how the parts will fit together. The drawing is not dimensioned. Proper location of center lines and datum surfaces is vital.

If a general assembly drawing is required, data from the check assembly can be used in its construction. A carefully made general assembly could serve as a check assembly.

## General Assembly Drawings

A general assembly drawing shows how the various parts of a product fit together. The parts are shown in outline form, and any movement of parts is

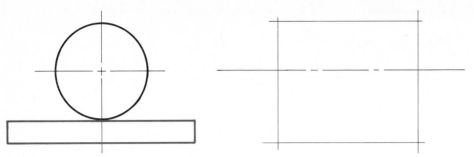

STEP I. LOCATE THE CENTER LINES AND THE MAIN OUTLINE OF THE ASSEMBLY.

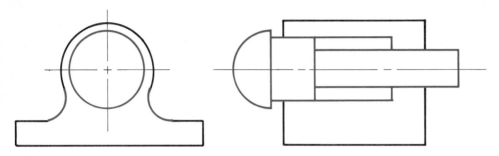

STEP 2. BLOCK IN THE DETAILS OF THE MAJOR PARTS.

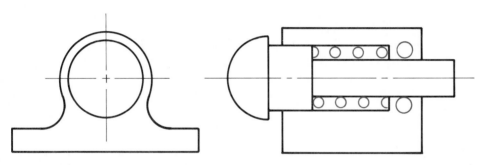

STEP 3. DRAW MINOR DETAILS, COMPLETE DETAILS ON MAJOR PARTS, AND REMOVE CONSTRUCTION LINES.

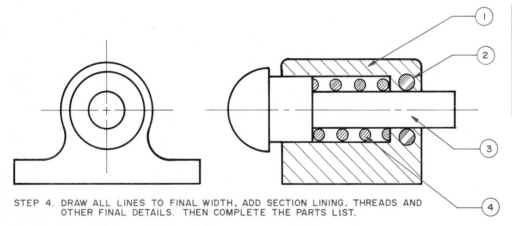

| PART NO. | NAME | MTL. | NO. REQD |
|---|---|---|---|
| I | SHELL | C.IRON | I |
| 2 | O-RING | PLAST | I |
| 3 | SHAFT | STEEL | I |
| 4 | SPRING | STEEL | I |
| | | | |
| | | | |
| | | | |

STEP 4. DRAW ALL LINES TO FINAL WIDTH, ADD SECTION LINING, THREADS AND OTHER FINAL DETAILS. THEN COMPLETE THE PARTS LIST.

FIGURE 18–33 Laying out an assembly drawing.

shown. Sections are required to show hidden details. These are useful in the assembly of the product in the manufacturing plant (Fig. 18–34).

The individual parts in a general assembly are not dimensioned. The only dimensions that may appear are those that are important to the assembled

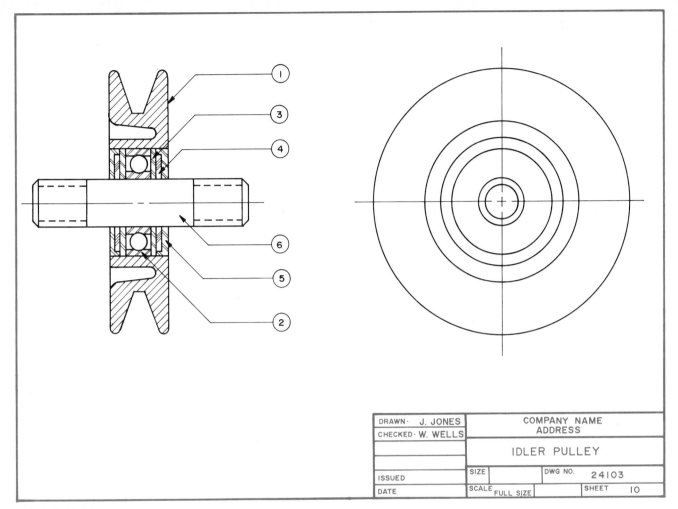

FIGURE 18-34  A general assembly drawing.

| IDLER PULLEY ASSEMBLY B-32 PARTS LIST | | | |
|---|---|---|---|
| KEY NO. | PART NAME | PART NO. | NO. REQD |
| 1 | PULLEY | 13-42 | 1 |
| 2 | BALL BEARING – ACE 1011 | 13-04 | 1 |
| 3 | BEARING SHIELD | 13-09 | 2 |
| 4 | PLASTIC WASHER | 13-17 | 2 |
| 5 | BEARING CAP | 13-25 | 2 |
| 6 | IDLER SHAFT | 10-03 | 1 |
|  |  |  |  |
|  |  |  |  |
|  |  |  |  |

FIGURE 18-35  A parts list.

condition, such as an overall assembled size. Each part is identified with a number. The numbers are often assigned in the order in which the parts are assembled. Each number is recorded inside a circle that is connected to the part with a leader, which may or may not have an arrowhead. These circles are placed on the drawing in an orderly fashion; whenever possible, they are lined up on the outside edges of the views in horizontal or vertical rows. Circle diameters of 10 mm (⅜ in.) or 11 mm (⁷⁄₁₆ in.) are commonly used, and the numbers are usually 5 mm (³⁄₁₆ in.) high.

The leaders can be at any convenient angle but should not be horizontal or vertical, nor should they cross another line or obscure a detail on the drawing. The leaders are drawn so that if they were extended into the circle, they would cross its center.

A parts list accompanies the general assembly drawing. It often includes the part number, name, material, and the quantity of each part needed to make one product. It is usually placed immediately above the title block (Fig. 18-35). If it is extensive, it is placed on a separate sheet.

## Working Drawing Assembly Drawings

A working drawing assembly drawing is often made when a product has only a few parts that are rather simple in structure. The drawing shows the assembled product, and each part is completely dimensioned. The necessary orthographic views are shown along with notes, sections, and dimensions. If several pieces cannot be clearly shown, they can be drawn on the same sheet as detail drawings (Fig. 18-36).

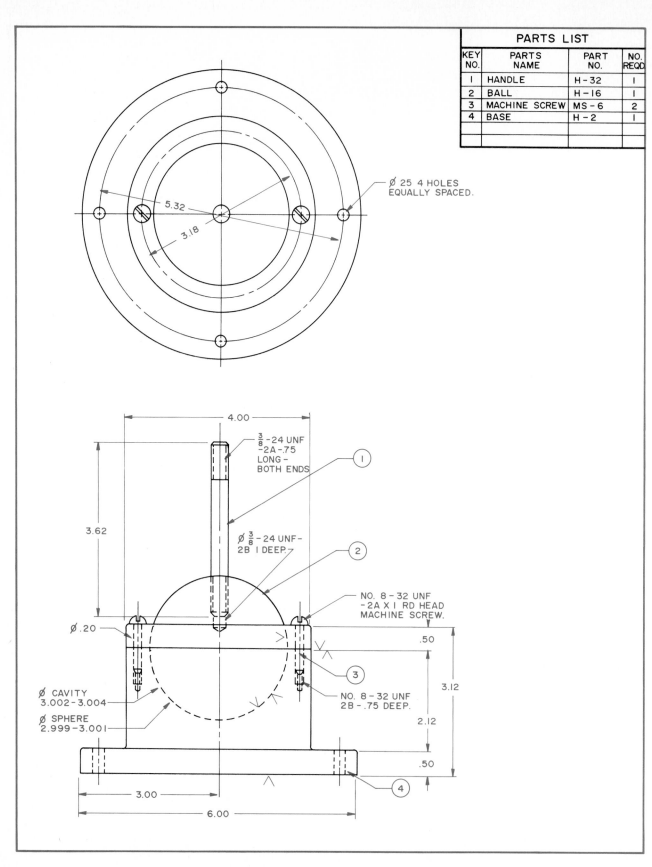

| KEY NO. | PARTS NAME | PART NO. | NO. REQD |
|---|---|---|---|
| 1 | HANDLE | H – 32 | 1 |
| 2 | BALL | H – 16 | 1 |
| 3 | MACHINE SCREW | MS – 6 | 2 |
| 4 | BASE | H – 2 | 1 |
| | | | |
| | | | |

Ø 25 4 HOLES
EQUALLY SPACED.

5.32

3.18

4.00

$\frac{3}{8}$ – 24 UNF
–2A –.75
LONG –
BOTH ENDS

3.62

Ø $\frac{3}{8}$ – 24 UNF–
2B 1 DEEP.

NO. 8 – 32 UNF
–2A X 1 RD HEAD
MACHINE SCREW.

.50

3.12

Ø .20

Ø CAVITY
3.002–3.004

Ø SPHERE
2.999–3.001

NO. 8 – 32 UNF
2B – .75 DEEP.

2.12

.50

3.00

6.00

FIGURE 18–36   A working drawing assembly drawing.

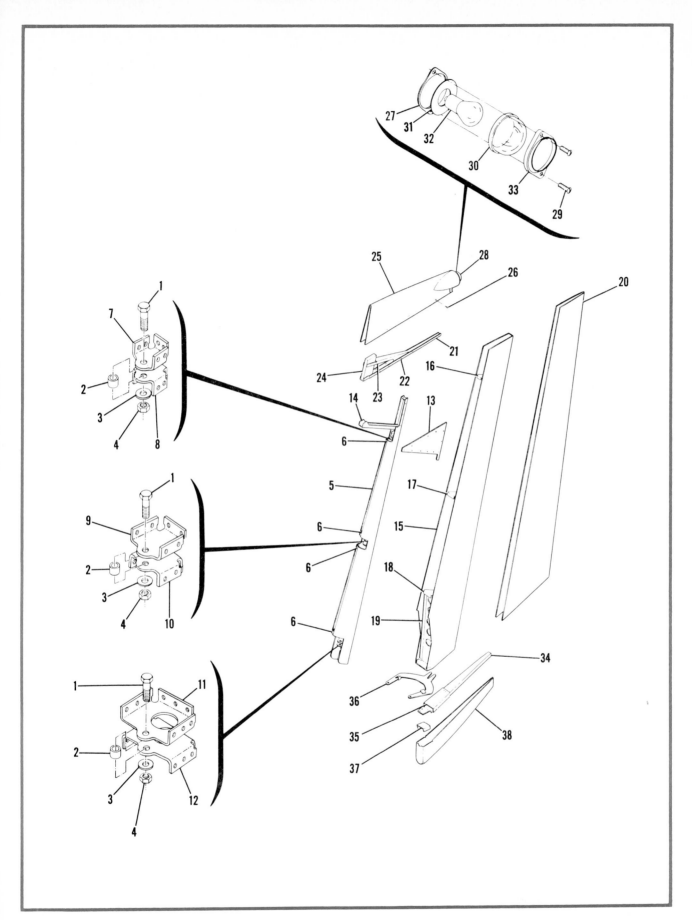

FIGURE 18-37 An installation assembly drawing of an aircraft rudder. (Courtesy of Cessna Aircraft Company.)

459

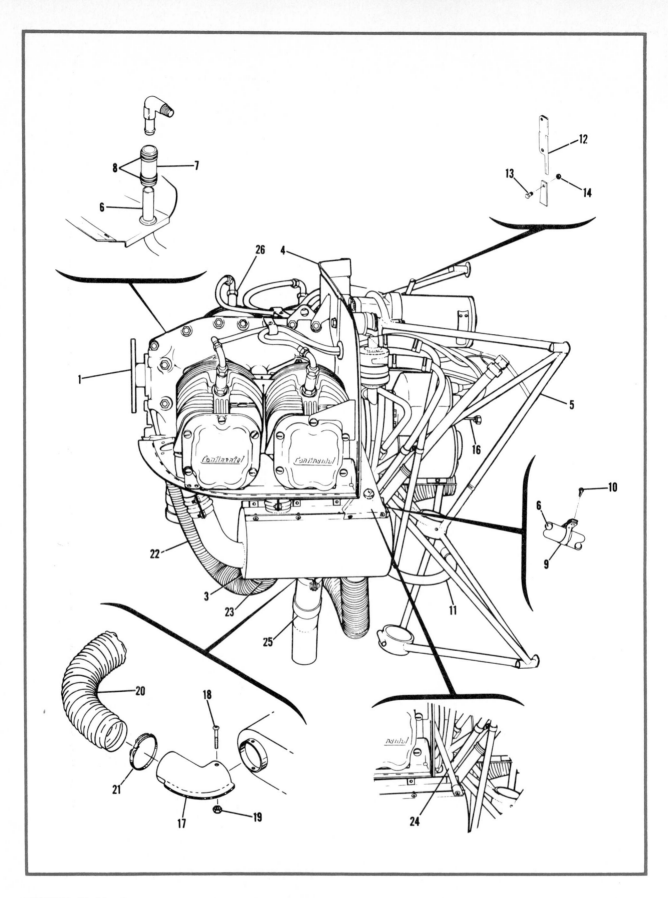

FIGURE 18–38  A pictorial assembly drawing of an
aircraft engine on its mounting structure. (Courtesy of
Cessna Aircraft Company.)

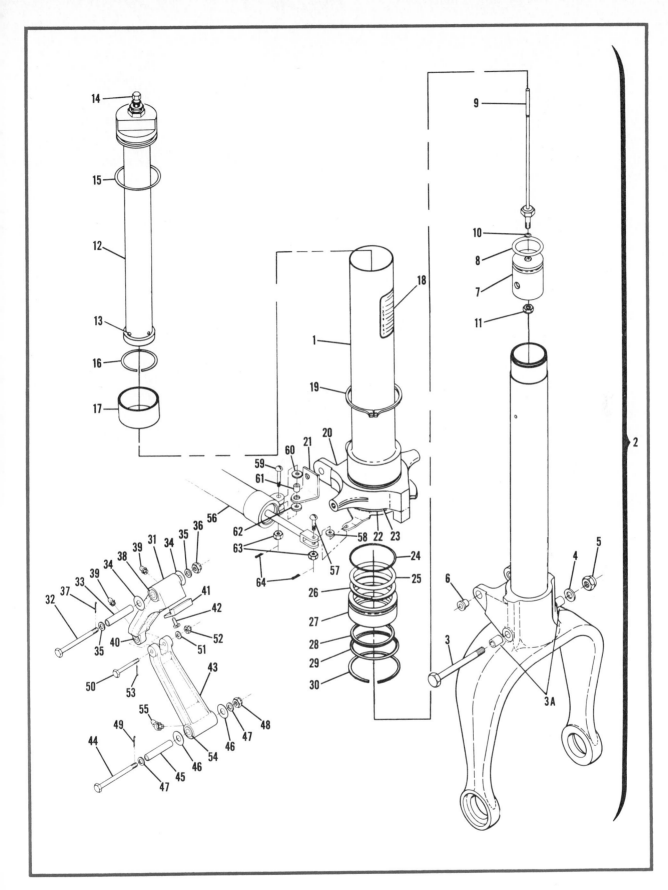

FIGURE 18–39 An exploded assembly of an aircraft front landing-wheel assembly. (Courtesy of Cessna Aircraft Company.)

## Subassembly Drawings

Many products, such as automobile engines, are too large and complex to draw in a single assembly drawing. In such cases, smaller, individually functioning units are drawn as separate assemblies, called subassemblies. The alternator on the automobile would be a subassembly. Subassemblies are drawn in the same manner as general assembly drawings and serve the same purpose.

## Installation Assembly Drawings

An *installation assembly drawing* provides information needed for installation or erection. It shows only the basic outline of the parts and those dimensions needed for installation. Overall dimensions are helpful when the space needed for installation is limited. Other dimensions, such as the distances moving parts will travel, are important. Often, one drawing can serve several identical machines of different sizes (Fig. 18–37).

## Maintenance Assembly Drawings

A maintenance assembly drawing can be a general assembly (with or without sections) or a pictorial assembly. The purpose is to provide information about lubrication, servicing procedures, and repair procedures for a machine. In some cases, the parts are identified by the part number and can be used when ordering replacement parts.

## Pictorial Assembly Drawings

*Pictorial assembly drawings* can serve as installation and maintenance assembly drawings. They are also used in parts catalogs and sales literature. They are useful in showing relationships between subassemblies. Because they are lifelike in appearance, they are understood by people who have little experience reading engineering drawings. Pictorial assembly drawings may consist of a single pictorial view (Fig. 18–38), an exploded view (Fig. 18–39), or a sectional view (Fig. 18–40). Usually, such drawings are not dimensioned, but the parts are identified by name or part number. They may be shaded to improve their appearance. The pictorial systems used depend on the desires of the company, but isometric, dimetric, trimetric, and perspective drawings are all used.

## Diagrammatic Assembly Drawings

There are other drawings that are really assembly drawings but use symbols and single lines to repre-

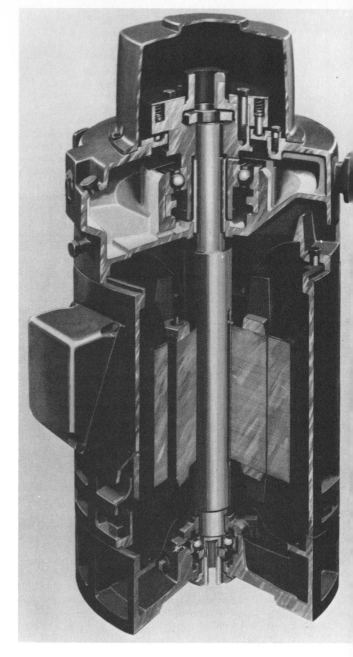

FIGURE 18–40 A section assembly of a 200-hp electric motor that has been shaded to give a more natural appearance.

sent the parts of the product. A schematic of an electrical product shows the components and connecting wires in an assembled view (Fig. 18–41). A piping drawing is also an assembled drawing of a circuit (Fig. 18–42). A diagrammatic assembly can be drawn as an orthographic projection or pictorially.

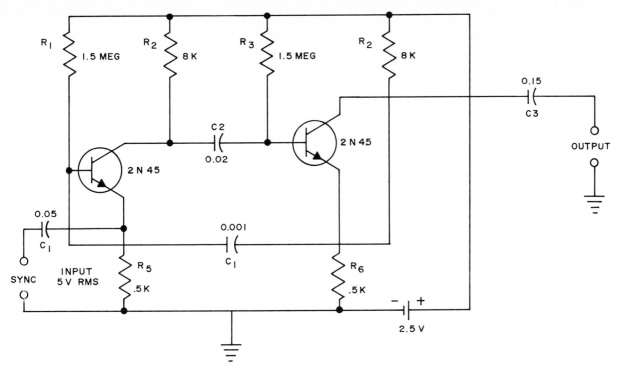

**FIGURE 18–41** An electronics schematic is a form of assembly drawing.

**FIGURE 18–42** Orthographic views, top and front, of a pipe diagram from an assembly drawing.

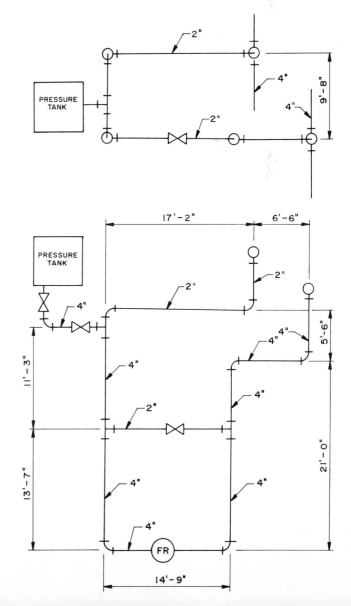

# Problems

The following are details of a variety of products to be used for the preparation of complete working drawings and parts lists. They include part production for castings, forgings, sheet metal, welding, and machining.

Draw on paper with a border that permits zoning the drawing. Use a standard title block. Decide whether you will use monodetail or multidetail drawings. Select a scale that will produce a drawing large enough to clearly show all the features. Assign each part a number. Identify each part with a descriptive title.

Make the necessary detail drawings (casting, machining, stamping, etc.) for each part. Use sections, auxiliary views, and other techniques to clearly describe each feature. Prepare an assembly drawing. Decide on the purpose of the assembly (general, check, working drawing, etc.) before proceeding to draw it. Then prepare a parts list to complete the project.

Some of the following projects are being used with the permission of the manufacturer. The design sizes shown have been changed to simplify or otherwise change the product to adapt it for use as a drawing experience. The following instructions give suggestions for the use of each product. The instructor can change or adapt these problems in many ways and may choose to prepare different or additional directions.

1. Prepare casting and machining drawings for each part of this serrated adjustable clamp. The draft angle is ± 7°, the casting tolerance is ± .03 in., and the machining tolerance is ± .01 in.; the material is steel. The surface finish data includes: roughness height 10 microinches, waviness height .001 in., waviness width .12 in., lay parallel with the long dimension, and roughness width .01 in.

2. Prepare a check assembly of this sliding bolt. It has some errors built into it. Find these errors with the check assembly, and correct the sizes and locations when you draw the detail drawings. When the handle is moved all the way to the right of the slot, the shaft must extend one in. beyond the base. When moved to the left the shaft must extend 1¼ in. beyond the base.

3. Prepare a casting drawing for the body of the clamp. It is aluminum. Prepare a complete working drawing of each of the other parts.

4. Prepare working drawings of each part with a tolerance of ± .10 in. Draw a general assembly of this straddle cam assembly.

5. Prepare detail drawings for each part of this shaft post. The shaft cover and post body are cast iron. The bearing liner is brass.

6. Prepare separate pattern and casting drawings for the handle using the same specifications as in Prob. 18–1. Prepare detail drawings of the other parts with a tolerance of ± .10 in. Then prepare an exploded pictorial assembly.

7. Prepare casting detail drawings of the base and handle and a detail drawing of the screw. Select tolerances suitable for the accuracy needed.

8. Prepare a machining drawing of the plate with a tolerance of ± .05. The base is cast with a spot face and threads to be machined into it. Prepare a casting detail of the base. Then make a general assembly drawing.

9. Prepare detail drawings for each part using a tolerance of ± .01 in. Then prepare a general assembly or exploded assembly as assigned by the instructor. Remember to give descriptive names to each part and prepare a parts list.

10. Prepare a machining detail for the chuck and collet using a machining tolerance of ± 0.25 mm. Surface finish data for both parts includes: roughness average 1.6 mm, waviness height 0.0012 mm, waviness width 3.0 mm, lay perpendicular to ends, and roughness width 0.25 mm.

11. Prepare stamping detail drawings of the 10 gauge steel parts and complete detail drawings of all the other parts. Then draw a general assembly showing how the unit is to be assembled.

12. Prepare a casting detail of the base and detail drawings of the other parts. Select casting tolerances that seem adequate for the accuracy needed.

13. Prepare a pattern detail drawing of the main casting. Core the two large diameter holes. The draft angle is 7°. Prepare detail drawings for the other parts using a tolerance of ± 0.25 mm. Draw a general assembly of the unit and give descriptive names to each part.

14. This problem is drawn with the main parts as castings. It could also be made by welding individual pieces together to form each part. Detail each part as either casting or welding drawings as assigned by your instructor. Show all machining information needed to complete each part.

15. The T-connection and head are cast parts. Prepare a casting detail for each. Draw machining details of the other parts. Then put them together in a general assembly drawing. If something does not fit, change the detail drawing to make it fit.

16. Prepare a pattern detail drawing of the handle and a casting detail of the main body. The draft angle is 7°, the casting tolerance is ± .03 in., and the machining tolerance is ± .01 in.; the material is cast iron. Prepare finished detail drawings of all other parts using a tolerance of ± .10.

17. Prepare detail drawings of each part. Make some decisions as to how you would make this stand and detail accordingly. For example, the base could be a large casting or made from separate steel parts welded together. Include a parts list.

18. The pulley is cast, so make a casting detail. The main arm and bottom shaft housing of the frame is cast. The housing on the top and the threaded shaft are welded in place. Prepare the necessary detail drawings to describe each part.

19. Prepare the necessary detail drawings to describe all the manufacturing operations necessary to make each part of this tool stop. Select tolerances suitable for the degree of accuracy required.

20. This is a large casting that serves as the base for an index table used in a machine shop. It has been simplified for this problem so it does not reflect all of the complexities of the design. Prepare pattern and casting detail drawings.

21. This disc sander is mounted on a heavy steel base and has a steel table. Prepare stamping detail drawings for each piece.

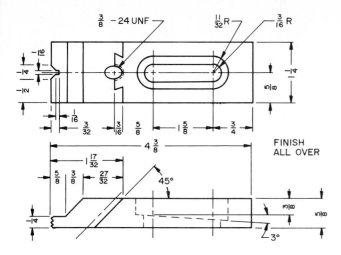

$\frac{3}{8}$ – 24 UNF $\quad$ $\frac{11}{32}$ R $\quad$ $\frac{3}{16}$ R

$\frac{1}{16}$

$\frac{1}{4}$

$\frac{1}{2}$

$\frac{1}{4}$

$\frac{5}{8}$

$\frac{1}{16}$ $\quad$ $\frac{3}{32}$ $\quad$ $\frac{3}{16}$ $\quad$ $\frac{5}{8}$ $\quad$ $1\frac{5}{8}$ $\quad$ $\frac{3}{4}$

FINISH ALL OVER

$4\frac{3}{8}$

$1\frac{17}{32}$

$\frac{5}{8}$ $\quad$ $\frac{3}{8}$ $\quad$ $\frac{27}{32}$ $\quad$ 45°

$\frac{3}{8}$

$\frac{5}{8}$

$\frac{1}{4}$

3°

**PROBLEM 18–1** Serrated adjustable clamp. (Courtesy Carr Lane Manufacturing Co.)

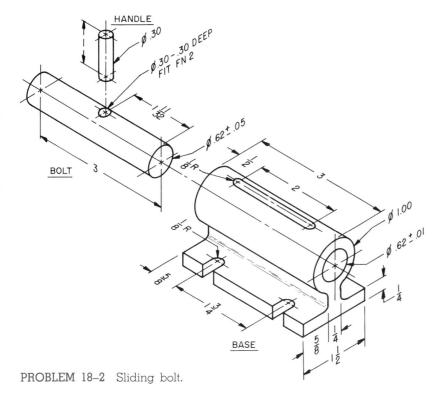

HANDLE

Ø .30

Ø .30 - .30 DEEP
FIT FN 2

Ø .62 $^+$ .05

$\frac{1}{32}$

BOLT

3

$\frac{1}{8}$ R

$\frac{1}{2}$

3

2

Ø 1.00

Ø .62 $^+$ .01

$\frac{1}{8}$ R

$\frac{5}{8}$

$\frac{3}{4}$

BASE

$\frac{5}{8}$

$\frac{1}{4}$

$\frac{1}{4}$

$\frac{1}{4}$

$1\frac{1}{2}$

**PROBLEM 18–2** Sliding bolt.

**PROBLEM 18–3** Clamp.

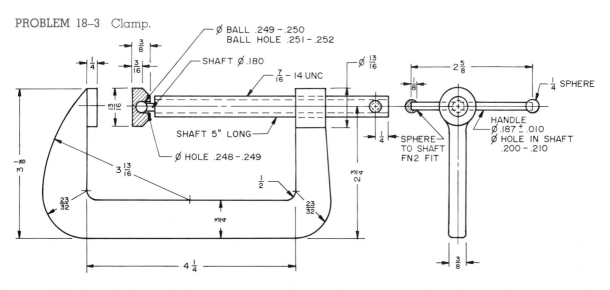

$\frac{3}{8}$

$\frac{1}{4}$

$\frac{3}{16}$

Ø BALL .249 – .250
BALL HOLE .251 – .252

SHAFT Ø .180

$\frac{7}{16}$ – 14 UNC

Ø $\frac{13}{16}$

$2\frac{5}{8}$

$\frac{1}{8}$

$\frac{1}{4}$ SPHERE

$\frac{13}{16}$

SHAFT 5" LONG

Ø HOLE .248 –.249

$\frac{1}{4}$

$2\frac{3}{4}$

SPHERE
TO SHAFT
FN2 FIT

HANDLE
Ø .187 $^+$ .010
Ø HOLE IN SHAFT
.200 – .210

$3\frac{1}{8}$

$3\frac{13}{16}$

$\frac{23}{32}$

$\frac{1}{2}$

$\frac{23}{32}$

$\frac{3}{4}$

$4\frac{1}{4}$

$\frac{3}{8}$

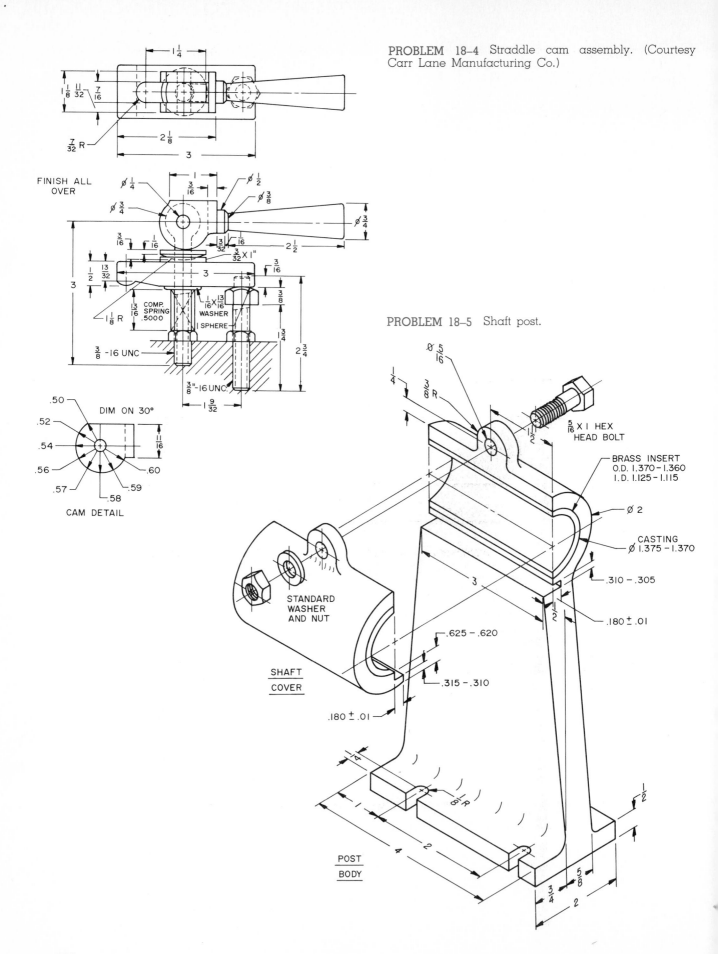

PROBLEM 18-4 Straddle cam assembly. (Courtesy Carr Lane Manufacturing Co.)

FINISH ALL OVER

COMP. SPRING .5000

WASHER SPHERE

$\frac{3}{8}$ -16 UNC

$\frac{3}{8}$"-16 UNC

DIM ON 30°

CAM DETAIL

PROBLEM 18-5 Shaft post.

$\frac{5}{16}$ X 1 HEX HEAD BOLT

BRASS INSERT
O.D. 1.370 - 1.360
I.D. 1.125 - 1.115

CASTING
Ø 1.375 - 1.370

STANDARD WASHER AND NUT

SHAFT COVER

POST BODY

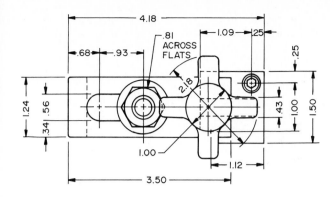

PROBLEM 18–7 Locking table clamp

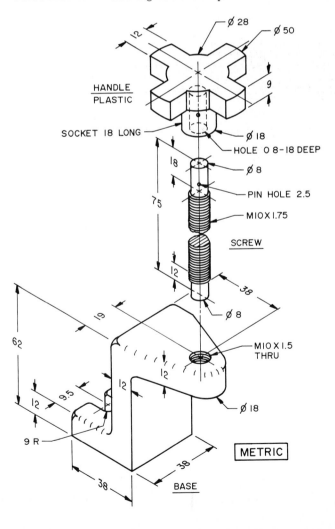

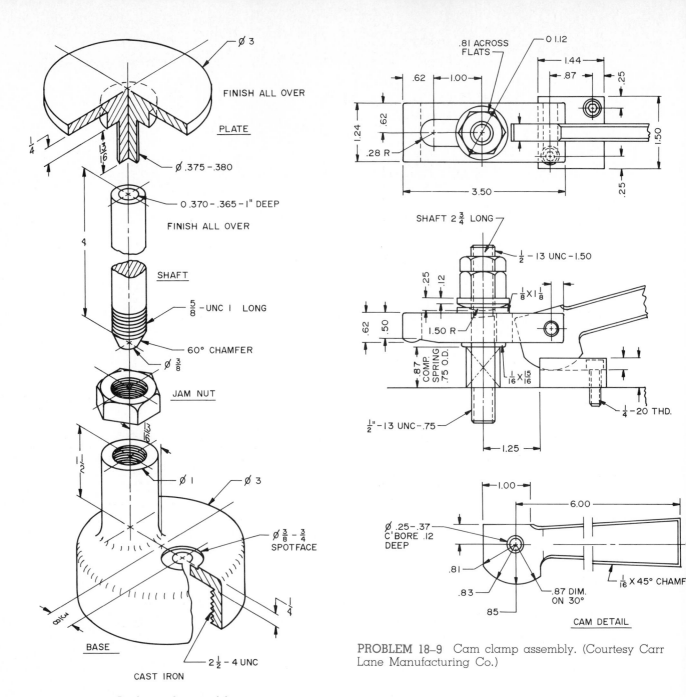

FINISH ALL OVER

Ø 3

PLATE

Ø .375 –.380

1/4

13/16

4

0 .370 –.365 –1" DEEP

FINISH ALL OVER

SHAFT

5/8 –UNC 1 LONG

60° CHAMFER

Ø 3/8

JAM NUT

13/16

1 1/2

Ø 1

Ø 3

Ø 3/8 – 3/4
SPOTFACE

1/4

3/16

BASE

2 1/2 – 4 UNC

CAST IRON

PROBLEM 18–8  Back-up plate and base.

.81 ACROSS
FLATS

0 1.12

1.44

.87

.25

.62

1.00

1.24

.62

.28 R

1.50

3.50

.25

SHAFT 2 3/4 LONG

1/2 – 13 UNC –1.50

.25

.12

1/8 X 1 1/8

.62

.50

1.50 R

.87
COMP.
SPRING
.75 O.D.

1/16 X 15/16

1/2" –13 UNC –.75

1.25

1/4 –20 THD.

CAM DETAIL

1.00

6.00

Ø .25 –.37
C'BORE .12
DEEP

.81

.83

85

.87 DIM.
ON 30°

1/16 X 45° CHAMF

PROBLEM 18–9  Cam clamp assembly. (Courtesy Carr
Lane Manufacturing Co.)

PROBLEM 18–10  Chuck with a collet.

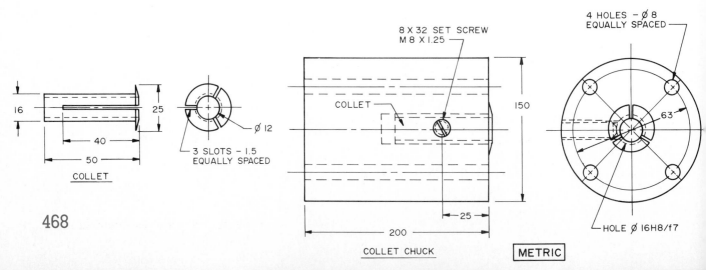

16

25

40

50

Ø 12

3 SLOTS – 1.5
EQUALLY SPACED

COLLET

8 X 32 SET SCREW
M 8 X 1.25

COLLET

150

25

200

COLLET CHUCK

4 HOLES – Ø 8
EQUALLY SPACED

63

HOLE Ø 16H8/f7

METRIC

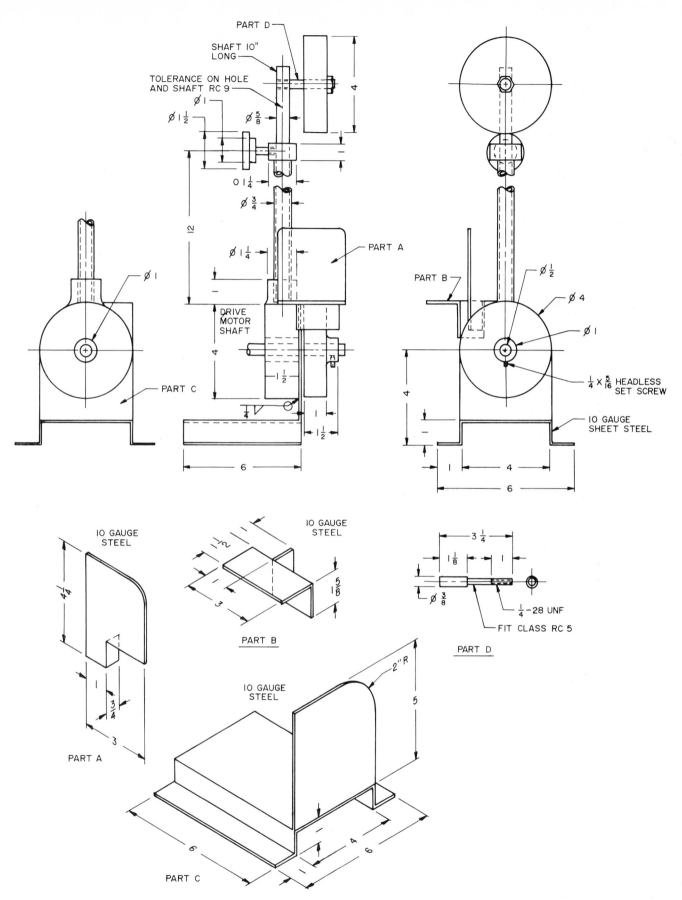

PART D

SHAFT 10" LONG

TOLERANCE ON HOLE AND SHAFT RC 9

$\phi 1$

$\phi 1\frac{1}{2}$

$\phi \frac{5}{8}$

$O 1\frac{1}{4}$

$\phi \frac{3}{4}$

12

$\phi 1\frac{1}{4}$

PART A

DRIVE MOTOR SHAFT

4

$1\frac{1}{2}$

$\frac{1}{4}$

1

$1\frac{1}{2}$

6

PART B

$\phi \frac{1}{2}$

$\phi 4$

$\phi 1$

$\frac{1}{4} \times \frac{5}{16}$ HEADLESS SET SCREW

IO GAUGE SHEET STEEL

4

1    4

6

$\phi 1$

PART C

IO GAUGE STEEL

$\frac{1}{2}$

$\frac{5}{8}$

3

PART B

IO GAUGE STEEL

$3\frac{1}{4}$

$1\frac{1}{8}$    1

$\phi \frac{3}{8}$

$\frac{1}{4}-28$ UNF

FIT CLASS RC 5

PART D

IO GAUGE STEEL

$4\frac{1}{4}$

1

$\frac{3}{4}$

3

PART A

IO GAUGE STEEL

2"R

5

6

4

6

PART C

PROBLEM 18–11   Belt sander. (Courtesy Kalamazoo Industries, Inc.)

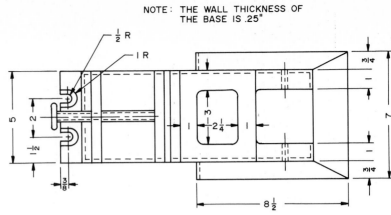

NOTE: THE WALL THICKNESS OF THE BASE IS .25"

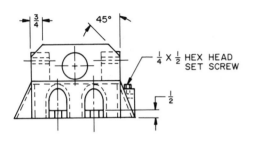

**PROBLEM 18–12** Base for a power cut-off saw. (Courtesy Kalamazoo Industries, Inc.)

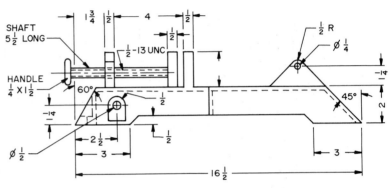

**PROBLEM 18–13** Swing clamp assembly. (Courtesy Carr Lane Manufacturing Co.)

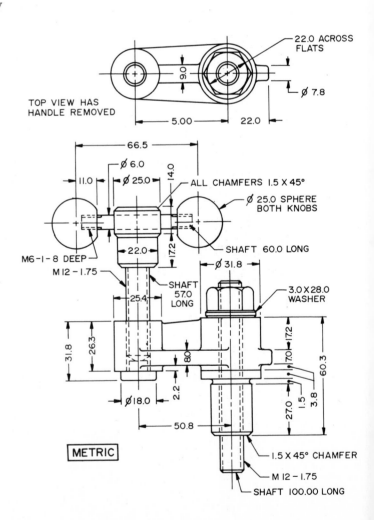

TOP VIEW HAS HANDLE REMOVED

22.0 ACROSS FLATS

METRIC

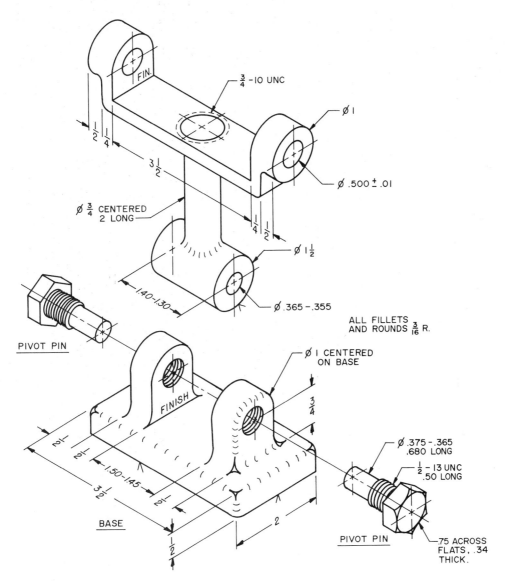

FIN.

$\frac{3}{4}$ -10 UNC

$\varnothing$ 1

$\varnothing$ .500 $\pm$ .01

$\frac{1}{2}$  $\frac{1}{4}$

$3\frac{1}{2}$

$\varnothing \frac{3}{4}$ CENTERED
2 LONG

$\frac{1}{4}$  $\frac{1}{2}$

$\varnothing 1\frac{1}{2}$

1.40-1.30

$\varnothing$ .365 – .355

ALL FILLETS
AND ROUNDS $\frac{3}{16}$ R.

PIVOT PIN

$\varnothing$ 1 CENTERED
ON BASE

FINISH

$\frac{3}{4}$

$\frac{1}{2}$  $\frac{1}{2}$

1.50-1.45

$\frac{1}{2}$

$3\frac{1}{2}$

BASE

$\frac{1}{2}$

2

$\varnothing$ .375 – .365
.680 LONG

$\frac{1}{2}$ – 13 UNC
.50 LONG

PIVOT PIN

.75 ACROSS
FLATS, .34
THICK.

PROBLEM 18–14   Idler yoke.

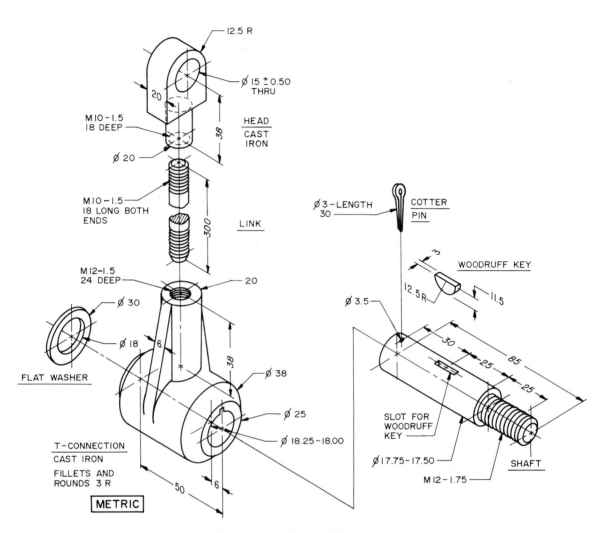

12.5 R

Ø 15 ± 0.50
THRU

20

M10-1.5
18 DEEP

HEAD
CAST
IRON

38

Ø 20

M10-1.5
18 LONG BOTH
ENDS

300

LINK

M12-1.5
24 DEEP

20

Ø 30

Ø 18

6

FLAT WASHER

38

Ø 38

Ø 25

Ø 18.25-18.00

T-CONNECTION
CAST IRON

FILLETS AND
ROUNDS 3 R

50

6

METRIC

Ø 3-LENGTH
30

COTTER
PIN

3

WOODRUFF KEY

12.5R

11.5

Ø 3.5

30

85

25

25

SLOT FOR
WOODRUFF
KEY

Ø 17.75-17.50

M 12-1.75

SHAFT

PROBLEM 18–15   Adjustable control arm.

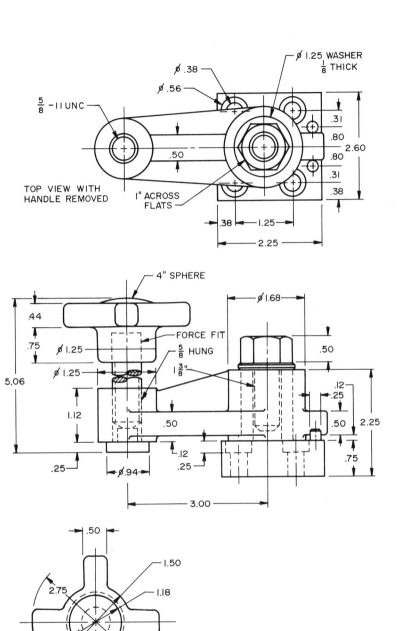

PROBLEM 18-16 Swing clamp assembly. (Courtesy
Carr Lane Manufacturing Co.)

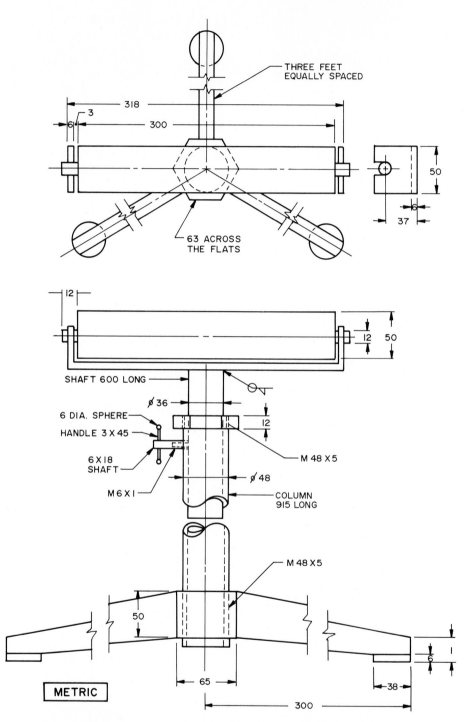

THREE FEET
EQUALLY SPACED

318

3

6

300

50

6

37

63 ACROSS
THE FLATS

12

12

50

SHAFT 600 LONG

⌀ 36

6 DIA. SPHERE

HANDLE 3 X 45

6 X 18
SHAFT

M 6 X 1

M 48 X 5

⌀ 48

12

COLUMN
915 LONG

M 48 X 5

50

6

65

300

38

METRIC

PROBLEM 18–17  Stock handling stand. (Courtesy Kalamazoo Industries, Inc.)

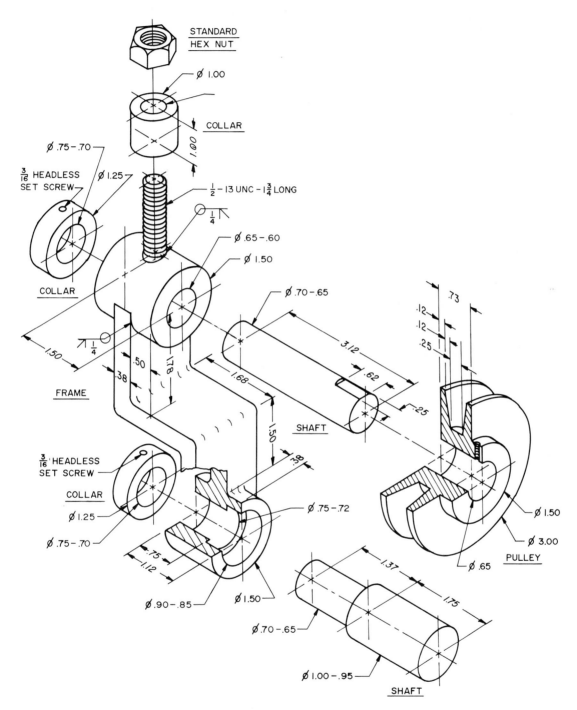

STANDARD
HEX NUT

Ø 1.00

COLLAR

1.00

Ø .75 - .70

3/16 HEADLESS
SET SCREW

Ø 1.25

1/2 - 13 UNC - 1 3/4 LONG

COLLAR

Ø .65 - .60

Ø 1.50

Ø .70 - .65

1.50

1/4

1.50

.73

.12

.12

.25

3.12

1/4

.50

.38

.87

FRAME

1.68

.62

.25

1.50

SHAFT

3/16 HEADLESS
SET SCREW

.38

COLLAR

Ø 1.25

Ø .75 - .70

Ø .75 - .72

Ø 1.50

Ø .90 - .85

Ø 1.50

Ø 3.00

PULLEY

.75

Ø .65

1.12

1.37

1.75

Ø .70 - .65

Ø 1.00 - .95

SHAFT

PROBLEM 18–18  Pulley idler arm.

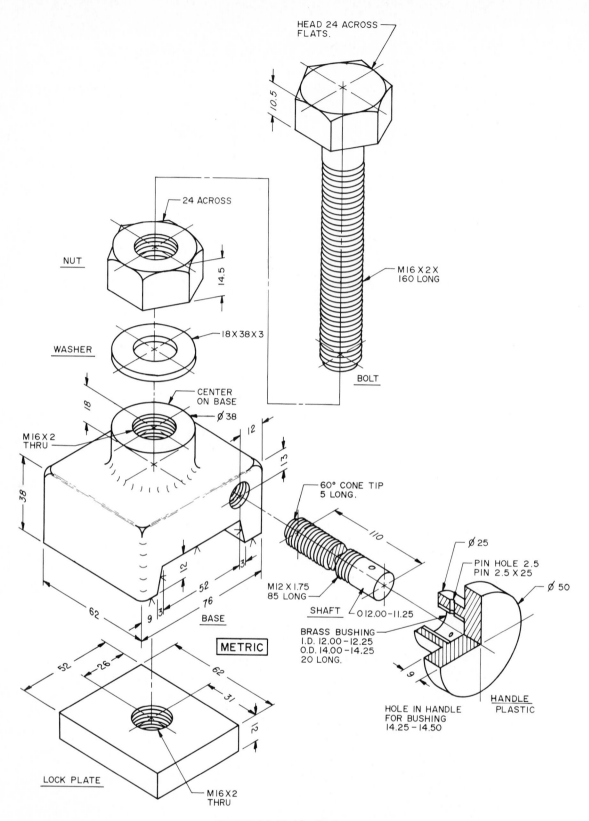

HEAD 24 ACROSS FLATS.

10.5

24 ACROSS

NUT

14.5

M16 X 2 X 160 LONG

WASHER

18 X 38 X 3

BOLT

CENTER ON BASE

Ø 38

18

M16 X 2 THRU

Ø 12

Ø 13

60° CONE TIP 5 LONG.

38

Ø 25

PIN HOLE 2.5
PIN 2.5 X 25

Ø 50

110

M12 X 1.75 85 LONG

12

3

52

76

9  3

BASE

SHAFT

Ø 12.00-11.25

BRASS BUSHING
I.D. 12.00-12.25
O.D. 14.00-14.25
20 LONG.

METRIC

9

HANDLE
PLASTIC

52

26

62

31

12

HOLE IN HANDLE
FOR BUSHING
14.25-14.50

LOCK PLATE

M16 X 2 THRU

PROBLEM 18–19  Tool stop.

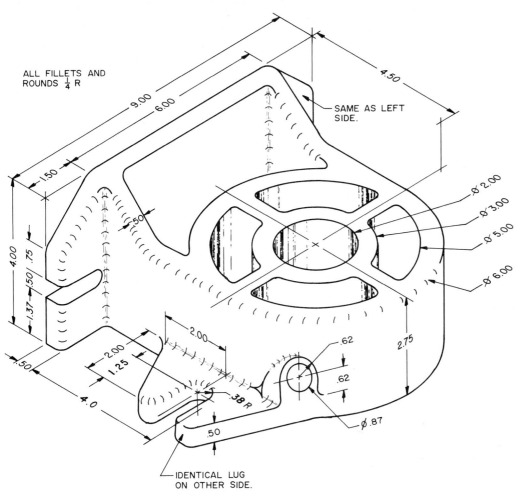

ALL FILLETS AND
ROUNDS $\frac{1}{4}$ R

9.00
6.00
4.50

SAME AS LEFT
SIDE.

1.50

4.00
1.37 .50 .75

.50

Ø 2.00
Ø 3.00
Ø 5.00
Ø 6.00

2.75

.50

2.00
2.00
1.25
4.0
.38 R

.62
.62

Ø .87

.50

IDENTICAL LUG
ON OTHER SIDE.

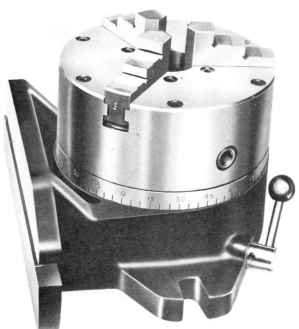

PROBLEM 18-20   Index table base. (Courtesy Kalamazoo Industries, Inc.)

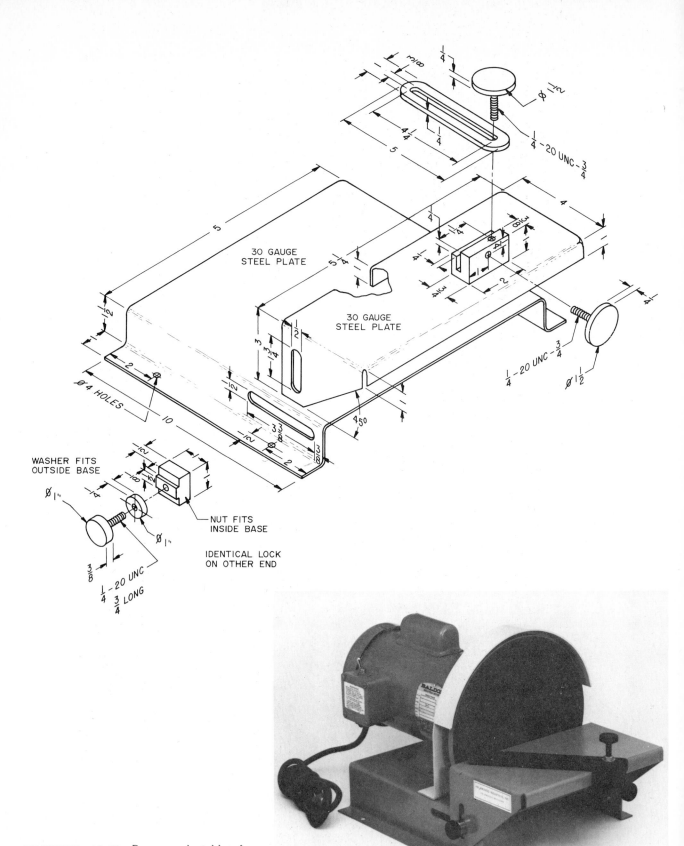

30 GAUGE STEEL PLATE

30 GAUGE STEEL PLATE

¼ - 20 UNC - ¾

¼ - 20 UNC - ¾

Ø 1½

Ø 4 HOLES

WASHER FITS OUTSIDE BASE

Ø 1"

Ø 1"

NUT FITS INSIDE BASE

IDENTICAL LOCK ON OTHER END

¼ - 20 UNC ¾ LONG

**PROBLEM** 18–21 Base and table for power disc sander. (Courtesy Kalamazoo Industries, Inc.)

# Developments

A development is the surface of an object unfolded and included in a plane. The development or pattern is designed so that when it is folded or rolled properly, it will form the object.

## GEOMETRIC SURFACES

A geometric surface is formed by the motion of a straight or curved line. One formed by a straight line is called a *ruled surface*. It may be a plane, single-curved, or warped surface (Fig. 19–1).

A *plane* is a ruled surface generated by a straight line moving with one point along a straight line and another point parallel with the straight line.

A *single-curved* surface is a surface that can be developed. It can be unrolled in a plane. A cylinder is a single-curved surface.

A *warped surface* is one that cannot be developed. No two adjacent positions of the generatrix lie in the same plane. An automobile hood is an example of a warped surface.

Surfaces formed by a curved line are called *double-curved surfaces*. They are formed by revolving a curved line about a straight line. A sphere is a double-curved surface.

The line generating the surface is called the

*generatrix*. Any position of the generatrix on a surface is called an *element*.

## GEOMETRIC SOLIDS

Solids are bounded by geometric surfaces. Those bounded by plane surfaces are called *polyhedra*. Examples include the pyramid and the prism. Those bounded by single-curved surfaces include a cone and a cylinder. Solids bounded by warped surfaces include the conoid and the hyperboloid. Those bounded by double-curved surfaces include the sphere and the torus (review Chapter 5).

## DEVELOPMENTS

If you look at the products you buy and use daily, you will notice an extensive use of parts formed from sheet materials. The exterior skin of light aircraft is generally formed from sheet material (Fig. 19–2). Even the lowly mailbox is a developed product. The designer must understand the types of developed surfaces and how they are designed. Many geometric shapes can be formed from flat stock. A pattern is developed that is used for laying out and cutting the part. The folds are made as indicated on the pattern (Fig. 19–3).

Developments can be laid out to produce either an inside or outside pattern. An inside pattern is one with the inside surfaces visible on the development. An outside pattern has the outside surfaces visible. Generally, patterns are inside because the fold lines marked on the surface are not visible when the product is folded. Also, the machines that fold the metal are generally designed to bend it inward. Always label the development inside or outside so there is no doubt when it is used.

Developments are built along a *stretchout line*. This is a long line from which measurements are taken.

The location of each plane at which the development is to be folded is shown with *fold lines*. Fold lines are shown as thin lines with an X across them.

The edges and fold lines on a development must be true length. If the drawing you are using to get the information to make the development does not show all the true lengths needed, you will have to find them by drawing auxiliary views or by revolution.

In addition to drawing the development, the designer must include some provision for holding the part together after it is formed. On many sheet metal products, this is some form of a tab. The de-

FIGURE 19–1 Geometric surfaces are formed by the motion of straight and curved lines.

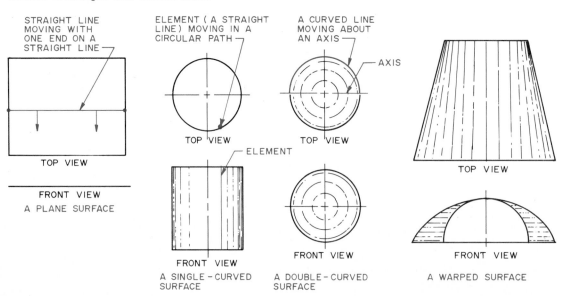

FIGURE 19–2 The skin of the aircraft is made of panels formed from sheet material. (Courtesy of Cessna Aircraft Company.)

signer should understand the use of sheet metal joints.

Usually, the pattern is developed so that the ends to be joined are the shortest. This enables the product be made with the shortest joint, reducing the amount of welding, adhesives, or rivets needed. Appearance sometimes influences where the joint is located. The pattern is laid out so that the joint is in a place where it is least likely to be seen on the finished product.

Some developed parts, such as those used on space vehicles, must be very accurate. Others, as in the mailbox, do not require much accuracy. The fol-

lowing material covers the procedures for drawing some of the more frequently used developments.

## TYPES OF DEVELOPMENTS

There are three basic types of developments—parallel line, radial line, and triangulation. *Parallel line development* is used for objects having parallel elements. Prisms and cylinders are examples. *Radial line* development has the fold lines coming from a point. Pyramids and cones are examples. *Triangulation* is a means of developing a surface by breaking it into a series of triangles. Objects made of a combination of curved and plane surfaces are developed by triangulation (Fig. 19–4).

*Warped surfaces* are those that cannot be laid out flat on a plane. They can be approximated using a series of triangles.

## PARALLEL LINE DEVELOPMENT

The geometric forms developed by parallel line development include prisms and cylinders (Fig. 19–5).

### Development of a Right Prism

A right prism has its base perpendicular to its sides. It produces square and rectangular forms (Fig. 19–6).

FIGURE 19–3 A pattern drawing for a sheet metal product.

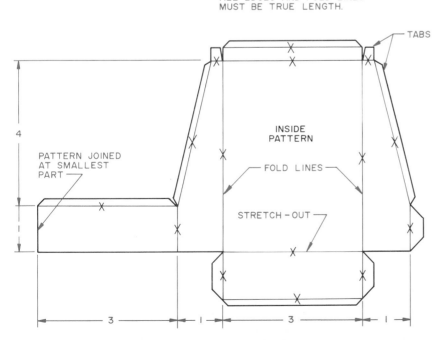

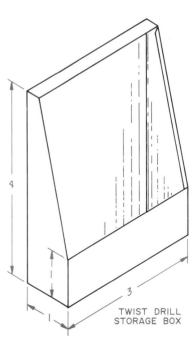

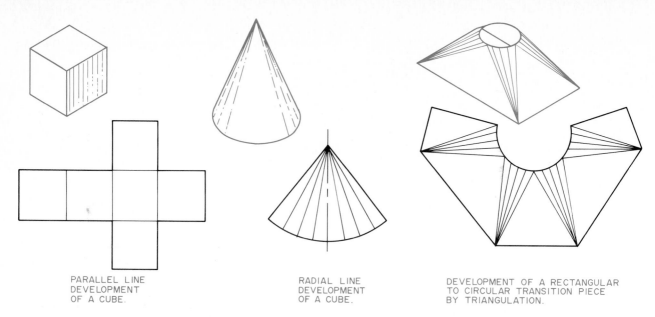

PARALLEL LINE
DEVELOPMENT
OF A CUBE.

RADIAL LINE
DEVELOPMENT
OF A CUBE.

DEVELOPMENT OF A RECTANGULAR
TO CIRCULAR TRANSITION PIECE
BY TRIANGULATION.

FIGURE 19-4  The three basic types of developments.

FIGURE 19-5  A cylinder is a type of parallel line development. (Courtesy of Phillips Petroleum Co.)

FIGURE 19-6  The basic shape of this display module is a rectangular prism. (Courtesy of Hewlett-Packard.)

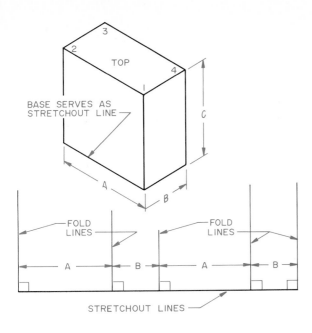

Assignment: Make a development of a right prism.

1. Draw the stretchout line. Locate each distance and draw the fold lines.

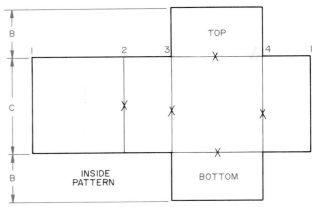

2. Measure the true length of the fold lines. Locate the top and bottom of the object. Connect the points to complete the pattern.

FIGURE 19–7   Drawing a development of a right prism.

Following are the steps for producing a development of a right prism (Fig. 19–7).

1. Draw the stretchout line. Its length is equal to the sum of the width of the four sides of the prism. Locate each distance on the stretchout line.
2. Draw fold lines perpendicular to the stretchout line at each mark. These are fold lines. Measure the true length of each corner of the prism on the fold lines.
3. Connect the marks to complete the pattern.

## Development of a Truncated Right Prism

A truncated geometric form has one of its surfaces cut on an angle (Fig. 19–8). The steps for developing a truncated right prism are the same as described for the right prism. The difference is that the lengths of the corners recorded on the fold lines will vary, depending on the corner. Notice that not all lines in the top view of the truncated right prism are true length. These can be found on other views, with no auxiliary view needed.

## Development of an Oblique Prism

An oblique prism has its sides at an angle other than 90° to the base. The development of an oblique prism is much the same as a right prism. The major difference is that the pattern does not unfold in a straight line. The stretchout line is located in the center of the prism.

Following are the steps for producing a development of an oblique prism (Fig. 19–9).

1. Draw the top and front views of the prism. Draw an auxiliary section cut perpendicular to the sides of the prism. This gives the true width of the sides of the prism.
2. To start the development, draw the stretchout line perpendicular to the side of the prism.
3. Measure the true width of each face on the auxiliary section. Draw fold lines at these points perpendicular to the stretchout line.
4. Lay out the true length of each corner on successive fold lines. These are projected from the front view of the prism. Connect the points to complete the development.

## Development of a Right Cylinder

A right cylinder has its axis perpendicular to the base. Following are the steps for producing a development of a right cylinder (Fig. 19–10).

1. Draw two views of the cylinder. One must show the true height and the other the true-size circular view.
2. Divide the circular view into an equal number of parts. The more parts, the more accurate the development. Number each of the parts.
3. Draw the stretchout line. On the left, draw a perpendicular representing the end of the development. Make it the same length as the true height of the cylinder.

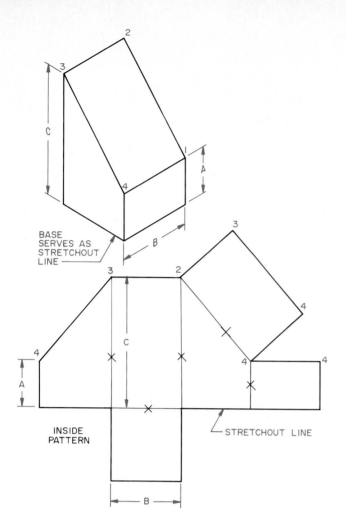

BASE
SERVES AS
STRETCHOUT
LINE

INSIDE
PATTERN

STRETCHOUT LINE

FIGURE 19–8 The development of a truncated right prism.

4. With dividers, measure the straight-line distance between two divisions on the circular view.

5. On the stretchout line, step off as many steps as there are divisions on the circular view. This gives the circumference.

6. Draw the true height at the last division. Connect the lines to complete the development.

A more accurate method for finding the circumference is to calculate it. The circumference of a circle is equal to the diameter multiplied by pi ($\pi$). The value of $\pi$ is equal to 3.1416. If a circle has a diameter of 10, the circumference is equal to 10 × 3.1416, or 31.416.

## Development of a Truncated Right Cylinder

A truncated right cylinder is one that is cut on an angle other than 90° with its axis. This will produce a curved line in the development.

Following are the steps for producing a development of a truncated right cylinder (Fig. 19–11).

FIGURE 19–9 Drawing a development of an oblique prism.

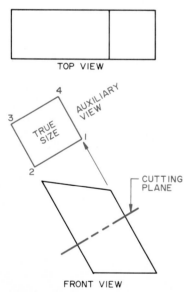

TOP VIEW

AUXILIARY VIEW

TRUE SIZE

CUTTING PLANE

FRONT VIEW

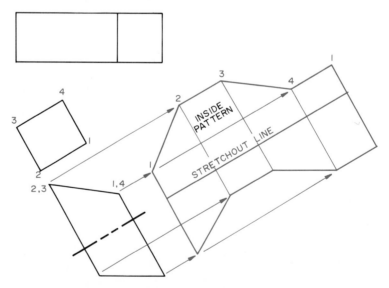

INSIDE PATTERN

STRETCHOUT LINE

Assignment: Make a development of an oblique prism.

1. Draw the top and front views of the prism. Draw an auxiliary section showing the true size of the prism.

2. Draw the stretchout line perpendicular to the sides of the prism. Project it from the cutting plane. Locate each fold line by securing the true distances from the auxiliary section. Lay out the true lengths of each fold by projecting them from the front view. Keep them in succession. Connect the points to complete the development.

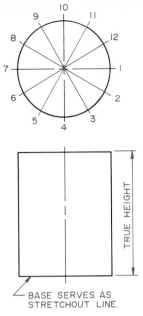

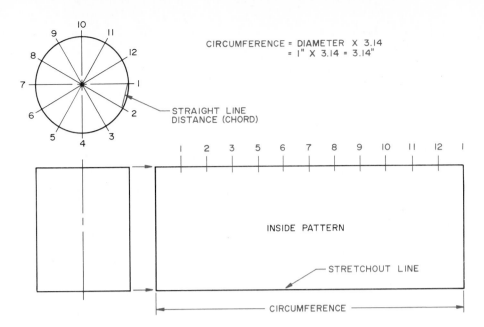

CIRCUMFERENCE = DIAMETER X 3.14
= 1" X 3.14 = 3.14"

STRAIGHT LINE
DISTANCE (CHORD)

INSIDE PATTERN

STRETCHOUT LINE

CIRCUMFERENCE

TRUE HEIGHT

BASE SERVES AS
STRETCHOUT LINE.

Assignment: Make a
development of a right
cylinder.

1. Draw the top and front
views. Divide the circular
view into a number of equal
parts.

2. Draw the stretchout line.
Lay out the straight-line
(chord) distance between
each point in the circular
view or compute the
circumference. Measure the
true height. Draw the edges
to complete the
development.

**FIGURE 19–10** Drawing a development of a right cylinder.

**FIGURE 19–11** Drawing a development of a truncated right cylinder.

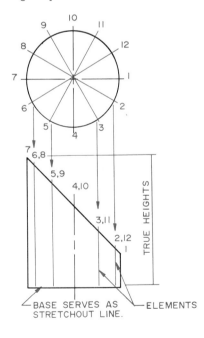

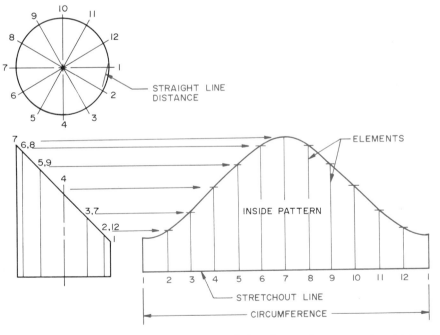

STRAIGHT LINE
DISTANCE

ELEMENTS

INSIDE PATTERN

STRETCHOUT LINE

CIRCUMFERENCE

TRUE HEIGHTS

BASE SERVES AS
STRETCHOUT LINE.

ELEMENTS

Assignment: Make a development
of a truncated right cylinder.

1. Draw the top and front views.
Divide the circular view into a
number of equal parts. Project
these to the front view. These form
elements on the surface of the
cylinder.

2. Draw the stretchout line. Lay out the straight-line distance between each point found
in the circular view. Draw lines through these perpendicular to the stretchout line.
Project the true height of each element in succession from the front view to the
development. Connect the points to form the development.

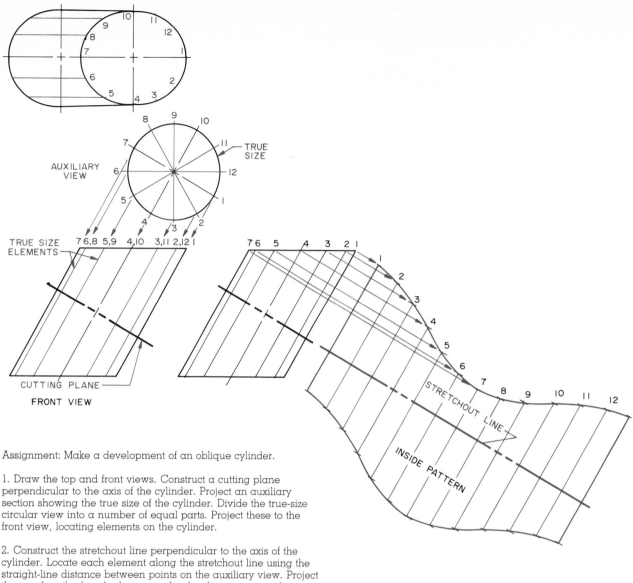

Assignment: Make a development of an oblique cylinder.

1. Draw the top and front views. Construct a cutting plane perpendicular to the axis of the cylinder. Project an auxiliary section showing the true size of the cylinder. Divide the true-size circular view into a number of equal parts. Project these to the front view, locating elements on the cylinder.

2. Construct the stretchout line perpendicular to the axis of the cylinder. Locate each element along the stretchout line using the straight-line distance between points on the auxiliary view. Project the true length of each element to the development, recording each in succession. Connect the points to form the development.

FIGURE 19–12   Drawing a development of an oblique cylinder.

1. Prepare the layout as described for a right cylinder. Project the divisions in the circular view to the front view, forming elements on the cylinder. Locate the stretchout line perpendicular to the axis of the cylinder and in line with the base.

2. Project each element from the circular view to the front view. Where they cross the inclined surface, project to development. Start the development with the shortest edge.

3. Connect the ends of the vertical layout lines with an irregular curve.

## Development of an Oblique Cylinder

An oblique cylinder has its axis on an angle other than 90° to its base. Following are the steps for pro-

ducing the development of an oblique cylinder (Fig. 19–12).

1. Draw the circular and front view of the oblique cylinder. The true-size circular view will be a primary auxiliary. The line of sight will be parallel with the axis of the cylinder.

2. The cutting plane for the auxiliary view can serve as the stretchout line. Draw the stretchout line equal to the circumference of the cylinder and perpendicular to the axis of the cylinder. Line it up with the stretchout line location on the cylinder.

3. Divide the true-size circular view into a number of equal parts. Project these to the front view.

4. Locate each division on the circular view on the

**486**   Chapter 19

stretchout line. Draw lines through each perpendicular to the stretchout line.

5. Project the length of each division to the development. Connect these with an irregular curve.

# RADIAL LINE DEVELOPMENT

The geometric forms developed by radial line developments include pyramids and cones.

## Development of a Right Rectangular Pyramid

A right rectangular pyramid has its axis perpendicular to its base. Following are the steps for producing a development of a right rectangular pyramid (Fig. 19–13).

1. Draw the top and front views.
2. Find the true length of the intersection between two surfaces by revolution.
3. Locate the point, A, to which the radial lines will converge. Set a compass on a radius equal to the true length of the intersection between two surfaces. Swing an arc from point A.
4. Draw a line from point A to one end of the arc. From this intersection, step off the lengths of the four

sides at the base. Start at one corner of the pyramid and move in sequence around the base.

5. Connect each of these points with the vertex at A. This locates the fold lines. Connect the points on the base to complete the development.

## Development of a Frustum of a Right Pyramid

A frustum of a right pyramid has its top cut away by a plane that is parallel with the base (Fig. 19–14). Following are the steps for producing a development of a frustum of a right pyramid (Fig. 19–15).

1. Lay out the pyramid as explained for a right pyramid.
2. Find the true length of each edge formed by the intersection of two surfaces by revolution. Rotate it from the vertex of the pyramid cut away from the frustum.
3. Start the development by locating the vertex of the pyramid, using the true length of the intersection as the radius. Locate each fold line on the arc. Draw the fold lines. Using the true lengths of the removed section of the intersections, locate the top end of each fold line.
4. Connect the points on each end of the fold lines to form the development.

FIGURE 19–13 Drawing a development of a right rectangular pyramid.

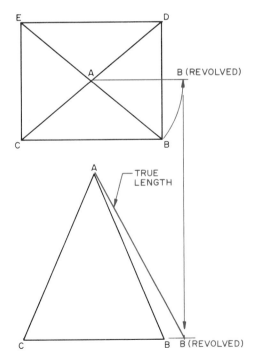

Assignment: Make a development of a right rectangular pyramid.

1. Draw the top and front views. Then, find the true length of the line of intersection between the surfaces by revolution.

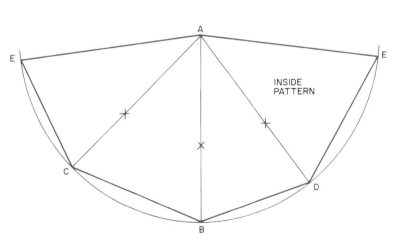

2. Locate the point of the pyramid, A. Using it as a center, swing an arc with a radius equal to the true length of the line of intersection. Connect point A with one end of the arc. Step off the lengths of the four sides along the arc. Connect these points to form the development.

FIGURE 19-15 Drawing a development of a frustum of a right pyramid.

## Development of a Truncated Right Pyramid

A truncated right pyramid is one that is cut on an angle other than 90° to its base. Following are the steps for producing a development of a truncated right pyramid (Fig. 19-16).

1. Lay out the pyramid as explained for a right frustum of a pyramid. The only difference is that the top plane slopes, so its corners are different distances from the vertex. Find the true lengths of these by revolution.
2. Locate these true-length edges on the fold lines.
3. Connect the ends to form the development.

## Development of an Oblique Pyramid

An oblique pyramid has its axis at an angle other than 90° to its base. Following are the steps for producing a development of an oblique pyramid (Fig. 19-17).

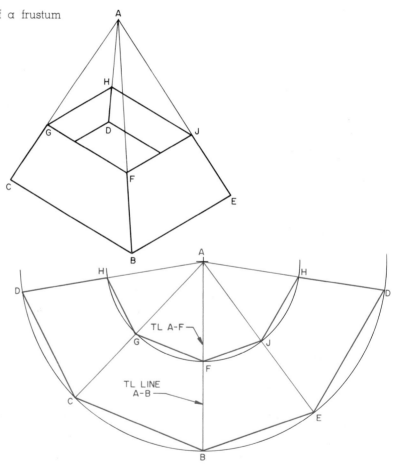

Assignment: Make a development of a frustum of a right pyramid.

1. Draw the top and front views. Find the true length of the line of intersection between two faces by revolution. This is line A–B. Then, find the true length of line A–F, which is cut away by the top plane.

2. Locate the vertex, A, of the pyramid. Using it as a center, swing an arc with a radius equal to the true length of line A–B. Step off the length of the sides of the base on the arc. This locates the fold lines, which converge on point A. Step off the true length of line A–F on each fold line. Connect the points to form the development.

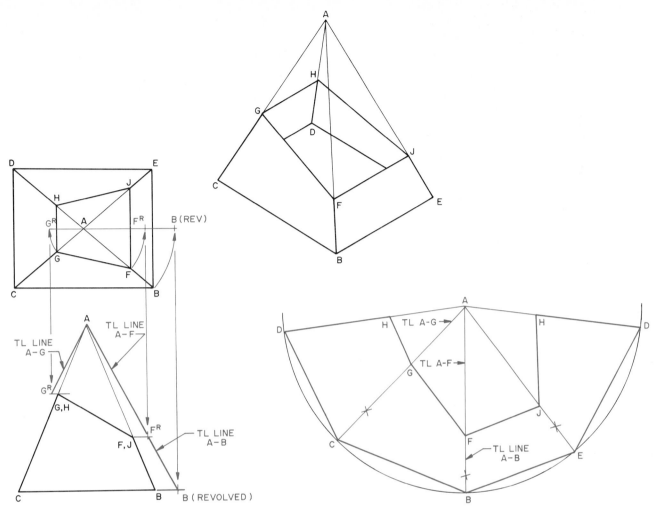

Assignment: Make a development of a truncated right pyramid.

1. Draw the top and front views. Find the true length of the line of intersection between two faces by revolution. This is line A–B. Also find the true length of each line removed by the cut forming the truncated top. These are lines A–F and A–G.

2. Locate the vertex of the pyramid. Using it as a center, swing an arc with a radius equal to the true length of line A–B. Step off the lengths of the base on the arc to locate the fold lines. Then, step off the true-length lines to locate the top end of each fold line. Connect the ends to form the development.

FIGURE 19–16 Drawing a development of a truncated right pyramid.

1. Draw the top and front views. Find the true lengths of the fold lines by revolution.
2. Start the development by locating the vertex of the pyramid and drawing one true-length fold line from it. The true length of the sides of the base are in the top view. Using as a radius the length of the base that touches the fold line drawn, swing a short arc. From the vertex, swing another arc whose length is equal to the next fold line. The point at which it crosses the arc just drawn is a point on the development. Continue in succession around the pyramid until all fold lines are located.
3. Connect the points to form the development. If the pyramid is truncated, find the top ends of the fold lines as explained in the development of a truncated pyramid.

## Development of a Right Cone

A right cone has its axis perpendicular to the base (Fig. 19–18). Following are the steps for producing a development of a right cone (Fig. 19–19).

1. Draw the top and front views of the cone.
2. Divide the circular view into a number of equal parts.
3. Project these parts to the base of the cone on the front view.
4. Draw elements from these points on the base to the vertex of the cone. The only elements that are true length are those that are parallel with the frontal plane.

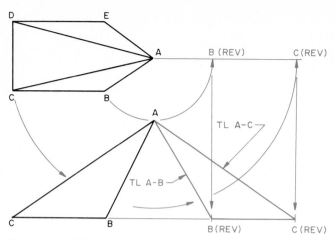

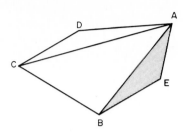

Assignment: Make a development of an oblique pyramid.

1. Draw the top and front views. Find the true length of each fold line by revolution.

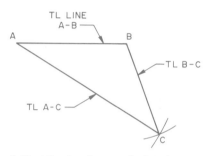

2. Start the development by locating the vertex, A. From it, draw the true length of one fold line, A–B. From B, swing a short arc with a radius equal to the length of the base, B–C. From A, swing a short arc with a radius equal to the next fold line, A–C, on the pyramid. The point at which the arcs cross is a point on the development.

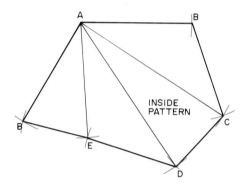

3. Continue in this manner until all fold lines are located in succession. Connect the points to form the development.

FIGURE 19–17  Drawing a development of an oblique pyramid.

FIGURE 19–18  This funnel is made of two right cones.

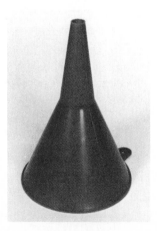

5. Locate the vertex of the cone in the area where the development will be drawn. Swing an arc from the vertex, using the true-length element as a radius.

6. Set dividers on the straight-line distance between two of the points located on the circular view. Step off this distance along the arc until there are the same number of points on the arc as there are on the circle. This produces an approximation of the circumference of the base of the cone. A more accurate method would be to calculate the actual circumference and divide it by the number of segments.

7. Connect the ends of the arc with point A, completing the development of the cone.

## Development of a Frustum of a Cone

A frustum of a cone is formed when the vertex of the cone is cut off by a plane parallel with the base.

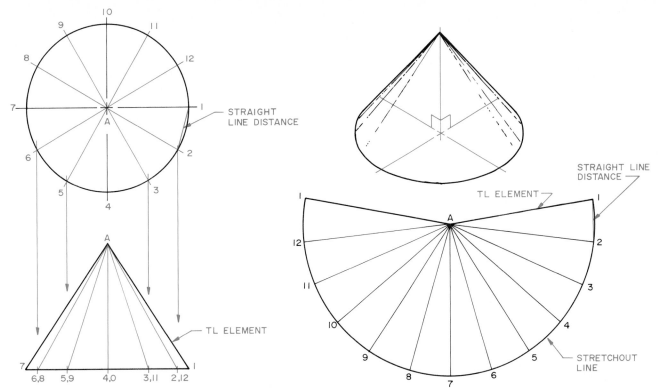

Assignment: Make a development of a right cone.

1. Draw the top and front views. Divide the circular view into a number of equal parts. Project these to the base of the cone in the front view. Then, connect them to the vertex to form elements on the surface of the cone. The elements parallel with the frontal plane are true length.

FIGURE 19–19 Drawing a development of a right cone.

2. Locate the vertex, A. Using it as a center, swing an arc with a radius equal to the true length of the element. This is the stretchout line. Lay off on the stretchout line the straight-line distance between the divisions on the circular view to locate the elements. Connect these to the vertex to form the development.

Following are the steps for producing the development of a frustum of a cone (Fig. 19–20).

1. Lay out the development for a right cone.
2. Using dividers, measure the true-length distance from the vertex of the cone to the top plane. Mark this on the development.
3. Draw the arc for the edge of the top plane.

### Development of a Truncated Right Cone

A truncated right cone has the top cut on an angle other than 90° to its axis. Following are the steps for producing the development of a truncated right cone (Fig. 19–21).

1. Lay out the development for a right cone.
2. Find the true length of each element between the vertex and the truncated surface by revolution.
3. Locate these true lengths of the elements on the development. Connect them with an irregular curve to complete the development.

## TRIANGULATION

Triangulation is used to develop surfaces that cannot be developed by the parallel line or radial line methods. Surfaces developed by triangulation are approximate.

### Development of an Oblique Cone

An oblique cone has an axis at an angle to its base other than 90°. Following are the steps for producing the development of an oblique cone (Fig. 19–22).

1. Draw the top and front views. Divide the base into a number of equal parts. Project these to the front view and then to the vertex to form elements on the surface. These form triangles on the surface. Now, find the true length of each element. Because the cone is oblique, each element in half the cone will have a different length. Find the true lengths by revolution.
2. Start the development by locating the vertex. Draw

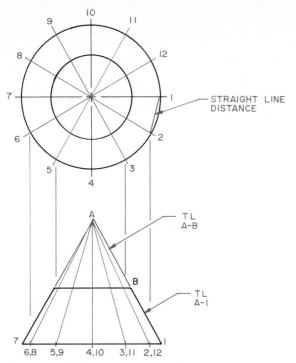

Assignment: Make a development of a frustum of a cone.

1. Draw the top and front views. Divide the circular view into a number of equal parts. Project these to the base of the cone in the front view. Connect them to the vertex to form elements on the surface of the cone. The elements parallel with the frontal plane are true length.

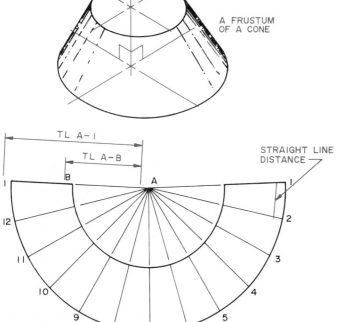

2. Locate the vertex, A. Using it as a center, swing an arc with a radius equal to the true length of the element from the vertex to the base, A–1. This is the stretchout line. Swing a second arc with a radius equal to the distance from the vertex to the plane cutting off the top of the cone, AB. Lay off on the stretchout line the straight-line distance between two divisions in the circular view to locate each element. Connect the elements to the vertex to form the development.

**FIGURE 19–20** Drawing a development of a frustum of a cone.

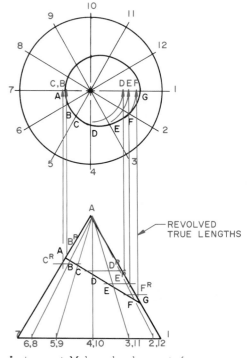

Assignment: Make a development of a truncated right cone.

1. Draw the development of the cone as described for a right cone. By revolution, find the true length of each of the elements forming the top of the cone, which is removed.

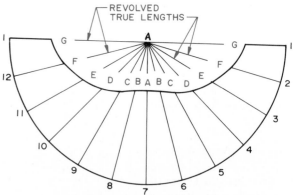

2. Measure these from the vertex on the development along the elements. Be certain to get them on the proper element. Connect the points to form the development.

**FIGURE 19–21** Drawing a development of a truncated right cone.

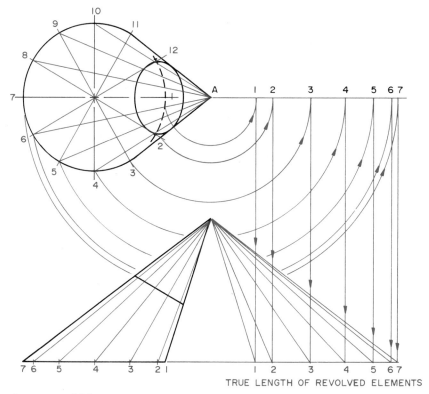

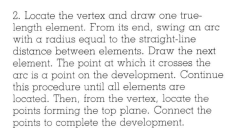

HALF THE DEVELOPMENT

TRUE LENGTH OF REVOLVED ELEMENTS

Assignment: Make a development of an oblique cone.

1. Draw the top and front views of the cone. Divide the base into a number of equal parts. Project these to the front view and then to the vertex. Find the true length of each element by revolution.

2. Locate the vertex and draw one true-length element. From its end, swing an arc with a radius equal to the straight-line distance between elements. Draw the next element. The point at which it crosses the arc is a point on the development. Continue this procedure until all elements are located. Then, from the vertex, locate the points forming the top plane. Connect the points to complete the development.

**FIGURE 19–22** Drawing a development of an oblique cone.

one true length from the vertex. From its end, swing an arc with a radius equal to the straight-line distance between two elements on the circular view. Then, draw the next element. The point at which it crosses the arc is a point on the development. Continue to locate each element in succession, forming a series of triangles. Connect the points to form the development.

## Development of a Warped Surface

A warped surface cannot be accurately developed on a flat sheet. The object in Fig. 19–23 has a warped surface. It can be developed approximately by triangulation.

Following are steps for producing the development of a warped surface by triangulation (Fig. 19–24).

**FIGURE 19–23** This technician is producing a full-size mock-up of an aircraft engine housing. It is a warped surface. (Courtesy of Cessna Aircraft Company.)

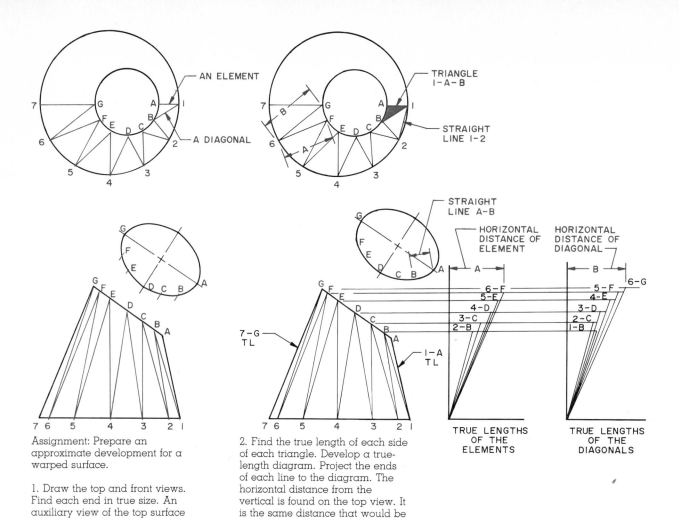

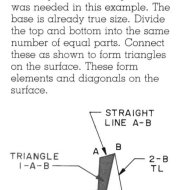

Assignment: Prepare an approximate development for a warped surface.

1. Draw the top and front views. Find each end in true size. An auxiliary view of the top surface was needed in this example. The base is already true size. Divide the top and bottom into the same number of equal parts. Connect these as shown to form triangles on the surface. These form elements and diagonals on the surface.

2. Find the true length of each side of each triangle. Develop a true-length diagram. Project the ends of each line to the diagram. The horizontal distance from the vertical is found on the top view. It is the same distance that would be found if the end of the line were rotated.

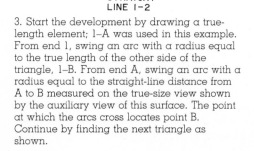

3. Start the development by drawing a true-length element; 1–A was used in this example. From end 1, swing an arc with a radius equal to the true length of the other side of the triangle, 1–B. From end A, swing an arc with a radius equal to the straight-line distance from A to B measured on the true-size view shown by the auxiliary view of this surface. The point at which the arcs cross locates point B. Continue by finding the next triangle as shown.

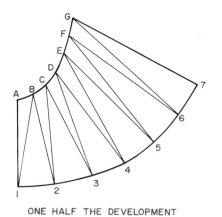

ONE HALF THE DEVELOPMENT

4. Repeat this procedure until all elements and diagonals are located. Connect their ends with an irregular curve to complete the development.

FIGURE 19–24 Drawing a development of a warped surface by triangulation.

494

1. Draw the top and front views.

2. Find the true size of the bottom and top openings.

3. Divide each true-size opening into the same number of equal parts. Join each division in the top with those on the bottom, forming triangles on the surface. The lines connecting alike points on each end of the object are elements, such as 1–A. The other lines, such as 1–B, are diagonals that form the triangles. They are actually curved but are drawn as though they are straight.

4. Find the true length of the sides of each triangle. This can be done by developing a true-length diagram. Locate one of the points on the base as the beginning of the diagram. Point 1 was used in this example. Then, in the top view, measure the horizontal distance from that base point to a point on the top to which it connects. Point B is an example. Locate point B on the diagram. Line B is true length. Repeat this for each side of each triangle. It is helpful to divide the sides into two diagrams to reduce the confusion of line. In this example, the true lengths of the elements were found in one diagram and the diagonals in another.

5. Start the development by drawing a true-length element. From one end, swing an arc having a radius equal to the true length of the next side of the triangle. From the other end, swing an arc having a radius equal to the straight-line distance between the other ends of the triangle. This must be measured on a true-size view. This forms one triangle. Repeat this procedure until all triangles are located. Connect the points with an irregular curve to form the development.

The development of warped (or compound) surfaces is more complex than presented by this method of approximating a surface. It is a critical design factor in aircraft and space vehicle development. Computers are used extensively to produce an accurate loft (layout) of the surface.

### Transition Pieces

A transition piece is a unit designed to connect pipes of different sizes or shapes. Figure 19–25 shows some typical examples. Transition pieces are developed by triangulation.

### Development of a Square-to-Square Transition Piece with the Same Axis

A square-to-square transition piece connects two rectangular pipes of different sizes. Actually, the transition piece is a frustum of a pyramid and can be developed as described earlier (Fig. 19–26).

### Development of a Circular-to-Rectangular Transition Piece with the Same Axis

The transition piece is developed by forming four conical surfaces and four isosceles triangles whose

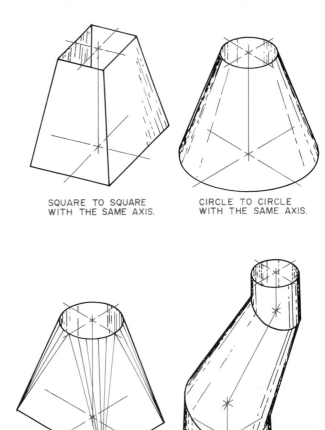

SQUARE TO SQUARE WITH THE SAME AXIS.

CIRCLE TO CIRCLE WITH THE SAME AXIS.

RECTANGLE TO CIRCLE WITH THE SAME AXIS.

CIRCLE TO CIRCLE WITH DIFFERENT AXES.

FIGURE 19–25 Transition pieces connect pipes of different sizes and shapes.

bases form the base of the transition piece (Fig. 19–27).

Following are the steps for developing this transition piece by triangulation (Fig. 19–28).

1. Draw the top and front views.

2. Divide the circular pipe in the top view into a number of equal parts. Connect these to the corners with elements.

3. Project the elements to the front view.

4. Find the true length of the elements by revolution. The true lengths of the sides of the base appear in the top view.

5. Start the development by locating the first true-length edge. Usually, this is an edge cutting an isosceles triangle in half. With a compass, swing an arc the length of half the base. From the other end of the line, swing the first true-length element until it crosses the base arc. Connect the base.

6. Swing an arc equal to the straight-line distance between two points on the circle. Swing an arc equal to the true length of the second element. The point at which they cross is a point on the development.

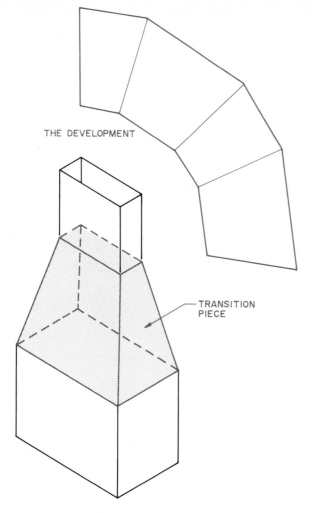

THE DEVELOPMENT

TRANSITION PIECE

**FIGURE 19-26** A rectangular transition piece.

**FIGURE 19-27** A circular-to-rectangular transition piece is made up of conical and triangular surfaces. (Courtesy of McDermott Incorporated).

7. Continue this procedure until all bases and elements are on the development. Connect the points to complete the development.

### Development of a Circle-to-Circle Transition Piece with the Same Axis

A circle-to-circle transition piece connects two pipes of different diameters. The transition piece is a frustum of a cone, which is developed as explained earlier (Fig. 19-29).

### Development of a Circle-to-Circle Transition Piece with Different Axes

This transition piece is actually a frustum of a cone. The vertex of the cone is found by extending the elements until they meet. The procedure for finding the development follows the plan given for the development of an oblique cone using triangulation.

Following are the steps for producing a circle-to-circle transition piece with different axes (Fig. 19-30).

1. Draw the top and front views.
2. Divide the circular view into a number of equal parts and project them to the front view. Since the circle is not the true size of the intersection, the true-size circle is found by revolution.
3. Revolve each element and find its true length.
4. Start the development by locating the vertex. Then, draw a true-length line representing the element chosen to start the development. With a compass, measure the true distance on the circular view between the elements on the base. Swing this arc from the end of the first element drawn. From the vertex, swing another arc the true length of the second element. The point at which it crosses the arc is a point on the development. Continue this until all the elements have been located on the development. Connect the points to complete the development.

### Development of a Sphere

The sphere is a strong shape often used where exterior pressures are great (Fig. 19-31). It forms an excellent storage tank and is used in various applications in the petroleum and chemical industries.

A sphere has a *double-curved surface* and is not developable on a flat surface. There are several methods for laying out an approximate sphere on a flat surface. One of these uses gores. A gore is a spherical element bounded by curved edges that run from one vertex to another.

Following are the steps for producing a development for an approximate sphere using gores (Fig. 19-32).

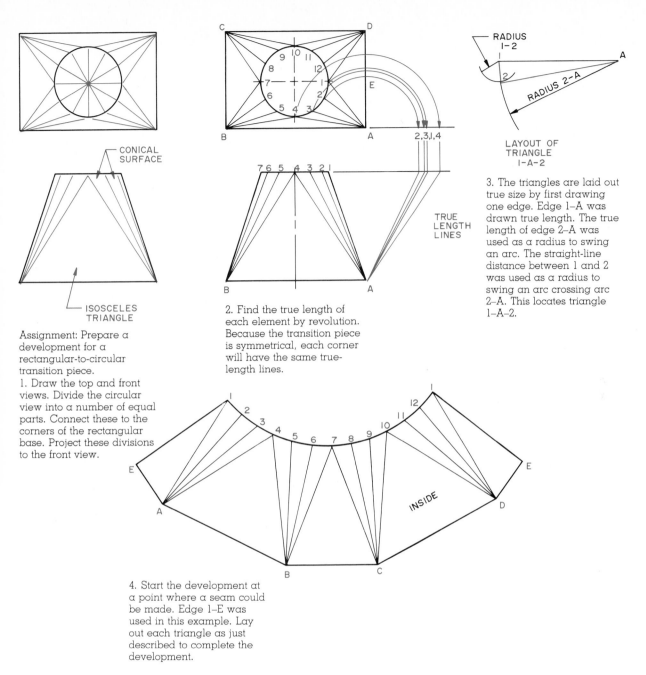

CONICAL
SURFACE

ISOSCELES
TRIANGLE

Assignment: Prepare a
development for a
rectangular-to-circular
transition piece.

1. Draw the top and front
views. Divide the circular
view into a number of equal
parts. Connect these to the
corners of the rectangular
base. Project these divisions
to the front view.

2. Find the true length of
each element by revolution.
Because the transition piece
is symmetrical, each corner
will have the same true-
length lines.

RADIUS
1-2

RADIUS 2-A

LAYOUT OF
TRIANGLE
1-A-2

3. The triangles are laid out
true size by first drawing
one edge. Edge 1-A was
drawn true length. The true
length of edge 2-A was
used as a radius to swing
an arc. The straight-line
distance between 1 and 2
was used as a radius to
swing an arc crossing arc
2-A. This locates triangle
1-A-2.

TRUE
LENGTH
LINES

INSIDE

4. Start the development at
a point where a seam could
be made. Edge 1-E was
used in this example. Lay
out each triangle as just
described to complete the
development.

FIGURE 19–28   Drawing a circular-to-rectangular transition piece by triangulation.

1. Draw the top and front views of the sphere.

2. Divide the top view into a number of equal parts (12
   parts are commonly used). The area between two
   elements is a gore.

3. Divide the front view into a number of horizontal
   planes.

4. Calculate the circumference of the sphere. The
   length of a gore is half the circumference.

5. Lay out a line representing the true length of the
   gore. Divide it into parts equal to the distances between the horizontal divisions. These are a part of

the circumference. If 8 divisions are used, the length
of each is equal to one eighth of half the circumference. Then, draw the width of each division. The
straight-line distance between the vertical divisions
can be measured on the top view. A more accurate
method would be to calculate the circumference of
each horizontal division and divide it by the number
of gores.

6. Connect the points to form the development of a
   gore. If the sphere is divided into 12 parts, it will
   take 12 identical gores to produce the approximate
   development.

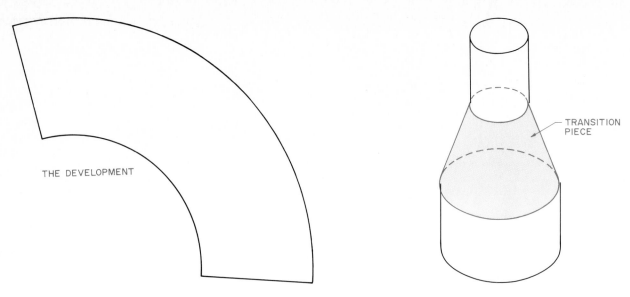

THE DEVELOPMENT

TRANSITION PIECE

FIGURE 19–29   A circle-to-circle transition piece.

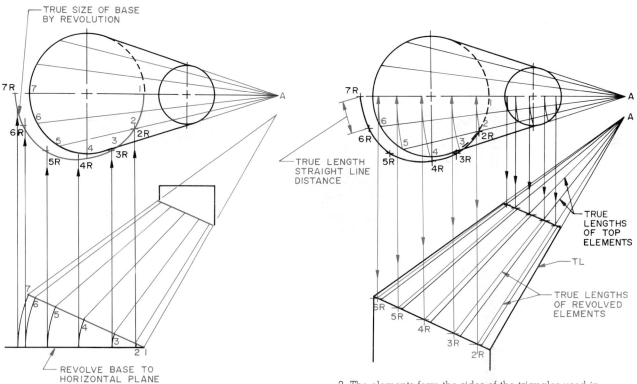

TRUE SIZE OF BASE BY REVOLUTION

7R

6R

5R   4R   3R

REVOLVE BASE TO HORIZONTAL PLANE

TRUE LENGTH STRAIGHT LINE DISTANCE

7R

6R

5R   4R   3R

TRUE LENGTHS OF TOP ELEMENTS

TL

TRUE LENGTHS OF REVOLVED ELEMENTS

5R   5R   4R   3R   2R

Assignment: Prepare a development for a transition piece to connect two circular pipes of different diameters having different axes.

1. Draw the top and front views. Extend the sides to find the vertex. Divide the circular base into a number of equal parts. Extend these to form elements. Project the elements to the front view.

The development is made by triangulation. The true-length straight-line distances between the divisions on the base are needed. Revolve the base parallel with the horizontal plane. Project the revolved points to the top view to form a true-size view of the base.

2. The elements form the sides of the triangles used in triangulation. They are found true length by revolving them parallel with the frontal plane. Project them to the front view to find their true lengths.

3. Start the development by laying off a true-length element. In this example, A–1 was used. Then, swing an arc for the next element, A–2. The straight-line distance between 1 and 2 locates point 2. Continue locating each element. The top of the transition piece is located by finding the true lengths of the top parts of each element, which are not part of the transition piece.

FIGURE 19–30   Drawing a transition piece to connect two circular pipes having different diameters.

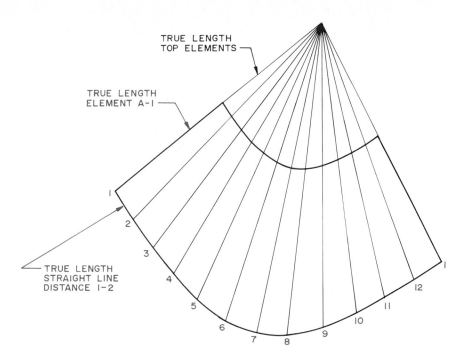

TRUE LENGTH
TOP ELEMENTS

TRUE LENGTH
ELEMENT A-1

TRUE LENGTH
STRAIGHT LINE
DISTANCE 1-2

FIGURE 19–30 *Cont.*

FIGURE 19–31 A sphere has many industrial applications. (Courtesy of Phillips Petroleum Co.)

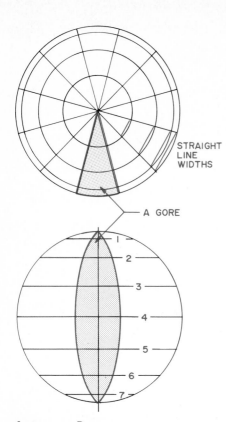

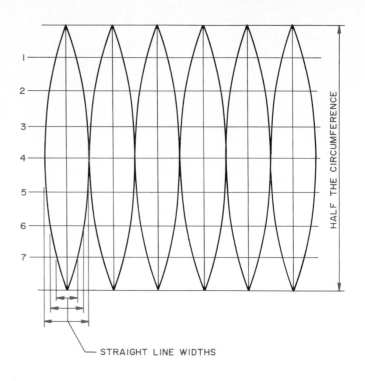

STRAIGHT
LINE
WIDTHS

A GORE

HALF THE CIRCUMFERENCE

STRAIGHT LINE WIDTHS

DEVELOPMENT FOR HALF A SPHERE

Assignment: Prepare an approximate development for a sphere using gores.

1. Draw the top and front views. Divide the top view into a number of equal parts. Divide the front view into a number of horizontal parts.

2. Lay out half the circumference. This is the center line of a gore. Divide it into the same number of equal parts as there are horizontal divisions. Number these horizontal parts to match those on the front view. On each horizontal line, lay out the straight-line width of the gore at each horizontal level. Connect these to form the gore. Repeat this for each gore.

FIGURE 19–32  Drawing an approximate development for a sphere using gores.

Another way to draw an approximate development of a sphere is to divide it into several horizontal layers that form a frustum of a cone. A development of each cone is made.

Following are the steps for drawing an approximate development of a sphere by dividing it into a series of cones (Fig. 19–33).

1. Draw the top and front views. Divide the top view into a number of equal parts. Divide the front view into a number of equal horizontal divisions. It is eas-

iest if the surface of the sphere is divided into equal parts. This divides the sphere into cones that are frustums.

2. The development involves making a development of each cone. The radius for each cone is found by extending the side forming the cone to a vertex. The circumference is found by stepping off the straight-line distances on each circumference as found in the top view. It is easiest to begin by drawing the development of the largest cone first. The width of the development is the straight-line width of the horizontal sections.

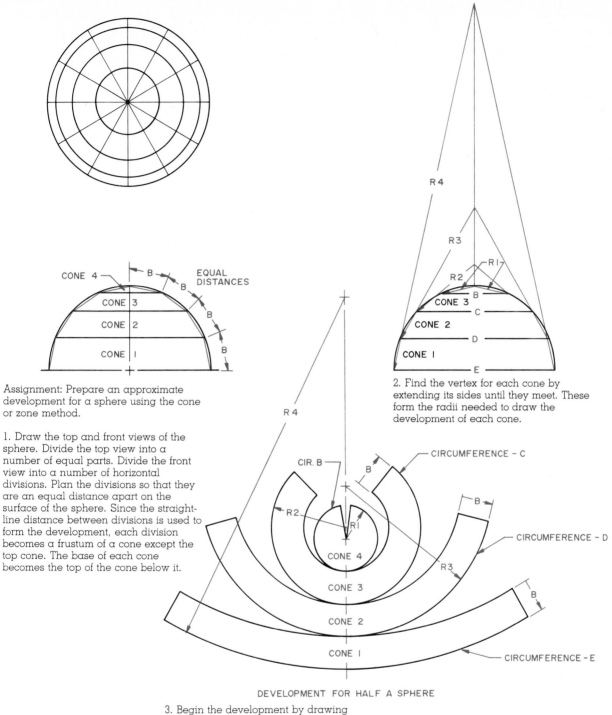

CONE 4
CONE 3
CONE 2
CONE 1
B
B
B
B
B
EQUAL
DISTANCES

Assignment: Prepare an approximate development for a sphere using the cone or zone method.

1. Draw the top and front views of the sphere. Divide the top view into a number of equal parts. Divide the front view into a number of horizontal divisions. Plan the divisions so that they are an equal distance apart on the surface of the sphere. Since the straight-line distance between divisions is used to form the development, each division becomes a frustum of a cone except the top cone. The base of each cone becomes the top of the cone below it.

R 4
R 3
R 1
R 2
CONE 3
CONE 2
CONE 1
B
B
C
D
E

2. Find the vertex for each cone by extending its sides until they meet. These form the radii needed to draw the development of each cone.

R 4
CIR. B
R2
R1
R3
B
B
B
CONE 4
CONE 3
CONE 2
CONE 1
CIRCUMFERENCE – C
CIRCUMFERENCE – D
CIRCUMFERENCE – E

DEVELOPMENT FOR HALF A SPHERE

3. Begin the development by drawing the largest cone first. This is cone 1 in the example. Then draw cone 2. The radius is shown in step 2. The circumference is found by stepping off the straight-line distance of its base. The top of the frustum of the cone is distance B from the base.

FIGURE 19–33  Drawing an approximate development for a sphere using the cone method.

# Problems

Use an A or A4 sheet for these problems unless directed otherwise.

1. Draw the views shown and develop the exterior surfaces of the prisms in Prob. 19–1.

2. Draw the views shown and develop the exterior surfaces of the prisms in Prob. 19–2.

3. Draw the views shown and develop the exterior surfaces of the cylinders in Prob. 19–3.

4. Draw the views shown and develop the exterior surfaces of the cylinders in Prob. 19–4.

5. Draw the views shown and develop the exterior surfaces of the pyramids in Prob. 19–5.

6. Draw the views shown and develop the exterior surfaces of the pyramids in Prob. 19–6.

7. Draw the views shown and develop the exterior surfaces of the cones in Prob. 19–7.

8. Draw the views shown and develop the exterior surfaces of the cones in Prob. 19–8.

9. Develop the transition pieces in Prob. 19–9. Select a scale that will enable them to fit on the paper. Cut out the drawing and tape it together to form a three-dimensional model.

10. Develop the transition pieces in Prob. 19–10. Select a scale that will enable them to fit on the paper. Cut out the drawing and tape it together to form a three-dimensional model.

11. Design a hood to completely cover the surface cooking unit in Prob. 19–11 and connect it to the vent pipe. Develop the surfaces to a suitable scale, cut them out, and tape them together to form a model of the hood.

12. Design a transition piece to connect the rectangular heat register opening in Prob. 19–12 with the round heat pipe below the floor. Select a suitable scale.

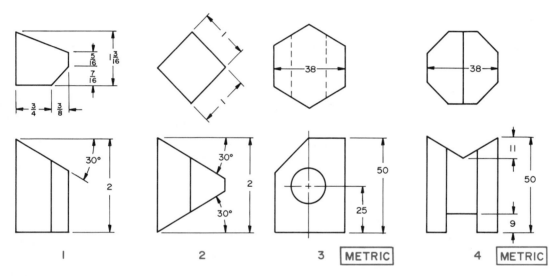

PROBLEM 19–1  Prisms.

PROBLEM 19–2  Prisms.

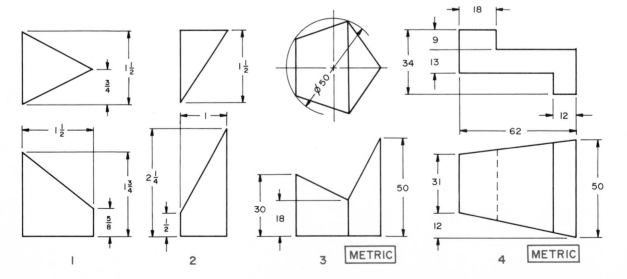

13. Develop the surfaces of the water tower in Prob. 19–13 on light cardboard. Cut out the surfaces and assemble the tower. Use the scale ¼ in. = 1 in.-0 ft.

14. Develop the surfaces of the products in Prob. 19–14. Select a suitable scale.

15. In Chapter 20 are a series of intersection problems. After the intersections are drawn, prepare developments for each surface.

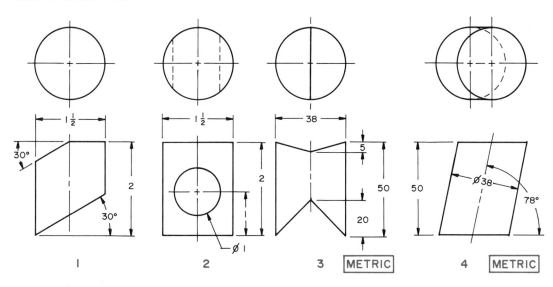

**PROBLEM 19–3** Cylinders.

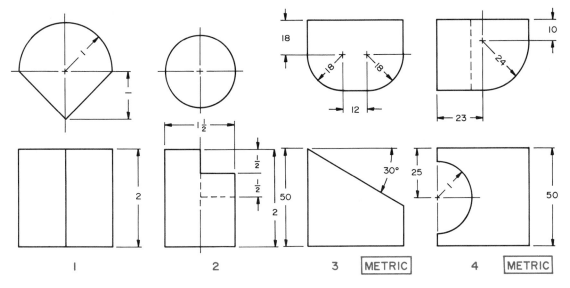

**PROBLEM 19–4** Cylinders.

**PROBLEM 19–5** Pyramids.

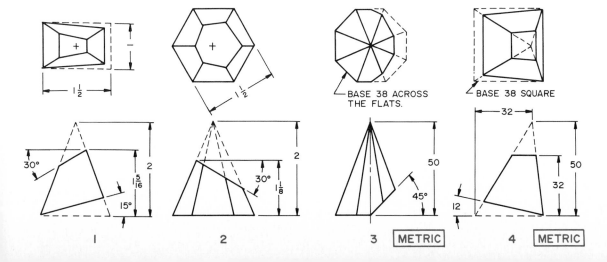

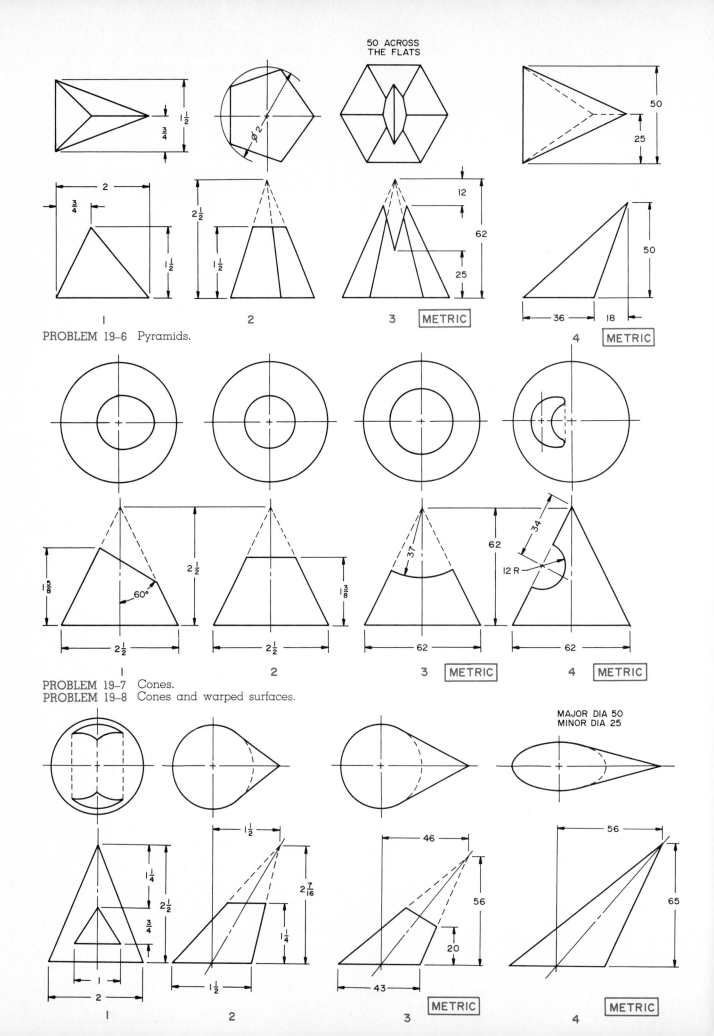

50 ACROSS
THE FLATS

1½
¾

2
¾
1½

2½
1½

12
62
25

50
25

36   18

PROBLEM 19–6  Pyramids.

1        2        3  [METRIC]        4  [METRIC]

1⅝
60°
2½
2½

⅜
2½
2½

37
62
62

34
12 R
62

PROBLEM 19–7  Cones.
PROBLEM 19–8  Cones and warped surfaces.

MAJOR DIA 50
MINOR DIA 25

1¼
2½
¾

1
2

1½
2 7/16
1¼
1½

46
56
20
43  [METRIC]

56
65  [METRIC]

1        2        3        4

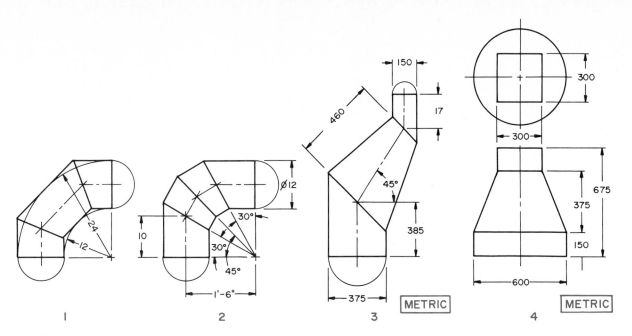

PROBLEM 19–9   Transition pieces.

PROBLEM 19–10   Transition pieces.

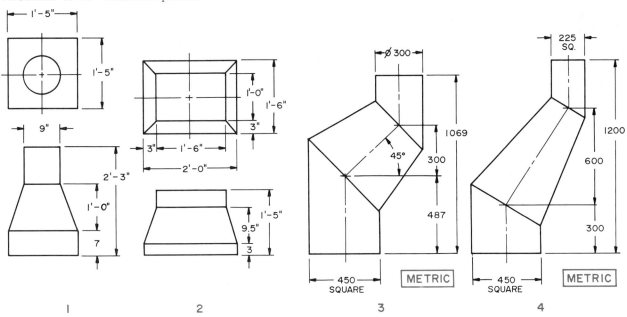

PROBLEM 19–11   Transition problem.   PROBLEM 19–12   Transition problem   PROBLEM 19–13   Development problem.

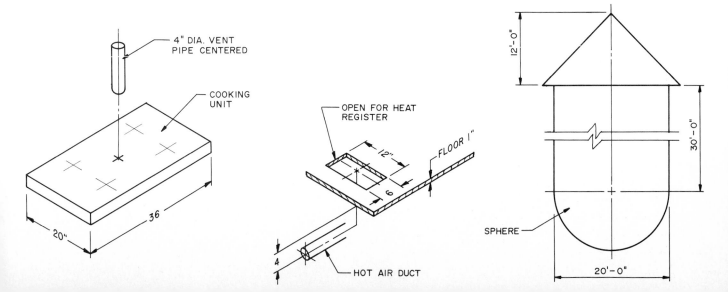

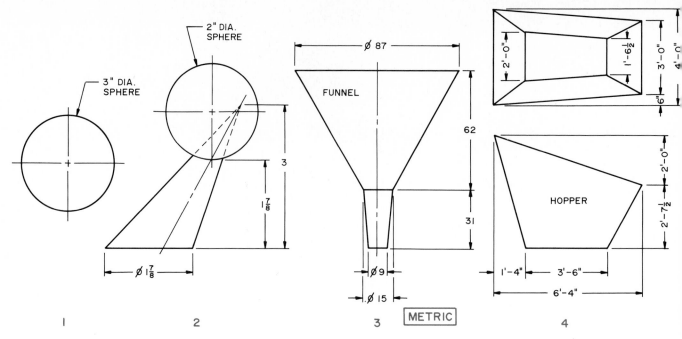

PROBLEM 19-14   Development problems.

# Intersections

When two surfaces, either plane or curved, meet, a line common to both called the *line of intersection* is formed. This occurs repeatedly on engineering drawings because almost every line on a drawing is the result of an intersection of planes. To draw orthographic views of the product, points on the line of intersection must be plotted and connected to locate it. Whether the line of intersection will be straight or curved depends on the surfaces involved. When two planes meet, the intersection is a straight line. When a plane and a curved surface or two curved surfaces meet, the intersection is a curved line. The more complex lines of intersection occur when geometric solids, such as cones, pyramids, and cylinders, intersect (Fig. 20–1). The careful and accurate plotting of intersections is essential to the development of a product. For example, sheet metal developments in which intersections are involved cannot be completed until the true location and size of the intersections have been plotted.

A thorough understanding of intersections is necessary to successful engineering design. As you look at products around you, you will notice that they contain intersections (Fig. 20–2). The type of intersections used on a design influences the cost and quality of a product. Intersections cost money to produce on a product. A simple straight-line inter-

## FIGURE 20–1  Typical lines of intersection.

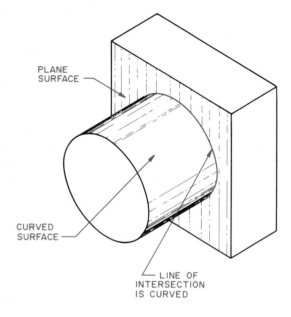

PLANE
SURFACE

CURVED
SURFACE

LINE OF
INTERSECTION
IS CURVED

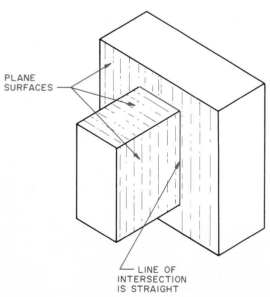

PLANE
SURFACES

LINE OF
INTERSECTION
IS STRAIGHT

FIGURE 20–2  This aircraft design shows some of the many intersections between planes that occur. (Courtesy of Boeing Co.)

section can usually be manufactured at less cost than one that has complex curves. This applies to all fields of engineering. An architect must constantly consider intersections. They are used to improve the exterior appearance of a building and become a factor in structural design (Fig. 20–3).

This chapter is concerned with intersections between solids and planes and between two solids; information related to these and other intersections can also be found in the principles illustrated in Chapters 8, 9, 10, and 19.

## INTERSECTION OF AN INCLINED PLANE AND A RIGHT PRISM

An inclined plane will appear in edge view in one of the normal views. In the example in Fig. 20–4, it is shown in the side view. The point at which the plane crosses a corner of the prism is a point where the prism pierces the plane. The line of intersection is found by connecting these piercing points.

Visibility is found by inspection. The top view shows the back corner, and the lines of intersection running to it are hidden.

FIGURE 20–3 Intersections occur in architectural design as well as in other areas of engineering. (Courtesy of American Plywood Association.)

## INTERSECTION OF AN OBLIQUE PLANE AND A RIGHT PRISM

The line of intersection is formed by connecting the points where the corners of the prism pierce the plane.

To find these points, pass a vertical plane through each corner in the horizontal view (Fig. 20–5). The vertical plane crosses the oblique plane on its edges. Project these intersections to the front view. The point at which they cross the corner of the prism is the piercing point on the plane.

To determine visibility, examine the top view. It shows that line 1–2 is behind the object and is therefore hidden.

## INTERSECTION OF AN OBLIQUE PLANE AND AN OBLIQUE PRISM

Neither the plane nor the prism appears in edge view or true size in the views given. The first step is to find the plane in edge view (Fig. 20–6). This can be done in a primary auxiliary view. The line of sight is parallel with the edges of the plane, giving a point view of those edges and an edge view of the plane.

The points at which the edge view of the oblique plane intersect the edges of the prism are the points where the prism pierces the plane. Project

FIGURE 20–4 The intersection of an inclined plane and a right prism.

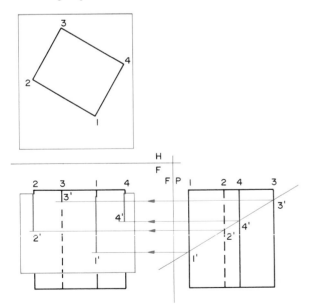

Assignment: Find the line of intersection of an inclined plane and a prism.
1. From the edge view of the plane, project the points at which it crosses the edges of the prism to the same edges in the front view.

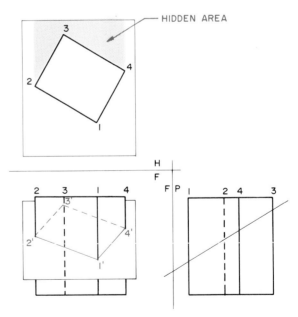

2. Connect the points in the proper order. Corners 1 and 2 are connected on the prism, so connect the projected points 1' and 2', which are on these corners. Visibility is found by examination. For example, corner 3 in the top view is clearly hidden by the plane and the prism, so it is hidden in the front view. The lines of intersection to this corner are also hidden.

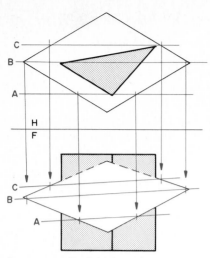

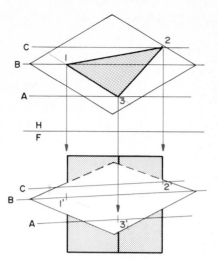

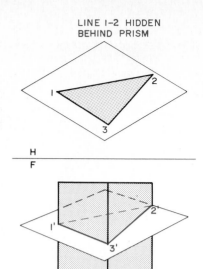

Assignment: Find the line of intersection of an oblique plane and a prism.

1. Pass a vertical plane through each corner of the prism and across the plane. Project the points where the vertical plane crosses the edge of the oblique plane to the front view. This transfers the line of intersection of the vertical and oblique planes to the front view.

2. Project each corner of the prism to its respective cutting plane in the front view. This locates points on the line of intersection.

3. Connect the points on the line of intersection in the same succession as the corners appear on the prism. Visibility is found by inspection of the top view.

FIGURE 20-5 The intersection of an oblique plane and a right prism.

FIGURE 20-6 The intersection of an oblique plane and an oblique prism.

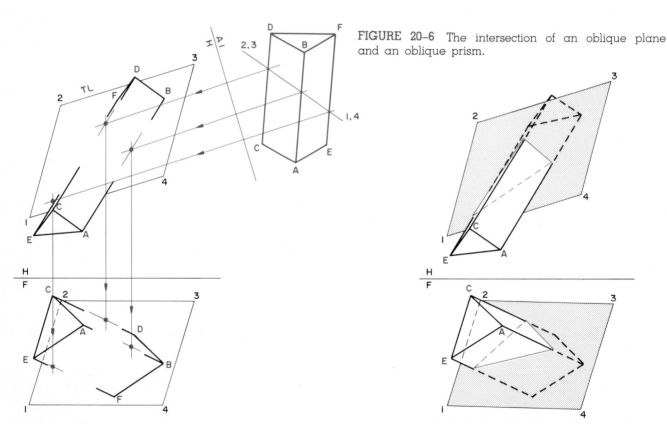

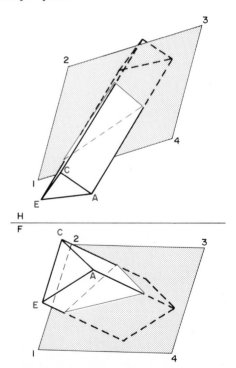

Assignment: Find the line of intersection of an oblique plane and an oblique prism.
1. Find the edge view of the plane and prism in a primary auxiliary view. In this example, edge AB is true length. It is the line of sight. Project the points located where the edge view of the plane crosses the edges of the prism to the top and front views.

2. Where they intersect their respective edges of the prism, locate points on the line of intersection. Connect these to locate the intersection. An examination of the auxiliary view will show that end BFD is behind the plane and is therefore hidden.

510

these from the auxiliary view to the given views. Connect the points to locate the line of intersection.

Visibility is determined by examining the auxiliary view. End BDF is behind plane 1–2–3–4 and is therefore hidden.

## INTERSECTION OF AN INCLINED PLANE AND A RIGHT CYLINDER

When an inclined plane intersects a cylinder, the plane is seen in edge view. To find the line of intersection, it is necessary to find a series of points where the elements of the cylinder pierce the plane (Fig. 20–7).

Locate points on the circular view of the cylinder. Project them to the side view. The line of intersection on the front view is found by projecting from the top and side views. The more points projected, the more accurate the drawing of the intersection.

Visibility is found by inspection.

## INTERSECTION OF AN OBLIQUE PLANE AND A RIGHT CYLINDER

Pass a series of vertical cutting planes through the circular view of the cylinder. The points at which they cross the cylinder are points where the cylinder

pierces the plane. These form the line of intersection. Project each vertical plane and piercing point to the front view. The series of points forms the line of intersection (Fig. 20–8).

Visibility is found by inspection.

## INTERSECTION OF AN OBLIQUE PLANE AND AN OBLIQUE CYLINDER

In this situation, the cylinder does not appear true length and the plane does not appear as an edge. The first step is to find the edge view of the plane (Fig. 20–9).

The edge view of the plane is found in a primary auxiliary view. The oblique plane has one edge that is true length. Sight parallel to this edge to construct the auxiliary view. Pass cutting planes through the cylinder parallel with its axis on the auxiliary view. Project to the profile view the points where the oblique plane crosses the cutting planes. These piercing points form the line of intersection on that view. Project the points to the front view. Locate them by measuring their distance from the profile plane.

Visibility is found by inspection.

**FIGURE 20–7** The intersection of an inclined plane and a right cylinder.

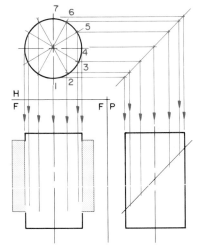

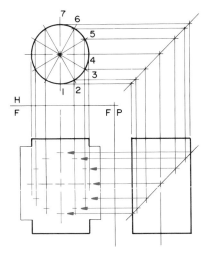

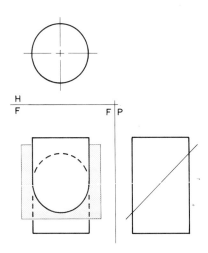

Assignment: Find the line of intersection of an inclined plane and a right cylinder.

1. Divide the circular view into a number of equal parts. Project these to the side and front views.

2. Project the points at which the divisions cross the edge view of the plane to the front view. Each point at which they cross their projection from the top view is a point on the line of intersection.

3. Project each point. Then, connect the points to form the intersection. Visibility is found by inspection.

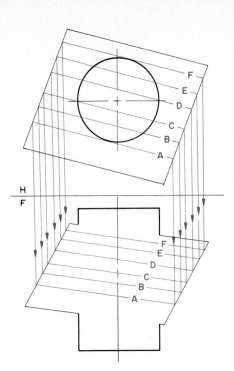

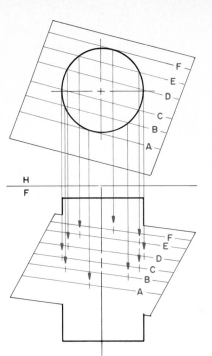

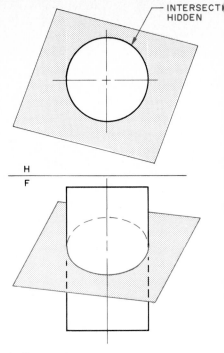

INTERSECT
HIDDEN

Assignment: Find the line of intersection of an oblique plane and a right cylinder.

1. Pass a number of vertical cutting planes parallel with the edge of the oblique plane through the circular view of the cylinder and cutting the plane. Project them to the front view.

2. Project the points at which the vertical planes cut the cylinder to the front view. This locates points on the line of intersection. The more points found, the more accurate the line of intersection.

3. Connect the points to find the line of intersection. Visibility is found by inspection.

FIGURE 20–8 The intersection of an oblique plane and a right cylinder.

FIGURE 20–9 The intersection of an oblique plane and an oblique cylinder.

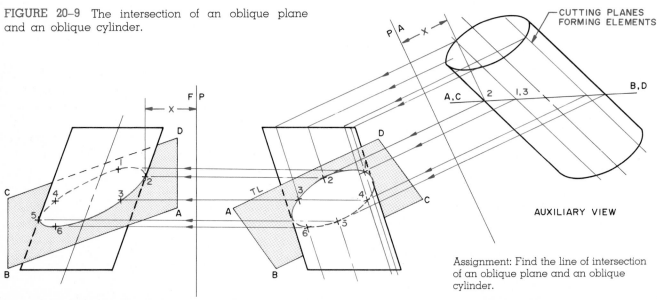

CUTTING PLANES
FORMING ELEMENTS

AUXILIARY VIEW

2. Pass cutting planes through the cylinder parallel with its axis in the auxiliary view. This locates elements on the surface of the cylinder that intersect the plane. Project these elements and points of intersection to the side and front views. The points on the line of intersection on the front view are located by measuring their distance from the profile plane on the auxiliary view. Visibility is found by inspection.

Assignment: Find the line of intersection of an oblique plane and an oblique cylinder.

1. First, find the edge view of the plane in a primary auxiliary view. Since the oblique plane used has a true-length edge, it can be used as the line of sight.

512

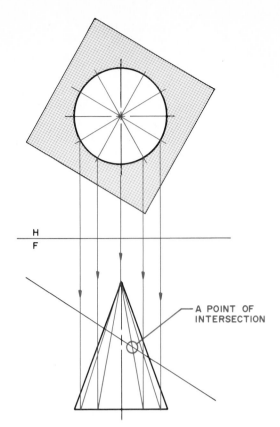

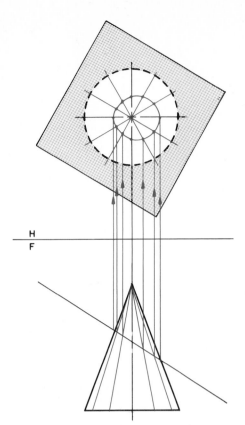

Assignment: Find the intersection of an inclined plane and a right cone.
1. In this example, the plane appears as an edge in the front view. Draw elements on the surface of the cone in the circular view. Project them to the front view. Each point at which they cross the edge view of the plane is a point on the line of intersection.

2. Project the points of intersection to their respective elements in the top view. Connect them to form the line of intersection.

FIGURE 20–10 The intersection of an inclined plane and a right cone.

## INTERSECTION OF AN INCLINED PLANE AND A RIGHT CONE

When an inclined plane intersects a right cone, the plane is seen in edge view in one of the principal planes. To find the line of intersection, draw a series of elements on the circular view of the cone (Fig. 20–10). Project these to the front view. The points at which the edge view of the plane intersects the elements in the front view are the points where the element pierces the plane. These are points on the line of intersection. Project these points to the top view. Connect them to form the line of intersection.

## INTERSECTION OF AN OBLIQUE PLANE AND A RIGHT CONE

Pass a series of horizontal cutting planes through the cone perpendicular to the axis (Fig. 20–11). Project these to the surface of the plane in the top view. The horizontal planes cut the cone into a series of circular sections. Draw these on the top view. The points

at which each circular section crosses the line on the plane formed by the horizontal cutting plane are points on the line of intersection. A plane may cross the line more than once. Repeat this for each of the horizontal cutting planes used to show the intersection. Then project the points to the front view.

Visibility is found by inspection.

## CONIC SECTIONS

When a plane intersects a cone, a conic section is formed. This can be either a circle, an ellipse, a triangle, a parabola, or a hyperbola.

A *circle* is formed when a right cone is cut by a plane perpendicular to the axis. The radius of the circle is found in the front view (Fig. 20–12).

An *ellipse* is formed when a right cone is cut by a plane at an angle other than 90° to the axis (Fig. 20–13). Its true size is found in a primary auxiliary view.

An ellipse is formed by the path of a point moving so that the sum of the distances from two

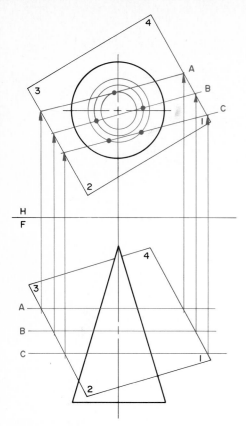

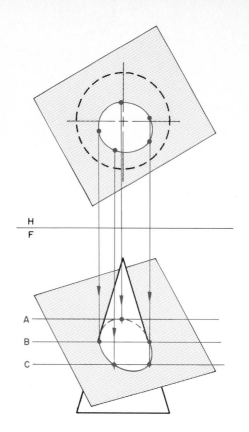

Assignment: Find the intersection of an oblique plane and a right cone.

1. First, pass several horizontal planes through the cone parallel with the axis. They cut the cone and the plane. Project these to the top view, where a series of circular sections are formed. The point at which each circular section crosses the line of the same cutting plane on the oblique plane locates a point on the line of intersection.

2. Project each point of intersection to its respective plane in the front view. Connect the points to form the line of intersection. Determine visibility by inspection.

**FIGURE 20–11** The intersection of an oblique plane and a right cone.

**FIGURE 20–13** An ellipse is formed when a right cone is cut by an inclined plane.

**FIGURE 20–12** A circle is formed when a right cone is cut by a plane that is perpendicular to the axis of the cone.

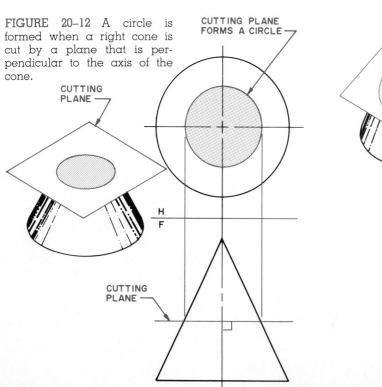

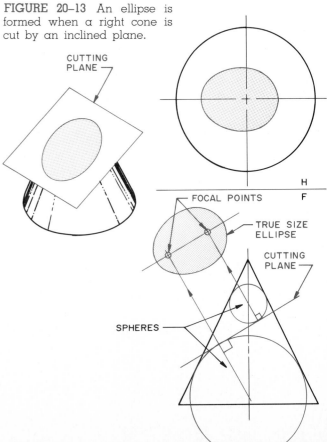

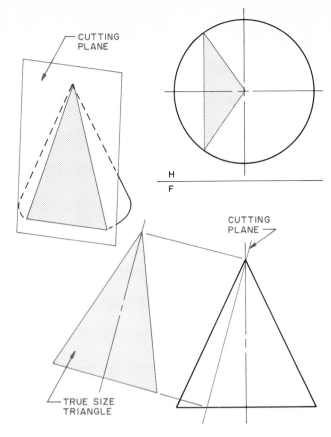

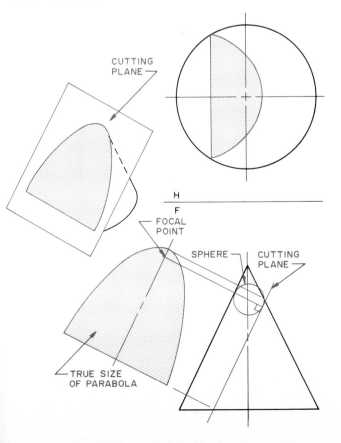

FIGURE 20–14 A triangle is formed when a right cone is cut by a plane passing through the apex of the cone.

FIGURE 20–15 A parabola is formed when a right cone is cut by a plane on the same angle with the base of the cone as the elements on the cone.

focal points is a constant. The focal points are found by drawing spheres on the front view that are tangent to the cone and the plane of the ellipse. The centers of the spheres are projected to the center line of the ellipse in the auxiliary view.

A *triangle* is formed when a cone is cut by a plane that passes through the apex and continues to the base. The true size is found by making a primary auxiliary view (Fig. 20–14).

A *parabola* is formed when a cutting plane passes through a cone at the same angle with the base of the cone as do the elements (Fig. 20–15).

To draw this intersection, locate the edge view of the plane on the front view (Fig. 20–16). Draw a series of horizontal planes on the front view. Project these to the top view. The points at which they cross the edge view of the plane are points on the line of intersection that form the parabola in the top view. The true size of the parabola is found in a primary auxiliary view projected perpendicular to the edge view.

All lines entering the open end of a parabola that are parallel to its axis are reflected to a common point called a focal point. This is used in lighting reflectors. If a bulb is placed at the focal point, its light will be reflected by the reflector (a parabola)

FIGURE 20–16 The procedure for drawing a parabola formed when a right cone is cut.

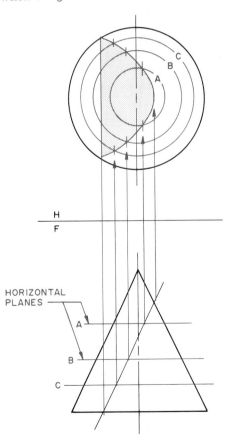

515

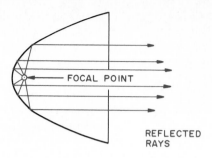

**FIGURE 20–17** A light placed at the focal point of a parabola will be reflected in parallel rays.

in parallel rays (Fig. 20–17). The focal point is found by drawing a sphere at the apex of the cone that is tangent to the cone and the edge view of the parabola. The point of tangency of the sphere with the edge view of the parabola locates the focal point (Fig. 20–15).

A *hyperbola* is formed when a cutting plane passes through a cone at an angle of 90° or less and makes an angle greater than the angle made by the elements with the base. When the angle is 90°, the true size of the hyperbola appears in the side view. When it is less than 90°, a primary auxiliary view is needed (Fig. 20–18).

**FIGURE 20–18** A hyperbola is formed when a right triangle is cut by a plane making an angle of 90° or less with the base and at an angle greater than the angle of the elements with the base.

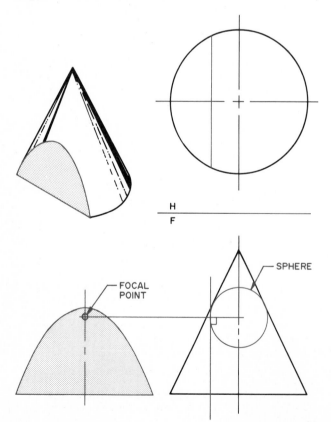

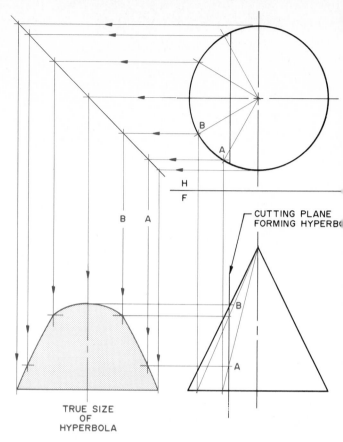

**FIGURE 20–19** The procedure for drawing a hyperbola produced by passing a plane through a right cone.

To draw this intersection, divide the circular view of the cone into equally spaced elements (Fig. 20–19). Project these to the front and side views. The points at which the elements cross the edge view of the hyperbola in the front view are points on the line of intersection. Project these to the side view to form the hyperbola.

The focal point is found by drawing a sphere at the apex of the cone tangent to the edge view of the hyperbola and the sides of the cone (Fig. 20–18).

## INTERSECTION OF AN INCLINED PLANE AND A SPHERE

When an inclined plane intersects a sphere, the plane is seen in edge view (Fig. 20–20).

To find the line of intersection, draw a series of horizontal planes on the view in which the plane appears as an edge. Project the circular intersections of the horizontal planes and the sphere to the top view. Project the points in the front view where the horizontal cutting plane intersects the edge view of the plane to their respective circular locations in the top view. The line of intersection in the top view appears elliptical because it is not parallel with a

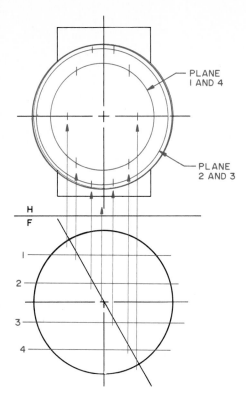

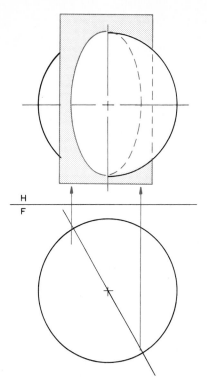

Assignment: Find the line of intersection of an inclined plane and a sphere.
1. Draw a number of horizontal planes on the front view. The plane appears here as an edge. Project the planes to the top view, where they appear as circular elements.

2. Project onto the front view the points at which the edge view of the plane crosses the horizontal planes. Connect the points to form the line of intersection.

**FIGURE 20–20** The intersection of an inclined plane and a sphere.

**FIGURE 20–21** Finding the line of intersection of two prisms at right angles to each other.

principal reference plane. If you look perpendicular to the intersection, it will be a circle.

## INTERSECTION OF TWO PRISMS AT RIGHT ANGLES TO EACH OTHER

The principles used to find the intersection of a plane and a prism are used when finding the intersection between two prisms. Since a prism is made of several planes, the intersection can be found by finding the intersection of each plane.

The steps for finding the intersection of two prisms are shown in Fig. 20–21.

1. Project the points in the top view where the corners of the horizontal plane pierce the planes of the second prism to the front view.
2. Draw the edges of the horizontal plane until they cross the projection from the top view. This is the piercing point, which is part of the line of intersection. Connect the points to form the intersection.

In Fig. 20–22, one prism is turned so that the edges do not pierce the planes of the second prism

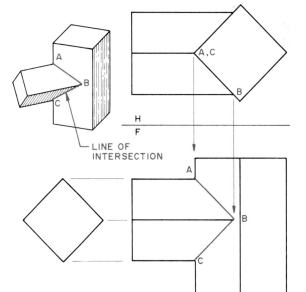

Assignment: Find the line of intersection of two prisms at right angles to each other.

1. The points at which the edges of the horizontal plane touch the surfaces of the vertical prism in the top view are points on the line of intersection. Project these to the same edge in the front view. Connect the points to form the intersection.

Intersections  **517**

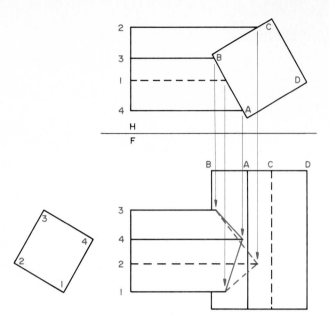

in a symmetrical manner. The piercing points of the horizontal prism on the vertical prism are found in the same manner as just described. Find the corners of the vertical prism that intersect the horizontal prism by drawing a side view of the horizontal prism. Then, project the piercing point in the top view to the side view and on to the front view. The piercing points must be connected in the same order as they occur on the horizontal cylinder.

## INTERSECTION OF TWO PRISMS OBLIQUE TO EACH OTHER

The steps are the same as described for prisms at right angles to each other. First, project the points of intersection in the top view to the front view (Fig. 20–23). Then, extend the corners of the oblique prism in the front view until they cross the projections from the top view. These are the ends of the lines of intersection of the various planes. Connect these to draw the intersection.

FIGURE 20–22 Finding the line of intersection of two prisms that form an asymmetrical union.

FIGURE 20–23 Finding the line of intersection of two prisms that are oblique to each other.

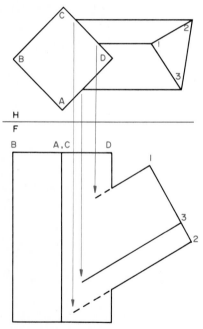

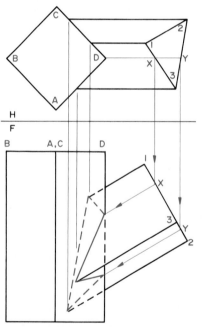

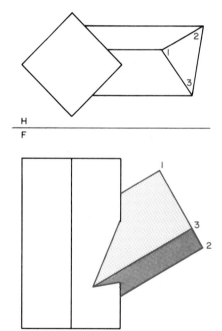

Assignment: Find the line of intersection of two prisms oblique to each other.
1. The points at which the edges of the triangular prism pierce the surface of the square prism are points on the line of intersection. Project these to the front view. This locates the piercing points of edges 1, 2, and 3. Establish visibility by inspection. This locates all the points except those needed to locate where corner D of the square prism intersects the triangular prism.

2. Corner D intersects the planes formed by edges 1–3 and 2–3. To find these points of intersection, pass a plane through corner D parallel with the edge of the triangular prism. Project the line of intersection of this plane DX and DY to the front view. This locates the intersection of X and Y on D. Connect the other points to these in proper succession to form the completed intersection.

3. Hidden lines are removed from the finished drawing. Match it up with the previous drawing to see how the line of intersection developed.

FIGURE 20–24 This tower has numerous intersections between cylindrical structural members. (Courtesy of Mc-Dermott Incorporated.)

## INTERSECTION OF TWO CYLINDERS AT RIGHT ANGLES TO EACH OTHER

One of the most common intersection problems is that of intersecting cylinders. Notice the intersection situations in the oil-drilling structure in Fig. 20–24.

Divide the end view of the smaller cylinder into a series of equally spaced elements (Fig. 20–25).

Project these elements to the top view and down to the front view. Project these elements from the side view to the front view. The point at which each element crosses itself on the front view is a point on the line of intersection. The more elements used, the more accurate the drawing of the line of intersection. Connect the points with an irregular curve.

FIGURE 20–25 Finding the line of intersection of two cylinders at right angles to each other.

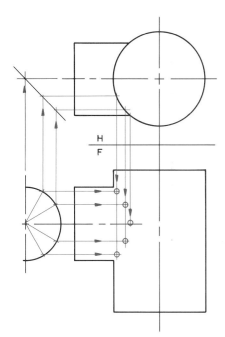

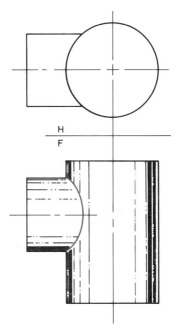

Assignment: Find the line of intersection of two cylinders at right angles to each other.
1. Divide the end view of the smaller cylinder into several equal parts. Project these to the top and front views. Project the points of intersection from the top view to the front view. The point at which each projection crosses itself is a point on the line of intersection.

2. Because the cylinders are symmetrical, the upper and lower halves of the intersection are the same. Connect the points to form the line of intersection.

## INTERSECTION OF TWO CYLINDERS OBLIQUE TO EACH OTHER

First, draw an auxiliary view of the oblique cylinder (Fig. 20–26). This shows its true size. Then, divide it into a number of equal parts by passing a number of cutting planes through it parallel with the frontal plane. Use these points to locate elements on the surface of the small cylinder. Locate these elements to the top view. Project from the top view down to the front view. The point at which each element crosses itself in the front view is a point on the line of intersection. Connect these points with an irregular curve. A typical industrial application is shown in Fig. 20–27.

## INTERSECTION OF A CYLINDER AND A PRISM

This intersection may be found using cutting planes or elements on the surface of the cylinder as used in Fig. 20–26. For this example, the cutting plane technique was chosen.

The cutting planes were drawn parallel with the frontal plane (Fig. 20–28). They were located on the end view of the cylinder. The piercing points

FIGURE 20–27 This manifold with downcomers forms a ring inside the suppression chamber of a nuclear plant. The intersections of the cylindrical parts must be completely tight. (Courtesy of Tennessee Valley Authority.)

FIGURE 20–26 Finding the line of intersection of two cylinders oblique to each other.

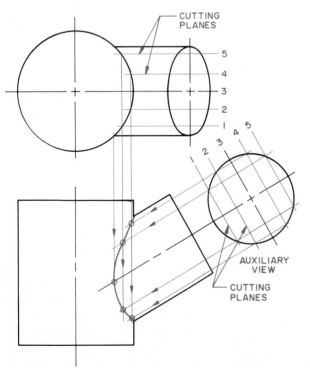

of the elements are projected to the front view from the top and side views. The point at which they cross locates a point on the line of intersection.

## INTERSECTION OF A CONE AND A PRISM

In this example, the axes of the cone and prism are parallel with a principal plane. The steps for finding the line of intersection are given in Fig. 20–29.

In the side view, draw planes through the axis of the cone and the corners of the prism. Project these to the top view and down to the front view. The intersection of the edges of the prism with these elements locates points on the line of intersection.

Other points on the curve are found by drawing additional planes on the cone that pass between those locating the sides. The point at which they cross the edge view of the prism locates a point on the line of intersection. These elements are actually cutting planes cutting the cone and the edge of the prism. The line of intersection is formed by connecting the piercing points with an irregular curve.

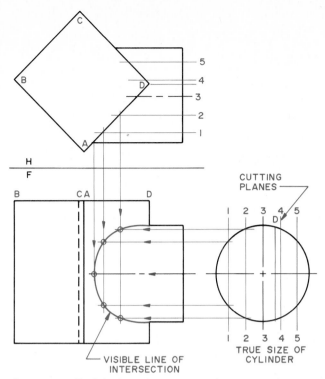

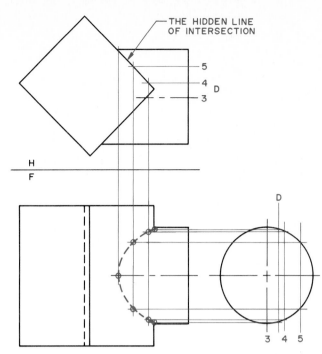

THE HIDDEN LINE OF INTERSECTION

CUTTING PLANES

TRUE SIZE OF CYLINDER

VISIBLE LINE OF INTERSECTION

Assignment: Find the line of intersection of a prism and a cylinder.

1. Find the true size of the cylinder. Draw cutting planes through the cylinder parallel with the frontal plane. Project these to the top view. The points at which they intersect the prism are points on the line of intersection.

This step shows how to locate the visible section of the line of intersection. Project the points at which the cutting planes cut the cylinder on the front half (1, 2) from the top and side views to the front view. Each such point is a point on the visible part of the line of intersection.

2. To find the hidden part of the line of intersection, project the points at which the cutting planes cut the back half of the cylinder (D, 4, 5) from the top and side views to the front view. Each is a point on the hidden part of the line of intersection.

FIGURE 20–28 Finding the line of intersection of a prism and a cylinder.

FIGURE 20–29 Finding the intersection of a cone and a prism.

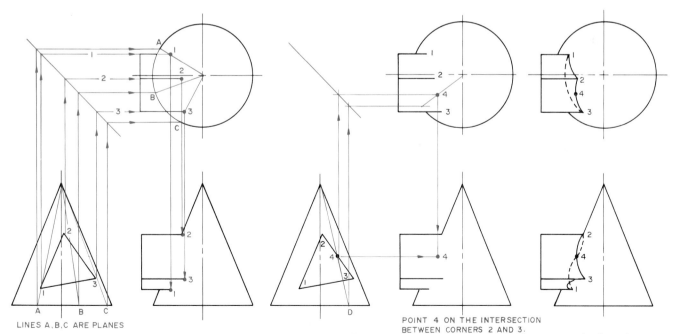

LINES A,B,C ARE PLANES

POINT 4 ON THE INTERSECTION BETWEEN CORNERS 2 AND 3.

Assignment: Find the line of intersection of a cone and prism.

1. Pass planes through the axis of the cone and the corners of the prism. Project these to the top and front views. The points at which they cross locate the corners of the prism on the surface of the cone.

2. To find additional points between the corners of the prism, pass additional planes through the sides of the prism. Project these points to the top and front views. One such point, 4, is projected to show how this is done. Additional points will help to locate the curve accurately. This procedure locates points on the surface of the cone forming the line of intersection of the corners of the prism and the cone.

3. The final solution gives the line of intersection, 1.

521

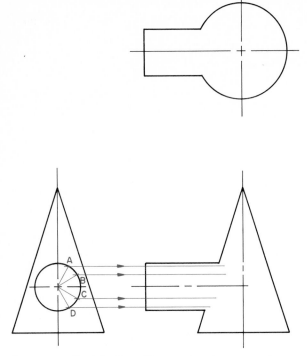

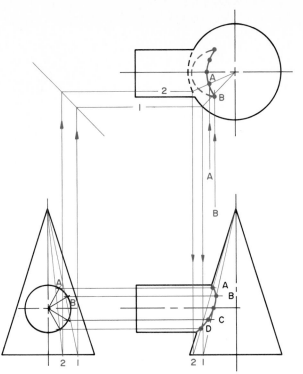

Assignment: Find the line of intersection of a cone and a cylinder.

1. Divide the end view of the cylinder into a number of equally spaced sections. Draw these as elements on the front view of the cylinder.

FIGURE 20–30   The intersection of a cone and a cylinder.

2. Then, draw elements on the cone, passing them through the divisions on the end of the cylinder. Points A and B are projected as examples. Project these elements to the top view and from the top view to the front view. The points at which they cross the elements on the cylinder are points on the line of intersection on the front view. Project these points to their respective elements on the cone in the top view. Because the object is symmetrical, the right and left halves of the intersection are the same.

## INTERSECTION OF A CONE AND A CYLINDER

The technique used to find this intersection involves dividing the cylinder into equally spaced elements in its circular view (Fig. 20–30). Then, draw elements on the cone through the elements on the cylinder. Project these to the top and front views. The points at which the elements of the cone and cylinder cross on the front view locate points on the line of intersection. Connect the points with an irregular curve.

## INTERSECTION OF A PYRAMID AND A PRISM

If the pyramid and prism are positioned so that the piercing points can be found by projection, the intersection is found as described for two right prisms. If they are positioned so that the edges are not true size, the cutting plane technique is used.

The steps for finding the intersection using cutting planes are given in Fig. 20–31.

In the top view, pass vertical cutting planes through each edge of the prism. The points at which the planes cross the edges of the pyramid locate points on the edge of that surface. Project these to the front view. In the front view, extend the edge of the prism that is in the plane until it crosses the edge of the cutting plane. This is a point of intersection. Project it to the same edge in the top view. Repeat these steps for each of the other edges.

Visibility is found by inspection.

## INTERSECTION OF A SPHERE AND A PRISM

The intersection is found by passing cutting planes parallel to the frontal plane (Fig. 20–32). These will appear in the top and side views. In the front view, these cutting planes appear as circles. Project from the side view to the front view the intersections between the prism and each cutting plane. The points at which they cross the respective cutting plane in the front view are points on the line of intersection. Project these to the top view to find the line of intersection there.

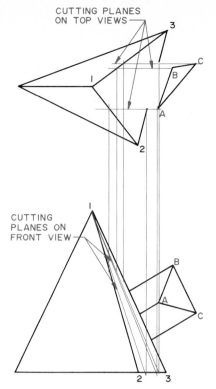

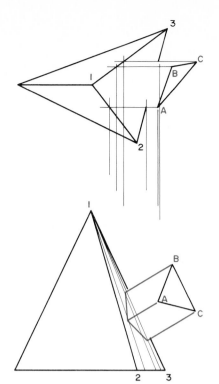

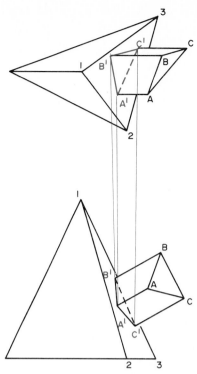

CUTTING PLANES ON TOP VIEWS

CUTTING PLANES ON FRONT VIEW

Assignment: Find the line of intersection of a pyramid and prism.

1. Pass a vertical cutting plane through each corner of the prism in the top view. These are planes A, B, and C. Project the points where they intersect the edges of surface 1–2–3 of the prism. Project these points to the front view. This locates the cutting plane on the front view.

2. On the front view, extend the edge of the prism until it crosses the cutting plane. This is where that edge pierces the plane.

3. Repeat this for each edge. Connect the points to complete the line of intersection. Project the points to their respective lines in the top view.

FIGURE 20–31 Finding the line of intersection of a pyramid and a prism.

FIGURE 20–32 Finding the line of intersection of a sphere and a prism.

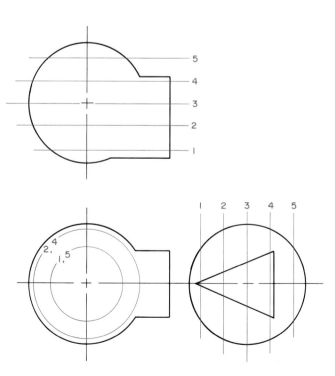

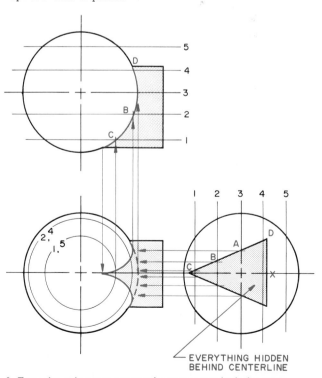

EVERYTHING HIDDEN BEHIND CENTERLINE

Assignment: Find the line of intersection of a sphere and a prism.

1. First, pass vertical cutting planes parallel with the frontal plane through the sphere in the side and top views. Then, project these to the front view, where they form circular elements.

2. From the side view, project the points at which the prism intersects the cutting planes to the respective planes in the front view. This locates points on the line of intersection. Project these points to the top view to locate the line of intersection in that view.

Visibility changes where the line of intersection of the prism and the axes of the sphere meet. This is cutting plane 3.

Because the object is a sphere, the line of intersection becomes hidden beyond the intersection of the prism and the axes of the sphere. These are found on the side view of the sphere. Cutting plane 3 is the point at which visibility changes on the front view. Point X is where the visibility changes on the top view.

## INTERSECTION OF A SPHERE AND A CYLINDER

The intersection is found by passing horizontal cutting planes through the front and side views (Fig. 20–33). These will appear in the top view as circles. Project from the side view to the top view the points where the cutting planes cross the cylinder. The points at which they cross the respective cutting plane in the top view are points on the line of intersection. Project these points from the top view to the front view to find the line of intersection there.

Visibility of the line of intersection is found by examining the axes of the cylinder. It is at A on the front view and B on the top view.

FIGURE 20–33 Finding the line of intersection of a sphere and a cylinder.

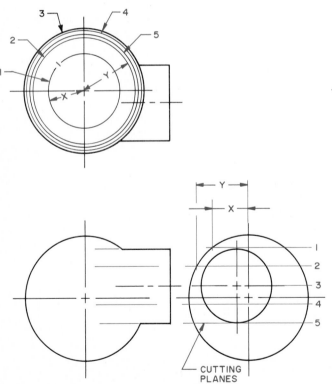

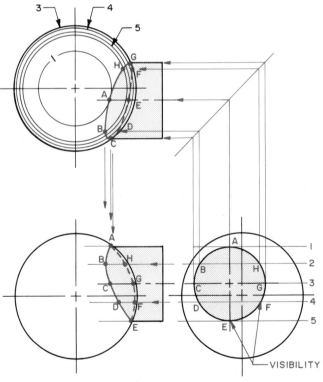

Assignment: Find the line of intersection of a sphere and a cylinder.

1. Pass horizontal cutting planes through the front and side views. Project them to the top view. They appear circular in this view.

2. Project the points at which the cylinder crosses each cutting plane in the side view to the front and top views. The points at which they cross the respective cutting planes in the top are points on the line of intersection. Project these points to their respective cutting planes in the front view to locate points on the line of intersection.

Visibility is found by examining the axes of the cylinder, point on the front view and point G on the top view.

# Problems

Use an A or A4 sheet for these drawings.

1. Draw the given views and the missing view of the right prisms and inclined planes in Prob. 20–1. Ascertain the visibility of the lines of intersection.

2. Draw the given views and find the line of intersection of the right prisms and oblique planes in Prob. 20–2. Ascertain the visibility of the lines in each view.

3. Draw the given views using the ¼-in. grid to locate the oblique prisms and oblique planes in Prob. 20–3. Then, find the lines of intersection between them and ascertain their visibility.

4. In part 1 of Prob. 20–4, find the line of intersection between the right cylinder and the inclined plane. In part 2, find the intersection with the oblique plane. In part 3, find the intersection between the oblique cylinder and oblique plane. Ascertain visibility for all lines. Use the squares in the grid to find the location of points.

5. In parts 1 and 2 of Prob. 20–5, find the line of intersection between the cone and the indicated plane. In part 3, construct planes to form conic sections, including a circle, ellipse, triangle, parabola, and hyperbola. Show these intersections on the top and front views and then produce a true size drawing of each. You will be able to fit two sections on one page.

6. In part 1 of Prob. 20–6, find the line of intersection between the plane and the sphere. In parts 2, 3, and 4, find the intersection between the objects and the given plane.

7. In Prob. 20–7, find the intersection of the cylinders and each of the intersecting solids. Use section A, B, or C as assigned by the instructor.

8. In Prob. 20–8, find the intersections of the rectangular prisms and each of the intersecting solids. Use section A, B, or C as assigned by the instructor.

9. In Prob. 20–9, find the intersections of the hexagonal prisms and each of the intersecting solids. Use section A, B, or C as assigned by the instructor.

10. In Prob. 20–10, find the intersections of the cone and each of the intersecting solids. Use section A, B, or C as assigned by the instructor.

11. In Prob. 20–11, find the intersections of the cones and each of the intersecting solids. Use section A, B, or C as assigned by the instructor.

12. In Prob. 20–12, find the intersections of the pyramids and each of the intersecting solids. Use section A, B, or C as assigned by the instructor.

13. In Prob. 20–13, find the intersections of the pyramids and each of the intersecting solids. Use section A, B, or C as assigned by the instructor.

14. In Prob. 20–14, find the intersections of the spheres and each of the intersecting solids. Add additional views if needed.

15. Find the intersections between the pipes in Prob. 20–15. See Fig. 20–27 for an example of this product. Use a suitable scale for the drawing. Prepare a development for each surface if assigned by your instructor.

16. Draw the intersections of the pipes forming the welded Y fitting in Prob. 20–16. Use a suitable scale. Prepare a development of each surface if assigned by your instructor.

PROBLEM 20–1 Intersections of right prisms and inclined planes.

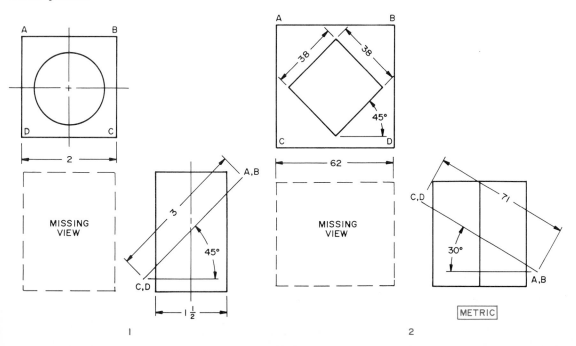

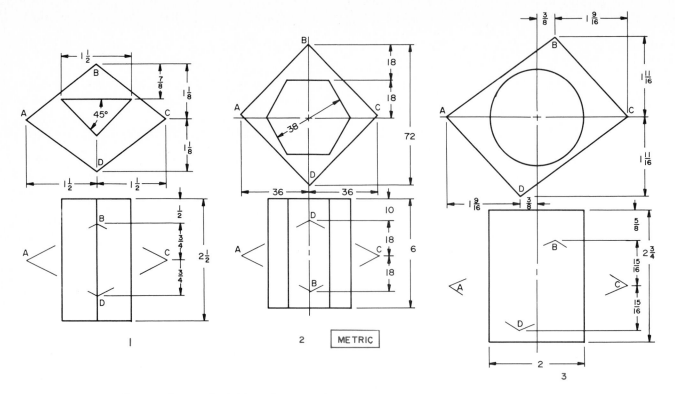

**PROBLEM 20-2** Intersections of right prisms and oblique planes.

**PROBLEM 20-3** Intersections of oblique prisms and oblique planes.

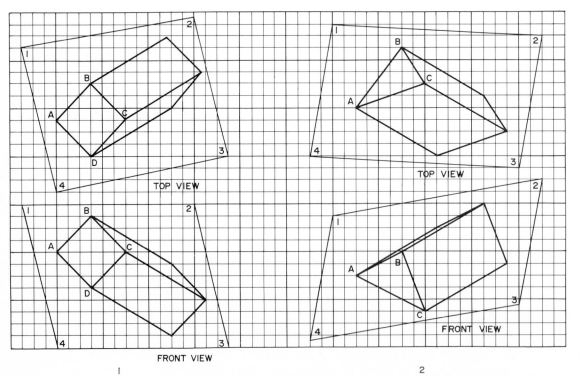

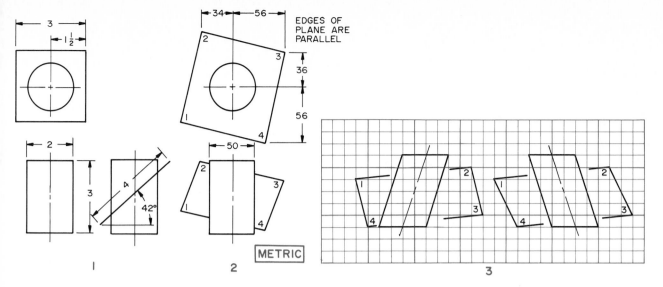

PROBLEM 20–4 Intersections of right cylinders and inclined and oblique planes.

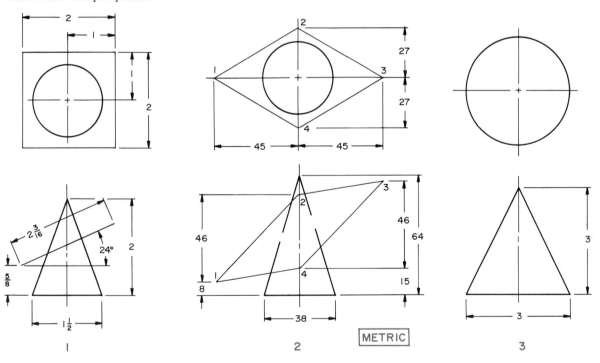

PROBLEM 20–5 Intersections of a right cone and various planes.

PROBLEM 20–6 Sphere and cylinder intersection problems.

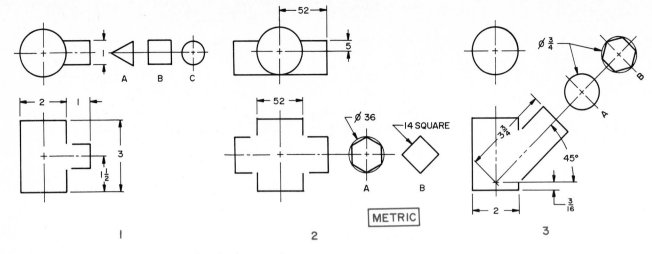

**PROBLEM 20–7** Intersections of cylinders and intersecting solids.

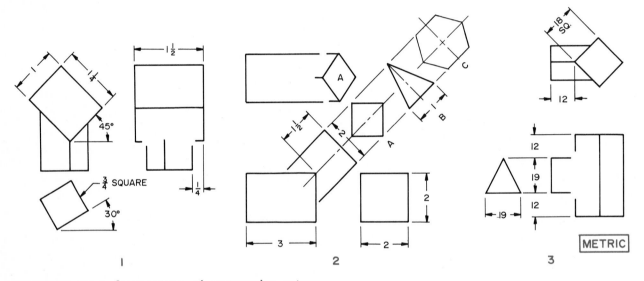

**PROBLEM 20–8** Intersections of rectangular prisms and intersecting solids.

**PROBLEM 20–9** Intersections of hexagonal prisms and intersecting solids.

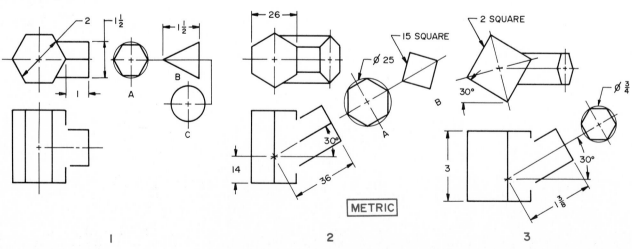

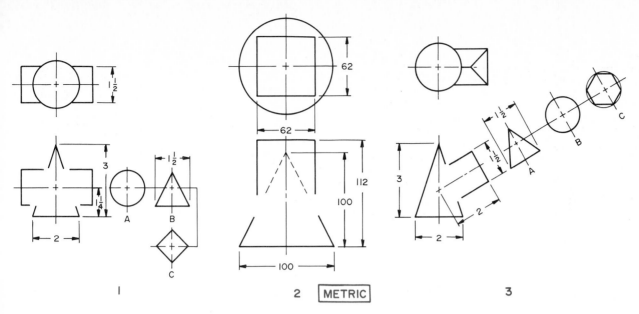

**PROBLEM 20–10** Intersections of right cones and intersecting solids.

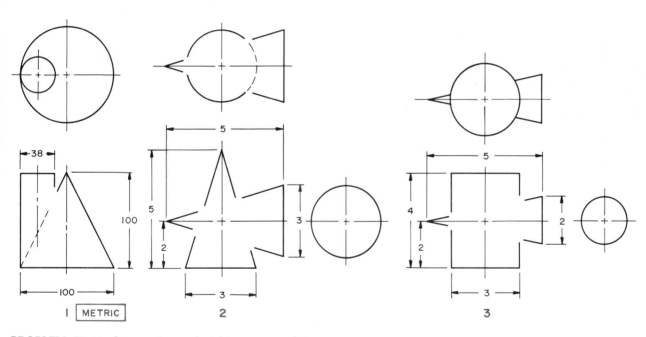

**PROBLEM 20–11** Intersections of right cones and intersecting solids.

**PROBLEM 20–12** Intersections of right pyramids and intersecting solids.

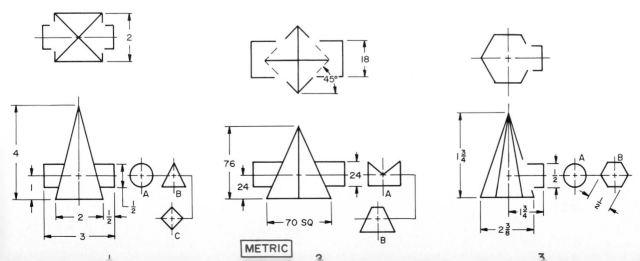

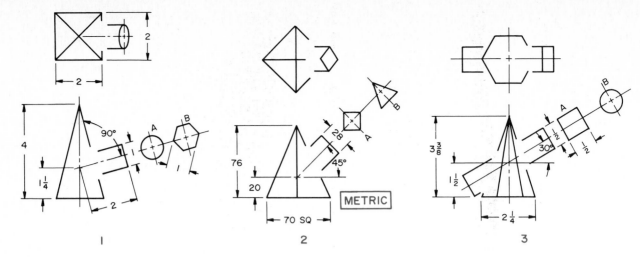

**PROBLEM 20–13** Intersections of right pyramids and intersecting solids.

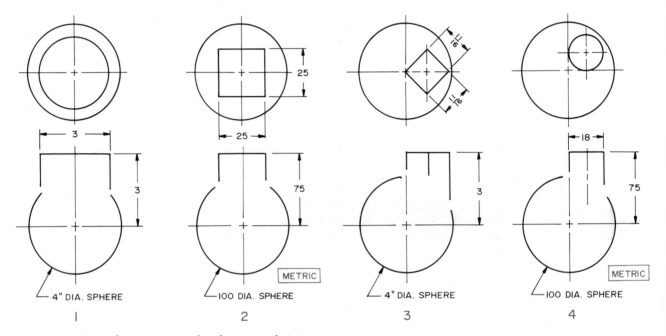

**PROBLEM 20–14** Intersections of spheres and intersecting solid.

**PROBLEM 20–15** Intersections of pipes.

**PROBLEM 20–16** Intersections of pipes in a Y fitting.

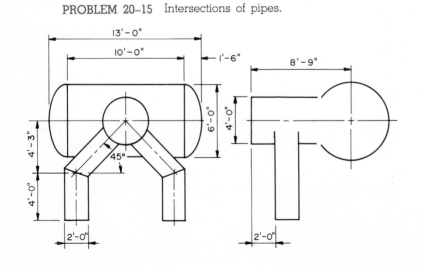

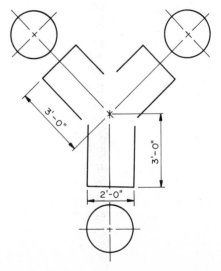

# Vector Analysis

After the structural system of a product has been designed, it must be analyzed to determine the stresses it must accommodate. The design must meet the original specifications set for it. Graphical vector analysis methods are most applicable to the preliminary analysis and refinement of the design. The design may be less than, equal to, or greater than the design specifications.

When analyzing a structure, the forces of tension, compression, shear, and moment must be considered (Fig. 21–1). Moment of a force is the rotation produced on a body when a force is applied. Shear is an applied force that tends to produce a shearing strain. Shearing strain is a condition in a body caused by forces that tend to produce an opposite but parallel sliding motion of the bodies' planes (Fig. 21–2). Tension and compression forces can be analyzed using vectors (Fig. 21–3). Vectors can also be used to represent other quantities such as velocity, distance, and electrical properties.

The graphical methods using vectors to make an analysis of some types of problems involving the action of forces on matter or mechanical systems provide a rapid and easy solution. Analysis can also be accomplished using conventional trigonometric and algebraic methods, but these are more difficult

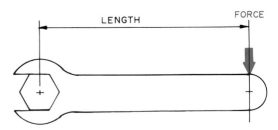

FIGURE 21-2 Typical moment and shear forces.

to use. The range of accuracy produced by graphical vector analysis is adequate for most purposes.

## TERMINOLOGY

Solving problems with vectors requires a knowledge of the terminology of graphical vectors. Following are descriptions of these terms. They are arranged in alphabetical order to make it easier to use as a reference.

> *Compression:* A condition in a member produced by the application of opposite pushing forces. A member under compression tends to be shortened. On structural analysis diagrams, compression is represented by a minus sign (–) (Fig. 21–3).

> *Components:* Individual forces that when combined result in a single force (Fig. 21–4). In this ex-

FIGURE 21-1 Before the actual structural members of this bridge can be selected, the stresses to be encountered must be calculated. (Courtesy of West Virginia Department of Highways.)

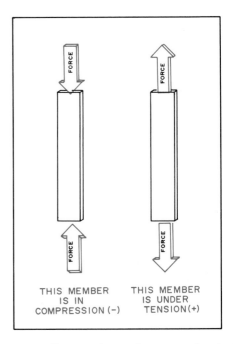

FIGURE 21-3 Structural members may be in tension or in compression.

ample, forces A and B produce a single force, C (called a resultant).

*Coplanar forces:* Forces that are acting in a single plane. This could include two or more forces. Those in Fig. 21–5 are coplanar.

*Concurrent forces:* Forces that are acting on a single point. This could include two or more forces. Those in Fig. 21–5 are concurrent. Concurrent forces may or may not be in the same plane (coplanar or noncoplanar).

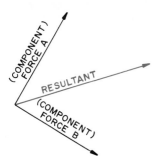

A FORCE SYSTEM

FIGURE 21-4 A component is an individual force. These components form a vector diagram.

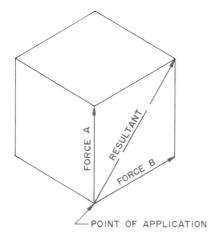

FIGURE 21-5 These are concurrent, coplanar forces.

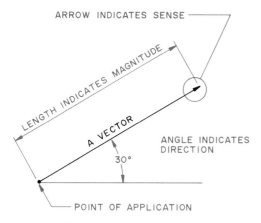

ARROW INDICATES SENSE

LENGTH INDICATES MAGNITUDE

A VECTOR

ANGLE INDICATES DIRECTION

30°

POINT OF APPLICATION

FIGURE 21-6 A vector represents the direction, point of application, sense, and magnitude of a force.

***Direction of a force:*** The inclination of a force in degrees, measured from a coordinate system such as horizontal and vertical planes (Fig. 21-6).

***Equilibrant:*** A force that is opposite and will counterbalance all the forces of a force system (the resultant). When forces acting on an object cancel each other, the object remains at rest. It is in a state of equilibrium.

***Force:*** An action (push or pull) that causes motion, a change in motion, or a change in shape. All forces have (1) magnitude, (2) direction, (3) sense, and (4) a point of application (Fig. 21-6).

***Force system:*** All of the forces acting on an object at a point of application (Fig. 21-4).

***Magnitude of a force:*** The amount of a force, expressed in pounds or newtons. Magnitude is shown on a vector diagram by the length of the vector (Fig. 21-6).

***Moment:*** The tendency of a force to cause the body to which it is applied to rotate about a point or axis.

***Nonconcurrent forces:*** Forces that do not act upon a single point of application (Fig. 21-7). They may be coplanar or noncoplanar.

***Noncoplanar forces:*** Forces that are not all acting in a single plane (Fig. 21-8).

***Point of application:*** The point at which a force is applied to an object (Fig. 21-6).

***Resultant:*** A single force that can replace all the forces in a force system. It has the same effect on the object as all the individual components of a force system (Fig. 21-4).

***Sense:*** A vector can point toward or away from the point of application. This is called the sense of the

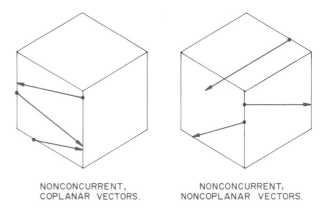

NONCONCURRENT, COPLANAR VECTORS.

NONCONCURRENT, NONCOPLANAR VECTORS.

FIGURE 21-7 Nonconcurrent vectors may lie in the same plane or in different planes.

FIGURE 21-8 These are concurrent, noncoplanar forces that are acting in several different planes.

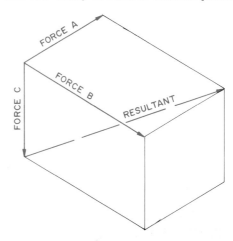

vector. It is shown by an arrowhead on the end of the vector line (Fig. 21–6).

*Space diagram:* A drawing showing the spatial relationship between the structural members of an object. A space diagram is shown in Fig. 21–4.

*Statics:* The study of forces and force systems that are in equilibrium.

*Tension:* A condition in a member produced by the application of opposite pulling forces. A member under tension tends to increase in length. On structural diagrams, tension is represented by a plus sign (+) (Fig. 21–3).

*Vector:* The representation of a quantity by a line drawn to scale to indicate magnitude, direction, sense, and point of application (Fig. 21–6).

*Vector diagram:* A diagram containing vectors drawn to scale that represent the forces involved in a particular situation (Fig. 21–4).

## UNITS USED ON VECTOR DIAGRAMS

The following metric and customary units are used on vector diagrams.

| QUANTITY | METRIC | CUSTOMARY |
|---|---|---|
| Velocity | Meters per second (m/s) | Miles per hour (MPH) |
| Angles | Degrees (°) | Degrees (°) |
| Force | Newtons (N) or kilonewtons (kN) | Pounds (LB) |
| Weight (mass) | Kilograms (kg) | Pounds (LB) |

A *newton* is the amount of force required to accelerate a mass of one kilogram one meter per second squared. Expressed as an equation, this reads $N = kg \cdot m/s^2$. When a product, such as a truss, is subject to a known load in kilograms, the forces along the members can be found in newtons by multiplying the loads in kilograms by the gravitational acceleration, $9.81 \text{ m/s}^2$. A *kilonewton* (1000 newtons) is often used because the newton is a small unit.

## DRAFTING SUGGESTIONS

Since the solution to vector analysis problems depends on measuring the length of the resultant found, it is essential that everything be drawn very accurately. Use a sharp, hard lead to get a thin line. Measure all distances carefully. Use as large a scale as is practical; the larger the scale, the more accurate the readings.

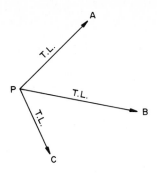

FIGURE 21–9 In a concurrent, coplanar force system, all vectors appear true length.

## CONCURRENT, COPLANAR FORCE SYSTEMS

When several forces are in the same plane and all act upon a common point of application, the system is coplanar and concurrent. All the vectors appear true length on one view (Fig. 21–9). This diagram can be analyzed to produce a resultant, which represents the effect of the three individual forces shown.

### Addition of Concurrent, Coplanar Forces Acting in Opposite Directions

Two farmers decided to have a tractor pull. They connected their tractors together with a chain and pulled in opposite directions (Fig. 21–10). Tractor A developed a total force of 5 kN, while tractor B could only develop 3 kN. The graphical solution shown by the vector diagram indicates a difference of 2 kN and

FIGURE 21–10 Concurrent, coplanar vectors can be added graphically.

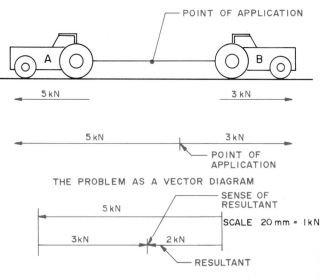

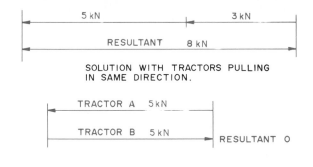

SOLUTION WITH TRACTORS PULLING
IN SAME DIRECTION.

SOLUTION IF TRACTORS PULL AGAINST
EACH OTHER BUT HAVE SAME POWER.

FIGURE 21–11  Graphical addition of concurrent, co-planar vectors acting opposite each other.

shows the sense of this resultant. If both tractors pulled in the same direction, the resultant force would have been 8 kN. If both tractors developed the same force, neither could move the other, and the vector diagram would be in equilibrium (Fig. 21–11).

Since most forces are not applied directly opposite or with each other, resultants can be found using a parallelogram or a polygon method as shown in the following examples.

## Addition of Concurrent, Coplanar Vectors

*Parallelogram method*  The steps for finding the resultant for two vectors are shown in Fig. 21–12. Since this is a concurrent, coplanar system, the vectors lie in the same plane and act upon a common point. The goal is to find the resultant that will replace the effects of vectors A (300 N) and B (400 N) and the direction of the resultant. This is found by drawing lines from the ends of each vector parallel with the other vector. Connect the point of application with the intersection of these parallel lines. This distance is the resultant, and its size and direction can be measured on the drawing.

An example of using vectors to solve a velocity problem appears in Fig. 21–13. The SI unit for velocity is meters per second (m/s). In this example, an airplane is flying at an air speed of 250 m/s heading 240° from north. A wind is blowing at 35 m/s in a direction 300° from north. What will be the actual air speed and direction of flight of the aircraft? This simple problem can be solved using the parallelogram technique described in Fig. 21–12.

The steps for finding the resultant for three or more concurrent, coplanar vectors using the parallelogram method are shown in Fig. 21–14. First, find the resultant of two of the vectors. Then, using this resultant and the third vector, find the resultant pro-

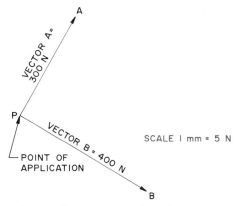

1. Lay out the vectors to scale in their known magnitude, sense, and direction.

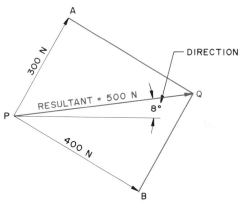

2. From the end of vector A, draw a line parallel with vector B. From the end of vector B, draw a line parallel with vector A. This locates point Q. Draw the resultant from P to Q. Measure the length of PQ to find the resultant force. The direction can be measured in degrees from a reference plane, as a horizontal plane.

FIGURE 21–12  Finding the resultant of two concurrent, coplanar vectors, using the parallelogram method.

FIGURE 21–13  Velocity problems can be solved using vector analysis.

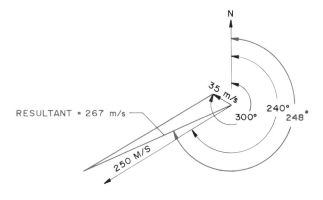

AIR SPEED 267 m/s
DIRECTION 248° FROM
NORTH.

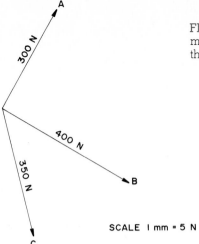

**FIGURE 21-14** Finding the resultant when three or more concurrent, coplanar vectors are known, using the parallelogram method.

SCALE 1 mm = 5 N

1. Lay out the vectors to scale in their known magnitude, sense, and direction.

duced by them. This is the resultant that produces the same effect as the three independent vectors.

A force triangle can be used to simplify the solution of concurrent, coplanar vector problems. Examine the problem in Fig. 21-15. When using the parallelogram method, the resultant is the diagonal of a parallelogram. The opposite sides of a parallelogram are equal; therefore, one vector can be moved so that its tail (without the arrow) touches the head (with the arrow) of the other vector. The vectors retain their magnitude and direction.

It is possible to resolve a known resultant into its components if the sense and direction of the components are known. Examine the forces in Fig. 12-16. The resultant is known, as are the direction and sense of each component. The size of the components can be found by placing them tail to tail with the resultant and constructing a parallelogram with the resultant as the diagonal. Measure each component to find its size.

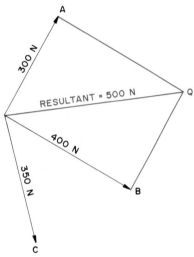

2. Find the resultant of two of the vectors. Vectors A and B were used in this example.

**FIGURE 21-15** How to use a force triangle to find a resultant of two concurrent, coplanar vectors.

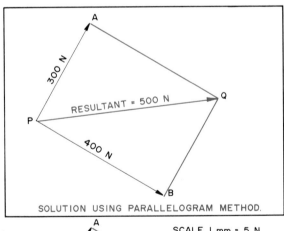

SOLUTION USING PARALLELOGRAM METHOD.

SCALE 1 mm = 5 N

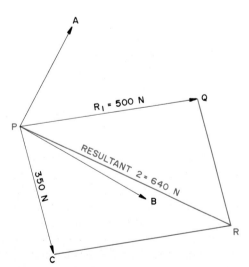

3. Using resultant 1 and the third vector, C, construct another parallelogram. The resultant, PR, is the resultant of all three vectors.

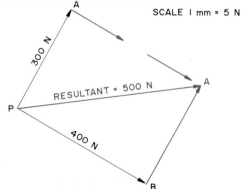

Move vector A so that its tail is touching the head of vector B and it has the same magnitude and direction as before it was moved. Connect the tail of vector B to the head of vector A to find the resultant.

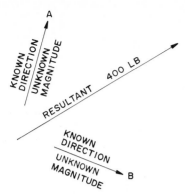

1. The problem: The direction, sense, and magnitude of the resultant are known. Find the magnitude of the two components shown.

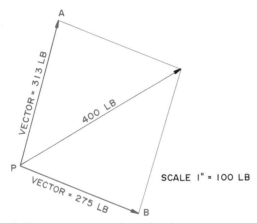

2. Place the vectors tail to tail with the resultant. Draw a parallelogram using the resultant as the diagonal. Measure the magnitude of each vector.

FIGURE 21-16 How to find the components for a known resultant for a concurrent, coplanar vector diagram.

## Addition of Concurrent, Coplanar Vectors

*Polygon method* The steps for finding the resultant for three or more vectors using the polygon method are shown in Fig. 12-17. A polygon is a closed geometric figure bounded on all sides by straight lines. The procedure involves connecting the vectors head to tail, keeping each in its original magnitude, sense, and direction. The resultant is the distance between the point of application and the head of the last vector.

## CONCURRENT, NONCOPLANAR FORCE SYSTEMS

When several forces lie in more than one plane yet all act upon a common point of application, the sys-

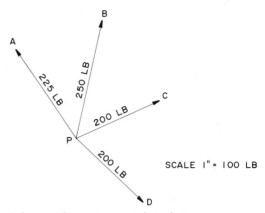

1. Lay out the vectors to scale in their known magnitude, sense, and direction.

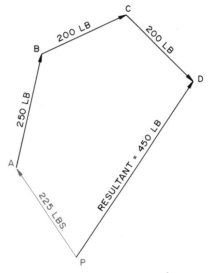

2. Leave one vector in its original position. Connect to its head the tail of the next vector, keeping its magnitude, sense and direction, the same as on the original vector diagram. Repeat this for each vector. Connect the point of application with the head of the last vector to find the resultant.

FIGURE 21-17 Finding the resultant for three or more vectors, using the polygon method.

tem is concurrent and noncoplanar. Because the vectors lie in several planes, more than one view is necessary to analyze the spatial relationships. The resultant can be found when the true projections are given in two adjacent orthographic views.

Concurrent, noncoplanar force systems can be analyzed using the parallelogram and polygon methods.

## Adding Concurrent, Noncoplanar Vectors

*Parallelogram method* The steps for finding the resultant using the parallelogram method are shown in Fig. 21–18. Three forces are acting from a single point, but they are not in the same plane. First, use two of the vectors to form a parallelogram and find its resultant. Rotate the resultant and find its true length. Then, using the true-length resultant and the third vector, form another parallelogram and find the second resultant. Revolve it to find its true length. This is the resultant that represents the effect of the three individual vectors.

*Polygon method* The steps for finding the resultant using the polygon method are shown in Fig. 21–19. The vectors are connected head to tail, with each maintaining its original magnitude, sense, and direction. The point of application is connected to the head of the last vector, producing the resultant. If the resultant is not parallel with a plane of projection, it must be revolved until it appears true size.

## EQUILIBRIUM

Equilibrium occurs when opposing forces are equal and the object subjected to them is at rest. If two forces are pulling in opposite directions and develop the same force, neither will be able to move the other (Fig. 21–20). The vector diagram represents them as being in equilibrium.

Three known vectors acting on a single point and in the same plane can be checked for equilibrium by drawing a vector triangle (Fig. 21–21). The space diagram shows the magnitude, sense, and direction. Connect the vectors head to tail and try to form a vector triangle. If a triangle is formed, the forces are in equilibrium.

In engineering design situations, the load to be carried is often known and the forces on the supporting structure must be found. When these forces are known, the proper type and size of structural members can be chosen (Fig. 21–22).

An example is given in Fig. 21–23. A weight is held away from a wall with a horizontal support. A

FIGURE 21–18   Finding the resultant of concurrent, noncoplanar vectors, using the parallelogram method.

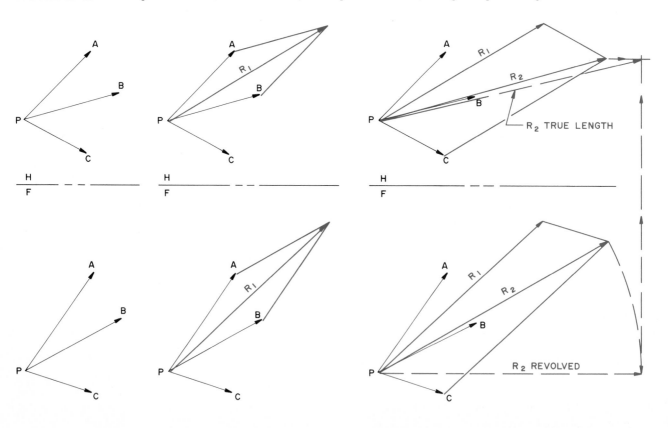

1. Locate the vectors in their proper magnitude, sense, and direction in two orthographic views.

2. Find the resultant, $R_1$, of any two of the vectors. Vectors A and B were used in this example.

3. Use resultant $R_1$ and the third vector to form a parallelogram and find resultant $R_2$. Revolve $R_2$ to find its true length. This is the magnitude of a single force that represents all three vectors.

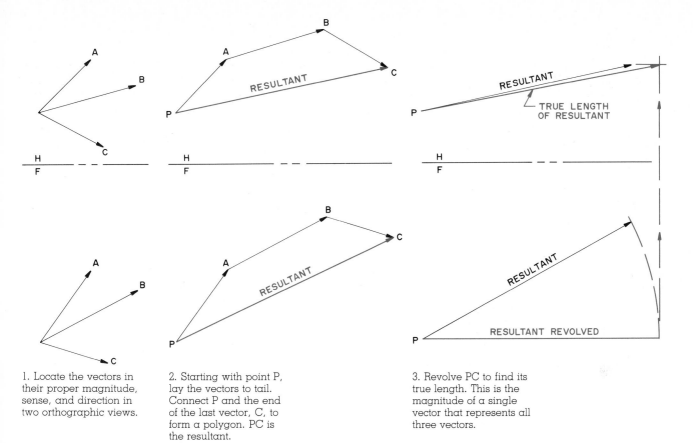

1. Locate the vectors in their proper magnitude, sense, and direction in two orthographic views.

2. Starting with point P, lay the vectors to tail. Connect P and the end of the last vector, C, to form a polygon. PC is the resultant.

3. Revolve PC to find its true length. This is the magnitude of a single vector that represents all three vectors.

**FIGURE 21–19** Finding the resultant of concurrent, noncoplanar vectors, using the polygon method.

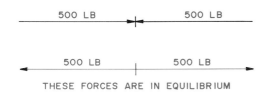

THESE FORCES ARE IN EQUILIBRIUM

**FIGURE 21–20** Opposite equal forces are in equilibrium.

cable is fastened to the weight, runs over the end of the support, and is fastened to the wall. The cable is in *tension*, and its sense is *away from* the point of application. The support is in *compression*, and its sense is *toward* the point of application. These principles hold true for all force diagrams and are used to determine sense.

To find the forces in the cable and the support needed to hold the weight in equilibrium, draw a vector triangle, using the known weight for one side. Draw the vectors for the cable and support in their known direction, connecting them head to tail. Measure their lengths to ascertain the needed forces.

**FIGURE 21–21** These concurrent, coplanar forces are in equilibrium.

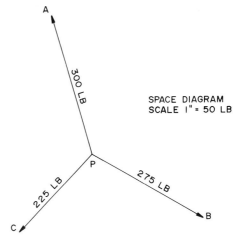

SPACE DIAGRAM
SCALE 1" = 50 LB

PROBLEM: ARE THESE FORCES IN EQUILIBRIUM?

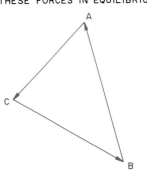

SOLUTION: LAY OUT THE FORCES IN A VECTOR TRIANGLE. IF IT CLOSES THE FORCES ARE IN EQUILIBRIUM.

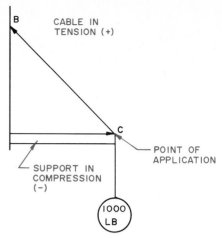

CABLE IN
TENSION (+)

POINT OF
APPLICATION

SUPPORT IN
COMPRESSION
(−)

1000
LB

1. The space diagram gives the direction
and sense of each force.

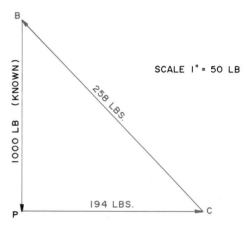

SCALE 1" = 50 LB

258 LBS.

1000 LB (KNOWN)

194 LBS.

2. Draw the known vector to scale in its
proper magnitude, sense, and direction.
Draw he other vectors head to tail in
their proper sense and direction to form
a vector triangle. Measure their lengths
to scale to determine their magnitude.

**FIGURE 21–23** The forces supporting a known load
can be found by using vector analysis.

**FIGURE 21–22** This electrical transmission line support is designed so that the combination of forces acting upon it are in equilibrium. (Courtesy of Kansas Gas and Electric Company.)

## BOW'S NOTATION

Bow's notation, a system for designation of coplanar forces, is widely used in engineering practice. With this system, a letter is placed in the spaces between the system of vectors (Fig. 21–24). One commonly used system uses capital letters placed outside the system in a clockwise direction between the external forces and lowercase letters between the interior forces (Fig. 21–25). A force is designated by the letters appearing on either side of it. The first letter identifies the tail, while the second letter identifies the head. The letters are read in a clockwise direction.

Bow's notation can also be applied to parallel forces (Fig. 21–26). The letters are recorded between forces in a clockwise direction. Force 1 is identified as AB, with A at the tail and B at the head. Force 2 is BC, and force 3 is CD. Reaction 1 is EA, and reaction 2 is DE.

## STRESS ANALYSIS OF TRUSSES

A *truss* is a rigid framework of structural members used to support a load while spanning a long distance. A typical example is a truss designed to support the roof of a building (Fig. 21–27). A truss must be analyzed to determine the stresses in each mem-

ber and which members are in tension and compression. To do this, the engineer needs to know the size of the truss, the location of each member, the loads to be carried, and the directions of these loads. Loads on a truss include dead and live loads. *Dead loads* are the weights of the materials in the truss plus other permanently installed material, such as roofing materials. *Live loads* are those that are occasionally applied to the truss, such as snow and wind loads.

Stress analysis requires a knowledge of statics. *Statics* is a branch of mechanics that deals with the equilibrium of forces. The application of the laws of statics makes it possible to find the stresses produced in trusses by the loads to which they are subjected. These stresses can be found mathematically

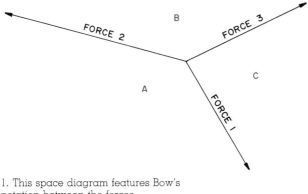

1. This space diagram features Bow's notation between the forces.

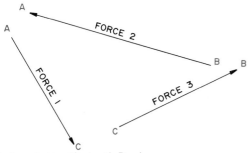

2. Each force is identified with Bow's notation, working in a clockwise direction. The first letter identifies the head and the second the tail.

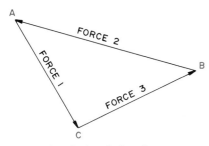

3. The forces are identified and placed head to tail to form a vector triangle.

FIGURE 21–24 A vector triangle with forces identified using Bow's notation.

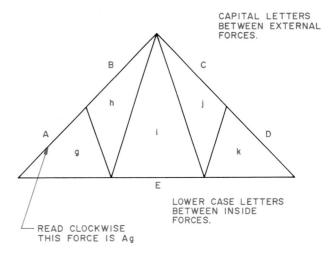

FIGURE 21–25 Identifying forces with Bow's notation when inside and outside forces are present.

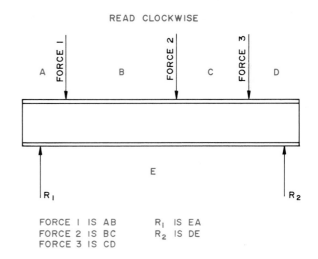

FORCE 1 IS AB    R₁ IS EA
FORCE 2 IS BC    R₂ IS DE
FORCE 3 IS CD

FIGURE 21–26 Using Bow's notation to identify forces on a beam.

FIGURE 21–27 A typical bow-type roof truss.

or graphically. In this chapter, graphical solutions are illustrated.

Each common point at which members of a truss meet can be treated as a separate stress problem. This joint is called a *free body*. Each joint in a truss must be held in static equilibrium by the forces in the members and any external loads or reactions applied to that joint. The forces are considered to pass along the center line of the structural member.

When considering stress analysis, there are three laws of static equilibrium to remember.

1. The algebraic sum of all the applied horizontal forces and the resulting horizontal reactions on a free body must equal 0. If it does not, the body will move horizontally.

2. The algebraic sum of all the applied vertical forces and the resulting vertical reactions acting on a free body must equal 0. If it does not, the body will move vertically.

3. The algebraic sum of the moments of all forces acting about any point in a free body must equal 0. If it does not, the body will *rotate* in the direction of the larger moment. *Moment* is the tendency of a force to rotate the body to which it is applied.

## DETERMINING TENSION AND COMPRESSION

The engineer must examine a drawing of the truss and determine for each joint which members are in tension and which are in compression. In a properly designed Fink truss, the forces appear as shown in Fig. 21–28. The various sections of each structural member are identified using Bow's notation. Even though loads may be distributed uniformly along a structural member for design purposes, they are considered to be concentrated at the joints.

Following are the steps for determining which members are in tension or compression.

1. Draw a free body diagram, separating each joint in the truss (Fig. 21–29). This establishes each joint and its members for analysis.

2. Place a letter of the alphabet between each joint and in each space using Bow's notation procedures. Capital letters indicate external forces. Lowercase letters indicate internal forces. Forces with two capital letters, such as AB and BC, are external forces. Forces noted as a capital and lowercase letter, such as Ag

FIGURE 21–28  The forces in a typical Fink truss.

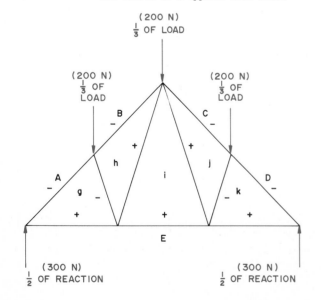

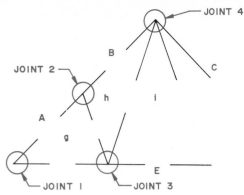

1. Draw a free body diagram and identify sections using Bow's notation.

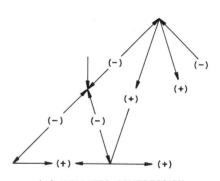

MAXWELL STRESS DIAGRAM FOR THE FINK TRUSS.    SCALE : 1 mm = 30 kg

2. Draw a Maxwell stress diagram for the Fink truss.

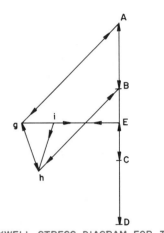

(−) INDICATES COMPRESSION
(+) INDICATES TENSION

3. Using the Maxwell diagram, determine the sense of each force at each joint.

FIGURE 21–29  Determining which members of a truss are in tension and which are in compression.

or Bh, are external truss-member forces. Forces noted with two lowercase letters, such as gh or hi, are internal truss-member forces.

3. Draw a Maxwell stress diagram. The procedure for drawing this is explained later in this chapter and is

shown in Fig. 21–31. It represents the loads on the truss drawn to scale.

4. Tension and compression are determined for each member by reading the Maxwell stress diagram and applying what is found to the free body diagram. When the stress sense has been determined for all members, those in compression (–) have arrows pointing *toward* the joint, while those in tension (+) have the arrows pointing *away from* the joint.

Begin the analysis with joint 1 and find whether the two members are in tension or compression. To do this, refer to the free body diagram. Move around joint 1 clockwise, beginning with member Ag, then proceeding to member gE, and so on.

The Maxwell stress diagram shows that Ag is going downhill to the left. Put an arrow on the downhill end g. This is the sense of the force. Superimpose the sense of Ag onto the free body diagram. Since the arrow points toward the joint, the member is in compression (–).

Next, move from g to E. On the Maxwell stress diagram, the arrow goes from g to E. Superimpose the sense of gE onto the free body diagram. Since the arrow points away from the joint, the member is in tension (+).

This analysis is repeated for each of the other joints, resulting in a free body diagram that displays the sense of each member at each joint shown.

## A SIMPLE TRUSS ANALYSIS

The simplest truss available is a triangle with no internal bracing having a load at one corner and support (reactions) at two corners. Because this truss is symmetrical, the resultant loads are each half the total load. Following are the steps for performing a stress analysis of this truss (Fig. 21–30).

1. Make a free body diagram of joint 1. The forces to be considered are the resultant, CA, and members Ad and dC.

2. Draw a vector diagram with the known side, the resultant, to scale. Because it is resisting the downward force, its sense is toward the joint. Put an arrow pointing up. The angular directions of Ad and dC are indicated on the truss drawing. Transfer these to the vector diagram. The sense of each is found by realizing that the vectors must go head to tail if the diagram is to achieve equilibrium. Therefore, Ad is in compression (toward the joint) and dC is in tension (away from the joint). The magnitude of Ad and dC can be found by measuring their length on the scaled vector diagram. These are the forces needed to keep the truss in equilibrium, and the structural members chosen must be able to withstand these stresses.

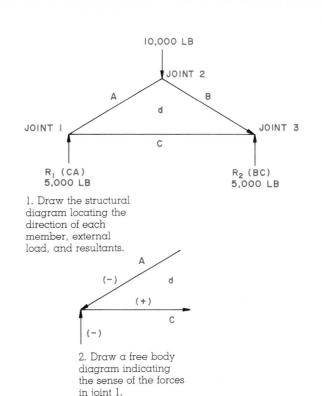

1. Draw the structural diagram locating the direction of each member, external load, and resultants.

2. Draw a free body diagram indicating the sense of the forces in joint 1.

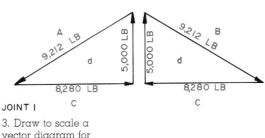

3. Draw to scale a vector diagram for joint 1. Measure the length of each side to find its magnitude.

FIGURE 21–30 Stress analysis of a simple truss.

## GRAPHICAL STRESS ANALYSIS OF A FINK TRUSS USING A MAXWELL STRESS DIAGRAM

The graphical analysis determines the sense of the force in each member at each joint and the magnitude of that force. First, make a scale structural diagram of the truss, being careful to make each member the correct length and at the correct angle. The line representing each member is actually its axis. Using Bow's notation, identify each section of the truss. Then, construct a free body diagram of each joint and indicate the sense of each member (Fig. 21–31). Because the truss is symmetrical and is symmetrically loaded, the resultants can be found by dividing the total force of 6000 kN by two, giving a reaction of 3000 kN at each end of the truss.

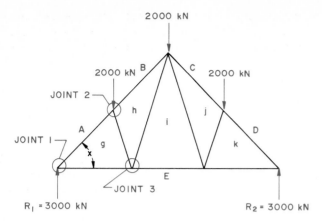

SCALE 1 mm = 60 mm

1. Make a scale structural diagram of the truss. Be certain to draw each member the proper length and at the correct angle. Apply Bow's notation and the forces to the diagram.

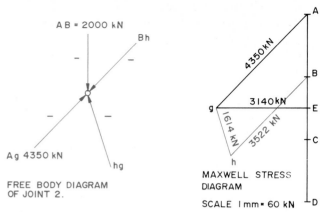

FREE BODY DIAGRAM OF JOINT 2.

MAXWELL STRESS DIAGRAM

SCALE 1 mm = 60 kN

3. Construct a free body diagram of joint 2. Then, prepare a Maxwell diagram to solve for hg and Bh. Draw these at the angle at which they appear on the free body diagram. Determine their magnitude by measuring their length on the Maxwell diagram. Determine their sense on the free body diagram. The sense of two of the vectors, AB and Ag, is known. The others can be found by forming a vector diagram, connecting the four forces head to tail.

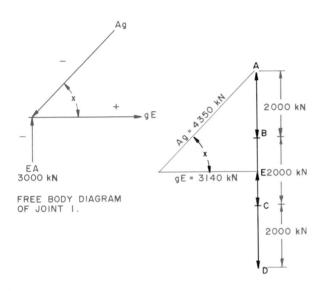

FREE BODY DIAGRAM OF JOINT 1.

2. Construct a free body diagram of joint 1. Then, prepare a Maxwell diagram for joint 1. Lay out to scale known forces AB, BC, and CD, giving their sense and direction. From E, the midpoint, draw Eg, which is horizontal. From A, draw gA at angle X. The point at which gA crosses Eg locates g. Measure each line to find its magnitude. The sense of vector EA is known. The others can be found by forming a vector diagram, connecting the three forces head to tail.

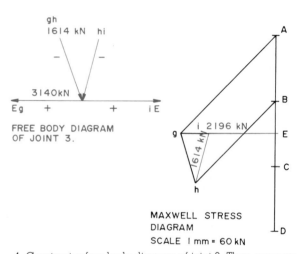

FREE BODY DIAGRAM OF JOINT 3.

MAXWELL STRESS DIAGRAM

SCALE 1 mm = 60 kN

4. Construct a free body diagram of joint 3. Then, prepare a Maxwell diagram to solve for hi and iE. Draw these at the angle at which they appear on the free body diagram. Determine their magnitude by measuring their length on the Maxwell diagram. Their sense is found on the free body diagram. The sense of two vectors, Eg and gh, is known. The others can be found by forming a vector diagram, connecting the four forces head to tail.

FIGURE 21–31 Graphical analysis of a Fink truss using a Maxwell stress diagram.

The actual analysis of each joint is made separately, using a Maxwell stress diagram. Beginning with joint 1, note that there are two unknowns, Ag and gE. The Maxwell stress diagram is laid out by connecting the three forces, AB, BC, and CD, head to tail, because they are parallel forces and can be added. E is the midpoint of these forces. Draw string Ag through A at the same angle as it is on the truss drawing. Construct a line perpendicular to the force line at E. Extend it until it crosses Ag at point g. Measure Ag and gE to find their true force. Their sense can be determined by examining the free body diagram and the stress diagram. Resultant EA is known to have an upward sense, so the arrow points toward the joint. Since EA is upward and the joint is in equilibrium, Ag must have its sense toward the joint. To bring the free body diagram into equilibrium, gE must have its sense away from the joint.

An examination of the second joint reveals that there are four forces. Two of the forces are already known: AB is the downward external force, and Ag was found when solving joint 1. The problem is to solve for the two unknowns, Bh and hg. On the Maxwell stress diagram, draw forces Bh and hg at the same angle as they appear on the truss diagram until they intersect. Their magnitudes can be measured on the drawing. Make a free body diagram of the joint. The directions of AB and Ag (toward the joint) are already known. The sense of the other vectors is found by laying them off head to tail with those known. Since the joint is in equilibrium, they both point toward the joint.

The third joint has four forces, two of which are already known, gh and Eg. To find the remaining two forces, hi and iE, draw them on the Maxwell

stress diagram until they cross. Measure their magnitude on this drawing. Their sense is determined by making a free body diagram as for the other joints.

## CONCURRENT, NONCOPLANAR VECTORS

Forces that have the same point of application but are not acting in the same plane are concurrent and noncoplanar (Fig. 21–32). In this illustration, three known forces are acting from point P, a common point of application. The problem is to find the resultant of these three forces, PA, PB, and PC. First, find the resultant for two of the forces (PA and PB were chosen) by constructing a parallelogram with the forces as sides. The diagonal is the resultant, $R_1$. Next, construct a parallelogram using $R_1$ and PC as sides. The resultant, $R_2$, is the resultant for all three forces. To find its magnitude, revolve it until it appears true length and measure its length.

Another typical problem is finding the magnitude of the components when the total force (resul-

FIGURE 21–32 Finding the resultant of three concurrent, coplanar forces.

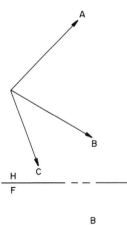
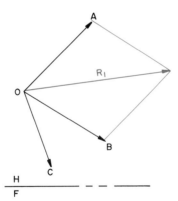
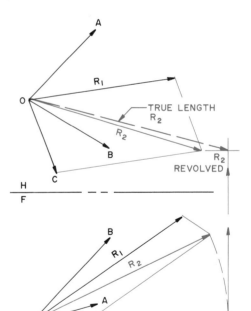
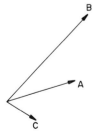
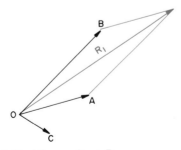

1. Given: Three concurrent, coplanar forces of known magnitude, sense, and direction. Find the resultant of these forces.

2. Find the resultant, $R_1$, of two of the forces. OA and OB were used here. Construct a paralllelogram in the top and front views using the two forces being considered. The diagonal is the resultant.

3. Construct another parallelogram using $R_1$ and the third side, OC. This produces the resultant for all three forces, $R_2$. Since it is not true length in any of the views, revolve it to find the true magnitude.

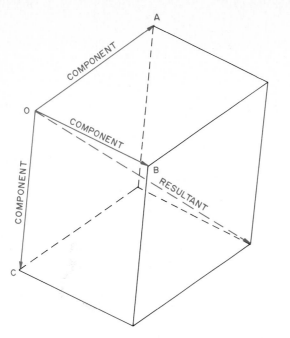

FIGURE 21–33 The components in a three-force diagram can form a parallelpiped, with the diagonal being the resultant.

FIGURE 21–34 Finding the magnitude of the components of a three-force diagram when their direction and sense are known and the resultant is known.

tant) and the direction and sense of the components are known. This can be accomplished using a parallelepiped (Fig. 21–33).

The steps for finding the magnitude of three components are shown in Fig. 21–34. Begin by drawing the opposite end of the parallelepiped. At the end of the resultant, draw two sides of the base, RE and RD, parallel to OB and OA. It is now necessary to find where force OC pierces the base. Using the two known sides, construct a plane and find where OC pierces the plane. (This is explained in Chapters 8 and 10.) The piercing point establishes the length of one of the sides of the base. Complete the base by constructing sides parallel with those shown. This locates the length (magnitude) of each of the three components, but they appear foreshortened. To find the true magnitude of the sides, find their true length by revolution and measure according to the scale for the drawing.

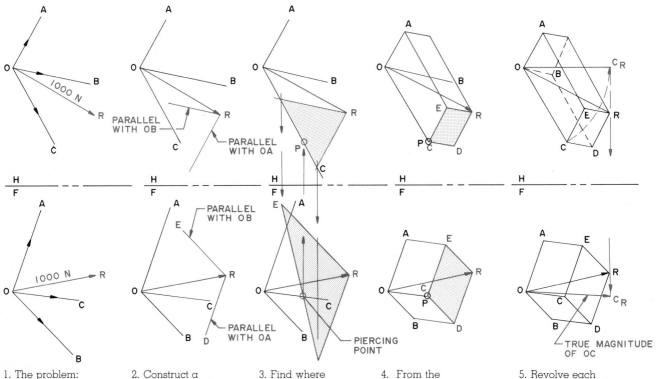

1. The problem: The resultant and the sense and direction of the components are known. Find the magnitude of the components.

2. Construct a parallelepiped using the components as sides and the resultant as the diagonal. Begin by drawing the sides of the other end of the parallelepiped parallel with the first end.

3. Find where component OC pierces the second end of the parallelepiped.

4. From the piercing point, draw the end of the parallelepiped with sides CD and CE parallel with components ER and RD. This establishes the length of the components. They are foreshortened in this view. You can complete the parallelepiped in these views.

5. Revolve each component to find its true length. Then, measure the length to scale to find the magnitude. In this example, only component OC was revolved to show the process. Components OA and OB will have to be revolved in the same manner.

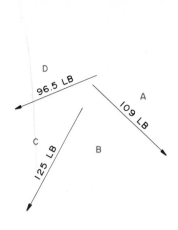

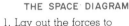

SCALE 1" = 50 LB

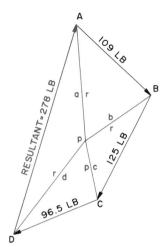

THE SPACE DIAGRAM

1. Lay out the forces to scale and identify them with Bow's notation.

THE VECTOR DIAGRAM

2. Draw a vector diagram to find the resultant. Locate point P in any convenient place and connect it to the ends of the vectors with strings. Identify each string.

THE FUNICULAR DIAGRAM

3. Transfer each string at the angle at which it appears on the vector diagram to the space diagram, forming a funicular diagram. Draw each between the two vectors with which it forms a common side. The point of intersection of the two strings falling off the diagram locates the resultant. Draw it at the same angle at which it appears on the vector diagram.

FIGURE 21–35 How to find the magnitude, direction, and line of action of a single force that will produce equilibrium of three nonconcurrent, coplanar vectors.

## NONCONCURRENT, COPLANAR VECTORS

An example of forces applied in such a way that they are nonconcurrent is given in Fig. 21–35. The problem is to find the magnitude, direction, and line of action of a single force (resultant) that will produce equilibrium.

1. On the space diagram, identify each vector using Bow's notation.
2. Draw a vector diagram to scale to find the resultant. The closing vector is the resultant, and its direction and magnitude can be measured. Then, locate a point, p, at a convenient place in the center of the diagram. Connect it with strings to the ends of the vectors and identify the strings. This resolves each vector into two components. These components (strings) are equal and opposite components of adjacent vectors. For example, component pc is common to vectors BC and CD. This actually forms a small vector diagram, one side being the resultant of the other two sides.
3. Transfer each string at its original angle to the space diagram to form a funicular diagram. The original angle is the angle each string has on the vector diagram. Draw it between the two vectors to which it

is related. For example, string pc is drawn in the area between BC and CD. The string crosses BC and CD at any convenient point. Then, draw the other strings on the funicular diagram, keeping them at their original angles and intersecting them with the previous string on a vector.

The point of intersection of the two strings that intersect off the three known vectors is the point through which the resultant will pass. The resultant can be drawn through this point parallel with the resultant in the vector diagram. This shows the magnitude, sense, direction, and point of application.

## NONCONCURRENT SYSTEMS: COUPLES

A *couple* is a system of parallel forces that are equal but opposite, are separated by some distance, and, when applied to a point on an object, cause it to rotate (Fig. 21–36). This rotational force is called a moment. A *moment* is the product of the *force* and

Vector Analysis   **547**

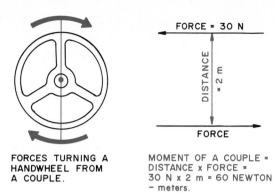

FORCES TURNING A
HANDWHEEL FROM
A COUPLE.

MOMENT OF A COUPLE =
DISTANCE x FORCE =
30 N x 2 m = 60 NEWTON
– meters.

FIGURE 21–36 A couple is a system of equal but opposite parallel forces that are separated by some distance and cause an object to rotate.

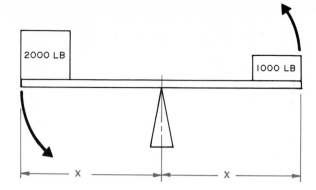

FIGURE 21–38 Because this system is not in equilibrium, it will rotate.

the perpendicular *distance* (F × D) from the point to the line of action of the force. This distance is called the *moment arm*. When the force is expressed in pounds and the moment arm in feet, the moment of the force about the point is given in foot-pounds (ft · lb). When the moment arm is expressed in meters and the force in newtons, the moment is in newton-meters (N · m). In Fig. 21–36, the moment is 30 N × 2 m, or 60 N · m.

A system of nonconcurrent, coplanar forces is in equilibrium when both the resultant force and the resultant moment equal 0.

## PARALLEL, NONCONCURRENT FORCES

Frequently, design situations involve parallel forces that are nonconcurrent. The forces on a beam are typical (Fig. 21–37). If the system is not in equilibrium, the forces could cause a rotational motion (Fig. 21–38). When the engineer knows the location and magnitude of the loads on a beam and the location of the resultant, the size of the resultant that will bring the situation into equilibrium can be found.

FIGURE 21–37 The forces on this beam are parallel and nonconcurrent.

A typical example is illustrated by a tower crane used to hoist materials during construction (Fig. 21–39). The location of the resultant is known, the counterbalance weight is known, and the length of the boom to the weight and the location of the hook on the boom are known. As the hook moves farther away from the resultant, the load it can carry will vary. The problem is to find the load the hook can safely carry.

FIGURE 21–39 A typical tower crane, with a predetermined weight on one end of the beam and a movable hook to lift a load on the other. (Courtesy of American Pecco Corporation.)

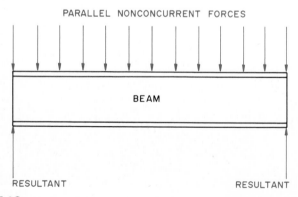

PARALLEL NONCONCURRENT FORCES

BEAM

RESULTANT

RESULTANT

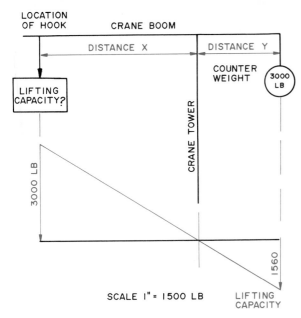

LIFTING CAPACITY WITH HOOK AT DISTANCE X

SCALE 1" = 1500 LB

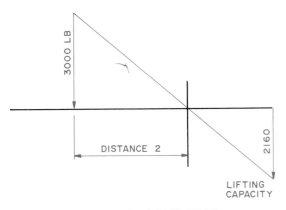

LIFTING CAPACITY WHEN HOOK IS MOVED
CLOSER TO TOWER SHORTENING BOOM.

FIGURE 21-40 The lifting capacity of the crane varies with the distance of the hook from the tower.

The solution involves the use of the law of moments, which means that the force is multiplied by the perpendicular distance to its line of action (weight × length of boom) (Fig. 21-40). In this example, the unit is in equilibrium (balance) when the effect of the moments is equal to 0, or force × boom X = weight × boom Y.

The weight can be found graphically by making a scale drawing. Draw to scale a horizontal line representing the boom length from the weight to the hook. Locate the point where the resultant meets the boom (the point of balance). Draw to scale the known vector and a line representing weight at each end of the line, *transposing* them to opposite ends of the beam. Then, draw a line from the end of the known force vector through the resultant point and

across the weight vector. Measure the length of the weight vector to scale to find its magnitude.

Notice in Fig. 21-40 what happens to the weight on the hook as its boom arm is shortened. Construction engineers know that they must reduce the weight of items lifted as they increase the length of the boom arm.

This same graphical technique can be used to find the balance point when the force and weight are known.

### Parallel, Nonconcurrent Forces on a Beam

Figure 21-41 shows a structural support for a bridge. Notice the separation in the center, which indicates that it is two separate structures. Among the design features to be considered are the parallel, nonconcurrent forces (the road bed) and the resultants (the columns).

Figure 21-42 shows a typical beam supported at two points that has three forces applied to it. The magnitude of the resultant on each end must be found so that adequate supports can be selected. The applied stresses are assumed to be working on the center line of the beam.

The problem is drawn as a space diagram, and the spaces between the vectors are identified in a

FIGURE 21-41 A poured concrete bridge structure is subject to parallel, nonconcurrent forces, while the columns represent the resultants. (Courtesy of Michigan Department of Transportation.)

Vector Analysis 549

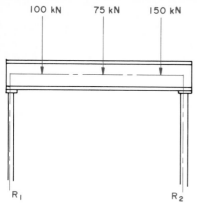

1. This beam is loaded with three parallel, unequal loads. Find reactions $R_1$ and $R_2$ and the location and size of the total resultant that will replace the three parallel loads.

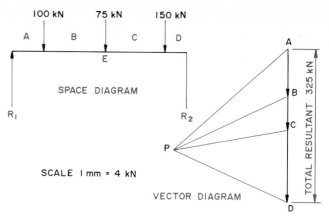

SPACE DIAGRAM

SCALE 1 mm = 4 kN

VECTOR DIAGRAM

2. Select a convenient scale and graphically add the vectors head to tail. Locate point P at any convenient location and draw strings to it from the ends of the vectors.

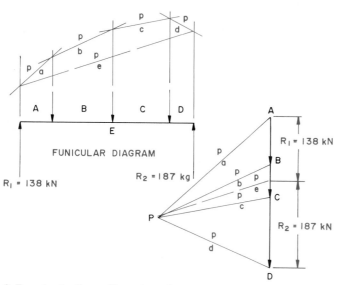

FUNICULAR DIAGRAM

$R_1$ = 138 kN

$R_2$ = 187 kg

$R_1$ = 138 kN

$R_2$ = 187 kN

3. Transfer the lines of force from the vector diagram to the space diagram. Draw a funicular diagram with pa in the A space, pb in the B space, and on through all the spaces. The last string, pe, closes the funicular diagram. Transfer pe to the vector diagram, locating point E. This divides the total force, AD, into two resultants, $R_1$ and $R_2$.

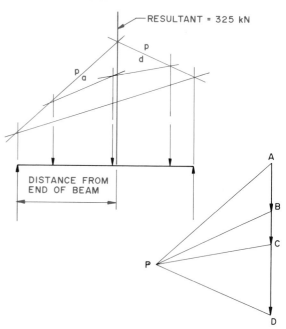

RESULTANT = 325 kN

DISTANCE FROM END OF BEAM

4. The magnitude of the resultant is the summation of the three vectors. The location of this resultant is found by extending strings pa and pd in the funicular diagram until they intersect. The resultant of the three forces, 325 kN, will act through this point at the location indicated.

**FIGURE 21–42** How to determine the resultants of a known load on a beam.

clockwise direction using Bow's notation. Next, graphically add the force vectors by joining them head to tail in a vector diagram. Since they are parallel forces, they form a single line. A point, P, is located a convenient distance and position from the vectors. Strings are drawn from P to the ends of each vector.

Next, a funicular diagram is drawn. The lines of force are extended, and strings from the vector diagram are drawn parallel to their direction on the vector diagram and located in their respective places. For example, string pa is drawn parallel to string OA in space A. The last string, oe, closes the funicular diagram. String pe is transferred at its original angle through p on the vector diagram. Its crossing the vectors divides them into two resultants. Vector AE is $R_1$ and ED is $R_2$. These are the resultants for the ends of the beam.

If it is desired to locate a third support to strengthen the beam, the best location is the point at which the summation of all downward forces is cen-

tered. The total of the downward force is found by adding the three individual forces. The location of this single force on the beam is found by extending the first and last strings, pa and pd, until they intersect. This point of intersection locates the resultant of the combined forces.

# Problems

The following problems represent the basic principles used in vector analysis. They offer the opportunity to apply these principles to a variety of situations involving the composition and resolution of forces and the determination of magnitude. It is recommended that A or A4 size paper be used. The scale selected should enable the drawing to fit the paper yet be as large as possible for maximum accuracy.

1. In Prob. 21-1, a force equal to 1100 lb is pulling east and a force equal to 650 lb is pulling west. What is the resultant?

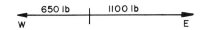

PROBLEM 21-1   Addition of vectors.

2. In Prob. 21-2, a force of 500 N is acting upward at an angle of 45° with the horizontal. Determine the horizontal and vertical components of this force.

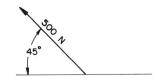

PROBLEM 21-2   A single force.

3. In Prob. 21-3 there are two concurrent, coplanar forces. Find the magnitude of the resultant and its angle from the horizontal.

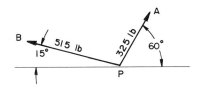

PROBLEM 21-3   Concurrent, coplanar forces.

4. In Prob. 21-4 there are three concurrent, coplanar forces. Find the magnitude of the resultant and its angle from the horizontal.

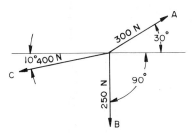

PROBLEM 21-4   A concurrent, coplanar force system.

5. Determine the resultant and its angle with the horizontal of the force system in Prob. 21-5. It is a concurrent, coplanar force system.

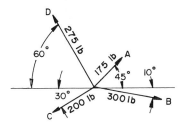

PROBLEM 21-5   Four concurrent, coplanar forces.

6. In Prob. 21-6, determine the angle of the resultant with the horizontal, and the magnitude of force PB, in this concurrent, coplanar force system.

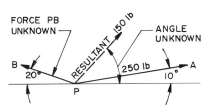

PROBLEM 21-6   Concurrent, coplanar forces with the resultant known.

7. In Prob. 21–7, determine the angle of the resultant with the horizontal, and the magnitude of force PC, in this concurrent, coplanar force system.

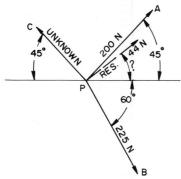

PROBLEM 21–7   A concurrent, coplanar force system.

8. In Prob. 21–8, an airplane is flying at an air speed of 275 m/s and the wind is blowing at an air speed of 25 m/s from the directions shown. What is the actual direction of flight in degrees from the north and the actual air speed?

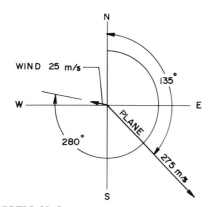

PROBLEM 21–8   Airspeed and wind diagram.

9. A boat is moving at a speed of 17 knots/hr and encounters a current from the direction shown in Prob. 21–9. This forces the boat on the course shown. What is the speed of the current?

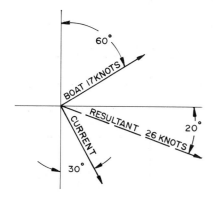

PROBLEM 21–9   Current changes the course of a boat.

10. Find the resultant of the four concurrent, noncoplanar forces in Prob. 21–10.

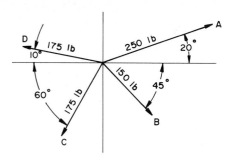

PROBLEM 21–10   Four concurrent, noncoplanar forces.

11. Find the true length of the resultant of the concurrent, noncoplanar forces in Prob. 21–11. Lay out the vectors four times as long as shown. Measure the resultant using the scale 1 mm = 10 mm.

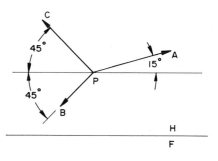

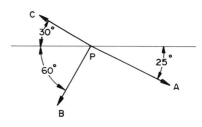

PROBLEM 21–11   Concurrent, noncoplanar forces.

12. Draw a diagram that places the force in Prob. 21–12 in equilibrium.

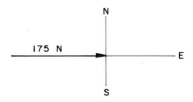

PROBLEM 21–12   A single force to be placed in equilibrium.

13. Determine if the concurrent, coplanar forces in Prob. 21–13 are in equilibrium. If they are not, indicate what change should be made in force PA to bring it into equilibrium.

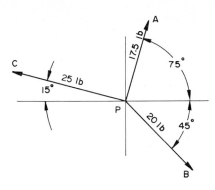

PROBLEM 21–13   Equilibrium.

14. Determine the magnitude and direction of the equilibrant for the concurrent, coplanar forces in Prob. 21–14.

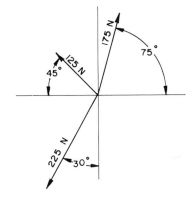

PROBLEM 21–14   Concurrent, coplanar forces in equilibrium.

15. In Prob. 21–15, a pipe frame fastened to a wall is to be designed to lift weights up to 500 lb. What are the forces in each member? Indicate whether they are in tension or compression. Then change the angle of the inclined pipe to 60°. How did this change the forces in each member?

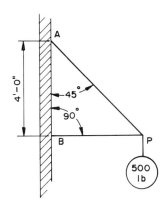

PROBLEM 21–15   A frame under tension and compression.

16. Prob. 21–16 shows a weight supported by two cables. Determine the magnitude of the tension in each cable.

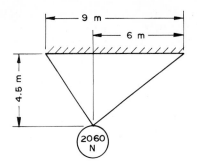

PROBLEM 21–16   Cables under tension.

17. In Prob. 21–17, a weight is supported by three cables. Determine the magnitude of the tension in each cable. Then place the ends of the cable so each is on a circle with a radius of 3 ft 0 in. from point P; find out how much this changes the forces.

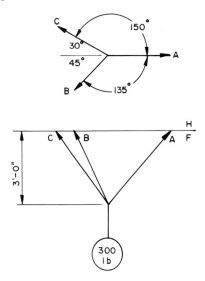

PROBLEM 21–17   Cables under tension.

18. Determine the magnitude of reactions $R_1$ and $R_2$ of the beam in Prob. 21–18.

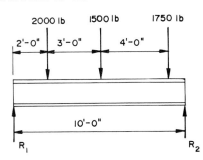

PROBLEM 21–18   A simple beam.

19. Determine the magnitude of reactions $R_1$ and $R_2$ of the beam in Prob. 21–19.

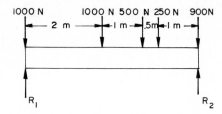

PROBLEM 21-19   A simple beam.

20. In Prob. 21-20, a crane boom has a 4000 lb counter weight located 25 ft from the balance point. What is the maximum weight it can lift if the boom is 50 ft from the hook to the balance point? If it is necessary to use a boom 75-ft long, what is the maximum weight that can be lifted?

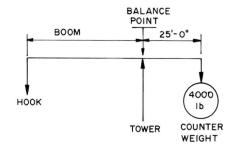

PROBLEM 21-20   A counterbalanced boom.

21. Determine the magnitude of reactions $R_1$ and $R_2$ for the truss in Prob. 21-21.

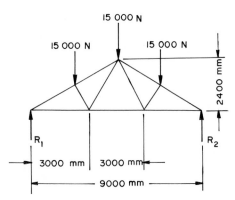

PROBLEM 21-21   A truss.

22. Determine the magnitude of the reactions $R_1$ and $R_2$ for the truss in Prob. 21-22.

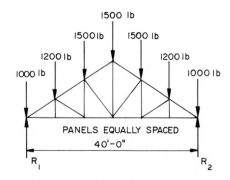

PROBLEM 21-22   A truss.

23. Determine the stresses in the members of the concurrent, noncoplanar space frame in Prob. 21-23.

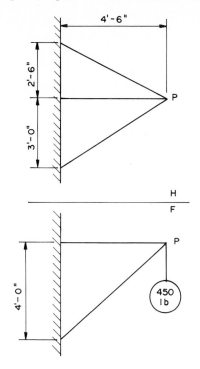

PROBLEM 21-23   A space frame.

24. Determine the stresses in the members of the concurrent, noncoplanar space frame in Prob. 21-24.

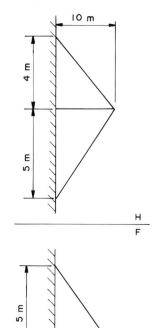

PROBLEM 21-24   A space frame.

25. Find the direction and sense of the components for the concurrent, noncoplanar force system in Prob. 21–25. The resultant is drawn to scale to represent a known force. Draw it four times as long as shown.

26. Find the resultant for the nonconcurrent, coplanar forces in Prob. 21–26. Then using a funicular diagram, indicate the magnitude, sense, direction, and point of application of the resultant.

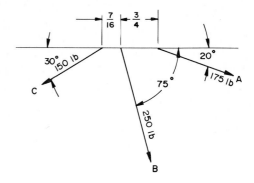

PROBLEM 21–26   Nonconcurrent, coplanar forces.

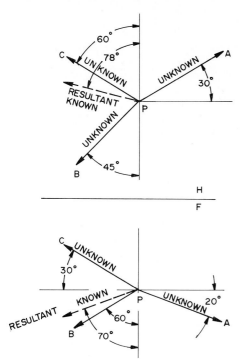

PROBLEM 21–25   A concurrent, noncoplanar force system.

27. Determine the moment of the couple in Prob. 21–27.

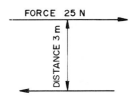

PROBLEM 21–27   A couple.

# Graphical Kinematics

The study of motion in machines is called *kinematics* or *mechanism*. It involves the design and analysis of linkages, cams, gears, and other components of machines used to produce motion. This includes an analysis of motion and of velocity and acceleration associated with motion.

One of the early phases of designing a machine involves determining the anticipated motion relationships between machine members. These are part of the scope statement of the design. For example, a part of the machine might require a constant speed input, but the output would be expected to be varied so as to offer five different output speeds. The velocity and acceleration of these members are also stated. Following the design statement, the engineer makes a graphical and mathematical analysis of these factors to determine if the proposed solution is satisfactory. This combination of graphical and mathematical analysis produces a viable design. The use of computers has greatly increased the ability of the engineer to test and meet kinematic design requirements.

As products are improved to meet competition or increase their productive capacity, motion analysis of the proposed new design is essential. While the redesigned product may look better, actual per-

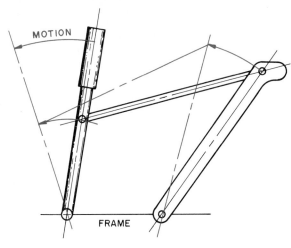

A LINKAGE MECHANISM

formance will depend on the kinematic analysis of the proposed design.

In this chapter are presented a few common linkages, gears, and cams that are used to produce motion.

## LINKAGES

The term *linkage* is generally used to identify a series of rigid links joined by pins or some other method. The symbols commonly used to represent parts of a linkage mechanism on kinematic drawings are shown in Fig. 22–1. Kinematic drawings are made with the assumption that links have no thickness, and all parts of a mechanism may be shown in one plane. A *rigid link* is a single solid part that is connected to other links or a fixed bearing. A *pin joint* is the connection between two links or a link and a fixed bearing. A *fixed bearing* is a means of fastening one end of a link to something other than another link, such as the frame of the machine. *Sliders* are moving parts in a mechanism that follow a predetermined path restricted by an internal or external guide as they slide.

FIGURE 22–1  Symbols used on linkage diagrams.

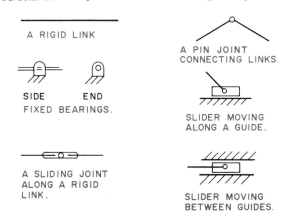

A RIGID LINK

SIDE    END
FIXED BEARINGS.

A SLIDING JOINT
ALONG A RIGID
LINK.

A PIN JOINT
CONNECTING LINKS.

SLIDER MOVING
ALONG A GUIDE.

SLIDER MOVING
BETWEEN GUIDES.

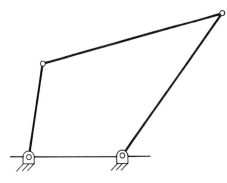

A KINEMATIC DRAWING OF THE
LINKAGE MECHANISM.

FIGURE 22–2  A typical mechanism and its linkage diagram.

The term *mechanism* refers to an assemblage of links in which, when one end is fixed, the motion of each link is related to the motion of every other link (Fig. 22–2). Examples of linkages that are not mechanisms appear in Fig. 22–3. One of these linkages is rigid; because it has no motion, it is not a mechanism. The other is not a mechanism because the motion is not constrained. When one link is moved, the motion of the other links cannot be specified.

When solving motion problems, deflection due to strains on the link are not considered because the major initial goal is to study the general motion of the mechanism. The initial analysis is usually done graphically, and the final analysis is mathematically resolved.

## MOTION OF LINKS

Links move in plane motion. *Plane motion* is when all parts of the links maintain a constant distance from a reference plane during motion (Fig. 22–4). *Rotation*

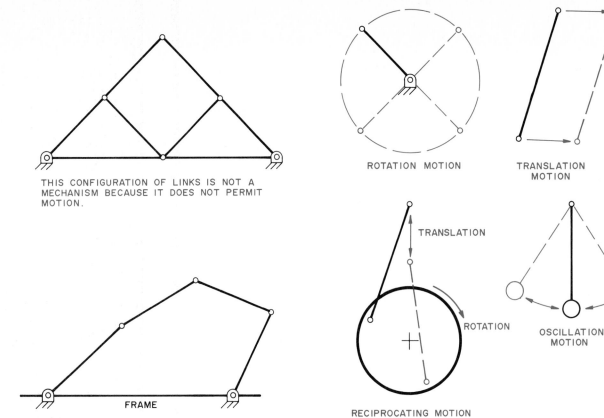

THIS CONFIGURATION OF LINKS IS NOT A MECHANISM BECAUSE IT DOES NOT PERMIT MOTION.

THIS LINKAGE PERMITS MOTION BUT THE MOTION OF THE LINKS CANNOT BE CONTROLLED THEREFORE IT IS NOT A MECHANISM.

FIGURE 22–3 Examples of linkages that are not mechanisms.

ROTATION MOTION

TRANSLATION MOTION

TRANSLATION

ROTATION

RECIPROCATING MOTION IS A COMBINATION OF TRANSLATION AND ROTATION.

OSCILLATION MOTION

FIGURE 22–4 Linkages move in plane motion. These are examples of various types of motion that occur in a plane.

*motion* occurs when all points on the link maintain constant distances from the axis of rotation. A hand crank utilizes rotation motion. *Translation* or *rectilinear motion* occurs when all points in the link move in the same direction at the same speed. *Oscillation* describes a back-and-forth rotation, such as the swinging of a pendulum. *Reciprocating motion* describes a combination of rotation and translation as illustrated by a piston and connecting rod.

## DISPLACEMENT, PATH, VELOCITY, AND ACCELERATION

*Displacement* refers to a change in location. For example, point A on the linkage in Fig. 22–5 moves to location A'. The straight-line distance is the displacement. As the point moves from A to A', it describes a circular path. This *path* is different from dis-

FIGURE 22–5 How to locate the path and displacement of parts of a linkage mechanism.

LINK 1

LINK 2

A

DISPLACEMENT OF POINT A

PATH OF POINT A

FRAME

LINK 1

LINK 2

A₁

PATH AND DISPLACEMENT OF POINT B

SLIDER B

B₁

placement because it is the actual route traveled by the point. Point B also moves, but the path is a straight-line route, so it is the same as the displacement.

Velocity is the speed at which some part of the linkage moves. Average velocity and instantaneous velocity are used in linkage design. The *average velocity* of a point on a linkage is found by dividing the total distance traveled by the time required. *Instantaneous velocity* is the velocity of a point on a linkage at a given instant. It is the limiting value of distance traveled divided by the time required *as the time interval approaches 0.*

## VECTOR NOTATION

Kinematics problems are solved using scalar and vector quantities. *Scalar quantities* are those having magnitude only and are represented by a number that stands for units such as time, weight, volume, or distance. *Vector quantities* possess magnitude and direction, such as velocity, acceleration, and displacement. The length of the vector (to scale) represents the magnitude, and the direction of the vector represents the direction of the force or motion. The arrow on the end of the vector indicates the sense of the force or the way the vector is acting. Review Chapter 21 for additional information.

## TRANSLATION MOTION

Translation or straight-line motion can be described by scalar rather than vector notation because the direction of the path remains constant (Fig. 22–6). The displacement and the path coincide. By drawing the crank in a series of positions representing successive instants in time, the linkage drawing can be made and the *slider displacement* can be found at each instant. Plotting the displacements against time produces a curve (Fig. 22–7). The change in displacement that occurs during a finite interval of time, $t_1$ to $t_2$, is $D_2 - D_1$. The average velocity of the slider during this interval is:

$$\text{Average velocity} = \frac{D_2 - D_1 \text{ (in inches or millimeters)}}{t_2 - t_1 \text{ (in seconds)}}$$
$$= \text{in./sec or mm/sec}$$

The average velocity is represented by the slope of the hypotenuse of the triangle.

## ROTATIONAL MOTION

Rotational motion is the movement in the plane of the mechanism about a fixed point. Displacement is angular because of the rotational motion. *Angular*

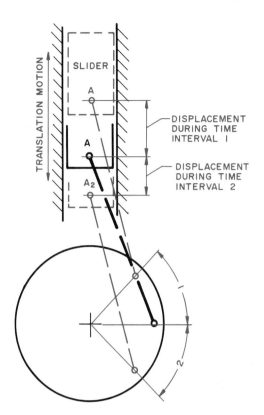

FIGURE 22–6 When translation motion exists, the path and displacement are the same.

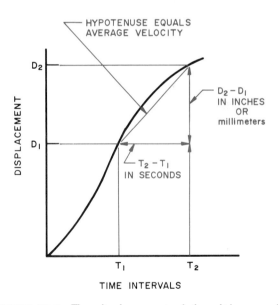

FIGURE 22–7 The displacement of the slider can be plotted against the time required for the displacement.

*displacement* is the change in the angular position of a point occuring during a time interval measured in radians. A radian is equivalent to the arc length subtending the angle divided by the radius of the arc (Fig. 22–8). One radian equals about 57.3°, and 6.28 radians equal one complete rotation (360°). Since the angle measured in radians is equivalent to the arc

$$\text{ANGULAR DISPLACEMENT} = \frac{\text{PATH (B TO B}^1\text{)(BOTH IN INCHES OR mm)}}{\text{RADIUS}}$$

$$\text{AVERAGE ANGULAR VELOCITY} = \frac{\text{ANGULAR DISPLACEMENT (RADIANS)}}{\text{TIME INTERVAL (SECONDS)}}$$

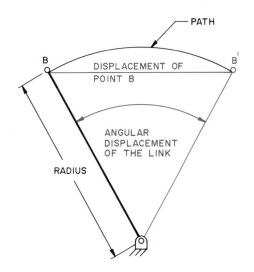

FIGURE 22–8 Angular displacement is the change in radians of the position of a point.

length subtending the angle divided by the radius of the arc, the following equation is used to find the length of the path the point moves (Fig. 22–8):

$$\frac{\text{Angular displacement}}{\text{Angle in radians}} = \frac{\text{Arc length}}{\text{Radius}} = \begin{array}{l}\text{(Both in}\\ \text{inches or}\\ \text{millimeters)}\end{array}$$

The displacement is the straight-line distance between a point and its location after motion.

Average angular velocity is the angular displacement in radians divided by the time interval in seconds. This is summarized in Fig. 22–8 as:

$$\frac{\text{Average angular}}{\text{velocity}} = \frac{\text{Angular displacement (radians)}}{\text{Time interval (seconds)}}$$

## TYPICAL APPLICATIONS

The design of linkages encompasses an almost infinite variety of situations. Some commonly used and commercially available components can be found in manufacturers' catalogs. A designer should be aware of these and utilize them when possible but should be able to develop and analyze new and possibly better solutions than those in traditional designs.

A *parallelogram linkage* is one type frequently used. A common application is in drafting machines (Fig. 22–9). In this device, rigid links are connected to discs with pins. The links move on the pins, but the discs do not rotate. This allows translation of the

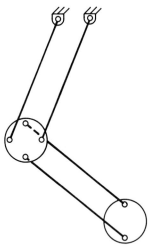

FIGURE 22–9 A typical parallelogram linkage.

straightedge used for drafting in any direction. A similar situation exists with a band-type drafting machine. The endless steel bands turn on pairs of discs. Disc A does not rotate. Discs B and C translate without rotation. The band between discs C and D permits D to translate the straightedge around the board without rotation.

The part of a belt between pulleys, or a chain between sprockets, is at a given instant a rigid link (Fig. 21–10).

An *in-line slider crank mechanism* is also frequently used. One application is in a reciprocating pump or air compressor. The input rotation is changed to reciprocating motion of the piston (Fig. 22–11). A drawing and linkage diagram of an in-line slider crank is shown in Fig. 22–12. Link A is the frame of the machine, link B is the crank, and link C is the connecting rod. The center of the crankshaft is point 1, and the crank revolves around this center.

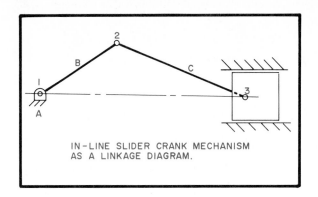

IN-LINE SLIDER CRANK MECHANISM
AS A LINKAGE DIAGRAM.

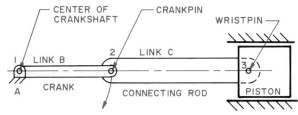

PISTON AT TOP OF STROKE

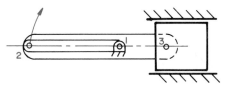

PISTON AT BOTTOM OF STROKE

FIGURE 22–12 A typical in-line slider crank mechanism.

the mechanism can be found if the magnitude and direction of the initial point are known.

In addition to mechanical links, fluid links using hydraulic or pneumatic cylinders are widely used. *Fluid links* are used to raise backhoe shovels, bulldozers, cherry picker baskets, and other such devices (Fig. 22–13). A typical linkage that includes

FIGURE 22–13 Hydraulic cylinders serve as fluid links on this self-propelled boom. (Courtesy of Mark Industries.)

FIGURE 22–10 The section of the belt between pulleys acts for an instant as a rigid link.

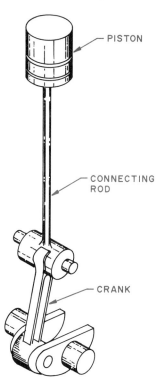

FIGURE 22–11 An application of an in-line slider crank mechanism.

Point 2 is the crankpin joining the crank and the connecting rod. Point 3 is the wristpin connecting the piston to the connecting rod. Using this diagram, the direction and magnitude of any point on

a fluid link appears in Fig. 22–14. Movement of the piston in the fluid link causes the mechanical link to pivot on a fixed connection and the end to oscillate. Figure 22–14 shows a typical kinematic drawing of this mechanism used for analysis.

A typical four-bar linkage is shown in Fig. 22–15. This is a crank rocker mechanism, one of several basic four-bar designs. Link A is the frame of the machine. Link B is the crank, which rotates about fixed center 1. Link C is the connecting rod, which joins the follower crank, link D. Connection C oscillates through a fixed angle as the crank rotates in one direction. This connection follows a cir-

FIGURE 22–14  A typical mechanism using a fluid link to produce motion.

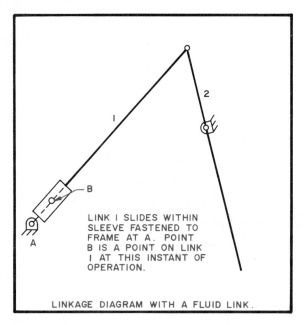

LINKAGE DIAGRAM WITH A FLUID LINK.

LINK I SLIDES WITHIN SLEEVE FASTENED TO FRAME AT A.  POINT B IS A POINT ON LINK I AT THIS INSTANT OF OPERATION.

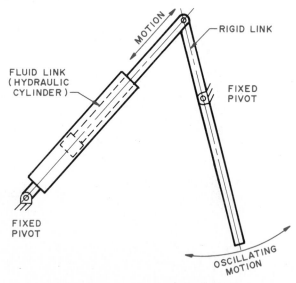

ILLUSTRATION OF LINKAGE MECHANISM WITH A HYDRAULIC CYLINDER.

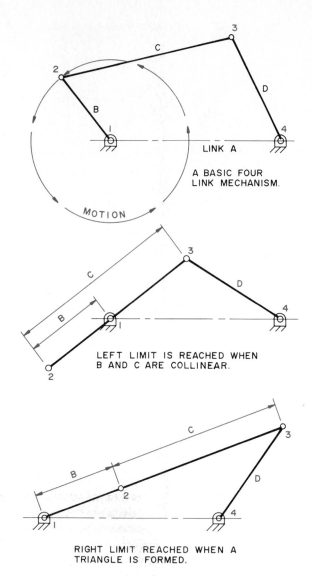

A BASIC FOUR LINK MECHANISM.

LEFT LIMIT IS REACHED WHEN B AND C ARE COLLINEAR.

RIGHT LIMIT REACHED WHEN A TRIANGLE IS FORMED.

FIGURE 22–15  A typical four-bar linkage.

cular path, the ends of which are the limits of the crank rocker mechanism. When B and C are colinear, the left limit is reached. When the linkage forms a triangle, the right limit is reached.

## CAMS

A cam is a machine component, usually a disc or a cylinder, having an irregularly shaped surface that upon rotation imparts a desired motion to another component, usually a follower, that is in contact with the cam. As the cam rotates, the follower moves in a single plane (Fig. 22–16).

The two basic types of cams are disc (or plate) and cylinder (Figs. 22–17 and 22–18). The disc cam has a follower moving in a plane perpendicular to the axis of the camshaft (Fig. 22–16). The surface of a disc cam is actually an inclined plane that causes a

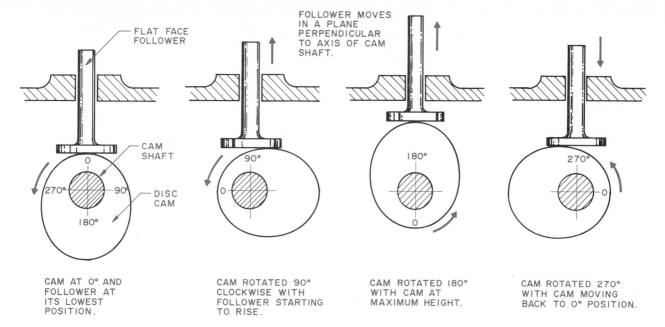

FLAT FACE FOLLOWER

FOLLOWER MOVES IN A PLANE PERPENDICULAR TO AXIS OF CAM SHAFT.

CAM SHAFT

DISC CAM

CAM AT 0° AND FOLLOWER AT ITS LOWEST POSITION.

CAM ROTATED 90° CLOCKWISE WITH FOLLOWER STARTING TO RISE.

CAM ROTATED 180° WITH CAM AT MAXIMUM HEIGHT.

CAM ROTATED 270° WITH CAM MOVING BACK TO 0° POSITION.

FIGURE 22–16 The disc cam converts the rotary motion of the camshaft to reciprocating motion in the follower. The follower moves in a plane perpendicular to the axis of the camshaft.

FIGURE 22–17 A standard disc or plate cam. (Courtesy of Ferguson Machine Company.)

FIGURE 22–18 A cylindrical external-barrel cam. (Courtesy of Ferguson Machine Corporation.)

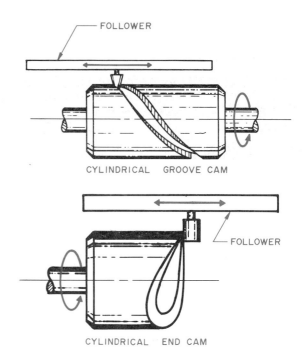

FOLLOWER

CYLINDRICAL GROOVE CAM

FOLLOWER

CYLINDRICAL END CAM

FIGURE 22–19 A cylindrical cam produces motion through a follower that is parallel with the axis of the camshaft.

change in the slope of the plane, thus producing the required motion in the follower. A cylindrical cam has a groove cut into its surface and moves a follower back and forth in a plane parallel with the axis of the camshaft (Fig. 22–19).

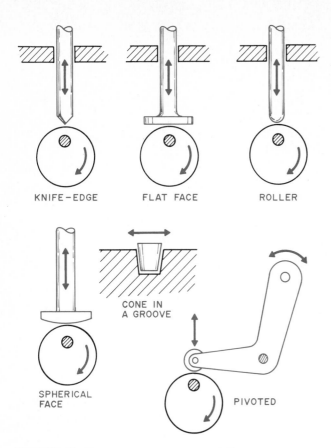

FIGURE 22–20    Basic types of cam followers.

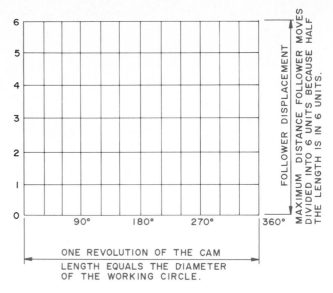

FIGURE 22–21    A grid layout for a displacement diagram drawn to a convenient scale.

## CAM FOLLOWERS

There are several basic types of followers (Fig. 22–20): knife-edge, flat-face, roller, spherical-face, and pivoted, as well as a cone follower used on cylindrical cams. The roller cam is extensively used because it has less friction and wear, operates effectively at higher speeds, and can transmit larger forces. The flat-face and knife-edge followers are limited to applications with low speeds and minimal forces. If the weight of the follower is not sufficient to keep it in contact with a cam, springs can be used.

## DISPLACEMENT DIAGRAMS

Displacement diagrams represent the motion of the follower during one revolution of the cam. The construction of the displacement diagram is the first step when designing a cam (Fig. 22–21). The *length* of the diagram represents one revolution of the cam. The length is usually drawn equal to the circumference of the working circle. The *working circle* has a radius equal to the distance from the center of the camshaft to the highest point in the cam rise (Fig. 22–22). The length of the diagram is subdivided into increments representing sections of the working cir-

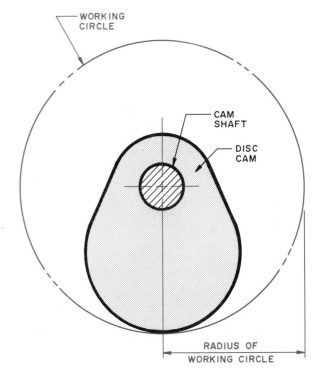

FIGURE 22–22    The working circle of a cam passes through the highest point of the cam rise.

cle. In Fig. 22–21, the length is divided into 12 equal units, each representing 30°.

The *height* of the displacement diagram represents the maximum length of travel of the cam follower in inches or millimeters. The height is called *follower displacement* and is divided into time intervals or angle of cam rotation. A *time interval* is the time it takes a cam to move the follower to a particular

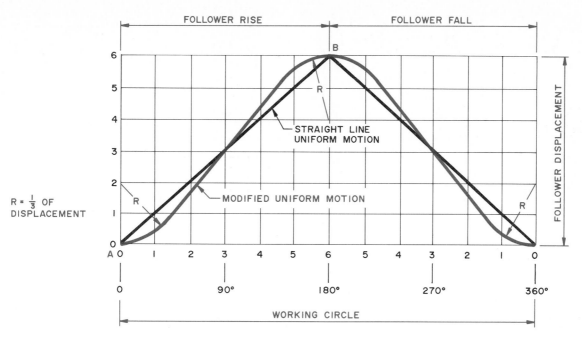

R = ⅓ OF
DISPLACEMENT

STRAIGHT LINE
UNIFORM MOTION

MODIFIED UNIFORM MOTION

FOLLOWER DISPLACEMENT

WORKING CIRCLE

**FIGURE 22–23** A displacement diagram for uniform motion and modified uniform motion.

height. The follower moves when the cam rotates through a specified number of degrees. *Rise* is that section of a cam profile, stated in degrees of rotation of the cam, during which the follower moves away from the cam axis of rotation. *Fall* is that section of a cam profile, stated in degrees of rotation of the cam, during which the follower moves toward the cam axis of rotation (Fig. 22–23).

## UNIFORM MOTION

When a cam rises and falls at a constant speed, it is said to be moving at a uniform motion. On a displacement diagram, this would appear as a straight line (Fig. 22–23). In this example, the follower rises from A to B, a height of 40 mm, through 180° revolution of the cam. Uniform motion is a straight line connecting A to B.

In actual practice, straight-line motion would produce abrupt changes of direction and velocity at two points every revolution. This would cause the follower to bounce and is therefore undesirable. The motion is *modified* by placing arcs at each point, thus producing a smooth change. The radius of these arcs is usually one-third to one-fourth the displacement. A straight line then connects these arcs.

## HARMONIC MOTION

Harmonic-motion cams are extensively used on applications requiring slow to medium speed ranges. The harmonic motion lifts the follower gradually

from the beginning position and increases its speed until halfway into the full rise. The speed then gradually decreases until the follower reaches its full height. A harmonic cam has uniformly changing follower velocity and acceleration. It results in a smooth, quiet operation.

A displacement diagram for a cam with harmonic motion appears in Fig. 22–24. The procedure for preparing the diagram is as follows.

1. Lay out the length equal to the circumference of the working circle.
2. Lay out the height (follower displacement).
3. Construct a semicircle on the height with a radius equal to half the height.
4. Divide the semicircle into the same number of parts as was half the length of the working circle.
5. Divide the length of the diagram into rise and fall. Divide each of these into the same number of parts as the semicircle. These divisions will be in degrees of rotation. Number each division to correspond with the numbers on the semicircle.
6. Project each point on the semicircle across the diagram until it crosses a line with the same number. This locates the points forming the curve. Connect these points to form a smooth curve.

## PARABOLIC MOTION

A cam with parabolic motion has uniform acceleration and deceleration. The follower has constant acceleration during the first half of the rise and constant deceleration during the second half of the rise.

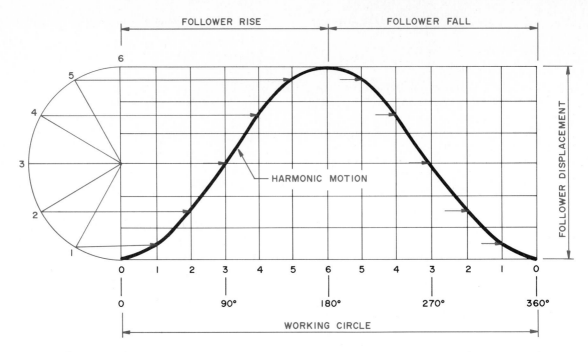

FIGURE 22–24 A typical harmonic motion displacement diagram.

This produces a smoother motion than harmonic motion.

A displacement diagram for a cam with parabolic motion appears in Fig. 22–25. The procedure for preparing the diagram is as follows.

1. Draw the length and height of the diagram.
2. Divide the diagram into several equal parts and number them.
3. Draw a line from the end of the length at any convenient angle. Using any convenient division, divide

FIGURE 22–25 A typical parabolic motion displacement diagram showing uniform acceleration and deceleration.

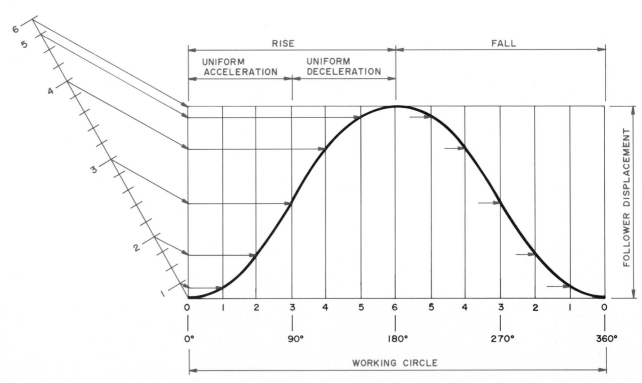

this line into equal parts. The number needed depends on the number of intervals used. The intervals are placed on the inclined line with the following number of divisions between them: 1, 3, 5, 7, 7, 5, 3, 1 for an eight-interval diagram, 1, 3, 5, 5, 3, 1 for one with six intervals. The number of intervals depends on the number of divisions used to divide up the rise on the length.

4. Connect the last division with the top of the diagram (height). Draw lines from the other divisions parallel with this until they cross the end of the diagram. Project them horizontally until they cross the vertical division with the same number.

5. Draw a smooth curve through these points to form the displacement curve.

## DWELL

*Dwell* is a period of time during the rotation of a cam when the follower does not move. It appears on a displacement diagram as a horizontal line. Dwell is shown on the multiple-motion diagram in Fig. 22–26.

## COMBINATION OF MOTIONS

The cams just discussed each produces a single type of motion. On many products, the engineer requires some combination of these to produce the desired results. When drawing a displacement diagram for a cam with a combination of motions, each is drawn separately as it occurs in the rise and fall (Fig. 22–26). In this example, the cam moves the follower half the

total height using harmonic motion through 90° of revolution. It then has a dwell for 60°, followed by a modified uniform motion to the total height through 90°. The remaining 120° are used to return the follower to the starting point using parabolic motion (uniform acceleration and deceleration).

## DRAWING A DISC CAM WITH A KNIFE-EDGE FOLLOWER ON CENTER

A knife-edge follower is used to produce harmonic motion from 0° to 180°, dwell from 180° to 210°, and modified uniform motion from 210° to 360°. The base circle is known to be 38 mm, and the follower displacement is 25 mm. Following are the steps for drawing the cam profile (Fig. 22–27).

1. Lay out the displacement diagram. In this example, 30° intervals were used. The displacement was drawn to scale. The same scale was used to draw the cam.

2. Draw the harmonic motion part of the cam by drawing the base circle and dividing it into 30° parts, numbered to correspond to the displacement diagram. Use the same scale as for the displacement diagram. Then, on each radial line, locate a point on the cam profile by extending the radial line beyond the base circle a distance equal to the length of the same line on the displacement diagram. This involves points 1 through 6 in this example.

FIGURE 22–26 A displacement diagram for a cam having a combination of motions.

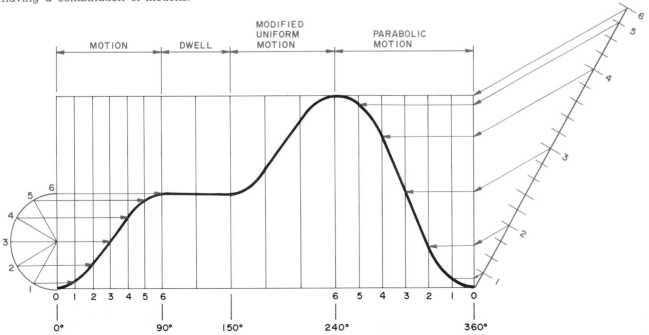

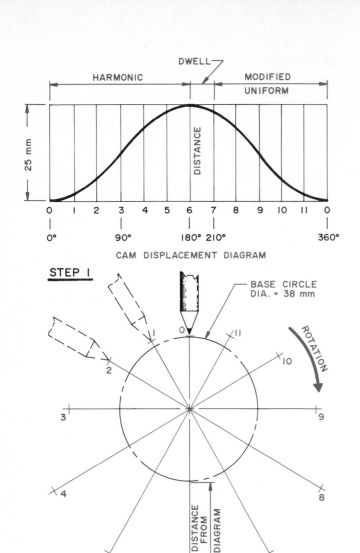

CAM DISPLACEMENT DIAGRAM

STEP I

BASE CIRCLE DIA. = 38 mm

ROTATION

DISTANCE FROM DIAGRAM

STEP 2-4

3. Locate the points for the dwell. Since the cam is not to move and the dwell is 30°, radial line 7 will have the same length as line 6. Dwell is from point 6 to point 7.

4. Now, draw the modified uniform motion part of the cam. It runs from radial line 7 back to 0 (150°). Measure the length of each numbered line on the displacement diagram and record it on the same numbered radial line on the cam drawing beyond the based circle. This motion is from 7 back to 0.

5. To complete the cam profile, draw a smooth curve between the points located.

## DRAWING A DISC CAM WITH AN OFFSET ROLLER FOLLOWER

The cam in this illustration will produce harmonic motion through 360° of rotation. Following are the steps for drawing the cam profile (Fig. 22–28).

1. Lay out the displacement diagram using the same scale as for the cam profile drawing.

2. Locate the center of the camshaft and draw the base circle from this center. Since the follower has a roller, the radius of the base circle equals the radius of the lowest part of the cam plus the radius of the follower roller.

3. Locate the center line of the offset follower. Draw the offset circle with a radius equal to the amount of offset.

4. Divide the offset circle into the same number of equal divisions as used on the displacement diagram.

5. Number the divisions on the offset circle, the same as they are on the displacement diagram. Begin with 0 and number in a direction opposite the direction of cam rotation. Point 0 is the point at which the center line of the follower is tangent to the offset circle.

6. Draw straight lines through each point on the offset circle tangent to the circle.

7. On each of these tangent lines, locate from the base circle the length of each point as it is found on the displacement diagram. Notice point 6 in the illustration. Locate each displacement distance in this manner. These points form the pitch curve, which is the path followed by the axis of the follower roller as it rolls on the cam.

8. Connect the pitch curve points to form a smooth curve.

9. Next, locate the cam profile. It will be the same shape as the pitch curve but smaller by a distance equal to the radius of the follower roller. Set a compass to the radius of the follower wheel. Using the pitch curve as a center, swing a series of arcs. The more arcs drawn, the more accurate will be the cam profile.

10. Draw a smooth curve tangent to the follower wheel arcs. This forms the finished cam profile.

11. Add the hub and follower to produce the finished layout.

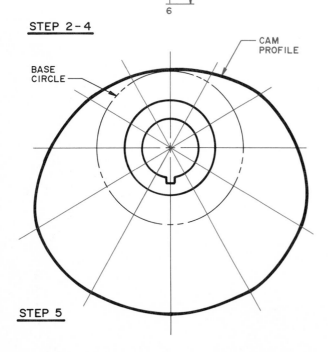

BASE CIRCLE

CAM PROFILE

STEP 5

FIGURE 22–27 Drawing a cam profile.

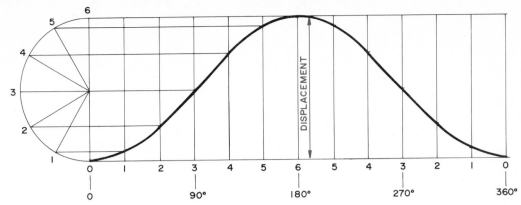

CAM DATA: HARMONIC MOTION 360°, FOLLOWER OFFSET 11 mm, DISPLACEMENT 33 mm, BACE CIRCLE DIA. 33 mm, HUB DIA. 16 mm, CAM SHAFT DIA. 8 mm, KEY 3 mm SQUARE, FOLLOWER ROLLER DIAMETER 10 mm.

STEP 1

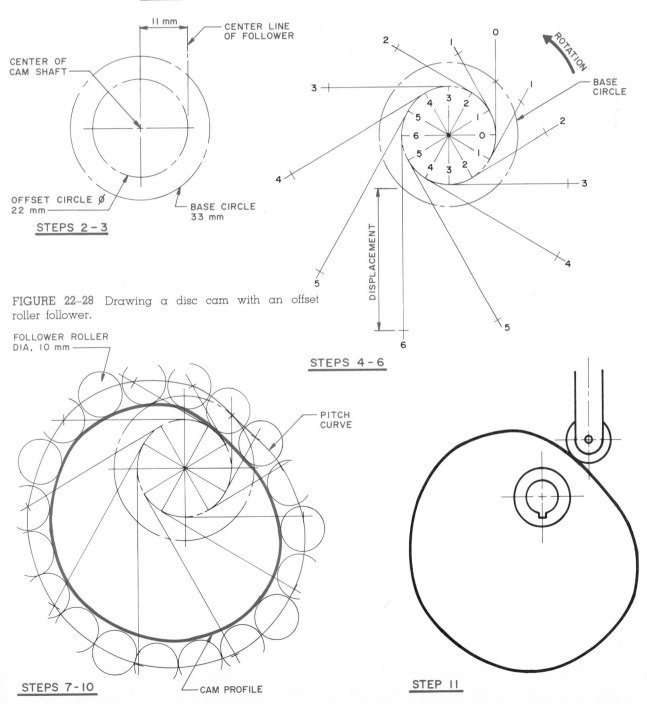

CENTER LINE OF FOLLOWER

11 mm

CENTER OF CAM SHAFT

OFFSET CIRCLE Ø 22 mm

BASE CIRCLE 33 mm

STEPS 2-3

ROTATION

BASE CIRCLE

DISPLACEMENT

STEPS 4-6

FIGURE 22–28 Drawing a disc cam with an offset roller follower.

FOLLOWER ROLLER DIA, 10 mm

PITCH CURVE

STEPS 7-10

CAM PROFILE

STEP 11

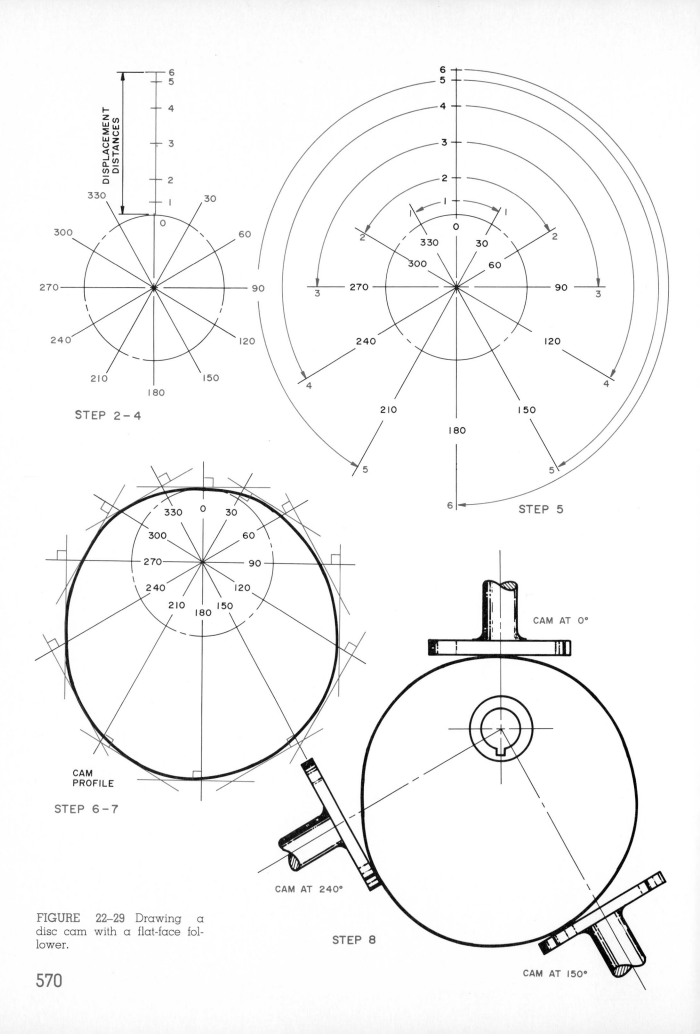

DISPLACEMENT DISTANCES

STEP 2-4

STEP 5

CAM AT 0°

CAM PROFILE

STEP 6-7

CAM AT 240°

STEP 8

CAM AT 150°

FIGURE 22–29 Drawing a disc cam with a flat-face follower.

570

# DRAWING A DISC CAM WITH A FLAT-FACE FOLLOWER

The cam in this illustration will produce harmonic motion through 360° rotation. Following are the steps for drawing the cam profile (Fig. 22–29).

1. Lay out the displacement diagram using the same scale as the cam profile drawing. In this example, the diagram in Fig. 22–28 was used.

2. Locate the center of the camshaft and draw the base circle. Its radius is equal to the distance from the center of the camshaft to the face of the follower at its lowest position. This is the point at which the base circle crosses the center line of the follower.

3. Lay out the follower displacement distances taken from the displacement diagram on the center line of the follower. Number in the same manner as they appear on the displacement diagram.

4. From the center of the camshaft, draw radial lines at the same degrees as on the displacement diagram. Number them as they appear on the diagram.

5. Using the distances from the displacement diagram, swing arcs from the center of the camshaft until they cross the radial line on which they were measured on the displacement diagram.

6. At each of these points, draw a line perpendicular to the radial line. Each represents the position of the follower at a specific point.

7. Draw the cam profile by drawing a smooth curve tangent to these perpendicular lines. The point of contact is not necessarily always on the center line of the cam. The face of the follower is made large enough to provide contact as needed.

8. Draw the hub, key, and follower to complete the layout.

## GEARS

The proper design of gears and gear trains is a major part of many engineering design projects. The selection of the best type of gear, the most desirable size, and the actual configuration of the gear itself are critical to the success of a design.

Gears are usually in the form of a cylinder or cone with teeth that engage to transfer motion or power from one shaft to another. They can be used to change the direction and speed of rotation. The gear shafts can be parallel or angular, and their axes may intersect or not.

Illustrating this discussion are a *spur gear* used for transmitting power from one shaft parallel with another, a *spur gear* with a *rack* that changes rotary motion into linear motion, a *bevel gear* for shafts having intersecting axes, and a *worm* and *worm gear* for nonintersecting shafts that are not parallel. *Helical gears* are another frequently used type. A single helical gear is used when one shaft is parallel with an-

other; a herringbone gear is used when shaft axes are nonintersecting and nonparallel (Fig. 22–30).

## GEAR TOOTH FORM

The profile of gear teeth is critical to the smoothness of motion transmitted by the gear. Satisfactory results may be had when the profile of the teeth is an involute. *Involute teeth* are those in which the active section of a transverse profile is the involute of the base circle of the teeth. An involute is a curve developed by a point on a line that is unwound like a string from the circumference of the base circle (Fig. 22–31). Detailed information on involutes can be found in Chapter 5.

A procedure for drawing an approximate involute that is satisfactory for detailing gear teeth is given in Fig. 22–32.

## GEAR NOMENCLATURE

Following are the terms commonly used when describing and specifying spur and helical gear data (Figs. 22–33, 22–34, and 22–35).

*Addendum:* The height of a tooth above the pitch circle.

*Addendum circle:* A circle formed by the outside surface of the teeth.

*Base circle:* A circle from which the tooth profile is generated. It is found by drawing a 14.5° or 20° line through the pitch point. A circle tangent to this line is the base circle. The angle used, 14.5° or 20°, depends on the gear pressure angle design.

*Base diameter:* The diameter of the base circle.

*Base pitch:* The distance on the base circle between corresponding sides of adjacent teeth.

*Center distance:* The distance between the parallel axes of spur gears and parallel helical gears or the crossed axes of crossed helical gears.

*Chordal addendum:* The distance from the top of the tooth to the chordal thickness.

*Chordal thickness:* The thickness of the tooth measured along the chord of the circular thickness arc.

*Circular pitch:* The circular distance along the pitch circle from a point on a tooth to the corresponding point on the next tooth.

*Circular thickness:* The thickness of a tooth measured along the arc of the pitch circle.

*Clearance:* The difference between the working depth and the whole depth.

*Dedendum:* The distance between the pitch circle and the minor diameter.

*Dedendum circle:* A circle formed by the minor diameter (sometimes called the *root circle*).

*Dimetral pitch:* The ratio between the pitch diameter in inches and the number of teeth. For example,

Spur Gear and Rack
Gear
Rack

Straight Bevel Gears

Single Helical Spur Gears

Internal Spur Gear

WORM

WORM
GEAR

Spiral Bevel Gears

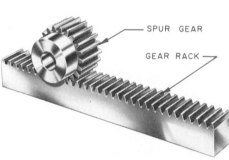

SPUR GEAR

GEAR RACK

Herringbone Helical Gear

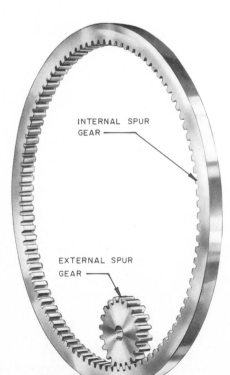

INTERNAL SPUR
GEAR

EXTERNAL SPUR
GEAR

Worm and Worm Gear
Worm
Worm Gear

External Spur Gear

FIGURE 22–30 Common types of gears. (Courtesy of Boston Gear Incom International Incorporated.)

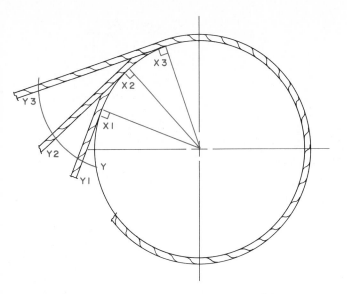

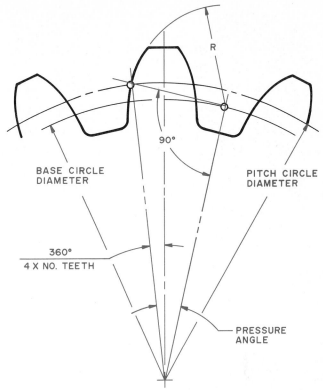

FIGURE 22–31 An involute curve is traced by a point on a string, Y, as the string is unwrapped from a cylinder. The involute is formed by point Y moving along a path indicated by $Y_1$, $Y_2$, and $Y_3$. The string is tangent to the cylinder at $X_1$, $X_2$, and $X_3$. These are instantaneous radii of the curvature of the involute.

$$R = \frac{\text{BASE CIRCLE DIAMETER}}{2} \times \text{TANGENT OF PRESSURE ANGLE.}$$

FIGURE 22–32 The procedure for drawing an approximate involute for purposes of detailing gear teeth.

FIGURE 22–33 Terms used to describe straight spur gear teeth. (Reproduced from ANSI Y14.7.1-1971 with the permission of the publisher, The American Society of Mechanical Engineers.)

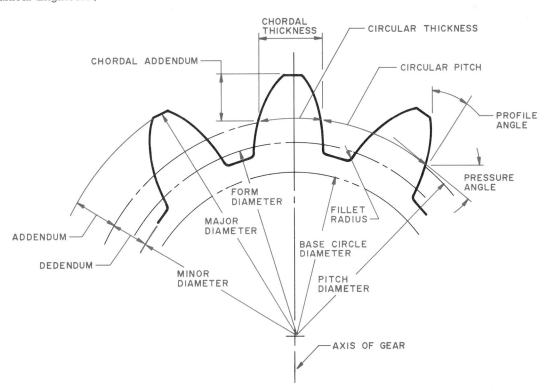

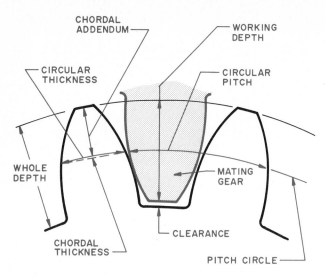

**FIGURE 22–34** Gear tooth terminology related to mating teeth of spur gears. (Reproduced from ANSI Y14.7.1-1971 with the permission of the publisher, The American Society of Mechanical Engineers.)

**FIGURE 22–35** Helical gear nomenclature. (Reproduced from ANSI Y14.7.1-1971 with the permission of the publisher, The American Society of Mechanical Engineers.)

a gear with 32 teeth and a pitch diameter of 8 would have a dimetral pitch of 4. For metric gears, the module is used instead of dimetral pitch.

*Face width:*   The width of the tooth measured from the front face of the gear to the back face.

*Fillet radius:*   The radius of the fillet curve at the base of the gear tooth.

*Form diameter:*   The diameter at which the involute tooth profile originates.

*Helix angle:*   The angle between the pitch helix and the axis of the gear.

*Lead angle:*   The angle between the pitch helix and the plane of rotation.

*Major diameter:*   The diameter of the circle generated by the top of the teeth in an external gear.

*Minor diameter:*   The diameter of the circle formed by the bottoms of the tooth spaces of an external gear (sometimes called the *root diameter*).

*Module:*   The ratio of the pitch diameter in millimeters to the number of teeth.

*Pitch circle:*   The reference circle from which all transverse tooth dimensions are taken. It is the point at which the teeth of mating gears are tangent.

*Pitch diameter:*   The diameter of the pitch circle.

*Pitch helix:*   The curve of intersection of a tooth surface and its pitch cylinder in a helical gear.

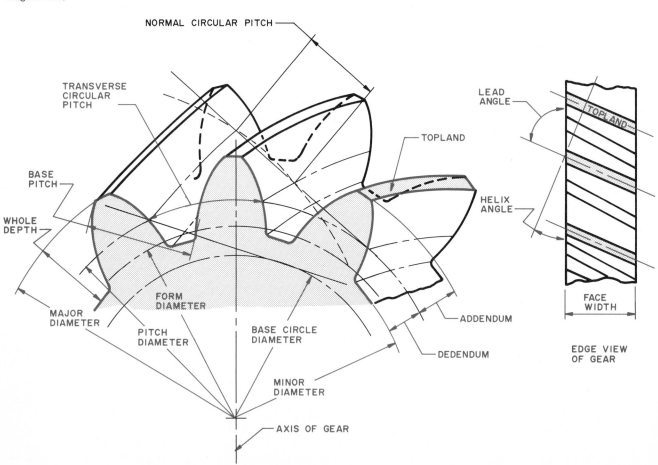

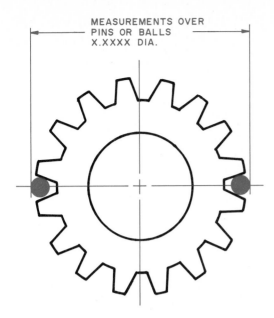

MEASUREMENTS OVER
PINS OR BALLS
X.XXXX DIA.

FIGURE 22–36 The distance between two measuring pins or balls gives the tooth-engagement diameter.

**Pitch point:** The point of tangency of two pitch circles on the line of centers. It is the intersection of the pitch circle and the tooth profile.

**Pressure angle:** The angle at which pressure from the tooth of one gear is passed to the tooth of another gear. One leg of the pressure angle is a line perpendicular to a line tangent to the involute at the pitch diameter, and the other leg is a line perpendicular to a radial line passing through the point of tangency.

**Tooth engagement diameter:** Measure between two diameter balls or pins. A method for measuring the tooth engagement diameter. Cylindrical gauge pins or spherical gauge balls are placed between diametrically opposite tooth spaces. The diameter over the pins or balls is measured and placed on the gear drawing (Fig. 22–36).

**Tooth space:** The space between two teeth into which the tooth from a mating gear will fall.

**Transverse:** The term used on gear drawings to indicate that the dimension applies in a transverse plane.

**Transverse plane:** A plane perpendicular to the axial plane and the pitch plane.

**Whole depth:** The total depth of a tooth space, equal to the addendum plus the dedendum. It also equals the working depth plus the clearance.

**Working depth:** The distance a tooth of one gear extends into the tooth space on a mating gear when they are fully engaged.

## DIMENSIONING GEAR DRAWINGS

The following publications are an important source of information pertaining to drawing gears: "Gear Drawing Standards—Part 1 for Spur, Helical, Double

Helical, and Rack," ANSI Y14.7.1-1971, and "Gear and Spline Drawing Standards—Part 2, Bevel and Hypoid Gears," ANSI Y14.7.2-1978. Both are published by the American Society of Mechanical Engineers, United Engineering Center, 345 East 47th Street, New York, N.Y. 10017.

The dimensioning practices in the following section relate to the gear teeth and their relation to mounting surfaces. Dimensions required on other parts of the gear are applied following standard dimensioning practices. The dimensional values are shown by the letter X. This indicates the number of decimal places recommended for each dimension. You will note that the metric dimensions have one less decimal place than the inch dimensions. Angular dimensions are given in degrees and decimal parts of a degree. The use of degrees, minutes, and seconds is optional.

## GEAR DRAWING PRACTICES

The general drafting practices recommended for engineering drawings are observed when making gear drawings. The gear drawing should be arranged clearly to show the general configuration. The gear drawing has extensive notes detailing the gear design data. These are grouped as shown in the examples in this chapter. The exact location on the drawing for design data is variable but must be clearly related to a particular gear drawing. This is essential when several gears are detailed on a single sheet.

When noting data for external gears, the major diameter may be specified as the outside diameter and the minor diameter as the root diameter. On internal gears, the major diameter may be specified as the root diameter and the minor diameter as the inside diameter.

A gear drawing should clearly show the design information but should not specify the method of manufacture or inspection. These decisions are made independent of the gear design process.

## SPUR AND HELICAL GEAR DRAWINGS

External spur and helical gears are drawn as shown in Fig. 22–37. The front view is not drawn unless it is required to show some feature, such as the relationship between teeth to another gear. Sections are used when there are internal features that need to be described. Gear tooth outlines are not drawn except when needed to orient the gear with another or when a feature of the tooth requires dimensioning, as when showing tip chamfer. If it is necessary to draw a tooth, it is drawn on a fragment or as a part

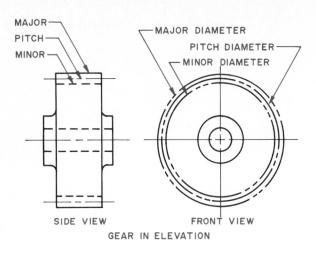

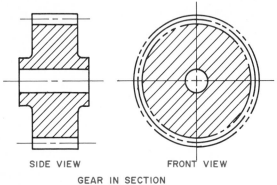

FIGURE 22–37 Recommended ways of drawing external gears. (Reproduced from ANSI Y14.7.1-1971 with the permission of the publisher, The American Society of Mechanical Engineers.)

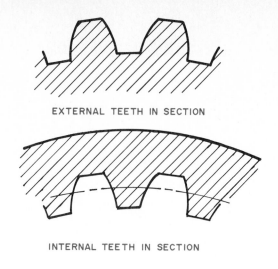

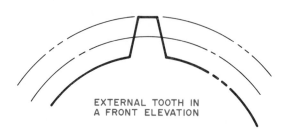

FIGURE 22–38 Representing individual teeth on external and internal gears.

FIGURE 22–39 How to represent individual teeth on internal and external gears. (Reproduced from ANSI Y14.7.1-1971 with the permission of the publisher, The American Society of Mechanical Engineers.)

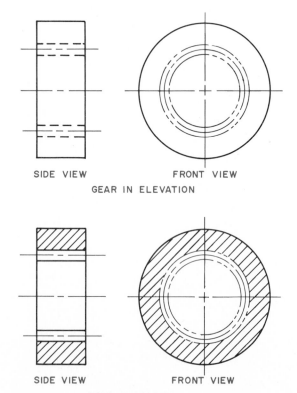

of a front view (Fig. 22–38). Internal gears are drawn following the same procedures (Fig. 22–39). Often, a gear will have more than one set of teeth. When there is no relationship between the different teeth on the gear, this is indicated in a note. When the sets of teeth do have a positional relationship, this must be clearly indicated with a note or a dimension (Fig. 22–40). In addition, the position of a set of teeth may have to be related to some other feature, such as a keyseat or a spline.

A typical drawing with design data for an *external spur gear* appears in Fig. 22–41. The basic design data shown are for a format A gear. Gear drawings are classified in four formats, A, B, C, and D. These relate to the degree of definitive specification. Format A is the lowest quality level and coarse pitch quality level.* A drawing of an *internal spur gear* appears in Fig. 22–42.

---

*All four levels are detailed in the "Gear Classification Manual" of the American Gear Manufacturers Association, 1330 Massachusetts Ave. N.W., Washington, D.C. 20005.

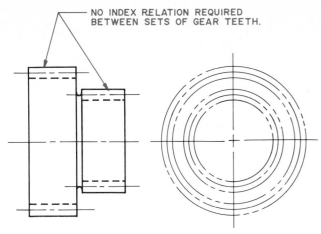

NO INDEX RELATION REQUIRED
BETWEEN SETS OF GEAR TEETH.

NOTE WHEN THERE IS NO RELATIONSHIP BETWEEN
SETS OF GEAR TEETH.

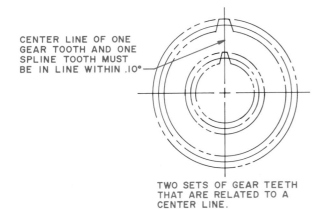

CENTER LINE OF ONE
GEAR TOOTH AND ONE
SPLINE TOOTH MUST
BE IN LINE WITHIN .10°

TWO SETS OF GEAR TEETH
THAT ARE RELATED TO A
CENTER LINE.

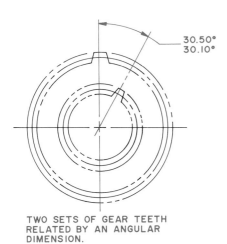

30.50°
30.10°

TWO SETS OF GEAR TEETH
RELATED BY AN ANGULAR
DIMENSION.

FIGURE 22–40 These details show how to indicate the relationship between sets of gear teeth. (Reproduced from ANSI Y14.7.1-1971 with the permission of the publisher, The American Society of Mechanical Engineers.)

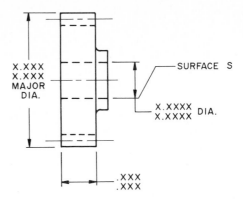

X.XXX
X.XXX
MAJOR
DIA.

SURFACE S

X.XXXX
X.XXXX DIA.

.XXX
.XXX

EXTERIOR INVOLUTE SPUR GEAR DATA (I)

| | ROOT |
(2)

DIAMETRAL PITCH XX (3)
OR
MODULE X (4)
PRESSURE ANGLE XX.XX°
PITCH DIA. X.XXXXXXX REF.
BASE DIA. X.XXXXXXX REF.
FORM DIA. X.XXX
MINOR DIA. X.XXX – X.XXX
ACTL CIRC TOOTH THK AT PITCH DIA. .XXXX–.XXXX
MEAS OVER TWO .XXXXX DIA. PINS X.XXXX–
  X.XXXX REF.
MAX RUNOUT .XXXX FIR.
ALL TOOTH ELEMENT SPECIFICATIONS ARE FROM
  DATUM ESTABLISHED BY THE AXIS OF
  SURFACE S.

1. DATA SHOWN ARE FOR INCH GEAR DRAWINGS,
   METRIC GEAR DRAWINGS HAVE ONE LESS
   DECIMAL PLACE TO THE RIGHT OF THE
   DECIMAL POINT.

2. SPECIFY WHETHER FULL FILLET OPTIONAL
   OR FLAT.

3. USED ON INCH DRAWING ONLY.

4. USED ON METRIC DRAWINGS INSTEAD OF
   DIAMETRAL PITCH.

FIGURE 22–41 A typical design drawing of an external spur gear showing tooth data only. Other dimensions are applied following standard dimensioning practices. (Reproduced from ANSI Y14.7.1-1971 with the permission of the publisher, The American Society of Mechanical Engineers.)

*External* and *internal* single helical gear drawings are shown in Figs. 22–43 and 22–44. If the gear is a double helical gear, it is drawn as shown in Fig. 22–45.

## RACK DRAWINGS

The terms used to identify parts of a rack are given in Fig. 22–46. They are the same as those found on a spur or helical gear. The pitch circle of the circular gear is a straight line called the *pitch line.* The gear tooth proportions are determined from the pitch line. A typical *spur gear rack* detail appears in Fig. 22–47. A helical gear rack is shown in Fig. 22–48. Since the racks are usually very long and all the teeth are identical along the entire length, a section is broken out to shorten it. Only one or two teeth are shown on each end. These are necessary for dimensioning.

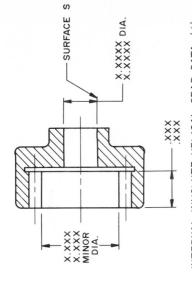

INTERNAL INVOLUTE SPUR GEAR DATA (1)

| (2) | ROOT |

NUMBER OF TEETH XX
DIAMETRAL PITCH XX (3)
    OR
MODULE X (4)
PRESSURE ANGLE  XX.XX° REF.
PITCH DIA.  X.XXXXXXXX REF.
BASE DIA.  X.XXXXXXXX REF.
FORM DIA.  X.XXX
MAJOR DIA. X.XXXX - X.XXX
ACTL CIRC TOOTH THK AT PITCH DIA. XXXX - XXXX
MEAS BETWEEN TWO .XXXXX DIA. PINS X.XXXX -
    X.XXXX REF.
MAX RUNOUT .XXXX FIR.
ALL TOOTH ELEMENT SPECIFICATIONS ARE
    FROM DATUM ESTABLISHED BY THE AXIS
    OF SURFACE S.

1. DATA SHOWN ARE FOR INCH GEAR DRAWINGS.
   METRIC GEAR DRAWINGS HAVE ONE LESS
   DECIMAL PLACE TO THE RIGHT OF THE
   DECIMAL POINT.
2. SPECIFY WHETHER FULL FILLET, OPTIONAL
   OR FLAT.
3. USED ON INCH DRAWINGS ONLY.
4. USED ON METRIC DRAWINGS INSTEAD OF
   DIAMETRAL PITCH.

FIGURE 22-42  A typical design drawing of an internal spur gear showing gear tooth data only. Other dimensions are applied following standard dimensioning practices. (Reproduced from ANSI Y14.7.1-1971 with the permission of the publisher, The American Society of Mechanical Engineers.)

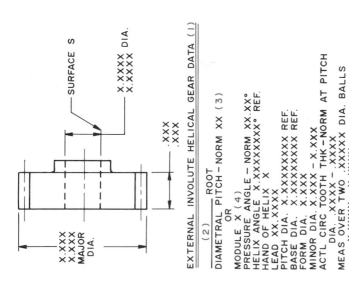

EXTERNAL INVOLUTE HELICAL GEAR DATA (1)

| (2) | ROOT |

DIAMETRAL PITCH - NORM XX (3)
    OR
MODULE X (4)
PRESSURE ANGLE - NORM XX.XX°
HELIX ANGLE X.XXXXXXXX° REF.
HAND OF HELIX X
LEAD XX.XXXX
PITCH DIA. X.XXXXXXXX REF.
BASE DIA. X.XXXXXXXX REF.
FORM DIA. X.XXX
MINOR DIA. X.XXX
ACTL CIRC TOOTH THK-NORM AT PITCH
    DIA.  .XXXX - .XXXX
MEAS OVER TWO .XXXXX DIA. BALLS
    X.XXXX - X.XXXX
MAX RUNOUT .XXXX FIR.
ALL TOOTH ELEMENT SPECIFICATIONS ARE
    FROM DATUM ESTABLISHED BY THE AXIS
    OF SURFACE S.

1. DATA SHOWN ARE FOR INCH GEAR DRAWINGS.
   METRIC GEAR DRAWINGS HAVE ONE LESS
   DECIMAL PLACE TO THE RIGHT OF THE
   DECIMAL POINT.
2. SPECIFY WHETHER FULL FILLET, OPTIONAL
   OR FLAT.
3. USED ON INCH DRAWINGS ONLY.
4. USED ON METRIC DRAWINGS INSTEAD OF
   DIAMETRAL PITCH.

FIGURE 22-43  A typical drawing of an external helix gear showing tooth data only. Other dimensions are applied following standard dimensioning practices. (Reproduced from ANSI Y14.7.1-1971 with the permission of the publisher, The American Society of Mechanical Engineers.)

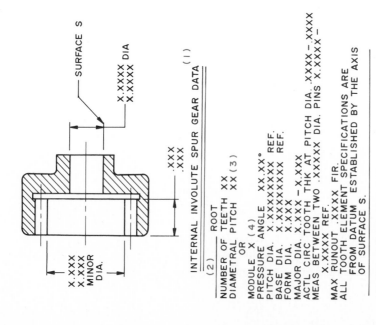

INTERNAL INVOLUTE HELICAL GEAR DATA. (1)

| (2) | ROOT |

NUMBER OF TEETH XX
DIAMETRAL PITCH - NORM XX (3)
    OR
MODULE X (4)
PRESSURE ANGLE - NORM XX.XX°
HELIX ANGLE X.XXXXXXXX REF.
HAND OF HELIX X
LEAD XX.XXXX
PITCH DIA. X.XXXXXXXXX REF.
BASE DIA. X.XXXXXXXXX REF.
FORM DIA. X.XXX
MAJOR DIA X.XXX - X.XXX
ACTL CIRC TOOTH THK-NORM AT PITCH DIA.
    .XXXX - .XXXX
MEAS BETWEEN TWO .XXXX DIA. BALLS
    X.XXXX - X.XXXX REF.
MAX RUNOUT .XXXX FIR.
ALL TOOTH ELEMENT SPECIFICATIONS ARE
    FROM DATUM ESTABLISHED BY THE AXIS
    OF SURFACE S.

1. DATA SHOWN ARE FOR INCH GEAR DRAWINGS.
   METRIC GEAR DRAWINGS HAVE ONE LESS
   DECIMAL PLACE TO THE RIGHT OF THE
   DECIMAL POINT.
2. SPECIFY WHETHER FULL FILLET, OPTIONAL
   OR FLAT.
3. USED ON INCH DRAWINGS ONLY.
4. USED ON METRIC DRAWINGS INSTEAD OF
   DIAMETRAL PITCH.

FIGURE 22-44  A typical design drawing of an internal helical gear showing tooth data only. Other dimensions are applied following standard dimensioning practices. (Reproduced from ANSI Y14.7.1-1971 with the permission of the publisher, The American Society of Mechanical Engineers.)

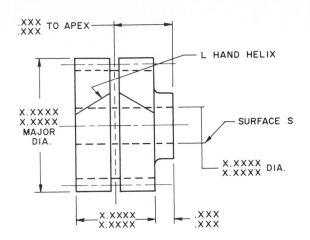

FIGURE 22–45 A typical drawing of an external involute double-helical (herringbone) gear. (Reproduced from ANSI Y14.7.1-1971 with the permission of the publisher, The American Society of Mechanical Engineers.)

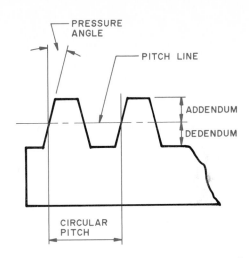

FIGURE 22–46 The basic parts of spur rack gear teeth.

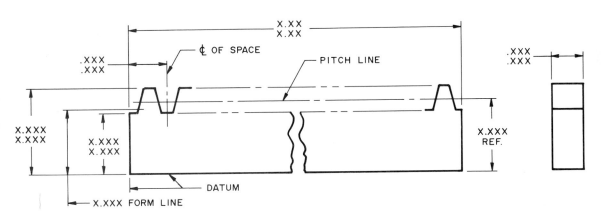

SPUR RACK DATA (1)

```
NUMBER OF TEETH  XX
DIAMETRAL PITCH   XX.XXXXXXX REF. (2)
MODULE X (3)
LINEAR PITCH        .XXXX
PRESSURE ANGLE   XX°
ACTUAL TOOTH THICKNESS AT PITCH LINE .XXXX-.XXXX
PITCH TOL .XXXX
INDEX TOL .XXXX
ALL TOOTH ELEMENT SPECIFICATIONS ARE
    FROM THE SPECIFIED DATUM.
```

1. DATA SHOWN ARE FOR INCH GEAR DRAWINGS. METRIC GEAR DRAWINGS HAVE ONE LESS DECIMAL PLACE, TO THE RIGHT OF THE DECIMAL POINT.
2. USE ON INCH DRAWINGS ONLY
3. USE ON METRIC DRAWINGS INSTEAD OF DIAMETRAL PITCH.

FIGURE 22–47 A typical design drawing for a spur rack. (Reproduced from ANSI Y14.7.1-1971 with the permission of the publisher, The American Society of Mechanical Engineers.)

## BEVEL GEAR DRAWINGS

The terms used to identify the parts of a bevel gear are given in Fig. 22–49. Most of the terms used for spur gears are also used on bevel gears. The termi-nology differences that exist are due to the taper of the bevel gear tooth and because it is cut on an angle.

Following are the terms used when designing a bevel gear (Figs. 22–50 and 22–51).

Graphical Kinematics  **579**

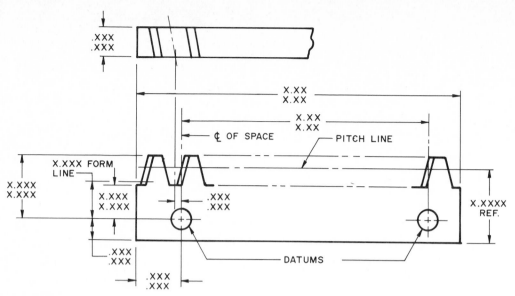

HELICAL RACK DATA

NUMBER OF TEETH  XX
DIAMETRAL PITCH—NORM  XX.XXXXXXX REF.
MODULE  X
LINEAR PITCH—NORM       .XXXX
PRESSURE ANGLE—NORM  XX°
ACTUAL TOOTH THICKNESS AT PITCH LINE—NORM  .XXXX—.XXXX
TOOTH ANGLE (HELIX ANGLE)  X.XXXXXXXX°
HAND OF HELIX  X
PITCH TOL.       .XXXX
INDEX TOL.       .XXXX
ALL TOOTH ELEMENT SPECIFICATIONS ARE FROM THE
SPECIFIED DATUM.

FIGURE 22–48  A typical design drawing for a helical
rack. (Reproduced from ANSI Y14.7.1-1971 with the
permission of the publisher, The American Society of
Mechanical Engineers.)

*AGMA quality class:*  A classification number established by the American Gear Manufacturers Association to designate the quality requirements of a gear.

*Backlash:*  The amount of play between gears that produces a jarring reaction.

*Backlash variation tolerance:*  Allowable variation in backlash measured at the tightest point of mesh.

*Circular thickness:*  The length of the arc between two sides of a gear tooth on the pitch circle.

*CONIFLEX®:*  A trade name applied to generated straight bevel gears whose teeth have lengthwise crowning.

*Driver member:*  The member to mate with the gear.

*Fillet radius:*  The radius of the fillet at the root of the tooth.

*Index tolerance:*  The allowable displacement of any tooth from its theoretical angular or linear position relative to a tooth datum.

*Mean measuring addendum:*  The height from the top of the tooth to the chord subtending the circular thickness arc in the normal plane.

*Mean measuring thickness:*  The length of the chord subtending a circular-thickness arc in the normal plane.

*Normal backlash with mate:*  The space between mating tooth surfaces.

*Pitch angle:*  The angle between the axis and the pitch element.

*Pitch tolerance:*  The allowable difference between the pitch and the measured distance between any adjacent teeth.

*Pressure angle:*  The angle at the pitch point between a line normal to the tooth profile and the pitch plane.

*REVACYCLE®:*  A trade name applied to straight bevel gears produced with circular-arc tooth profiles.

*Root angle:*  The angle between the axis and the root.

*Runout tolerance:*  The total allowable variation of the distance between a surface of revolution and an indicated surface measured perpendicular to the surface of revolution.

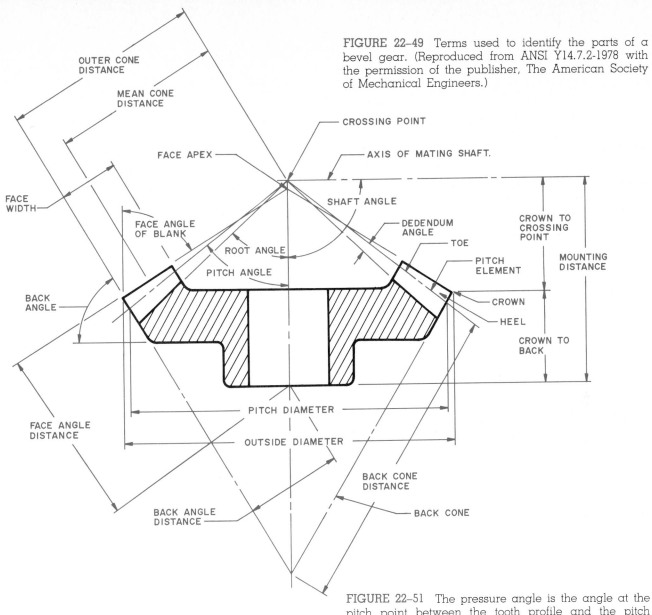

FIGURE 22–49 Terms used to identify the parts of a bevel gear. (Reproduced from ANSI Y14.7.2-1978 with the permission of the publisher, The American Society of Mechanical Engineers.)

FIGURE 22–50 Bevel gear nomenclature related to the design of teeth.

FIGURE 22–51 The pressure angle is the angle at the pitch point between the tooth profile and the pitch plane. (Reproduced from ANSI Y14.7.2-1978 with the permission of the publisher, The American Society of Mechanical Engineers.)

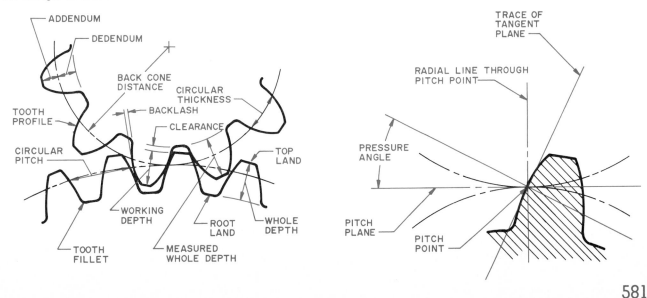

*Shaft angle:* The angle between the shafts of mating gears.

*Theoretical crown to back:* The distance from the crown to the back of the gear measured parallel with the gear axis.

*Theoretical outside diameter:* The overall diameter of a gear measured perpendicular to the axis.

*Tooth form:* The shape of the tooth profile.

*Tooth surface texture:* The texture of the finish on the working tooth surface.

*Working depth:* The distance the tooth penetrates a mating gear.

A typical bevel gear drawing appears in Fig. 22–52. If the gear is to be metric, the dimetral pitch

FIGURE 22–52 A typical design drawing for a straight bevel gear or pinion. (Reproduced from ANSI Y14.7.2-1978 with the permission of the publisher, The American Society of Mechanical Engineers.)

data is replaced by the module. Generally, the back of the gear is the datum to which dimensions are related.

Often, mating gears are drawn together because they are designed in pairs. They are drawn and dimensioned as for a single gear. A typical layout is shown in Fig. 22–53.

## WORM AND WORM GEAR DRAWINGS

A worm is a form of screw or helix. It mates with a worm gear, which permits the transmission of power between two shafts whose axes are at right angles but do not intersect.

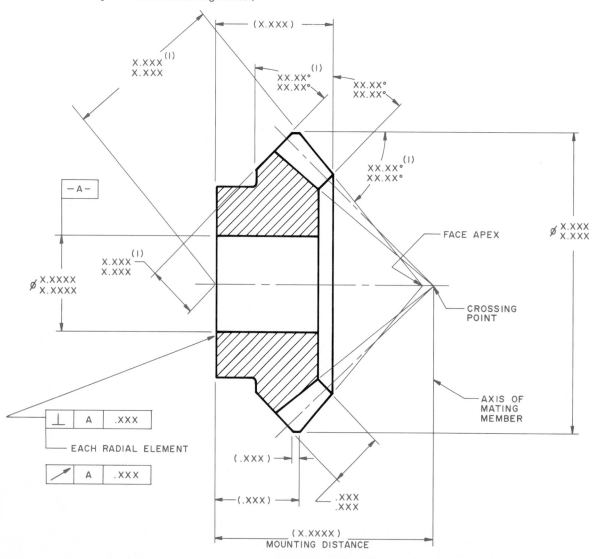

1. WHEN FACE ANGLE DISTANCE AND BACK ANGLE DISTANCE ARE USED FOR DIMENSIONING THE GEAR BLANK, THE FACE ANGLE AND BACK ANGLE SHOULD BE GIVEN AS REFERENCE DIMENSIONS ON THE DRAWING, WITHOUT A TOLERANCE.

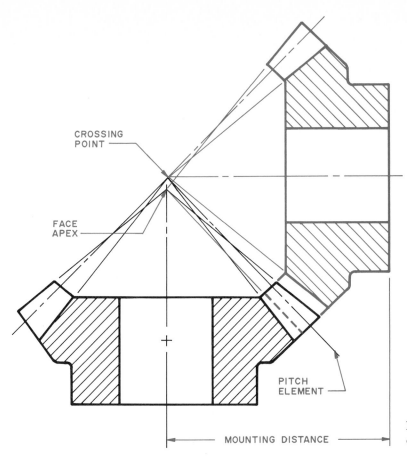

CROSSING
POINT

FACE
APEX

PITCH
ELEMENT

MOUNTING DISTANCE

FIGURE 22–53  A typical layout for mating bevel gears.

The *worm gear* is much like a spur gear but has
a curved tooth form that enables it to mate with the
thread of the worm. Some worm gears have teeth
that are straight across and have concave faces.
These are called single-enveloping worm gears.
They provide straight-line contact with the worm.
Another type has a concave rather than straight top-
land. This is called a double-enveloping worm gear
and provides additional contact with the worm. It
transmits more power than the single-enveloping
gear (Fig. 22–54).

The terms commonly used to describe a worm
and worm gear are given in Fig. 22–55. The *lead angle* is the angle between the thread helix and the
plane of rotation. The *pitch diameter* of the worm is
the average diameter of the working depth. The *pitch*
is the distance between corresponding points on adjacent threads measured along the axis of the worm.
The *lead* is the distance a point on the worm moves
parallel with the axis in one revolution. In this figure, a single thread worm is used, so the lead and
pitch are the same. If a double thread were used, the
lead would be twice the pitch.

A drawing of a worm appears in Fig. 22–56.
Because the worm is often long and the thread is the
same the entire length, a few threads are drawn on
each end and the threads in the center are repre-

FIGURE 22–54  Two types of worm gear teeth.

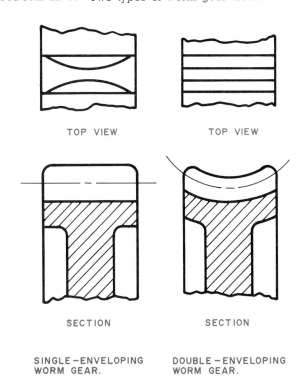

TOP VIEW                    TOP VIEW

SECTION                     SECTION

SINGLE–ENVELOPING           DOUBLE–ENVELOPING
WORM GEAR.                  WORM GEAR.

Graphical Kinematics  583

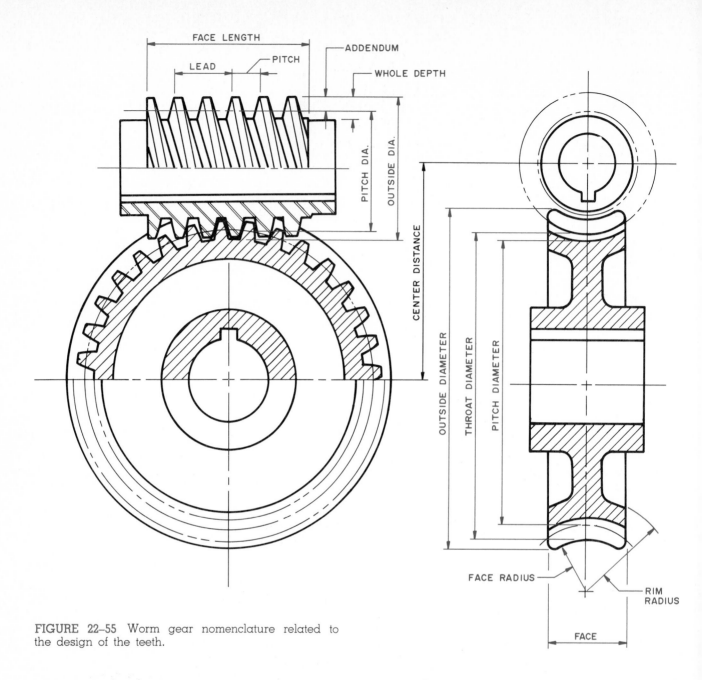

FIGURE 22–55 Worm gear nomenclature related to the design of the teeth.

sented with phantom lines. The design data are presented in note form.

A drawing of a worm gear appears in Fig. 22–57. This example is drawn as a half section, which exposes interior details and the tooth form. The design data are given as a note where the recommended decimal places for each is shown. This example is a double-enveloping worm gear.

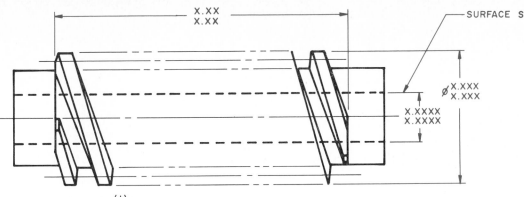

WORM TOOTH DATA (1)

NUMBER OF THREADS        X
PITCH DIAMETER           X.XXX
AXIAL PITCH              .XXX
LEAD (RIGHT OR LEFT)     X.XXX
LEAD ANGLE               XX.XX°
PRESSURE ANGLE           XX.XX°
ADDENDUM                 .XXX
WHOLE DEPTH              .XXX
ALL TOOTH ELEMENT SPECIFICATIONS ARE FROM
DATUM ESTABLISHED BY THE AXIS OF SURFACE S.

1. DATA SHOWN ARE FOR INCH GEAR DRAWINGS.
   METRIC GEAR DRAWINGS HAVE ONE LESS
   DECIMAL PLACE TO THE RIGHT OF THE DECIMAL
   POINT.

FIGURE 22–56 A typical design drawing for a worm showing tooth data only. Other dimensions are applied following standard dimensioning practices.

FIGURE 22–57 A typical design drawing for a worm gear showing tooth data only. Other dimensions are applied following standard dimensioning practices.

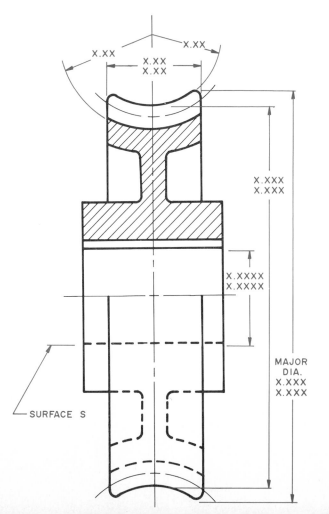

WORM GEAR TOOTH DATA (1)

NUMBER OF TEETH         XX
PITCH DIAMETER          X.XXX
ADDENDUM                X.XX
WHOLE DEPTH             X.XXX
BACKLASH ASSEMBLED      .XXX–.XXX

WORM DATA – REFERENCE (1)

NUMBER OF THREADS       X
AXIAL PITCH             .XXX
LEAD (RIGHT OR LEFT)    X.XXX
PITCH DIAMETER          X.XXX
LEAD ANGLE              XX.XX°
PRESSURE ANGLE          XX.XX°
ALL TOOTH ELEMENT SPECIFICATIONS ARE FROM
DATUM ESTABLISHED BY THE AXIS OF SURFACE S.

1. DATA SHOWN ARE FOR INCH GEAR DRAWINGS,
   METRIC GEAR DRAWINGS HAVE ONE LESS
   DECIMAL PLACE TO THE RIGHT OF THE
   DECIMAL POINT.

# Problems

The following problems offer a variety of experiences in the principles of linkages, gears, and cams. The problems are designed to be solved on A or A4 size sheets.

## Linkage Problems

1. Indicate the displacement and the path of point B as point C on the slider reaches the maximum and minimum distances from point A. At the maximum distance label the points $B_1$ and $C_1$. At the minimum distance label them $B_2$ and $C_2$. If slider C moves 1.5 in. in 2 seconds what is its average velocity? Scale full size.

2. Rotate point D clockwise through 360° and plot the location of points D and C every 30°. Label the rotated points in succession as $D_1$, $D_2$, $D_3$, and $C_1$, $C_2$, $C_3$, etc. Scale full size.

3. Move the slider, point E, 2 in. up and 2 in. down. Draw the path and displacement of point D. Measure the length of the displacement. Scale full size.

4. Move point C to trace the shape formed by triangle 1–2–3. Locate the position of point A when C is at each corner. Label these locations $A_1$, $A_2$, and $A_3$. Scale full size.

5. Rotate point B 60° to the right and left of the position shown. Draw the location, displacement, and path of rotation of points C and D. Label the points to the right $C_1$ and $D_1$ and those to the left $C_2$ and $D_2$. Measure the length of the displacement. Scale full size.

6. Rotate point B 360° clockwise. Locate point C when point B has rotated to the 0°, 90°, 180°, and 270° positions. If it takes 10 seconds to complete a 360° rotation what is the angular velocity to point C? Scale full size.

7. Find the displacement and path when the fluid link moves point C from its given location upward 2 in. What type of motion is imparted to points C and E? Scale full size.

8. Design a linkage mechanism that will enable a point to oscillate through an arc of 3 in.

9. Design a linkage mechanism that will produce translation motion of 50 mm to a point 75 mm above a horizontal surface.

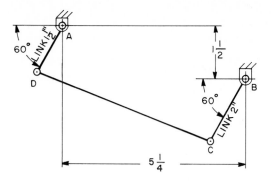

PROBLEM 22–2    A four-bar linkage.

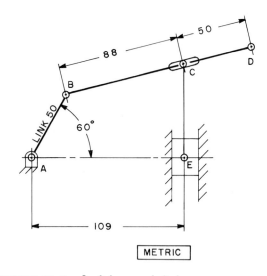

PROBLEM 22–3    A slider-crank linkage.

PROBLEM 22–4    A pantagraph linkage.

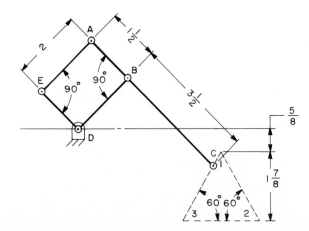

PROBLEM 22–1    A slider-crank linkage.

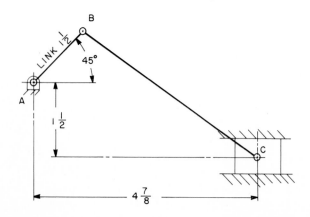

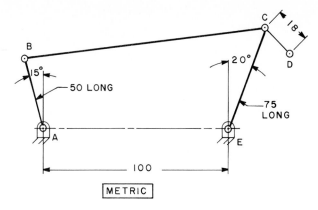

PROBLEM 22–5  A four-bar linkage.

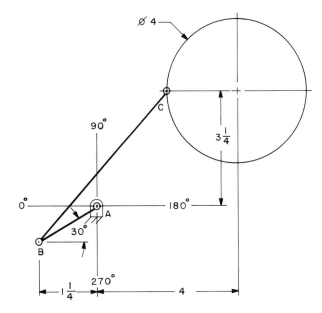

PROBLEM 22–6  A crank rocker mechanism.

PROBLEM 22–7  A linkage with a fluid link.

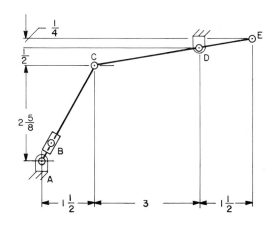

## Cam Problems

Using the cam data that follows solve problems 1 through 10.

### Cam Data

|  | INCH | METRIC |
|---|---|---|
| Diameter of cam shaft | 1.0 in. | 25.0 |
| Diameter of cam hub | 2.0 in. | 50.0 |
| Diameter of working circle | 3.25 in. | 80.0 |
| Keyway | .25 × .12 | 6 × 3 |
| Follower displacement | 2.5 in. | 62.0 |
| **Scale: Full size** | | |

1. Using the cam data provided, draw a displacement diagram for the following motions:
    a. Uniform motion
    b. Rise 2.5 in. (62 mm) in 120°
    c. Dwell for 60°
    d. Fall for 90°
    e. Dwell for 90°

2. Using the cam data provided, draw a displacement diagram for the following motions:
    a. Modified uniform motion
    b. Rise 1.5 in. (37 mm) for 120°
    c. Dwell 60°
    d. Rise 1.0 in. (25 mm) for 60°
    e. Fall for 120°

3. Using the cam data provided, draw a displacement diagram for the following motions:
    a. Harmonic motion
    b. Rise 1.75 in. (43 mm) 180°
    c. Dwell for 45°
    d. Fall for 90°
    e. Dwell for 45°

4. Using the cam data provided, draw a displacement diagram for the following motions:
    a. Parabolic motion
    b. Rise 2.5 in. (62 mm) in 120°
    c. Dwell for 50°
    d. Fall for 60°
    e. Dwell for 130°

5. Using the cam data provided, draw a displacement diagram for the following motions:
    a. Rise 2.0 in. (50 mm) in 120° with modified uniform motion
    b. Dwell 30°
    c. Rise to 0.5 in. (12 mm) in 30° with modified uniform motion
    d. Dwell 30°
    e. Fall 150° with harmonic motion

6. Using the cam data provided, draw a displacement diagram for the following motions:
    a. Rise 1.5 in. (37 mm) in 90° with parabolic motion
    b. Dwell 30°
    c. Rise 1.0 in. (25 mm) in 120° with harmonic motion
    d. Dwell 30°
    e. Fall 90° with parabolic motion

7. Using the design data and information in problem 1, lay out a disc cam that is to move a knife-edge follower in the position shown in the drawing for problem 7. The cam will rotate clockwise.

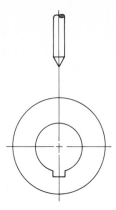

CAM PROBLEM 22-7 A cam with a knife-edge follower.

8. Using the design data and information in problem 2, lay out a disc cam that is to have a roller follower with a diameter of .75 in. (18 mm) located as shown in the drawing for problem 8. The cam will rotate counterclockwise.

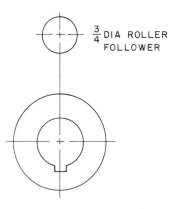

$\frac{3}{4}$ DIA ROLLER FOLLOWER

CAM PROBLEM 22-8 A cam with a roller follower.

9. Using the design data and information in problem 3, lay out a disc cam that is to move a flat-face follower located as shown in the drawing for problem 9. Find the length of the follower face by making it .25 in. (6 mm) wider than the surface of contact. The cam will rotate clockwise.

CAM PROBLEM 22-9 A cam with a flat-face follower.

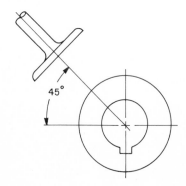

45°

10. Using the cam data and information in problem 4, lay out a disc cam that is to have an offset-roller follower in the position shown in the drawing for problem 10. The cam will rotate counterclockwise.

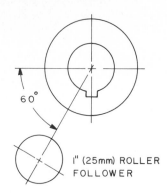

60°

1" (25mm) ROLLER FOLLOWER

CAM PROBLEM 22-10 A cam with an off-set roller follower.

## Gear Problems

1. Draw and dimension a working drawing of an external spur gear having the following specifications:

| | |
|---|---|
| Face width | 1.2500 in. |
| Length of hub | 2.00 in. |
| Diameter of hub | 1.50 in. |
| Shaft diameter | .85 in. |
| Keyway | $\frac{1}{4} \times \frac{1}{8}$ in. |
| Full-root fillet | |
| Number of teeth | 49 |
| Diametral pitch | 7 |
| Pressure angle | 20.00° |
| Major diameter | 7.2762 in. |
| Pitch diameter | 7.00125000 in. |
| Base diameter | 6.2130000 in. |
| Form diameter | 6.704 in. |
| Minor diameter | 6.660–6.670 in. |
| Actual circular tooth thickness at pitch diameter | .2440–.2470 in. |
| Measurement over two .2500 diameter pins | 7.2500–7.3500 in. |
| Maximum runout | .0010 in. |

2. Draw and dimension an internal spur gear having the same specifications as in problem 1.

3. Draw and dimension two sets of spur gears on a single axis as shown in Fig. 22–40. The large gear has the same specifications as in problem 1. The data for the small gear are as follows. The center line of one tooth on the large gear must be within .10° of the center line of the tooth on the second gear. The hub extends .50 in. on each side of the gear.

| | |
|---|---|
| Face width | 2.50 |
| Full-fillet root | |
| Number of teeth | 35 |
| Dimetral pitch | 5 |
| Pressure angle | 20.00° |
| Major diameter | 7.4000 in. |
| Pitch diameter | 7.00000000 in. |
| Base diameter | 6.47440000 in. |
| Form diameter | 6.600 in. |
| Minor diameter | 6.539–6.569 in. |
| Actual circular tooth thickness at pitch diameter | .3141–.3191 in. |

Measurement over two  
.2500 diameter pins        4.1250–4.1350 in.  
Maximum runout        .0010 in.

4. Draw a spur rack using the data given below. Dimension it completely. The rack is 31.42 in. long and .750 in. wide and the bottom to top tooth height is 1.5750–1.5950 in.

| | |
|---|---|
| Number of teeth | 50 |
| Diametral pitch | 5 |
| Linear pitch | .6284 in. |
| Pressure angle | 20.00° |
| Actual tooth thickness at pitch line | .3142–.3242 in. |
| Pitch tolerance | .0100 in. |
| Index tolerance | .0100 in. |
| Form line | 1.118 in. |
| Pitch line | 1.344 in. |
| Whole depth | .4314–.4364 in. |

5. Draw an external involute double-helical (herringbone) gear having the following specifications:

Full-fillet root

| | |
|---|---|
| Major diameter | 3.7460 in. |
| Number of teeth | 25 |
| Diametral pitch | 8 |
| Pressure angle | 20.00° |
| Helix angle | 26°–46' |
| Lead angle | 30.00° |
| Lead | 21.7987 in. |
| Tooth depth | .2696 in. |
| Pitch diameter | 3.49610000 in. |
| Base diameter | 2.99601200 in. |
| Form diameter | 3.5340 in. |
| Minor diameter | 3.2460–3.2470 in. |
| Actual circular tooth thickness | .1963–.1983 in. |
| Total face width | .7500 in. |
| Hub diameter | .7500 in. |
| Hub extends | .2500 in. from right side of gear |

Left gear, left hand  
Right gear, right hand  

| | |
|---|---|
| Groove between gears | .1500 in. |

6. Draw a section through a spur as shown in Fig. 22–49 and dimension it completely. Use the following data. Draw full size or larger.

| | |
|---|---|
| Outer cone distance | 4.0938 in. |
| Mean cone distance | 3.554 in. |

| | |
|---|---|
| Face width | 1.250 in. |
| Face angle of blank | 58° |
| Root angle | 48.5° |
| Pitch angle | 53.5° |
| Book angle | 58° |
| Face angle distance | 3.5625 in. |
| Pitch diameter | 6.8125 in. |
| Outside diameter | 7.1250 in. |
| Back angle distance | 2.1406 in. |
| Back cone distance | 6.5000 in. |
| Crown to back | 1.8750 in. |
| Crown to crossing point | 2.3750 in. |
| Dedendum angle | 5.5° |
| Shaft angle | 90.00° |

7. Make a working drawing of a worm as shown in Fig. 22–56. Dimension the teeth and all parts. Draw twice normal size. Use the following data:

| | |
|---|---|
| Face length of threads | 4.80 in. |
| Length of entire worm | 5.55 in. |
| Shaft diameter | .5050 in. |
| Major diameter | 2.0000 in. |
| Number of threads | 12 |
| Pitch diameter | 1.841 in. |
| Axial pitch | .400 in. |
| Lead to the right | .250 in. |
| Lead angle | 22.0° |
| Pressure angle | 14.5° |
| Addendum | .076 in. |
| Whole depth | .185 in. |
| Keyway | ⅛ × 1/16 in. |

8. Make a working drawing of a worm gear as shown in Fig. 22–57. Dimension the teeth and all parts. Draw full size using the data that follows.

| | |
|---|---|
| Number of teeth | 32 |
| Pitch diameter | 6.937 in. |
| Major diameter | 8.500 in. |
| Throat diameter | 7.937 in. |
| Shaft diameter | 2.000 in. |
| Hub diameter | 4.000 in. |
| Hub length, overall | 3.00 in. |
| Keyway | ½ × ¼ in. |
| Whole depth | .5050 in. |
| Backlash assembled | ±.010 in. |
| Face of teeth | 2.010 in. |
| Face radius | 1.505 in. |
| Rim radius | 2.100 in. |
| Thickness of tooth at pitch circle | .3125 in. |

# Graphical Presentation and Analysis

There are many types of graphs or charts used to present engineering data and for engineering analysis. They are used extensively by engineers, mathematicians, scientists, and marketing staff.

Engineers use them to present information to explain and support design proposals. Visual, oral, and written presentations of data enable the engineer to set forth clearly the necessary information. Graphical techniques make it possible to make engineering analyses of data including situations with several variables. Empirical equations, nomographs, and graphical calculus can be used for solutions. Much of these data are derived from laboratory experiments or from physical relationships that can be expressed as mathematical equations. The equation form of empirical data can be determined if an equation exists. If it is possible to graph a function, it should be equally possible to express the same information with some type of mathematical relation. The engineer can use graphical methods to analyze performance data from prototypes before the product goes into full production. As the product enters the market and use, performance data can be analyzed. Improvements in the design might be indicated as a result of these analyses. A product could be compared with others on such criteria as the potentials

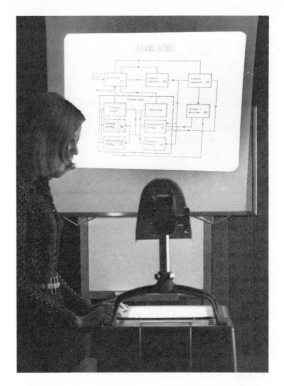

FIGURE 23-1 Graphics are used for engineering, marketing, and sales presentations. (Courtesy of Chartpak.)

of various fuels, materials, power requirements, the effect of altitude, or other such factors.

The marketing staff also makes extensive use of graphical presentations. Engineers could be called upon to assist with the technical aspects of such presentations. Marketing would use graphics to analyze data from field studies to determine the market potential of a product. The staff could compare the results regionally or with existing products. These data may determine if the design project continues, is dropped, or is redirected. Marketing also represents graphically information about potential purchasers, such as their income, where they live, and the size of their population. Graphics are used extensively to inform the sales force about all aspects of the product and how it compares with competitive products. Advertising also relies on these data and uses graphical presentations in ads in magazines, newspapers, and technical brochures and on television. When all factors are considered, the use of graphics for presentation and analysis is a major graphic activity (Fig. 23–1).

## THE COMPUTER

The computer has dramatically changed the production of charts for use as presentation graphics and engineering analysis. The basic data from tests or product design are usually already stored in the computer. See Chapter 24 for information about computers. The engineer has to decide what would be the best form in which to present the data. The computer programmer will instruct the computer to develop the graphics. It can be seen on a CRT or produced as hard copy in a large range of colors (Fig. 23–2). Computer graphics techniques are finding increasing use in the production of the charts and graphs needed. It should be emphasized, however, that the engineer and drafter must still under-

stand the types of charts and how they are designed so that the data in the computer are properly displayed. The computer cannot display the graphics unless it has been properly programmed (Fig. 23–3).

FIGURE 23-2 This computer graphics installation produces graphics for analysis, displays them on a CRT, and produces hard copy on command. (Courtesy of Tektronix, Inc.)

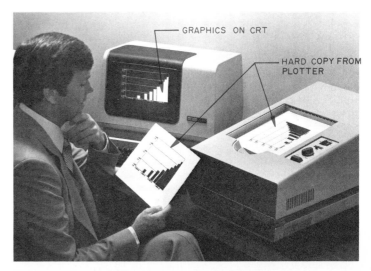

FIGURE 23–3 A digital plotter producing a bar chart. The pen tips are changed to produce graphics in several colors. The pens can be ballpoint, felt-tip, or India-ink types. (Courtesy of Bausch and Lomb.)

## DRAFTING AIDS

There are a wide variety of pressure-sensitive tapes, shading materials, and symbols that greatly facilitate the production of charts.

*Tapes* range in width from $\frac{1}{16}$ to 2 in. and are available in assorted colors, densities, and surfaces (Fig. 23–4). They can be used for grid lines, curves, and bars. Curves can be laid on by using a special tape dispenser.

Pressure-sensitive screening is used to shade the bars and other parts of a chart. The various dot and line patterns can be used to identify different parts of the chart (Fig. 23–5).

Symbols also are available in pressure-sensitive form. They are removed from the backing and pressed into place on the chart. Hundreds of symbols and pictures are available (Fig. 23–6).

FIGURE 23–4 Examples of pressure-sensitive tapes. (Courtesy of Chartpak.)

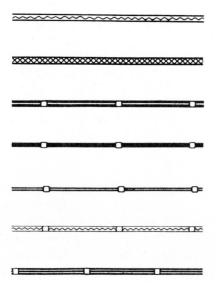

FIGURE 23–5 Pressure-sensitive screening is used to shade large areas. (Courtesy of Chartpak.)

FIGURE 23–6 A few of the symbols available on pressure-sensitive material. (Courtesy of Chartpak.)

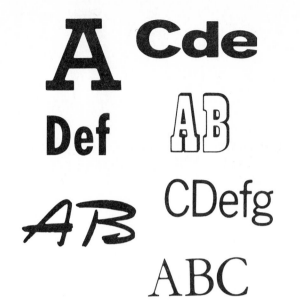

FIGURE 23-7  Many styles of lettering are available on pressure-sensitive tape. (Courtesy of Chartpak.)

Pressure-sensitive lettering speeds up chart production and results in a professional-looking job. The letters adhere directly to the chart. They are available in a wide range of sizes and styles (Fig. 23-7).

## BAR CHARTS

Bar charts are widely used for making comparisons. The numerical values are shown by rectangular bars drawn to scale. The length of the bar represents the numerical value. The bars can be shaded or colored for emphasis.

### Single-Column Bar Chart

A *single-column bar chart* has one column that represents the total, which could represent a quantity such as the cost of a product (Fig. 23-8). The bar represents 100%. It is subdivided into the elements

FIGURE 23-8  A single-column bar chart.

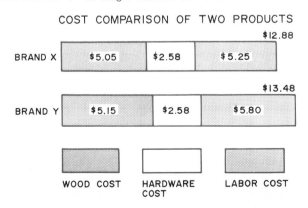

COST COMPARISON OF TWO PRODUCTS

| | | | $12.88 |
|---|---|---|---|
| BRAND X | $5.05 | $2.58 | $5.25 |

| | | | $13.48 |
|---|---|---|---|
| BRAND Y | $5.15 | $2.58 | $5.80 |

WOOD COST    HARDWARE COST    LABOR COST

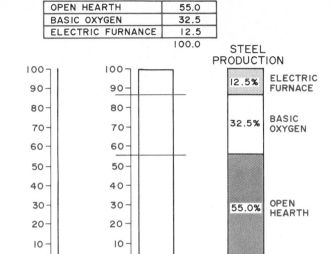

| | PERCENT |
|---|---|
| OPEN HEARTH | 55.0 |
| BASIC OXYGEN | 32.5 |
| ELECTRIC FURNANCE | 12.5 |
| | 100.0 |

FIGURE 23-9  The steps for drawing a single-column bar chart.

that make up the total. These are identified by shading and titles. To draw a single-column bar chart (Fig. 23-9):

1. Select a scale that produces the size desired. Lay out the horizontal and vertical scales.
2. Divide the vertical distance into equal divisions. Lay off the vertical divisions.
3. Draw the bar to width and divide it into the required divisions.
4. Shade and letter on or near the parts as required.

### Multiple-Bar Chart

A *multiple-bar chart* can be drawn vertically or horizontally (Fig. 23-10). A grid is established to show the value represented by each bar. The value can be lettered on the bar if desired (Fig. 23-11). The vertical and horizontal axes have identifying titles and numbers indicating the grid data and bar identification. The vertical axis title is lettered so as to be read from the right side of the drawing. The entire chart should have an overall identifying title. It should be lettered larger than the other words and placed above the chart. Keep the title as short as possible.

### Divided-Bar Chart

A *divided-bar chart* makes comparisons of two or more factors. Each factor is represented by a bar shaded with an identifying pattern. A key indicating the meaning of the shading patterns used is included (Fig. 23-12). Be certain to allow enough space between the elements of the chart for letter-

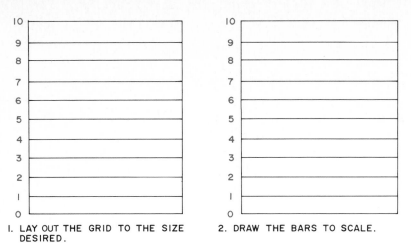

1. LAY OUT THE GRID TO THE SIZE DESIRED.

2. DRAW THE BARS TO SCALE.

3. SHADE THE BARS AND LETTER NEEDED DATA.

FIGURE 23–10 The steps for drawing a vertical multiple-bar chart.

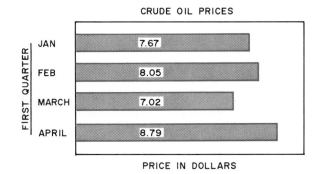

FIGURE 23–11 Multiple-bar charts can be drawn horizontally and the data lettered on the bar.

FIGURE 23–12 A divided-bar chart is used to make comparisons.

ing. Identification of each part should be clear. The overall layout is the same as for a single-bar chart.

## Over-and-Under Bar Chart

An *over-and-under bar chart* shows data ranging above and below a zero point. Various shading patterns focus attention on each element (Fig. 23–13).

## Progressive Bar Chart

A *progressive bar chart* does not have a zero starting point because it shows variables that change. When one variable changes, the other does also, giving a

FIGURE 23–13 An over-and-under bar chart.

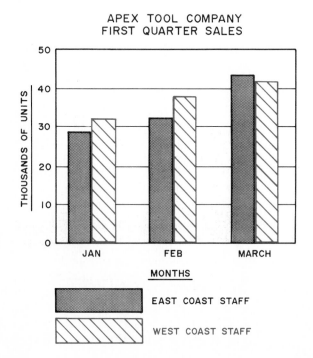

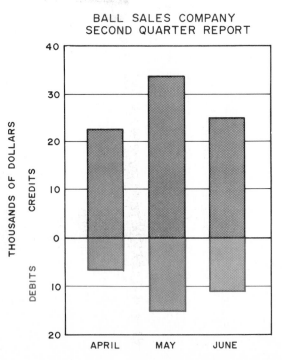

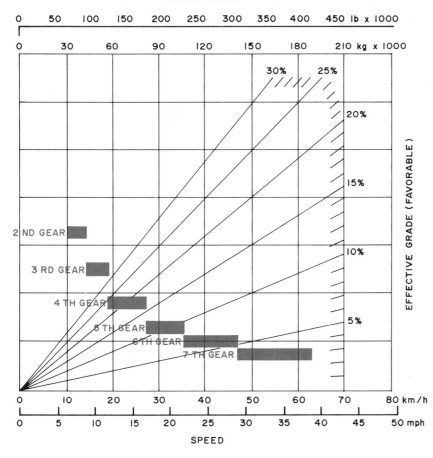

BRAKE PERFORMANCE—
CONTINUOUS GRADE RETARDING

FIGURE 23–14 This progressive bar chart is used to determine brake performance of off-highway trucks on a continuous grade. (Courtesy of Caterpillar Tractor Co.)

progressive range of data. In Fig. 23–14, the weight and grade of downhill slope influence the maximum safe speed an off-road tractor can reach and not exceed the braking capacity. The chart shows the speed limits for each gear.

## Range-Bar Chart

A *range-bar chart* shows the high and low figures over a predetermined period of time. The bar therefore represents the range from high to low. Comparisons can be made by examining the ranges over various time frames (Fig. 23–15).

## Planning Considerations

Some things to consider when drawing bar charts include:

1. Keep the bars the same width.
2. Use a scale on the horizontal and vertical axes to show the values.
3. Numerical data and identifying words can be lettered on or beside the bar.
4. Bars representing different features should have different surface indications. The use of color is excellent for this purpose.

FIGURE 23–15 A bar chart showing a range of stock values. The highest to lowest averages per month are charted.

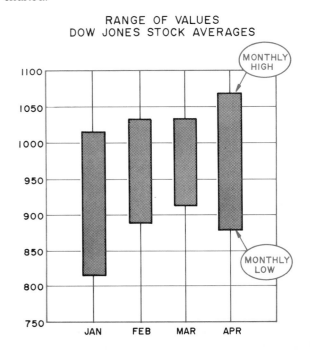

595

5. Each chart should have a short, descriptive title. It is usually lettered above the chart. The lettering is usually slightly larger than that used on the chart.

6. The meaning of the chart should be clear even when it is separated from supporting text.

7. Keys are often used to indicate what each bar represents.

## PIE CHARTS

A pie chart uses a circle to represent the whole or 100%. It is divided into sectors, which represent parts of the whole. They are most effective when there are no more than eight divisions. Small divisions are difficult to identify. The circle should be large enough to permit the needed lettering to be done without crowding. Generally, all lettering is horizontal so that it can be read from the bottom of the chart. If it is necessary to identify a small segment, letter it off the area and connect it with a leader, if necessary, for clarity (Fig. 23–16).

To convert a percentage to degrees, multiply the 360° in the circle by the percent. Then, lay off the number of degrees found with a protractor. Circular percentage charts are commercially available. The circle is divided into 100 parts and serves as a guide when developing pie charts (Fig. 23–17). The various sections can be shaded or colored to provide a more effective differentiation. A title is placed above the chart. A key identifying the sectors can be below it, if necessary. Pie charts can be drawn pictorially (Fig. 23–18).

## PICTORIAL CHARTS

Pictorial charts, or pictographs, are a variety of a bar chart that use pictures or symbols to form the bar. They are a visual way of presenting technical information to nontechnical readers. Each picture or symbol represents specific unit or quantity (Fig. 23–19). The value of each symbol is stated on the chart along with the necessary titles and numerical data. In Fig. 23–19, each figure represents 1000 wells. Pictorial charts have a limited degree of accuracy and are not suitable for reporting engineering data.

Some forms of pictorial charts vary the size of the image as a means of representing data. The viewer makes the comparison by noting the difference in the sizes of the symbols. If improperly drawn, this type of chart can be misleading because the reader does not know if the drafter emphasized area or linear size to make the comparison (Fig. 23–20).

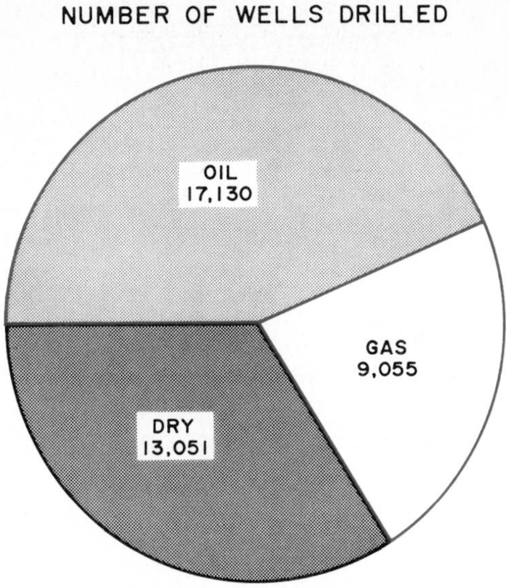

FIGURE 23–16 A pie chart.

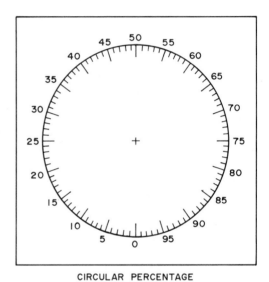

CIRCULAR PERCENTAGE

FIGURE 23–17 A circular percentage guide.

FIGURE 23–18 A pictorial pie chart.

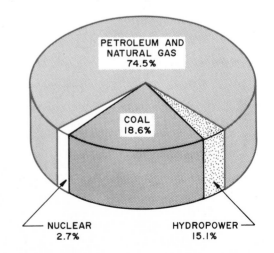

## NUMBER OF WELLS DRILLED

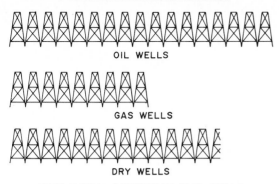

OIL WELLS

GAS WELLS

DRY WELLS

EACH SYMBOL REPRESENTS 1000 WELLS

FIGURE 23–19 Pictorial charts give general information.

EMPLOYMENT IN THREE
MAJOR CONSTRUCTION TRADES

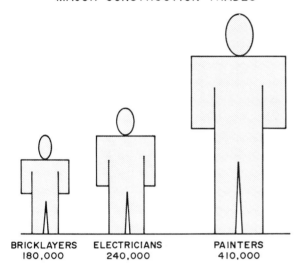

| BRICKLAYERS | ELECTRICIANS | PAINTERS |
| 180,000 | 240,000 | 410,000 |

FIGURE 23–20 The sizes of the symbols can be used to make comparisons.

FIGURE 23–21 This flow chart shows the processes used to make glass.

## FLOW CHARTS

A flow chart shows a schematic representation of the sequence involved in some process such as a manufacturing operation. In Fig. 23–21, pictorials are used to show the float process for making glass. Flow charts can also represent units in the sequence with rectangular, circular, or other appropriate shapes (Fig. 23–22).

## ORGANIZATION CHARTS

Organization charts show the line of authority or responsibility, beginning with the chief executive of the division. They show the interrelated and interlocking activities of the organization by connecting them with lines. Usually, the various areas are identified and lettered inside rectangular boxes, though other shapes can be used (Fig. 23–23).

## LINEAR CHARTS

Linear charts are used to show change or trends. They are also used to plot experimental data and in engineering problem solving including mathematical relationships. The basic types of linear charts use rectangular, semilogarithmic, and logarithmic grids.

Linear charts are built around two axes—a vertical, or Y, axis and a horizontal, or X, axis. A scale is laid out on each axis, representing some quantity, such as pounds, pressure, or time. This produces a grid pattern. Each quantity is located on the grid and connected with a straight or curved line, depending upon the factors being charted. Finally, the numerical data and identifying words are lettered (Fig. 23–24).

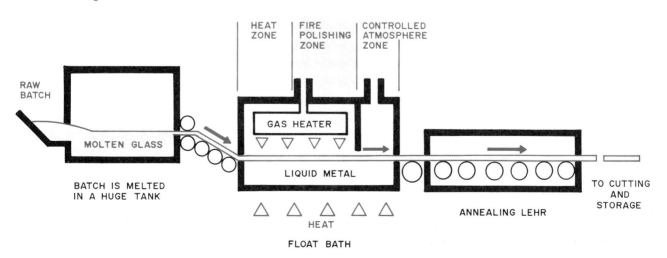

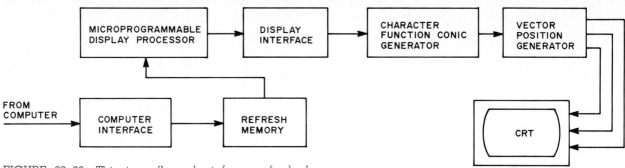

FIGURE 23–22 This is a flow chart for a refreshed stroke-writing computer-aided design system.

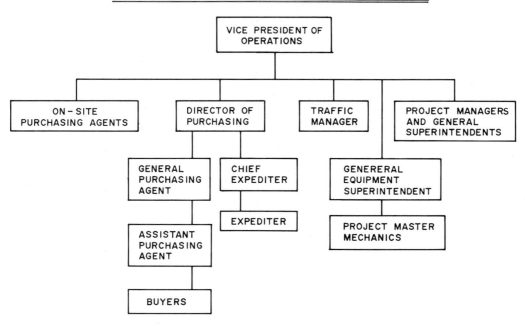

A CONSTRUCTION COMPANY OPERATIONS DIVISION

FIGURE 23–23 An organization chart shows the flow of responsibility through the various levels of an organization.

FIGURE 23–24 The steps for drawing a linear chart.

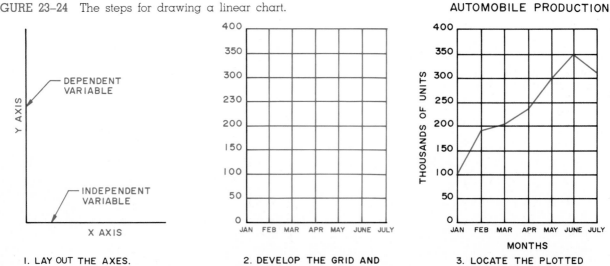

1. LAY OUT THE AXES.

2. DEVELOP THE GRID AND IDENTIFY THE VALUES.

3. LOCATE THE PLOTTED POINTS AND CONNECT. ADD THE NECESSARY DESCRIPTIVE WORDS.

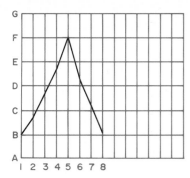

1. HORIZONTAL SCALE TOO SMALL MAKING THE RISE APPEAR SUDDEN.

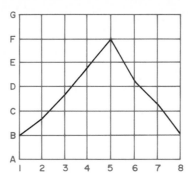

2. GRID PRODUCES A NORMAL WELL PROPORTIONED SCALE.

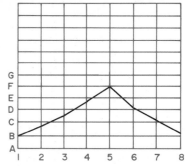

3. VERTICAL SCALE TOO SMALL MAKING THE CHART APPEAR FLAT AND THE RISE MINIMAL.

FIGURE 23–25 The proportions of the grid influence the appearance of the chart.

## The Variables

The independent variable is located on the horizontal or X axis. An independent variable is one that is controlled. For example, in a study of temperature over a period of a year, the months and days are controlled. The temperature varies, so the dependent variable, the temperature, is placed on the vertical or Y axis.

## Grid Design

The selection of the grid pattern is critical to the proper presentation of data. The size of the grid directly influences the appearance of the charted data (Fig. 23–25). When the horizontal scale is too small, rises and falls appear steep. When the vertical scale is too small, rises and falls appear flat and unimportant. After examining the data and selecting the variables to be charted, use a grid that most accurately portrays what the data reveal.

Grids involving time are usually rectangular, having a proportion of 2:3 or 3:4. A grid that is wider than it is high is best for charts that have a large number of points to plot (Fig. 23–26). A grid

FIGURE 23–26 Time charts and charts that have many plotted points often use grids wider than they are high.

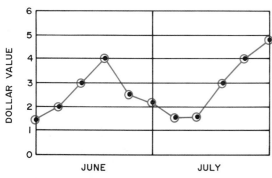

PRICE OF APEX COMPANY STOCK

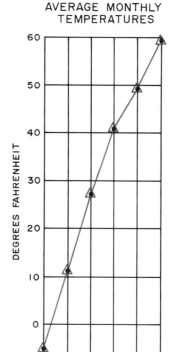

AVERAGE MONTHLY TEMPERATURES

FIGURE 23–27 Charts with plotted data having a rapid rise use a grid that is higher than it is wide.

that is higher than it is wide is often used for time charts covering short periods of time or those having a rapid change (Fig. 23–27). Logarithmic grids have scales that are graduated with logarithmic divisions along both axes. Semilogarithmic grids have a vertical logarithmic scale and an arithmetic horizontal scale. These are shown later in this chapter.

## Graph Sheets

Printed graph sheets are widely used for developing linear charts. The printed grid forms the basis for

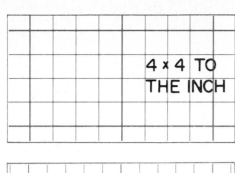

4 × 4 TO
THE INCH

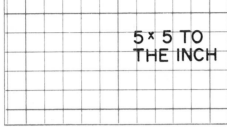

5 × 5 TO
THE INCH

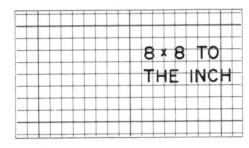

8 × 8 TO
THE INCH

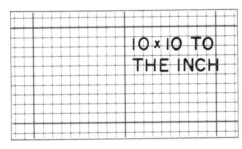

10 × 10 TO
THE INCH

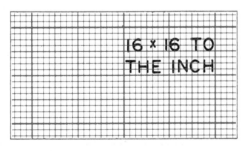

16 × 16 TO
THE INCH

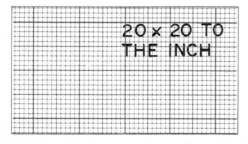

20 × 20 TO
THE INCH

FIGURE 23–28  Inch graph sheets.

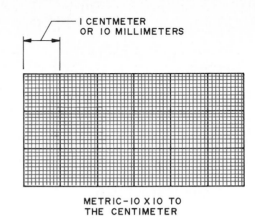

I CENTMETER
OR 10 MILLIMETERS

METRIC–10 X 10 TO
THE CENTIMETER

FIGURE 23–29  A metric graph sheet.

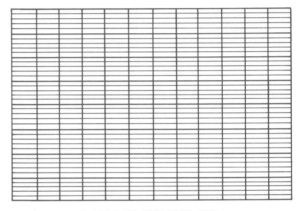

A 4X20 TO THE INCH GRID

FIGURE 23–30  One type of rectangular grid.

selecting the grid lines of the chart. A variety of grids are available. Inch-type grids are available in a variety of divisions, such as 4, 5, 6, 8, 10, 12, 16, and 20 divisions per inch (Fig. 23–28). Metric paper is divided into millimeters, with a heavy accent line every tenth millimeter (Fig. 23–29). These are available in specially prepared forms for drum and flatbed plotters.

Also available are sheets with a variety of rectangular grids for use in making economic and time-series charts (Fig. 23–30). Semilogarithmic and full-logarithmetic grids will be shown later in this chapter.

## Charted Lines

Charted lines may fall in a continuous straight line, but more often they take other patterns. They may form a curve, be broken, jagged, or stepped, or be adjusted to match the trend of a series of plotted points (Fig. 23–31). They can be drawn as solid lines or formed by any variety of dashed lines. The line is usually drawn thick and may be in color or formed with pressure-sensitive tape of various colors.

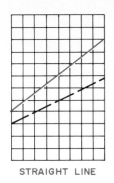  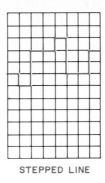 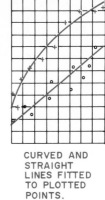 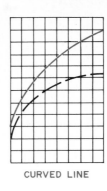

STRAIGHT LINE     BROKEN LINE     STEPPED LINE     CURVED LINE     CURVED AND STRAIGHT LINES FITTED TO PLOTTED POINTS.

FIGURE 23–31   Charted lines may take several forms.

## Plotted Points

Linear charts used for general information rather than engineering analysis usually plot the beginning and end point of each section to be recorded. If the location of each point is important, it can be indicated within a small circle or triangle, as may be seen in Figs. 23–27 and 23–34.

## STEPPED-LINE CHART

A stepped-line chart is used to show data that remain fixed over a period of time and then abruptly go higher or lower. Figure 23–32 shows that the price of a barrel of oil in January remained at $41.00. In February, it was raised to $43.00. It remained at this price until May, when it went to $44.00. Such a chart is also useful when data have been accumulated or averaged over a period of time, such as the Dow-Jones Index.

## LINEAR CHARTS USING RECTANGULAR GRIDS

Curves appearing on linear charts may be in one of three forms. If the data collected are not backed up by a definite theory or mathematical law, the plotted

points are usually connected with straight lines. In some cases an empirical relationship (a relationship derived by observation or experiment) between the curve and plotted points may be used when the engineer believes that the curve should pass through some points but pass beside others. If the curve should follow a known true theoretical curve, it can be drawn and related as closely as possible to the plotted points (Fig. 23–33).

FIGURE 23–33   Methods used to draw curves of plotted data.

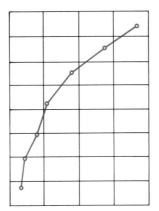 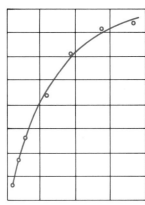

DATA NOT SUPPORTED BY A THEORY OR MATHEMATICAL LAW.

DATA PLOTTED WHEN AN EMPIRICAL RELATIONSHIP EXISTS.

FIGURE 23–32   A stepped-line chart.

### PRICE OF OIL PER BARREL

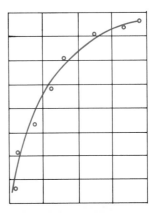

DATA PLOTTED WHEN CURVE FITS A KNOWN TRUE THEORETICAL CURVE.

There are many forms of line charts. The form a chart takes depends on the data to be presented and the ingenuity of the person preparing the chart. Several sets of data can be presented on one chart. The various curves can be identified with a key (Fig. 23–34), or the identification can be lettered on the curve (Fig. 23–35). When line charts are used to compare several sets of data, the area below the line can be shaded. This is called a *surface chart,* and the shading helps to visualize the comparison (Fig. 23–36) and gives added emphasis to the data. Surface charts can also show the net difference between comparative data. Figure 23–37 shows graphically the gain and loss due to the sale of a product at prices differing from the manufacturer's list price.

When the points to be plotted have positive and negative values, the line chart can be built using the four quadrants (Fig. 23–38). Most line charts are drawn in the first quadrant because all points have positive values. When plotting mathematical data, several quadrants may be needed. Each quadrant represents a positive and/or negative value of X and Y. The X and Y axes intersect at point 0. The values to the right of the Y axis are positive. Those to the left are negative. The Y values above the X axis are positive and those below are negative. Since each

point has an X and Y value, it can be located in one of the quadrants. The plotting of selected data forming a curve is shown in Fig. 23–39. The positive and negative values of each point are essential to correct plotting.

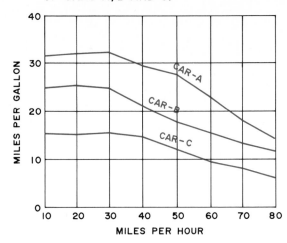

FIGURE 23–35  Curves can be identified by lettering on the curve.

FIGURE 23–36  Several sets of data can be compared on a surface chart.

FIGURE 23–34  Several sets of data can be presented on one chart.

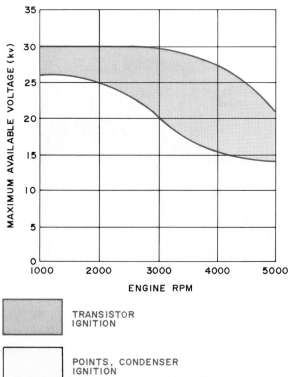

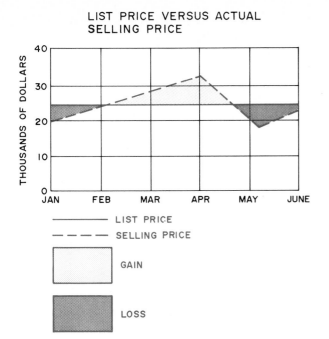

LIST PRICE VERSUS ACTUAL
SELLING PRICE

— — — LIST PRICE

– – – – SELLING PRICE

GAIN

LOSS

FIGURE 23-37 Surface charts can show gain and loss data.

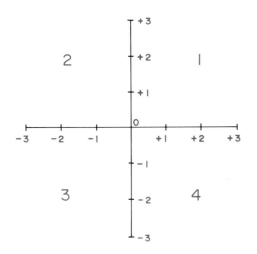

FIGURE 23-38 The four quadrants in which line charts may be drawn.

## TRILINEAR CHARTS

Trilinear charts are commonly used in chemistry and metallurgy because comparisons of metallurgical and chemical compositions with three related variables occur frequently. Such charts are useful in comparing three related variables relative to their total composition, which equals 100 percent.

The principle upon which this chart is based is the geometric principle that the sum of the perpendiculars to the three sides of an equilateral triangle from any point within the triangle is equal to the altitude of the triangle. In the equilateral triangle in

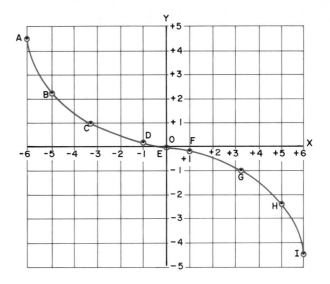

DATA TO BE
PLOTTED

|   | X | Y |
|---|------|------|
| A | -6.0 | +4.5 |
| B | -5.0 | +2.3 |
| C | -3.2 | +1.0 |
| D | -1.0 | +0.2 |
| E | 0.0 | 0.0 |
| F | +1.0 | -0.2 |
| G | +3.2 | -1.0 |
| H | +5.0 | -2.3 |
| I | +6.0 | -4.5 |

FIGURE 23-39 The positive and negative values of the points on this curve have been plotted in the second and fourth quadrants.

Fig. 23-40, point X is located within the triangle. The sum of distances WX, YX, and ZX is equal to the altitude, A2, B3, or C1. For example, if distances WX, YX, and ZX were 20, 30, and 50, respectively, the altitude would be 100.

Trilinear charts are drawn on commercially available triangular coordinate grids that have the altitude divided into 100 equal parts and numbered every fifth division. The coordinates for these points are drawn parallel to the sides of the triangle (Fig. 23-41). They can easily be drawn on plain paper, but the coordinates must be carefully located. The designation of the factor on each of the three scales must be clearly lettered by each scale but off the grid area. The curves developed can be recorded as solid or dashed lines as desired.

In Fig. 23-42 is a trilinear chart showing a plot for the composition of an alloy made up of three variables, steel 50%, nickel 20%, and chromium 30%. Vertex A of the diagram represents 100% of one component, steel. The base of the diagram at C represents 0% for steel. Thus, lines parallel to the base indicate varying percentages of steel. The point representing alloy 1 lies 50% of the distance between A

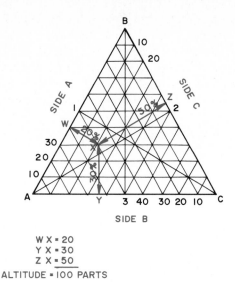

WX = 20
YX = 30
ZX = 50
ALTITUDE = 100 PARTS

FIGURE 23-40 The sum of the distances from a point inside the chart perpendicular to each side of the trilinear chart is equal to the altitude of the triangle.

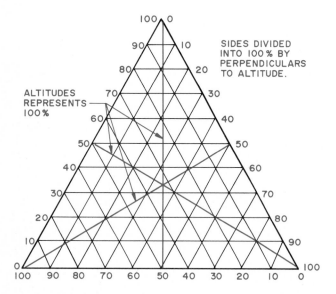

SIDES DIVIDED INTO 100% BY PERPENDICULARS TO ALTITUDE.

ALTITUDES REPRESENTS 100%

FIGURE 23-41 A trilinear chart showing the division of the sides into 100 equal parts.

and C, so alloy 1 contains 50% steel. The same procedure is used for the other sides of the chart to locate values for the nickel, 20%, and chromium, 30%. The percentage values lie on the lines parallel with the respective sides.

## Laying Out Trilinear Charts

Commercially available triangular coordinate grids are generally used for trilinear charts; however, they can be drawn on plain paper if necessary. It is important that the grid spacing be accurately drawn. The identification of each of the three scales is nec-

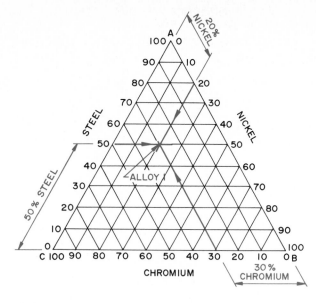

FIGURE 23-42 A trilinear chart showing a plot for the composition of an alloy made up of 50% steel, 20% nickel, and 30% chromium.

essary. Generally, this is lettered parallel with the edge of the triangle. Locate all notations outside the grid area.

Generally, the individual points plotted are not shown and the curves are drawn as continuous lines. They are usually solid lines, but dashed lines or other types can be used.

One application is to develop a chart that can be used to find data about the three variables. For example, if three materials when combined in various percentages have a common weight of 250 kg, the contour for this weight could be drawn on the chart. Data developed for three metals are shown in Fig. 23-43. They are plotted, forming the given curve. This can be used to find the percentage of each metal required when the percentage of one of them is altered.

FIGURE 23-43 A typical plot on a trilinear chart.

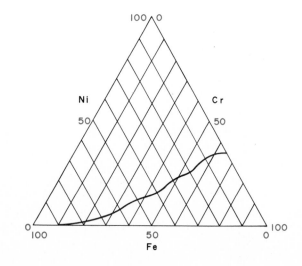

# POLAR CHARTS

Polar charts are used to plot technical data having two variables, one a linear quantity (called a radius vector) and the other its angular quantity. They are drawn on polar coordinate paper that consists of concentric circles—representing linear distances from the pole—that are divided into degrees (Fig. 23–44). Polar charts are drawn by self-recording devices attached to scientific equipment (Fig. 23–45) or by drafters plotting known points to form a continuous curve that represents scientific data.

The 0° polar axis is located on the right horizontal axis or the vertical axis. The degrees are rotated either clockwise or counterclockwise. The radius vector values are lettered on the vertical or horizontal axis. The angular designations are placed off the grid along its edges.

FIGURE 23–44 A polar chart containing data for a 24-hr period.

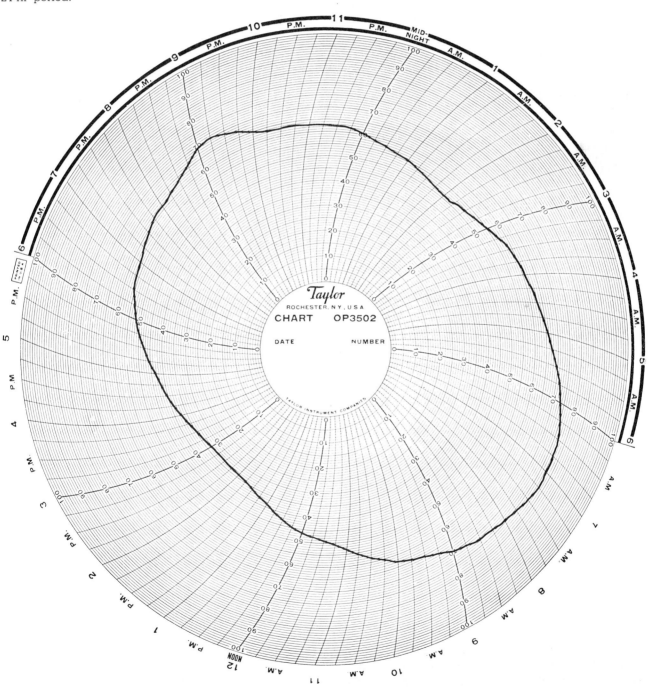

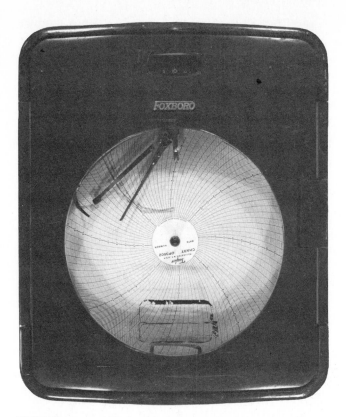

FIGURE 23-45 This instrument is recording data on a polar chart.

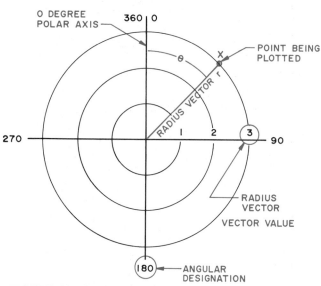

FIGURE 23-46 To plot a curve on a polar chart, the radius vector and the angle of that vector should be known.

To plot the curve, it is necessary to know the radius vector, $r$, and the angle of that vector, $\theta$, for each point. In Fig. 23-46, point X has an $r$ of 3 and $\theta$ is 45°. A completed plot for the decibels produced by an automobile turbine engine is shown in Fig. 23-47. The 0° axis represents the front of the auto, while the 180° axis is the rear. When you are about

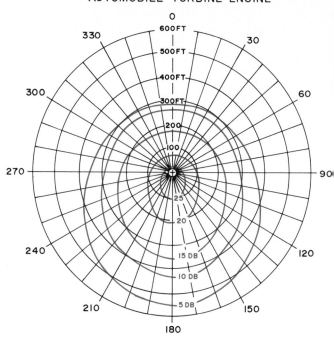

FIGURE 23-47 A typical polar chart.

300 ft in front of the vehicle, the decibel rating is 5. When you are 300 ft behind the vehicle, the decibel rating is about 17. You have to get approximately 550 ft from the rear before the decibel rating drops to 5.

## NOMOGRAPHS

The engineer uses mathematics to solve design problems involving equations. Solutions to equations can be found rapidly by using nomographs. Once developed, a nomograph can be used repeatedly for solving the equation from which it was derived.

Nomographs provide solutions to problems containing three or more variables. They have a series of calibrated scales, each representing one of the variables. The solution is derived by placing a straight line called an *isopleth* joining the known values on two of the scales and crossing the third scale. The unknown value is read off the third scale. For example, the nomograph in Fig. 23-48 relates to production rates of a track-type loader. The variables are cycles per hour, hourly production, and required payload volume. If you locate the cycles per hour at 100 and are moving 150 yd³/hr of soil and connect these with a straight line that intersects the third scale, you will see the required payload volume is 1.5 yd³/cycle. In other words, it is necessary to dig 1.5 yd³ for each of the 100 digging cycles to produce a total hourly production of 150 yd³/hr. Notice that

1) Enter Scale A cycles per hour (100) and B hourly production 115 m³/hr (150 yd³/hr).
2) Connect A and B and extend to C to find required payload 1.2 m³ (1.5 yd³).

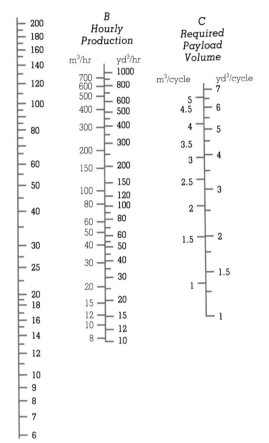

FIGURE 23–48 A nomograph. (Courtesy of Caterpillar Tractor Co.)

this nomograph is scaled for customary and metric units. A five-variable nomograph related to the earth loader is shown in Fig. 23–49. Following this example in metric, we have a required payload of 1.2 m³/cycle and are digging soil with a density of 1958 kg/m³, giving a bucket payload of 2250 kg/cycle. If the loader operates at 100 cycles/hr, it will dig 225 metric tons/hr.

## LOGARITHMIC SCALES

A logarithm is the exponent indicating the power to which a fixed number, the base, must be raised to produce a given number. For example, if $M^x = b$, the logarithm of $b$ with $M$ as the base is $X$, or $\log_m b = X$.

The divisions on logarithmic charts are made proportional to the logarithms of the numbers instead of the numbers themselves. This is why they are not equally spaced. Logarithmic grids are commercially available in several combinations of ruling. *Semilogarithmic grids* have *logarithmic spacing* on one axis and *equal spacing* on the other axis. The grids are available with one, two, three, four, five, and seven log cycles per page. The length of each cycle varies. For example, three-cycle grids have 3½ in. per cycle. Seven-cycle grids allow 1¼ in. per cycle (Fig. 23–50). *Full-logarithmic grids* have *logarithmic spacing* on both axes. Typical grids available include 1 × 1 cycle, 2 × 1 cycle, 2 × 2 cycles, 2 × 3 cycles, 3 × 3 cycles, and 3 × 5 cycles (Fig. 23–51).

## SEMILOGARITHMIC CHARTS

The logarithmically divided scale is usually positioned as the Y axis (called the ordinate), and the equally divided scale is the X axis (called the *abscissa*)—though they can be reversed if it is helpful. The logarithmic scales are positioned so that the largest division is at the bottom, with the 10 decreasing sizes above it forming a log cycle.

The locations of the grid divisions are determined by taking the logarithm to the base 10 of each value to be plotted. The layout of a logarithmic scale is shown in Fig. 23–52. The desired length of one cycle is established. Then, the log of each subdivision up to 10 is multiplied by the length. This establishes the location of each grid line on the chart. For example, if the length is to be 3.5 in., the number 5 grid line is 3.5 in. × log of 5, or 3.5 × 0.699 = 2.4 in. from the origin. Using these logarithmic proportions, additional cycles can be added, forming a grid with several cycles.

The selection of the number of cycles to use depends on the range of data to be plotted. Log cycle designations must start with a power of 10 and end with the next power of 10. Typically used designations include 0.01 to 0.1, 0.1 to 1, 1 to 10, 10 to 100, and 100 to 1000. These are expressed on the chart as powers of 10. The preceding designations would be $10^{-2}$ to $10^{-1}$, $10^{-1}$ to $10^0$, $10^0$ to $10^1$, $10^1$ to $10^2$, and $10^2$ to $10^3$ (Fig. 23–53). A three-cycle grid starting with 1 could therefore accommodate a range of values from 1 to 1000. Remember that the logarithms of numbers between 0 and 1 are negative and above 1 are positive. When the positive variable approaches zero, its logarithm becomes negatively infinite; therefore, semilogarithmic charts never have a zero value on the logarithmic scale.

When deciding whether to use a grid with rectilinear ruling or semilogarithmic ruling, consider what the chart is expected to show. If it is to show

## Track-Type Loaders

Connect C 1.2 m³ (1.5 yd³) to F 1958 kg/m³ (3000 lb/yd³) and
extend to G to find payload weight 2250 kg (5000 lb.)
Extend Scale G reading 2250 kg (5000 lb) through Scale A (100)
to Scale I to find tons per hour 225 metric ton/hr (250 U.S. ton/hr).

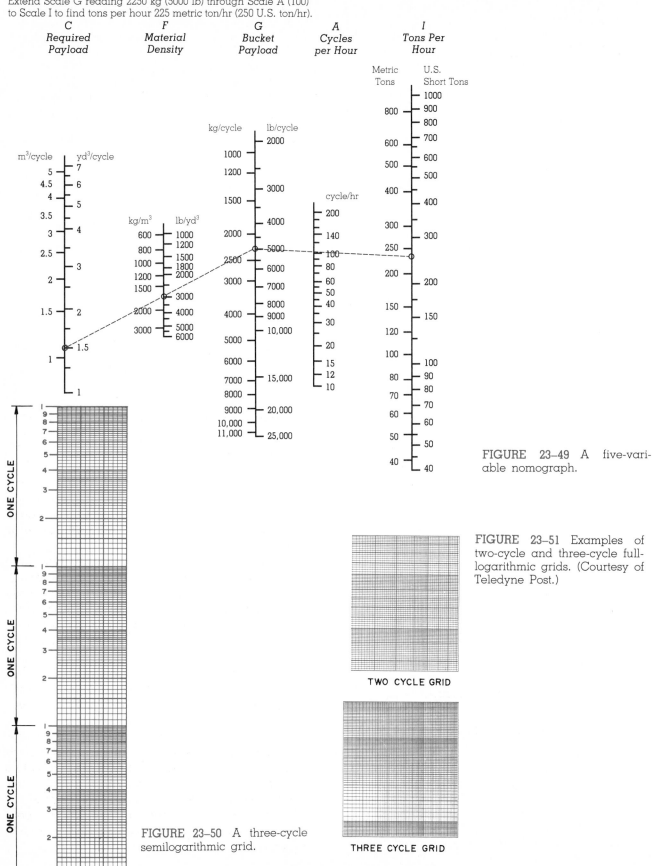

**FIGURE 23–49** A five-variable nomograph.

**FIGURE 23–51** Examples of two-cycle and three-cycle full-logarithmic grids. (Courtesy of Teledyne Post.)

TWO CYCLE GRID

THREE CYCLE GRID

**FIGURE 23–50** A three-cycle semilogarithmic grid.

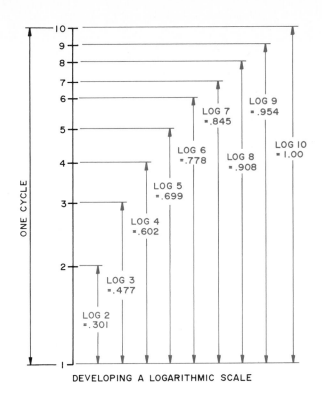

DEVELOPING A LOGARITHMIC SCALE

FIGURE 23–52 Logarithmic grid divisions are determined by taking the logarithm to the base 10 of each value in the cycle.

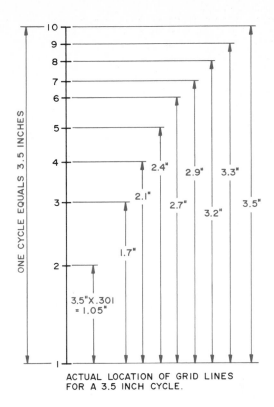

ACTUAL LOCATION OF GRID LINES
FOR A 3.5 INCH CYCLE.

FIGURE 23–53 Several logarithmic cycles are used to increase the range of values in the scale.

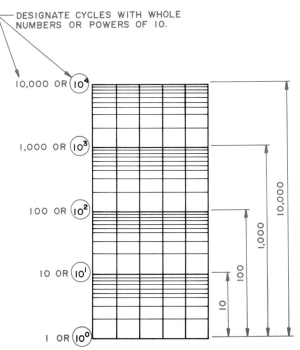

*numerical* increases or decreases, a rectilinear ruling should be used. If it is to show percentages or *rate of change,* the semilogarithmic grid is used.

In Fig. 23–54, the same data are plotted on both types of grids. The arithmetic chart shows the quantitative change, which appears steeper than the actual rate of change shown on the logarithmic chart. It is the true picture of the price increase regardless of the size of the increase.

Semilogarithmic charts represent data in a manner that is not possible on rectangular arithmetic charts. They display the data so that a relative comparison can be made visually and show whether the data follow a consistent relative-change pattern or whether no forces are in effect to produce a consistent relative change.

Semilogarithmic charts cannot be used for data that have negative or zero values. They also should not be mistaken as arithmetic charts or the data read as they would be on such a chart.

The procedures for constructing logarithmic charts are the same as those discussed for rectilinear charts. The plotted points are usually located with a circle or triangle. The titles, scale values, and curves drawn are identified in the same manner. A semilogarithmic chart is shown in Fig. 23–55. Notice that the logarithmic scale is located in this case along the abscissa.

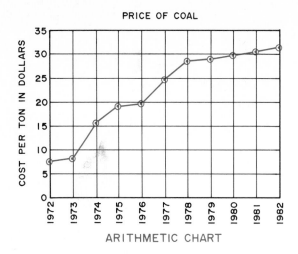

PRICE OF COAL

ARITHMETIC CHART

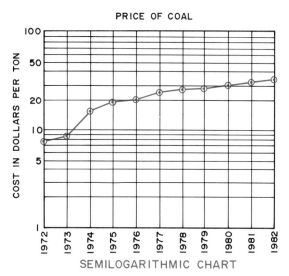

PRICE OF COAL

SEMILOGARITHMIC CHART

FIGURE 23–54 A comparison of data plotted on arithmetic and semilogarithmic charts.

FIGURE 23–55 A semilogarithmic chart.

## LOGARITHMIC CHARTS

Logarithmic charts contain logarithmic scales on the ordinate and abscissa. Full log grids are shown in Fig. 23–51.

Logarithmic charts are used in engineering analysis where two variables must be plotted. They are especially useful when a large number of points are to be plotted and can be used to compare trends shown by several curves drawn on the chart. Since the chart has log cycles on both axes, an extensive range of values can be plotted on a rather small chart. Logarithmic charts are frequently used to determine empirical equations.

Logarithmic charts are developed in the same manner as semilogarithmic charts. The scales need to be fully and carefully identified. A typical logarithmic chart is shown in Fig. 23–56.

## EMPIRICAL EQUATIONS

Engineers deal constantly with equations. An equation is a linear array of mathematical symbols separated into left and right sides that are initially indicated as being equal by the inclusion of an equal sign (Fig. 23–57). Empirical data are those obtained usually by measurement when conducting field tests or laboratory tests. In many situations, the data can be set up in equations that are then used to evaluate mathematically the characteristics of the data within the limits of the tests made. The empirical data can be used to develop equations using one of the three types of equations covered in this unit.

The development of the equation begins by plotting the data on rectangular, semilogarithmic,

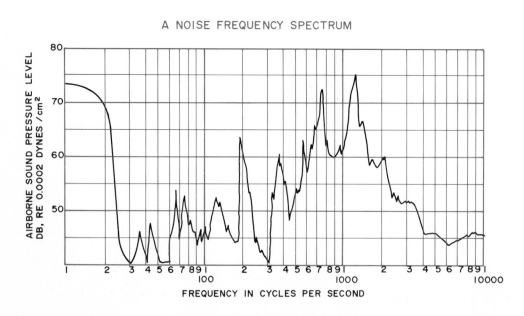

A NOISE FREQUENCY SPECTRUM

FREQUENCY IN CYCLES PER SECOND

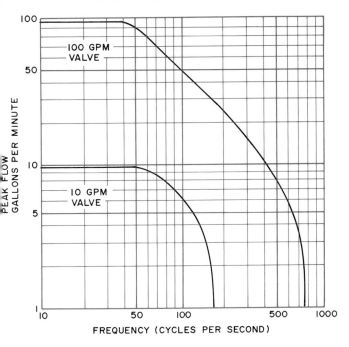

VALVE RESPONSE TO A PRESSURE DROP ACROSS THE VALVE

FIGURE 23–56  A logarithmic chart with two cycles on each side.

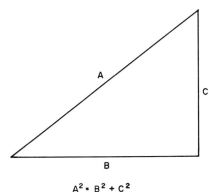

THE PYTHAGOREAN THEOREM

FIGURE 23–57  This equation means that the square of the hypotenuse is equal to the sum of the squares of the other two sides.

and logarithmic grids. If they appear as a straight line on one of the grids, that straight-line curve is used to develop the equation. Empirical data are plotted on the grids in Fig. 23–58. On each grid, one of the three sets of data plot as a straight line. This determines the form of equation to be used.

The three forms of empirical equations to be considered include:

$y = mx + b$ (straight line on a uniform rectangular grid)

$y = bx^m$ (straight line on a rectangular log-log grid)

$y = bm^x$ (straight line on a rectangular semilog grid)

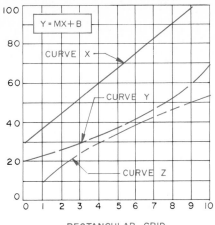

RECTANGULAR GRID

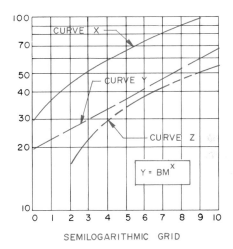

SEMILOGARITHMIC GRID

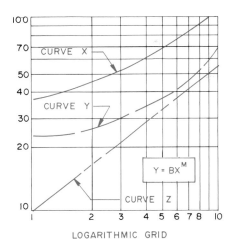

LOGARITHMIC GRID

FIGURE 23–58  Empirical data shown plotted on three types of grids. This helps determine which grid will produce a straight-line plot of the data. If the data plot as a straight line on one of the grids, the equation for the data can be found.

## PLOTTING TECHNIQUES

As was mentioned in the discussion of graphic presentation, the independent variable is usually placed on the X axis and the dependent variable on the Y axis. The plotting must be done with great accuracy. Locate the plotted points with a sharp x or a point rather than the large circle or triangle used on graphical presentations. When the points are plotted, draw a single straight line through as many of them as is possible. Since there may be inaccuracies both in reading the instruments used in the experiment and also in plotting the points, it is not usually possible to get all the points on the straight line. Try to get about the same number of points that miss the line above it as below it. Ignore any points that are far from those forming the line. They are probably due to a recording error (Fig. 23–59).

If the data are rather scattered, draw a straight line on each side to locate the maximum and minimum readings. Then, locate a line between these that is near the midpoint of the space. This represents an average of the readings (Fig. 23–60).

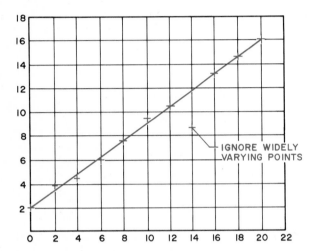

FIGURE 23–59 Locate the straight line so that it runs through as many points as possible.

FIGURE 23–60 When plotted points are scattered, an average can be used to locate the line representing the data.

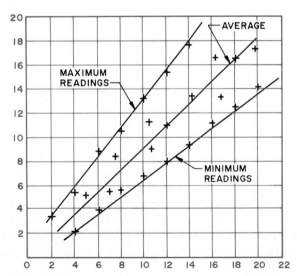

Remember, the empirical equation developed in this manner is an approximation of the data presented. Since there are many chances for deviation, it must be understood that this is not absolute. Also, the equation is valid only for the range of data developed by the experiment. It is not acceptable to use the equation developed for extensions of the curve beyond that developed in the experiment.

## METHODS FOR DETERMINING THE EQUATION

The methods used for finding the equation for a straight line on any of the three types of grids include:

1. Slope and intercept.
2. Selected points.
3. Averages.
4. Least squares.

The slope-and-intercept and selected-points methods are adequate for data accurate to two significant digits. The averages method is accurate for almost any situation. The least-squares method is more accurate than the others but is more complex and time-consuming and will not be covered in this unit. These methods are used in the following examples.

## THE LINEAR EQUATION
$y = mx + b$

When the data plot as a straight line on a rectangular coordinate grid, the data are linear and can be used to develop a *linear equation*. This means that each measurement on the Y axis is directly proportional to the values on the X axis. The linear equation is $y = mx + b$. The constant $m$ is the tangent of the angle between the straight line and the horizontal. The constant $b$ is the Y intercept, which is the value of $y$ when $x = 0$. The values $x$ and $y$ are the variables.

The equation can be written using any of the three methods mentioned earlier. Each is explained in the following examples.

### Slope-and-Intercept Method

To develop the equation, plot the data from the experiment and draw the straight lines connecting the points (Fig. 23–61). Find the value of $m$ as follows:

1. Select two known points on the line that are a reasonable distance apart. These are A and B in this example.

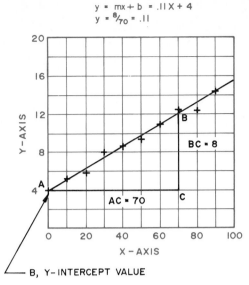

$$y = mx + b = .11 X + 4$$
$$y = {}^8\!/_{70} = .11$$

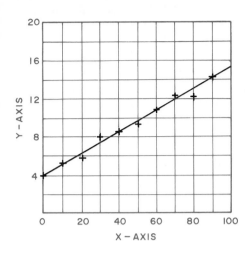

1. LOCATE THE POINTS AND DRAW THE STRAIGHT LINE.

2. DETERMINE THE VALUES FOR AC AND BC. FIND M AND Y-INTERCEPT. SUBSTITUTE FOR THE CONSTANTS IN THE EQUATION.

B, Y-INTERCEPT VALUE

BC = 8

AC = 70

FIGURE 23–61 How to write a linear equation for data that plot as a straight line on a rectangular grid using the slope-and-intercept method.

2. Draw a right triangle, ABC. The values of the sides are found by counting the units on the vertical and horizontal scales.

3. The value of $m$ is the tangent of the angle at corner A. Therefore,

$$m = \frac{\text{side BC}}{\text{side AC}} = \frac{8}{70} = 0.11$$

4. The Y intercept (point A on the Y axis) can be read off the vertical scale and, in this case, is 4, the value at $b$ in the equation $y = mx + b$.

5. The substitution of these constants in the equation produces $y = 0.11x + 4$. Substitute values for $x$ and find the value of $y$ for the data on this chart. For example, when $x$ equals 40, $y$ equals 8.40.

## Selected-Points Method

1. Select two points on the line, A and B (Fig. 23–62).

2. Find the coordinates of each point. The coordinates of A are 10($x$), and 5($y$). The coordinates of B are 70($x$) and 12($y$).

3. Enter the coordinates into the equation $y = mx + b$. Solve the simultaneous equation for $m$.

$$
\begin{aligned}
12 &= 70m + b \\
5.5 &= 10m + b \\
\hline
\text{subtract} \quad 6.5 &= 60m \\
m &= 0.11
\end{aligned}
$$

4. Substitute $m$ into one of the equations to find $b$.

$$
\begin{aligned}
12 &= 70m + b \\
12 &= 70(0.11) + b \\
b &= 4.3
\end{aligned}
$$

5. The equation is therefore

$$y = 0.11x + 4.3$$

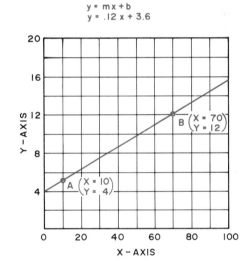

$$y = mx + b$$
$$y = .12 x + 3.6$$

B $\begin{pmatrix} X = 70 \\ Y = 12 \end{pmatrix}$

A $\begin{pmatrix} X = 10 \\ Y = 4 \end{pmatrix}$

FIGURE 23–62 Writing a linear equation using the selected-points method.

## Averages Method

The method of averages is based on the numerical data and not on the plot itself. A plot is drawn to verify that the curve is straight. The accuracy of the plot is not important since it is not used to generate the data. The straight-line curve is based on the assumption that if the data values of $y$ are added together and the same number of computed values of $y$ derived from an equation are added, the difference between these two resultants would be 0. Therefore, the equation $y = xm + nb$ is applied to two equal groups of data values of $x$ and $y$, and these are solved as simultaneous equations. This provides the

Graphical Presentation and Analysis **613**

| X | Y |
|---|---|
| 10 | 5.5 |
| 30 | 7.5 |
| 50 | 9.7 |
| 70 | 12 |
| 90 | 14.5 |
| 100 | 15.5 |

**FIGURE 23–63** Coordinates on the straight-line curve shown in Fig. 23–61.

slope ($m$) and intercept ($b$) values. The $n$ in the equation is the number of data values added.

The data are presented in Fig. 23–63. The top three and bottom three sets of data are added, substituted in the equation, and solved as simultaneous equations.

$$
\begin{aligned}
y &= mx + b \\
42.0 &= 260m + 3b \\
22.7 &= 90m + 3b \\
\hline
\text{subtract } 19.3 &= 170m \\
m &= 0.11
\end{aligned}
$$

The value for $b$ is found by inserting the value of $m$ in one of the equations.

$$
42.0 = 260(0.11) + 3b
$$
$$
b = \frac{13.4}{3} = 4.46
$$

## THE EQUATION $y = bm^x$

The exponential equation for a straight line on a semilogarithmic grid is $y = bm^x$. The experimental data in Fig. 23–64 plotted on a rectangular grid do

not form a straight line; therefore, we know they cannot be expressed in the form of a linear equation. When plotted on a semilogarithmic grid, they produced an approximately straight line for which the equation $y = bm^x$ will apply. The slope of the curve is $m$, and the Y intercept is $b$. The procedure for writing the equation uses the slope-and-intercept method, though either of the others discussed earlier could be used.

To develop the equation:

1. Select two points on the line, A and B.
2. Draw a right triangle using these points. The sides of the triangle are the differences between the coordinates of points A, B, and C.
3. Solve for $m$ as follows:

$$
\log m = \frac{\log 44 - \log 5}{8 - 1.8} = \frac{1.6435 - 0.6990}{5} = 0.1889
$$

Find the antilog of 0.1889
$$
m = 1.5
$$

4. The value of $b$ is 3.
5. Substitute the values of $m$ and $b$ in the equation: $y = 3(1.5)^x$. In this equation, $x$ is the variable that can be substituted into the equation to give additional values for $y$.
6. If desired, write the equation in logarithm form so that it can easily be used to solve for $y$ values. Rewrite the equation as follows:

$$
\begin{aligned}
\log y &= \log b + x \log m \\
&= \log b + x \log 1.5
\end{aligned}
$$

## THE EQUATION $y = bx^m$

The power equation for a straight line on a logarithmic grid is $y = bx^m$. The experimental data in Fig. 23–65 plotted on a rectangular grid do not form a straight line; therefore, we know they cannot be ex-

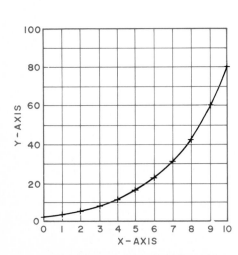

EXPERIMENTAL DATA PLOTTED ON
RECTANGULAR COORDINATE GRID.

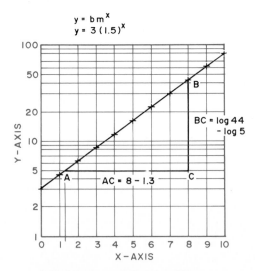

EQUATION IN LOGARITHM FORM
log y = log b + X log 1.5

**FIGURE 23–64** Developing the exponential equation for a straight line on a semilogarithmic grid.

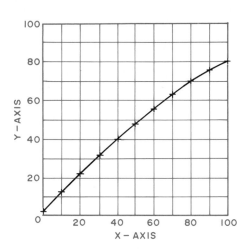

EXPERIMENTAL DATA PLOTTED ON
RECTANGULAR COORDINATE GRID.

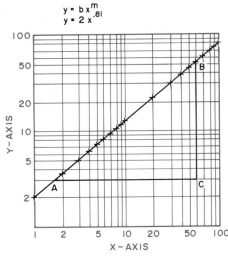

$y = bx^m$
$y = 2x^{.81}$

EQUATION IN LOGARITHM FORM
$\log y = \log 2 + .81 \log x$

FIGURE 23–65 Developing the power equation for a straight line on a logarithmic grid.

pressed in the form of a linear equation. When the data were plotted on a logarithmic grid, an approximate straight line was produced. The data can be expressed in the form of a *power equation* in which $y$ is a function of $x$ raised to a given power, or $y = bx^m$. The equation for these data is obtained much the same as explained for exponential equations, where $m$ is the slope of the line (tangent of the angle) and $b$ is the Y intercept. Following is the procedure using the slope-and-intercept method.

To develop the equation:

1. Select two points on the line, A and B.

2. Draw a right triangle using these points. The sides of the triangle are the differences between the coordinates of points A, B, and C.

3. Solve for $m$ as follows:

$$m = \frac{\text{side } y}{\text{side } x} = \frac{\log 53 - \log 3}{\log 60 - \log 1.7} =$$
$$\frac{1.724 - 0.477}{1.778 - 0.230} = \frac{1.247}{1.548} = 0.81$$

4. The value of $b$ is 2 on the Y axis.

5. Substituting the values of $m$ and $b$ in the equation $y = bx^m$ gives the equation $y = 2x^{0.81}$. Converting this to logarithmic form produces the equation:

$$\log y = \log b + m \log x$$
$$\log y = \log 2 + 0.81 \log x$$

## GRAPHICAL ALGEBRA

Engineers will sometimes find it is faster to use a graphical procedure for solving a mathematical problem than using the algebraic procedure. This is especially useful when working with equations that require lengthy or burdensome algebraic processes.

It also is of some help in visualizing relationships between variables and can also serve as a check on solutions reached using algebraic procedures.

Graphic techniques give approximate solutions. The accuracy of the solution depends on the scale of the drawing and the skill of the person producing it. The larger the scale, the more accurate the answer. Since the data for many problems result from observations or experimental procedures, the original data may not be absolutely accurate, and graphical techniques will often be more accurate than the data plotted. In all cases, the charts should be carefully drawn, using fine lines and accurate measurements. If great accuracy is required, traditional algebraic procedures should be used.

Graphics can be used to solve equations having one or two variables. The following discussion will show how to solve a one-variable equation such as

$$\tan x = 2 + \cos x$$

This could be solved by trial and error by examining trigonometric function tables, isolating an approximate answer and using trial and error interpolation to get the result, or it could be solved graphically (Fig. 23–66). To solve the equation graphically:

1. Develop a table that contains values of $x$ within which the solution will fall. These can be found by examining trigonometric function tables. Record the trigonometric values for each side of the equation as found in the trigonometric tables.

2. Develop a chart using as large a scale as is practical. In this case, the degrees were placed on the X axis and the trigonometric functions on the Y axis.

3. Plot these data on the chart. Where they cross is the value of $x$.

Graphical Presentation and Analysis  **615**

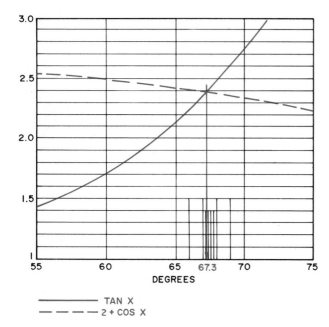

| FIND THE VALUE OF X WHEN TAN X = 2 + COS X | | |
|---|---|---|
| X | TAN X | 2 + COS X |
| 55 | 1.42 | 2.57 |
| 60 | 1.73 | 2.50 |
| 65 | 2.14 | 2.42 |
| 70 | 2.75 | 2.34 |
| 75 | 3.73 | 2.26 |

——————— TAN X
— — — — — 2 + COS X

FIGURE 23–66 *Using graphics to solve an equation having one variable.*

While there are many algebraic applications, one that is easily solved graphically is situations involving relative velocities. Following is a production problem that can be best solved graphically.

A manufacturer has an automated assembly line and wants to know how far a part will travel before it intersects an assembly upon which it will be mounted. The part moves from storage on a conveyor running at 2 mph to a machining center. It takes 20 min to reach the center. Here, it is removed from the conveyor and machined for 20 min. It is then placed back on the conveyor and moves toward the assembly.

The assembly leaves storage 4000 ft from the part storage area on another conveyor running at 3 mph parallel with the part conveyor. The assembly leaves 40 min after the part left its storage area. How many feet will each travel before they intersect, and how many minutes will this take?

To solve this problem graphically, follow these steps (Fig. 23–67).

1. Lay out a grid as large as possible, placing the distances traveled on the X axis and the time on the Y

axis. The total distance between the moving items is 4000 ft and the total time is 80 min.

2. First, plot the travel of the part on the grid. The slope of the plot depends on its velocity. It travels at 2 mph (10,560 ft/hr) for 20 min. In the machining center, it consumes time but no travel distance. After 20 min here, it resumes its trip on the conveyor at 2 mph.

3. Next, plot the travel of the assembly. Since it starts at the opposite side of the factory, it is plotted from right to left. It does not move for 40 min after the part leaves storage, so the first plot is at the 40-min grid. It travels at a uniform speed until it intersects the part.

4. The coordinates intersect at the 2875-ft mark, and this occurs 47 min after the part left storage. These values are found by projecting the point of intersection to the time and distance-traveled scales.

## GRAPHICAL CALCULUS

When solving engineering problems, the engineer has to work with relationships between variables that can be solved only by using the principles of calculus. There are many known equations of curves, and these can be solved using analytical calculus. However, when engineering data cannot be expressed easily in algebraic form, graphical techniques can be used to produce approximate solutions.

The two forms of calculus are differential and integral calculus. *Differential calculus* is used to determine the rate of change of one variable when compared with another. For example, notice in Fig. 23–68 that as the X variable (temperature) increases, the Y variable (pressure) also increases. The rate of change would be an important bit of information needed by the engineer as a product is designed.

*Integral calculus* is the process of finding the area under a given curve by summation of a series of areas below the curve. It is the reverse of differential calculus. The area below a curve is found by dividing one of the variables into small intervals forming rectangular areas. These small areas are located so that about as much of the end of the area increment is above the curve as is below it (Fig. 23–69). The average height, called the *mean ordinate,* of the increment is near its center line. These small areas are added together to find the approximate area under the curve.

## GRAPHICAL DIFFERENTIATION

Graphical differentiation is the determination of the rate of change of one variable with respect to a related variable at a given point. *Differentiation* is finding or calculating the limit of the ratio of an incre-

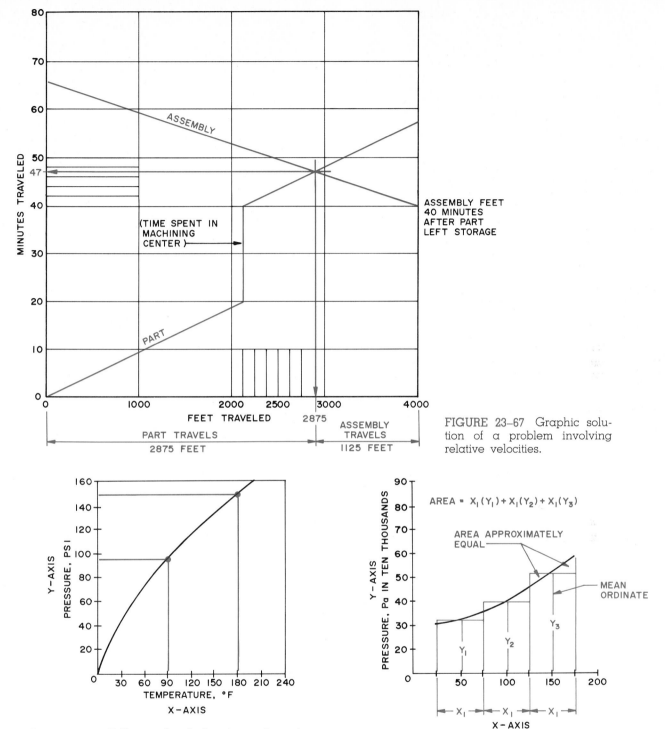

FIGURE 23-67 Graphic solution of a problem involving relative velocities.

FIGURE 23-68 Differential calculus is used to determine the rate of change of variables in respect to each other.

FIGURE 23-69 Integral calculus is the process of finding the area under a given curve.

ment in a function of a variable to the corresponding increment of the variable as the increment of the variable approaches zero as the limit. The process of differentiation produces a derivative curve. The curve represents the derivative of the data for the given curve. This can be illustrated by Fig. 23-70, in which the relationship between two variables is

shown. When an increment of time, called $\Delta X$ (or $\Delta t$), is known, there is a corresponding increment of velocity, called $\Delta Y$ (or $\Delta v$). As $\Delta X$ becomes smaller (approaches zero as the limit), $\Delta Y$ also becomes smaller and the ratio $\dfrac{\Delta Y}{\Delta X}$ approaches the value of the tangent function of the angle the tangent to the

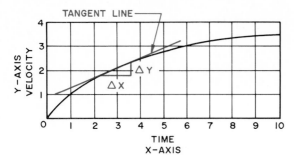

FIGURE 23–70  The numerical value of the derivative at a point on a curve is the same as the tangent function of the angle that a tangent to the curve at the point makes with the positive axis.

curve of that point makes with the X axis. The tangent of the angle that a tangent to the curve makes with X axis is the derivative of the curve at that point. Using the derivatives for selected points on the given curve, it is possible to draw another curve, the derivative curve, using the same X-axis values and using the derivative for each point (the tangent of curve at that point) as the Y-axis value. The tangent represents the slope of the original curve or the rate of change of the function of the variable with respect to that variable at that point.

## Constructing the Derivative Grid

The given curve is drawn using convenient scales, as discussed earlier in this chapter. The grid for the derivative curve is drawn below the given curve because it is a function of a lower order. Use a scale that is as large as possible. Accuracy is of great importance, and a larger scale increases the accuracy of the results.

The scale on the X axis is the same as that used on the grid for the given curve. The maximum scale that is needed on the Y axis of the derivative grid is equal to the maximum slope of the given curve. To find the maximum slope, draw chords on the given curve and estimate the maximum slope. In Fig. 23–71, the greatest slope appears at the beginning of the curve and, by projection to the Y axis, indicates a slope of 2. Generally, this scale is made a little longer, just in case the actual slope is slightly greater than that estimated.

## Differentiation Using the String Polygon Method

There are a number of methods of differentiation. This method is relatively simple and lends itself to graphical techniques. It divides the curve up into

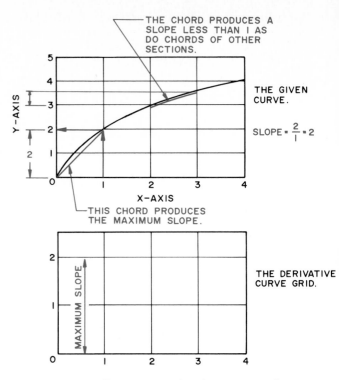

FIGURE 23–71  Constructing the derivative grid.

segments by using chords. A perpendicular bisector of a chord locates the tangent point on the curve (Fig. 23–72).

Following are the steps using the string polygon method (Fig. 23–73):

1. Lay out the grid for the given curve.
2. Lay out the grid for the derivative curve below the grid for the given curve.
3. Divide the given curve into segments by chords, bisect each, and label them.

FIGURE 23–72  Differentiation using the string polygon method.

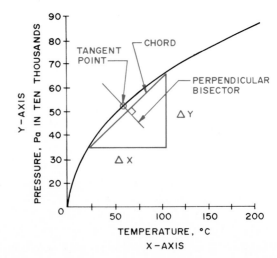

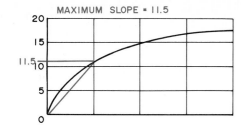

MAXIMUM SLOPE = 11.5

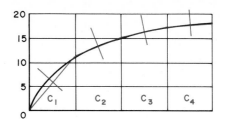

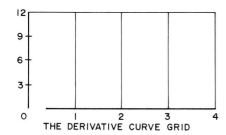

THE DERIVATIVE CURVE GRID

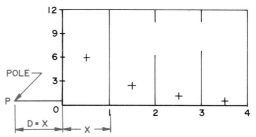

POLE

D = X

X

1. Lay out the given curve. Develop the scale and lay out the grid for the derivative curve. The maximum slope is 11.5 so the Y-axis on the derivative curve grid should equal or exceed this.

2. Divide the given curve into segments with chords. Bisect each chord. Locate the pole position, P, off the derivative curve grid. In this case it was decided to locate it two x units from the grid.

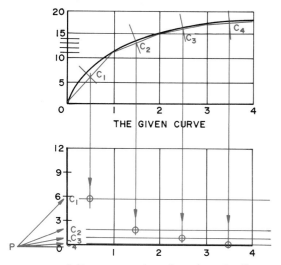

THE GIVEN CURVE

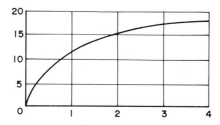

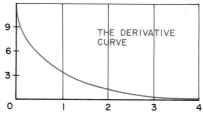

THE DERIVATIVE CURVE

3. Draw strings from the pole to the Y-axis parallel to the cords. Then draw them parallel to the X-axis. Project the perpendicular bisectors of the chords down to the derivative curve grid. Where they cross their respective horizontal projections locates a point on the curve.

4. Connect the points to form the derivative curve.

FIGURE 23–73 The steps for developing a derivative curve.

4. Locate a pole position, P, so that the pole distance, D, from the grid is equal to one or more units on the X axis. If one unit is used, the vertical scale of the derivative curve will be the same size as the vertical scale on the given curve. In this example, a pole distance of one X unit was used. If it had been made to equal two X units, the vertical scale of the derivative curve would have been twice as large as the given curve.

5. From the pole, P, draw strings, $C_1$, $C_2$, $C_3$, parallel to the chords, $C_1$, $C_2$, $C_3$, until they meet the Y axis.

6. Where the strings meet the Y axis, draw horizontal lines across the grid.

7. Project the perpendicular bisectors of the chords (the tangent points) on the given curve to the derivative curve grid until they intersect their respective horizontal projections.

8. Connect these points of intersection to form the derivative curve. This curve shows the relationship between the two variables.

## GRAPHICAL INTEGRATION

*Integration* is a process of determining the area under a given curve by adding a series of increments. Another important application of integral calculus is the determination of a function from a given derivative, which is the inverse of differentiation. The resulting function is called the *integral*.

Graphically, integration is a process of finding the area between a given curve and the X axis by dividing the area into a series of increments and drawing an integral curve. The difference between any two ordinates on the integral curve is equal to the area of the increment found under the given curve (Fig. 23–74). For example, the area in the shaded increment is 13 mm². The difference in value of the two ordinates of that portion of the integral curve equals 13 mm. From this it can be seen that the difference between the first and last ordinates of the integral curve equals the area below the given curve.

There are several graphical methods of integration. The one to be shown, the string polygon method, lends itself well to graphical techniques (Fig. 23–75).

### Integration Using the String Polygon Method

1. Plot the given curve as described earlier in this chapter.

2. Divide the area under the curve into small increments. The narrower the increments, the more accurate the solution. The increments should be narrower on steep curves than on flatter curves. The increments need not be the same size.

3. Lay out the equivalent-area rectangle for each increment. This is done by drawing a horizontal line through the curve as it crosses each increment so that the areas above and below appear to be equal in area. The shaded areas shown represent approximately equal areas. Notice that the first two increments were made half the width of the others because the curve is very steep at that location.

4. Project the top of each increment to the Y axis.

5. Locate a pole on the X axis a distance of one or more unit values on the X axis. In this example, three X units were used. The distance of the pole from the grid influences the slope of the integral curve. It is

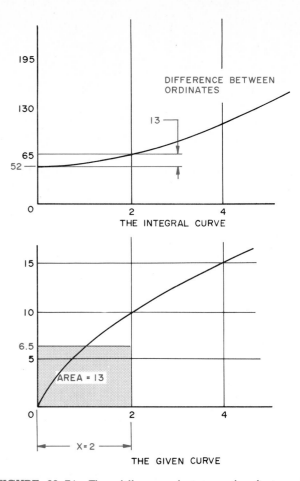

FIGURE 23–74 The difference between the first and last ordinates of an integral curve equals the area below the given curve.

best to make this distance long enough to avoid having the integral curve become too flat. Larger units on the Y axis will produce a steeper curve. The larger the distance to the pole, the larger will be the scale of the Y axis.

6. Draw rays from each of the points located on the Y axis to the pole and label them.

7. Lay out the grid for the integral curve above the grid for the given curve. It is usually located here because the integral will be an equation raised to a higher order. The scale on the X axis is the same as that used with the given curve. The scale on the Y axis is equal to the scale on the Y axis on the given curve multiplied by the number of X units used to locate the pole, P, which in this case is 3.

8. Now, begin to locate the integral curve by drawing strings parallel to the rays on the given curve drawing. For example, string $R_1$ is parallel to ray $R_1$ and crosses the first two divisions on the X axis. Repeat this for each ray.

9. Now, draw a smooth curve through the points at which the strings cross the divisions of the X axis. The strings become chords. The area below the given curve is equal to the difference between the first ordinate, 0, and the last ordinate, 26.5, which is 26.5 units.

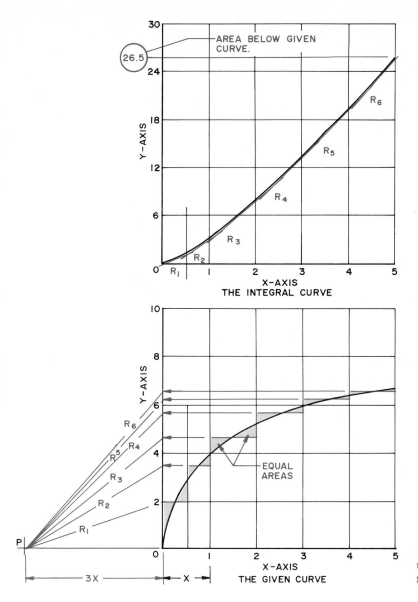

FIGURE 23–75 Integration using the string polygon method.

# Problems

Draw the solutions to the following problems on A or A4 size sheets. Select sizes, scales, and grids that will produce the most accurate description of the data. The very best drafting quality and lettering are expected.

1. Bring to class data from various sources, such as automobile or motorcycle performance reports, population data, or financial data from a newspaper. Present these data in three different graphical forms. Prepare a report citing the limitations and merits of each chart.

2. Select five models of used automobiles. Choose automobiles made in the same year and by different manufacturers. Look up the prices in the want ads of the local paper. Call auto dealers for additional prices on these

models. Then prepare a chart comparing the prices and showing the range in price for each model.

3. Prepare a flow chart for structural clay products. There are three main subdivisions of these products: solid masonry units, hollow masonry units, and architectural terra cotta. The solid masonry units include bricks. The hollow units include structural clay tile and structural facing tile. The terra cotta units include anchored ceramic veneer, adhesion ceramic veneer, and ornamental units.

4. Make copying machine reproductions of four different types of charts found in technical and professional journals. Present these data by developing charts using a different method of graphically reporting the data.

5. Secure data from the registrar of your school on the number of graduates in each area of engineering or technology for the last five years. Construct a pictorial chart to present these data.

6. Develop a pie chart that presents the enrollment data for your school. This could be reported by colleges or schools within a university, by departments within a college or school, or by majors within a department.

7. Have everyone in class report their height. Prepare a chart that represents the class in terms of their height.

8. Prepare an organizational chart showing the administrative structure of your school. Begin with the governing board and work it through to the teaching faculty and physical plant staff.

9. Degrees Fahrenheit are converted to degrees Celsius by the formula $C° = 5/9 (F° - 32°)$. Plot a chart that will permit a given Fahrenheit reading between 0° and 220° to be converted to an equivalent Celsius reading.

10. Pounds are converted to grams by multiplying pounds by the factor 453.592. Plot a chart that will permit a given number of pounds between 0.1 and 3.0 to be converted to an equivalent kilogram reading. Break the first pound into tenths and use one-half pound graduations from one to three pounds.

11. Construct a pie chart that will show the occupational groups that make up the construction industry. Use the following data: skilled trades and supervisors (51%), laborers (17%), managers and officials (12%), semiskilled workers (10%), professional and technical workers (5%), and clerical workers (5%).

12. Construct a chart to report the following employment data for engineers and drafters: chemical engineers (55,000), aerospace engineers (68,000), surveyors (61,000), architects (79,500), electrical engineers (230,000), civil engineers (165,000), mechanical engineers (213,000), ceramic engineers (15,000), biomedical engineers (4,000), industrial engineers (115,000), and petroleum engineers (18,000).

13. Prepare a chart to clearly report the procedural steps in handling a worker grievance. The process is as follows: A worker has a grievance and discusses it with the shop steward. Together they talk to the shift supervisor and try to resolve the grievance. If this fails, the union business agent talks to the supervisor's superior and tries to reach an agreement. If this fails, a formal grievance is sent to the company's grievance committee. If the committee fails to resolve the grievance, the formal grievance is sent to the local or national union officer and the local labor relations director of the company. If agreement is not reached at this level, the formal grievance is sent to arbitration where a decision is made that is binding on both sides.

14. Hardwood plywood is classified into two classes: specialty and standard. Uses of the specialty types include v-grooved wall panel, architectural custom panels, and miscellaneous uses. Standard types include stock panels, door skins, furniture, and other uses. Prepare a flow chart to show these relationships.

15. Prepare a diagrammatic representation and a flow chart of the process of manufacturing brick. The steps are as follows:
1. Winning (mining the clay)
2. Crushing and storing the mined material
3. Pulverizing the material
4. Screening the material

5. Forming and cutting the bricks
6. Glazing the green bricks
7. Drying the bricks on carts
8. Burning the bricks in a kiln
9. Store and ship the finished bricks

16. Prepare a chart to show the length of the following U.S. rivers:

| RIVER | mi | km |
|-------|-----|------|
| Mississippi | 2,348 | 3778 |
| Missouri | 2,315 | 3725 |
| Arkansas | 1,450 | 2333 |
| Snake | 1,000 | 1609 |
| Cimarron | 698 | 1123 |
| Wisconsin | 430 | 692 |

17. Prepare a pie chart to show the major elements found in seawater.

| ELEMENT | mg · kg |
|---------|---------|
| Chlorine | 18,980 |
| Sodium | 10,561 |
| Magnesium | 1,272 |
| Sulfur | 884 |
| Calcium | 400 |
| Potassium | 380 |
| Bromine | 65 |
| Other | 48 |

18. Plot a chart giving a comparison of the heights of selected buildings.

| BUILDING | HEIGHT (ft) | HEIGHT (m) |
|----------|-------------|------------|
| Sears Tower | 1454 | 443 |
| United California Bank | 859 | 262 |
| John Hancock Tower | 790 | 241 |
| Bank of America | 700 | 213 |
| Irving Trust Building | 654 | 199 |
| Waldorf Astoria Hotel | 625 | 190 |

19. Make a graphic comparison of the height of the following: hydroelectric plant dams:

| | m | ft |
|--|-----|------|
| Rogunsky, U.S.S.R. | 325 | 1,066 |
| Vaiont, Italy | 262 | 858 |
| Oroville, U.S.A. | 235 | 770 |
| Daniel Johnson, Canada | 214 | 703 |
| Pah Lavi, India | 203 | 666 |

20. Compare the capacities of the following hydroelectric plants:

| | MEGAWATTS |
|---|---|
| Italpu, Brazil | 12,870 |
| Sayanskaya, U.S.S.R. | 6,400 |
| Churchill Falls, Canada | 5,225 |
| Inga, Zaire | 2,820 |
| Blue Ridge, U.S.A. | 1,600 |
| San Carlos, Colombia | 1,550 |

21. Draw an interesting chart that shows the average monthly temperatures in St. Louis, Missouri in degrees Fahrenheit.

| | | | |
|---|---|---|---|
| January | 31.7 | July | 79.3 |
| February | 34.8 | August | 77.5 |
| March | 44.3 | September | 70.1 |
| April | 56.1 | October | 59.0 |
| May | 65.9 | November | 45.3 |
| June | 75.1 | December | 35.3 |

22. Prepare a chart that compares males and females on income and education which could be used in a color magazine article.

| | MALE | FEMALE |
|---|---|---|
| Elementary | | |
| Less than 8 years | $ 8,991 | $ 5,644 |
| 8 years | 11,312 | 6,433 |
| High School | | |
| 1 to 3 years | 12,301 | 6,800 |
| 4 years | 13,703 | 8,069 |
| College | | |
| 1 or more years | 17,327 | 10,813 |
| 5 or more years | 20,597 | 13,569 |

23. Make a chart that compares various types of glass by their softening temperatures.

| TYPE | TEMPERATURE °F | °C |
|---|---|---|
| Soda-lime silica | 1350 | 738 |
| Borosilicate | 1500 | 822 |
| Lead-alkali silicate | 1160 | 632 |
| Fused silica | 3000 | 1662 |
| 96% silica | 2700 | 1494 |
| Alumina silicate | 1675 | 920 |

24. Compare graphically the types of woods listed below on their shear strength parallel to grain.

| SPECIES OF WOOD | SHEAR STRENGTH (PSI) 12% MOISTURE CONTENT |
|---|---|
| Ash, white | 1,950 |
| Birch, yellow | 1,880 |
| Ebony, African | 2,473 |
| Elm, American | 1,510 |
| Cedar, Western Red | 860 |
| Fir, white | 930 |
| Pine, red | 1,210 |

25. Make a range-bar chart showing the heating value of the gaseous fuels below.

| FUEL | BTU/CU FT RANGE |
|---|---|
| Natural gas | 850 to 1950 |
| Coal gases | 500 to 600 |
| Blast furnace gas | 80 to 150 |
| Producer gases | 120 to 180 |
| Water gas | 525 to 555 |
| Liquefied petroleum gases | 1000 to 4000 |

26. Design a pie chart to show the composition of soda-lime-silica glass.

| | |
|---|---|
| $SiO_2$ (silica) | 72% |
| $Na_2O$ (soda) | 15% |
| $CaO$ (lime) | 9% |
| $MgO$ (magnesia) | 3% |
| $Al_2O_3$ (alumina) | 1% |
| | 100% |

27. Compare the light transmittance values of the plastic materials listed below.

| PLASTIC | TRANSMITTANCE (%) |
|---|---|
| Polystyrene | 90 |
| High-impact polystyrene | 45 |
| Acrylics | 92 |
| Polycarbonate | 88 |
| A.B.S. | 30 |
| Polyethylene (high density) | 40 |
| Polyethylene (low density) | 75 |

28. Compare these gray cast irons on their range of hardness.

| ASTM CLASS OF IRON | BRINELL HARDNESS |
|---|---|
| 20 | 60–180 |
| 25 | 170–210 |
| 30 | 210–230 |
| 35 | 210–230 |
| 40 | 210–240 |
| 50 | 230–270 |
| 60 | 250–290 |

29. Design a graphical means of telling the homeowner how much moisture evaporates from the soil in the crawl space under a house. Consider using some type of special pictorial symbol.

| CONDITION | MOISTURE |
|---|---|
| Bare earth with high moisture content | 16.25 gal |
| Standing water | 14.8 gal |
| Bare earth with average moisture content | 10.3 gal |
| Vapor barrier over earth | 0.24 gal |

30. Draw a range-bar chart comparing selected materials on their noise reduction capacity from outside to inside a building.

| MATERIAL | DIFFERENCE (dB) | |
|---|---|---|
| | Inside | Outside |
| ⅛-in. glass | 11 | 16 |
| ¼-in. plateglass | 20 | 26 |
| ¼-in. glass and ⁷⁄₃₂-in. glass spaced 3¾-in. apart | 35 | 42 |
| 2 × 4 wall, wood sheathing, ½-in. plaster board | 23 | 28 |
| 4½-in. brick wall | 33 | 37 |
| 9-in. brick wall | 42 | 46 |
| Builtup roof on 1-in. wood deck with ½-in. gypsum board and 2 × 8 joists | 18 | 23 |

31. Plot the following points and find the nearest straight line value for the sound transmission loss data recorded from tests on 24 concrete masonry walls.

| SOUND TRANSMISSION CLASS (dB) | SOUND TRANSMISSION LOSS (dB) |
|---|---|
| 25 | 26 |
| 26 | 25 |
| 28 | 25 |
| 30 | 27 |
| 30 | 32 |
| 31 | 31 |
| 32 | 29 |
| 32 | 30 |
| 35 | 32 |
| 28 | 36 |
| 35 | 38 |
| 40 | 35 |
| 40 | 41 |
| 41 | 38 |
| 43 | 45 |
| 44 | 41 |
| 47 | 41 |
| 49 | 42 |
| 49 | 51 |
| 50 | 51 |
| 50 | 56 |
| 51 | 57 |
| 52 | 50 |
| 53 | 50 |

32. Prepare a chart to make a comparison of the yield strengths of selected metals and plastics.

| MATERIAL | STRENGTH (1000 psi) |
|---|---|
| Steel | 22–160 |
| Aluminum alloys | 2– 38 |
| Copper alloys | 4– 42 |
| Plastics | 1– 4 |

33. Show the relationship between these materials and their service temperatures.

| MATERIAL | SERVICE TEMPERATURES (°F) |
|---|---|
| Plastics | 95– 500 |
| Wood | 225– 415 |
| Iron and steel | 880–1340 |
| Aluminum alloys | 300– 580 |
| Copper alloys | 400–1110 |
| Glass | 360–1640 |
| Concrete | 460–1420 |

34. Illustrate graphically the chemical composition of cast monel metal which is: Ni (67%), Cu (29%), Fe (1.5%), Mn (.9%), Si (1.25%), and C (.35%).

35. Prepare two charts showing the Annual Production of Steel Products. Use net tons on one chart and the percent of the total on the other.

|  | NET TONS (THOUSANDS) | % |
|---|---|---|
| Sheets and strip | 35,260 | 42.6 |
| Bars | 12,165 | 14.7 |
| Plates | 8,180 | 9.9 |
| Tubular products | 7,225 | 8.7 |
| Tin mill products | 6,000 | 7.3 |
| Structural shapes | 5,580 | 6.5 |
| Wire and wire products | 4,625 | 5.6 |
| Semifinished shapes | 1,275 | 1.2 |
| Other mill products | 1,135 | 1.4 |
| Cast products | 1,748 | 2.1 |
| Total steel products | 82,993 | 100 |

36. Plot the mechanical properties of full-annealed, high-carbon cast steel as related to stress and the percent of carbon in the steel.

| TENSILE STRENGTH | | BRINELL HARDNESS | | YIELD POINT | |
|---|---|---|---|---|---|
| Stress (psi) | Carbon (%) | Stress (psi) | Carbon (%) | Stress (psi) | Carbon (%) |
| 98000 | .50 | 77000 | .50 | 51000 | .50 |
| 104000 | .60 | 82000 | .60 | 56000 | .60 |
| 112000 | .70 | 90000 | .70 | 60000 | .70 |
| 120000 | .80 | 97000 | .80 | 62000 | .80 |
| 125000 | .90 | 101000 | .90 | 63000 | .90 |
| 135000 | 1.0 | 103000 | 1.0 | 64000 | 1.00 |

## Linear Charts

37. Plot the curves of creep strain for the selected stress levels shown below. Creep is a term used to describe slow deformation of a material when under stress.

| STRESS 40,000 LB/IN.$^2$ | | STRESS 30,000 LB/IN.$^2$ | | STRESS 20,000 LB/IN.$^2$ | |
|---|---|---|---|---|---|
| Creep Stain (in./in.) | Time (hr) | Creep strain (in./in.) | Time (hr) | Creep Strain (in./in.) | Time (hr) |
| .0018 | 100 | .0010 | 100 | .0005 | 100 |
| .0028 | 200 | .0014 | 200 | .0008 | 200 |
| .0034 | 400 | .0019 | 400 | .0010 | 400 |
| .0040 | 600 | .0021 | 600 | .0011 | 600 |
| .0045 | 800 | .0024 | 800 | .0012 | 800 |
| .0053 | 1000 | .0025 | 1000 | .0013 | 1000 |

38. Plot the curves for the data for an in-line crank mechanism for a crank through an angle of 0 to 180°. The data represent velocity vs. the crank angle for various ratios of crank length, C, and rod length, R.

| R/C RATIO (1.5) | | R/C RATIO (2.0) | | R/C RATIO (8.0) | |
|---|---|---|---|---|---|
| Crank Angle (Deg) | Slider Velocity/ Crank Pin Velocity | Crank Angle (Deg) | Slider Velocity/ Crank Pin Velocity | Crank Angle (Deg) | Slider Velocity/ Crank Pin Velocity |
| 0 | 0 | 0 | 0 | 0 | 0 |
| 20 | .50 | 20 | .45 | 20 | .32 |
| 40 | 1.0 | 40 | .95 | 40 | .65 |
| 60 | 1.30 | 60 | 1.12 | 60 | .86 |
| 80 | 1.15 | 80 | 1.08 | 80 | .98 |
| 85 | 1.0 | 85 | 1.0 | 85 | 1.0 |
| 100 | .73 | 100 | .88 | 100 | .96 |
| 120 | .40 | 120 | .63 | 120 | .84 |
| 140 | 1.96 | 140 | .58 | 140 | .64 |
| 160 | .50 | 160 | .15 | 160 | .36 |
| 180 | 0 | 180 | 0 | 180 | 0 |

39. Plot the curves for the following data, which give yield strengths of cast aluminum alloys at various temperatures.

| ALLOY IDENTIFICATION | TEMPERATURES °F | | | | |
|---|---|---|---|---|---|
| | 75 | 300 | 400 | 500 | 600 |
| SAE 321 | 28000* | 22000 | 13500 | 9500 | 5000 |
| SAE 39 | 18000 | 18000 | 14000 | 8000 | 4000 |
| SAE 38 | 10000 | 13000 | 9000 | 6000 | 3000 |
| SAE 380 | 26000 | 21000 | 9000 | 5000 | 2000 |

*Yield strength in lb/in.$^2$

40. Plot the speed-torque curves for the general-purpose small electric motors given below.

| SINGLE PHASE CAPACITOR START | | THREE PHASE | |
|---|---|---|---|
| Torque (% Of Full Load) | Speed (% Synchronous Speed) | Torque (% Of Full Load) | Speed (% Synchronous Speed) |
| 425 | 0 | 300 | 0 |
| 435 | 10 | 315 | 10 |
| 450 | 20 | 320 | 20 |
| 475 | 30 | 330 | 30 |
| 500 | 40 | 345 | 40 |
| 500 | 50 | 350 | 50 |
| 475 | 60 | 348 | 60 |
| 425 | 70 | 340 | 70 |
| — | 80 | 300 | 80 |
| — | 100 | 210 | 90 |
| | | 0 | 100 |

41. Chart the strength of plain and beveled lap joints.

| PLAIN | | BEVELED | |
|---|---|---|---|
| Breaking Load (100 lb/in. Width) | Overlap (in.) | Breaking Load (100 lb/in. Width) | Overlap (in.) |
| 3.0 | .5 | 3.0 | .5 |
| 4.1 | 1.0 | 5.2 | 1.0 |
| 4.5 | 1.5 | 6.2 | 1.5 |
| 4.7 | 2.0 | 8.3 | 2.0 |
| 4.8 | 2.5 | — | 2.5 |
| 4.9 | 3.0 | — | 3.0 |
| 5.0 | 3.5 | — | 3.5 |
| 5.1 | 4.0 | — | 4.0 |

42. Make a linear chart comparing these sound transmission values for sealed and porous masonry walls according to the weight of the walls in lb/ft$^2$

| UNSEALED SURFACE | | SEALED SURFACE | |
|---|---|---|---|
| Sound Transmission Class (db) | Wall Weight (lb/ft$^2$) | Sound Transmission Class (db) | Wall Weight (lb/ft$^2$) |
| 23 | 15 | 40 | 15 |
| 27 | 20 | 42 | 20 |
| 30 | 25 | 44 | 25 |
| 34 | 30 | 46 | 30 |
| 37 | 35 | 47 | 35 |
| 41 | 40 | 49 | 40 |
| 45 | 45 | 50 | 45 |
| 47 | 50 | 52 | 50 |
| 49 | 55 | 54 | 55 |

43. Compare these three materials on their stress-strain behavior using a linear chart.

| COPPER | | STEEL | | POLYETHYLENE | |
|---|---|---|---|---|---|
| Stress (Thousand psi) | Strain (%) | Stress (Thousand Psi) | Strain (%) | Stress (Thousand psi) | Strain (%) |
| 0 | 0 | 0 | 0 | 0 | 0 |
| 10 | 2 | 10 | .3 | .1 | 10 |
| 20 | 7 | 20 | .7 | .2 | 20 |
| 30 | 18.5 | 30 | 1.1 | .3 | 30 |
| 37 | 35 | 40 | 1.2 | .4 | 40 |
| 32 | 46.5 | 45 | 2.25 | .5 | 50 |
| | | 50 | 5 | | |
| | | 60 | 9.5 | | |
| | | 70 | 13.5 | | |
| | | 80 | 20.3 | | |
| | | 75 | 33 | | |

44. Moisture condensation on windows is an important factor in design. Prepare a linear chart making a comparison of the effect of glazing and relative humidity on interior moisture condensation.

| | RELATIVE HUMIDITY % | | |
|---|---|---|---|
| Temperature, °F | Triple Glazing | Double Glazing | Single Glazing |
| −30 | 42 | 31 | 5 |
| −20 | 44 | 35 | 8 |
| −10 | 48 | 40 | 12 |
| 0 | 55 | 45 | 15 |
| 10 | 60 | 51 | 21 |
| 20 | 65 | 56 | 28 |
| 30 | 71 | 64 | 37 |
| 40 | 75 | 71 | 47 |

45. Plot the curve representing the average velocity of a slider having rectilinear displacement. The displacement is plotted against the time required for the displacement. The following times were noted during one cycle of the slider.

| RECTILINEAR DISPLACEMENT (in.) | TIME (s) |
|---|---|
| 0 | 0 |
| .20 | 5 |
| .65 | 10 |
| 1.30 | 15 |
| 1.90 | 20 |
| 2.20 | 25 |
| 2.30 | 28 |
| 2.20 | 30 |

## Semilogarithmic Charts

46. Plot the data below on a semilogarithmic chart. Place the log scale on the y-axis.

| y | 2 | 5 | 9 | 14 | 18 | 24 | 30 | 36 | 42 | 51 |
|---|---|---|---|---|---|---|---|---|---|---|
| x | 10 | 20 | 30 | 40 | 50 | 60 | 70 | 80 | 90 | 100 |

47. Plot the thermal conductivity of selected materials using a semilogarithmic scale for the thermal conductivity (y-axis) and temperature (x-axis).

| MATERIAL | TEMPERATURE, °F | | | | | | | |
|---|---|---|---|---|---|---|---|---|
| | | 200 | 400 | 600 | 800 | 1000 | 1200 | 1400 |
| Brick | 1 | 0 | 315 | 3.2 | 3.0 | 2.8 | 2.6 | 2.5 |
| Stainless Steel | 10 | 10* | 10 | 10 | 10 | 11 | 13 | 14 |
| Low Carbon Steel | 26 | 25.5 | 24.8 | 23 | 22.5 | 21.0 | 19 | 18 |
| Aluminum | 125 | 130 | 140 | 150 | 160 | | | |
| Copper | 220 | 210 | 205 | 205 | 207 | 210 | | |

*Thermal conductivity in BTU/hr ft, °F

48. Plot the tripping characteristics of an electric motor temperature overload relay as related to the percent of motor full-load current and tripping time on a semilogarithmic chart. Place the current data on the y-axis and tripping time on the x-axis. The x-axis should be divided into a three-cycle log scale and the y-axis into a 10 unit linear scale.

| CURRENT IN PERCENT OF MOTOR FULL-LOAD CURRENT | TRIPPING TIME (MIN) |
|---|---|
| 100 | 1.00 |
| 150 | 1.5 |
| 200 | 3.8 |
| 250 | 1.6 |
| 300 | 1.0 |
| 350 | .70 |
| 400 | .52 |
| 450 | .42 |
| 500 | .35 |
| 550 | .29 |
| 600 | .24 |
| 650 | .20 |
| 700 | .19 |
| 750 | .18 |
| 800 | 0.00 |

49. Plot the data in problem 36 on a semilogarithmic chart.

50. Plot the data in problem 43 on a semilogarithmic chart.

## Logarithmic Charts

51. Prepare a chart showing the relationship between particle size and terminal velocity of solid spherical particles of selected specific gravities. These data are used when designing dust collecting systems. Plot as a logarithmic chart.

| PARTICLE DIAMETER microns | TERMINAL VELOCITY in./min | |
|---|---|---|
| | Particles With Specific Gravity of 1.0 | Particles With Specific Gravity of 12.0 |
| 1.0 | | .9 |
| 1.5 | — | 2.0 |
| 2.0 | .3 | 3.5 |
| 3 | .6 | 8.0 |
| 4 | 1.2 | 15.0 |
| 5 | 1.7 | 22.0 |
| 6 | 2.7 | 32.0 |
| 7 | 3.7 | 42.0 |
| 8 | 4.7 | 60.0 |
| 9 | 5.9 | 70.0 |
| 10 | 7.0 | 90.0 |
| 15 | 15.2 | 200.0 |
| 20 | 20.9 | 350.0 |
| 30 | 60.8 | — |
| 40 | 130 | — |
| 50 | 170 | — |
| 60 | 270 | — |
| 70 | 360 | — |

52. Plot the data below on a two-cycle logarithmic chart showing the relationship between humidifier output and the revolutions per minute of the fan motor.

| HUMIDIFIER OUTPUT 1 GAL/DAY | FAN (RPM) |
|---|---|
| 10.0 | 500 |
| 17.0 | 600 |
| 20.0 | 700 |
| 24.0 | 800 |
| 28.5 | 1000 |
| 30.5 | 1200 |
| 33.5 | 1400 |
| 38.0 | 1750 |

53. Prepare a logarithmic chart showing the temperature versus the pressure of one type of Freon gas. The data are as follows:

| TEMPERATURE (°C) | PRESSURE OF FREON REFRIGERANT (PSI) |
|---|---|
| 0 | 30.1 |
| 10.0 | 36.7 |
| 15.5 | 57.7 |
| 20.0 | 67.6 |
| 25.5 | 81.3 |
| 30.0 | 93.3 |
| 35.5 | 110.0 |
| 40.0 | 124.6 |
| 45.5 | 144.7 |
| 50 | 162.2 |

## Empirical Equations: Linear

54. Construct a linear graph to determine the equation for the temperature-pressure relationship on a specific gas. The data are as follows: 0°F, 0 psi; 100°F, 600 psi; 200°F, 1200 psi; 300°F, 1800 psi; 400°F, 2400 psi; 500°F, 3000 psi; 600°F, 3600 psi.

55. Construct a linear graph to determine the equation for the cost of fuel for operating a tractor at various engine speeds. The following data relate engine revolutions per minute to the amount of fuel consumed per hour. The data are as follows: 0 rpm, 0 gal/hr; 500 rpm, 1.15 gal/hr; 1000 rpm, 2.30 gal/hr; 1500 rpm, 3.45 gal/hr; 2000 rpm, 4.60 gal/hr; 2500 rpm, 5.75 gal/hr; 3000 rpm, 7 gal/hr.

56. Plot the data in the table below on a linear graph and determine its equation. These empirical data show linear measurements in inches and millimeters.

| in. | mm |
|---|---|
| .5 | 12.7 |
| 1.0 | 25.4 |
| 1.7 | 43.2 |
| 2.3 | 58.4 |
| 3.9 | 99.1 |
| 5.1 | 129.5 |
| 7.8 | 198.1 |
| 8.4 | 213.4 |
| 9.6 | 243.8 |

## Empirical Equations: Semilogarithmic

57. Construct a semilog chart showing the rate of change of pressure as the temperature increases. Then determine the equation. Plot the pressure on the logarithmically divided scale on the y-axis and the temperature on the linear scale on the x-axis. The data are shown in the chart below.

| TEMPERATURE (°F) | PRESSURE (LB/IN.$^2$) |
|---|---|
| 0 | 10.0 |
| 10 | 14.0 |
| 15 | 16.0 |
| 20 | 18.5 |
| 25 | 21.0 |
| 30 | 26.0 |
| 35 | 31.5 |
| 40 | 45.5 |
| 45 | 41.0 |
| 50 | 50.0 |

58. Construct a two-cycle semilogarithmic chart of the following data and determine their equation. Place the two-cycle log scale on the y-axis and give it a range of 10–1000. Make the x-axis a 10 unit linear scale.

| X-AXIS | Y-AXIS |
|---|---|
| 0 | 25.0 |
| 5 | 36.0 |
| 10 | 51.0 |
| 15 | 72.0 |
| 20 | 110.0 |
| 25 | 170.0 |
| 30 | 230.0 |
| 35 | 350.0 |
| 40 | 500.0 |
| 45 | 710.0 |
| 50 | 1000.0 |

59. Construct a single-cycle semilogarithmic chart showing the relationship between temperature in a room and the percent relative humidity. Plot the relative humidity on the log scale on the y-axis. The data are as follows:

| TEMPERATURE (°F) | RELATIVE HUMIDITY (%) |
|---|---|
| 0 | 15.0 |
| 10 | 18.0 |
| 20 | 25.0 |
| 30 | 31.5 |
| 40 | 39.5 |
| 50 | 50.0 |
| 60 | 62.0 |
| 70 | 77.0 |

## Empirical Equations: Logarithmic

60. Construct a logarithmic chart of the following data and find the equation.

| X-AXIS | Y-AXIS |
|---|---|
| 10 | 2.0 |
| 20 | 3.7 |
| 40 | 6.4 |
| 60 | 8.5 |
| 80 | 11.5 |
| 100 | 14.5 |
| 200 | 25.0 |
| 400 | 44.5 |
| 600 | 60.0 |
| 800 | 75.5 |
| 1000 | 90.0 |

61. Construct the logarithmic chart of the data shown below. These data show the relationship between the chromium content of alloy steels and corrosion resistance. Determine the equation for these data.

| CORROSION RESISTANCE | CHROMIUM CONTENT (%) |
|---|---|
| 1.0 | 1 |
| 2.4 | 2 |
| 6.2 | 4 |
| 9.5 | 6 |
| 15.5 | 8 |
| 20.5 | 10 |

62. Construct the logarithmic chart of the data shown below. These data show the relationship between the Brinell hardness of hot-worked steel and the percent of carbon content. Determine the equation.

| BRINELL HARDNESS | CARBON (%) |
|---|---|
| 100 | .1 |
| 145 | .2 |
| 190 | .4 |
| 220 | .6 |
| 267 | .8 |
| 300 | 1.0 |
| 325 | 1.2 |
| 340 | 1.4 |
| 362 | 1.6 |

## Algebraic Problems

63. Plot the following equations:
a. $y = 2 + 2x$
b. $y = 5 - x$
c. $4y = 4 + 3x$
d. $y = 2x - 2$
e. $y = x^2 + 2x + 3$
f. $y = x^2 - 3x + 4$

64. Find the value of $x$ for the following equations:
a. $\tan x = 1 + \cos x$, using 45 through 65 degrees
b. $\cos x + 1 = 1 + \sin x$, using 45 through 70 degrees

## Calculus: Integration

65. Determine the area under the following acceleration-time diagram for a cam.

| ACCELERATION (IN./S²) | TIME (S) |
|---|---|
| 0 | 0 |
| −.20 | 1 |
| −.37 | 2 |
| −.56 | 3 |
| −.72 | 4 |
| −.78 | 5 |
| −.81 | 6 |
| −.78 | 7 |
| −.72 | 8 |
| −.56 | 9 |
| −.37 | 10 |
| −.20 | 11 |
| 0 | 12 |

66. Determine the area under the following velocity-time diagram for a cam.

| VELOCITY (IN./S) | TIME (S) |
|---|---|
| 0 | 0 |
| 2.95 | 1 |
| 2.60 | 2 |
| 2.15 | 3 |
| 1.50 | 4 |
| .75 | 5 |
| 0 | 6 |
| −.75 | 7 |
| −1.50 | 8 |
| −2.15 | 9 |
| −2.60 | 10 |
| −2.95 | 11 |
| 0 | 12 |

67. Determine the area under the integral curve $y = x^2$. Plot for a range of $x$-values 0 through 6.

## Calculus: Differentiation

68. Plot the equation $y = \dfrac{x^2}{4}$ as a rectangular graph. Then find the derivative curve on a graph placed below the plotted equation.

69. Plot the equation $y = x^3$ as a rectangular graph. Then find the derivative curve on a graph placed below the plotted equation.

70. Plot the equation $2y = x^2$ as a rectangular graph. Then find the derivative curve on a graph placed below the plotted equation.

71. Plot the equation $2y = 3x + 6$ as a rectangular graph. Then find the derivative curve on a graph placed below the plotted equation.

# Computers in Engineering and Graphics

Computers have revolutionized engineering design, engineering graphics, and manufacturing. It is essential that the engineer and drafter be thoroughly prepared to use computers in their daily work. In computer-aided design (CAD), design problems are solved using the mathematical analysis and graphics produced by a computer. The designer can rapidly analyze and design a product and store, recall, and modify data for later use.

Computer-aided design drafting (CADD) uses computers, software, and graphics devices, such as a plotter, to produce preliminary and final drawings. The drawings can be viewed on a cathode-ray tube (CRT) and revised before the final hard copy of the drawing is produced.

Computer-aided manufacturing (CAM) uses computer-controlled machinery to produce identical parts with a minimum of human assistance in the operation. Computer-controlled robots with multiple-joint arms can perform delicate assembly operations or work in harsh conditions where humans could not work effectively or safely.

This chapter provides an introduction to computers and their use in design, drafting, and manufacturing. This field uses a unique vocabulary. To get the maximum benefit from this material, it is

## Types of Computers

Computers are classified into two major types—analog and digital. An *analog computer* represents the variables in a situation as an analogy, such as an equation, or by proportional physical quantities, such as the mechanical positions of parts or electrical voltages. A *digital computer* uses sequences of the binary digits 1 and 0 to represent the quantities of a problem or calculation entered into it through a computer program. It performs most mathematical operations by adding and subtracting. For example, multiplication may be performed by adding the number to be multiplied, as 150, as many times as there are units in the second number, as 3. So, 150 × 3 is actually calculated as 150 + 150 + 150 = 450.

Since computer graphics units generally use digital computers, the following discussion centers on them.

### The Digital Computer

A digital computer is an assembly of mechanical and electrical parts that perform functions such as storage of data, processing control, input, and output (Fig. 24–2). Basically, it is made up of a central processing unit, which includes an arithmetic-logic unit, controlling circuits, storage, input/output devices, and interface units (Fig. 24–3).

The *central processing unit* (CPU) is the heart of the computer. Generally, it is subdivided into three major areas—control circuits, the arithmetic-logic unit, and part of the storage memory.

The *storage area* contains a large number of

---

advisable to study the glossary at the end of the chapter.

## COMPUTERS

The computer has revolutionized the work of design engineers. The time formerly spent on slow, laborious calculations can now be spent on creative design and problem solving. Solutions to problems that formerly might have taken months can now be found in hours. Many of the dramatic engineering accomplishments, as in our space program, are due largely to the development and constant improvement of our computers. Almost instant decisions in complex situations can be made when a computer analyzes the data and communicates the situation and possible alternatives to the decision makers (Fig. 24–1).

FIGURE 24–1 Computers aid in monitoring processes and making decisions that would not be possible without their assistance. (Courtesy of National Aeronautics and Space Administration.)

FIGURE 24–2 This is a large, fast asynchronous processing system including a central processor module, input/output subsystem module, host data unit, universe I/O subsystem, auxiliary processor, memory subsystem module, and a system console with dual displays and in-built maintenance terminal and maintenance interface processor. (Courtesy of Burroughs Corporation.)

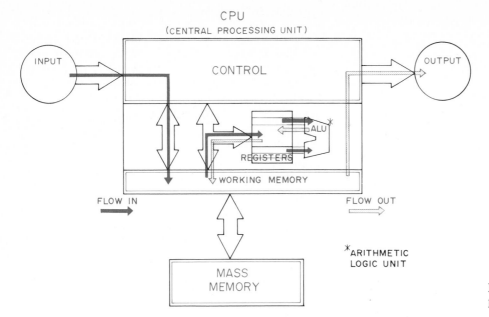

CPU
(CENTRAL PROCESSING UNIT)

INPUT

OUTPUT

CONTROL

ALU *

REGISTERS

WORKING MEMORY

FLOW IN

FLOW OUT

*ARITHMETIC
LOGIC UNIT

MASS
MEMORY

FIGURE 24–3 A typical computer system.

"bins" in which information is saved for processing and to send to the input/output units. The most commonly used system stores information formed by combinations of the binary numbers 1 and 0. Other large storage devices, such as disks, are external.

The *arithmetic-logic unit* has the processing circuits that perform basic manipulations such as adding. It is associated directly with a register or registers called an accumulator that holds the current operand or number to be operated on; after the operation, it may contain the answer.

The *controlling circuits* are the connections provided between storage, the processing unit, and input/output devices. This unit maintains the proper timing pulse cycles needed to queue data flow. Cycle time is in terms of nanoseconds or microseconds. The registers associated closely with timing and data flow are the current instruction register and program counter.

*Input/output (I/O) devices* vary considerably with the design of the computer system. Input devices may include a card reader, ASCII keyboard, light pen, digitizer board mouse, and analog-to-digital converters. Output devices include line printers, plotters, CRTs, magnetic tapes, and disks. Tape and disks store data and can be used to reintroduce programs into the computer at some future time.

The *interface* is the hardware and electronics (switches, relays, buses, circuits, and software) used to connect the peripheral units via the I/O ports to the CPU. Several devices can be wired, or multiplexed, to one channel, but only one can be operational at any time on a single channel.

## Computer Numerics

A digital computer stores binary digits in its memory storage. Each 1 or 0 digit, called a *bit*, has a location in the memory. A *byte* is a group of eight adjacent binary bits that form a sequence in the computer and is usually operated on as a unit. Numeric characters have a precise length for each computer. Commonly available lengths include 4, 8, 16, and 32 bits per computer word.

A computer word contains an exact number of bits for each machine. For example, if a word is composed of four bytes and a byte contains eight bits, the word will occupy 32 bits of memory. The size of the storage of a computer memory is usually given in terms of the number of bytes or megabytes (million bytes).

Computers use several different number bases—binary, octal, decimal, and hexadecimal. The binary (base 2) system uses two digits, 0 and 1. The octal uses eight digits, 0 through 7. The hexadecimal uses 16 digits: 10 are decimal digits, 0 through 9, and six alphabetic, A, B, C, D, E, and F (Fig. 24–4).

Most digital computers use the binary code, using the digits 0 and 1 to represent numbers, letters, symbols, and codes. When recorded on magnetic tape or disk, the presence or absence of a sensitized spot or its magnetic polarity may represent the 0 and 1. These two conditions correspond to "on" and "off" in electrical switching devices. For example, a switch "on" (a sensitized spot) may represent a 1 and a switch "off" (no spot) may represent a 0. A word might be represented as 0110110111. Examples of the binary code for selected decimal numbers ap-

| Binary | Decimal | Hexadecimal | Octal |
|--------|---------|-------------|-------|
| 0 | 0 | 0 | 0 |
| 1 | 1 | 1 | 1 |
| 10 | 2 | 2 | 2 |
| 11 | 3 | 3 | 3 |
| 100 | 4 | 4 | 4 |
| 101 | 5 | 5 | 5 |
| 110 | 6 | 6 | 6 |
| 111 | 7 | 7 | 7 |
| 1000 | 8 | 8 | |
| 1001 | 9 | 9 | |
| 1010 | 10 | A | |
| 1011 | 11 | B | |
| 1100 | 12 | C | |
| 1101 | 13 | D | |
| 1110 | 14 | E | |
| 1111 | 15 | F | |

FIGURE 24–4   Binary and decimal codes.

pear in Fig. 24–4. The binary code is generated by expressing a number in terms of powers of 2. For example, 15 is

$$(1 \times 2^3) + (1 \times 2^2) + (1 \times 2) + 1 = 15$$
$$8 + 4 \qquad + 2 \qquad + 1 = 15$$

In binary code, this would be

$$1000 + 0100 + 0010 + 0001 = 1111$$

When the digital computer is solving a problem, its circuit switches go on and off in response to the 1s and 0s of a specific signal. This occurs so rapidly that the computer appears to operate continuously. The faster computers make more than 1 million computations per second.

## Computer Languages

As computers become more complex and are used to solve more difficult problems and find new applications and uses, the task of programming becomes more demanding. There are a variety of different kinds of computers, and provision has to be made to transfer programs between computers. The computer program is written in an *algorithmic language,* and the computer is used to translate the algorithmic language to a *machine-language* program. Machine language is the language of a particular computer. There are many algorithmic languages in use in the computer industry. The following are some of the most widely used.

*FORTRAN* (Formula Translative) is used for scientific computation.
*ALGOL* is an algorithmic language for computer scientists.

*COBOL* (Common Business-Oriented Language) is a business-oriented algorithmic language. It was designed specifically for business data processing.

*PL/1* (Programming Language I) is a language combining many of the features of FORTRAN, ALGOL, and COBOL into a single language. Its major use is in business data processing.

*Pascal* (named for Blase Pascal) is a well-structured, easy-to-use modern language.

*LISP* is a list-processing language used in research in the area of machine intelligence, such as the development of robots.

*BASIC* (Beginner's All-Purpose Symbolic Instruction Code) enables the operator to communicate with the computer through keyboard terminals. It is a FORTRAN-like language developed as an introductory programming tool. It is an interactive language.

*APL* (Associative Programming Language) enables the operator to communicate with the computer through keyboard terminals. Like BASIC, it is an interactive language permitting interactive computation.

*Interactive computation* is one in which a sequence of intermediate results are made available to the user and additional inputs are provided by the user as needed in the solution of the program. Examples of current applications of interactive languages include computer-aided instruction (CAI), which presents students information and asks questions. The student gets immediate responses and can communicate with the computer. Airline reservation systems use an interactive language to communicate with the central computer and receive immediate responses. Engineering designers can communicate with the computer and calculate the feasibility of a specific design. Changes can be made and new results made available rapidly. A variety of standards can be stored in the computer for instant recall.

There is and will continue to be constant development in the technological aspects of the computer and of programming concepts. The engineer will have to read constantly and attend seminars to keep current in computer-language developments.

## The Digital Computer Program

A machine-language program is a sequence of instructions written in symbolic form using the language of the computer. They are written line after line in the sequence in which they are to be performed. Each line is a single instruction to the computer. A high-level-language computer program is written in symbols more common to the individual and is translated by the machine into its own language. Examples of high-level languages follow.

These programs, written in BASIC and in FOR-TRAN, calculate the average of four numbers.

## BASIC program

```
10 REM THIS PROGRAM AVERAGES FOUR NUMBERS PROVIDED BY THE
OPERATOR
20 PRINT ''TYPE IN THE FOUR VALUES TO BE AVERAGED WHEN YOU
SEE THE QUESTION MARK. SEPARATE YOUR VALUES WITH
A COMMA''
30 INPUT A,B,C,D
40 LET A1={A+B+C+D}/4
50 PRINT ''THE AVERAGE OF'' A;B;C;D ''IS='' A1
60 END


LINE 10 IS A REMARK STATEMENT TO THE PROGRAMMER WILL NOT BE
RUN OR SEEN BY THE OPERATOR UNLESS THE PROGRAM IS LISTED.
LINE 20 IS A STATEMENT TO BE PRINTED TO THE OPERATOR
TELLING WHEN AND HOW TO ENTER THE FOUR VALUES.
LINE 30 IS THE INPUT STATEMENT WHICH WILL CAUSE THE
COMPUTER TO ASK FOR THE VALUES A,B,C,D.
LINE 40 TELLS THE COMPUTER TO ADD THE FOUR VALUES THEN TO
DIVIDE BY THE NUMBER FOUR AND MAKE ANSWER EQUAL TO A1
LINE 50 TELLS THE COMPUTER TO PRINT THE ANSWER.
LINE 60 TELLS THE COMPUTER THAT THE PROGRAM IS FINISHED.
```

## FORTRAN program

```
READ A,B,C,D    (insert numbers to be averaged)
AVG = {A + B + C + D}/4    (divide the sum by 4)
WRITE {6, 1} AVG    (print the answer)
1 FORMAT { '', 'AVERAGE OF THE NUMBERS
&4F13.6, ' IS = ', F13.6} {
STOP
END    (end of program)
```

## Computer Software

*Software* is a term used to describe the variety of programs a computer system has in its library. These programs may be purchased from a software developing company, be provided by the firm selling the computer, or be written by programmers on the staff of the company that uses them. Frequently used programs are often stored in a direct-access device, such as a disk pack, where they are available to the user without the need for an operator to move the program. The engineer should be familiar with the programs available in the computer "library" so that maximum value can be gained from existing resources; this saves considerable time in problem solving.

There are a variety of application package programs designed for engineers who have little knowledge of programming. For example, COGO and STRESS are program packages available for use in civil engineering; the ECAP package contains programs for electrical engineering. If the engineer or computer staff knows how to write programs in the language used in the package, additional subroutines can be added to the library.

The algorithmic languages and the compiler to process them into machine language are important parts of computer software. When a program is written in algorithmic language, it must be translated into the language of the machine. A *compiler* translates the algorithmic language rapidly and at a lower cost. An *optimized compiler* takes longer to do the translation but tries to produce a very efficient machine program. Some users run a program through the compiler to debug it and then through the optimized compiler for improvement.

An *interpreter* is another form of language translator. It translates one line at a time and runs the line on the computer as soon as it is translated. This detects errors immediately, permitting correc-

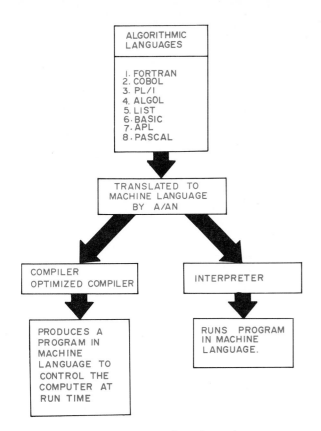

```
ALGORITHMIC
LANGUAGES

  1. FORTRAN
  2. COBOL
  3. PL/I
  4. ALGOL
  5. LIST
  6. BASIC
  7. APL
  8. PASCAL

        ↓

TRANSLATED TO
MACHINE LANGUAGE
   BY A/AN

   ↓          ↓

COMPILER      INTERPRETER
OPTIMIZED COMPILER

   ↓             ↓

PRODUCES A    RUNS PROGRAM
PROGRAM IN    IN MACHINE
MACHINE       LANGUAGE.
LANGUAGE TO
CONTROL THE
COMPUTER AT
RUN TIME
```

FIGURE 24–5 The various algorithmic languages are translated to machine language, producing a program to control the computer.

tions to be made. However, it interrupts program operation when an error is found. The compiler runs the entire program before a trial run to seek errors (Fig. 24–5).

## Basic Arithmetic Operations

The following is a simplified explanation of the relationship between the input, central processing unit (CPU), memory or storage, and output of a computer.

Tracing the flow of information through a digital computer is a complex task if an attempt is made to follow the movement of each character or value through each of the temporary storage locations known as registers. However, the general operation can be seen by following simple arithmetic manipulations such as adding and subtracting through the major structural blocks in the machine.

Before an operation such as addition can start, the individual operating the computer must place a program or set of instructions in the memory telling it exactly what to do. In the case of addition, the numbers must be entered by the operator at a keyboard or some other input device. The numbers then flow into memory via the CPU, where they can be retrieved by the CPU when called by the appropriate program step. In a simple addition, the numbers would be pulled up one at a time and placed in the CPU's registers to be added by the arithmetic-logic unit (ALU), from there to be placed back in memory or sent via the CPU to the output device (Fig. 24–6).

Subtraction is handled in much the same fashion. After the subtract program is placed in memory, the data (numbers) are entered into the input device and are placed in memory by the CPU's control section. At the appropriate program step, the numbers are placed in the registers associated with the ALU and subtracted. They then flow back to memory or output under CPU control (Fig. 24–7). Note that the CPU contains two sections in most computers: the control, which routes data and sets up special subsections such as the ALU, and the register, which contains the temporary storage locations known as registers where the actual data manipulation takes place. Some computers use memory for register operations rather than special register locations in the CPU.

## COMPUTER-AIDED DESIGN DRAFTING (CADD)

As you begin a study of computer-aided design drafting, first realize that the human operator is intelligent, can reason, is creative, and is slow in personally handling data. The computer has no intel-

FIGURE 24–6 The arithmetic-logic unit processes numbers sent to it and performs the indicated action.

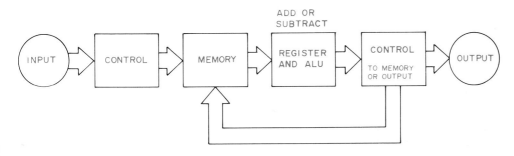

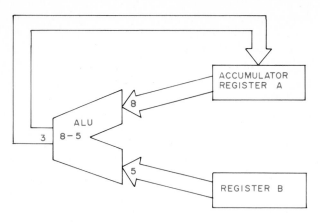

FIGURE 24-7 The subtract program is placed in the memory; the numbers to be subtracted are placed in the registers associated with the arithmetic-logic unit and are subtracted.

lect, is not creative, but is extremely fast in processing data placed there by a human. The success of the total concept is the ability of the human and the machine to complement each other. Each has a different language. Man thinks in pictures and symbols; the computer understands only electrical impulses.

Computer-aided design drafting replaces the traditional drawing board and design tools with a video display terminal in a computer-based electronics system. One such system is shown in Fig. 24-8.

CADD is rapidly changing the way engineers and architects approach design problems. Most major engineering and architectural firms have some form of in-house computer and have progressed from using it only for engineering analysis and financial management to complex applications such as architectural programming, design studies, and en-

gineering drawing. In addition to enabling the architect and engineer to handle increasingly complex design problems more rapidly and accurately, the computer is used to produce drawings and other graphics more rapidly and at a lower cost. Manufacturers of printed and integrated circuits were the first to make use of this system. CADD has found wide use in electrical and mechanical drafting applications such as schematics, logic and wiring diagrams, mechanical drawings of all types, and detailing. In addition, it has many applications in other areas, such as cartography, architectural drawings, charts, graphs, and almost any other type of drawing.

Computer-aided graphic systems provide drawings of improved and uniform quality, produced in ink at the same or less cost than conventional pencil drawings. The drawings are in compliance with standards for parts, notes, symbols, and drafting practices, because these are all stored in the software for instant retrieval. Because the data are stored and instantly retrievable, any sector of a company can call upon the computer and the automated drafting equipment to produce a copy of the drawing that incorporates the latest corrections and revisions. There is no longer any need ever to be working with an obsolete drawing. The system enables the manufacturer to serve the customer better by giving faster response turnaround time on modifications requested.

Most systems now in use are stand-alone systems (systems that do not depend on a central computer). These systems include a minicomputer and all hardware and software to perform computer-aided design and drafting assignments. In addition to the minicomputer, the CADD system would have

FIGURE 24-8 A computer-aided design drafting system. (Courtesy of Calma Company.)

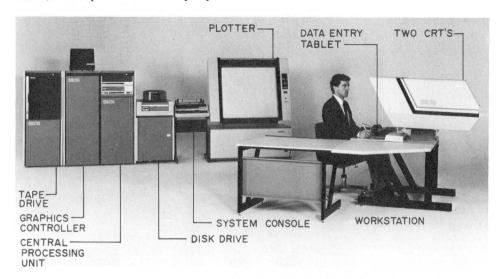

hardware and software to produce new drawings, convert existing drawings into a digital form with a digitizer, edit an existing digitized drawing, store many drawings, and produce hard-copy drawings using a device such as a plotter or computer-output microfilm (COM) equipment.

CADD has had a big impact on the engineering operation. The traditional files for drawings are gradually being replaced because the drawing is stored in digital form. Corrections and revisions are made on this record, so it is always ready to print out an up-to-date copy on a plotter. The file of drawings might be in a central digital store where all users would have access through remote drawing and editing work stations. Computer-output microfilm-produced aperture cards and microfiche have also become significant media for hard-copy storage of drawings. Thousands of drawings can be stored this way in a small drawer, providing a hard-copy backup to the digitized data stored in the digital store. While a variety of skill levels will continue to be required by design and drafting personnel, such as detailers, drafters, and designers, they will work at computer-controlled work stations rather than at conventional drafting tables. They need a sound knowledge of drafting standards and procedures, but the application of these is different.

Computer-aided design drafting involves the conversion of the electrical impulses of the computer into a form readable by humans and the conversion of the instructions of a human into a form that the computer understands. In some sophisticated systems, the engineer needs to know very little about programming in order to control the computer. CADD therefore includes any device that converts the language of the computer to a language recognized by humans and vice versa.

Computer graphics devices can be classified as graphic or nongraphic and are either interactive or passive (Fig. 24–9). A *passive* type has a significant lag between the time when the information is entered into the system and the time when output is received. This often requires a delay of hours for an initial response and a similar delay if other inquiries about the response are entered. In a passive system, the commands are recorded on some device, as punched or magnetic tape, which is transferred to the unit to perform the drawing, such as a plotter. This can provide a response in a reasonable time but does not provide immediate results. *Interactive* systems give the user almost immediate response (within seconds or minutes) to any inquiries entered into the system. For example, a design can be changed or new data for a calculation can be entered and the result of these will be known immediately.

A block diagram of an interactive CADD station is shown in Fig. 24–10. The human operator is the key to the mechanical design situation. The operator has experience, design knowledge, creativity, decision-making capacity, and, using both eyes and hands, the capability to input data and receive output. The design decisions of the human operator are input to the *interactive device* through the use of functional buttons, a tablet digitizer, keyboard, thumb wheel, joystick, or light pen. The output of the interactive device is input to a computer in the form of commands or data. The *computer* has registers for calculating, memory for storing programs and data, and other capabilities. The graphics software in the computer gives the system capabilities to augment the designer. The *display* provides feedback of infor-

|  | NON-GRAPHIC | GRAPHIC |
|---|---|---|
| Passive | Line printer | Incremental pen plotter<br><br>Computer-output microfilm (COM) |
| Interactive | Teletypewriter CRT A/N Consoles | CRT Graphic Consoles |

FIGURE 24–9 The spectrum of computer graphics devices.

FIGURE 24–10 A typical interactive computer-aided design drafting station.

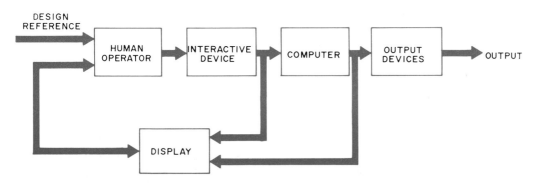

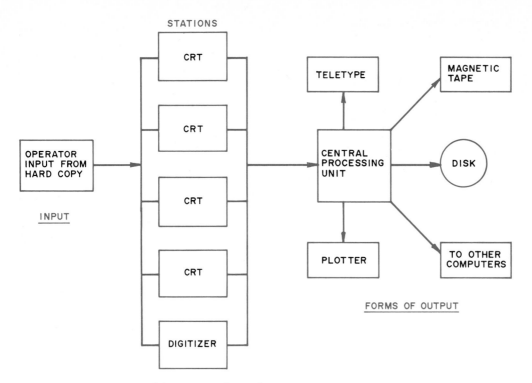

FIGURE 24–11 A typical four-terminal graphics system.

mation (commands, drawing data) to the designer. The types of display devices include the CRT and raster-scan tube. The designer interacts directly with the display and continues the design process by accepting or changing the displayed solution. Once the solution is acceptable, some type of *output device* is activated. A common design output is a plotter. This may be a pen plotter or an electrostatic plotter. These devices produce hard copies of the design solution, such as an engineering drawing or a land survey. Other forms of output include storage of the solution on paper or magnetic tape and communication to other systems. A typical four-terminal graphics system is shown in Fig. 24–11.

Graphic devices use lines, circles, curves, and characters to build a picture. Nongraphic devices produce alphanumeric output.

A wide variety of equipment, called hardware, is manufactured for use in computer-aided design drafting systems. The easiest way to get into operation is to acquire a turnkey system. A *turnkey system* is a complete hardware installation plus software purchased from a single manufacturer.

Graphics systems can be broadly classified into three categories: graphics input devices, graphics display devices, and computer systems.

### Input Devices

Input devices are used to make data available to the graphics program. The program processes the input data and controls the output on the display device.

It should be noted that the input and output are separated by the graphics program (software). The control provided by the graphics program makes it possible for the computer to assist the operator by straightening lines or performing calculations.

There are four major categories of input devices found at a computer graphics work station: light pens, analog-to-digital devices, keyboard devices, and voice data-entry systems.

*Light pens*  Light pens are used to "write" on a CRT screen (Fig. 24–12). They do not emit light but detect the presence of light on the screen. They are used only on sophisticated graphics systems because they require extensive programming support. The pen points to graphic objects on a refreshed CRT. It is also used to locate positions on the display area.

The light pen requires a refresh CRT be used because the pen responds to the instantaneous peak in brightness that occurs when the electron beam excites the CRT phosphor at the peak of brightness. The operator is assisted in aiming the pen at the screen by a small circle of light projected from inside the pen. This light does not produce the image on the screen.

The operator controls the light pen by turning on a switch located on the tip of the pen. The switch may be held on for drawing an image or momentarily turned on to identify a single element in the image. The switch actually controls a shutter that covers the opening in the tip of the pen. This controls the flow of light from the CRT into the high-speed

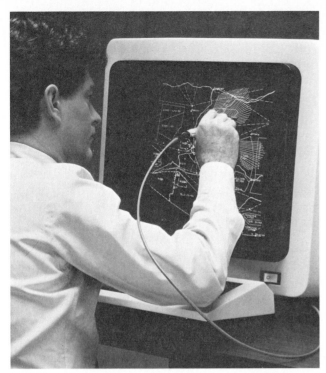

FIGURE 24–12 Light pens are used to produce images on a CRT. (Courtesy of California Computer Products, Inc.)

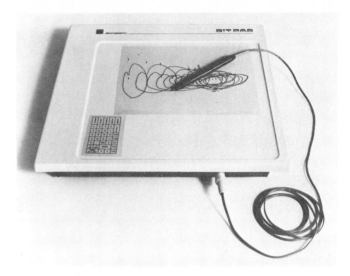

FIGURE 24–13 A digitizer tablet with a pen. To use it, lay a drawing on the tablet and touch the pertinent items on it with the pen. This information is entered into the data processing system. (Courtesy of Summagraphics Corp.)

light detector in the pen. The light is carried to an auxiliary electronic unit by a fiber-optics conductor. Here, the pen output is amplified and filtered. A digital pulse generator responds to the light by setting a one-bit digital code that indicates a hit, and this is sent to the computer. The response time for a light pen is less than one-millionth of a second.

A light pen has advantages and disadvantages when compared with other input devices. It depends heavily upon software. Elaborate programming is necessary to enable it to track across the CRT. It is difficult to aim the pen accurately, and resolution is poor compared to other members. Poor resolution makes it difficult to distinguish between several closely spaced lines. The pen must be moved at an almost constant speed or it will lose contact with the cross-hair image. This image is displayed at the point where the tracking stopped. It is the target used by the operator in aiming the pen to resume tracking.

The pen offers the advantage of permitting an operator to interact directly with the displayed image by pointing to the image.

*Analog-to-digital devices* The most frequently used input devices are analog-to-digital converters. They sense the quantity, such as speed, acceleration, force, position, direction, distance, and rotation, and convert it into a numeric quantity acceptable to the computer. These analog converters include digitizers, joysticks, trackballs, and dials.

A *digitizer* input device has a stationary flat surface and some form of hand-held tool used to position points on the surface. A flat surface is called a *digitizer tablet* (Fig. 24–13). It is about the size of a tablet of note paper, whereas a *digitizer table* is about the size of a drafting table, 40 × 60 in. (Fig. 24–14). The *positioning tool* is either a flat box with a crosshair sight, called a *cursor* or *mouse*, or a *stylus* or *pen*

FIGURE 24–14 A digitizer table with a cursor that is used to enter numeric data and program commands. (Courtesy of Summagraphics Corp.)

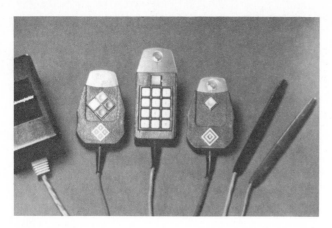

FIGURE 24-15 Various cursors and pens used on digitizer tablets and tables. (Courtesy of Summagraphics Corp.)

which has a sharp point like a pencil that contacts the flat surface. The point on the surface of the tablet, which locates the active point (a symbol) in a design, is called a *cursor point*. Often the term *cursor* is used in general to describe both the mouse and the point, and this causes some confusion. Some cursors have up to 12 buttons that enable numeric data and program commands to be entered (Fig. 24-15). A cursor point is recorded on the tablet by positioning the cross hairs on the stylus at the location and pressing a switch.

Tables are available with translucent tops that make it possible to project 35-mm slides or microfilm images on the back of the top. X-rays can be shown by backlighting the table (Fig. 24-16).

FIGURE 24-16 A large black-lighted digitizer table with a 12-button cursor. Maximum system flexibility is attained by simultaneous communication with several host computers and/or peripherals through multiple interface.

The tablet digitizer uses a pen. It has a resolution of 400 lines per inch and a selectable output rate of up to 240 coordinate points per second. (Courtesy of California Computer Products, Inc.)

The digitizer permits the operator to maintain control over the interactive graphics system. Common uses for the digitizer are to change or add to an existing image on the table or to enter commands or nongraphic data through a tablet menu and convert an existing document, such as a map or engineering drawing, into computer data.

To digitize an existing drawing so it is recorded in computer language, the drawing is taped to the table. The operator locates the beginnings and ends of lines and any corners with a mouse or stylus. It is much like making a tracing. The operator must develop a high level of eye and hand coordination to control the mouse or stylus on the table.

Many digitizers generate an electronic or magnetic field in one component (for example, the stylus). That is detected by the other component (the table). An electronic control unit sends a field-producing signal and receives the feedback signal, which determines the location. The flat surface commonly has within it a grid of tiny wires. There may be as many as 1000 wires per inch from top to bottom and across the tablet forming a grid with a resolution of 0.001 in. The control unit sends digitally coded pulses to the conductors that are picked up by the mouse or stylus from the wires closest to its location on the tablet. They are transmitted through the wire to the mouse or stylus and to the control unit, where they are decoded, thus locating the position of the mouse or stylus on the tablet.

One type of digitizer uses an ordinary pen or pencil as the stylus. The surface is *touch-sensitive* and senses the pen movements as *xy* analog values. The control unit converts these to digital *xy* coordinates, thus locating the points of contact (Fig. 24-17).

A *sonic digitizer* has a pen that produces sound

FIGURE 24-17 A touch-sensitive digitizer. (Courtesy of Elographics, Inc.)

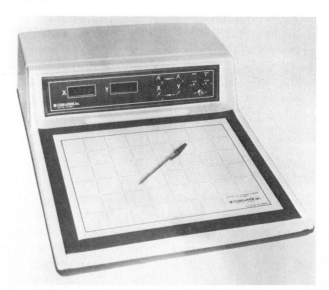

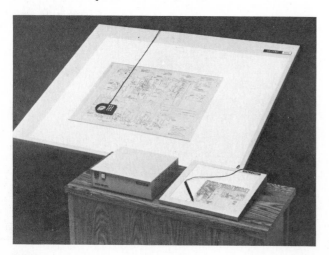

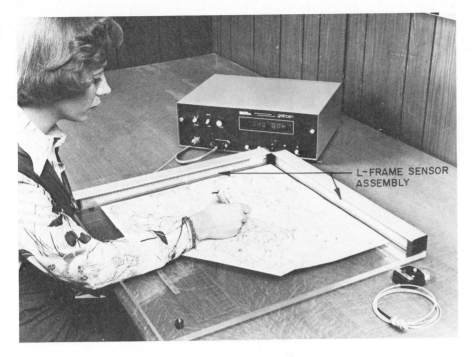

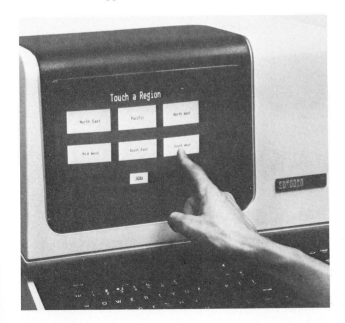

FIGURE 24–18 This sonic digitizer can interact with CRT or plasma displays and project images from x-rays and films and maps and drawings on drafting tables. (Courtesy of Science Accessories Corporation.)

impulses generated by an electric spark. These are detected by an L-frame sensor assembly on the border of the assembly area that establishes the location of a point. The time it takes for the sound to reach the two microphones is converted into distance measurements, giving the *xy* coordinates of the pen point (Fig. 24–18). It uses either a stylus or a cursor.

A *touch panel* or *light detector* is a type of digitizer that utilizes light detection (Fig. 24–19). The

FIGURE 24–19 This touch-sensitive digitizer requires no special skills. It eliminates a great deal of keyboard operation. Almost 1000 discrete touch points are available on a 12-in. display; more than 3000 discrete points are on the 19-in. display. (Courtesy of Carroll Touch Technology.)

display area is in the center of the screen, and the optical devices are located around the frame. Small, light-emitting diodes (LEDs) are located on one horizontal and one vertical edge of the frame. Phototransistor detectors are located on the other two sides, opposite the light-emitting diodes. This creates a grid of infrared light beams which are broken when the operator places a pen (or finger) on the digitizer area. The photo detectors send this interruption to the electronic logic section of the digitizer, which determines the coordinates of the point touched. This type of unit has a resolution of 0.25 in., which is adequate for many uses.

*Dials, joysticks,* and *trackballs* are other types of interactive graphics input devices. They are used to control *cursor points* in the same manner as the light pen. They move the cursor on the screen. The cursor moves only when the operator manipulates the device. The cursor-point movements are accumulated by the digitizer graphics program, and the coordinates of the cursor point are always stored in the computer. Resolution is not great for these devices because it is measured in terms of units of resolution on the screen.

A *joystick* is a vertical shaft mounted on a square base (Fig. 24–20). It is pulled or pushed by the operator in any direction, causing forward-backward and right-left movements of the cursor point.

A *trackball* is a sphere positioned inside a box with part of the top of the sphere exposed (Fig. 24–21). As the sphere is rotated, data-carrying pulses are converted into digital data that show the distance and direction of the trackball rotation. The distance the cursor point moves on the screen is proportional to the number of revolutions of the sphere.

Computers in Engineering and Graphics **643**

A three-axis joystick.

A two-axis joystick with a hand grip.

FIGURE 24–20 Two of the many variations of joysticks available. (Courtesy of Measurement Systems, Inc.)

FIGURE 24–21 A trackball with a 3-in. ball. (Courtesy of Measurement Systems, Inc.)

FIGURE 24–22 A dial-type cursor point control. (Courtesy of Measurement Systems, Inc.)

A *dial* is a one-axis cursor-point control device (Fig. 24–22). It is a knob that, when turned, moves the cursor point in a specific direction. The speed of cursor-point movement is determined by the speed at which the dial is rotated.

*Keyboards* Most computer graphics terminals have a full alphanumeric keyboard. It is an efficient method for entering data into the graphics system data base. The shift key on the keyboard enables the operator to choose between an alphabet of uppercase letters and a set of graphic-symbol characters. The keyboard has the same spacing between keys as a standard typewriter, which helps an operator who can type to adapt quickly to this keyboard (Fig. 24–23).

Some keyboards have five keys used to control the cursor point. Each key has an arrow on it showing the direction in which the cursor will move when it is pressed. When the two perpendicular direction keys are pressed simultaneously, the cursor moves on a diagonal. The fifth key is the "home" key, which causes the cursor point to return to its original programmed position.

A set of program function keys provides a rapid and more accurate way to enter commands than using the standard keyboard. A keyboard may contain 16 or 32 keys. When a single key is pressed, a complete command statement is entered.

*Voice data-entry systems* Voice data entry (VDE) permits the operator to input information using normal oral commands. Equipment that recognizes the spoken word can replace other forms of input devices (Fig. 24–24). It can also be used to supplement other forms of input, enabling the operator to give oral commands. When digitizing, this system permits the operator to use a cursor and oral commands; this is more efficient and does not pull the attention away from the movement of the cursor when it is necessary to input additional data. The

FIGURE 24–23 An intelligent graphics terminal. (Courtesy of Hewlett-Packard.)

system has a predefined vocabulary that, spoken aloud, activates the command. For example, the word *cancel* could be the code word that the result is incorrect.

The store vocabulary is created using the VDE equipment and a standard keyboard. The command is typed and then the word is spoken 10 times by

FIGURE 24–24 A voice-entry data terminal permits the operator to input using oral commands. (Courtesy of Threshold Technology, Inc.)

the operator. The system uses these repetitions to determine the characteristics of the word and transform it into a pattern. The patterns are stored in a vocabulary table. Storage can be on tape cassettes or disks so that each operator can have a table. The table is entered into the microcomputer when needed.

## Interactive Display Devices

There are a variety of devices used to display the output of a computer graphics system. Some produce visual displays on a CRT, while others produce hard copy.

Interactive display devices are those that give instant visual feedback to the input from the operator. The two major approaches are stroke-writing and raster.

Stroke-writing devices draw lines to create an image. Raster devices use a series of closely spaced dots to produce a visual image. Both systems may use a refresh CRT or a storage CRT. The refresh CRT continuously regenerates (refreshes) the image on the surface of the tube; the storage type retains the image until it is erased.

*Stroke-writing screens* *Stroke-writing* devices produce a picture by drawing individual lines one at a time at any desired length and oriented on any desired angle. It has an associated unit called a *vector generator*. When drawing a line, the end-point coordinates are given to the vector generator by the computer. The vector generator converts them into *xy* analog voltages that drive the deflection systems in the CRT. The voltages control the direction of the electron beam on the phosphor screen of the CRT and produce a line on it.

*Storage displays using stroke-writing screens* are made up of a standard CRT plus a secondary cathode located between the electron gun and the screen. The electron gun generates the writing beam that produces the lines on the phosphor-coated screen. The secondary cathode emits a stream of low-energy electrons, called the flood beam, that cause the lines on the screen to glow. The lines on the screen are erased by raising the energy of the flood beam until it excites the entire screen and then reducing the voltage to zero. This allows the phosphors to return to normal. The erasure cycle takes about 1 sec.

The resolution of *storage-type* CRTs is equal to any other type of display device. The image on the storage tube is displayed with the same intensity in a single color. Selected lines can be emphasized by drawing them wider. This is done by the graphics program, which can direct that two or more lines be stroked close together to form a single, wider line. When an operator adds something new to a design displayed on a storage tube, the change is immediately stroked on the CRT. When the design is changed, the screen is completely erased and an entirely new image with the changes is stroked on the CRT.

*Refresh displays on stroke-writing screens* provide a rapid response. They are therefore especially useful when the projects involved require considerable operator and computer interaction. A refresh display provides simulated motion for applications in chemistry, physics, and engineering research. It also provides excellent viewing capabilities for three-dimensional computer models.

A stroke-writing refresh display system operates by first building a *data base* containing a complete description of the object being drawn using interactive input devices. A *display file* holds the most recent design. The display file is kept in a *refresh memory*, which is a storage area in the computer. The *display processor* reads the refresh memory and sends instructions to the vector (line) and circle generators. These take the geometric data and convert them into *xy* analog voltages, which control the electron beam. The entire image on the screen is stroked during each refresh cycle. Refresh displays do not have an erase cycle like storage displays. The image on the screen is changed by altering the contents of the display file.

Most refresh displays have a *fixed refresh rate*. Normal speeds are 30 or 60 cycles/sec. The rate used is matched to the phosphor persistence to eliminate or reduce flicker in the image. The amount of information that can be on the screen is limited to the lines the CRT can trace during one refresh cycle. If there is a very long display file, the display processor may miss the last of it because it returns to the start of the file at the beginning of each refresh cycle.

*Multicolor stroke-writing displays* are produced by means of a process called *beam penetration*. The CRT has a single electron gun and a screen coated with layers of two or more different phosphors separated by a thin layer of dielectric. The color of each stroke depends on the velocity of the beam when the vector is stroked. A screen with two phosphors, red and green, produces a display in four colors— red, orange, yellow, and green. The digital-to-analog converter receives the usual description of a vector plus a color code. The color code indicates which of the four voltages are to be used to stroke that vector, thus determining its color.

The lowest velocity penetrates the first layer of phosphor and produces red. The second velocity penetrates the first layer and partially penetrates the second layer (green), mixing red and some green to produce orange. The third-highest velocity produces yellow, and the highest fully penetrates the second

phosphor layer to produce green. The green phosphor is much more efficient than the red, so the green predominates when both layers are fully penetrated.

*Raster screens* A raster CRT screen has a large number of small phosphor dots of varying degrees of brightness. These dots are called *picture elements* or *pixels*. They are in rows of parallel, horizontal lines covering the entire area of the screen. This pattern of lines is called the *scanning raster*. The image is a display made up of pixels excited by the electron beam.

Here is how a raster scan produces an image. The graphic information is stored in the computer data base. Because the raster display method requires a video signal, like a television signal, to drive the CRT, an extensive array of electronic equipment is needed to handle the conversion from the computer data base to the video signal. Once the video signal is produced, an electron beam is magnetically deflected to trace one horizontal line of pixels on the screen. The voltage applied to the electron gun is constantly changing due to the demands of the incoming video signal. As the beam traces a single horizontal line, it excites a pixel as it passes points on the drawing that are to appear on the screen. The scanning starts at the upper left corner of the screen and proceeds at a constant speed from left to right and from top to bottom, one line of pixels at a time (Fig. 24–25). As the beam reaches the right side of the screen, it shuts off and returns to the left side to scan the next line. When the bottom line has been scanned, the beam turns off and returns to the top left corner of the screen (Fig. 24–26).

Raster scans use sequential scanning or interlaced scanning. *Sequential scanning* is tracing each horizontal line from top to bottom as just described. The picture does flicker when the refresh rate is not

FIGURE 24–25 Sequential raster scanning.

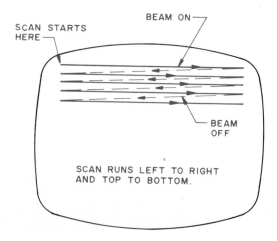

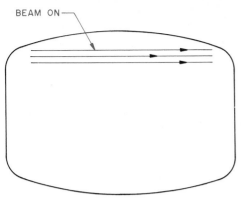

1. THE BEAM STARTS TRACING THE ROWS OF PIXELS.

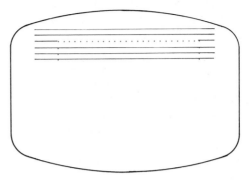

2. THE BEAM EXCITES PIXELS ON THE DISPLAY THAT ARE PART OF THE IMAGE.

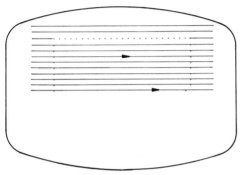

3. THE BEAM CONTINUES THE SCAN FROM TOP TO BOTTOM.

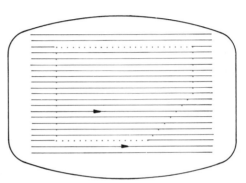

4. THE FINISHED IMAGE IS DISPLAYED BY THE PIXELS EXCITED ON EACH HORIZONTAL ROW.

FIGURE 24–26 The raster display develops an image by tracing horizontal rows of pixels.

high enough to match the persistence of the phosphor pixels. The upper part of the image will start to fade while the lower is refreshed. *Interlaced scanning* traces all the odd-numbered horizontal lines first and then traces the even-numbered lines. This helps prevent flicker because the screen is refreshed twice during each cycle.

*Full-color monitors* use raster patterns, but each pixel is made of three layers of phosphor in the three primary colors—red, green, and blue. The CRT receives three separate video signals that control the voltages applied to the three primary-color beams. The digital data input that identifies each raster point is placed in the refresh memory. Each raster point is described by means of a short address that references it to a color lookup table. The entries in the color lookup table contain the instructions needed to generate the hue, saturation, and intensity required for each pixel. These instructions, which are digital, are sent to digital-to-analog converters that generate the voltages to control the in-

tensity of each beam for each primary color. These video signals are sent to the color monitor and produce the color display.

*Plasma panel displays*  A *plasma panel* is an interactive display device but not a CRT (Fig. 24–27). The panel is made up of several layers of material sandwiched between glass panels. The void between two of the glass panels is filled with a neon gas. Thin conductive strips serving as horizontal and vertical electrodes are attached to the inner faces of the glass panels. These form the X axis and Y axis electrodes. The point at which the elements cross locates a display point. The electrodes are covered with a dielectric layer, a sheet of glass coated with magnesium oxide. These two glass layers are sealed, and the space between them is filled with neon gas (Fig. 24–28).

Images are displayed by exciting a series of points. To display a point, a voltage pulse is applied to the horizontal and vertical electrodes correspond-

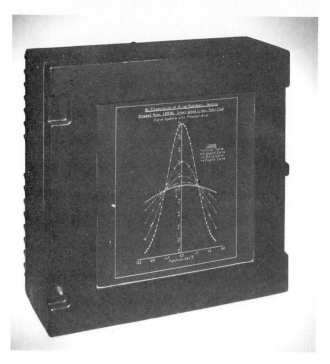

FIGURE 24–27 A flat-panel plasma display assembly. (Courtesy of Interstate Electronics, Corp.)

FIGURE 24–28 A typical plasma display panel.

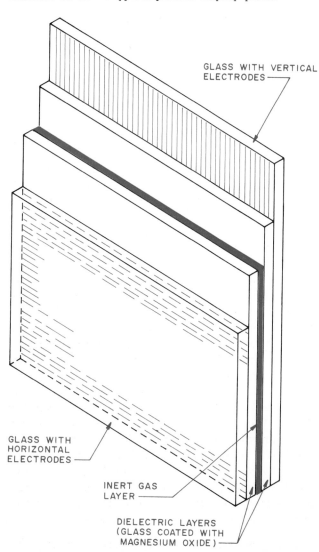

GLASS WITH VERTICAL ELECTRODES

GLASS WITH HORIZONTAL ELECTRODES

INERT GAS LAYER

DIELECTRIC LAYERS (GLASS COATED WITH MAGNESIUM OXIDE)

ing to the $xy$ coordinates of the point. This ionizes the gas at this point, causing it to produce an orange glow. Points are erased by removing the electrical charge from the dielectric layer. This display is stable as long as power is supplied to the panel. Since the dot pattern on the screen resembles the dots on a printed halftone picture, surfaces can be displayed shaded. Because the panel is transparent, images from 35-mm slides or microfilm can be projected from the rear. This enhances the displayed image.

## Plotters

When all the design work is finished and decisions and data are stored in a computer, it is often necessary to produce a hard copy of the final design. Hard copy is some form of reproduction, such as a drawing on paper. A plotter is used to produce these drawings.

*Pen plotters* The most widely used plotters use servomechanisms to drive a carriage which contains a pen that draws on a paper (Fig. 24–29). There are three types of pen plotters—flatbed, drum, and belt-bed.

The *flatbed plotter* consists of an accurately man-

FIGURE 24–29 A plotter drawing-head carriage with pens. (Courtesy of Data Technology, Inc.)

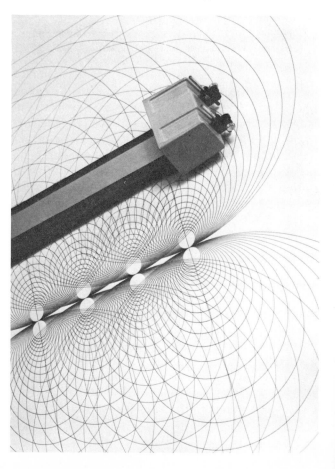

FIGURE 24-30 A flatbed plotter that has a pressurized inking system and four pens on the head carriage. (Courtesy of California Computer Products, Inc.)

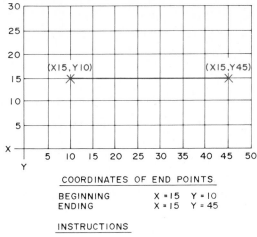

COORDINATES OF END POINTS

BEGINNING      X = 15   Y = 10
ENDING          X = 15   Y = 45

INSTRUCTIONS

RAISE PEN, MOVE TO X 15, Y 10
LOWER PEN, MOVE TO X 15, Y 45
RAISE PEN

FIGURE 24-31 How a plotter draws a horizontal line.

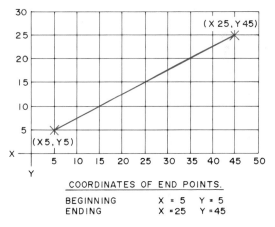

COORDINATES OF END POINTS.

BEGINNING      X = 5   Y = 5
ENDING          X = 25   Y = 45

FIGURE 24-32 An inclined line is drawn from the point of beginning to the end point using the x and y coordinates. The computer calculates xy coordinates for very small increments, and the line is drawn in small steps. This is done so rapidly that the pen appears to move in a straight line.

ufactured flatbed resembling a drafting table (Fig. 24-30). It is usually made of steel or aluminum. The drawing surface is often covered with a precision-ground neoprene platen. On small models, the paper is taped to the table. On larger models, it is held electrostatically or by a vacuum system.

The pen is held in a drawing head carriage. The writing device can be a ballpoint pen, felt-tip pen, or wet-tip pen. The ballpoint is used when rapid display of information on paper is needed because it dries rapidly. Felt-tips produce a rather heavy line but have good speed. Wet-tip pens are especially useful for high-quality drawings on plastic drafting film. Some machines will also hold a scriber that is used to cut fine lines in scribe coat. A drawing head will hold several pens, permitting a variety of line widths and colors to be used.

The control system reads the coded coordinate information that is stored in the computer, on tape, or on a deck of cards. It converts these into electronic signals that move the drafting machine head to the locations of the coordinates. After moving the pen to the x and y coordinates, it sends signals to the step motors that drive the drawing head so that they operate at a rate that permits a line to be drawn. The signal also raises and lowers the pen as needed to draw a line. In Fig. 24-31, the drawing head is moved to beginning coordinates x 15, y 10, and the pen is then lowered to the paper. Next, it moves to new x and y coordinates at the end of the line and raises the pen. A straight horizontal line has been drawn.

An inclined line is drawn in Fig. 24-32. Basically, the pen is moved to the coordinates of the point of beginning, x 5 and y 5. It is lowered and instructed to move to the end point at coordinates x 25 and y 45.

To draw curved lines, the computer plots points on the curve by interpolation and directs the

pen on the drafting machine to draw the curve. Interpolation is finding the value of a function between two known values of that function. There are three methods of interpolation used to find the points on a curve—linear, circular, and parabolic.

*Linear interpolation* involves straight-line motion between closely spaced points on a curve or circle (Fig. 24-33). This is the most common method. The chord length is the distance used to locate the points. The chord height is the distance between the curve and the chord. Chord height is very small, so the curve, even though it is a series of short, straight lines, appears as a smooth curve. The computer calculates coordinates for hundreds of points to generate a curve at such a speed that the pen appears to move smoothly when drawing the curve.

Computers in Engineering and Graphics   **649**

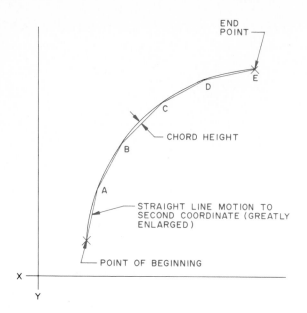

FIGURE 24-33 A curve plotted by linear interpolation locates coordinates of points on the curve using the chord height.

FIGURE 24-34 A four-pen drum plotter that has a plot width of 1320 mm (51.2 in.) and a velocity of 762 mm (30.0 in.) per second. (Courtesy of California Computer Products, Inc.)

*Circular interpolation* requires as its input the beginning and end points of the curve or the position of a circle along with coordinate points for the location of the center and the radius. This information is substituted into the general equation for a circle. The computer calculates intermediate points on the circle.

*Parabolic interpolation* requires the beginning and end points of the curve plus coordinate values of several intermediate points on the curve. These data are substituted into the general equation for a parabola, and the computer calculates intermediate points on the curve.

*Drum plotters* have the paper attached to a revolving drum similar to paper on a typewriter. The pen is in a drawing head and changes position in the $x$ and $y$ directions in relation to the paper on the drum. The drum rotates, thus providing the Y axis movement. The pen moves parallel with the length of the drum, providing the X axis movement.

Drum plotters are usually driven by stepping motors, which rotate the drum through angular steps. This makes it more difficult to maintain the accuracy of a drawing when compared to a flatbed plotter. Drum plotters usually use the same programming as flatbed plotters. Since the plot paper is often up to 120 ft long, a very long drawing can be made, or several smaller drawings can be made without stopping the machine to replace the paper (Fig. 24-34).

*Belt-bed plotters* use a wide, endless belt that moves over a flatbed surface and drums. A pen carriage moves to produce a plot similar to that of the drum plotter. This is really a hybrid flatbed and drum-type machine (Fig. 24-35).

*Electrostatic plotters* *Electrostatic plotters* use either the matrix-writing technique or the photoconductive-plate method. The *matrix-writing technique* places an invisible image on the paper in the form of a static electric charge. To this is applied a liquid toner and carbon particles, which cling to the areas electrically charged. The image is fused to the paper (Fig. 24-36).

The plot is placed on the paper as electric charges by needles that form lines by a series of overlapping dots. As the paper rolls under these needles, computer data decide which needles give electric charges to the paper. The image is developed one line at a time, in a similar manner to the raster-scan method described for graphics display. The *photoconductive-plate method* has the image developed on an internal CRT. The visible light from the CRT is accepted by a photoconductive plate, which transfers it into a static electric charge on the paper. The toner with carbon particles is applied and fused to the paper.

*Photoplotters* A *photoplotter*, sometimes called a *graphic film recorder*, is a camera that photographs

FIGURE 24–35 A beltbed plotter with a plotting area of 856 mm (33.7 in.) × 1518 mm (59.7 in.). (Courtesy of California Computer Products, Inc.)

FIGURE 24–36 An electrostatic plotter-printer that produces an image up to 894 mm (35.2 in.) wide. (Courtesy of California Computer Products, Inc.)

the image on an internal CRT. The CRT is a black-and-white, single-line raster-scan type. As each line is scanned, the film moves forward a very small distance until the entire image has been scanned and recorded on film. This system can be used for direct color photoplotting by using appropriate color filters.

*Modes of operation* Plotters can be operated on-line, off-line, or remote. *On-line operation* has the plotter connected directly to the host computer. It provides the fastest results because of the immediate response from the computer to which it is connected. The plotter usually has to be near the computer due to difficulties in connections from a distance.

An *off-line plotter* has the graphics data transferred from the computer to the plotter through some form of data storage device, such as magnetic tape, a hard disk pack, or a floppy diskette. The computer records the data to storage; they are physically carried to the plotter, which reads the stored data and produces the plot.

*Remote plotter* operation uses telecommunications to transfer the graphics data from the host computer to magnetic tape at the new location, or in some cases, it may go directly into the plotter. This enables data to be transferred rapidly over considerable distances. It also enables a company to run graphics programs when using a time-shared centralized computer. It is not possible to provide interactive response with a remote system, so its use in graphics is somewhat limited.

## Languages

*Language* is a set of rules followed by the user and the computer to communicate with each other. Computer graphics devices use languages generally developed from FORTRAN, though BASIC and AGL are used. Even though most of the languages used are developed from FORTRAN, they are not easily transferred from one computer graphics system to that made by another company. Many are proprietary languages tailored by a manufacturer to a specific system. Someone who has learned to operate a computer graphics system made by one company will have to have special instruction to learn to operate a system made by another company. These languages are given names by the various companies, such as PLOT-10 and GINO-F.

Language is part of the program (software) written to direct the computer and related graphics devices to produce an engineering drawing or perform calculations, analysis, or some other function.

The lines of a program used for computer-aided design will be one of the following:

1. Arithmetic: gives the computation to be completed.
2. Input/Output: controls the movement of data between the machine and the operator.
3. Control: specifies the flow of machine operations.
4. Declaration: gives descriptive information about programs.

## Software

*Software* is a term used to describe the computer programs used in computer-aided design drafting. Software for a simple drawing includes the information needed to generate a picture of the object. Areas of computer-aided design, simulation, or analysis require extensive software that associates the graphic data with additional facts or mathematical equations that define nonvisual characteristics of the object under consideration. The software therefore creates the data-based model, using operator input.

Graphics software provides the intelligence of the graphics system. It processes the commands and data input to produce displays for the designer and output of the finished design (such as drawings). The essential features of graphics software include points, lines, surfaces, data-base organization, and chaining commands. Points provide references for operations. Creation, deletion, movement, and extraction of points are part of the software program. Lines are entities shown on the display or on drawings. Creation, deletion, movement, extraction, trimming, and extending lines are part of the program. Surfaces such as ruled or free-form, with intersections of surfaces and projection of lines onto surfaces, are common parts of a software program. The software program will have some level of data-base organization. It will also use not only basic commands but chain a group of commands as a single command.

## BASIC PROGRAMMING

Following is a brief explanation of the general principles of programming. This explanation is based on the Tektronix 4051 computer graphics system. While some of the comments and procedures relate only to this system, the principles illustrated are applicable to other systems. The sample programs presented are based on the BASIC computer language.

### Entering the Program

The program is entered by typing the data on the keyboard. If the computer is not processing a program the input is displayed on the CRT. Each statement must have a line number preceding it to tell the computer the order in which the instructions are to be executed. When a statement with a line number is entered the computer accepts it. If line numbers are not used the computer considers this a command and will execute the operation immediately. For example, the operation PRINT 2*2 RETURN will produce an immediate answer of 4 on the screen; the program will be lost. When program line numbers are used, the program is retained in the computer and can be run many times.

Line numbers are usually entered with gaps of 5 to 10 units (100, 110, 120, etc.) to allow for program modification and flexibility. A simple series of instructions with line numbers follows. This set of instructions demonstrates statement structure and order of execution.

```
100 FOR I = 1 TO 10

110 PRINT ''COMPUTERS ARE FUN''

120 NEXT I

130 END
```

This program will print COMPUTERS ARE FUN ten times on the CRT and then return to a ready and waiting state. To see how the program does this, examine each line. Line 100 tells the computer to keep this instruction for later use and not to forget it. This is called a simple FOR statement and it tells the computer to do something 10 times in increments of 1. In other words, we want the computer to count the things we want done from 1 to 10. The letter I is the place in the memory where we want the computer to keep track of the count. Line 110 tells the computer to print something on the CRT and that it is to be the same information as is displayed in the statement between the quotation marks. After the computer prints this information it will move to line 120. Line 120 tells the computer to return to line 100 and increase I by 1. This process will continue until I = 10. After it prints the tenth line of COMPUTERS ARE FUN it will continue to line 130, which tells it that the program has been completed and it should cease execution until it receives additional instructions. The computer will signify that it has completed the program by showing a flashing cursor on the screen.

Some instructions to the computer are made with *reserved* or *key words*. The computer understands to do something special when a key word appears. The following are key words used by some computer systems (such as the Tektronix 4051).

*LIST.* Print program instructions on the CRT or plotter.
*PRINT.* Execute math operations or print information on the CRT or plotter.

*SAVE.* Store a program on tape or disk.

*MEM.* Show available space in computer memory.

*TLIST.* List the files stored on tape on the CRT.

*RUN.* Command to execute a program.

*AXIS.* Sets the number of spaces in the *x* and *y* direction.

*DRAW.* Drops plotter pen to table to draw a line.

*MOVE.* Directs the cursor to a new position.

*STEP.* Sets the increment between data points.

*READ.* Remove from memory needed data.

*DATA.* Numerical values to be used in a program.

*DIM.* Reserves a set of memory slots in the computer.

*REM.* Used to enter remarks.

*RDRAW.* A command to draw a line to a given coordinate.

*ROTATE.* Move a geometric shape to another position.

*PAGE.* Clear the CRT of all displayed information.

*HOME.* Move the cursor to the upper left hand corner of the screen.

*OLD.* Copy a program stored on tape into the computer memory.

*NEW.* Clear the memory of all stored data.

## Executing the Program

All computer programs written in BASIC are executed by typing the key word RUN and then depressing the execution key called RETURN. The computer will then execute the desired problem solution using the stored series of commands in the program.

## Applications

The following series of programs are examples of special commands used by computers to execute graphic solutions either on the CRT or a plotter. Each program will be described in detail to provide a basic understanding of special statements and their logic.

*Locate coordinates* Figure 24–37 is an example using a program to describe coordinate locations on the CRT or plotter. These will be used in later pro-

```
100  INIT
110  PRINT  ''ENTER MACHINE ADDRESS  32 CRT  2=PLOTTER''
120  INPUT P
130  PAGE
140  REM  *** 10X10 COORDINATES ***
150  REM  *** GRID PLOTTING ***
160  AXIS @P:10,10
170  AXIS @P:10,10,150,100
180  FOR I=10 TO 140 STEP 10
190  FOR J=10 TO 90 STEP 10
200  MOVE @P:I,J
210  DRAW @P:I,J
220  NEXT J
230  NEXT I
240  HOME
250  END
```

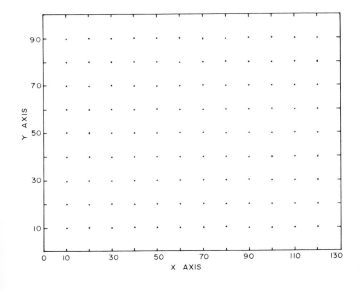

**FIGURE 24–37** This program establishes screen and plotter coordinate locations.

grams in this chapter. This program allows selection of the output device, CRT or plotter, in line 110. The computer stops at line 120 and allows the user to choose which output device is to be used. Line 130 clears the CRT for drawing.

There are several key words in this program: AXIS, MOVE, DRAW, HOME, STEP, and PAGE.

Lines 140 and 150 are ignored by the computer but are printed out to remind the user what the program is about. The AXIS command in lines 160 and 170 is used to set the number of spaces in the $x$ and $y$ directions of the grid, and set the size of the finished grid. Lines 180 and 190 determine the number of times this part of the loop will be executed to print the desired dots on the grid. The STEP function in lines 180 and 190 sets the increment within each pass of the loop for spacing the dots. They are 10 units apart in this example. The MOVE command in line 200 moves the cursor to a new position, while the DRAW statement causes the pen to lower, creating a dot at the desired coordinate location. Lines 220 and 230 repeat this process until the grid is complete. The HOME command in line 240 returns the cursor to the upper left corner of the screen, thereby indicating that the program execution will be complete upon the execution of line 250.

*Line construction* Figure 24–38 describes line construction on a CRT at a specified coordinate location. At line 125 the user is instructed to select the output device. This decision is indicated at line 127. Line 130 moves the cursor to a coordinate point of 30

units above and 30 units to the right of the lower left corner of the grid. Then line 140 instructs the computer to draw a line from the $x$, $y$ coordinate of (30, 30) to an $x$, $y$ coordinate of (30, 60). The movement is 30 units along the $y$ axis while the $x$ axis remains at 30. Line 150 raises the pen on the plotter and moves it to a coordinate location of $x = 70$ and $y = 30$. Line 160 tells the computer to draw a line from (70, 30) to $x = 100$, $y = 30$. Line 170 ends the execution of the program. Two dimensional objects, for example, a square, can be drawn by using these move and draw statements to provide the proper coordinates for the ends of each line.

*Drawing two dimensional objects* Figure 24–39 uses MOVE and DRAW statements to draw a square with sides 30 units long. The procedure is the same as in Fig. 24–38 except that the coordinates for four lines of equal length are needed. Lines 130 through 170 give the MOVE and DRAW commands and the coordinates for each line.

Figure 24–40 demonstrates a different approach to this graphic solution using DIM, DATA, and READ statements. Line 120, a DIM statement, reserves a set of memory storage slots for numeric $x$ and $y$ data. This is called an array and has four data slots for $x$ and four for $y$. The number of reserved slots is shown in parentheses. Line 130 shows a DATA statement that indicates the figures placed in these slots with the first four values being $x$ and the last four being $y$ values. The first $x$, 105, and the first $y$, 50, are the $x$ and $y$ values required for the coor-

```
100  REM  *** THE BASIC DRAW INSTRUCTION ***
120  REM  THIS PROGRAM ILLUSTRATES HOW TO DRAW A SIMPLE LINE
125  PRINT  ''SELECT PLOTTER = 2  CRT = 32''
127  INPUT M
130  MOVE @M:30,30
140  DRAW @M:30,60
150  MOVE @M:70,30
160  DRAW @M:100,30
170  END
```

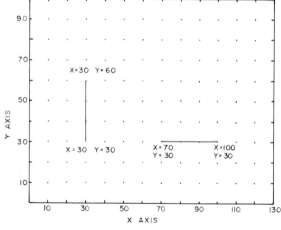

FIGURE 24–38  This program draws vertical and horizontal lines at specific coordinates on the grid.

```
100  REM *** THIS PROGRAM USES THE BASIC DRAW INSTRUCTION ***
120  REM *** WE WILL DRAW A SQUARE USING THIS INSTRUCTION ***
125  PRINT  ''SELECT PLOTTER = 2 CRT = 32''
127  INPUT M
130  MOVE @M:30,30
140  DRAW @M:60,30
150  DRAW @M:60,60
160  DRAW @M:30,60
170  DRAW @M:30,30
180  END
```

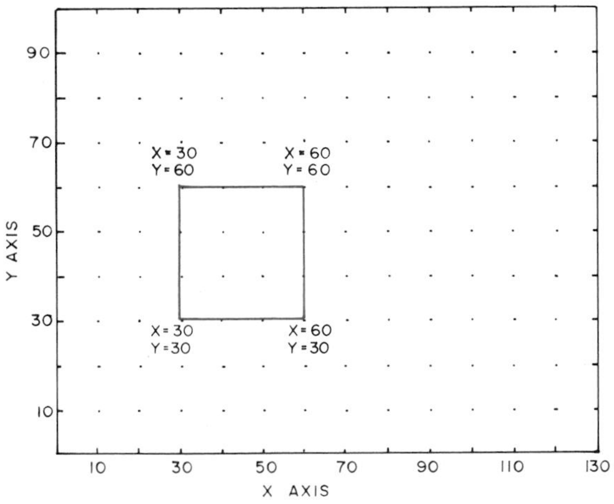

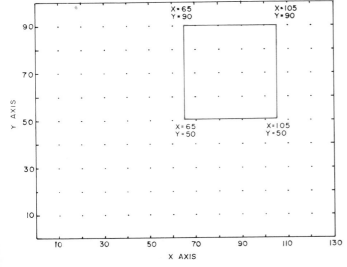

**FIGURE 24-39**  This program draws a square using MOVE and DRAW statements.

```
100  INIT
110  PAGE
120  DIM X{4}, Y{4}
130  DATA 105,105,65,65,50,90,90,50
140  PRINT ''SELECT PLOTTER=2   CRT=32''
150  INPUT M
160  READ X,Y
170  MOVE @M:65,50
180  DRAW @M:X,Y
190  HOME @M:
210  END
```

**FIGURE 24-40**  This program shows another way to draw two dimensional objects using DIM, DATA and READ statements.

```
100 INIT
105 REM *** THIS PROGRAM USES 'MOVE', 'RDRAW', AND 'REMOVE' COMMANDS ***
110 VIEWPORT 0,130,0,100
120 WINDOW 0,13,0,10
123 PRINT "SELECT PLOTTER = 2   CRT = 32"
125 INPUT M
130 MOVE @M:3,3
140 RDRAW @M:3,0
150 RDRAW @M:0,4
160 RDRAW @M:-3,0
170 RDRAW @M:0,-4
180 RMOVE @M:1.5,2
190 RDRAW @M:1.5,2
200 END
```

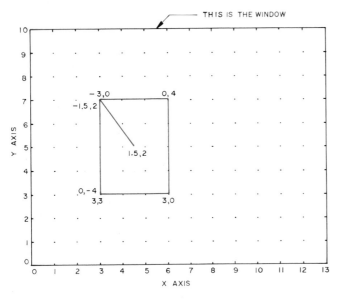

FIGURE 24–41 This program shows the use of WINDOW, VIEWPORT and RDRAM commands. The window is established by line 120.

dinates for the point to start the drawing process. The other coordinates are (105, 90), (65, 90), and (65, 50). These matched pairs of data must be pulled from the memory to be used for execution. This is done by line 160, a READ statement. After the data are read the cursor moves to the lower left corner of the square to be drawn as shown by line 170. Line 180 uses the data read to draw all four sides of the square. Line 190 returns the cursor to the upper left corner of the screen and the program ends with line 210.

*Using window, viewport, and rdraw commands*
Figure 24–41 uses WINDOW, VIEWPORT, and RDRAW commands. The VIEWPORT command controls the viewing area, which is $x = 130$ and $y = 100$ units in this program. The WINDOW command establishes a grid on the viewing area, which in this example is $x = 13$ and $y = 10$. The illustration shown is the window. Line 123 gives the user the choice of output on a CRT or plotter; line 125 is where this choice is indicated by the user. Line 130 moves the cursor over $3x$ units and up $3y$ units from the lower left corner of the grid. Line 140 is a

RDRAW statement that tells the computer to draw from this $xy$ of 3.3 to an $x$ of 3 more units to the right, but the $y$ is 0 so the line does not move vertically. In other words the line was drawn from (3, 3) to (6, 3). Line 150 tells the computer to move up on the $y$ axis 4 units, but to not vary the $x$. Pay particular attention to the minus value in line 160 as this is necessary to draw lines to the left, such as the top line of the square. Line 170 also has a minus value that is needed to draw down the $y$ axis to close the square. Line 180 moves the cursor to the center of the square by moving right an $x$ value of 1.5 and up a $y$ value of 2. Line 190 draws the diagonal line with a negative $1.5x$ and a positive $y$ of 2. The program is completed at line 200.

*Drawing circles*   Figure 24–42 will draw a circle of a specified radius and position. The beginning lines establish the type of output, the size of the window, and the viewport. Lines 170 through 200 allow the user to choose the $xy$ coordinate location for the center of the circle. Lines 210 and 220 establish the radius of the circle. Line 225 clears the CRT and line 230 moves the cursor to the center $x$ and $y$ coordi-

```
100  INIT
110  PRINT ''SELECT PLOTTER = 2 CRT = 32''
120  INPUT M
130  REM *** THIS PROGRAM DRAWS A CIRCLE OF SELECTABLE RADIUS ***
140  REM *** THE CIRCLE CENTER IS ALSO SELECTABLE ***
150  VIEWPOINT 0,100,0,100
160  WINDOW 0,10,0,10
170  PRINT ''SELECT X AXIS COORDINATE FOR CIRCLE CENTER''
180  INPUT X
190  PRINT ''SELECT Y AXIS COORDINATE FOR CIRCLE CENTER''
200  INPUT Y
210  PRINT  ''ENTER CIRCLE RADIUS {IN GRAPHIC DISPLAY UNITS}''
220  INPUT R
225  PAGE
230  MOVE @M:X,Y
240  SCALE 1,1
250  SET DEGREES
260  MOVE @M:R,0
270  FOR I-0 TO 360 STEP 5
280  DRAW @M:R*COS{I},R*SIN{I}
290  NEXT I
295  HOME @2:
310  PRINT @M:''X='',X
315  PRINT @M:''Y='',Y
320  PRINT @M:''R='',R
330  END
```

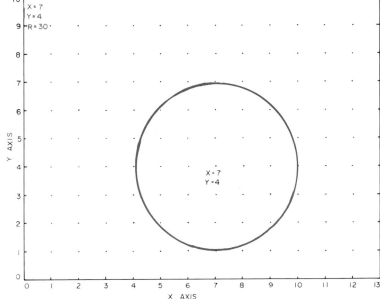

FIGURE 24-42 A circle is drawn using the computer to calculate the line segments.

nates. Line 240 sets the scale of the finished drawing while line 250 programs the computer to do the calculations in degrees. If this command is omitted this system will do the calculations in radians. Line 260 moves the cursor to the edge of the circle by the radius value. The computer calculates the line segment to be drawn (a truer circle can be drawn if many calculations are made); therefore line 270 sets up a loop providing calculations every 5° up to 360°. Line 280 draws the small segment of the circle that was

calculated. Line 290, NEXT I, tells the computer to return to line 270 and repeat the process until I = 360°. Line 295 returns the cursor to the upper left corner of the screen. Lines 310 and 320 print out the center *x* and *y* values and the radius used. This is not needed to draw the circle but does record the data involved.

*Drawing angles* Figure 24–43 draws angles using MOVE and DRAW statements as discussed in the

```
100  REM *** THIS PROGRAM DRAWS AN ANGLE USING DRAW INSTRUCTIONS ***
120  PRINT  ''SELECT PLOTTER = 2  CRT = 32''
130  INPUT M
140  MOVE @M:50,30
150  DRAW @M:110,30
160  MOVE @M:110,90
170  DRAW @M:50,30
180  END
```

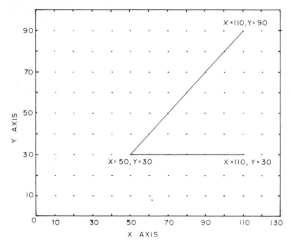

**FIGURE 24–43** This program draws an angle by locating the ends of the lines forming the angle.

earlier program examples. It simply draws two lines forming an angle. The number of degrees in the angle is not known because the lines are drawn using their $x$ and $y$ coordinates. Line 140 moves the cursor to the starting point, (50, 30). Line 150 draws the horizontal line. The pen is lifted and moved by line 160 to the top end of the inclined line. The angle is completed by line 170, and the program execution ceases at line 180.

Figure 24–44 is more versatile because the angle can be entered in degrees and the computer will complete the process with no input from the keyboard. The angle is computed using trigonometry and then plotted. Lines 195 and 200 do the trigonometric computations while side length and vertex position are handled by lines 230–260. The angle drawing is completed by lines 270–310. Lines 330–350 print on the drawing the numeric values used in the process of constructing the angle. This program could be stored as a subroutine and used when necessary to draw an angle on a drawing problem.

*Geometric figure rotation* Figure 24–45 introduces the ROTATE command used to move an object to many different positions. A computer generated design is shown using a triangle rotated through 355° in increments of 12°. Lines 130–180 allow the user to select the number of rotation degrees, the increment of rotation, and the device where it is to be printed. Line 190 clears the screen, and line 200 moves to the center or starting point. Line 220 sets up the condition for drawing the triangle the specified number of

times. Notice the rotation is done in line 230 inside the drawing loop, so that every time the object is drawn line 310 returns the computer to line 220 to draw the object again but rotated the desired increment from the last drawing. The completed drawing provides an interesting and complex looking computer generated design. Lines 330 to 350 print out the data used to create the design.

These programs are only a small set of examples illustrating the power of the computer in graphic presentation. The programs will vary from system to system.

*Pictorial drawings* Combinations of these examples can be used to draw pictorial illustrations. Extensive instructions are available in computer graphics books.

## COMPUTER OUTPUT MICROFILM (COM)

Computer output microfilm involves the use of microfilm processing units to store the output from a computer. The impact printers normally used for computer output do not keep up with the increased speed of the computer. Computer output microfilm is microfilm produced by a computer output microfilmer from computer-generated electrical signals. The microfilmer is a recorder that converts data from a computer into a human-readable language and records it on microfilm.

```
100  INIT
120  REM *** THIS PROGRAM DRAWS AN ANGLE USING DRAW INSTRUCTIONS ***
130  PRINT  ''SELECT PLOTTER=2  CRT=32''
140  INPUT M
145  SET DEGREES
150  PRINT  ''ENTER THE ANGLE IN DEGREES TO BE DRAWN''
160  INPUT D
170  PRINT  ''ENTER THE ADJACENT SIDE LENGTH IN GRAPHIC DISPLAY UNITS''
180  INPUT A
190  REM *** COMPUTE THE OPPOSITE SIDE LENGTH ***
195  T=TAN{D}
200  O=A*T
230  PRINT  ''ENTER X COORDINATE FOR ANGLE VERTEX''
240  INPUT X
250  PRINT  ''ENTER Y COORDINATE FOR ANGLE VERTEX''
260  INPUT Y
265  PAGE
270  MOVE @M:X,Y
280  DRAW @M:X+A,Y
290  MOVE @M:X+A,Y+O
300  MOVE @M:X,Y
310  DRAW @M:X+A,Y+O
325  MOVE @M:10,10
330  PRINT @M: ''ANGLE IN DEGREES = '';D
335  MOVE @M:10,6
340  PRINT @M:ADJACENT SIDE = '';A
345  MOVE @M:10,2
350  PRINT @M: ''OPPOSITE SIDE = '';O
360  END
```

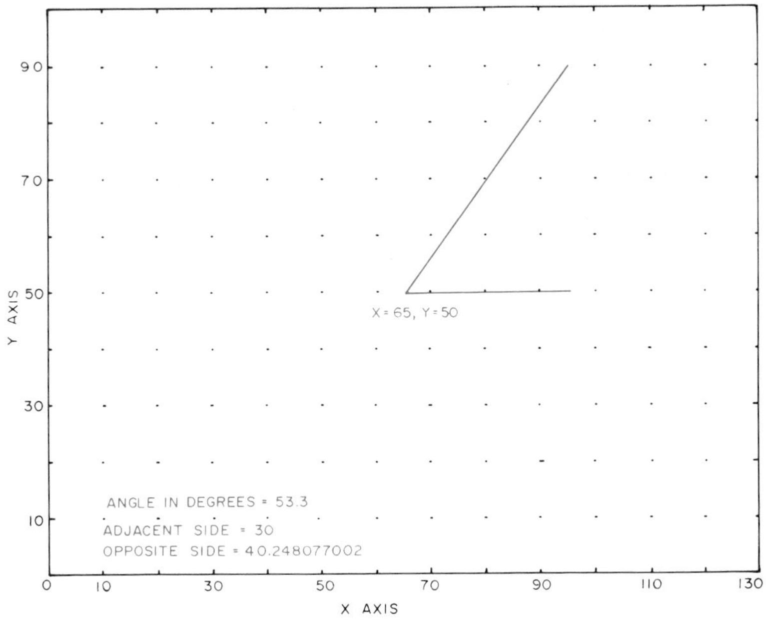

FIGURE 24–44 This program uses the number of degrees in the angle to plot its sides.

The COM recorder may be connected directly to the computer for on-line operation, or it may be operated by a magnetic tape unit as an off-line operation. The information from the computer is recorded on magnetic tape and is read by the COM recorder, which produces the microfilm.

There are three basic types of COM recorders—alphanumeric printers, alphanumeric printers and plotters, and special alphanumeric printers and plotters. For business operations, the *alphanumeric printer* is used to produce jobs such as parts catalogs, financial records, mailing lists, and inventories (Fig. 24–46). *Alphanumeric printers and plotters* combine the alphanumeric printing and graphic printing of a

```
100  INIT
110  PAGE
120  REM *** THIS PROGRAM ROTATES A GEOMETRIC FIGURE ***
130  PRINT  ''ENTER ROTATION INCREMENT IN DEGREES''
140  INPUT X
150  PRINT  ''ENTER THE NUMBER OF DEGREES OBJECT IS TO BE ROTATED''
160  INPUT D
170  PRINT  ''SELECT PLOTTER = 2 CRT = 32''
180  INPUT M
190  PAGE
200  MOVE @M:65,50
210  SET DEGREES
220  FOR I=0 TO D STEP X
230  ROTATE I
240  RDRAW @M:20,0
250  RDRAW @M:20,20
260  RDRAW @M:-40,-20
270  RDRAW @M:-20,20
280  RDRAW @M:40,-20
290  RDRAW @M:-10,30
300  RDRAW @M:-10,-30
310  NEXT I
320  HOME @M:
330  PRINT @M: ''NUMBER OF DEGREES PER ROTATION INCREMENT IS '';X
340  PRINT @M: ''THE NUMBER OF DEGREES OBJECT IS ROTATED IS '';D
350  PRINT @M: ''THE FIGURE ROTATED IS A TRIANGLE''
```

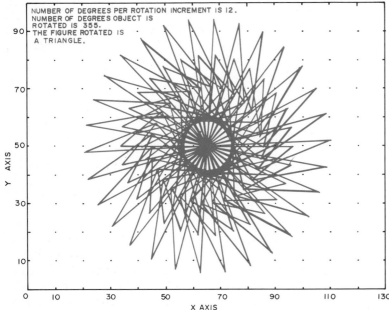

FIGURE 24–45 This program uses the ROTATE command to move a triangle to many different positions.

plotter to produce images requiring lines, vectors, and plots. These are heavily used in scientific, engineering, and business graphics (Fig. 24–47). *Special alphanumeric printers and plotters* produce alphanumerics and graphics of a very high quality. They are capable of exposing to microfilm any character that can be generated by a computer. Typical applications include circuit diagrams, three-dimensional drawings, and precise scientific data (Fig. 24–48).

A COM recorder is shown in Fig. 24–49. The *input section* can be a magnetic tape reader or electronic interface with a computer. The *logic section* interprets the code and subjects it to a logic conversion required to generate the required signals, such as setting the format. The *conversion section* converts the binary data from the logic section to the analog signals necessary to convert the data to human-readable alphanumeric or graphic information. The *deflection section* controls the positioning of the images on the CRT or microfilm. The *display section* converts the

| Code: 42-2-793-AB |||||
|---|---|---|---|
| **Inventory Report** ||||
| **Drafting Sets—Coleman Manufacturing Co.** ||||
| *STOCK NO.* | *QUANTITY* | *COST PER UNIT* | *TOTAL VALUE* |
| 107531 | 100 | 20.00 | 2000.00 |
| 177512 | 75 | 10.50 | 787.50 |
| 197731 | 300 | 19.75 | 5925.00 |
| 199754 | 151 | 15.25 | 2302.75 |
| 201371 | 20 | 40.00 | 800.00 |
| 213076 | 507 | 10.50 | 5323.50 |
| 231794 | 206 | 12.50 | 2575.00 |
| 251371 | 355 | 9.00 | 3195.00 |
| 270001 | 25 | 35.00 | 875.00 |
| 291071 | 6 | 50.00 | 300.00 |
| | | | 24083.25 |

FIGURE 24–46 Inventories, financial records, mailing lists, and other business records can be stored on computer output microfilm.

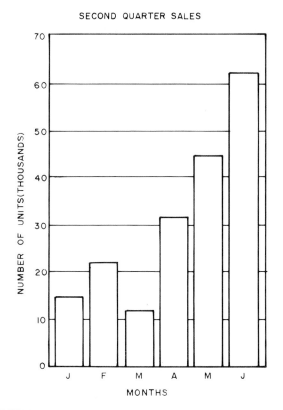

FIGURE 24–47 Business, scientific, and engineering graphics can be stored on computer output microfilm.

computer-generated data into human-readable form using cathode-ray-tube recording, electron-beam recording, light-emitting diodes, or laser-beam recording. The *film-handling section* consists of a lens and the exposure and film-handling mechanisms (Fig. 24–50).

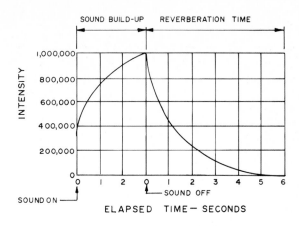

FIGURE 24–48 Scientific data and drawings can be stored on computer output microfilm.

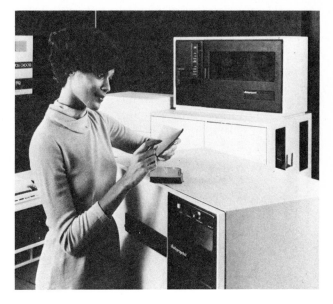

FIGURE 24–49 A COM recorder. (Courtesy of Datagraphix, Inc.)

COM systems use 16-mm, 35-mm, and 105-mm microfilm. It is packaged in 16-mm and 35-mm reels, cartridges, cassettes, and microfiche (Fig. 24–51). Aperture cards can be used to store one or more 35-mm microfilms. Jackets have channels that hold 16-mm or 35-mm film. Images may be copied directly from the jacket without removing the film (Fig. 24–52).

A summary of COM is shown in Fig. 24–53. The input is on-line .or off-line. On-line it goes directly to the COM recorder. Off-line data are recorded on magnetic tape, which goes to a tape drive that drives the COM recorder. The exposed film produced by the COM recorder goes to an automatic film processor that produces finished microfilm. At this point, a duplicator produces copies of the microfilm in the form of reels, cartridges, cassettes, mi-

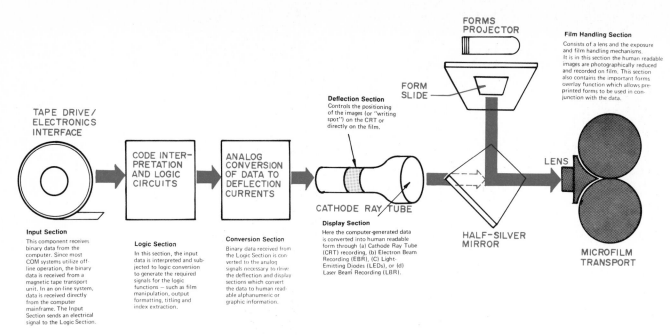

**FORMS PROJECTOR**

**Film Handling Section**
Consists of a lens and the exposure and film handling mechanisms. It is in this section the human readable images are photographically reduced and recorded on film. This section also contains the important forms overlay function which allows pre-printed forms to be used in conjunction with the data.

**TAPE DRIVE/ ELECTRONICS INTERFACE**

CODE INTERPRETATION AND LOGIC CIRCUITS

ANALOG CONVERSION OF DATA TO DEFLECTION CURRENTS

**FORM SLIDE**

**Deflection Section**
Controls the positioning of the images (or "writing spot") on the CRT or directly on the film.

**CATHODE RAY TUBE**

**LENS**

**HALF-SILVER MIRROR**

**MICROFILM TRANSPORT**

**Input Section**
This component receives binary data from the computer. Since most COM systems utilize off-line operation, the binary data is received from a magnetic tape transport unit. In an on-line system, data is received directly from the computer mainframe. The Input Section sends an electrical signal to the Logic Section.

**Logic Section**
In this section, the input data is interpreted and subjected to logic conversion to generate the required signals for the logic functions — such as film manipulation, output formatting, titling and index extraction.

**Conversion Section**
Binary data received from the Logic Section is converted to the analog signals necessary to drive the deflection and display sections which convert the data to human readable alphanumeric or graphic information.

**Display Section**
Here the computer-generated data is converted into human readable form through (a) Cathode Ray Tube (CRT) recording, (b) Electron Beam Recording (EBR), (C) Light-Emitting Diodes (LEDs), or (d) Laser Beam Recording (LBR).

FIGURE 24–50 A typical COM system. (Courtesy of National Micrographics Association.)

FIGURE 24–51 COM systems use microfilm packaged in reels, cartridges, cassettes, and as microfiche. (Courtesy of National Micrographics Association.)

crofiche, or microfilm in aperture cards. The image can be reproduced for use by a reader; hard copy can be produced by a reader-printer or an automatic retrieval system.

## COMPUTER-AIDED MANUFACTURING (CAM)

CAM is the integration of computers into the production process. CAM systems store, manipulate, retrieve, and display graphical information. They use data developed during the design phase to assist in the production of the product. The computers are connected to numerically controlled machine tools and control their operation. Therefore, the design data are used in the manufacturing process. CAM can use part numbers to produce a bill of material and keep track of other items such as the number of parts made, setup time, down time, and machine failures. It is used to control inventories and schedule work. A summary can be produced at the end of each shift, providing management with timely and much needed information.

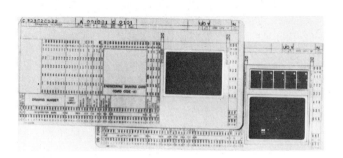

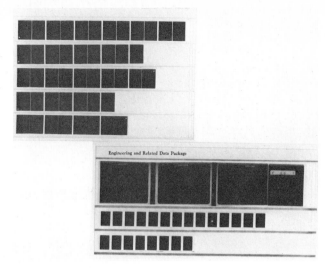

FIGURE 24–52 Microfilm is stored in jackets and aperture cards. (Courtesy of National Micrographics Association.)

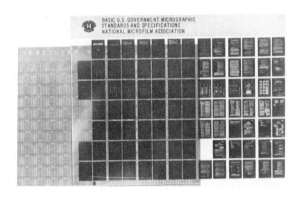

FIGURE 24–53 A flow diagram showing the steps in a COM system. The image is generated at the computer and retrieved on a reader or as hard copy. (Courtesy of National Micrographics Association.)

Simplified flow diagram of the steps in a COM program — from its origin at the computer to the image retrieved on a reader or reproduced in hardcopy form.

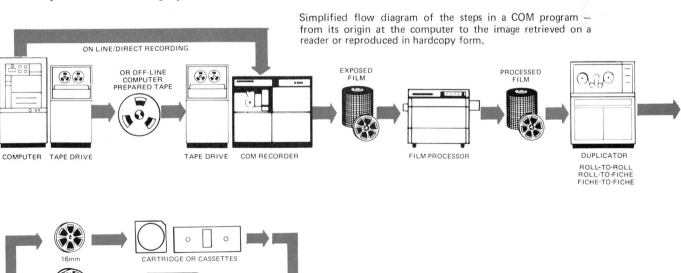

Computers in Engineering and Graphics    663

## Numerical Control
## and Computer Numerical Control

*Numerical control* (N/C) involves the control of a machine tool by use of a punched-paper tape or magnetic tape. The program is similar to a computer program. The tape contains the program that directs the machine to perform certain operations. The tape is read by the machine as it goes through the operations. For example, the tape can cause the machine to drill a series of holes of various diameters in precisely located positions on a part clamped to the table of the machine. This is performed in much the same way as if a human operator were doing the work, but the operations are performed without human intervention. This permits a series of operations to be performed rapidly and accurately, and they are the same on every part produced.

The most commonly used numerically controlled machine tools are mills, lathes, and drilling machines. There are many other adaptations to other machines and special-purpose machines. A two-axis machine is one that moves in a right-to-left (or X-axis) direction and a forward-to-backward (or Y-axis) direction. A three-axis machine moves in the X and Y axes as well as vertically (or Z axis) (Fig. 24–54). If a three-axis drill had a drilling head capable of tilting left to right and forward to backward, it would be a five-axis machine. This would enable the machine to drill a hole at any angle from vertical to horizontal at any location on the face of the part. Milling operations can also be performed on five-axis machines. The five-axis machine can produce parts that could not be machined before numerical control came into use (Fig. 24–55).

The program directs all of the operations to be performed. These include actions such as locating points, moving the cutting tool on the desired path, choosing the proper cutting tool for each operation from a turret (Fig. 24–56), setting the rotational speed of the tool, turning coolant on and off as needed, feeding various speeds, and stopping the machine when the operation is complete. The machining axis movements are driven by hydraulic pis-

FIGURE 24–54 A three-axis vertical machining center. The travel of the table across the spindle is the X axis. The travel of the spindle up and down on the column is the Z axis. The Y axis is the movement of the column carrying the spindle in and out on its own base. (Courtesy of Fritz Werner Machine Tool Corporation.)

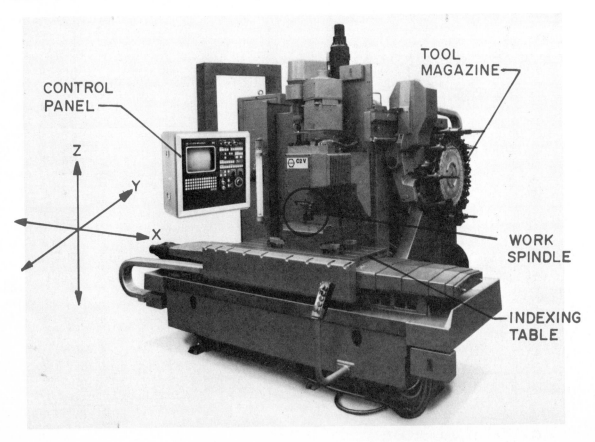

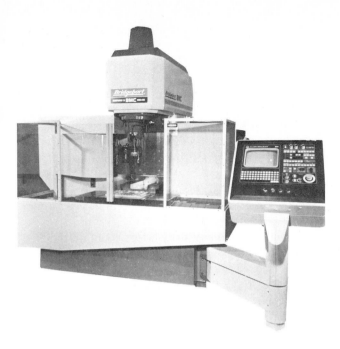

FIGURE 24–55 A five-axis machining center with an automatic tool changer that holds 20 tools up to 4 in. in diameter. It has a powerful CNC that uses distributed microprocessor technology where the executive is stored in nonvolatile bubble memory. It gives three-axis linear and two-axis circular interpolation in switchable planes. (Courtesy of Bridgeport Machines Division of Textron, Inc.)

FIGURE 24–56 The tool magazine. (Courtesy of Fritz Werner Machine Tool Corporation.)

tons or mechanical linkages that are activated by electric motors.

The writing of the program to produce the proper motions to shape the part as designed can become very complex. This plus the other information, as cutting speeds, complicate the problem. The use of a computer to handle the often large number of calculations is essential for complex parts. Several special languages for numerical control have been developed. These include APT, SPLIT, UNIAPT, and COMPACT. The program describes the geometry of the part in the language and, with the help of a computer, produces the tape to control the machine tool.

*Computer numerical control* (CNC) involves using a dedicated computer as part of a numerical control unit to input data to operate the machine tool (Fig. 24–57). The computer is applied to stand-alone or dedicated service to a single N/C machine. The CNC control is often very simple and may have storage for information on 15 m (50 ft) of tape or may have much greater capabilities.

A CNC system has memory storage that can run the tape and then put it in memory. The machine is then controlled by the computer memory, so it is not necessary to rerun the tape. On some units, it is possible to edit the tape right at the machine and enter corrections at that time. Using a special unit, the edited computer memory can produce a new correct tape.

N/C programs can be produced using a voice numerical control (VNC) system, a complete system used to generate N/C tapes via voice input from a parts programmer (Fig. 24–58).

FIGURE 24–57 This CNC lathe features quick tool changing, fast programming, a diagnostic system designed to prevent down time, and an input/output status panel made up of LEDs that monitor all kinds of machine functions and provide detailed information to pinpoint any malfunction. It has a maximum swing up to 533 mm (21 in.) and turning lengths up to 3048 mm (120 in.). (Courtesy of LeBlond Machine Tool Co.)

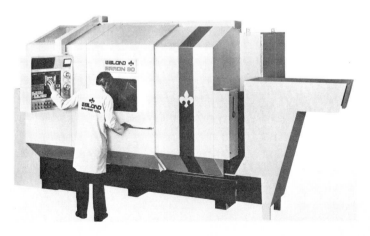

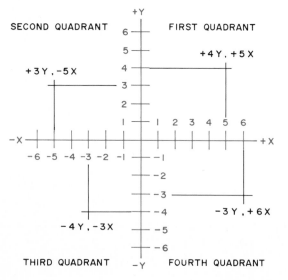

## Drafting for Numerical Control

The N/C programmer prepares the tapes from information given on engineering drawings. For these drawings to be most useful, they must be dimensioned using quadrants. The quadrant used in the cartesian coordinate system is a quarter of a circle. The N/C machine can be programmed in terms of four quadrants. The coordinates of features on the part must be programmed into the tape so that the machine can perform its work in relation to these quadrants (Fig. 24-59). The quadrants are numbered

FIGURE 24-59 The cartesian coordinate system.

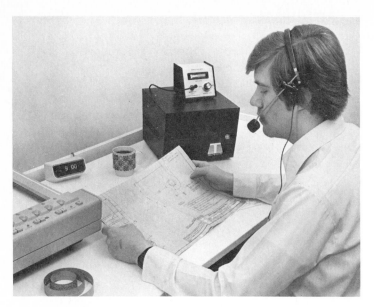

SCALE: EACH DIVISION = 1 mm

in a counterclockwise direction following the same system used in mathematics and drafting. The plus and minus signs indicate a direction from the zero point along the X and Y axis. The Z axis is indicated by a plus ($+$) if it is above the XY plane and a minus ($-$) if below the XY plane. These are used to indicate the direction of travel of a cutter on an N/C machine.

It is easiest if a drawing can be located totally in the first quadrant because all X and Y values are positive ($+$). Any of the four quadrants can be used, and the plus and minus signs become very important.

Confusion sometimes arises over the location of the axes. On a machine with a vertical spindle, such as a drill press, the Z axis is vertical through the spindle. On a machine with a horizontal spindle, such as a horizontal boring machine, the Z axis is horizontal (through the spindle). A line through the center of the spindle is used as the Z axis (Fig. 24-60).

FIGURE 24-60 The positions of the X, Y, and Z axes in relation to the machine spindle.

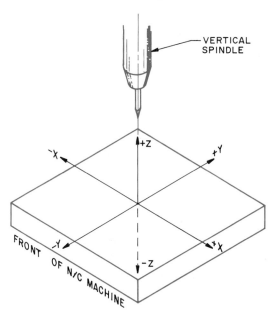

AXES WITH A VERTICAL SPINDLE.

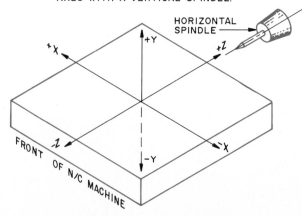

AXIS WITH A HORIZONTAL SPINDLE.

The *zero point* is the point where the X, Y, and Z axes intersect. It is from this point that all coordinate dimensions are taken. This point is often fixed by the manufacturer of the N/C machine, while on some machines it can be located as part of the program. The coordinate dimensions are always from the zero point. The dimensions from the zero point to the work piece must be given on the drawing. A typical detail drawing is shown in Fig. 24–61. It is dimensioned as is normal for a machine drawing, but the zero point is shown. To enable this to be programmed for use in N/C, it is redrawn using baseline or datum dimensioning with these related to the zero point (Fig. 24–62). This is a general example, and special considerations must be made depending on the operation (drilling, milling, punching, etc.) and the make of the N/C machine.

A *setup point* is a point on the work piece or the fixture to hold the piece as it is machined. It is often the intersection of two previously finished edges that form the baseline (datum). It could be the center of a hole already formed in the work piece. The setup point is accurately related to the zero point. The person setting up the machine knows where to locate the fixture on the machine table so that the machining operations on the tape will occur at the proper places on the work piece.

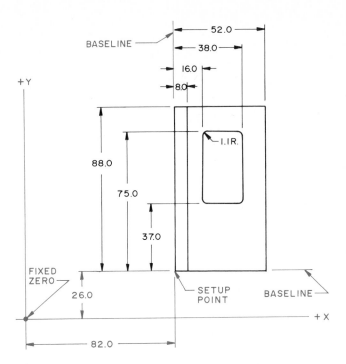

FIGURE 24–62   The part from Fig. 24–52 redrawn using baseline dimensions and relating the setup location to fixed zero.

FIGURE 24–61   A standard machine detail drawing with the fixed zero and setup point indicated.

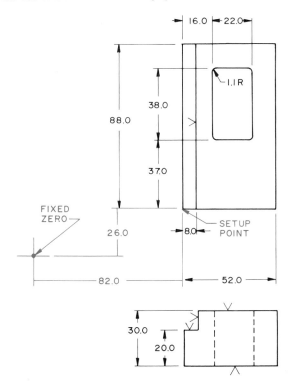

New developments in numerical control occur frequently, and the engineer should attempt to keep informed. These have a definite impact on drafting practices and production techniques.

Drawings for N/C can be used for other functions, such as quality control checks, inspection, and ordering materials. In some cases, it is possible to combine detail, assembly, and installation drawings into a single drawing. This requires that complete design information be given, such as tolerances, surface finish, and type of materials.

## GLOSSARY

*Address:*   The location in computer storage that contains a computer word.

*Algorithm:*   A special method of solving a certain kind of mathematical problem, as finding the greatest common divisor of two numbers.

*Alphanumeric:*   Numbers (0–9), alphabetic characters (A–Z), and special symbols (such as *) that are used to form computer words.

*Analog:*   Representing a variable using physical quantities such as electrical voltage or mechanical pressure.

*Analog computer:* An automatic computing device that operates in terms of continuous variations in some physical quantity such as electrical voltage or mechanical positions of parts. A major use of analog computers is to solve differential equations.

*Analog-to-digital converter:* Hardware that senses some physical quality, such as speed or force, and converts it into a numeric quantity that the computer can accept. Typical devices include digitizers, joysticks, trackballs, and dials.

*Arithmetic-logic unit:* The section of the central processing unit that performs the mathematical operations (addition, subtraction, etc.).

*Batch processing:* Running computer programs in groups or batches.

*Binary:* Of a base 2 number system that uses only 0s and 1s.

*Bit:* A single binary digit that can take the value of 0 or 1.

*Byte:* A group of adjacent bits that form one alphanumeric character.

*CAD (computer-aided design):* Using computers in the design process to speed the flow of information, store data, increase analytical capabilities, and speed the decision-making process in the creation or modification of a design.

*CAM (computer-aided manufacturing):* The utilization of computer technology in the management, control, and operations of a manufacturing facility.

*Circle generator:* An electronic device that is part of the display device (CRT). It receives the data supplied by the computer, such as the center-point coordinates and the numeric length of the radius, and calculates the coordinates for the points on the circumference of the circle.

*CNC (computer numerical control):* Using a dedicated computer within a numerical control unit to input data.

*COM (computer output microfilm):* The automated drawing is stored on microfilm by feeding the computer-produced signals through a COM recorder. This produces a microfilm for permanent storage of the drawing or report.

*Command:* The part of a computer program that specifies an operation, such as addition.

*Computer graphics:* A human-oriented system that uses the capabilities of a computer to create, modify, and display pictorial and symbolic data.

*Core:* The main storage area in a computer.

*CPU (central processing unit):* The nerve center of the computer system, consisting of the controlling circuits, main storage, and the arithmetic-logic unit. It receives information, acts according to commands, and either stores or displays the results.

*CRT (cathode-ray tube):* A display device resembling a television tube.

*Cursor:* A movable symbol, such as an arrow, that shows the current grid position on a CRT.

*Cycle:* The amount of time associated with a fundamental sequence of computer executions measured in terms of the number of beats from an internal crystal clock.

*Digital:* Representing information with discrete numerical values.

*Digital computer:* An automatic computing device that operates on discrete numerical data by performing arithmetic and logic manipulations.

*Digitize:* To transform a graphical shape into a digital signal so that it can be stored in the computer.

*Digitizer board:* A flat surface that is used to convert positional locations, such as points on a drawing, to *xy* coordinates.

*Disk:* A coated circular disk upon which information is recorded using a magnetic read-write head.

*Floppy disk:* A flexible disk used to store information from a computer.

*Hard copy:* A computer printout or drawings produced on a plotter.

*Hardware:* Computer equipment.

*Hexadecimal:* A numbering system using the base 16.

*IC (integrated circuit):* A solid-state device that contains several electronic circuits on one chip.

*I-CAD (interactive computer-aided design):* The designer works directly with the computer design system to design, draft, detail, and document a drawing, in two or three dimensions. The designer works at a digitizer station much the same as at a drafting board. Working at a computer terminal with a video display screen and a digitizing table, the designer moves a cursor from point to point on an engineer's drawing to pick up the details of the drawing and feed them into a computer. Drafting instructions are also typed into the system on a terminal keyboard.

Once the information is in the computer system, the designer can recall it onto the screen for updating and editing. The system is linked to plotters for producing hard copies. The information is stored on computer disks or tapes and can be used by others for reports and documentation.

*Intelligent terminal:* A terminal that has some computing and storage capabilities of its own.

*Interactive system:* A system that permits the user to receive output immediately and edit or change the original input.

*Interface:* Electronic circuitry that changes the output of one piece of hardware so that the signal can be received by another.

*I/O (input/output):* Information that is put into or received from a computer.

*Language:* A set of symbols, conventions, and rules used to communicate information.

*Library:* A collection of programs and subroutines in a form that is readily usable.

*Location:* An area of computer storage that is identified by an address.

*Loop:* A series of instructions that can be executed repetitively.

*Machine language:* A sequence of binary instructions that can be used directly by the machine. The instructions are in a form the machine can read directly and require no translation.

*Magnetic tape:* A plastic tape upon which data can be stored by means of magnetized spots on its surface.

*Megabyte:* One million bytes.

*Memory:* The storage in a computer.

*Menu:* A set of symbols on display, as on a CRT, that permits the operator to select the next operation when using an interactive system.

*Microprocessor:* The electronics of a minicomputer miniturized onto a single integrated-circuit (IC) chip.

*Microsecond:* One-millionth of a second.

*Minicomputer:* A small special-purpose computer.

*Monitor:* A program used to verify the correct operation of another program while the latter is being run.

*N/C (numerical control):* The control of machine tools using punched-paper tape.

*Off-line:* Not connected directly to the central processor.

*On-line:* Connected directly to the central processor.

*Operand:* A quantity upon which a mathematical operation is performed.

*Passive graphics:* Computer-controlled graphics produced by an off-line method.

*Peripheral:* Hardware that may be connected to a computer.

*Plotter:* An output device used to present data in graphic form, such as a line drawing.

*Program:* Instructions to a computer that enable it to perform a specified task.

*Programmer:* A person who sets out computer instructions and numbers in their correct sequence and allocates addresses to each instruction and each number.

*Program register:* The storage area in a computer that receives and stores the program instructions.

*Raster scan:* A scanning system on a CRT that uses a beam that moves horizontally across the screen locating all X values at each Y level. The beam moves down the screen (each Y level) until the entire screen has been scanned.

*Refresh CRT:* A CRT upon which the image fades and must be regenerated constantly so that the image appears steady.

*Register:* Temporary storage in a computer for control, arithmetic operations, and logic.

*Software:* Computer programs.

*Storage CRT:* A CRT that maintains a steady image until the image is erased.

*Subroutine:* A part of a program designed for a specific operation. Sometimes called a *subprogram.*

*Terminal:* A point in a system where data may be put into it or received from it.

*Translator:* A program that converts a program from one language to another language.

*VDE (voice data entry):* A system whereby the operator can input design information or commands using normal speech.

*Vector generator:* An electronic device that draws lines. The end-point coordinates are supplied by the computer to the vector generator, which converts them into $xy$ analog voltages that drive the deflection system in the CRT.

*Word:* A series of bytes operated on as a unit.

*Word length:* The number of bits in a word.

# Engineering Design Problems

The engineering design process was described in Chapter 2, and references to design have been made in the other chapters in this book. The principles of engineering graphics and descriptive geometry presented are constantly used as a product proceeds from an idea through the design process until it is ready for delivery to a user.

The *identification of the problem* is a difficult thing to do and in industry is often ambiguous or unclear. Research is needed to finally get the need or problem defined and parameters set so that it can be attacked by the designer. The problems in this chapter are designed for the academic setting and have a clearer delineation. They will not give the student the opportunity to work to find out what really is the problem. These problems are general in nature and permit a wide variation in potential solutions. As solutions are proposed, the student designer will often need to record decisions and assumptions made; these will be used to limit and further define the problem. Questions that are raised should be recorded. For example, questions such as the following are typical:

1. What is the maximum weight?
2. How much should it cost?
3. Where will it be used?

4. How will temperature affect the device?
5. What materials might be used?

In defining the problem, the student designer may have to study related devices or read in related areas to gain the knowledge needed to clearly prepare a statement. For example, if the problem involves some device to measure blood pressure, knowledge about blood, how it flows, how the heart works, how current blood pressure devices work, and other such information will need to be studied. The alert designer will want to know why it is desired to measure blood pressure. Perhaps there is some other device that can measure something else that will give better results with less difficulty and meet the original need for wanting to know the blood pressure.

As the student designer proceeds to collect *technical information,* a record should be kept of what references were consulted and if anything of value was found. This prevents going back to useless sources when you forget what information they held. Consider any source and also use the faculty as consultants. Their experience can be an invaluable resource.

As *ideas* are generated, keep a record of them. Even the most insignificant or far-out idea could be useful, not necessarily as the solution itself but as a device to trigger other ideas. If you do not write them down, you will forget them. Hold brainstorming sessions and talk with others about the project. You never know when someone, even a person unrelated to the project, will come up with a helpful idea.

As the *statement of the problem* becomes clear, review the qualities of the design and list the alternatives. The qualities might include weight, color, cost, shape, texture, and size. Suppose you have decided that the qualities of a newly designed lawn clipper will include a horizontal cutting action, battery operation, operation from a standing position,

and a selling price of under $50. Now review these to see if a 120-volt electric motor would be better. Could the market bear a $75 price tag? Would revolving cutters be more efficient? These and other questions will suggest ways in which the product might be improved.

Begin by preparing a written statement of the problem. Make it as complete as possible at the time it is written. Keep each factor short and easy to understand. Write a justification for the need to study this problem. Develop a list of *design requirements* to be met. These may be positive statements or questions to be resolved later. List the *limitations* to be imposed on the design. These should be quite specific and clearly written. These statements could be clarified with sketches, technical reports, or other supporting documents (Fig. 25–1).

In Fig. 25–1, the problem is stated as being to develop a lock for sliding glass patio doors. The justification cites an observation that many doors on the market have very poorly designed locks that are virtually useless in keeping out intruders. It is a simple problem that can be stated briefly and quite clearly. The scope is limited by providing a lock to metal-framed doors with metal tracks. Perhaps the lock developed would also work on wood sliding doors, but they have been excluded from consideration at this time. The design requirements show that cost and ease of installation are major considerations. The design requirements record all of the things the designer can cite about the project, such as cost, weight, material, and use. Since every project is limited in some way, these must be carefully noted. The acceptance of the solution depends on an understanding of how far the project goes and what it does not include.

Now, begin to identify and record the significant aspects of the problem. This includes reviewing the design requirements and limitations and securing factual data about them. The situation can be studied, and specific data can be collected and recorded. The problem begins to come into focus and is more clearly understood (Fig. 25–2). A second work sheet can be used to elaborate on the problem, on which known or possible answers are recorded. Suggestion for solutions or factors influencing solutions are noted. For example, it is pointed out that the glass in the door is likely to be broken unless special care is taken to avoid all contact with it. It suggests that materials be carefully considered and other locks examined for material use and mechanism design. The suggestion that a plunger or lever be considered is only a suggestion at this point.

Now that the problem is identified and defined, prepare a work plan. If it is an individual project, this will involve breaking the project down into units, arranging them in a workable sequence,

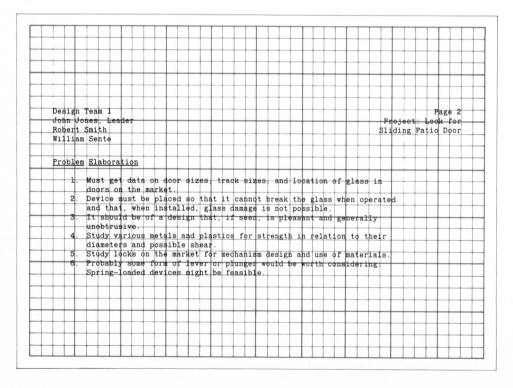

```
Design Team 1
John Jones, Leader                                  Project: Lock for
Robert Smith                                        Sliding Patio Door
William Sente

Statement of the Problem

Design a device to lock sliding glass patio doors from the inside.

Justification

Locks on existing doors are held by a small hook lever which can be easily
broken with a pry bar.  This type of door is a major source of intrusion into
homes, and
positive locking devices are needed.

Scope

This problem involves designing a device to lock metal-framed sliding patio
doors running in metal tracks.

Design Requirements
        1.  The device must be low in cost, probably selling retail for less than
            $3.00.
        2.  It must be easily installed by the average person.
        3.  It should be installed with common tools, such as screwdrivers,
            pliers, and wrenches.
        4.  It must enable the door to be easily opened from the inside yet must
            positively lock out intruders.

Limitation on Design
Applies to metal-framed doors sliding in metal tracks.
```

FIGURE 25-1   A statement of the problem.

and assigning time values to them. In this way, a projection can be made of the project completion date. If it is a team effort, the work units need not only a time schedule but should also be assigned to the various team members. Times for the team to meet to report progress should also be scheduled. The form for recording progress can vary, but some kind of progress chart is commonly used (Fig. 25–3).

FIGURE 25–2   Identify significant aspects of the problem.

```
Design Team 1                                                Page 2
John Jones, Leader                                  Project: Lock for
Robert Smith                                        Sliding Patio Door
William Sente

Problem Elaboration

        1.  Must get data on door sizes, track sizes, and location of glass in
            doors on the market.
        2.  Device must be placed so that it cannot break the glass when operated
            and that, when installed, glass damage is not possible.
        3.  It should be of a design that, if seen, is pleasant and generally
            unobtrusive.
        4.  Study various metals and plastics for strength in relation to their
            diameters and possible shear.
        5.  Study locks on the market for mechanism design and use of materials.
        6.  Probably some form of lever or plunger would be worth considering.
            Spring-loaded devices might be feasible.
```

**PROJECT SCHEDULE**

PROJECT: TOOTHBRUSH FOR THE HANDICAPPED
BEGINNING DATE: JAN 10    COMPLETION DATE: MAR 4
PROGRESS REPORTS: JAN 14, 21, 28, FEB 4, 11, 18, 28
FINAL REPORT: MARCH 4

DESIGN TEAM: #1
JONES, LEADER
SMITH
SENTE

**WEEKS**

| ASSIGNMENT | JAN 10 | JAN 17 | JAN 24 | JAN 31 | FEB 7 | FEB 14 | FEB 21 | FEB 28 |
|---|---|---|---|---|---|---|---|---|
| WRITE PROBLEM STATEMENT, LIMITATIONS, JUSTIFICATION | JONES | | | | | | | |
| GATHER AND PRESENT DATA | SMITH/SENTE | | | | | | | |
| BRAINSTORM | | | ALL | | | | | |
| PRESENT PRELIMINARY IDEAS | | | | ALL | | | | |
| REFINE PROBLEM, COMPLETE ANALYSIS, MAKE DECISION, GIVE A REPORT | | | | | ALL | | | |
| PREPARE WORKING DRAWINGS | | | | | | SMITH/SENTE | | |
| MAKE FINAL COST ANALYSIS | | | | | | | JONES | |
| MAKE PROTOTYPE AND TEST | | | | | | | JONES | |
| PREPARE FINAL REPORT, REPORT SOLUTION TO CLASS | | | | | | | | ALL |

FIGURE 25–3  Prepare a schedule.

If you were working for a client, the client would expect periodic progress reports.

If the project begins to require funds for such things as telephone calls, postage, and materials for model construction, some agreement needs to be reached and recorded as to how these expenses will be met. One way is for the team to agree to share costs equally regardless of which member incurs the expense. No expenses should be incurred without the prior approval of the team leader. If the project appears to be becoming excessively expensive, the class instructor should be consulted before proceeding. Sometimes, industrial support can be made available through the college or university.

## WORKING ALONE

When a problem is handled by an individual, that person assumes responsibility for planning the project and following the plan. He or she must perform the activity scheduled for each step. Such an experience gives, the person the opportunity to experi-

ence each step of the design process and to proceed without having to coordinate activities and secure input from others. On a complex problem, an individual is at a disadvantage, because some of the factors arising may be beyond the person's educational preparation or experience. It is possible for an individual to overlook possible solutions or relationships that may be obvious to someone else.

## WORKING ON A TEAM

Most industrial design projects use the team approach because of the complexity of the problems and a need for people with a variety of educational backgrounds and industrial experiences. This introduces the student to a new dimension in the design process, that of interaction with others on a common venture. Individuals will learn that they must overcome tendencies to dominate the situation and learn to accept ideas from others yet be able to present their ideas clearly and concisely and subjugate their individual desires to the good of the project.

A team must have a leader. This person can be appointed by the instructor, or the team can select one of its own. The leader must function or the team will most likely fail. If a leader fails to function, the team or the instructor should appoint a replacement.

The size of the team depends somewhat on the scope of the project. Seldom should a student team be larger than three or four persons. If it is larger, it becomes increasingly difficult for team members to arrange times to meet and work as a team. Larger teams also limit the amount of participation available for each student and thus reduce the learning experience.

Members of the team should be selected by the instructor. Often, friends want to group themselves into a team. This reduces the value of the experience because they need to learn to work with anyone who may be assigned to a project. In industry, those assigned have a contribution to make and may not be acquainted with the other team members. The team members could be selected by a random drawing of names, if desired.

## ASSIGNING A PROBLEM

One factor to be considered when assigning design problems is the amount of time available and the number of projects a student is expected to experience. Some instructors want each student to solve a rather simple problem and carry the entire project individually. Thereafter, the class can be divided into teams and approach more complex projects. In this way, the experience of each student would be broader. In some cases, students are given an entire semester to work on a rather large problem that becomes a senior design project. In any case, the methods for assigning problems can vary.

The instructor could assign a problem that each person solves individually. The class can then see a variety of solutions to the same problem. Team projects could also be assigned by the instructor. This is more like industry, because the team is given a problem the company wants solved.

The problem could be selected by students from several proposed by the instructor. Those in this chapter could serve in this way.

Finally, the students, individually or as teams, could be required to find a problem, define it, and seek approval of the instructor. The class could have a brainstorming session and develop a list of possible problems from which to choose. Again, the problem selected should be carefully examined to see if it is within the ability of the class, if facilities are available, if the cost is reasonable, and if it can be accomplished in the time available.

## PRELIMINARY IDEAS

After the problem has been described and the research assignments made, the preliminary design ideas are recorded. This record can consist of a series of freehand drawings and accompanying notes. Usually, several proposals are recorded for consideration. These become part of the file being accumulated on the project. Some partial design sheets are shown in Fig. 25–4. The number of sheets will vary with the complexity of the problem. The sketches should be made in a professional manner and dimensioned as needed to describe the solution.

## REFINING THE DESIGN

The preliminary design proposals are examined, changed, and restudied, and the solution with the greatest potential is selected for additional development. Several proposals are often carried on to the refinement stage before the final one is selected. This could be followed in the classroom if time is available. These drawings are made with instruments to scale, with the dimensions and notes needed to continue project assessment and the eventual production from them of finished working drawings. They usually are assemblies of the product with accompanying details, such as sections or developments. Several ideas may be carried into this refinement phase or may be even combined into a single solution. The drawings use all the types of engineering drawings discussed in this text, such as orthographic projection, sections, auxiliary views, and revolutions. Materials to be used are noted, as are the necessary size descriptions (Fig. 25–5). After these are completed, the refined proposed solution is carefully analyzed before the preparation of the final working drawings. Refinement process drawings may be of any appropriate size, depending on the material being developed.

## ANALYSIS OF THE DESIGN

Now the students must begin one of the most interesting and challenging parts of the design project, analysis of the proposed design. They must apply known conditions to the design to see if it measures up to the need. This could include building a model or prototype, a mathematical analysis, or actual testing of certain materials to see if they will fill the need. The anticipated cost must be developed and a decision made as to whether it is realistic or should be reduced (Fig. 25–6).

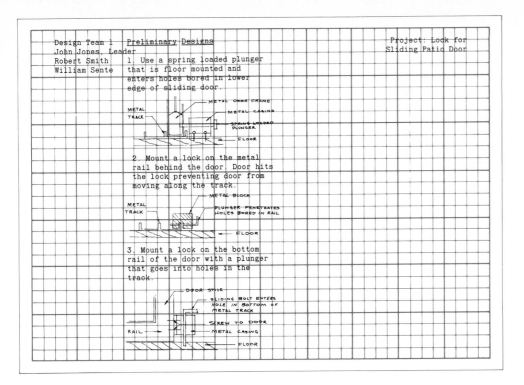

FIGURE 25–4  Preliminary designs.

Design Team 1
John Jones, Leader
Robert Smith
William Sente

Project: Lock for
Sliding Patio Door

REFINEMENT OF THE DESIGN

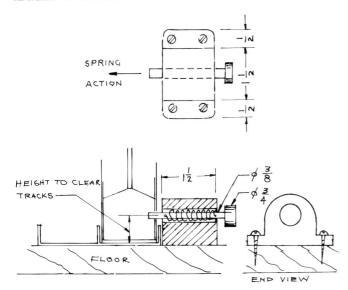

FIGURE 25–5  Refine the design.

## FINALIZING THE DESIGN

The final proposed design will have to be presented for acceptance. The student designers should prepare a presentation of their design in much the same way as a design is presented in industry. They should prepare charts, visuals, drawings, and a carefully written script for an oral presentation. This

FIGURE 25–6  An analysis might include models and prototypes.

could be made to the instructor and the class. The class can react, and the instructor can reject the proposal and explain why or indicate acceptance and give approval to proceed with the project.

## THE ENGINEERING DRAWINGS

The final working drawings can now be prepared, using the graphical methods explained in this book. National standards, proper drafting procedures, tolerancing and dimensioning practices, and other requirements for working drawings should be adhered to closely. If possible, some or all of the drawings could be plotted and drawn using computer graphics techniques. The final drawings should all be on the same size paper, with a standard border and title block.

## THE FINAL REPORT

The final report is written after a project has been completed. Although the organization of this report will vary depending on company practices, it could include the following:

> Title page
> Abstract
> Table of contents
> List of figures
> List of tables
> Body of the report
>> Statement of the problem
>> Research methods used
>> Proposals that were discarded
>> Proposal selected
> Technical analysis
>> Physical analysis
>> Engineering analysis
>> Human engineering analysis
>> Performance analysis
>> Environmental analysis
> Findings
> Conclusions
> Bibliography
> Appendix

The *title page* includes the title of the report, the name or names of the persons and company involved in the study, and the reporting date. The abstract is written concisely and gives in one page or less a summary of the entire project. The table of contents lists the name of each section and gives the page number. Since most technical reports rely heavily on *charts*, *graphs*, and *tables* for reporting

data, these must be numbered and identified following the table of contents. The body of the report includes the *statement of the problem*, which presents the problem under study as clearly as possible. The statement may present the actions that led to the isolation and study of the problem. If others have worked on related problems and reported their research, this could be reported to help the reader understand the current situation as it pertains to the problem. The *methods of research* section explains the procedures used in developing solutions and how these were tested or analyzed. This gives the reader an understanding of how the data were generated and enables a judgment to be made on the value of the study. If poor or questionable techniques are used, the report will be suspect. Next, report on the design proposals that were discarded and go into detail about the one selected, including the working drawings. The results of the technical analyses are reported and usually make extensive use of charts and tables for presenting data. The *findings* are a report of the factual outcomes of the study. They do not include opinions or subjective material but are a clear-cut statement of what was actually learned. The *conclusions* are statements based on the findings. They sum up the findings and cite the decisions that can be made after careful deliberation of the facts. For example, a finding would be that three competitive products sell for $3.50, $3.75, and $4.15. A conclusion would be that since your product will sell for $3.95, it is competitive with existing products. The *bibliography* is a list of publications used. They are often listed alphabetically and are organized in groups such as books, periodicals, and government reports. A typical entry for a book includes, in order, the author, title, publisher, date, and page references. Here is an example:

> Jones, Robert C. *Theory of Semiconductor Electronics.* Chicago: Ward Publishing Co., 1989, pp. 99–135.

Periodical entries include the author, title of the article, title of the periodical, volume and number, date of publication, and page references. Here is an example:

> Sites, James R. "Recent Research in Semiconductor Electronics." *Modern Electronics* 1:3 (1990), pp. 35–37.

## INDIVIDUAL DESIGN PROBLEMS

The problems that follow are rather limited in scope and could be useful as individual research problems. It is suggested students prepare carefully made reports as explained in this chapter and in Chapter 2. The extent of the project will depend on the instructor. It could logically end after the student makes a final design proposal but has not prepared final working drawings. If time is available, it would be of benefit to prepare carefully dimensioned drawings. The writing of the final report is a logical conclusion to the experience if time is available.

*Design project 1. Travel lamp* Design a small lamp that can be packed in a suitcase and used in hotel and motel rooms for reading. It should be able to be mounted for reading in bed, at a table, or in a chair.

*Design project 2. Shoe rack* Design a rack to store shoes. It should be possible to remove any pair of shoes from the rack without moving shoes above, below, or next to them. It could stand on the floor, hang on the wall, or be placed in any other position in a clothes closet (Fig. 25–7).

*Design project 3. Tie rack* Design a tie rack to hold 15 ties. It should be possible to remove any tie without removing other ties. The rack should hold the ties so that they do not constantly fall on the floor (Fig. 25–8).

*Design project 4. Ski carrier* Design a carrier to hold water skis outside the passenger area of a boat. Skis are long and always in the way inside the boat.

FIGURE 25–7  Design of a better shoe rack.

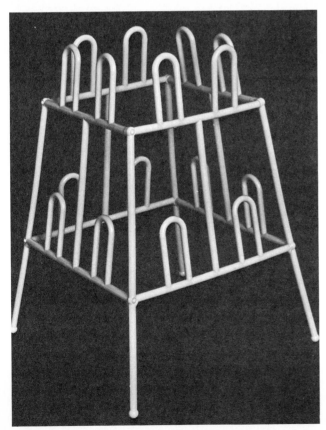

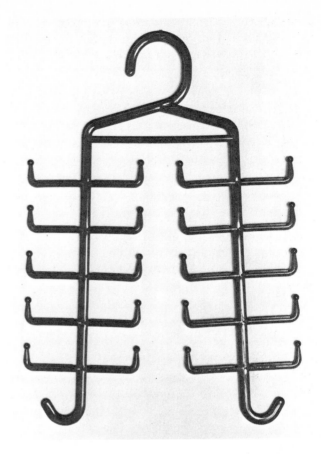

FIGURE 25–8  One possible tie rack.

Since they are lightweight, they could be carried outside the boat. The carrier should be adjustable to fit many brands of boats.

*Design project 5. Fan carrier* Design a carrier to suspend the fan in Fig. 25–9 from the ceiling. The

FIGURE 25–9  Design a system to hang this fan.

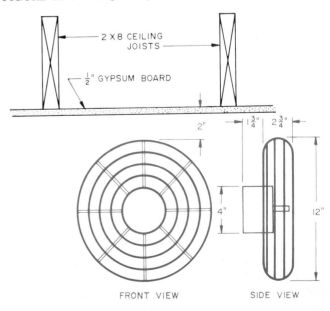

FRONT VIEW          SIDE VIEW

Engineering Design Problems  **677**

carrier should permit the fan to rotate 360° and be held tightly at various angles to the floor. The wire guard can be redesigned if necessary.

*Design project 6. Carton*  Design a corrugated-board carton to store and protect from damage glass products measuring 3 in. in diameter by 5 in. tall. It should hold 10 of these objects. Draw the developments for each piece of the carton.

*Design project 7. Screw holder*  Design a device that can be used on a screwdriver to hold a screw on the tip so that it can be driven in or removed from places your hand cannot reach.

*Design project 8. Auxiliary table*  Design a collapsible, lightweight table, standing 36 in. high and with a top 12 × 18 in., that when folded occupies a space 2 × 12 × 18 in.

*Design project 9. Doorstop*  Design a doorstop that will keep a house door from hitting the wall yet hold the door back against the wall so that the wind will not slam it closed.

*Design project 10. Horizontal thrust lever*  Design a lever and, if necessary, a linkage system that will enable lever A to push horizontally the piston in hydraulic cylinder B (Fig. 25–10). The cylinder should remain fixed and not have to twist or turn when the lever moves the piston. The length, diameter, and location of the lever can be changed. The piston must travel 35 mm horizontally.

*Design project 11. Toothbrush*  Design a toothbrush to be used by persons who have little finger movement and no gripping ability in their hands.

*Design project 12. Pipe carrier*  Design a pipe carrier that will suspend pipe 2 to 3 in. in diameter 3 to

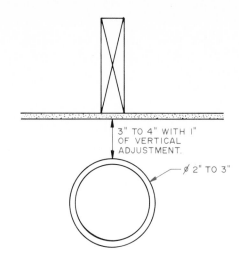

FIGURE 25–11  Design a pipe carrier.

4 in. from the ceiling and permit 1 in. of vertical adjustment in the pipe (Fig. 25–11).

*Design project 13. Porch railing kit*  It is desired to market a wooden porch railing kit as shown in Fig. 25–12. Design fastening devices to help the unskilled homeowner assemble the railing at the points circled.

*Design project 14. Center locater tool*  Design a tool to be used to find the center of cylindrical objects up to 6 in. in diameter.

*Design project 15. Angle layout tool*  Design a tool to be used by carpenters to lay out angles from 0° to 90°.

*Design project 16. Cabinet lock*  Design a mechanism to move the bolt horizontally to lock a cabinet

FIGURE 25–12  This product needs a device that permits easy assembly.

FIGURE 25–10  Design the linkage to move the piston.

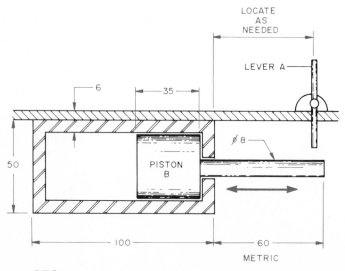

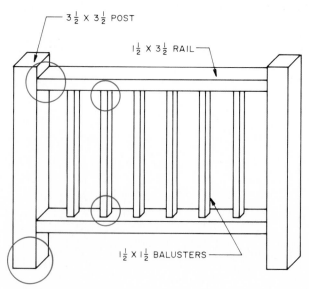

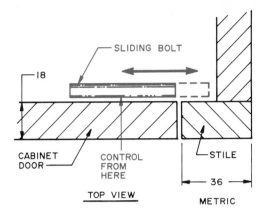

FIGURE 25-13  A slide-bolt cabinet-lock design.

door. The bolt is on the inside of the cabinet. It should be able to be moved from the outside of the door. The bolt can be round, square, or rectangular. It should move horizontally 18 mm (Fig. 25-13).

*Design project 17. Tamperproof bottle*  Design a tamperproof bottle for products such as aspirin and cold tablets.

*Design project 18. Feeding mechanism*  Figure 25-14 shows a loading chute on a machine that carries ball bearings to a finishing operation. Design a mechanism that will release one ball at a time, hold the others back for a predetermined number of seconds, and then release another ball. You may alter the chute, perforate it, or change it in any way needed.

*Design project 19. Measuring device*  Design an automatic device that will measure and drop into a container 2 pt of granular feed from a hopper. The more pints per hour that can be dropped, the better the solution (Fig. 25-15).

*Design project 20. Self-loading gripper*  Design a hook or gripper device to be used on the end of a cable from a crane that will close around the object

FIGURE 25-14  Can you devise an automatic feeding system for this ball bearing–loading machine?

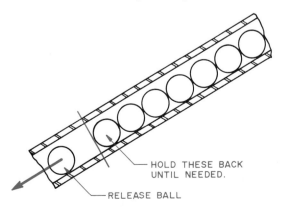

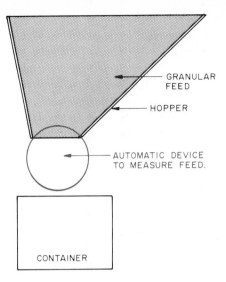

FIGURE 25-15  This hopper should release premeasured quantities.

to be lifted, as logs, when the crane begins to lift the hook.

*Design project 21. One-way valve*  Design a valve that will permit liquid to flow in one direction but will close to prevent liquid from flowing in the opposite direction.

*Design project 22. Soap dispenser*  Design a liquid-soap dispenser that will distribute a single teaspoonful of soap each time it is activated.

*Design project 23. Welding clamp*  Design a clamp that can be fastened to a welding bench and used to hold a part being welded. It should be a fast-action clamp that can be tightened and loosened with one hand in just a few seconds.

*Design project 24. Heat salvage device*  Gas- and oil-fired furnaces lose considerable heat up the chimney. If the furnace has a metal vent pipe to the chimney, it gets hot. Design a device that will salvage some of that heat and distribute it into the room or into the cold-air-return duct found on hot-air heating systems.

*Design project 25. Angle plate*  Design a pivot and positive indexing system so the movable bar will stop on 0°, 45°, 90°, 135°, and 180° and be locked in that position (Fig. 25-16).

*Design project 26. Drafting triangle*  Design a drafting triangle that will permit you to directly lay out 15°, 30°, 45°, 60°, 75°, and 90° angles.

*Design project 27. Indexing mechanism*  Design an indexing mechanism for the shaft in Fig. 25-17. The shaft should have a stop every 45°. Design the stop so that it will release after a certain torque is applied and will set itself at the next indexing point.

Engineering Design Problems  **679**

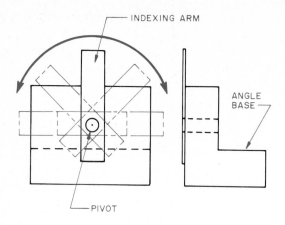

FIGURE 25–16 This indexing arm needs a positive indexing system.

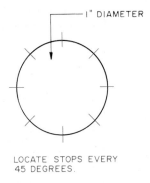

FIGURE 25–17 Prepare a device to index this shaft every 45°.

You may add something on the shaft and build apparatus off the shaft.

*Design project 28. Clamping mechanism* Design a mechanism that will raise and lower the jaw of the clamp in Fig. 25–18. The jaw should move vertically at least 8 mm and have a design that will permit it to clamp thin work to the table. Alter the base in any

FIGURE 25–18 Design a mechanism to raise and lower this clamp.

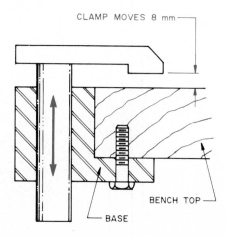

way necessary to produce the action and clamping force.

*Design project 29. Grinder hanger* Design a bracket to hold a tool grinder on a wall. The bracket must be able to carry the weight of the machine and resist downward pressures exerted during the grinding operation (Fig. 25–19).

*Design project 30. Parallel jaws* Design a mechanism to make the jaws in Fig. 25–20 both move toward or away from each other with a single action on the part of the operator.

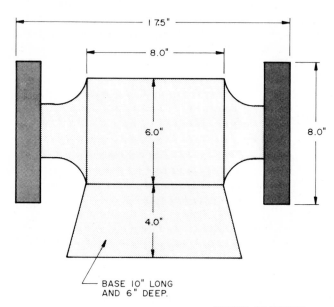

FIGURE 25–19 This grinder is to be mounted on a wall-hung bracket.

FIGURE 25–20 Design a mechanism to open and close these clamp jaws.

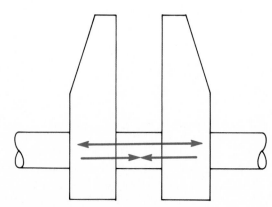

# COMPREHENSIVE PROBLEMS

These problems are rather comprehensive and might be better solved using a team approach. They involve two basic types of problems, product development and systems development. They could be carried through the final proposal stage, working drawing stage, or all the way through the final report. The construction of a model or prototype is helpful in evaluating the proposed solution.

## Product Development Problems

*Design project 31. Camping stove*  The stove will burn propane and heat two containers at the same time. It should be able to be folded into a small, compact unit. Safety from fire and explosion are vital considerations. Consider using some form of remote ignition so that matches will not be needed.

*Design project 32. Portable boat*  Design a 12-ft boat that can be collapsed for storage and fit in an automobile trunk, yet quickly be assembled with a minimum of tools. Do not use an inflatable type. It will be powered with oars and seat four people.

*Design project 33. Precast panel fasteners*  Design a method for fastening precast concrete panels to the exterior walls of commercial buildings. Alterations can be made in the exterior surface of the building as it is being built to accommodate the fasteners (Fig. 25–21).

FIGURE 25–21 This precast concrete panel must be firmly mounted on the wall.

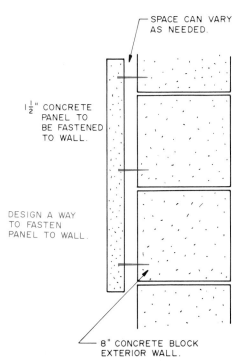

*Design project 34. Jogging exerciser*  Design a device that will let you jog indoors. It should reduce the common injuries suffered by joggers from running on concrete and asphalt. Look into these injuries and see what actions are recommended by research in this area.

*Design project 35. Bleachers*  Design small bleachers that will be four rows high and seat 40 people per section. Ease of erection and speedy dismantling are essential. Try to design it to have as few pieces as possible and no small, loose parts that can get lost. Be certain to test the design for load-carrying capacity.

*Design project 36. Wheelbarrow*  Design a wheelbarrow-type earth-moving device that will not get easily stuck at muddy building sites. It should carry about 4 ft³ of soil and be balanced so that one person can handle it.

*Design project 37. Lifting device*  Design a device that can be operated with one hand that will enable a person in a wheelchair to pick up items, such as clothing or a newspaper, from the floor.

*Design project 38. Trash can*  Design a trash can with a lid that is easy to open but will not blow off in a windstorm.

*Design project 39. Gripping tool*  Figure 25–22 shows a gripping type of pliers that uses a lever and a cam to provide gripping action. Design a similar tool to compete with this on the market that uses a different principle.

*Design project 40. Modelworker's vise*  In Fig. 25–23 is a typical modelworker's vise. It is adjusted by turning a long screw, which is time-consuming. Redesign the product so that it can be opened,

FIGURE 25–22 Can you design a better gripping tool?

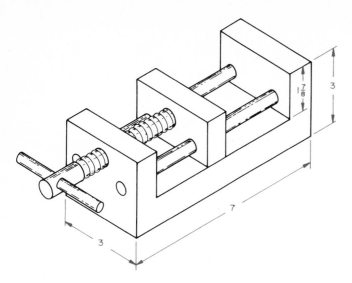

FIGURE 25–23 This vise needs a rapid-action closing and opening mechanism.

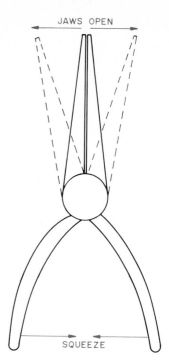

FIGURE 25–24 Expanding pliers require a unique design solution.

closed, and locked rapidly, such as by using a simple 90° turn or with a short throw-lever action.

*Design project 41. Water dispenser*  Design a device that can be used to keep a pet's water dish filled. Keep the cost as low as possible so that it can be sold in large quantities by pet stores.

*Design project 42. Portable shower*  Design a portable shower stall that can be taken on camping trips to ensure privacy while bathing. It must be lightweight and can use a local well, lake, or river as a source of water.

*Design project 43. Workbench*  Design a workbench that can be folded up and moved around on a construction site. The top can be 2 in. thick, 24 in. wide, and 72 in. long. It must be able to withstand hard use and support 400 lb.

*Design project 44. Expanding pliers*  Design a pair of pliers that will exert an expanding force when the handles are squeezed together. Allow the jaws to expand 2 in. from the closed position (Fig. 25–24).

*Design project 45. Solar collector*  Design an air-to-air solar collector that could be mounted in a window much like a window air conditioner and provide some heat for one room. A low-velocity blower can be used to move the air. It could be controlled by a thermostat inside the solar collector.

*Design project 46. Transmission jack*  Design a jack that can be used to remove and install automotive transmissions. Consider the weight of transmissions and the height they must be raised. The transmission should be able to be moved in the shop while on the jack. The cradle should be designed to fit most transmissions.

*Design project 47. Auto parts cleaning tank*  Design a tank containing a solvent used to clean greasy parts removed from automobiles. Include a way to keep the solvent moving in the tank and a system for filtering out the sludge.

*Design project 48. Metal punch*  Design a punch that will produce holes in metal up to 18 gauge thick. Make the punch so that it can be adjusted to produce round, triangular, and square holes.

*Design project 49. Band-saw guard*  Design a guard to cover the exposed portion of a band-saw blade in the area immediately above the table (Fig. 25–25).

*Design project 50. Nose protector*  Design a strong, lightweight device to protect the broken nose of an athlete. It must be attached firmly so it will not dislodge during the sports activity, such as basketball.

*Design project 51. Anchor*  Design an anchor for use with a canoe or small aluminum fishing boat. It should be easy to store, not excessively heavy, yet grip the bottom of the lake firmly, holding the boat steady.

*Design project 52. Slide projector*  Design an inexpensive 35-mm slide projector that can load and show up to 10 slides in a single cartridge. The key to success is a high-quality projected image at a very

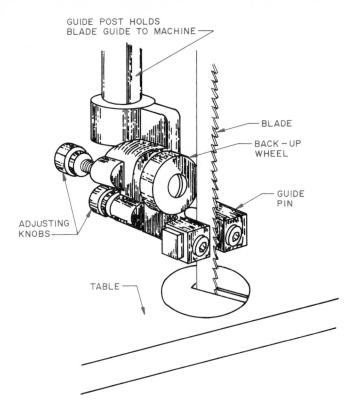

FIGURE 25–25 This band saw needs a blade guard.

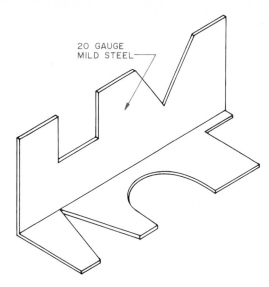

FIGURE 25–27 Can you design a device to notch sheet metal?

low cost. Simplicity of operation plus few moving parts will contribute to its acceptance.

*Design project 53. Auto antitheft device* Design a device you could secure inside an automobile that would prevent its being driven by unauthorized persons. Consider the material used and the possibility of removing the device by breaking the part of the auto to which it is attached.

*Design project 54. Swivel base* Design a base for the vise in Fig. 25–23 that will clamp to the table and permit the vise to rotate 360° yet be locked in position and tilted up to 30° with the horizontal and remain locked in that position.

*Design project 55. Retractable-blade utility knife* Design a utility knife similar to the one in Fig. 25–26, with a mechanism that retracts the blade into the handle. Be certain it has a positive stop so that it does not close accidentally while being used.

FIGURE 25–26 Design a mechanism that will allow the blade to retract when not in use.

*Design project 56. Sheet-metal notcher* Design a hand-held and hand-operated tool that will cut square, round, and V-shaped notches in mild steel up to 20 gauge (Fig. 25–27).

*Design project 57. Hand-powered punch* Design a hand-powered punch that will produce round holes up to 6 mm in diameter through 16-gauge mild steel. It should be able to punch a hole at least 75 mm from the edge of the sheet. The punches and dies for the various size holes must be easily mounted on the device and able to be quickly changed. The device must have some type of stops or guides that, once they are set, automatically locate the piece of metal so that the hole is punched in the correct location. The device must be small enough to carry in a normal toolbox and be operated by fastening it to the top of a workbench.

*Design project 58. Foot-powered punch* Design a punch as described in Design Project 57 that can be mounted on a floor stand and operated with your foot (Fig. 25–28). This frees both hands to hold the metal as it is punched.

*Design project 59. Punched-hole locater* Design a device that will accurately measure the distance from the center of a hole to be punched to the edge of the sheet (Fig. 25–29) and permit a punch mark to be made to locate the center.

*Design project 60. Pipe crimper* It is easier to slip one sheet-metal pipe inside another if the one to go inside is crimped. Crimping means producing indentations around the circumference of the pipe at its end (Fig. 25–30). Design a hand-held tool that will crimp sheet-metal pipes 1⅛ in. from their ends.

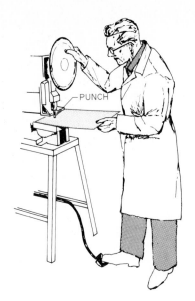

FIGURE 25–28 Design a foot-powered system for the metal punch.

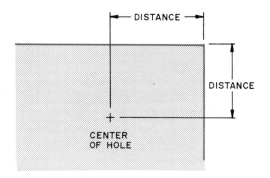

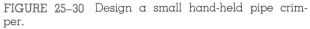

FIGURE 25–29 Prepare a simple location device for punched holes.

FIGURE 25–30 Design a small hand-held pipe crimper.

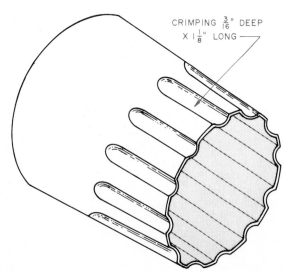

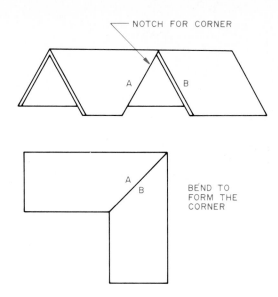

FIGURE 25–31 Design a device that will bend angle iron up to 90°.

*Design project 61. Angle iron bender* Design a device that will bend angle iron from 0° to 90°. The maximum size angle it will handle will be 2 × 2 × ¼ in. Design it so that it can be securely fastened to a workbench. Remember, to bend angle iron and get a sharp corner, one of the legs must be notched (Fig. 25–31).

*Design project 62. Metal rod bender* The bending of solid metal rods is a common process. Design a device that will enable you to hand-bend mild steel rods up to 12 mm in diameter (Fig. 25–32).

*Design project 63. Clip punch* Sheet metal can be joined by punching a clip through both pieces simultaneously and bending the clip to hold them together (Fig. 25–33). Design a device to punch the clip through the two pieces of 22-gauge mild steel.

*Design project 64. Rivet squeezer* Design a device to set aluminum rivets up to ⁵⁄₃₂ in. in diameter, as-

FIGURE 25–32 A hand-held device to bend this round rod is needed.

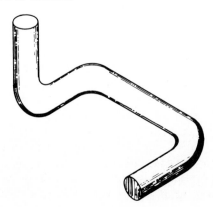

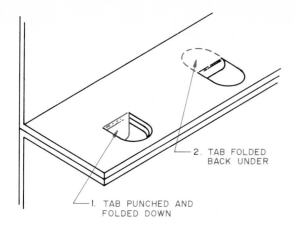

FIGURE 25–33 Design a punch to join sheet metal with folded clips.

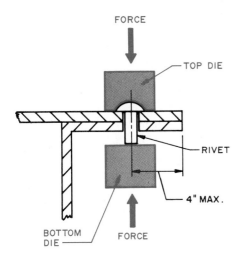

FIGURE 25–34 Prepare a design for a tool to form heads on small rivets.

suming the tool has a throat permitting it to extend 4 in. into a metal sheet from the edge (Fig. 25–34). The rivets have a preformed head on one side. Therefore, the die on one jaw of the device will have to be concave to accommodate the head. Since there are several types of rivet heads, the dies in the jaws must be replaceable. They must also be adjusted for rivets of different lengths.

*Design project 65. Measuring microscope stand*
A measuring microscope is used in tool and die work. It must be held securely on a base yet be able to be raised so that the end of the lens is 8 in. above the base and move horizontally 360° at a maximum of 8 in. from the center line of the base (Fig. 25–35). Design the base and the necessary columns and supports to carry the microscope. The microscope weighs 3 lb, so remember to provide some means to counterbalance this weight when the microscope is at its maximum distance from the base.

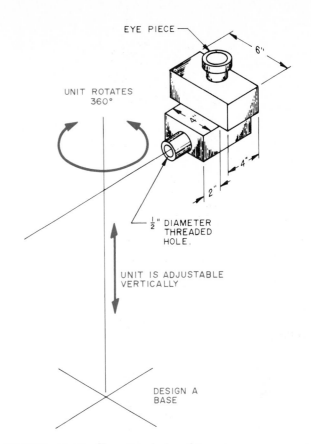

FIGURE 25–35 The stand for this microscope must provide vertical and rotational movement.

*Design project 66. Leveling base* A microscope is mounted on a rectangular base and slides in slots on it (Fig. 25–36). Design a way the base can be ad-

FIGURE 25–36 Come up with a way to level this microscope base.

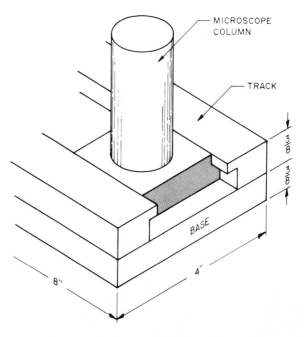

justed so that it sits absolutely level. Since the base is small, the leveling adjustment must be very fine. The base may be altered in any way desired.

*Design project 67. Hole depth gauge* It is often necessary to find the depth of holes in metal products. Design a device that will find the depth of holes up to 3 in. deep and having a minimum diameter of 0.25 in. Devise a scale that will measure the depth to the nearest 0.10 in.

*Design project 68. Field magnifier* When inspecting a job for surface finish, it is helpful to view it through a magnifier. Design a base to hold a 3-in. round magnifying lens that focuses on the surface from a distance of 3.5 in. Have the unit provide illumination on the surface being inspected.

*Design project 69. Swivel plate* A swivel plate has a horizontal table that can be adjusted to 45° left or right. It must be able to be locked at any angle and not slip. Design the base and the tilting mechanism for a 5 × 6 × ½ in. plate (Fig. 25–37). It should have a scale in degrees to indicate the angle of slope.

*Design project 70. Trolley* Design a trolley to ride on the steel beam in Fig. 25–38. It should carry 1000 lb. Make provision for a way to hang a cable from it or, with alterations, to hang a permanent unit, such as a bucket.

*Design project 71. Drill-press table adjuster* Design a device to enable the table on the drill press in Fig. 25–39 to be easily moved vertically. A locking device is needed to hold it firmly in the desired position.

*Design project 72. Small parts pickup device* Design a device to enable you rapidly to pick up quantities of loose, small metal parts, such as nails or screws, and deposit them in a container.

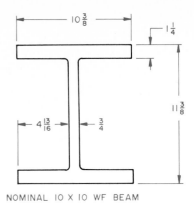

NOMINAL 10 X 10 WF BEAM

FIGURE 25–38 Design a trolley to roll along this steel overhead beam.

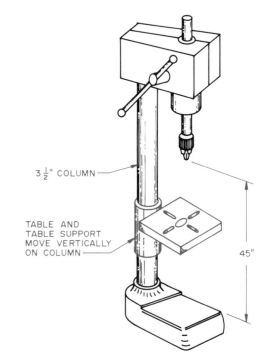

FIGURE 25–39 This drill-press table must be easily raised and lowered.

*Design project 73. Sheet pickup device* Design a device that will enable you to pick up a single metal sheet from a stack, move the sheet to another location, and release it. Assume that the sheets are approximately 12 × 12 in.

*Design project 74. Shelf storage unit* Design a metal shelf storage unit 36 in. wide, 72 in. tall, and 12 in. deep. The shelves should be able to be located every 1 in. Design vertical dividers to break up every shelf with divisions every 6 in. Design a shelf box that could be used to store small parts on the shelves.

FIGURE 25–37 Design a tilting mechanism for this base.

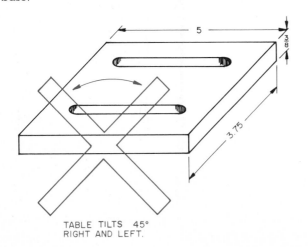

TABLE TILTS 45°
RIGHT AND LEFT.

*Design project 75. Safety shield* Design an eye safety shield to be used on machine tools as lathes and milling machines to keep the hot metal chips from flying in the face of the operator. The device should be portable and designed so that it can be used on any machine tool. The shield itself can be approximately 250 × 300 mm.

*Design project 76. In-duct fan* Design a mounting system for installing a low-velocity electric fan in an 8 × 10 in. rectangular heating duct. The fan is used to boost the amount of air flowing through the duct. A typical fan design is shown in Fig. 25–40. Use it or any other brand you may find for this design. This fan has an 8-in. outside diameter, uses 120 V, requires 14 W and 0.23 A, and moves 150 ft$^3$/min.

*Design project 77. Fireplace insert* Design an insert to be installed in a fireplace to increase the heat flow from the fireplace into the room. You may use electric fans, natural convection, or any other means to induce flow. Be certain the materials selected will withstand the effects of exposure to the fire.

*Design project 78. Weatherstripping* Most weatherstripping around doors works well until the door or the frame warps or twists slightly due to moisture or temperature changes. Design a weatherstripping material that can be installed by the homeowner that will allow for a change in the space between a door and its frame (Fig. 25–41).

*Design project 79. Window insulation* Windows, even when covered with storm windows, are a major factor in heat loss in a building. Design some unit that could be used to cover the windows to reduce heat loss. It might be used only when the room is unoccupied, as at night. It should be possible to remove the device easily so that the window is uncovered. If it is installed on the inside of the window, it

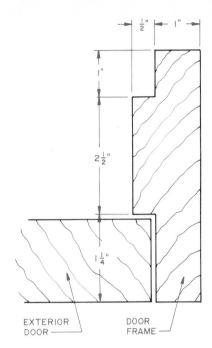

FIGURE 25–41 Design easily installed weatherstripping for this door.

should be attractive and blend with the walls, curtains, and furnishings.

*Design project 80. Cabinet* Design a wall cabinet that will be used by a barber to hold the tools and supplies used frequently including a facility for washing and tinting hair. Give special attention to the location of each item in relation to the barber chair so that a minimum of reaching and walking is required. Consider some innovative chair location, such as inside a U- or V-shaped counter.

## System Development Problems

*Design project 81. Go-cart braking system* Design a breaking system for all four wheels on a go-cart. Make provision for easy adjustment so that all four wheels stop uniformly. The system can be hand- or foot-operated (Fig. 25–42).

*Design project 82. Go-cart brake light* Design a go-cart electrical system that includes one headlight and two stoplights.

*Design project 83. Go-cart steering system* Design a go-cart steering system that permits the alignment of the front wheels.

*Design project 84. Moving sidewalk* Design a moving sidewalk for use indoors, such as in a mall or airport. Observe safety features and maximum safe speeds. It should be able to be used by the handicapped. Locate several emergency buttons that can stop it if a problem occurs.

FIGURE 25–40 Mount this fan in a sheet-metal duct.

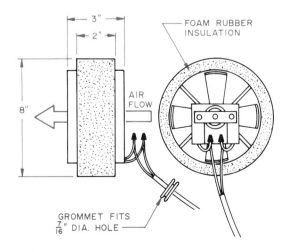

FIGURE 25–42 Design a braking system for this go-cart. (Courtesy of Montgomery Ward and Co.)

*Design project 85. Door answering system*   Design a system that could be installed on private homes that would permit someone at the door to leave a recorded message if no one is at home.

*Design project 86. Solar water heater*   Design a portable solar water heater that could be used on camping trips to produce small quantities of hot water for bathing. Include some means of storing the heated water and keeping it hot until it is needed.

*Design project 87. Tent electrical system*   Design an electric lighting system that could be installed in a large camping tent and provide at least four interior lights plus one outside the door of the tent. It must have a self-sustaining unit that can be used where normal 120-V electrical service is not available.

*Design project 88. Water system*   Design a portable water system that will accommodate at least 5 gal of fresh drinking water and deliver it when needed under 20 psi pressure.

*Design project 89. Burglar alarm*   Design an inexpensive burglar-alarm system that would be useful in a home or small business. It could be electrical or mechanical but should protect against all forms of entry and alert those outside the building that it has been entered.

*Design project 90. Air filtration system*   Design an air filtration system for a below-grade area such as a storm shelter or fallout shelter. The shelter will accommodate six people. Plan the system to remove the stale interior air and bring in filtered fresh outside air. Have provision to immediately seal the ducts to the outside if necessary during a nuclear attack. Also make provision for some means of operating the system, even if it is on a reduced basis, if the electrical power fails.

*Design project 91. Rear-seat auto heater*   The rear floor of most automobiles is much colder than the other areas. Design a system to provide additional heat to this area. It could be a part of the existing auto heating system or a totally independent source of heat.

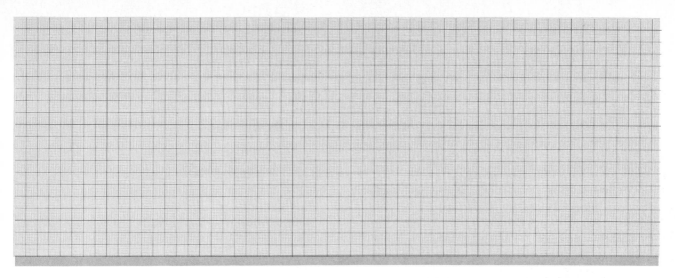

# APPENDICES

# A ABBREVIATIONS

Abbreviations are conventional ways to represent words or names in a particular language. Letter symbols represent quantities or units, not names, and are independent of a language such as MA for milliampere. The use of letter symbols is preferred over abbreviations because they can be understood by all, regardless of the language.

Abbreviations are used in place of spelling out a word only when necessary to save time and space. Abbreviations should be used only where their meaning is unquestionably clear to the person using the drawing. If there is any doubt spell out the word.

The following abbreviations were abstracted from ANSI Y1.1–1972, *Abbreviations For Use on Drawings and in Text*, with permission of the American Society of Mechanical Engineers.

| WORD | ABBREVIATION | WORD | ABBREVIATION | WORD | ABBREVIATION |
|---|---|---|---|---|---|
| Abbreviate | ABBR | Base plate | BP | Centigram | CG |
| Absolute | ABS | Battery | BAT. | Centiliter | CL |
| Accelerate | ACCEL | Bearing | BRG | Centimeter | cm |
| Access panel | AP | Bent | BT | Centrifugal | CENT. |
| Accessory | ACCESS. | Between | BET. | Centrifugal force | CF |
| Actual | ACT. | Between centers | BC | Ceramic | CER |
| Adapter | ADPT | Between perpendiculars | BP | Chain | CH |
| Addendum | ADD. | | | Chamfer | CHAM |
| Adjust | ADJ | Bevel | BEV | Change | CHG |
| Advance | ADV | Bill of material | B/M | Channel | CHAN |
| After | AFT. | Birmingham Wire Gage | BWG | Check | CHK |
| Aggregate | AGGR | | | Check valve | CV |
| Aileron | AIL | Blank | BLK | Chemical | CHEM |
| Aircraft | ACFT | Block | BLK | Chord | CHD |
| Airplane | APL | Blueprint | BP | Chrome molybdenum | CR MOLY |
| Airtight | AT | Bolt circle | BC | Chromium plate | CR PL |
| Alarm | ALM | Bottom | BOT | Chrome vanadium | CR VAN |
| Allowance | ALLOW | Bottom chord | BC | Circle | CIR |
| Alloy | ALY | Brake | BK | Circuit | CKT |
| Alteration | ALT | Brass | BRS | Circular | CIR |
| Alternate | ALT | Brazing | BRZG | Circular pitch | CP |
| Alternating current | AC | Break | BRK | Circulate | CIRC |
| Alternator | ALT | Breaker | BKR | Circumference | CIRC |
| Altitude | ALT | Brinnell hardness | BH | Clamp | CLP |
| Aluminum | AL | British Standard | BR STD | Class | CL |
| American Standard | AMER STD | British thermal units | BTU | Clear | CLR |
| American Wire Gage | AWG | Broach | BRO | Clearance | CL |
| Ammeter | AM | Bronze | BRZ | Clockwise | CW |
| Amount | AMT | Brown & Sharp | B & S | Closing | CL |
| Ampere | AMP | Brush | BR | Clutch | CL |
| Ampere hour | AMP HR | Burnish | BNH | Coated | CTD |
| Amplifer | AMPL | Bushing | BUSH. | Coaxial | COAX |
| Anneal | ANL | Bypass | BYP | Coefficient | COEF |
| Antenna | ANT. | | | Cold drawn | CD |
| Apparatus | APP | Cadmium plate | CD PL | Cold-drawn steel | CDS |
| Approved | APPD | Calculate | CALC | Cold rolled | CR |
| Approximate | APPROX | Calibrate | CAL | Cold-rolled steel | CRS |
| Arc weld | ARC/W | Calking | CLKG | Column | COL |
| Area | A | Capacitor | CAP | Combination | COMB. |
| Armature | ARM. | Capacity | CAP | Combustion | COMB |
| Arrange | ARR. | Cap screw | CAP. SCR | Communication | COMM |
| Arrester | ARR. | Carburize | CARB | Commutator | COMM |
| Asbestos | ASB | Caseharden | CH | Complete | COMPL |
| Assemble | ASSEM | Casing | CSG | Composite | CX |
| Assembly | ASSY | Cast (used with other materials) | C | Composition | COMP |
| Atomic | AT | | | Compressor | COMPR |
| Attach | ATT | Cast iron | CI | Concentric | CONC |
| Automatic | AUTO | Cast-iron pipe | CIP | Concrete | CONC |
| Auxiliary | AUX | Cast steel | CS | Condition | COND |
| Average | AVG | Casting | CSTG | Conduct | COND |
| | | Castle nut | CAS NUT | Conductor | COND |
| Babbitt | BAB | Cement | CEM | Conduit | CND |
| Back to back | B to B | Center | CTR | Connect | CONN |
| Baffle | BAF | Center line | CL | Constant | CONST |
| Balance | BAL | Center to center | C to C | Contact | CONT |
| Ball bearing | BB | Centering | CTR | Container | CNTR |
| Base line | BL | Centigrade | C | Continue | CONT |

| WORD | ABBREVIATION | WORD | ABBREVIATION | WORD | ABBREVIATION |
|---|---|---|---|---|---|
| Control | CONT | Drawing | DWG | Foot | (') FT |
| Control relay | CR | Drawing list | DL | Force | F |
| Control switch | CS | Drill | DR | Forging | FORG |
| Controller | CONT | Drill rod | DR | Forward | FWD |
| Convert | CONV | Drive | DR | Foundation | FDN |
| Conveyor | CNVR | Drive fit | DF | Foundry | FDRY |
| Cooled | CLD | Drop | D | Fractional | FRAC |
| Copper oxide | CUO | Drop forge | DF | Frame | FR |
| Copper plate | COP. PL | Duplex | DX | Freezing point | FP |
| Cord | CD | Duplicate | DUP | Frequency | FREQ |
| Correct | CORR | Dynamic | DYN | Frequency, high | HF |
| Corrosion resistant | CRE | Dynamo | DYN | Frequency, low | LF |
| Corrosion-resistant | CRES | | | Frequency, medium | MF |
| steel | | Each | EA | Frequency modulation | FM |
| Corrugate | CORR | Eccentric | ECC | Frequency, super high | SHF |
| Cotter | COT | Effective | EFF | Frequency, ultra high | UHF |
| Counter | CTR | Electric | ELEC | Frequency, very high | VHF |
| Counterclockwise | CCW | Elevation | EL | Frequency, very low | VLF |
| Counterbalance | CBAL | Enclose | ENCL | Friction horsepower | FHP |
| Counterbore | CBORE | End to end | E to E | From below | FR BEL |
| Counterdrill | CDRILL | Envelope | ENV | Front | FR |
| Counterpunch | CPUNCH | Equal | EQ | Fuel | F |
| Countersink | CSK | Equation | EQ | Furnish | FURN |
| Countersink other side | CSK-O | Equipment | EQUIP. | Fusible | FSBL |
| Coupling | CPLG | Equivalent | EQUIV | Fusion point | FNP |
| Cover | COV | Estimate | EST | | |
| Crank | CRK | Evaporate | EVAP | Gage or Gauge | GA |
| Cross connection | XCONN | Excavate | EXC | Gallon | GAL |
| Cross section | XSECT | Exhaust | EXH | Galvanize | GALV |
| Cubic | CU | Expand | EXP | Galvanized iron | GI |
| Current | CUR | Exterior | EXT | Galvanized steel | GS |
| Cycle | CY | External | EXT | Galvanized steel wire | GSWR |
| Cycles per minute | CPM | Extra heavy | X HVY | rope | |
| Cycles per second | CPS | Extra strong | X STR | Gas | G |
| Cylinder | CYL | Extrude | EXTR | Gasket | GSKT |
| | | | | Gasoline | GASO |
| Decibel | DB | Fabricate | FAB | General | GEN |
| Decimal | DEC | Face to face | F to F | Glaze | GL |
| Dedendum | DED | Fahrenheit | F | Governor | GOV |
| Deep drawn | DD | Fairing | FAIR. | Grade | GR |
| Deflect | DEFL | Farad | F | Graduation | GRAD |
| Degree | (°) DEG | Far side | FS | Gram | g |
| Density | D | Feed | FD | Graphic | GRAPH. |
| Describe | DESCR | Feeder | FDR | Graphite | GPH |
| Design | DSGN | Feet | (') FT | Grating | GRTG |
| Designation | DESIG | Feet per minute | FPM | Gravity | G |
| Detail | DET | Feet per second | FPS | Grid | G |
| Detector | DET | Female | FEM | Grind | GRD |
| Detonator | DET | Fiber | FBR | Groove | GRV |
| Develop | DEV | Field | FLD | Ground | GRD |
| Diagonal | DIAG | Figure | FIG. | | |
| Diagram | DIAG | Filament | FIL | Half hard | ½H |
| Diameter | DIA | Fillet | FIL | Half round | ½RD |
| Diametral pitch | DP | Filling | FILL. | Handle | HDL |
| Diaphragm | DIAPH | Fillister | FIL | Hanger | HGR |
| Differential | DIFF | Filter | FLT | Hard | H |
| Dimension | DIM. | Finish | FIN. | Hard-drawn | HD |
| Diode | DIO | Finish all over | FAO | Harden | HDN |
| Direct current | DC | Fireproof | FPRF | Hardware | HDW |
| Directional | DIR | Fitting | FTG | Head | HD |
| Discharge | DISCH | Fixture | FIX. | Headless | HDLS |
| Disconnect | DISC. | Flange | FLG | Heat | HT |
| Distance | DIST | Flashing | FL | Heat treat | HT TR |
| Distribute | DISTR | Flat | F | Heavy | HVY |
| Ditto | DO. | Flat head | FH | Height | HGT |
| Double | DBL | Flexible | FLEX. | Henry | H |
| Dovetail | DVTL | Float | FLT | Hexagon | HEX |
| Dowel | DWL | Floor | FL | High | H |
| Down | DN | Fluid | FL | High frequency | HF |
| Drafting | DFTG | Fluorescent | FLUOR | High point | H PT |
| Draftsman | DFTSMN | Flush | FL | High pressure | HP |
| Drain | DR | Focus | FOC | High speed | HS |

| WORDS | ABBREVIATION | WORDS | ABBREVIATION | WORDS | ABBREVIATION |
|---|---|---|---|---|---|
| High-speed steel | HSS | Kilovolt-ampere | kva | Microhenry | μH or UH |
| High tension | HT | Kilovolt-ampere hour | kvah | Micro-inch-root-mean | μ-IN-RMS or |
| High voltage | HV | Kilowatt | kW | square | U-IN-RMS |
| Highway | HWY | Kilowatt-hour | kWh | Micrometer | MIC |
| Holder | HLR | Kip (1000 lb) | K | Micron | μ or U |
| Hollow | HOL | Knots | KN | Microvolt | μV or UV |
| Horizontal | HOR | | | Microwatt | μW or UW |
| Horsepower | HP | Laboratory | LAB | Miles | MI |
| Hot rolled | HR | Lacquer | LAQ | Miles per gallon | MPG |
| Hot-rolled steel | HRS | Laminate | LAM | Miles per hour | MPH |
| Hour | HR | Lateral | LAT | Milli | m |
| Hydraulic | HYD | Lead-coated metal | LCM | Milliampere | ma |
| | | Lead covered | LC | Milligram | mg |
| Identify | IDENT | Leading edge | LE | Millihenry | mH |
| Ignition | IGN | Left | L | Millimeter | mm |
| Illuminate | ILLUM | Left hand | LH | Milliseconds | ms |
| Illustrate | ILLUS | Length | LG | Millivolt | mv |
| Impact | IMP | Length over all | LOA | Milliwatt | mW |
| Impedance | IMP. | Letter | LTR | Minimum | MIN |
| Inch | (″) IN. | Light | LT | Minute | MIN |
| Inches per second | IPS | Limit | LIM | Miscellaneous | MISC |
| Include | INCL | Line | L | Mixture | MIX. |
| Increase | INCR | Linear | LIN | Modify | MOD |
| Indicate | IND | Link | LK | Morse taper | MOR T |
| Inductance or induc- | IND | Liquid | LIQ | Motor | MOT |
| tion | | Liter | L | Mounted | MTD |
| Industrial | IND | Locate | LOC | Mounting | MTG |
| Information | INFO | Long | LG | Multiple | MULT |
| Injection | INJ | Longitude | LONG. | Multiple contact | MC |
| Inlet | IN | Low explosive | LE | | |
| Inspect | INSP | Low frequency | LF | National | NATL |
| Install | INSTL | Low pressure | LP | Natural | NAT |
| Instantaneous | INST | Low tension | LT | Near face | NF |
| Instruct | INST | Low voltage | LV | Near side | NS |
| Instrument | INST | Low speed | LS | Negative | NEG |
| Insulate | INS | Low torque | LT | Network | NET |
| Interchangeable | INTCHG | Lubricate | LUB | Neutral | NEUT |
| Interior | INT | Lubricating oil | LO | Nickel-silver | NI-SIL |
| Interlock | INTLK | Lumen | L | Nipple | NIP. |
| Intermediate | INTER | Lumens per watt | LPW | Nominal | NOM |
| Intermittent | INTMT | | | Normal | NOR |
| Internal | INT | Machine | MACH | Normally closed | NC |
| Interrupt | INTER | Magnet | MAG | Normally open | NO |
| Interrupted continu- | ICW | Main | MN | Not to scale | NTS |
| ous wave | | Male and female | M&F | Number | NO. |
| Interruptions per | IPM | Malleable | MALL | | |
| minute | | Malleable iron | MI | Obsolete | OBS |
| Interruptions per sec- | IPS | Manual | MAN. | Octagon | OCT |
| ond | | Manufacture | MFR | Ohm | Ω |
| Intersect | INT | Manufactured | MFD | On center | OC |
| Inverse | INV | Manufacturing | MFG | One pole | 1 P |
| Invert | INV | Material | MATL | Opening | OPNG |
| Iron | I | Material list | ML | Operate | OPR |
| Iron-pipe size | IPS | Maximum | MAX | Opposite | OPP |
| Irregular | IRREG | Maximum working | MWP | Optical | OPT |
| Issue | ISS | pressure | | Orifice | ORF |
| | | Mean effective pres- | MEP | Original | ORIG |
| Jack | J | sure | | Oscillate | OSC |
| Joint | JT | Mechanical | MECH | Ounce | OZ |
| Junction | JCT | Mechanism | MECH | Out to Out | O to O |
| | | Medium | MED | Outlet | OUT. |
| Kelvin | K | Mega | M | Output | OUT. |
| Key | K | Megacycles | MC | Outside diameter | OD |
| Keyseat | KST | Megwatt | MW | Outside face | OF |
| Keyway | KWY | Megohm | MEG | Outside radius | OR |
| Kilo | k | Melting point | MP | Over-all | OA |
| Kilocycle | kc | Metal | MET | Overhead | OVHD |
| Kilocycles per sec- | kc/s | Meter (instrument) | M | Overload | OVLD |
| ond | | Meter (linear mea- | m | Overvoltage | OVV |
| Kilogram | kg | sure) | | Oxidized | OXD |
| Kiloliter | kL | Micro | μ or U | | |
| Kilometer | km | Microampere | μA or UA | Pack | PK |
| Kilovolt | kv | Microfarad | μF or UF | Packing | PKG |

| WORD | ABBREVIATION | WORD | ABBREVIATION | WORD | ABBREVIATION |
|---|---|---|---|---|---|
| Painted | PTD | Quart | QT | Segment | SEG |
| Pair | PR | Quarter | QTR | Select | SEL |
| Panel | PNL | Quarter hard | ¼ H | Semifinished | SF |
| Parallel | PAR. | Quarter round | ¼ RD | Semifixed | SFXD |
| Part | PT | Quartz | QTZ | Semisteel | SS |
| Pattern | PATT | | | Separate | SEP |
| Perforate | PERF | Radial | RAD | Sequence | SEQ |
| Permanent | PERM | Radio frequency | RF | Serial | SER |
| Permanent magnet | PM | Radius | R | Series | SER |
| Perpendicular | PERP | Reactor | REAC | Serrate | SERR |
| Phase | PH | Ream | RM | Service | SERV |
| Phosphor bronze | PH BRZ | Reassemble | REASM | Set screw | SS |
| Photograph | PHOTO | Received | RECD | Shaft | SFT |
| Physical | PHYS | Receiver | REC | Shield | SHLD |
| Piece | PC | Receptacle | RECP | Short wave | SW |
| Piece mark | PC MK | Recriprocate | RECIP | Shunt | SH |
| Pierce | PRC | Recirculate | RECIRC | Side | S |
| Pipe Tap | PT | Reclosing | RECL | Signal | SIG |
| Pitch | P | Record | REC | Sink | SK |
| Pitch circle | PC | Rectangle | RECT | Sketch | SK |
| Pitch diameter | PD | Rectifier | RECT | Sleeve | SLV |
| Plastic | PLSTC | Reduce | RED. | Slide | SL |
| Plate | PL | Reference | REF | Slotted | SLOT. |
| Plotting | PLOT. | Reference line | REF L | Small | SM |
| Pneumatic | PNEU | Regulator | REG | Smoke | SMK |
| Point | PT | Reinforce | REINF | Smokeless | SMKLS |
| Point of compound curve | PCC | Relay | REL | Socket | SOC |
| | | Release | REL | Soft | S |
| Point of curve | PC | Relief | REL | Solder | SLD |
| Point of intersection | PI | Remove | REM | Solenoid | SOL |
| Point of reverse curve | PRC | Repair | REP | Sound | SND |
| Point of switch | PS | Replace | REPL | South | S |
| Point of tangent | PT | Reproduce | REPRO | Space | SP |
| Polar | POL | Require | REQ | Spare | SP |
| Pole | P | Required | REQD | Speaker | SPKR |
| Polish | POL | Resistance | RES | Special | SPL |
| Port | P | Resistor | RES | Specific | SP |
| Position | POS | Retainer | RET. | Specific gravity | SP GR |
| Positive | POS | Retard | RET. | Specific heat | SP HT |
| Potential | POT. | Return | RET. | Specification | SPEC |
| Pound | LB | Reverse | REV | Speed | SP |
| Pounds per cubic foot | PCF | Revise | REV | Spherical | SPHER |
| Pounds per square foot | PSF | Revolution | REV | Spindle | SPDL |
| | | Revolutions per minute | RPM | Split phase | SP PH |
| Pounds per square inch | PSI | | | Spot-faced | SF |
| | | Revolutions per second | RPS | Spring | SPG |
| Pounds per square inch absolute | PSIA | Rheostat | RHEO | Square | SQ |
| | | Right | R | Stabilize | STAB |
| Power | PWR | Right hand | RH | Stainless | STN |
| Power amplifier | PA | Ring | R | Standard | STD |
| Power direction relay | PDR | Rivet | RIV | Static pressure | SP |
| Power factor | PF | Rockwell hardness | RH | Station | STA |
| Preamplifier | PREAMP | Roller bearing | RB | Stationary | STA |
| Precast | PRCST | Root diameter | RD | Steel | STL |
| Prefabricated | PREFAB | Root mean square | RMS | Stiffener | STIFF |
| Preferred | PFD | Rotary | ROT. | Stock | STK |
| Premolded | PRMLD | Rotate | ROT. | Storage | STG |
| Prepare | PREP | Rough | RGH | Straight | STR |
| Press | PRS | Round | RD | Strip | STR |
| Pressure | PRESS. | Rubber | RUB. | Structural | STR |
| Pressure angle | PA. | | | Substitute | SUB |
| Primary | PRI | Safe working pressure | SWP | Suction | SUCT |
| Process | PROC | Safety | SAF | Summary | SUM. |
| Production | PROD | Sand blast | SD BL | Supervise | SUPV |
| Profile | PF | Saturate | SAT. | Supply | SUP |
| Project | PROJ | Schedule | SCH | Surface | SUR |
| Punch | PCH | Schematic | SCHEM | Survey | SURV |
| Purchase | PUR | Screen | SCRN | Switch | SW |
| Push-pull | P-P | Screw | SCR | Symbol | SYM |
| | | Second | SEC | Symmetrical | SYM |
| Quadrant | QUAD | Section | SECT | Synchronous | SYN |
| Quality | QUAL | | | Synthetic | SYN |
| Quantity | QTY | | | System | SYS |

| WORD | ABBREVIATION | WORD | ABBREVIATION | WORD | ABBREVIATION |
|------|--------------|------|--------------|------|--------------|
| Tabulate | TAB. | Tracer | TCR | Video-frequency | VDF |
| Tachometer | TACH | Transfer | TRANS | Vibrate | VIB |
| Tandem | TDM | Transformer | TRANS | Viscosity | VISC |
| Tangent | TAN. | Transmission | XMSN | Vitreous | VIT |
| Taper | TPR | Transmitter | XMTR | Volt | V |
| Technical | TECH | Transmitting | XMTG | Volt-ampere | VA |
| Tee | T | Transportation | TRANS | Voltmeter | VM |
| Teeth per inch | TPI | Transverse | TRANSV | Volts per mil | VPM |
| Television | TV | Trimmer | TRIM. | Volume | VOL |
| Temperature | TEMP | Triode | TRI | | |
| Template | TEMP | Truss | T | Washer | WASH |
| Tensile strength | TS | Tubing | TUB | Water | W |
| Tension | TENS. | Turbine | TURB | Water line | WL |
| Terminal | TERM. | Typical | TYP | Watertight | WT |
| Terminal board | TB | | | Watt | W |
| That is | IE | Ultimate | ULT | Watt-hour | WHR |
| Theoretical | THEO | Ultra-high frequency | UHF | Watt-hour meter | WHM |
| Thermal | THRM | Under voltage | UV | Wattmeter | WM |
| Thermostat | THERMO | Unit | U | Weight | WT |
| Thick | THK | United States Gage | USG | West | W |
| Thousand | M | United States Stan- | USS | Wet bulb | WB |
| Thread | THD | dard | | Width | W |
| Throttle | THROT | Universal | UNIV | Wind | WD |
| Through | THRU | | | Winding | WDG |
| Time | T | Vacuum | VAC | Wire | W |
| Time delay | TD | Vacuum tube | VT | With | W/ |
| Time-delay closing | TDC | Valve | V | Without | W/O |
| Time-delay opening | TDO | Vapor proof | VAP PRF | Wood | WD |
| Tinned | TD | Variable | VAR | Woodruff | WDF |
| Toggle | TGL | Variable-frequency | VFO | Working point | WP |
| Tolerance | TOL | oscillator | | Working pressure | WP |
| Tongue and groove | T&G | Velocity | V | Wrought | WRT |
| Tool steel | TS | Ventilate | VENT. | Wrought iron | WI |
| Tooth | T | Versed sine | VERS | | |
| Total | TOT | Versus | VS | Yard | YD |
| Total indicator read- | TIR | Vertical | VERT | Year | YR |
| ing | | Very-high frequency | VHF | Yield point | YP |
| Trace | TR | Very-low frequency | VLF | Yield strength | YS |

## ABBREVIATIONS FOR COLORS

| | | | | | |
|------|------|------|------|------|------|
| Amber | AMB | Brown | BRN | White | WHT |
| Black | BLK | Green | GRN | Yellow | YEL |
| Blue | BLU | Orange | ORN | | |

# B INCH MILLIMETER EQUIVALENCY TABLES

## Inch–millimeter equivalents

| 4ths | 8ths | 16ths | 32nds | 64ths | Decimal To 4 places | To 3 places | To 2 places | Millimeters To 4 places |
|---|---|---|---|---|---|---|---|---|
| | | | | 1/64 | .0156 | .016 | .02 | 0.3969 |
| | | | 1/32 | | .0312 | .031 | .03 | 0.7938 |
| | | | | 3/64 | .0469 | .047 | .05 | 1.1906 |
| | | 1/16 | | | .0625 | .062 | .06 | 1.5875 |
| | | | | 5/64 | .0781 | .078 | .08 | 1.9844 |
| | | | 3/32 | | .0938 | .094 | .09 | 2.3813 |
| | | | | 7/64 | .1094 | .109 | .11 | 2.7781 |
| | 1/8 | | | | .1250 | .125 | .12 | 3.1750 |
| | | | | 9/64 | .1406 | .141 | .14 | 3.5719 |
| | | | 5/32 | | .1562 | .156 | .16 | 3.9688 |
| | | | | 11/64 | .1719 | .172 | .17 | 4.3656 |
| | | 3/16 | | | .1875 | .188 | .19 | 4.7625 |
| | | | | 13/64 | .2031 | .203 | .20 | 5.1594 |
| | | | 7/32 | | .2188 | .219 | .22 | 5.5563 |
| | | | | 15/64 | .2344 | .234 | .23 | 5.9531 |
| 1/4 | | | | | .2500 | .250 | .25 | 6.3500 |
| | | | | 17/64 | .2656 | .266 | .27 | 6.7469 |
| | | | 9/32 | | .2812 | .281 | .28 | 7.1438 |
| | | | | 19/64 | .2969 | .297 | .30 | 7.5406 |
| | | 5/16 | | | .3125 | .312 | .31 | 7.9375 |
| | | | | 21/64 | .3281 | .328 | .33 | 8.3344 |
| | | | 11/32 | | .3438 | .344 | .34 | 8.7313 |
| | | | | 23/64 | .3594 | .359 | .36 | 9.1281 |
| | 3/8 | | | | .3750 | .375 | .38 | 9.5250 |
| | | | | 25/64 | .3906 | .391 | .39 | 9.9219 |
| | | | 13/32 | | .4062 | .406 | .41 | 10.3188 |
| | | | | 27/64 | .4219 | .422 | .42 | 10.7156 |
| | | 7/16 | | | .4375 | .438 | .44 | 11.1125 |
| | | | | 29/64 | .4531 | .453 | .45 | 11.5094 |
| | | | 15/32 | | .4688 | .469 | .47 | 11.9063 |
| | | | | 31/64 | .4844 | .484 | .48 | 12.3031 |
| | | | | | .5000 | .500 | .50 | 12.7000 |
| | | | | 33/64 | .5156 | .516 | .52 | 13.0969 |
| | | | 17/32 | | .5312 | .531 | .53 | 13.4938 |
| | | | | 35/64 | .5469 | .547 | .55 | 13.8906 |
| | | 9/16 | | | .5625 | .562 | .56 | 14.2875 |
| | | | | 37/64 | .5781 | .578 | .58 | 14.6844 |
| | | | 19/32 | | .5938 | .594 | .59 | 15.0813 |
| | | | | 39/64 | .6094 | .609 | .61 | 15.4781 |
| | 5/8 | | | | .6250 | .625 | .62 | 15.8750 |
| | | | | 41/64 | .6406 | .641 | .64 | 16.2719 |
| | | | 21/32 | | .6562 | .656 | .66 | 16.6688 |
| | | | | 43/64 | .6719 | .672 | .67 | 17.0656 |
| | | 11/16 | | | .6875 | .688 | .69 | 17.4625 |
| | | | | 45/64 | .7031 | .703 | .70 | 17.8594 |
| | | | 23/32 | | .7188 | .719 | .72 | 18.2563 |
| | | | | 47/64 | .7344 | .734 | .73 | 18.6531 |
| 3/4 | | | | | .7500 | .750 | .75 | 19.0500 |
| | | | | 49/64 | .7656 | .766 | .77 | 19.4469 |
| | | | 25/32 | | .7812 | .781 | .78 | 19.8438 |
| | | | | 51/64 | .7969 | .797 | .80 | 20.2406 |
| | | 13/16 | | | .8125 | .812 | .81 | 20.6375 |
| | | | | 53/64 | .8281 | .828 | .83 | 21.0344 |
| | | | 27/32 | | .8438 | .844 | .84 | 21.4313 |
| | | | | 55/64 | .8594 | .859 | .86 | 21.8281 |
| | 7/8 | | | | .8750 | .875 | .88 | 22.2250 |
| | | | | 57/64 | .8906 | .891 | .89 | 22.6219 |
| | | | 29/32 | | .9062 | .906 | .91 | 23.0188 |
| | | | | 59/64 | .9219 | .922 | .92 | 23.4156 |
| | | 15/16 | | | .9375 | .938 | .94 | 23.8125 |
| | | | | 61/64 | .9531 | .953 | .95 | 24.2094 |
| | | | 31/32 | | .9688 | .969 | .97 | 24.6063 |
| | | | | 63/64 | .9844 | .984 | .98 | 25.0031 |
| | | | | | 1.0000 | 1.000 | 1.00 | 25.4000 |

## Common Fractional Inches to Millimeters

| In. | Mm | | In. | Mm | | In. | Mm | | In. | Mm |
|---|---|---|---|---|---|---|---|---|---|---|
| 1/64 | 0.397 | | 17/64 | 6.747 | | 33/64 | 13.097 | | 49/64 | 19.447 |
| 1/32 | 0.794 | | 9/32 | 7.144 | | 17/32 | 13.494 | | 25/32 | 19.844 |
| 3/64 | 1.191 | | 19/64 | 7.541 | | 35/64 | 13.890 | | 51/64 | 20.240 |
| 1/16 | 1.587 | | 5/16 | 7.937 | | 9/16 | 14.287 | | 13/16 | 20.637 |
| 5/64 | 1.984 | | 21/64 | 8.334 | | 37/64 | 14.684 | | 53/64 | 21.034 |
| 3/32 | 2.381 | | 11/32 | 8.731 | | 19/32 | 15.081 | | 27/32 | 21.431 |
| 7/64 | 2.778 | | 23/64 | 9.128 | | 39/64 | 15.478 | | 55/64 | 21.828 |
| 1/8 | 3.175 | | 3/8 | 9.525 | | 5/8 | 15.875 | | 7/8 | 22.225 |
| 9/64 | 3.572 | | 25/64 | 9.922 | | 41/64 | 16.272 | | 57/64 | 22.622 |
| 5/32 | 3.969 | | 13/32 | 10.319 | | 21/32 | 16.669 | | 29/32 | 23.019 |
| 11/64 | 4.366 | | 27/64 | 10.716 | | 43/64 | 17.065 | | 59/64 | 23.415 |
| 3/16 | 4.762 | | 7/16 | 11.113 | | 11/16 | 17.462 | | 15/16 | 23.812 |
| 13/64 | 5.159 | | 29/64 | 11.509 | | 45/64 | 17.859 | | 61/64 | 24.209 |
| 7/32 | 5.556 | | 15/32 | 11.906 | | 23/32 | 18.256 | | 31/32 | 24.606 |
| 15/64 | 5.953 | | 31/64 | 12.303 | | 47/64 | 18.653 | | 63/64 | 25.003 |
| 1/4 | 6.350 | | 1/2 | 12.700 | | 3/4 | 19.050 | | 1 | 25.400 |

From Drafting: Technology and Practice, copyright 1980 by Spence. Used with permission of the publisher, Bennett Publishing Company, Peoria, IL., 61615.

## Decimal Inches to Millimeters

| DECIMAL INCHES | MILLIMETER EQUIVALENT | DECIMAL INCHES | MILLIMETER EQUIVALENT |
|---|---|---|---|
| .02 | .508 | .53 | 13.462 |
| .03 | .762 | .54 | 13.716 |
| .04 | 1.016 | .55 | 13.970 |
| .05 | 1.270 | .56 | 14.224 |
| .06 | 1.524 | | |
| .08 | 2.032 | .58 | 14.732 |
| .09 | 2.286 | .59 | 14.986 |
| .10 | 2.540 | .60 | 15.240 |
| .12 | 3.048 | .62 | 15.748 |
| .14 | 3.556 | .64 | 16.256 |
| .15 | 3.810 | .65 | 16.510 |
| .16 | 4.064 | .66 | 16.764 |
| .18 | 4.572 | .68 | 17.272 |
| .19 | 4.826 | .69 | 17.526 |
| .20 | 5.080 | .70 | 17.780 |
| .22 | 5.588 | .72 | 18.288 |
| .24 | 6.096 | .74 | 18.796 |
| .25 | 6.350 | .75 | 19.050 |
| .26 | 6.604 | .76 | 19.304 |
| .28 | 7.112 | .78 | 19.812 |
| .30 | 7.620 | .80 | 20.320 |
| .31 | 7.874 | .81 | 20.574 |
| .32 | 8.128 | .82 | 20.828 |
| .34 | 8.636 | .84 | 21.336 |
| .35 | 8.890 | .85 | 21.590 |
| .36 | 9.144 | .86 | 21.844 |
| .37 | 9.398 | .87 | 22.098 |
| .38 | 9.652 | .88 | 22.352 |
| .40 | 10.160 | .90 | 22.860 |
| .41 | 10.414 | .91 | 23.114 |
| .42 | 10.668 | .92 | 23.368 |
| .44 | 11.176 | .94 | 23.876 |
| .45 | 11.430 | .95 | 24.130 |
| .46 | 11.684 | .96 | 24.384 |
| .47 | 11.938 | .97 | 24.638 |
| .48 | 12.192 | .98 | 24.892 |
| .50 | 12.700 | 1.00 | 25.400 |
| .52 | 13.208 | | |

## Millimeters to Decimal Inches

| Mm | In. | Mm | In. | Mm | In. | Mm | In. | Mm | In. |
|---|---|---|---|---|---|---|---|---|---|
| 1 | 0.0394 | 21 | 0.8268 | 41 | 1.6142 | 61 | 2.4016 | 81 | 3.1890 |
| 2 | 0.0787 | 22 | 0.8662 | 42 | 1.6536 | 62 | 2.4410 | 82 | 3.2284 |
| 3 | 0.1181 | 23 | 0.9055 | 43 | 1.6929 | 63 | 2.4804 | 83 | 3.2678 |
| 4 | 0.1575 | 24 | 0.9449 | 44 | 1.7323 | 64 | 2.5197 | 84 | 3.3071 |
| 5 | 0.1969 | 25 | 0.9843 | 45 | 1.7717 | 65 | 2.5591 | 85 | 3.3465 |
| 6 | 0.2362 | 26 | 1.0236 | 46 | 1.8111 | 66 | 2.5985 | 86 | 3.3859 |
| 7 | 0.2756 | 27 | 1.0630 | 47 | 1.8504 | 67 | 2.6378 | 87 | 3.4253 |
| 8 | 0.3150 | 28 | 1.1024 | 48 | 1.8898 | 68 | 2.6772 | 88 | 3.4646 |
| 9 | 0.3543 | 29 | 1.1418 | 49 | 1.9292 | 69 | 2.7166 | 89 | 3.5040 |
| 10 | 0.3937 | 30 | 1.1811 | 50 | 1.9685 | 70 | 2.7560 | 90 | 3.5434 |
| 11 | 0.4331 | 31 | 1.2205 | 51 | 2.0079 | 71 | 2.7953 | 91 | 3.5827 |
| 12 | 0.4724 | 32 | 1.2599 | 52 | 2.0473 | 72 | 2.8247 | 92 | 3.6221 |
| 13 | 0.5118 | 33 | 1.2992 | 53 | 2.0867 | 73 | 2.8741 | 93 | 3.6615 |
| 14 | 0.5512 | 34 | 1.3386 | 54 | 2.1260 | 74 | 2.9134 | 94 | 3.7009 |
| 15 | 0.5906 | 35 | 1.3780 | 55 | 2.1654 | 75 | 2.9528 | 95 | 3.7402 |
| 16 | 0.6299 | 36 | 1.4173 | 56 | 2.2048 | 76 | 2.9922 | 96 | 3.7796 |
| 17 | 0.6693 | 37 | 1.4567 | 57 | 2.2441 | 77 | 3.0316 | 97 | 3.8190 |
| 18 | 0.7087 | 38 | 1.4961 | 58 | 2.2835 | 78 | 3.0709 | 98 | 3.8583 |
| 19 | 0.7480 | 39 | 1.5355 | 59 | 2.3229 | 79 | 3.1103 | 99 | 3.8977 |
| 20 | 0.7874 | 40 | 1.5748 | 60 | 2.3622 | 80 | 3.1497 | 100 | 3.9371 |

From *Fundamentals of Engineering Drawing: For Design, Product Development, and Numerical Control*, 8th ed., copyright 1981 by Prentice-Hall, Inc. Used with permission of the publisher, Prentice-Hall, Inc., Englewood Cliffs NJ 07632.

## Decimal Inch Design Tolerances to Millimeters

| Design practice | | Calculated equivalents of decimal-inch tolerances† | Design practice | | Calculated equivalents of decimal-inch tolerances° |
|---|---|---|---|---|---|
| Decimal, in. | Metric, mm | | Decimal, in. | Metric, mm | |
| 0.0001 | 0.003 | 0.00254 | 0.004 | 0.1 | 0.1016 |
| 0.0002 | 0.005 | 0.00508 | 0.005 | 0.13 | 0.1270 |
| 0.0003 | 0.008 | 0.00762 | 0.006 | 0.15 | 0.1524 |
| 0.0004 | 0.01 | 0.01016 | 0.007 | 0.18 | 0.1778 |
| 0.0005 | 0.013 | 0.01270 | 0.008 | 0.2 | 0.2032 |
| 0.0006 | 0.015 | 0.01524 | 0.009 | 0.23 | 0.2286 |
| 0.0007 | 0.018 | 0.01778 | 0.010 | 0.25 | 0.254 |
| 0.0008 | 0.02 | 0.02032 | 0.015 | 0.4 | 0.381 |
| 0.0009 | 0.023 | 0.02286 | 0.02 | 0.5 | 0.508 |
| 0.001 | 0.025 | 0.0254 | 0.03 | 0.8 | 0.762 |
| 0.0015 | 0.04 | 0.0381 | 0.04 | 1 | 1.016 |
| 0.002 | 0.05 | 0.0508 | 0.06 | 1.5 | 1.524 |
| 0.0025 | 0.06 | 0.0635 | 0.08 | 2 | 2.032 |
| 0.003 | 0.08 | 0.0762 | 0.10 | 2.5 | 2.540 |

* To be used for tolerance conversions only. Table gives commonly used inch tolerances.
† Calculated on the basis of 1 in. = 25.4 mm.
Additional information relating to the conversion of tolerances may be obtained from the SAE Standard-Dual Dimensioning-SAE J390.

From *Fundamentals of Engineering Drawing: For Design, Product Development, and Numerical Control*, 8th ed., copyright 1981 by Prentice-Hall, Inc. Used with permission of the publisher, Prentice-Hall, Inc., Englewood Cliffs, NJ 07632.

# C  METRIC CONVERSION FACTORS

Conversion Factors—Current Practices to Metric

| FROM CURRENT | TO METRIC | MULTIPLY BY* |
|---|---|---|
| Fahrenheit | Celsius | $(t(°F) - 32) \times 5/9$ |
| miles | kilometers | 1.609 344 |
| ounces | grams | 28.349 52 |
| pounds | grams | 453.5924 |
| pounds | kilograms | 0.453 592 4 |
| pounds | metric tons | 0.000 453 592 4 |
| tons of 2000 pounds | metric tons | 0.907 184 7 |
| tons of 2240 pounds | metric tons | 1.016 047 |
| pounds per gallon | kilograms per liter | 0.119 826 4 |
| pounds per gallon (LPG only) | kilograms per liter | 0.120 066 5 |
| cents per pound | cents per kilogram | 2.204 623 |
| cents per pound | cents per metric ton | 2204.623 |
| cents per 100 pounds | cents per 100 kilograms | 2.204 623 |
| cents per 100 pounds | cents per metric ton | 22.046 23 |
| cents per ton of 2000 pounds | cents per metric ton | 1.102 311 |
| cents per ton of 2240 | cents per metric ton | 0.984 206 5 |
| cents per gallon | cents per liter | 0.264 172 0 |
| cents per foot or per linear foot | cents per meter | 3.280 840 |
| mills per mile | mills per kilometer | 0.621 371 2 |
| cents per ton mile | cents per kilometer | 0.684 944 5 |
| cents per mile or cents per car mile | cents per kilometer or cents per car kilometer | 0.621 371 2 |
| mils (0.001 inches) | millimeters | 0.0254 |
| inches | centimeters | 2.54 |
| inches | millimeters | 25.5 |
| fractional or decimal inches | millimeters | 25.4 |
| square inches | square centimeters | 6.4516 |
| square inches | square millimeters | 645.16 |
| cubic inches | cubic centimeters | 16.387 06 |
| cubic inches | cubic millimeters | 163 87.06 |
| inches | meters | 0.0254 |
| feet or linear feet | meters | 0.3048 |
| feet | centimeters | 30.48 |
| feet | millimeters | 304.8 |
| square feet | square meters | 0.092 903 04 |
| square feet | square centimeters | 929.0304 |
| square feet | square millimeters | 92 903.04 |
| cubic feet | cubic meters | 0.028 316 85 |
| cubic feet | cubic centimeters | 28 316.85 |
| yards | meters | 0.9144 |
| square yards | square meters | 0.836 127 4 |
| cubic yards | cubic meters | 0.764 554 9 |
| fluid ounces | milliliters | 29.573.53 |
| pints (liquid) | liters | 0.473 176 5 |
| quarts (liquid) | liters | 0.946 352 9 |
| gallons | liters | 3.785 412 |
| cord (128 cubic feet) | cubic meters | 3.624 557 |
| 1000 board feet (1000 fbm) | cubic meters | 2.359 737/1000 fbm |
| cunits (100 cubic feet) | cubic meters | 2.831 685 |
| pounds per inch or inch of width | kilograms per meter | 17.857 97 |
| pounds per square inch | kilograms per square meter | 703.0696 |
| pounds per square foot | kilograms per square meter | 4.882 428 |
| pounds per cubic foot | kilograms per cubic meter | 16.018 46 |
| pounds-force per square inch or pounds per square inch (psi) | kilopascals | 6.894 757 |
| pounds-force per square foot or pounds per square foot (psf) | pascals | 47.880 26 |
| foot pounds force | joules | 1.355 818 |
| cents per cubic yard | cents per cubic meter | 1.307 951 |
| cents per unit of 200 cubic feet (pulpwood chips) | cents per cubic meter | 0.176 573 6 |
| cents per cord of 128 cubic feet (pulpwood chips) | cents per cubic meter | 0.275 896 3 |
| cents per cubic foot | cents per cubic meter | 35.31466 |

*Round product to nearest whole number, or to number of decimal places required, in accordance with Item 10.

## Conversion Factors—Metric (SI) to Current Practices

| FROM METRIC | TO CURRENT | MULTIPLY BY* |
|---|---|---|
| Celsius | Fahreheit | 1.8 t(°C) + 32 |
| kilometers | miles | 0.621 371 2 |
| grams | ounces | 0:035 273 96 |
| grams | pounds | 0.002 204 623 |
| kilograms | pounds | 2.204 623 |
| metric tons | pounds | 2204.623 |
| metric tons | tons of 2000 pounds | 1.102 311 |
| metric tons | tons of 2240 pounds | 0.984 206 5 |
| kilograms per liter | pounds per gallon | 8.345 405 |
| cents per kilogram | cents per pound | 0.453 592 4 |
| cents per metric ton | cents per pound | 0.000 453 592 4 |
| cents per 100 kilograms | cents per 100 pounds | 0.453 592 4 |
| cents per metric ton | cents per 100 pounds | 0.045 359 24 |
| cents per metric ton | cents per ton of 2000 pounds | 0.907 184 7 |
| cents per metric ton | cents per ton of 2240 pounds | 1.016 047 |
| cents per liter | cents per gallon | 3.785 412 |
| cents per meter | cents per foot or per linear foot | 0.3048 |
| mills per kilometer | mills per mile | 1.609 344 |
| cents per kilometer | cents per ton mile | 1.459 972 |
| cents per kilometer or per car kilometer | cents per mile or cents per car mile | 1.609 344 |
| millimeters | mils (0.001 inches) | 39.370 08 |
| centimeters | inches | 0.393 700 8 |
| millimeters | inches | 0.039 370 08 |
| millimeters | fractional or decimal inches | 0.039 370 08 |
| square centimeters | square inches | 0.155 000 3 |
| square millimeters | square inches | 0.001 550 003 |
| cubic centimeters | cubic inches | 0.061 023 74 |
| cubic millimeters | cubic inches | 0.000 061 023 74 |
| meters | inches | 39.370 08 |
| meters | feet or linear feet | 3.280 840 |
| centimeters | feet | 0.032 808 40 |
| millimeters | feet | 0.003 280 840 |
| square meters | square feet | 10.763 91 |
| square centimeters | square feet | 0.001 076 391 |
| square millimeters | square feet | 0.000 010 763 91 |
| cubic meters | cubic feet | 35.314 6 |
| cubic centimeters | cubic feet | 0.000 035 314 66 |
| meters | yards | 1.093 613 |
| square meters | square yards | 1.195 990 |
| cubic meters | cubic yards | 1.307 951 |
| milliliters | fluid ounces | 0.033 814 02 |
| liters | pints (liquid) | 2.113 376 |
| liters | quarts (liquid) | 1.056 688 |
| liters | gallons | 0.264 172 0 |
| cubic meters | cords (128 cubic feet) | 0.275 895 8 |
| cubic meters | 1000 board feet (1000 fbm) | 0.423 776 0 |
| cubic meters | cunits (100 cubic feet) | 0.353 146 7 |
| kilograms per meter | pounds per inch or inch of width | 0.055 997 41 |
| kilograms per square meter | pounds per square inch | 0.001 422 334 |
| kilograms per square meter | pounds per square foot | 0.204 816 1 |
| kilograms per cubic meter | pounds per cubic foot | 0.062 427 96 |
| kilopascals | pounds-force per square inch or pounds per square inch (psi) | 0.145 037 7 |
| pascals | pounds-force per square foot or pounds per square foot (psf) | 0.020 885 43 |
| joules | foots pounds force | 0.737 562 1 |
| cents per cubic meter | cents per cubic yard | 0.764 554 9 |

*Round product to nearest whole number, or to number of decimal places required, in accordance with Item 10.

## Current Measures—Metric Equivalents

### *MEASURE OF LENGTH*

1 Meter = 39.37 Inches = 3.281 Feet = 1.094 Yards
1 Centimeter = 0.3937 Inch
1 Millimeter = ¹⁄₂₅ Inch (approximately)
1 Kilometer = 0.621 Mile
1 Inch = 2.540 Centimeters = 25.400 Millimeters
1 Foot = 0.305 Meter
1 Mile = 1.609 Kilometers

### *MEASURES OF SURFACE*

1 Square Meter = 10.764 Square Feet = 1.196 Square Yards
1 Square Centimeter = 0.155 Square Inch
1 Square Millimeter = 0.00155 Square Inch
1 Square Yard = 0.836 Square Meter
1 Square Foot = 0.0929 Square Meter
1 Square Inch = 6.452 Square Centimeters = 645.2 Square Millimeters

### *MEASURES OF VOLUME*

1 Cubic Meter = 35.314 Cubic Feet = 1.308 Cubic Yards
1 Cubic Decimeter = 61.023 Cubic Inches = 0.0353 Cubic Foot
1 Cubic Centimeter = 0.061 Cubic Inch
1 Liter = 61.223 Cubic Inches = 0.0353 Cubic Foot = 0.2642 (U.S.) Gallon
1 Cubic Foot = 28.317 Cubic Decimeters = 28.317 Liters
1 Cubic Inch = 16.387 Cubic Centimeters

### *MEASURES OF WEIGHT*

1 Kilogram = 2.2046 Pounds
1 Metric Ton = 2204.6 Pounds = .9842 Ton (2240 pounds)
1 Ounce (avoirdupois) = 28.35 Grams
1 Pound = 0.4536 Kilogram

### *SPEED MEASUREMENTS*

1 Kilometer per Hour = 0.621 Mile per Hour

# D INCH AND METRIC DRILL SIZES

Number and Letter Size Inch Drills With Metric Equivalents*

| | Number sizes | | | | | | | Letter sizes | | | |
|---|---|---|---|---|---|---|---|---|---|---|---|
| No. size | Decimal equivalent | Metric equivalent | Closest metric drill (mm) | No. size | Decimal equivalent | Metric equivalent | Closest metric drill (mm) | Size letter | Decimal equivalent | Metric equivalent | Closest metric drill (mm) |
| 1 | .2280 | 5.791 | 5.80 | 41 | .0960 | 2.438 | 2.45 | A | .234 | 5.944 | 5.90 |
| 2 | .2210 | 5.613 | 5.60 | 42 | .0935 | 2.362 | 2.35 | B | .238 | 6.045 | 6.00 |
| 3 | .2130 | 5.410 | 5.40 | 43 | .0890 | 2.261 | 2.25 | C | .242 | 6.147 | 6.10 |
| 4 | .2090 | 5.309 | 5.30 | 44 | .0860 | 2.184 | 2.20 | D | .246 | 6.248 | 6.25 |
| 5 | .2055 | 5.220 | 5.20 | 45 | .0820 | 2.083 | 2.10 | E | .250 | 6.350 | 6.40 |
| 6 | .2040 | 5.182 | 5.20 | 46 | .0810 | 2.057 | 2.05 | F | .257 | 6.528 | 6.50 |
| 7 | .2010 | 5.105 | 5.10 | 47 | .0785 | 1.994 | 2.00 | G | .261 | 6.629 | 6.60 |
| 8 | .1990 | 5.055 | 5.10 | 48 | .0760 | 1.930 | 1.95 | H | .266 | 6.756 | 6.75 |
| 9 | .1960 | 4.978 | 5.00 | 49 | .0730 | 1.854 | 1.85 | I | .272 | 6.909 | 6.90 |
| 10 | .1935 | 4.915 | 4.90 | 50 | .0700 | 1.778 | 1.80 | J | .277 | 7.036 | 7.00 |
| 11 | .1910 | 4.851 | 4.90 | 51 | .0670 | 1.702 | 1.70 | K | .281 | 7.137 | 7.10 |
| 12 | .1890 | 4.801 | 4.80 | 52 | .0635 | 1.613 | 1.60 | L | .290 | 7.366 | 7.40 |
| 13 | .1850 | 4.699 | 4.70 | 53 | .0595 | 1.511 | 1.50 | M | .295 | 7.493 | 7.50 |
| 14 | .1820 | 4.623 | 4.60 | 54 | .0550 | 1.397 | 1.40 | N | .302 | 7.671 | 7.70 |
| 15 | .1800 | 4.572 | 4.60 | 55 | .0520 | 1.321 | 1.30 | O | .316 | 8.026 | 8.00 |
| 16 | .1770 | 4.496 | 4.50 | 56 | .0465 | 1.181 | 1.20 | P | .323 | 8.204 | 8.20 |
| 17 | .1730 | 4.394 | 4.40 | 57 | .0430 | 1.092 | 1.10 | Q | .332 | 8.433 | 8.40 |
| 18 | .1695 | 4.305 | 4.30 | 58 | .0420 | 1.067 | 1.05 | R | .339 | 8.611 | 8.60 |
| 19 | .1660 | 4.216 | 4.20 | 59 | .0410 | 1.041 | 1.05 | S | .348 | 8.839 | 8.80 |
| 19 | .1610 | 4.089 | 4.10 | 60 | .0400 | 1.016 | 1.00 | T | .358 | 9.093 | 9.10 |
| 21 | .1590 | 4.039 | 4.00 | 61 | .0390 | 0.991 | 1.00 | U | .368 | 9.347 | 9.30 |
| 22 | .1570 | 3.988 | 4.00 | 62 | .0380 | 0.965 | 0.95 | V | .377 | 9.576 | 9.60 |
| 23 | .1540 | 3.912 | 3.90 | 63 | .0370 | 0.940 | 0.95 | W | .386 | 9.804 | 9.80 |
| 24 | .1520 | 3.861 | 3.90 | 64 | .0360 | 0.914 | 0.90 | X | .397 | 10.084 | 10.00 |
| 25 | .1495 | 3.797 | 3.80 | 65 | .0350 | 0.889 | 0.90 | Y | .404 | 10.262 | 10.50 |
| 26 | .1470 | 3.734 | 3.75 | 66 | .0330 | 0.838 | 0.85 | Z | .413 | 10.491 | 10.50 |
| 27 | .1440 | 3.658 | 3.70 | 67 | .0320 | 0.813 | 0.80 | | | | |
| 28 | .1405 | 3.569 | 3.60 | 68 | .0310 | 0.787 | 0.80 | | | | |
| 29 | .1360 | 3.454 | 3.50 | 69 | .0292 | 0.742 | 0.75 | | | | |
| 30 | .1285 | 3.264 | 3.25 | 70 | .0280 | 0.711 | 0.70 | | | | |
| 31 | .1200 | 3.048 | 3.00 | 71 | .0260 | 0.660 | 0.65 | | | | |
| 32 | .1160 | 2.946 | 2.90 | 72 | .0250 | 0.635 | 0.65 | | | | |
| 33 | .1130 | 2.870 | 2.90 | 73 | .0240 | 0.610 | 0.60 | | | | |
| 34 | .1110 | 2.819 | 2.80 | 74 | .0225 | 0.572 | 0.55 | | | | |
| 35 | .1100 | 2.794 | 2.80 | 75 | .0210 | 0.533 | 0.55 | | | | |
| 36 | .1065 | 2.705 | 2.70 | 76 | .0200 | 0.508 | 0.50 | | | | |
| 37 | .1040 | 2.642 | 2.60 | 77 | .0180 | 0.457 | 0.45 | | | | |
| 38 | .1015 | 2.578 | 2.60 | 78 | .0160 | 0.406 | 0.40 | | | | |
| 39 | .0995 | 2.527 | 2.50 | 79 | .0145 | 0.368 | 0.35 | | | | |
| 40 | .0980 | 2.489 | 2.50 | 80 | .0135 | 0.343 | 0.35 | | | | |

*Fraction-size drills range in size from one-sixteenth—4 in. and over in diameter—by sixty-fourths.

From *Fundamentals of Engineering Drawing: For Design, Product Development, and Numerical Control,* 8th ed., copyright 1981 by Prentice-Hall, Inc. Used with permission of the publisher, Prentice-Hall, Inc., Englewood Cliffs, NJ 07632.

Metric Twist Drill Sizes With Decimal Inch Equivalents*

| Metric drill diameter | Diameter (in.) | Metric drill diameter | Diameter (in.) | Metric drill diameter | Diameter (in.) | Metric drill diameter | Diameter (in.) |
|---|---|---|---|---|---|---|---|
| 0.200 | .0078 | 2.300 | .0905 | 5.70 | .2244 | 9.20 | .3622 |
| 0.250 | .0098 | 2.35 | .0925 | 5.75 | .2264 | 9.25 | .3642 |
| 0.300 | .0118 | 2.400 | .0945 | 5.80 | .2283 | 9.30 | .3661 |
| 0.350 | .0138 | 2.45 | .0964 | 5.90 | .2323 | 9.40 | .3701 |
| 0.400 | .0157 | 2.50 | .0984 | 6.00 | .2362 | 9.50 | .3740 |
| 0.450 | .0177 | 2.60 | .1024 | 6.10 | .2401 | 9.60 | .3779 |
| 0.500 | .0197 | 2.70 | .1063 | 6.20 | .2441 | 9.70 | .3819 |
| 0.550 | .0216 | 2.75 | .1083 | 6.25 | .2461 | 9.75 | .3838 |
| 0.600 | .0236 | 2.80 | .1102 | 6.30 | .2480 | 9.80 | .3858 |
| 0.650 | .0256 | 2.90 | .1142 | 6.40 | .2520 | 9.90 | .3898 |
| 0.700 | .0275 | 3.00 | .1181 | 6.50 | .2559 | 10.00 | .3937 |
| 0.750 | .0295 | 3.10 | .1220 | 6.60 | .2598 | 10.50 | .4134 |
| 0.800 | .0315 | 3.20 | .1260 | 6.70 | .2638 | 11.00 | .4331 |
| 0.850 | .0335 | 3.25 | .1279 | 6.75 | .2657 | 11.50 | .4527 |
| 0.900 | .0354 | 3.30 | .1299 | 6.80 | .2677 | 12.00 | .4724 |
| 0.950 | .0374 | 3.40 | .1338 | 6.90 | .2716 | 12.50 | .4921 |
| 1.000 | .0394 | 3.50 | .1378 | 7.00 | .2756 | 13.00 | .5118 |
| 1.050 | .0413 | 3.60 | .1417 | 7.10 | .2795 | 13.50 | .5315 |
| 1.100 | .0433 | 3.70 | .1457 | 7.20 | .2835 | 14.00 | .5512 |
| 1.150 | .0453 | 3.75 | .1476 | 7.25 | .2854 | 14.50 | .5709 |
| 1.200 | .0472 | 3.80 | .1496 | 7.30 | .2874 | 15.00 | .5905 |
| 1.250 | .0492 | 3.90 | .1535 | 7.40 | .2913 | 15.50 | .6102 |
| 1.300 | .0512 | 4.00 | .1575 | 7.50 | .2953 | 16.00 | .6299 |
| 1.350 | .0531 | 4.10 | .1614 | 7.60 | .2992 | 16.50 | .6496 |
| 1.400 | .0551 | 4.20 | .1653 | 7.70 | .3031 | 17.00 | .6693 |
| 1.450 | .0571 | 4.25 | .1673 | 7.75 | .3051 | 17.50 | .6890 |
| 1.500 | .0590 | 4.30 | .1693 | 7.80 | .3071 | 18.00 | .7087 |
| 1.550 | .0610 | 4.40 | .1732 | 7.90 | .3110 | 18.50 | .7283 |
| 1.600 | .0630 | 4.50 | .1772 | 8.00 | .3150 | 19.00 | .7480 |
| 1.650 | .0650 | 4.60 | .1811 | 8.10 | .3189 | 19.50 | .7677 |
| 1.700 | .0669 | 4.70 | .1850 | 8.20 | .3228 | 20.00 | .7874 |
| 1.750 | .0689 | 4.75 | .1870 | 8.25 | .3248 | 20.50 | .8071 |
| 1.800 | .0709 | 4.80 | .1890 | 8.30 | .3268 | 21.00 | .8268 |
| 1.850 | .0728 | 4.90 | .1929 | 8.40 | .3307 | 21.50 | .8464 |
| 1.900 | .0748 | 5.00 | .1968 | 8.50 | .3346 | 22.00 | .8661 |
| 1.950 | .0768 | 5.10 | .2008 | 8.60 | .3386 | 22.50 | .8858 |
| 2.000 | .0787 | 5.20 | .2047 | 8.70 | .3425 | 23.00 | .9055 |
| 2.050 | .0807 | 5.25 | .2067 | 8.75 | .3445 | 23.50 | .9252 |
| 2.100 | .0827 | 5.30 | .2087 | 8.80 | .3464 | 24.00 | .9449 |
| 2.150 | .0846 | 5.40 | .2126 | 8.90 | .3504 | 24.50 | .9646 |
| 2.200 | .0866 | 5.50 | .2165 | 9.00 | .3543 | 25.00 | .9842 |
| 2.250 | .0886 | 5.60 | .2205 | 9.10 | .3583 | 25.50 | 1.0039 |

* Drills beyond the range of this table increase in diameter by increments of 0.50 mm—26.00, 26.50, 27.00, etc.

From *Fundamentals of Engineering Drawing: For Design, Product Development, and Numerical Control*, 8th ed., copyright 1981 by Prentice-Hall, Inc. Used with permission of the publisher, Prentice-Hall, Inc., Englewood Cliffs, NJ 07632.

# APPENDIX E: METRIC THREADS AND FASTENERS

## Metric Screw Thread Series

The standard M profile diameter/pitch combinations are shown in boldface.)
ISO 261–1973 Diameter/Pitch Combination.

| Nominal Diameters | | | Pitches | | | | | | | | | | |
|---|---|---|---|---|---|---|---|---|---|---|---|---|---|
| Col. 1 1st Choice | Col. 2 2nd Choice | Col. 3 3rd Choice | Coarse | Fine | | | | | | | | | |
| | | | | 3 | 2 | 1.5 | 1.25 | 1 | 0.75 | 0.5 | 0.35 | 0.25 | 0.2 |
| **1.6** | | | **0.35** | ... | ... | ... | ... | ... | ... | ... | ... | ... | 0.2 |
| | 1.8 | | 0.35 | ... | ... | ... | ... | ... | ... | ... | ... | ... | 0.2 |
| **2** | | | **0.4** | ... | ... | ... | ... | ... | ... | ... | ... | 0.25 | ... |
| | 2.2 | | 0.45 | ... | ... | ... | ... | ... | ... | ... | ... | 0.25 | ... |
| **2.5** | | | **0.45** | ... | ... | ... | ... | ... | ... | ... | 0.35 | ... | ... |
| **3** | | | **0.5** | ... | ... | ... | ... | ... | ... | ... | 0.35 | ... | ... |
| | **3.5** | | **0.6** | ... | ... | ... | ... | ... | ... | ... | 0.35 | ... | ... |
| **4** | | | **0.7** | ... | ... | ... | ... | ... | ... | 0.5 | ... | ... | ... |
| | 4.5 | | 0.75 | ... | ... | ... | ... | ... | ... | 0.5 | ... | ... | ... |
| **5** | | | **0.8** | ... | ... | ... | ... | ... | ... | 0.5 | ... | ... | ... |
| | | 5.5 | ... | ... | ... | ... | ... | ... | ... | 0.5 | ... | ... | ... |
| **6** | | | **1** | ... | ... | ... | ... | ... | 0.75 | ... | ... | ... | ... |
| | | 7 | 1 | ... | ... | ... | ... | ... | 0.75 | ... | ... | ... | ... |
| **8** | | | **1.25** | ... | ... | ... | ... | **1** | 0.75 | ... | ... | ... | ... |
| | | 9 | 1.25 | ... | ... | ... | ... | 1 | 0.75 | ... | ... | ... | ... |
| **10** | | | **1.5** | ... | ... | ... | **1.25** | 1 | **0.75** | ... | ... | ... | ... |
| | | 11 | 1.5 | ... | ... | ... | ... | 1 | 0.75 | ... | ... | ... | ... |
| **12** | | | **1.75** | ... | ... | **1.5**[d] | **1.25** | 1 | ... | ... | ... | ... | ... |
| | 14 | | 2 | ... | ... | **1.5** | **1.25**[a] | 1 | ... | ... | ... | ... | ... |
| | | 15 | ... | ... | ... | 1.5 | ... | 1 | ... | ... | ... | ... | ... |
| **16** | | | **2** | ... | ... | **1.5** | ... | 1 | ... | ... | ... | ... | ... |
| | | 17 | ... | ... | ... | 1.5 | ... | 1 | ... | ... | ... | ... | ... |
| | 18 | | 2.5 | ... | 2 | **1.5** | ... | 1 | ... | ... | ... | ... | ... |
| **20** | | | **2.5** | ... | 2 | **1.5** | ... | 1 | ... | ... | ... | ... | ... |
| | 22 | | 2.5[c] | ... | 2 | **1.5** | ... | 1 | ... | ... | ... | ... | ... |
| **24** | | | **3** | ... | **2** | 1.5 | ... | 1 | ... | ... | ... | ... | ... |
| | | 25 | ... | ... | 2 | **1.5** | ... | 1 | ... | ... | ... | ... | ... |
| | | 26 | ... | ... | ... | 1.5 | ... | ... | ... | ... | ... | ... | ... |
| | 27 | | 3[c] | ... | **2** | 1.5 | ... | 1 | ... | ... | ... | ... | ... |
| | | 28 | ... | ... | 2 | 1.5 | ... | 1 | ... | ... | ... | ... | ... |
| **30** | | | **3.5** | (3)[e] | **2** | 1.5 | ... | 1 | ... | ... | ... | ... | ... |
| | | 32 | ... | ... | 2 | 1.5 | ... | ... | ... | ... | ... | ... | ... |
| | 33 | | 3.5 | (3)[e] | **2** | 1.5 | ... | ... | ... | ... | ... | ... | ... |
| | | 35[b] | ... | ... | ... | 1.5 | ... | ... | ... | ... | ... | ... | ... |
| **36** | | | **4** | 3 | **2** | 1.5 | ... | ... | ... | ... | ... | ... | ... |
| | | 38 | ... | ... | ... | 1.5 | ... | ... | ... | ... | ... | ... | ... |
| | 39 | | 4 | 3 | **2** | 1.5 | ... | ... | ... | ... | ... | ... | ... |

(a) Only for spark plugs for engines.
(b) Only for nuts for bearings.
(c) Only for high strength structural steel bolts.

(d) Only for wheel studs and nuts.
(e) Pitches shown in brackets are to be avoided as far as possible.

Reproduced from ANSI B1.13M–1979 with the permission of the publisher, the American Society of Mechanical Engineers.

# Metric Screw Thread Series

## ISO 261–1973 Diameter/Pitch Combinations (con't.)

| Col. 1<br>1st Choice | Col. 2<br>2nd Choice | Col. 3<br>3rd Choice | Coarse | Fine | | | | |
|---|---|---|---|---|---|---|---|---|
| | | | | 6 | 4 | 3 | 2 | 1.5 |
| | | 40 | ... | ... | ... | 3 | 2 | 1.5 |
| 42 | | | 4.5 | ... | 4 | 3 | 2 | 1.5 |
| | 45 | | 4.5 | ... | 4 | 3 | 2 | 1.5 |
| 48 | | | 5 | ... | 4 | 3 | 2 | 1.5 |
| | | 50 | ... | ... | ... | 3 | 2 | 1.5 |
| | 52 | | 5 | ... | 4 | 3 | 2 | 1.5 |
| | | 55 | ... | ... | 4 | 3 | 2 | 1.5 |
| 56 | | | 5.5 | ... | 4 | 3 | 2 | 1.5 |
| | | 58 | ... | ... | 4 | 3 | 2 | 1.5 |
| | 60 | | 5.5 | ... | 4 | 3 | 2 | 1.5 |
| | | 62 | ... | ... | 4 | 3 | 2 | 1.5 |
| 64 | | | 6 | ... | 4 | 3 | 2 | 1.5 |
| | | 65 | ... | ... | 4 | 3 | 2 | 1.5 |
| | 68 | | 6 | ... | 4 | 3 | 2 | 1.5 |
| | | 70 | ... | 6 | 4 | 3 | 2 | 1.5 |
| 72 | | | 6 | 6 | 4 | 3 | 2 | 1.5 |
| | | 75 | ... | ... | 4 | 3 | 2 | 1.5 |
| | 76 | | ... | 6 | 4 | 3 | 2 | 1.5 |
| | | 78 | ... | ... | ... | ... | 2 | ... |
| 80 | | | 6 | 6 | 4 | 3 | 2 | 1.5 |
| | | 82 | ... | ... | ... | ... | 2 | ... |
| | 85 | | ... | 6 | 4 | 3 | 2 | ... |
| 90 | | | 6 | 6 | 4 | 3 | 2 | ... |
| | 95 | | ... | 6 | 4 | 3 | 2 | ... |
| 100 | | | 6 | 6 | 4 | 3 | 2 | ... |
| | 105 | | ... | 6 | 4 | 3 | 2 | ... |
| 110 | | | ... | 6 | 4 | 3 | 2 | ... |
| | 115 | | ... | 6 | 4 | 3 | 2 | ... |
| | 120 | | ... | 6 | 4 | 3 | 2 | ... |
| 125 | | | ... | 6 | 4 | 3 | 2 | ... |
| | 130 | | ... | 6 | 4 | 3 | 2 | ... |
| | | 135 | ... | 6 | 4 | 3 | 2 | ... |
| 140 | | | ... | 6 | 4 | 3 | 2 | ... |
| | | 145 | ... | 6 | 4 | 3 | 2 | ... |
| | 150 | | ... | 6 | 4 | 3 | 2 | ... |
| | | 155 | ... | 6 | 4 | 3 | ... | ... |
| 160 | | | ... | 6 | 4 | 3 | ... | ... |
| | | 165 | ... | 6 | 4 | 3 | ... | ... |
| | 170 | | ... | 6 | 4 | 3 | ... | ... |
| | | 175 | ... | 6 | 4 | 3 | ... | ... |
| 180 | | | ... | 6 | 4 | 3 | ... | ... |
| | | 185 | ... | 6 | 4 | 3 | ... | ... |
| | 190 | | ... | 6 | 4 | 3 | ... | ... |
| | | 195 | ... | 6 | 4 | 3 | ... | ... |
| 200 | | | ... | 6 | 4 | 3 | ... | ... |

Reproduced from ANSI B1.13M–1979 with the permission of the publisher, the American Society of Mechanical Engineers.

Metric Finished Hexagon Head Bolts and Cap Screws

| Nominal Screw Size & Thread Pitch | E Body Diameter Max | E Body Diameter Min | F Width Across Flats Max | F Width Across Flats Min | G Width Across Corners Max | G Width Across Corners Min | H Head Height Max | H Head Height Min | J Wrenching Height Min | K Washer Face Thickness Max | K Washer Face Thickness Min | M Washer Face Dia Min | E_a Fillet Transition Dia Max | L_a Fillet Transition Length Max | R Radius of Fillet Min | L_T Screw Lengths ≤125 | L_T Screw Lengths >125 and ≤200 | L_T Screw Lengths >200 | Y Transition Thread Length Max |
|---|---|---|---|---|---|---|---|---|---|---|---|---|---|---|---|---|---|---|---|
| M5x0.8 | 5.00 | 4.82 | 8.00 | 7.78 | 9.24 | 8.79 | 3.65 | 3.35 | 2.4 | 0.5 | 0.2 | 7.0 | 5.7 | 1.2 | 0.2 | 16 | 22 | 35 | 4.0 |
| M6x1 | 6.00 | 5.82 | 10.00 | 9.78 | 11.55 | 11.05 | 4.15 | 3.85 | 2.8 | 0.5 | 0.2 | 8.9 | 7.0 | 1.8 | 0.3 | 18 | 24 | 37 | 5.0 |
| M8x1.25 | 8.00 | 7.78 | 13.00 | 12.73 | 15.01 | 14.38 | 5.50 | 5.10 | 3.7 | 0.6 | 0.3 | 11.6 | 9.2 | 2.0 | 0.4 | 22 | 28 | 41 | 6.2 |
| M10x1.5 | 10.00 | 9.78 | 16.00 | 15.73 | 18.48 | 17.77 | 6.63 | 6.17 | 4.5 | 0.6 | 0.3 | 14.6 | 11.2 | 2.0 | 0.4 | 26 | 32 | 45 | 7.5 |
| M12x1.75 | 12.00 | 11.73 | 18.00 | 17.73 | 20.78 | 20.03 | 7.76 | 7.24 | 5.2 | 0.6 | 0.3 | 16.6 | 13.2 | 3.0 | 0.4 | 30 | 36 | 49 | 8.8 |
| M14x2 | 14.00 | 13.73 | 21.00 | 20.67 | 24.25 | 23.35 | 9.09 | 8.51 | 6.2 | 0.6 | 0.3 | 19.6 | 15.2 | 3.0 | 0.4 | 34 | 40 | 53 | 10.0 |
| M16x2 | 16.00 | 15.73 | 24.00 | 23.67 | 27.71 | 26.75 | 10.32 | 9.68 | 7.0 | 0.8 | 0.4 | 22.5 | 17.7 | 3.0 | 0.6 | 38 | 44 | 57 | 10.0 |
| M20x2.5 | 20.00 | 19.67 | 30.00 | 29.16 | 34.64 | 32.95 | 12.88 | 12.12 | 8.8 | 0.8 | 0.4 | 27.7 | 22.4 | 4.0 | 0.8 | 46 | 52 | 65 | 12.5 |
| M24x3 | 24.00 | 23.67 | 36.00 | 35.00 | 41.57 | 39.55 | 15.44 | 14.56 | 10.5 | 0.8 | 0.4 | 33.2 | 26.4 | 4.0 | 0.8 | 54 | 60 | 73 | 15.0 |
| M30x3.5 | 30.00 | 29.67 | 46.00 | 45.00 | 53.12 | 50.85 | 19.48 | 17.92 | 13.1 | 0.8 | 0.4 | 42.7 | 33.4 | 6.0 | 1.0 | 66 | 72 | 85 | 17.5 |
| M36x4 | 36.00 | 35.61 | 55.00 | 53.80 | 63.51 | 60.78 | 23.38 | 21.62 | 15.8 | 0.8 | 0.4 | 51.1 | 39.4 | 6.0 | 1.0 | 78 | 84 | 97 | 20.0 |
| M42x4.5 | 42.00 | 41.38 | 65.00 | 62.90 | 75.06 | 71.71 | 26.97 | 25.03 | 18.2 | 1.0 | 0.5 | 59.8 | 45.6 | 6.3 | 1.2 | 90 | 96 | 109 | 22.5 |
| M48x5 | 48.00 | 47.38 | 75.00 | 72.60 | 86.60 | 82.76 | 31.07 | 28.93 | 21.0 | 1.0 | 0.5 | 69.0 | 52.6 | 8.0 | 1.5 | 102 | 108 | 121 | 25.0 |
| M56x5.5 | 56.00 | 55.26 | 85.00 | 82.20 | 98.15 | 93.71 | 36.20 | 33.80 | 24.5 | 1.0 | 0.5 | 78.1 | 62.0 | 10.5 | 2.0 | — | 124 | 137 | 27.5 |
| M64x6 | 64.00 | 63.26 | 95.00 | 91.80 | 109.70 | 104.65 | 41.32 | 38.68 | 28.0 | 1.0 | 0.5 | 87.2 | 70.0 | 10.5 | 2.0 | — | 140 | 153 | 30.0 |
| M72x6 | 72.00 | 71.26 | 105.00 | 101.40 | 121.24 | 115.60 | 46.45 | 43.55 | 31.5 | 1.2 | 0.6 | 96.3 | 78.0 | 10.5 | 2.0 | — | 156 | 169 | 30.0 |
| M80x6 | 80.00 | 79.26 | 115.00 | 111.00 | 132.79 | 126.54 | 51.58 | 48.42 | 35.0 | 1.2 | 0.6 | 105.4 | 86.0 | 10.5 | 2.0 | — | 172 | 185 | 30.0 |
| M90x6 | 90.00 | 89.13 | 130.00 | 125.50 | 150.11 | 143.07 | 57.74 | 54.26 | 39.2 | 1.2 | 0.6 | 119.2 | 96.0 | 10.5 | 2.0 | — | 192 | 205 | 30.0 |
| M100x6 | 100.00 | 99.13 | 145.00 | 140.00 | 167.43 | 159.60 | 63.90 | 60.10 | 43.4 | 1.2 | 0.6 | 133.0 | 107.0 | 12.2 | 2.5 | — | 212 | 225 | 30.0 |

All dimensions are in millimetres.

SHADED SECTION REPRESENTS ENVELOPE OF FILLET LIMITS

FILLET

ENLARGED DETAIL OF FILLET

OPTIONAL POINT CONSTRUCTIONS

PROPERTY CLASS AND MANUFACTURER'S IDENTIFICATION TO APPEAR ON TOP OF HEAD

FILLET SEE DETAIL

15°–30°

Reproduced from ANSI B18.2.3.5M–1979 with the permission of the publisher, the American Society of Mechanical Engineers.
From *Technical Drafting*, copyright 1980 by Spence and Atkins. Used with permission of the publisher, Bennett Publishing Company, Peoria, IL 61615.

Metric Unfinished Hexagon Head Bolts

| Nominal Bolt Size & Thread Pitch | E Body Diameter | | F Width Across Flats | | G Width Across Corners | | H Head Height | | $E_a$ Fillet Transition Dia | R Radius of Fillet | $L_T$ (Ref) Thread Length (Basic) | | |
|---|---|---|---|---|---|---|---|---|---|---|---|---|---|
| | Max | Min | Max | Min | Max | Min | Max | Min | Max | Min | Bolt Lengths ⩽125 | Bolt Lengths >125 and ⩽200 | Bolt Lengths >200 |
| M5x0.8 | 5.48 | 4.52 | 8.00 | 7.64 | 9.24 | 8.63 | 3.88 | 3.35 | 5.7 | 0.2 | 16 | 22 | 35 |
| M6x1 | 6.48 | 5.52 | 10.00 | 9.64 | 11.55 | 10.89 | 4.38 | 3.85 | 6.8 | 0.3 | 18 | 24 | 37 |
| M8x1.25 | 8.58 | 7.42 | 13.00 | 12.57 | 15.01 | 14.20 | 5.68 | 5.10 | 9.2 | 0.4 | 22 | 28 | 41 |
| M10x1.5 | 10.58 | 9.42 | 16.00 | 15.57 | 18.48 | 17.59 | 6.85 | 6.17 | 11.2 | 0.4 | 26 | 32 | 45 |
| M12x1.75 | 12.70 | 11.30 | 18.00 | 17.57 | 20.78 | 19.85 | 7.95 | 7.24 | 13.7 | 0.4 | 30 | 36 | 49 |
| M14x2 | 14.70 | 13.30 | 21.00 | 20.16 | 24.25 | 22.78 | 9.25 | 8.51 | 15.7 | 0.6 | 34 | 40 | 53 |
| M16x2 | 16.70 | 15.30 | 24.00 | 23.16 | 27.71 | 26.17 | 10.75 | 9.68 | 17.7 | 0.6 | 38 | 44 | 57 |
| M20x2.5 | 20.84 | 19.16 | 30.00 | 29.16 | 34.64 | 32.95 | 13.40 | 12.12 | 22.4 | 0.8 | 46 | 52 | 65 |
| M24x3 | 24.84 | 23.16 | 36.00 | 35.00 | 41.57 | 39.55 | 15.90 | 14.56 | 26.4 | 0.8 | 54 | 60 | 73 |
| M30x3.5 | 30.84 | 29.16 | 46.00 | 45.00 | 53.12 | 50.85 | 19.75 | 17.92 | 33.4 | 1.0 | 66 | 72 | 85 |
| M36x4 | 37.00 | 35.00 | 55.00 | 53.80 | 63.51 | 60.79 | 23.55 | 21.72 | 39.4 | 1.0 | 78 | 84 | 97 |
| M42x4.5 | 43.00 | 41.00 | 65.00 | 62.90 | 75.06 | 71.71 | 27.05 | 25.03 | 45.4 | 1.2 | 90 | 96 | 109 |
| M48x5 | 49.00 | 47.00 | 75.00 | 72.60 | 86.60 | 82.76 | 31.07 | 28.93 | 52.0 | 1.5 | 102 | 108 | 121 |
| M56x5.5 | 57.20 | 54.80 | 85.00 | 82.20 | 98.15 | 93.71 | 36.20 | 33.80 | 62.0 | 2.0 | — | 124 | 137 |
| M64x6 | 65.52 | 62.80 | 95.00 | 91.80 | 109.70 | 104.65 | 41.32 | 38.68 | 70.0 | 2.0 | — | 140 | 153 |
| M72x6 | 73.84 | 70.80 | 105.00 | 101.40 | 121.24 | 115.60 | 46.45 | 43.55 | 78.0 | 2.0 | — | 156 | 169 |
| M80x6 | 82.16 | 78.80 | 115.00 | 111.00 | 132.79 | 126.54 | 51.58 | 48.42 | 86.0 | 2.0 | — | 172 | 185 |
| M90x6 | 92.48 | 88.60 | 130.00 | 125.50 | 150.11 | 143.07 | 57.74 | 54.26 | 96.0 | 2.0 | — | 192 | 205 |
| M100x6 | 102.80 | 98.60 | 145.00 | 140.00 | 167.43 | 159.60 | 63.90 | 60.10 | 107.0 | 2.5 | — | 212 | 225 |

All dimensions are in millimetres.

Reproduced from ANSI B18.2.3.5M–1979 with the permission of the publisher, the American Society of Mechanical Engineers.

From *Technical Drafting*, copyright 1980 by Spence and Atkins. Used with permission of the publisher, Bennett Publishing Company, Peoria, IL 61615.

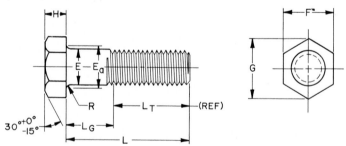

## Metric Hex Nuts

| Nominal Nut Size and Thread Pitch | F Width Across Flats | | G Width Across Corners | | O Bearing Face Dia | H Nut Thickness Style 1 | | H₁ Nut Thickness Style 2 | |
|---|---|---|---|---|---|---|---|---|---|
| | Max | Min | Max | Min | Min | Max | Min | Max | Min |
| M1.6x0.35 | 3.20 | 3.08 | 3.69 | 3.48 | 2.5 | — | — | 1.3 | 1.1 |
| M2x0.4 | 4.00 | 3.88 | 4.61 | 4.38 | 3.1 | — | — | 1.6 | 1.3 |
| M2.5x0.45 | 5.00 | 4.88 | 5.77 | 5.51 | 4.1 | — | — | 2.0 | 1.7 |
| M3x0.5 | 5.50 | 5.38 | 6.35 | 6.08 | 4.6 | — | — | 2.4 | 2.1 |
| M3.5x0.6 | 6.00 | 5.88 | 6.92 | 6.64 | 5.1 | — | — | 2.8 | 2.5 |
| M4x0.7 | 7.00 | 6.85 | 8.08 | 7.74 | 6.0 | — | — | 3.2 | 2.9 |
| M5x0.8 | 8.00 | 7.85 | 9.23 | 8.87 | 7.0 | 4.7 | 4.4 | 5.1 | 4.8 |
| M6x1 | 10.00 | 9.78 | 11.54 | 11.05 | 8.9 | 5.2 | 4.9 | 5.7 | 5.4 |
| M8x1.25 | 13.00 | 12.73 | 15.01 | 14.38 | 11.6 | 6.8 | 6.4 | 7.5 | 7.1 |
| M10x1.5 | 16.00 | 15.73 | 18.47 | 17.77 | 14.6 | 8.4 | 8.0 | 9.3 | 8.9 |
| M12x1.75 | 18.00 | 17.73 | 20.78 | 20.03 | 16.6 | 10.8 | 10.4 | 12.0 | 11.6 |
| M14x2 | 21.00 | 20.67 | 24.24 | 23.35 | 19.4 | 12.8 | 12.1 | 14.1 | 13.4 |
| M16x2 | 24.00 | 23.67 | 27.71 | 26.75 | 22.4 | 14.8 | 14.1 | 16.4 | 15.7 |
| M20x2.5 | 30.00 | 29.16 | 34.64 | 32.95 | 27.9 | 18.0 | 16.9 | 20.1 | 19.0 |
| M24x3 | 36.00 | 35.00 | 41.56 | 39.55 | 32.5 | 21.5 | 20.2 | 23.9 | 22.6 |
| M30x3.5 | 46.00 | 45.00 | 53.11 | 50.85 | 42.5 | 25.6 | 24.3 | 28.6 | 27.3 |
| M36x4 | 55.00 | 53.80 | 63.50 | 60.79 | 50.8 | 31.0 | 29.4 | 34.7 | 33.1 |

All dimensions are in millimetres.

Reproduced from ANSI 18.2.4.1M–1979 with the permission of the publisher, the American Society of Mechanical Engineers.

From *Technical Drafting*, copyright 1980 by Spence and Atkins. Used with permission of the publisher, Bennett Publishing Company, Peoria, IL 61615.

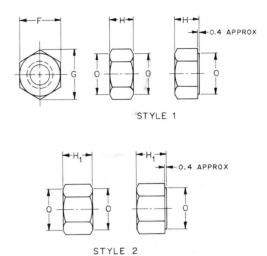

STYLE 1

STYLE 2

## Metric Hex Slotted Nuts

| Nominal Nut Size and Thread Pitch | F Width Across Flats | | G Width Across Corners | | O Bearing Face Dia | H Nut Thickness Style 1 | | H₁ Nut Thickness Style 2 | | T Unslotted Thickness Style 1 | | T₁ Unslotted Thickness Style 2 | | S Width of Slot | |
|---|---|---|---|---|---|---|---|---|---|---|---|---|---|---|---|
| | Max | Min | Max | Min | Min | Max | Min | Max | Min | Max | Min | Max | Min | Max | Min |
| M5x0.8 | 8.00 | 7.78 | 9.24 | 8.87 | 7.0 | 4.5 | 4.2 | 5.3 | 5.0 | 3.2 | 2.7 | 3.7 | 3.2 | 2.2 | 1.4 |
| M6.3x1 | 10.00 | 9.76 | 11.55 | 11.13 | 8.9 | 5.6 | 5.3 | 6.5 | 6.2 | 3.9 | 3.4 | 4.5 | 4.0 | 2.8 | 2.0 |
| M8x1.25 | 13.00 | 12.73 | 15.01 | 14.51 | 11.6 | 6.6 | 6.2 | 7.8 | 7.4 | 4.5 | 4.0 | 5.3 | 4.8 | 3.3 | 2.5 |
| M10x1.5 | 15.00 | 14.70 | 17.32 | 16.76 | 13.6 | 9.0 | 8.5 | 10.7 | 10.2 | 6.0 | 5.5 | 7.1 | 6.6 | 3.6 | 2.8 |
| M12x1.75 | 18.00 | 17.67 | 20.78 | 20.14 | 16.6 | 10.7 | 10.2 | 12.8 | 12.3 | 7.1 | 6.6 | 8.5 | 8.0 | 4.3 | 3.5 |
| M14x2 | 21.00 | 20.64 | 24.25 | 23.53 | 19.4 | 12.5 | 11.9 | 14.9 | 14.3 | 8.2 | 7.7 | 9.8 | 9.3 | 4.3 | 3.5 |
| M16x2 | 24.00 | 23.61 | 27.71 | 26.92 | 22.4 | 14.5 | 13.9 | 17.4 | 16.8 | 9.5 | 9.0 | 11.4 | 10.9 | 6.0 | 4.5 |
| M20x2.5 | 30.00 | 29.00 | 34.64 | 33.06 | 27.6 | 18.4 | 17.4 | 21.2 | 20.2 | 12.1 | 11.3 | 13.9 | 13.1 | 6.0 | 4.5 |
| M24x3 | 36.00 | 34.80 | 41.57 | 39.67 | 32.9 | 22.0 | 20.9 | 25.4 | 24.3 | 14.4 | 13.6 | 16.6 | 15.8 | 7.0 | 5.5 |
| M30x3.5 | 46.00 | 44.50 | 53.12 | 50.73 | 42.5 | 26.7 | 25.4 | 31.0 | 29.7 | 17.3 | 16.5 | 20.1 | 19.3 | 9.3 | 7.0 |
| M36x4 | 55.00 | 53.20 | 63.51 | 60.65 | 50.8 | 32.0 | 30.5 | 37.6 | 36.1 | 20.8 | 19.8 | 24.5 | 23.5 | 9.3 | 7.0 |

Reproduced from ANSI B18.2.4.3M–1979 with the permission of the publisher, the American Society of Mechanical Engineers.

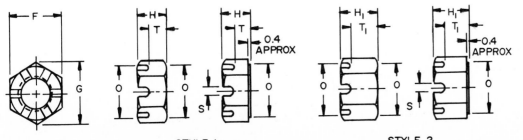

STYLE 1                     STYLE 2

707

Metric 12-Spline Flange Bolts

| Nominal Screw Size & Thread Pitch | E Body Dia | | F Flange Dia | N Bearing Circle Dia | K Flange Edge Thickness | M Flange Height | J Wrenching Height | H Head Height | S Chamfer Height | R Spline Junction Radius | L_T (Ref) Thread Length, Basic | | | Y (Ref) Transition Thread Length |
|---|---|---|---|---|---|---|---|---|---|---|---|---|---|---|
| | Max | Min | Max | Min | Min | Min | Min | Max | Max | Min | For Screw Lengths ≤125 | For Screw Lengths >125 and ≤200 | For Screw Lengths >200 | Max |
| M5x0.8 | 5.00 | 4.82 | 9.4 | 8.4 | 1.0 | 1.7 | 1.8 | 5.0 | 0.6 | 0.4 | 16.0 | 22.0 | 35.0 | 4.0 |
| M6x1 | 6.00 | 5.78 | 11.8 | 10.7 | 1.2 | 2.2 | 2.3 | 6.3 | 0.8 | 0.5 | 18.0 | 24.0 | 37.0 | 5.0 |
| M8x1.25 | 8.00 | 7.78 | 15.0 | 13.7 | 1.5 | 2.7 | 3.0 | 8.0 | 1.0 | 0.6 | 22.0 | 28.0 | 41.0 | 6.2 |
| M10x1.5 | 10.00 | 9.78 | 18.6 | 17.1 | 2.0 | 3.4 | 3.8 | 10.0 | 1.2 | 0.7 | 26.0 | 32.0 | 45.0 | 7.5 |
| M12x1.75 | 12.00 | 11.73 | 22.8 | 21.1 | 2.3 | 4.1 | 4.5 | 12.0 | 1.5 | 0.8 | 30.0 | 36.0 | 49.0 | 8.8 |
| M14x2 | 14.00 | 13.73 | 26.4 | 24.5 | 2.7 | 4.8 | 5.4 | 14.0 | 1.8 | 0.9 | 34.0 | 40.0 | 53.0 | 10.0 |
| M16x2 | 16.00 | 15.73 | 30.3 | 28.1 | 3.2 | 5.7 | 5.8 | 16.0 | 2.1 | 1.0 | 38.0 | 44.0 | 57.0 | 10.0 |
| M20x2.5 | 20.00 | 19.67 | 37.4 | 34.9 | 4.1 | 7.2 | 7.2 | 20.0 | 2.5 | 1.2 | 46.0 | 52.0 | 65.0 | 12.5 |

All dimensions are in millimetres.

From *Technical Drafting*, copyright 1980 by Spence and Atkins. Used with permission of the publisher, Bennett Publishing Company, Peoria, IL 61615.

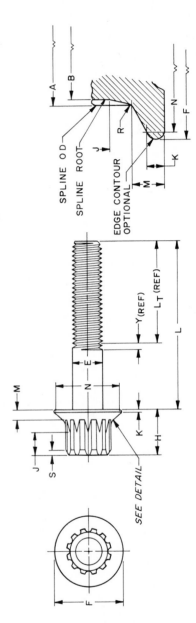

# Metric Hexagon Head and Hexagon Washer Head Machine Screws

| Nominal Screw Size and Thread Pitch | E Body Diameter | | A Width across Flats | | W Width across Corners | H Head Height | | B Washer Diameter | | U Washer Thickness | | R Fillet Radius | | $R_1$ Head Fillet Radius |
|---|---|---|---|---|---|---|---|---|---|---|---|---|---|---|
| | Max | Min | Max | Min | Min | Max | Min | Max | Min | Max | Min | Max | Min | Max |
| M2x0.4 | 2.00 | 1.65 | 3.20 | 3.02 | 3.36 | 1.27 | 1.02 | 4.2 | 3.9 | 0.4 | 0.2 | 0.3 | 0.1 | 0.4 |
| M2.5x0.45 | 2.50 | 2.12 | 4.00 | 3.82 | 4.25 | 1.40 | 1.12 | 5.3 | 5.0 | 0.5 | 0.3 | 0.4 | 0.1 | 0.5 |
| M3x0.5 | 3.00 | 2.58 | 5.00 | 4.82 | 5.36 | 1.52 | 1.24 | 6.2 | 5.8 | 0.5 | 0.3 | 0.5 | 0.2 | 0.5 |
| M3.5x0.6 | 3.50 | 3.00 | 5.50 | 5.32 | 5.92 | 2.36 | 2.03 | 7.5 | 6.8 | 0.6 | 0.4 | 0.5 | 0.2 | 0.6 |
| M4x0.7 | 4.00 | 3.43 | 7.00 | 6.78 | 7.55 | 2.79 | 2.44 | 9.2 | 8.5 | 0.7 | 0.5 | 0.6 | 0.2 | 0.7 |
| M5x0.8 | 5.00 | 4.36 | 8.00 | 7.78 | 8.66 | 3.05 | 2.67 | 10.5 | 9.8 | 0.8 | 0.5 | 0.8 | 0.3 | 0.8 |
| M6x1 | 6.00 | 5.21 | 10.00 | 9.78 | 10.89 | 4.83 | 4.37 | 13.2 | 12.2 | 1.3 | 0.8 | 1.0 | 0.3 | 1.0 |
| M8x1.25 | 8.00 | 7.04 | 13.00 | 12.73 | 14.17 | 5.84 | 5.28 | 17.2 | 15.9 | 1.4 | 0.9 | 1.2 | 0.4 | 1.2 |
| M10x1.5 | 10.00 | 8.86 | 15.00 | 14.73 | 16.41 | 7.49 | 6.86 | 19.8 | 18.3 | 1.6 | 0.9 | 1.5 | 0.5 | 1.5 |
| M12x1.75 | 12.00 | 10.68 | 18.00 | 17.73 | 19.75 | 9.50 | 8.66 | 23.8 | 22.0 | 1.8 | 1.1 | 1.8 | 0.6 | 1.8 |

Note: For lengths less than 30 mm, completely threaded or $L_T = 25$ mm MIN. For lengths greater than 30 mm, completely threaded or $L_T = 38$ mm MIN.
All dimensions are in millimetres.

Reproduced from ANSI B18.3.6M–1979 with the permission of the publisher, the American Society of Mechanical Engineers.

From *Technical Drafting*, copyright 1980 by Spence and Atkins. Used with permission of the publisher, Bennett Publishing Company, Peoria, IL 61615.

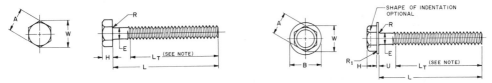

# Metric Slotted Flat and Oval Head Machine Screws

| Nominal Screw Size and Thread Pitch | E Body Diameter | | A Head Diameter | | | H Head Side Height | F Raised Head Height | $R_1$ Head Radius | R Fillet Radius | | J Slot Width | | T Slot Depth | |
|---|---|---|---|---|---|---|---|---|---|---|---|---|---|---|
| | | | Theoretical Sharp | | Actual | | | | | | | | | |
| | Max | Min | Max | Min | Min | Max Ref | Max | Approx | Max | Min | Max | Min | Max | Min |
| M2x0.4 | 2.00 | 1.65 | 4.40 | 3.90 | 3.60 | 1.20 | 0.50 | 3.8 | 0.8 | 0.2 | 0.7 | 0.5 | 1.0 | 0.8 |
| M2.5x0.45 | 2.50 | 2.12 | 5.50 | 4.90 | 4.60 | 1.50 | 0.60 | 5.0 | 1.0 | 0.3 | 0.8 | 0.6 | 1.2 | 1.0 |
| M3x0.5 | 3.00 | 2.58 | 6.60 | 5.80 | 5.50 | 1.80 | 0.75 | 5.7 | 1.2 | 0.3 | 1.0 | 0.8 | 1.5 | 1.2 |
| M3.5x0.6 | 3.50 | 3.00 | 7.70 | 6.80 | 6.44 | 2.10 | 0.90 | 6.5 | 1.4 | 0.4 | 1.2 | 1.0 | 1.7 | 1.4 |
| M4x0.7 | 4.00 | 3.43 | 8.65 | 7.80 | 7.44 | 2.32 | 1.00 | 7.8 | 1.6 | 0.4 | 1.4 | 1.2 | 1.9 | 1.6 |
| M5x0.8 | 5.00 | 4.36 | 10.70 | 9.80 | 9.44 | 2.85 | 1.25 | 9.9 | 2.0 | 0.5 | 1.5 | 1.2 | 2.3 | 2.0 |
| M6x1 | 6.00 | 5.21 | 13.50 | 12.30 | 11.87 | 3.60 | 1.60 | 12.2 | 2.5 | 0.6 | 1.9 | 1.6 | 3.0 | 2.6 |
| M8x1.25 | 8.00 | 7.04 | 16.80 | 15.60 | 15.17 | 4.40 | 2.00 | 15.8 | 3.2 | 0.8 | 2.3 | 2.0 | 3.7 | 3.2 |
| M10x1.5 | 10.00 | 8.86 | 20.70 | 19.50 | 18.98 | 5.35 | 2.50 | 19.8 | 4.0 | 1.0 | 2.8 | 2.5 | 4.5 | 4.0 |
| M12x1.75 | 12.00 | 10.68 | 24.70 | 23.50 | 22.88 | 6.35 | 3.00 | 23.8 | 4.8 | 1.2 | 2.8 | 2.5 | 5.3 | 4.8 |

Note: For lengths less than 30 mm, completely threaded or $L_T = 25$ mm MIN. For lengths greater than 30 mm, completely threaded or $L_T = 38$ mm MIN. All dimensions are in millimetres.

Reproduced from ANSI B18.3.4M–1979 with the permission of the publisher, the American Society of Mechanical Engineers.

From *Technical Drafting*, copyright 1980 by Spence and Atkins. Used with permission of the publisher, Bennett Publishing Company, Peoria, IL 61615.

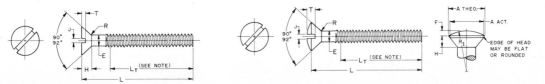

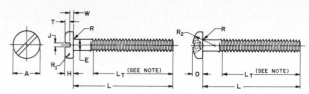

## Metric Slotted and Recessed Pan Head Machine Screws

| Nominal Screw Size and Thread Pitch | E Body Diameter | | A Head Diameter | | H Head Height Slotted Head | | O Head Height Recessed Head | | R₁ Head Radius (Slttd) | R₂ Head Radius (Rcssd) | R Fillet Radius | | J Slot Width | | T Slot Depth | W Un-slotted Thickness |
|---|---|---|---|---|---|---|---|---|---|---|---|---|---|---|---|---|
| | Max | Min | Max | Min | Max | Min | Max | Min | Max | Ref | Max | Min | Max | Min | Min | Min |
| M2x0.4 | 2.00 | 1.65 | 3.90 | 3.60 | 1.35 | 1.15 | 1.60 | 1.40 | 0.8 | 4 | 0.3 | 0.1 | 0.7 | 0.5 | 0.55 | 0.44 |
| M2.5x0.45 | 2.50 | 2.12 | 4.90 | 4.60 | 1.65 | 1.45 | 1.95 | 1.75 | 1.0 | 5 | 0.4 | 0.1 | 0.8 | 0.6 | 0.73 | 0.55 |
| M3x0.5 | 3.00 | 2.58 | 5.80 | 5.50 | 1.90 | 1.65 | 2.30 | 2.05 | 1.2 | 6 | 0.5 | 0.2 | 1.0 | 0.8 | 0.80 | 0.66 |
| M3.5x0.6 | 3.50 | 3.00 | 6.80 | 6.44 | 2.25 | 2.00 | 2.50 | 2.25 | 1.4 | 7 | 0.5 | 0.2 | 1.2 | 1.0 | 0.95 | 0.77 |
| M4x0.7 | 4.00 | 3.43 | 7.80 | 7.44 | 2.55 | 2.30 | 2.80 | 2.55 | 1.6 | 8 | 0.6 | 0.2 | 1.4 | 1.2 | 1.15 | 0.88 |
| M5x0.8 | 5.00 | 4.36 | 9.80 | 9.44 | 3.10 | 2.85 | 3.50 | 3.25 | 2.0 | 10 | 0.8 | 0.3 | 1.5 | 1.2 | 1.35 | 1.10 |
| M6x1 | 6.00 | 5.21 | 12.00 | 11.57 | 3.90 | 3.50 | 4.30 | 4.00 | 2.5 | 13 | 1.0 | 0.3 | 1.9 | 1.6 | 1.70 | 1.36 |
| M8x1.25 | 8.00 | 7.04 | 15.60 | 15.17 | 5.00 | 4.60 | 5.60 | 5.20 | 3.2 | 16 | 1.2 | 0.4 | 2.3 | 2.0 | 2.20 | 1.76 |
| M10x1.5 | 10.00 | 8.86 | 19.50 | 18.98 | 6.20 | 5.70 | 7.00 | 6.50 | 4.0 | 20 | 1.5 | 0.5 | 2.8 | 2.5 | 2.70 | 2.20 |
| M12x1.75 | 12.00 | 10.68 | 23.40 | 22.88 | 7.50 | 6.90 | 8.30 | 7.80 | 4.8 | 24 | 1.8 | 0.6 | 2.8 | 2.5 | 3.20 | 2.70 |

Note: For lengths less than 30 mm, completely threaded or $L_T$ = 25 mm MIN. For lengths greater than 30 mm, completely threaded or $L_T$ = 38 mm MIN.
All dimensions are in millimetres.

Reproduced from ANSI B18.3.4M–1979 with the permission of the publisher, the American Society of Mechanical Engineers.

From *Technical Drafting*, copyright 1980 by Spence and Atkins. Used with permission of the publisher, Bennett Publishing Company, Peoria, IL 61615.

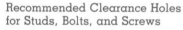

## Recommended Clearance Holes for Studs, Bolts, and Screws

| Nominal Fastener Size | D—Clearance Hole Diameter, Basic | | |
|---|---|---|---|
| | Close Clearance | Normal Clearance (Preferred) | Loose Clearance |
| 1.6 | 1.75 | 1.9 | 2.1 |
| 2 | 2.2 | 2.4 | 2.6 |
| 2.5 | 2.7 | 2.9 | 3.1 |
| 3 | 3.2 | 3.4 | 3.6 |
| 3.5 | 3.7 | 4.0 | 4.2 |
| 4 | 4.2 | 4.5 | 4.8 |
| 5 | 5.3 | 5.5 | 5.8 |
| 6 | 6.4 | 6.6 | 7.0 |
| 8 | 8.4 | 9.0 | 10.0 |
| 10 | 10.5 | 11.0 | 12.0 |
| 12 | 13.0 | 13.5 | 14.5 |
| 14 | 15.0 | 15.5 | 16.5 |
| 16 | 17.0 | 17.5 | 18.5 |
| 20 | 21.0 | 22.0 | 24.0 |
| 24 | 25.0 | 26.0 | 28.0 |
| 30 | 31.0 | 33.0 | 35.0 |
| 36 | 37.0 | 39.0 | 42.0 |
| 42 | 43.0 | 45.0 | 48.0 |
| 48 | 50.0 | 52.0 | 56.0 |
| 56 | 58.0 | 62.0 | 66.0 |
| 64 | 66.0 | 70.0 | 74.0 |
| 72 | 74.0 | 78.0 | 82.0 |
| 80 | 82.0 | 86.0 | 91.0 |
| 90 | 93.0 | 96.0 | 101.0 |
| 100 | 104.0 | 107.0 | 112.0 |

All dimensions are in millimetres.

From *Technical Drafting*, copyright 1980 by Spence and Atkins. Used with permission of the publisher, Bennett Publishing Company, Peoria, IL 61615.

## Metric Tapping Screws Type D, F, and T

| Nominal Screw Size and Thread Pitch | D Major Diameter | | P Point Diameter | | S Point Taper Length | | Body Diameter | L Minimum Practical Screw Lengths Pan, Hex and Hex Washer Heads | L Minimum Practical Screw Lengths Flat and Oval CTSK Heads |
|---|---|---|---|---|---|---|---|---|---|
| | Max | Min | Max | Min | Max | Min | Min | | |
| 2x0.4 | 2.00 | 1.88 | 1.61 | 1.54 | 1.4 | 1.0 | 1.65 | 3 | 5 |
| 2.5x0.45 | 2.50 | 2.37 | 2.07 | 2.00 | 1.6 | 1.1 | 2.12 | 4 | 6 |
| 3x0.5 | 3.00 | 2.87 | 2.52 | 2.44 | 1.8 | 1.3 | 2.58 | 5 | 7 |
| 3.5x0.6 | 3.50 | 3.35 | 2.92 | 2.83 | 2.1 | 1.5 | 3.00 | 6 | 8 |
| 4x0.7 | 4.00 | 3.83 | 3.33 | 3.24 | 2.5 | 1.8 | 3.43 | 6 | 9 |
| 5x0.8 | 5.00 | 4.82 | 4.22 | 4.10 | 2.8 | 2.0 | 4.36 | 8 | 11 |
| 6x1 | 6.00 | 5.78 | 5.08 | 4.96 | 3.5 | 2.5 | 5.21 | 10 | 13 |
| 8x1.25 | 8.00 | 7.76 | 6.79 | 6.67 | 4.4 | 3.1 | 7.04 | 12 | 16 |
| 10x1.5 | 10.00 | 9.73 | 8.55 | 8.42 | 5.2 | 3.8 | 8.86 | 15 | 20 |
| 12x1.75 | 12.00 | 11.70 | 10.31 | 10.16 | 6.1 | 4.4 | 10.68 | 18 | 24 |

All dimensions are in millimetres.

From *Technical Drafting*, copyright 1980 by Spence and Atkins. Used with permission of the publisher, Bennett Publishing Company, Peoria, IL 61615.

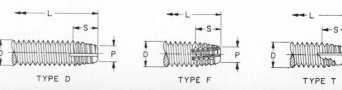

# Metric Tapping Screws Type B, AB, BF, and BT

| Nominal Screw Size and Thread Pitch | Basic Screw Dia | D Major Diameter | | d Minor Diameter | | V Point Diameter | | S Point Taper Length for Types B, BF, BT | | Z Point Length for Type AB | L Minimum Practical Screw Lengths — Type AB | | L Types B, BF, BT | |
|---|---|---|---|---|---|---|---|---|---|---|---|---|---|---|
| | Ref | Max | Min | Max | Min | Max | Min | Max | Min | Ref | Pan, Hex and Hex Washer Heads | Flat and Oval CTSK Heads | Pan, Hex and Hex Washer Heads | Flat and Oval CTSK Heads |
| 2.2x0.79 | 2.184 | 2.24 | 2.13 | 1.63 | 1.52 | 1.47 | 1.37 | 1.57 | 1.19 | 3.0 | 4 | 6 | 4 | 5 |
| 2.9x1.06 | 2.845 | 2.90 | 2.79 | 2.18 | 2.08 | 2.01 | 1.88 | 2.11 | 1.60 | 4.0 | 6 | 8 | 5 | 7 |
| 3.5x1.27 | 3.505 | 3.53 | 3.43 | 2.64 | 2.51 | 2.41 | 2.26 | 2.54 | 1.90 | 4.9 | 7 | 9 | 6 | 8 |
| 4.2x1.41 | 4.166 | 4.22 | 4.09 | 3.10 | 2.95 | 2.84 | 2.69 | 2.82 | 2.11 | 5.7 | 8 | 10 | 7 | 9 |
| 4.8x1.59 | 4.826 | 4.80 | 4.65 | 3.58 | 3.43 | 3.30 | 3.12 | 3.18 | 2.39 | 6.5 | 9 | 12 | 8 | 11 |
| 5.5x1.81 | 5.486 | 5.46 | 5.31 | 4.17 | 3.99 | 3.86 | 3.68 | 3.63 | 2.72 | 7.5 | 11 | 14 | 9 | 12 |
| 6.3x1.81 | 6.350 | 6.25 | 6.10 | 4.88 | 4.70 | 4.55 | 4.34 | 3.63 | 2.72 | 8.5 | 12 | 16 | 10 | 13 |
| 8.0x2.12 | 7.938 | 8.00 | 7.82 | 6.20 | 5.99 | 5.84 | 5.64 | 4.24 | 3.18 | 10.6 | 16 | 20 | 12 | 16 |
| 9.5x2.12 | 9.525 | 9.65 | 9.42 | 7.85 | 7.59 | 7.44 | 7.24 | 4.24 | 3.18 | 12.9 | 19 | 25 | 14 | 19 |

All dimensions are in millimetres.

From *Technical Drafting*, copyright 1980 by Spence and Atkins. Used with permission of the publisher, Bennett Publishing Company, Peoria, IL 61615.

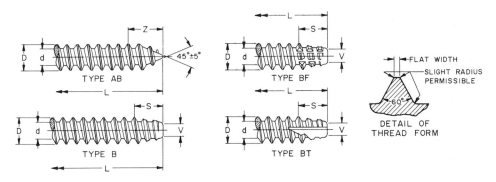

# Metric Studs

| Nominal Stud Size & Thread Pitch | Body Diameter E | | Nut Ends | | | | | |
|---|---|---|---|---|---|---|---|---|
| | | | F | G | F | G | F | G |
| | | | For Stud Lengths ⩽125 | | For Stud Lengths >125 and ⩽200 | | For Stud Lengths >200 | |
| | | | Thread Length | | Thread Length | | Thread Length | |
| | | | Full Thread | Total Thread | Full Thread | Total Thread | Full Thread | Total Thread |
| | Max | Min | Min | Max | Min | Max | Min | Max |
| M6x1 | 6.00 | 5.78 | 18 | 22 | 24 | 28 | 37 | 41 |
| M8x1.25 | 8.00 | 7.76 | 22 | 27 | 28 | 33 | 41 | 46 |
| M10x1.5 | 10.00 | 9.73 | 26 | 32 | 32 | 38 | 45 | 51 |
| M12x1.75 | 12.00 | 11.70 | 30 | 37 | 36 | 43 | 49 | 56 |
| M14x2 | 14.00 | 13.68 | 34 | 42 | 40 | 48 | 53 | 61 |
| M16x2 | 16.00 | 15.68 | 38 | 46 | 44 | 52 | 57 | 65 |
| M20x2.5 | 20.00 | 19.62 | 46 | 56 | 52 | 62 | 65 | 75 |
| M24x3 | 24.00 | 23.58 | — | — | 60 | 72 | 73 | 85 |
| M30x3.5 | 30.00 | 29.52 | — | — | 72 | 86 | 85 | 99 |
| M36x4 | 36.00 | 35.46 | — | — | — | — | 97 | 113 |

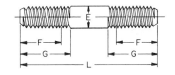

All dimensions are in millimetres.

From *Technical Drafting*, copyright 1980 by Spence and Atkins. Used with permission of the publisher, Bennett Publishing Company, Peoria, IL 61615.

## Metric Countersunk External Tooth Lock Washers

| Nominal Washer Size (1) | A Inside Diameter | | B Outside Diameter | C Material Thickness | | D Length | |
|---|---|---|---|---|---|---|---|
| | Max | Min | Nominal | Max | Min | Max | Min |
| 2 and 2.2 | 2.40 | 2.25 | 3.75 | 0.25 | 0.17 | 1.05 | 0.75 |
| 2.5 | 2.80 | 2.65 | 4.50 | 0.25 | 0.17 | 1.70 | 1.30 |
| 3 and 2.9 | 3.30 | 3.15 | 5.85 | 0.45 | 0.30 | 2.10 | 1.70 |
| 3.5 | 3.80 | 3.65 | 6.60 | 0.45 | 0.30 | 2.30 | 1.80 |
| 4 and 4.2 | 4.47 | 4.27 | 7.50 | 0.55 | 0.40 | 2.40 | 1.80 |
| 5 and 4.8 | 5.30 | 5.15 | 9.60 | 0.65 | 0.50 | 3.40 | 2.80 |
| 5.5 | 5.87 | 5.61 | 10.90 | 0.65 | 0.50 | 3.90 | 3.30 |
| 6.3 | 6.80 | 6.50 | 12.00 | 0.65 | 0.50 | 4.20 | 3.40 |
| 8 | 8.50 | 8.20 | 15.60 | 0.70 | 0.55 | 5.50 | 4.75 |
| 10 and 9.5 | 10.60 | 10.20 | 19.50 | 0.85 | 0.70 | 6.80 | 5.80 |
| 12 | 12.70 | 12.30 | 23.00 | 1.00 | 0.80 | 7.80 | 6.80 |

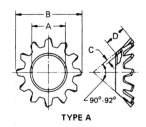
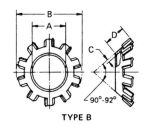

**TYPE A**   **TYPE B**

## Metric External Tooth Lock Washers

| Nominal Washer Size (1) | A Inside Diameter | | B Outside Diameter | | C Material Thickness | |
|---|---|---|---|---|---|---|
| | Max | Min | Max | Min | Max | Min |
| 2.2 | 2.40 | 2.25 | — | — | 0.35 | 0.25 |
| 2.5 | 2.80 | 2.65 | 5.85 | 5.45 | 0.35 | 0.25 |
| 2.9 | 3.12 | 2.92 | 6.60 | 6.20 | 0.45 | 0.30 |
| 3 | 3.30 | 3.15 | 7.35 | 6.85 | 0.50 | 0.35 |
| 3.5 | 3.80 | 3.65 | 8.05 | 7.55 | 0.55 | 0.40 |
| 4 | 4.30 | 4.15 | 9.00 | 8.50 | 0.60 | 0.45 |
| 4.2 | 4.47 | 4.27 | 9.70 | 9.20 | 0.60 | 0.45 |
| 4.8 | 5.18 | 4.95 | 10.30 | 9.80 | 0.65 | 0.50 |
| 5 | 5.35 | 5.15 | 10.50 | 10.00 | 0.65 | 0.50 |
| 5.5 | 5.87 | 5.61 | 12.00 | 11.50 | 0.70 | 0.55 |
| 6.3 | 6.80 | 6.50 | 12.95 | 12.20 | 0.70 | 0.55 |
| 8 | 8.50 | 8.20 | 15.50 | 14.75 | 0.85 | 0.70 |
| 9.5 | 10.11 | 9.75 | 17.75 | 17.00 | 1.00 | 0.80 |
| 10 | 10.60 | 10.20 | 17.75 | 17.00 | 1.00 | 0.80 |
| 12 | 12.70 | 12.30 | 20.25 | 19.50 | 1.00 | 0.80 |
| 14 | 14.80 | 14.30 | 23.00 | 22.00 | 1.15 | 0.95 |
| 16 | 17.00 | 16.40 | 27.50 | 26.50 | 1.25 | 1.05 |
| 20 | 21.20 | 20.50 | 32.00 | 31.00 | 1.40 | 1.20 |
| 24 | 25.30 | 24.50 | 35.80 | 34.50 | 1.50 | 1.30 |
| 30 | 31.40 | 30.60 | 46.80 | 45.10 | 1.70 | 1.50 |

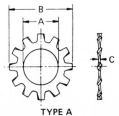

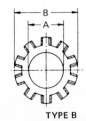

**TYPE A**   **TYPE B**

Metric Internal Tooth Lock Washers

| Nominal Washer Size (1) | A | | B | | C | |
|---|---|---|---|---|---|---|
| | Inside Diameter | | Outside Diameter | | Material Thickness | |
| | Max | Min | Max | Min | Max | Min |
| 2.2◆ | 2.40 | 2.25 | 4.70 | 4.40 | 0.35 | 0.25 |
| 2.5 | 2.80 | 2.65 | 5.35 | 5.45 | 0.35 | 0.25 |
| 2.9◆ | 3.12 | 2.92 | 6.75 | 6.25 | 0.45 | 0.30 |
| 3 | 3.30 | 3.15 | 7.35 | 6.85 | 0.50 | 0.35 |
| 3.5 | 3.80 | 3.65 | 8.05 | 7.55 | 0.55 | 0.40 |
| 4 | 4.30 | 4.15 | 8.75 | 8.25 | 0.60 | 0.45 |
| 4.2◆ | 4.47 | 4.27 | 8.75 | 8.25 | 0.60 | 0.45 |
| 4.8◆ | 5.18 | 4.95 | 9.70 | 9.20 | 0.65 | 0.50 |
| 5 | 5.30 | 5.15 | 10.50 | 10.00 | 0.70 | 0.55 |
| 5.5◆ | 5.87 | 5.61 | 10.50 | 10.00 | 0.70 | 0.55 |
| 6.3 | 6.80 | 6.50 | 12.95 | 12.20 | 0.70 | 0.55 |
| 5 | 8.50 | 8.20 | 15.50 | 14.75 | 0.85 | 0.70 |
| 9.5◆ | 10.11 | 9.75 | 17.60 | 16.85 | 1.00 | 0.80 |
| 10 | 10.60 | 10.20 | 17.60 | 16.85 | 1.00 | 0.80 |
| 12 | 12.70 | 12.30 | 20.25 | 19.50 | 1.00 | 0.80 |
| 14 | 14.80 | 14.30 | 22.90 | 21.90 | 1.15 | 0.95 |
| 16 | 17.00 | 16.40 | 27.20 | 26.20 | 1.25 | 1.05 |
| 20 | 21.20 | 20.50 | 32.00 | 31.00 | 1.40 | 1.20 |
| 24 | 25.30 | 24.50 | 35.30 | 34.50 | 1.50 | 1.30 |
| 30 | 31.40 | 30.60 | 46.30 | 45.10 | 1.70 | 1.50 |

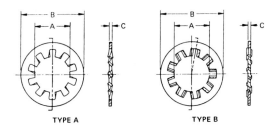

TYPE A          TYPE B

Metric Flat Washers, Lockwashers, and Spring Lockwashers

| BOLT SIZE | FLAT WASHER | | | LOCKWASHERS | | | SPRING LOCKWASHERS | | |
|---|---|---|---|---|---|---|---|---|---|
| | ID | OD | THICK | ID | OD | THICK | ID | OD | THICK |
| 2 | 2.2 | 5.5 | 0.5 | 2.1 | 3.3 | 0.5 | | | |
| 3 | 3.2 | 7 | 0.5 | 3.1 | 5.7 | 0.8 | | | |
| 4 | 4.3 | 9 | 0.8 | 4.1 | 7.1 | 0.9 | 4.2 | 8 | 0.3 0.4 |
| 5 | 5.3 | 11 | 1 | 5.1 | 8.7 | 1.2 | 5.2 | 10 | 0.4 0.5 |
| 6 | 6.4 | 12 | 1.5 | 6.1 | 11.1 | 1.6 | 6.2 | 12.5 | 0.5 0.7 |
| 7 | 7.4 | 14 | 1.5 | 7.1 | 12.1 | 1.6 | 7.2 | 14 | 0.5 0.8 |
| 8 | 8.4 | 17 | 2 | 8.2 | 14.2 | 2 | 8.2 | 16 | 0.6 0.9 |
| 10 | 10.5 | 21 | 2.5 | 10.2 | 17.2 | 2.2 | 10.2 | 20 | 0.8 1.1 |
| 12 | 13 | 24 | 2.5 | 12.3 | 20.2 | 2.5 | 12.2 | 25 | 0.9 1.5 |
| 14 | 15 | 28 | 2.5 | 14.2 | 23.2 | 3 | 14.2 | 28 | 1.0 1.5 |
| 16 | 17 | 30 | 3 | 16.2 | 26.2 | 3.5 | 16.3 | 31.5 | 1.2 1.7 |
| 18 | 19 | 34 | 3 | 18.2 | 28.2 | 3.5 | 18.3 | 35.5 | 1.2 2.0 |
| 20 | 21 | 36 | 3 | 20.2 | 32.2 | 4 | 20.4 | 40 | 1.5 2.25 |

Metric Coiled Spring Pins

| Nominal Pin Size | D | | | | | | B | C | Recommended Hole Size | |
| | Diameter | | | | | | Chamfer | Chamfer | | |
| | Standard Duty | | Heavy Duty | | Light Duty | | | | | |
| | Max | Min | Max | Min | Max | Min | Max | Ref | Max | Min |
|---|---|---|---|---|---|---|---|---|---|---|
| 0.8 | 0.89 | 0.84 | — | — | — | — | 0.74 | 0.4 | 0.83 | 0.80 |
| 1 | 1.12 | 1.04 | — | — | — | — | 0.94 | 0.4 | 1.03 | 1.00 |
| 1.2 | 1.32 | 1.24 | — | — | — | — | 1.14 | 0.4 | 1.23 | 1.20 |
| 1.5 | 1.73 | 1.61 | 1.68 | 1.58 | 1.75 | 1.61 | 1.44 | 0.5 | 1.57 | 1.50 |
| 2 | 2.24 | 2.11 | 2.19 | 2.08 | 2.26 | 2.11 | 1.90 | 0.6 | 2.07 | 1.97 |
| 2.5 | 2.77 | 2.63 | 2.72 | 2.60 | 2.79 | 2.62 | 2.40 | 0.6 | 2.58 | 2.47 |
| 3 | 3.32 | 3.15 | 3.25 | 3.12 | 3.35 | 3.15 | 2.90 | 0.8 | 3.10 | 2.97 |
| 4 | 4.35 | 4.15 | 4.27 | 4.12 | 4.38 | 4.15 | 3.90 | 1.0 | 4.10 | 3.97 |
| 5 | 5.45 | 5.20 | 5.36 | 5.16 | 5.50 | 5.20 | 4.85 | 1.2 | 5.12 | 4.95 |
| 6.3 | 6.83 | 6.55 | 6.75 | 6.50 | 6.88 | 6.55 | 6.15 | 1.4 | 6.45 | 6.25 |
| 8 | 8.60 | 8.27 | 8.52 | 8.22 | 8.65 | 8.27 | 7.80 | 1.7 | 8.17 | 7.93 |
| 10 | 10.73 | 10.33 | 10.66 | 10.28 | — | — | 9.75 | 2.0 | 10.20 | 9.93 |
| 12 | 12.82 | 12.37 | 12.72 | 12.32 | — | — | 11.70 | 2.4 | 12.22 | 11.90 |
| 14 | 14.89 | 14.41 | 14.82 | 14.36 | — | — | 13.65 | 2.9 | 14.25 | 13.85 |
| 16 | 16.95 | 16.43 | 16.86 | 16.38 | — | — | 15.65 | 3.2 | 16.25 | 15.85 |
| 20 | 21.00 | 20.45 | 20.90 | 20.40 | — | — | 19.65 | 3.8 | 20.25 | 19.85 |

Preferred Lengths of Metric Keys

| Length | Type of Key | | | |
| | SQ | RECT | SQ-Taper | RECT-Taper |
|---|---|---|---|---|
| 6 | X | | X | |
| 8 | X | | X | |
| 10 | X | | X | |
| 12 | X | | X | |
| 14 | X | | X | |
| 16 | X | | X | |
| 18 | X | X | X | X- |
| 20 | X | X | X | X |
| 22 | X | X | X | X |
| 25 | X | X | X | X |
| 28 | X | X | X | X |
| 32 | X | X | X | X |
| 40 | X | X | X | X |
| 45 | X | X | X | X |
| 50 | X | X | X | X |
| 56 | X | X | X | X |
| 63 | X | X | X | X |
| 70 | X | X | X | X |
| 80 | | X | | X |
| 90 | | X | | X |
| 100 | | X | | X |

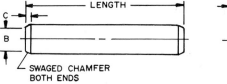

LENGTH

SWAGED CHAMFER BOTH ENDS

All dimensions are in millimetres.

From *Technical Drafting*, copyright 1980 by Spence and Atkins. Used with permission of the publisher, Bennett Publishing Company, Peoria, IL 61615.

Metric Square and Rectangular Parallel, Taper, and Gib Head Keys

| Nominal Ø Shaft | | Nominal Key Size | | Nominal Seat Depth | | | Nominal Gib Head Height |
|---|---|---|---|---|---|---|---|
| Over | to Include | Width W | Height H | Shaft $D_1$ | Hub $D_2$ | Hub (Taper) $D_2$ | Height A |
| 6.0 | 8.0 | 2.0 | 2.0 | 1.2 | 1.0 | 0.5 | — |
| 8.0 | 10.0 | 3.0 | 3.0 | 1.8 | 1.4 | 0.9 | — |
| 10.0 | 12.0 | 4.0 | 4.0 | 2.5 | 1.8 | 1.2 | 7.0 |
| 12.0 | 17.0 | 5.0 | 5.0 | 3.0 | 2.3 | 1.7 | 8.0 |
| 17.0 | 22.0 | 6.0 | 6.0 | 3.5 | 2.8 | 2.2 | 10.0 |
| 22.0 | 30.0 | 8.0 | 7.0 | 4.0 | 3.3 | 2.4 | 11.0 |
| 30.0 | 38.0 | 10.0 | 8.0 | 5.0 | 3.3 | 2.4 | 12.0 |
| 38.0 | 44.0 | 12.0 | 8.0 | 5.0 | 3.3 | 2.4 | 12.0 |
| 44.0 | 50.0 | 14.0 | 9.0 | 5.5 | 3.8 | 2.9 | 14.0 |
| 50.0 | 58.0 | 16.0 | 10.0 | 6.0 | 4.3 | 3.4 | 16.0 |
| 58.0 | 65.0 | 18.0 | 11.0 | 7.0 | 4.4 | 3.4 | 18.0 |
| 65.0 | 75.0 | 20.0 | 12.0 | 7.5 | 4.9 | 3.9 | 20.0 |
| 75.0 | 85.0 | 22.0 | 14.0 | 9.0 | 5.4 | 4.4 | 22.0 |
| 85.0 | 95.0 | 25.0 | 14.0 | 9.0 | 5.4 | 4.4 | 22.0 |
| 95.0 | 110.0 | 28.0 | 16.0 | 10.0 | 6.4 | 5.4 | 25.0 |
| 110.0 | 130.0 | 32.0 | 18.0 | 11.0 | 7.4 | 6.4 | 28.0 |
| 130.0 | 150.0 | 36.0 | 20.0 | 12.0 | 8.4 | 7.1 | 32.0 |
| 150.0 | 170.0 | 40.0 | 22.0 | 13.0 | 9.4 | 8.1 | 36.0 |
| 170.0 | 200.0 | 45.0 | 25.0 | 15.0 | 10.4 | 9.1 | 40.0 |
| 200.0 | 230.0 | 50.0 | 28.0 | 17.0 | 11.4 | 10.1 | 45.0 |
| 230.0 | 260.0 | 56.0 | 32.0 | 20.0 | 12.4 | 11.1 | 50.0 |
| 260.0 | 290.0 | 63.0 | 32.0 | 20.0 | 12.4 | 11.1 | 50.0 |

All dimensions are in millimetres.

From *Technical Drafting*, copyright 1980 by Spence and Atkins. Used with permission of the publisher, Bennett Publishing Company, Peoria, IL 61615.

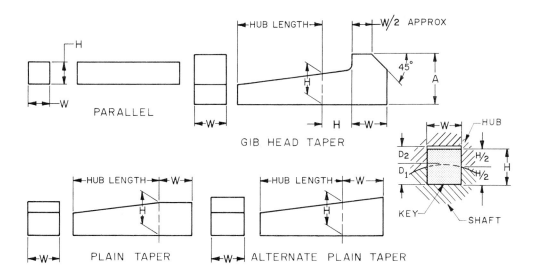

Unified Inch Screw Threads*

| Sizes (Primary) | Sizes (Secondary) | Basic Major Diameter | Coarse UNC | Fine UNF | Extra fine UNEF | 4UN | 6UN | 8UN | 12UN | 16UN | 20UN | 28UN | 32UN | Sizes |
|---|---|---|---|---|---|---|---|---|---|---|---|---|---|---|
| 0 | | 0.0600 | — | 80 | — | — | — | — | — | — | — | — | — | 0 |
| | 1 | 0.0730 | 64 | 72 | — | — | — | — | — | — | — | — | — | 1 |
| 2 | | 0.0860 | 56 | 64 | — | — | — | — | — | — | — | — | — | 2 |
| | 3 | 0.0990 | 48 | 56 | — | — | — | — | — | — | — | — | — | 3 |
| 4 | | 0.1120 | 40 | 48 | — | — | — | — | — | — | — | — | — | 4 |
| 5 | | 0.1250 | 40 | 44 | — | — | — | — | — | — | — | — | — | 5 |
| 6 | | 0.1380 | 32 | 40 | — | — | — | — | — | — | — | — | UNC | 6 |
| 8 | | 0.1640 | 32 | 36 | — | — | — | — | — | — | — | — | UNC | 8 |
| 10 | | 0.1900 | 24 | 32 | — | — | — | — | — | — | — | — | UNF | 10 |
| | 12 | 0.2160 | 24 | 28 | 32 | — | — | — | — | — | — | UNF | UNEF | 12 |
| 1/4 | | 0.2500 | 20 | 28 | 32 | — | — | — | — | — | UNC | UNF | UNEF | 1/4 |
| 5/16 | | 0.3125 | 18 | 24 | 32 | — | — | — | — | — | 20 | 28 | UNEF | 5/16 |
| 3/8 | | 0.3750 | 16 | 24 | 32 | — | — | — | — | UNC | 20 | 28 | UNEF | 3/8 |
| 7/16 | | 0.4375 | 14 | 20 | 28 | — | — | — | — | 16 | UNF | UNEF | 32 | 7/16 |
| 1/2 | | 0.5000 | 13 | 20 | 28 | — | — | — | — | 16 | UNF | UNEF | 32 | 1/2 |
| 9/16 | | 0.5625 | 12 | 18 | 24 | — | — | — | UNC | 16 | 20 | 28 | 32 | 9/16 |
| 5/8 | | 0.6250 | 11 | 18 | 24 | — | — | — | 12 | 16 | 20 | 28 | 32 | 5/8 |
| | 11/16 | 0.6875 | — | — | 24 | — | — | — | 12 | 16 | 20 | 28 | 32 | 11/16 |
| 3/4 | | 0.7500 | 10 | 16 | 20 | — | — | — | 12 | UNF | UNEF | 28 | 32 | 3/4 |
| | 13/16 | 0.8125 | — | — | 20 | — | — | — | 12 | 16 | UNEF | 28 | 32 | 13/16 |
| 7/8 | | 0.8750 | 9 | 14 | 20 | — | — | — | 12 | 16 | UNEF | 28 | 32 | 7/8 |
| | 15/16 | 0.9375 | — | — | 20 | — | — | — | 12 | 16 | UNEF | 28 | 32 | 15/16 |
| 1 | | 1.0000 | 8 | 12 | 20 | — | — | UNC | UNF | 16 | UNEF | 28 | 32 | 1 |
| | 1 1/16 | 1.0625 | — | — | 18 | — | — | 8 | 12 | 16 | 20 | 28 | — | 1 1/16 |
| 1 1/8 | | 1.1250 | 7 | 12 | 18 | — | — | 8 | UNF | 16 | 20 | 28 | — | 1 1/8 |
| | 1 3/16 | 1.1875 | — | — | 18 | — | — | 8 | 12 | 16 | 20 | 28 | — | 1 3/16 |
| 1 1/4 | | 1.2500 | 7 | 12 | 18 | — | — | 8 | UNF | 16 | 20 | 28 | — | 1 1/4 |
| | 1 5/16 | 1.3125 | — | — | 18 | — | — | 8 | 12 | 16 | 20 | 28 | — | 1 5/16 |
| 1 3/8 | | 1.3750 | 6 | 12 | 18 | — | UNC | 8 | UNF | 16 | 20 | 28 | — | 1 3/8 |
| | 1 7/16 | 1.4375 | — | — | 18 | — | 6 | 8 | 12 | 16 | 20 | 28 | — | 1 7/16 |
| 1 1/2 | | 1.5000 | 6 | 12 | 18 | — | UNC | 8 | UNF | 16 | 20 | 28 | — | 1 1/2 |
| | 1 9/16 | 1.5625 | — | — | 18 | — | 6 | 8 | 12 | 16 | 20 | — | — | 1 9/16 |
| 1 5/8 | | 1.6250 | — | — | 18 | — | 6 | 8 | 12 | 16 | 20 | — | — | 1 5/8 |
| | 1 11/16 | 1.6875 | — | — | 18 | — | 6 | 8 | 12 | 16 | 20 | — | — | 1 11/16 |
| 1 3/4 | | 1.7500 | 5 | — | — | — | 6 | 8 | 12 | 16 | 20 | — | — | 1 3/4 |
| | 1 13/16 | 1.8125 | — | — | — | — | 6 | 8 | 12 | 16 | 20 | — | — | 1 13/16 |
| 1 7/8 | | 1.8750 | — | — | — | — | 6 | 8 | 12 | 16 | 20 | — | — | 1 7/8 |
| | 1 15/16 | 1.9375 | — | — | — | — | 6 | 8 | 12 | 16 | 20 | — | — | 1 15/16 |
| 2 | | 2.0000 | 4 1/2 | — | — | — | 6 | 8 | 12 | 16 | 20 | — | — | 2 |
| | 2 1/8 | 2.1250 | — | — | — | — | 6 | 8 | 12 | 16 | 20 | — | — | 2 1/8 |
| 2 1/4 | | 2.2500 | 4 1/2 | — | — | — | 6 | 8 | 12 | 16 | 20 | — | — | 2 1/4 |
| | 2 3/8 | 2.3750 | — | — | — | — | 6 | 8 | 12 | 16 | 20 | — | — | 2 3/8 |
| 2 1/2 | | 2.5000 | 4 | — | — | UNC | 6 | 8 | 12 | 16 | 20 | — | — | 2 1/2 |
| | 2 5/8 | 2.6250 | — | — | — | 4 | 6 | 8 | 12 | 16 | 20 | — | — | 2 5/8 |

Reproduced from ANSI B1.1–1974 with the permission of the publisher, the American Society of Mechanical Engineers. From *Technical Drafting*, copyright 1980 by Spence and Atkins. Used with permission of the publisher, Bennett Publishing Company, Peoria, IL 61615.

## Inch Hexagon, Heavy Hexagon, and Square Bolts

| Nominal Size or Major Thread Dia. | Regular Hexagon | | Heavy Hexagon | | Regular Square | |
|---|---|---|---|---|---|---|
| | F Across Flats | H | F Across Flats | H | F Across Flats | H |
| $1/4$ | $7/16$ | $11/64$ | | | $3/8$ | $11/64$ |
| $5/16$ | $1/2$ | $7/32$ | | | $1/2$ | $13/64$ |
| $3/8$ | $9/16$ | $1/4$ | | | $9/16$ | $1/4$ |
| $7/16$ | $5/8$ | $19/64$ | | | $5/8$ | $19/64$ |
| $1/2$ | $3/4$ | $11/32$ | $7/8$ | $11/32$ | $3/4$ | $21/64$ |
| $5/8$ | $15/16$ | $27/64$ | $1\,1/16$ | $27/64$ | $15/16$ | $27/64$ |
| $3/4$ | $1\,1/8$ | $1/2$ | $1\,1/4$ | $1/2$ | $1\,1/8$ | $1/2$ |
| $7/8$ | $1\,5/16$ | $37/64$ | $1\,7/16$ | $37/64$ | $1\,5/16$ | $19/32$ |
| $1$ | $1\,1/2$ | $43/64$ | $1\,5/8$ | $43/64$ | $1\,1/2$ | $21/32$ |
| $1\,1/8$ | $1\,11/16$ | $3/4$ | $1\,13/16$ | $3/4$ | $1\,11/16$ | $3/4$ |
| $1\,1/4$ | $1\,7/8$ | $27/64$ | $2$ | $27/32$ | $1\,7/8$ | $27/32$ |
| $1\,3/8$ | $2\,1/16$ | $29/32$ | $2\,3/16$ | $29/32$ | $2\,1/16$ | $29/32$ |

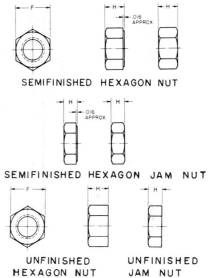

Reproduced from ANSI B18.2.1–1972 with the permission of the publisher, the American Society of Mechanical Engineers.

From *Drafting: Technology and Practice*, copyright 1980 by Spence. Used with permission of the publisher, Bennett Publishing Company, Peoria, IL 61615.

## Inch Hexagon Nuts and Jam Nuts

| Diameter | Regular | Nut | Jam Nut |
|---|---|---|---|
| | F | H | H |
| $1/4$ | $7/16$ | $7/32$ | $5/32$ |
| $5/16$ | $1/2$ | $17/64$ | $3/16$ |
| $3/8$ | $9/16$ | $21/64$ | $7/32$ |
| $7/16$ | $11/16$ | $3/8$ | $1/4$ |
| $1/2$ | $3/4$ | $7/16$ | $5/16$ |
| $9/16$ | $7/8$ | $31/64$ | $5/16$ |
| $5/8$ | $15/16$ | $35/64$ | $3/8$ |
| $3/4$ | $1\,1/8$ | $41/64$ | $27/64$ |
| $7/8$ | $1\,5/16$ | $3/4$ | $31/64$ |
| $1$ | $1\,1/2$ | $55/64$ | $35/64$ |
| $1\,1/8$ | $1\,11/16$ | $31/32$ | $5/8$ |
| $1\,1/4$ | $1\,7/8$ | $1\,1/16$ | $3/4$ |
| $1\,3/8$ | $2\,1/16$ | $1\,11/64$ | $13/16$ |
| $1\,1/2$ | $2\,1/4$ | $1\,9/32$ | $7/8$ |

Reproduced from ANSI B18.2.1–1972 with permission of the publisher, the American Society of Mechanical Engineers.

From *Drafting: Technology and Practice*, copyright 1980 by Spence. Used with permission of the publisher, Bennett Publishing Company, Peoria, IL 61615.

## Inch Standard Regular and Heavy Square Nuts

| Diameter | | Regular | | Heavy | |
|---|---|---|---|---|---|
| | | F | H | F | H |
| $1/4$ | 0.2500 | $7/16$ | $7/32$ | $1/2$ | $1/4$ |
| $5/16$ | 0.3125 | $9/16$ | $17/64$ | $9/16$ | $5/16$ |
| $3/8$ | 0.3750 | $5/8$ | $21/64$ | $11/16$ | $3/8$ |
| $7/16$ | 0.4375 | $3/4$ | $3/8$ | $3/4$ | $7/16$ |
| $1/2$ | 0.5000 | $13/16$ | $7/16$ | $7/8$ | $1/2$ |
| $5/8$ | 0.6250 | $1$ | $35/64$ | $1\,1/16$ | $5/8$ |
| $3/4$ | 0.7500 | $1\,1/8$ | $21/32$ | $1\,1/4$ | $3/4$ |
| $7/8$ | 0.8750 | $1\,5/16$ | $49/64$ | $1\,7/16$ | $7/8$ |
| $1$ | 1.0000 | $1\,1/2$ | $7/8$ | $1\,5/8$ | $1$ |

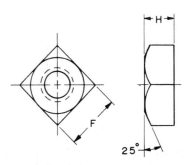

Reproduced from ANSI B18.2.1–1972 with the permission of the publisher, the American Society of Mechanical Engineers.

From *Drafting: Technology and Practice*, copyright 1980 by Spence. Used with permission of the publisher, Bennett Publishing Company, Peoria, IL 61615.

## Inch Acme and Stub Acme Threads

| Size | Threads per Inch | Size | Threads per Inch | Size | Threads per Inch | Size | Threads per Inch |
|---|---|---|---|---|---|---|---|
| 1/4 | 16 | 3/4 | 6 | 1 1/2 | 4 | 3 | 2 |
| 5/16 | 14 | 7/8 | 6 | 1 3/4 | 4 | 3 1/2 | 2 |
| 3/8 | 10 | 1 | 5 | 2 | 4 | 4 | 2 |
| 7/16 | 12 | 1 1/8 | 5 | 2 1/4 | 3 | 4 1/2 | 2 |
| 1/2 | 10 | 1 1/4 | 5 | 2 1/2 | 3 | 5 | 2 |
| 5/8 | 8 | 1 3/8 | 4 | 2 3/4 | 3 | — | — |

Reproduced from ANSI B1.5–1977 with the permission of the publisher, the American Society of Mechanical Engineers.

From *Drafting: Technology and Practice*, copyright 1980 by Spence. Used with permission of the publisher, Bennett Publishing Company, Peoria, IL 61615.

## Diameter-Pitch Combinations for 7°/45° Buttress Threads

| Major Diameter Range | Preferred Nominal Major Diameters | Threads per Inch — Preferred TPI Between Heavy Lines | | | | | | | | | | | | | |
|---|---|---|---|---|---|---|---|---|---|---|---|---|---|---|---|
| From 0.5 thru 0.75 | 0.5, 0.625, 0.75 | 20[a] | 16 | 12 | | | | | | | | | | | |
| Over 0.75 thru 1.0 | 0.875, 1.0 | | 16[a] | 12 | 10 | | | | | | | | | | |
| Over 1.0 thru 1.5 | 1.25, 1.375, 1.5 | | 16 | 12[a] | 10 | 8 | 6 | | | | | | | | |
| Over 1.5 thru 2.5 | 1.75, 2, 2.25, 2.5 | | 16 | 12 | 10[a] | 8 | 6 | 5 | 4 | | | | | | |
| Over 2.5 thru 4 | 2.75, 3, 3.5, 4 | | 16 | 12 | 10 | 8 | 6 | 5 | 4 | | | | | | |
| Over 4 thru 6 | 4.5, 5, 5.5, 6 | | | 12 | 10 | 8 | 6 | 5 | 4 | 3 | | | | | |
| Over 6 thru 10 | 7, 8, 9, 10 | | | | 10 | 8 | 6 | 5 | 4 | 3 | 2.5 | 2 | | | |
| Over 10 thru 16 | 11, 12, 14, 16 | | | | 10 | 8 | 6 | 5 | 4 | 3 | 2.5 | 2 | 1.5 | 1.25 | |
| Over 16 thru 24 | 18, 20, 22, 24 | | | | | 8 | 6 | 5 | 4 | 3 | 2.5 | 2 | 1.5 | 1.25 | 1 |

[a]When the pitch diameter is measured with "best-size" wires the measurement may be incorrect due to the double contact of the wire on the 7° flank because the lead angle exceeds 2°.

Reproduced from ANSI with the permission of the publisher, the American Society of Mechanical Engineers.

## Inch Hexagon Head Cap Screws*

| Nominal Size or Basic Major Diameter of Thread | | F | H |
|---|---|---|---|
| | | Width Across Flats | Height |
| 1/4 | 0.2500 | 7/16 | 5/32 |
| 5/16 | 0.3125 | 1/2 | 13/64 |
| 3/8 | 0.3750 | 9/16 | 15/64 |
| 7/16 | 0.4375 | 5/8 | 9/32 |
| 1/2 | 0.5000 | 3/4 | 5/16 |
| 9/16 | 0.5625 | 13/16 | 23/64 |
| 5/8 | 0.6250 | 15/16 | 25/64 |
| 3/4 | 0.7500 | 1 1/8 | 15/32 |
| 7/8 | 0.8750 | 1 5/16 | 35/64 |
| 1 | 1.0000 | 1 1/2 | 39/64 |
| 1 1/8 | 1.1250 | 1 11/16 | 11/16 |
| 1 1/4 | 1.2500 | 1 7/8 | 25/32 |
| 1 3/8 | 1.3750 | 2 1/16 | 27/32 |
| 1 1/2 | 1.5000 | 2 1/4 | 15/16 |

*Thread length
Screw lengths up to 6 inches have thread length equal to 2D + 1/4 inch.

Threads may be National coarse, National fine, or 8-Thread series, class 2A fit.

Reproduced from ANSI B18.2.1–1972 with the permission of the publisher, the American Society of Mechanical Engineers.

From *Drafting: Technology and Practice*, copyright 1980 by Spence. Used with permission of the publisher, Bennett Publishing Company, Peoria, IL 61615.

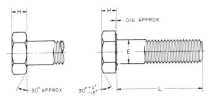

## Inch Hexagon and Spline Socket Head Cap Screws

| Nominal Size | D | A | H | M | J | T |
|---|---|---|---|---|---|---|
| | Body Diameter | Head Diameter | Head Height | Spline Socket Size | Hexagon Socket Size | Key Engagement |
| | Max | Max | Max | Nom | Nom | Min |
| 0 | 0.0600 | 0.096 | 0.060 | 0.060 | 0.050 | 0.025 |
| 1 | 0.0730 | 0.118 | 0.073 | 0.072 | 1/16 | 0.031 |
| 2 | 0.0860 | 0.140 | 0.086 | 0.096 | 5/64 | 0.038 |
| 3 | 0.0990 | 0.161 | 0.099 | 0.096 | 5/64 | 0.044 |
| 4 | 0.1120 | 0.183 | 0.112 | 0.111 | 3/32 | 0.051 |
| 5 | 0.1250 | 0.205 | 0.125 | 0.111 | 3/32 | 0.057 |
| 6 | 0.1380 | 0.226 | 0.138 | 0.133 | 7/64 | 0.064 |
| 8 | 0.1640 | 0.270 | 0.164 | 0.168 | 9/64 | 0.077 |
| 10 | 0.1900 | 0.312 | 0.190 | 0.183 | 5/32 | 0.090 |
| 1/4 | 0.2500 | 0.375 | 0.250 | 0.216 | 3/16 | 0.120 |
| 5/16 | 0.3125 | 0.469 | 0.312 | 0.291 | 1/4 | 0.151 |
| 3/8 | 0.3750 | 0.562 | 0.375 | 0.372 | 5/16 | 0.182 |
| 7/16 | 0.4375 | 0.656 | 0.438 | 0.454 | 3/8 | 0.213 |
| 1/2 | 0.5000 | 0.750 | 0.500 | 0.454 | 3/8 | 0.245 |
| 5/8 | 0.6250 | 0.938 | 0.625 | 0.595 | 1/2 | 0.307 |
| 3/4 | 0.7500 | 1.125 | 0.750 | 0.620 | 5/8 | 0.370 |
| 7/8 | 0.8750 | 1.312 | 0.875 | 0.698 | 3/4 | 0.432 |
| 1 | 1.0000 | 1.500 | 1.000 | 0.790 | 3/4 | 0.495 |
| 1 1/8 | 1.1250 | 1.688 | 1.125 | ....... | 7/8 | 0.557 |
| 1 1/4 | 1.2500 | 1.875 | 1.250 | ....... | 7/8 | 0.620 |
| 1 3/8 | 1.3750 | 2.062 | 1.375 | ....... | 1 | 0.682 |
| 1 1/2 | 1.5000 | 2.250 | 1.500 | ....... | 1 | 0.745 |

Screw head shall be flat and chamfered. Chamfer E shall be at an angle of 30° ± with surface of the flat.

*Screw lengths:*
Screw lengths 1/8 to 1 inch available in 1/8 inch increments.
Screw lengths 1 to 3 1/2 inches available in 1/4 increments.
Screw lengths 3 1/2 to 6 inches available in 1/2 increments.

*Thread lengths:*
National Coarse threads—thread length, $L_T$, equal 2D + 1/2 inch.
National Fine Threads—thread length, $L_T$, equal 1 1/2 D + 1/8 inch.
Thread class of fit is 3A.

Reproduced from ANSI B18.3–1972 with the permission of the publisher, the American Society of Mechanical Engineers.

From *Drafting: Technology and Practice*, copyright 1980 by Spence. Used with permission of the publisher, Bennett Publishing Company, Peoria, IL 61615.

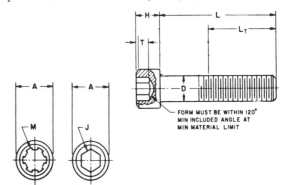

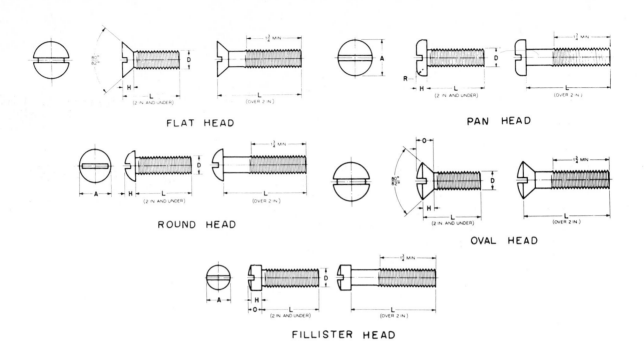

FLAT HEAD

PAN HEAD

ROUND HEAD

OVAL HEAD

FILLISTER HEAD

Inch Slotted Head Machine Screws

| Nominal Size | D Diameter of Screw | Flat Head | | Round Head | | Pan Head | | Oval Head | | | Fillister Head | | |
|---|---|---|---|---|---|---|---|---|---|---|---|---|---|
| | | A | H | A | H | A | H | A | H | O | A | H | O |
| 0 | 0.0600 | 0.119 | 0.035 | 0.113 | 0.053 | 0.116 | 0.039 | 0.119 | 0.035 | 0.056 | 0.096 | 0.043 | 0.055 |
| 1 | 0.0730 | 0.146 | 0.043 | 0.138 | 0.061 | 0.142 | 0.046 | 0.146 | 0.043 | 0.068 | 0.118 | 0.053 | 0.066 |
| 2 | 0.0860 | 0.172 | 0.051 | 0.162 | 0.069 | 0.167 | 0.053 | 0.172 | 0.051 | 0.080 | 0.140 | 0.062 | 0.083 |
| 3 | 0.0990 | 0.199 | 0.059 | 0.187 | 0.078 | 0.193 | 0.060 | 0.199 | 0.059 | 0.092 | 0.161 | 0.070 | 0.095 |
| 4 | 0.1120 | 0.225 | 0.067 | 0.211 | 0.086 | 0.219 | 0.068 | 0.225 | 0.067 | 0.104 | 0.183 | 0.079 | 0.107 |
| 5 | 0.1250 | 0.252 | 0.075 | 0.236 | 0.095 | 0.245 | 0.075 | 0.252 | 0.075 | 0.116 | 0.205 | 0.088 | 0.120 |
| 6 | 0.1380 | 0.279 | 0.083 | 0.260 | 0.103 | 0.270 | 0.082 | 0.279 | 0.083 | 0.128 | 0.226 | 0.096 | 0.132 |
| 8 | 0.1640 | 0.332 | 0.100 | 0.309 | 0.120 | 0.322 | 0.096 | 0.332 | 0.100 | 0.152 | 0.270 | 0.113 | 0.156 |
| 10 | 0.1900 | 0.385 | 0.116 | 0.359 | 0.137 | 0.373 | 0.110 | 0.385 | 0.116 | 0.176 | 0.313 | 0.130 | 0.180 |
| 12 | 0.2160 | 0.438 | 0.132 | 0.408 | 0.153 | 0.425 | 0.125 | 0.438 | 0.132 | 0.200 | 0.357 | 0.148 | 0.205 |
| 1/4 | 0.2500 | 0.507 | 0.153 | 0.472 | 0.175 | 0.492 | 0.144 | 0.507 | 0.153 | 0.232 | 0.414 | 0.170 | 0.237 |
| 5/16 | 0.3125 | 0.635 | 0.191 | 0.590 | 0.216 | 0.615 | 0.178 | 0.635 | 0.191 | 0.290 | 0.518 | 0.211 | 0.295 |
| 3/8 | 0.3750 | 0.762 | 0.230 | 0.708 | 0.256 | 0.740 | 0.212 | 0.762 | 0.230 | 0.347 | 0.622 | 0.253 | 0.355 |
| 7/16 | 0.4375 | 0.812 | 0.223 | 0.750 | 0.328 | 0.863 | 0.247 | 0.812 | 0.223 | 0.345 | 0.625 | 0.265 | 0.368 |
| 1/2 | 0.5000 | 0.875 | 0.223 | 0.813 | 0.355 | 0.987 | 0.281 | 0.875 | 0.223 | 0.354 | 0.750 | 0.297 | 0.412 |
| 9/16 | 0.5625 | 1.000 | 0.260 | 0.938 | 0.410 | 1.041 | 0.315 | 1.000 | 0.260 | 0.410 | 0.812 | 0.336 | 0.466 |
| 5/8 | 0.6250 | 1.125 | 0.298 | 1.000 | 0.438 | 1.172 | 0.350 | 1.125 | 0.298 | 0.467 | 0.875 | 0.375 | 0.521 |
| 3/4 | 0.7500 | 1.375 | 0.372 | 1.250 | 0.547 | 1.435 | 0.419 | 1.375 | 0.372 | 0.578 | 1.000 | 0.441 | 0.612 |

Reproduced from ANSI B18.6.3–1972 with the permission of the publisher, the American Society of Mechanical Engineers.

From *Drafting: Technology and Practice*, copyright 1980 by Spence. Used with permission of the publisher, Bennett Publishing Company, Peoria, IL 61615.

## Inch Slotted Head Cap Screws

| Nominal Size | D — Diameter of Screw | Flat Head A | Flat Head H | Round Head A | Round Head H | Fillister Head A | Fillister Head H | Fillister Head O |
|---|---|---|---|---|---|---|---|---|
| ¼ | 0.250 | 0.500 | 0.140 | 0.437 | 0.191 | 0.375 | 0.172 | 0.216 |
| ⁵⁄₁₆ | 0.3125 | 0.625 | 0.177 | 0.562 | 0.245 | 0.437 | 0.203 | 0.253 |
| ³⁄₈ | 0.375 | 0.750 | 0.210 | 0.625 | 0.273 | 0.562 | 0.250 | 0.314 |
| ⁷⁄₁₆ | 0.4375 | 0.8125 | 0.210 | 0.750 | 0.328 | 0.625 | 0.297 | 0.368 |
| ½ | 0.500 | 0.875 | 0.210 | 0.812 | 0.354 | 0.750 | 0.328 | 0.413 |
| ⁹⁄₁₆ | 0.5625 | 1.000 | 0.244 | 0.937 | 0.409 | 0.812 | 0.375 | 0.467 |
| ⅝ | 0.625 | 1.125 | 0.281 | 1.000 | 0.437 | 0.875 | 0.422 | 0.521 |
| ¾ | 0.750 | 1.375 | 0.352 | 1.250 | 0.546 | 1.000 | 0.500 | 0.612 |
| ⅞ | 0.875 | 1.625 | 0.423 | | | 1.125 | 0.594 | 0.720 |
| 1 | 1.000 | 1.875 | 0.494 | | | 1.312 | 0.656 | 0.803 |
| 1⅛ | 1.125 | 2.062 | 0.529 | | | | | |
| 1¼ | 1.250 | 2.312 | 0.600 | | | | | |
| 1⅜ | 1.375 | 2.562 | 0.665 | | | | | |
| 1½ | 1.500 | 2.812 | 0.742 | | | | | |

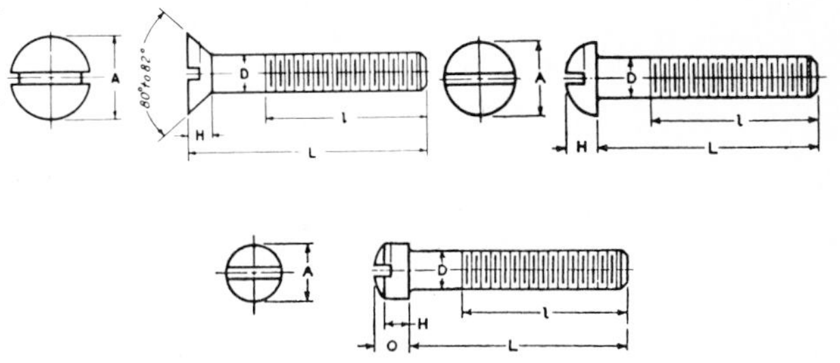

Reproduced from ANSI B18.6.2–1972 with the permission of the publisher, the American Society of Mechanical Engineers.

From *Drafting: Technology and Practice*, copyright 1980 by Spence. Used with permission of the publisher, Bennett Publishing Company, Peoria, IL 61615.

## Inch Square Head Set Screws

| Nominal Size | | F — Width Across Flats | H — Nom. | H — Max | G — Width Across Corners |
|---|---|---|---|---|---|
| #10 | 0.190 | 0.188 | ⁹⁄₆₄ | 0.148 | 0.247 |
| ¼ | 0.250 | 0.250 | ³⁄₁₆ | 0.196 | 0.331 |
| ⁵⁄₁₆ | 0.3125 | 0.312 | ¹⁵⁄₆₄ | 0.245 | 0.415 |
| ³⁄₈ | 0.3750 | 0.375 | ⁹⁄₃₂ | 0.293 | 0.497 |
| ⁷⁄₁₆ | 0.4375 | 0.438 | ²¹⁄₆₄ | 0.341 | 0.581 |
| ½ | 0.500 | 0.500 | 3.8 | 0.389 | 0.665 |

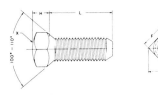

Square head set screw points are the same types and sizes as shown for slotted headless set screws.

Reproduced from ANSI B 18.6.2–1972 with the permission of the publisher, the American Society of Mechanical Engineers.

From *Drafting: Technology and Practice*, copyright 1980 by Spence. Used with permission of the publisher, Bennett Publishing Company, Peoria, IL 61615.

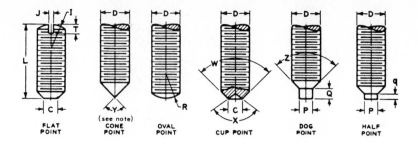

FLAT POINT   CONE POINT (see note)   OVAL POINT   CUP POINT   DOG POINT   HALF POINT

Inch Slotted Headless Set Screws

| D | | I | J | T | R | C | P | Q | q |
|---|---|---|---|---|---|---|---|---|---|
| Nominal Size | | Radius of Headless Crown | Width of Slot | Depth of Slot | Oval Point Radius | Diameter of Cup and Flat Points | Diameter of Dog Point | Length of Dog Point (see note) | |
| | | | | | | Max | Max | Full | Half |
| 5 | 0.125 | 0.125 | 0.026 | 0.036 | 0.094 | 0.067 | 0.083 | 0.063 | 0.033 |
| 6 | 0.138 | 0.138 | 0.028 | 0.040 | 0.104 | 0.074 | 0.092 | 0.073 | 0.038 |
| 8 | 0.164 | 0.164 | 0.032 | 0.046 | 0.123 | 0.087 | 0.109 | 0.083 | 0.043 |
| 10 | 0.190 | 0.190 | 0.035 | 0.053 | 0.142 | 0.102 | 0.127 | 0.095 | 0.050 |
| 12 | 0.216 | 0.216 | 0.042 | 0.061 | 0.162 | 0.115 | 0.144 | 0.115 | 0.060 |
| 1/4 | | 0.250 | 0.049 | 0.068 | 0.188 | 0.132 | 0.156 | 0.130 | 0.068 |
| 5/16 | | 0.312 | 0.055 | 0.083 | 0.234 | 0.172 | 0.203 | 0.161 | 0.083 |
| 3/8 | | 0.375 | 0.068 | 0.099 | 0.281 | 0.212 | 0.250 | 0.193 | 0.099 |

All dimensions given in inches.

Where usable length of thread is less than the nominal diameter, half-dog point shall be used.

When $L$ (length of screw) equals nominal diameter or less, = 118 deg $\pm$ 2 deg; when $L$ exceeds nominal diameter, $Y$ = 90 deg $\pm$ 2 deg.

Point Angles. $W$ = 80 deg to 90 deg; $X$ = 118 deg $\pm$ 5 deg; $Z$ = 100 deg to 110 deg.

Reproduced from ANSI B18.6.2–1972 with the permission of the publisher, the American Society of Mechanical Engineers.

From *Drafting: Technology and Practice*, copyright 1980 by Spence. Used with permission of the publisher, Bennett Publishing Company, Peoria, IL 61615.

Inch Studs

| E | B | D |
|---|---|---|
| Nominal Diameter -Inches- | Threads per Inch | Maximum Thread Length |
| 1/4 | 20 | .3750 |
| 5/15 | 18 | .5000 |
| 3/8 | 16 | .5625 |
| 7/16 | 14 | .6875 |
| 1/2 | 13 | .7500 |
| 9/16 | 12 | .8750 |
| 5/8 | 11 | .9375 |
| 3/4 | 10 | 1.1250 |

From *Drafting: Technology and Practice*, copyright 1980 by Spence. Used with permission of the publisher, Bennett Publishing Company, Peoria, IL 61615.

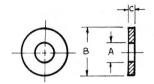

## Inch Type A Plain Washers[1]

| Nominal Washer Size | | | Inside Dia | Outside Dia. | Nominal Thickness |
|---|---|---|---|---|---|
| | | | A | B | C |
| ...... | ...... | | 0.078 | 0.188 | 0.020 |
| ...... | ...... | | 0.094 | 0.250 | 0.020 |
| ...... | ...... | | 0.125 | 0.312 | 0.032 |
| No. 6 | 0.138 | | 0.156 | 0.375 | 0.049 |
| No. 8 | 0.164 | | 0.188 | 0.438 | 0.049 |
| No. 10 | 0.190 | | 0.219 | 0.500 | 0.049 |
| | | | | | |
| 3/16 | 0.188 | | 0.250 | 0.562 | 0.049 |
| No. 12 | 0.216 | | 0.250 | 0.562 | 0.065 |
| 1/4 | 0.250 | N | 0.281 | 0.625 | 0.065 |
| 1/4 | 0.250 | W | 0.312 | 0.734 | 0.065 |
| 5/16 | 0.312 | N | 0.344 | 0.688 | 0.065 |
| 5/16 | 0.312 | W | 0.375 | 0.875 | 0.083 |
| | | | | | |
| 3/8 | 0.375 | N | 0.406 | 0.812 | 0.065 |
| 3/8 | 0.375 | W | 0.438 | 1.000 | 0.083 |
| 7/16 | 0.438 | N | 0.469 | 0.922 | 0.065 |
| 7/16 | 0.438 | W | 0.500 | 1.250 | 0.083 |
| 1/2 | 0.500 | N | 0.531 | 1.062 | 0.095 |
| 1/2 | 0.500 | W | 0.562 | 1.375 | 0.109 |
| | | | | | |
| 9/16 | 0.562 | N | 0.594 | 1.156 | 0.095 |
| 9/16 | 0.562 | W | 0.625 | 1.469 | 0.109 |
| 5/8 | 0.625 | N | 0.656 | 1.312 | 0.095 |
| 5/8 | 0.625 | W | 0.688 | 1.750 | 0.134 |
| 3/4 | 0.750 | N | 0.812 | 1.469 | 0.134 |
| 3/4 | 0.750 | W | 0.812 | 2.000 | 0.148 |
| | | | | | |
| 7/8 | 0.875 | N | 0.938 | 1.750 | 0.134 |
| 7/8 | 0.875 | W | 0.938 | 2.250 | 0.165 |
| 1 | 1.000 | N | 1.062 | 2.000 | 0.134 |
| 1 | 1.000 | W | 1.062 | 2.500 | 0.165 |
| 1 1/8 | 1.125 | N | 1.250 | 2.250 | 0.134 |
| 1 1/8 | 1.125 | W | 1.250 | 2.750 | 0.165 |
| | | | | | |
| 1 1/4 | 1.250 | N | 1.375 | 2.500 | 0.165 |
| 1 1/4 | 1.250 | W | 1.375 | 3.000 | 0.165 |
| 1 3/8 | 1.375 | N | 1.500 | 2.750 | 0.165 |
| 1 3/8 | 1.375 | W | 1.500 | 3.250 | 0.180 |
| 1 1/2 | 1.500 | N | 1.625 | 3.000 | 0.165 |

[1]Additional sizes in standards.

[2]Preferred sizes are for the most part from series previously designated "Standard Plate" and "SAE." Where common sizes existed in the two series the SAE is designated "N" (narrow) and the Standard Plate "W" (wide).

Reproduced from ANSI B18.22.1–1965(R1975) with the permission of the publisher, the American Society of Mechanical Engineers.

From *Drafting: Technology and Practice*, copyright 1980 by Spence. Used with permission of the publisher, Bennett Publishing Company, Peoria, IL 61615.

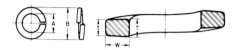

Inch Helical Spring Lock Washers

| Nominal Washer Size | | Inside Diameter A | | Outside Diameter B | Washer Section | |
|---|---|---|---|---|---|---|
| | | | | | Width W | Thickness $\frac{T + t}{2}$ |
| | | Min | Max | Max** | Min | Min |
| No.  2 | 0.086 | 0.088 | 0.094 | 0.172 | 0.035 | 0.020 |
| No.  3 | 0.099 | 0.101 | 0.107 | 0.195 | 0.040 | 0.025 |
| No.  4 | 0.112 | 0.115 | 0.121 | 0.209 | 0.040 | 0.025 |
| No.  5 | 0.125 | 0.128 | 0.134 | 0.236 | 0.047 | 0.031 |
| No.  6 | 0.138 | 0.141 | 0.148 | 0.250 | 0.047 | 0.031 |
| No.  8 | 0.164 | 0.168 | 0.175 | 0.293 | 0.055 | 0.040 |
| No. 10 | 0.190 | 0.194 | 0.202 | 0.334 | 0.062 | 0.047 |
| No. 12 | 0.216 | 0.221 | 0.229 | 0.377 | 0.070 | 0.056 |
| 1/4 | 0.250 | 0.255 | 0.263 | 0.489 | 0.109 | 0.062 |
| 5/16 | 0.312 | 0.318 | 0.328 | 0.586 | 0.125 | 0.078 |
| 3/8 | 0.375 | 0.382 | 0.393 | 0.683 | 0.141 | 0.094 |
| 7/16 | 0.438 | 0.446 | 0.459 | 0.779 | 0.156 | 0.109 |
| 1/2 | 0.500 | 0.509 | 0.523 | 0.873 | 0.171 | 0.125 |
| 9/16 | 0.562 | 0.572 | 0.587 | 0.971 | 0.188 | 0.141 |
| 5/8 | 0.625 | 0.636 | 0.653 | 1.079 | 0.203 | 0.156 |
| 11/16 | 0.688 | 0.700 | 0.718 | 1.176 | 0.219 | 0.172 |
| 3/4 | 0.750 | 0.763 | 0.783 | 1.271 | 0.234 | 0.188 |
| 13/16 | 0.812 | 0.826 | 0.847 | 1.367 | 0.250 | 0.203 |
| 7/8 | 0.875 | 0.890 | 0.912 | 1.464 | 0.266 | 0.219 |
| 15/16 | 0.938 | 0.954 | 0.978 | 1.560 | 0.281 | 0.234 |
| 1 | 1.000 | 1.017 | 1.042 | 1.661 | 0.297 | 0.250 |
| 1 1/16 | 1.062 | 1.080 | 1.107 | 1.756 | 0.312 | 0.266 |
| 1 1/8 | 1.125 | 1.144 | 1.172 | 1.853 | 0.328 | 0.281 |
| 1 3/16 | 1.188 | 1.208 | 1.237 | 1.950 | 0.344 | 0.297 |
| 1 1/4 | 1.250 | 1.271 | 1.302 | 2.045 | 0.359 | 0.312 |
| 1 5/16 | 1.312 | 1.334 | 1.366 | 2.141 | 0.375 | 0.328 |
| 1 3/8 | 1.375 | 1.398 | 1.432 | 2.239 | 0.391 | 0.344 |
| 1 7/16 | 1.438 | 1.462 | 1.497 | 2.334 | 0.406 | 0.359 |
| 1 1/2 | 1.500 | 1.525 | 1.561 | 2.430 | 0.422 | 0.375 |

Reproduced from ANSI B18.21.1–1972 with the permission of the publisher, the American Society of Mechanical Engineers.

From *Drafting: Technology and Practice*, copyright 1980 by Spence. Used with permission of the publisher, Bennett Publishing Company, Peoria, IL 61615.

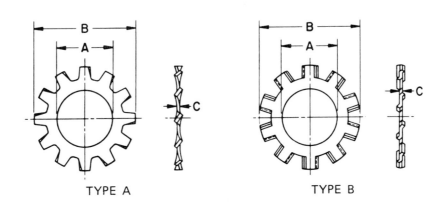

TYPE A          TYPE B

Dimensions of External Tooth Lock Washers

| Nominal Washer Size | | A | | B | | C | |
|---|---|---|---|---|---|---|---|
| | | Inside Diameter | | Outside Diameter | | Thickness | |
| | | Max | Min | Max | Min | Max | Min |
| No. 3 | 0.099 | 0.109 | 0.102 | 0.235 | 0.220 | 0.015 | 0.012 |
| No. 4 | 0.112 | 0.123 | 0.115 | 0.260 | 0.245 | 0.019 | 0.015 |
| No. 5 | 0.125 | 0.136 | 0.129 | 0.285 | 0.270 | 0.019 | 0.014 |
| No. 6 | 0.138 | 0.150 | 0.141 | 0.320 | 0.305 | 0.022 | 0.016 |
| No. 8 | 0.164 | 0.176 | 0.168 | 0.381 | 0.365 | 0.023 | 0.018 |
| No. 10 | 0.190 | 0.204 | 0.195 | 0.410 | 0.395 | 0.025 | 0.020 |
| No. 12 | 0.216 | 0.231 | 0.221 | 0.475 | 0.460 | 0.028 | 0.023 |
| $\frac{1}{4}$ | 0.250 | 0.267 | 0.256 | 0.510 | 0.494 | 0.028 | 0.023 |
| $\frac{5}{16}$ | 0.312 | 0.332 | 0.320 | 0.610 | 0.588 | 0.034 | 0.028 |
| $\frac{3}{8}$ | 0.375 | 0.398 | 0.384 | 0.694 | 0.670 | 0.040 | 0.032 |
| $\frac{7}{16}$ | 0.438 | 0.464 | 0.448 | 0.760 | 0.740 | 0.040 | 0.032 |
| $\frac{1}{2}$ | 0.500 | 0.530 | 0.513 | 0.900 | 0.880 | 0.045 | 0.037 |
| $\frac{9}{16}$ | 0.562 | 0.596 | 0.576 | 0.985 | 0.960 | 0.045 | 0.037 |
| $\frac{5}{8}$ | 0.625 | 0.663 | 0.641 | 1.070 | 1.045 | 0.050 | 0.042 |
| $\frac{11}{16}$ | 0.688 | 0.728 | 0.704 | 1.155 | 1.130 | 0.050 | 0.042 |
| $\frac{3}{4}$ | 0.750 | 0.795 | 0.768 | 1.260 | 1.220 | 0.055 | 0.047 |
| $\frac{13}{16}$ | 0.812 | 0.861 | 0.833 | 1.315 | 1.290 | 0.055 | 0.047 |
| $\frac{7}{8}$ | 0.875 | 0.927 | 0.897 | 1.410 | 1.380 | 0.060 | 0.052 |
| 1 | 1.000 | 1.060 | 1.025 | 1.620 | 1.590 | 0.067 | 0.059 |

Reproduced from ANSI B18.21.1–1972 with the permission of the publisher, the American Society of Mechanical Engineers.

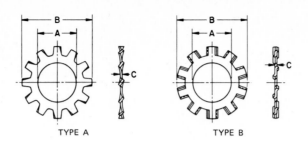

TYPE A　　　　　　　　　　　TYPE B

Dimensions of Internal Tooth Lock Washers

| Nominal Washer Size | | A | | B | | C | |
|---|---|---|---|---|---|---|---|
| | | Inside Diameter | | Outside Diameter | | Thickness | |
| | | Max | Min | Max | Min | Max | Min |
| No. 2 | 0.086 | 0.095 | 0.089 | 0.200 | 0.175 | 0.015 | 0.010 |
| No. 3 | 0.099 | 0.109 | 0.102 | 0.232 | 0.215 | 0.019 | 0.012 |
| No. 4 | 0.112 | 0.123 | 0.115 | 0.270 | 0.255 | 0.019 | 0.015 |
| No. 5 | 0.125 | 0.136 | 0.129 | 0.280 | 0.245 | 0.021 | 0.017 |
| No. 6 | 0.138 | 0.150 | 0.141 | 0.295 | 0.275 | 0.021 | 0.017 |
| No. 8 | 0.164 | 0.176 | 0.168 | 0.340 | 0.325 | 0.023 | 0.018 |
| No. 10 | 0.190 | 0.204 | 0.195 | 0.381 | 0.365 | 0.025 | 0.020 |
| No. 12 | 0.216 | 0.231 | 0.221 | 0.410 | 0.394 | 0.025 | 0.020 |
| ¼ | 0.250 | 0.267 | 0.256 | 0.478 | 0.460 | 0.028 | 0.023 |
| ⁵⁄₁₆ | 0.312 | 0.332 | 0.320 | 0.610 | 0.594 | 0.034 | 0.028 |
| ³⁄₈ | 0.375 | 0.398 | 0.384 | 0.692 | 0.670 | 0.040 | 0.032 |
| ⁷⁄₁₆ | 0.438 | 0.464 | 0.448 | 0.789 | 0.740 | 0.040 | 0.032 |
| ½ | 0.500 | 0.530 | 0.512 | 0.900 | 0.867 | 0.045 | 0.037 |
| ⁹⁄₁₆ | 0.562 | 0.596 | 0.576 | 0.985 | 0.957 | 0.045 | 0.037 |
| ⅝ | 0.625 | 0.663 | 0.640 | 1.071 | 1.045 | 0.050 | 0.042 |
| ¹¹⁄₁₆ | 0.688 | 0.728 | 0.704 | 1.166 | 1.130 | 0.050 | 0.042 |
| ¾ | 0.750 | 0.795 | 0.769 | 1.245 | 1.220 | 0.055 | 0.047 |
| ¹³⁄₁₆ | 0.812 | 0.861 | 0.832 | 1.315 | 1.290 | 0.055 | 0.047 |
| ⅞ | 0.875 | 0.927 | 0.894 | 1.410 | 1.364 | 0.060 | 0.052 |
| 1 | 1.000 | 1.060 | 1.019 | 1.637 | 1.590 | 0.067 | 0.059 |
| 1⅛ | 1.125 | 1.192 | 1.144 | 1.830 | 1.799 | 0.067 | 0.059 |
| 1¼ | 1.250 | 1.325 | 1.275 | 1.975 | 1.921 | 0.067 | 0.059 |

Dimensions of Heavy Internal Tooth Lock Washers

| Nominal Washer Size | | A | | B | | C | |
|---|---|---|---|---|---|---|---|
| | | Inside Diameter | | Outside Diameter | | Thickness | |
| | | Max | Min | Max | Min | Max | Min |
| ¼ | 0.250 | 0.267 | 0.256 | 0.536 | 0.500 | 0.045 | 0.035 |
| ⁵⁄₁₆ | 0.312 | 0.332 | 0.320 | 0.607 | 0.590 | 0.050 | 0.040 |
| ³⁄₈ | 0.375 | 0.398 | 0.384 | 0.748 | 0.700 | 0.050 | 0.042 |
| ⁷⁄₁₆ | 0.438 | 0.464 | 0.448 | 0.858 | 0.800 | 0.067 | 0.050 |
| ½ | 0.500 | 0.530 | 0.512 | 0.924 | 0.880 | 0.067 | 0.055 |
| ⁹⁄₁₆ | 0.562 | 0.596 | 0.576 | 1.034 | 0.990 | 0.067 | 0.055 |
| ⅝ | 0.625 | 0.663 | 0.640 | 1.135 | 1.100 | 0.067 | 0.059 |
| ¾ | 0.750 | 0.795 | 0.768 | 1.265 | 1.240 | 0.084 | 0.070 |
| ⅞ | 0.875 | 0.927 | 0.894 | 1.447 | 1.400 | 0.084 | 0.075 |

Reproduced from ANSI B18.21.1–1972 with the permission of the publisher, the American Society of Mechanical Engineers.

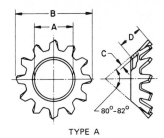
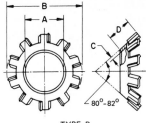

TYPE A            TYPE B

Dimensions of Countersunk External Tooth Lock Washers

| Nominal Washer Size | | A | | $B^1$ | C | | D | |
|---|---|---|---|---|---|---|---|---|
| | | Inside Diameter | | Outside Diameter | Thickness | | Length | |
| | | Max | Min | Approx | Max | Min | Max | Min |
| No. 4 | 0.112 | 0.123 | 0.113 | 0.213 | 0.019 | 0.015 | 0.065 | 0.050 |
| No. 6 | 0.138 | 0.150 | 0.140 | 0.289 | 0.021 | 0.017 | 0.092 | 0.082 |
| No. 8 | 0.164 | 0.177 | 0.167 | 0.322 | 0.021 | 0.017 | 0.105 | 0.088 |
| No. 10 | 0.190 | 0.205 | 0.195 | 0.354 | 0.025 | 0.020 | 0.099 | 0.083 |
| No. 12 | 0.216 | 0.231 | 0.220 | 0.421 | 0.025 | 0.020 | 0.128 | 0.118 |
| 1/4 | 0.250 | 0.267 | 0.255 | 0.454 | 0.025 | 0.020 | 0.128 | 0.113 |
| No. 16 | 0.268 | 0.287 | 0.273 | 0.505 | 0.028 | 0.023 | 0.147 | 0.137 |
| 5/16 | 0.312 | 0.333 | 0.318 | 0.599 | 0.028 | 0.023 | 0.192 | 0.165 |
| 3/8 | 0.375 | 0.398 | 0.383 | 0.765 | 0.034 | 0.028 | 0.255 | 0.242 |
| 7/16 | 0.438 | 0.463 | 0.448 | 0.867 | 0.045 | 0.037 | 0.270 | 0.260 |
| 1/2 | 0.500 | 0.529 | 0.512 | 0.976 | 0.045 | 0.037 | 0.304 | 0.294 |

Reproduced from ANSI B18.21.1–1972 with the permission of the publisher, the American Society of Mechanical Engineers.

Inch Straight Pins

| Nominal Diameter | Diameter A | | Chamfer B |
|---|---|---|---|
| | Max | Min | |
| 0.062 | 0.0625 | 0.0605 | 0.015 |
| 0.094 | 0.0937 | 0.0917 | 0.015 |
| 0.109 | 0.1094 | 0.1074 | 0.015 |
| 0.125 | 0.1250 | 0.1230 | 0.015 |
| 0.156 | 0.1562 | 0.1542 | 0.015 |
| 0.188 | 0.1875 | 0.1855 | 0.015 |
| 0.219 | 0.2187 | 0.2167 | 0.015 |
| 0.250 | 0.2500 | 0.2480 | 0.015 |
| 0.312 | 0.3125 | 0.3095 | 0.030 |
| 0.375 | 0.3750 | 0.3720 | 0.030 |
| 0.438 | 0.4375 | 0.4345 | 0.030 |
| 0.500 | 0.500 | 0.4970 | 0.030 |

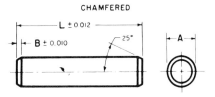

CHAMFERED

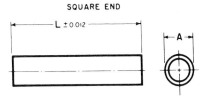

SQUARE END

All dimensions are given in inches.

These pins must be straight and free from burrs or any other defects that will affect their serviceability.

Reproduced from ANSI B18.8.2–1978 with the permission of the publisher, the American Society of Mechanical Engineers.

From *Drafting: Technology and Practice*, copyright 1980 by Spence. Used with permission of the publisher, Bennett Publishing Company, Peoria, IL 61615.

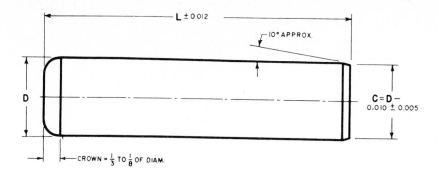

CROWN = $\frac{1}{3}$ TO $\frac{1}{8}$ OF DIAM.

Inch Hardened and Ground Dowel Pins

| Length, L | Nominal Diameter D | | | | | | | | | |
|---|---|---|---|---|---|---|---|---|---|---|
| | $\frac{1}{8}$ | $\frac{3}{16}$ | $\frac{1}{4}$ | $\frac{5}{16}$ | $\frac{3}{8}$ | $\frac{7}{16}$ | $\frac{1}{2}$ | $\frac{5}{8}$ | $\frac{3}{4}$ | $\frac{7}{8}$ |
| | Diameter of Standard Pins ±0.0001 | | | | | | | | | |
| | 0.1252 | 0.1877 | 0.2502 | 0.3127 | 0.3752 | 0.4377 | 0.5002 | 0.6252 | 0.7502 | 0.8752 |
| | Diameter Oversize Pins ±0.0001 | | | | | | | | | |
| | 0.1260 | 0.1885 | 0.2510 | 0.3135 | 0.3760 | 0.4385 | 0.5010 | 0.6260 | 0.7510 | 0.8760 |
| $\frac{1}{2}$ | X | X | X | X | | | | | | |
| $\frac{5}{8}$ | X | X | X | X | | | | | | |
| $\frac{3}{4}$ | X | X | X | X | X | | | | | |
| $\frac{7}{8}$ | X | X | X | X | X | X | | | | |
| 1 | X | X | X | X | X | X | | | | |
| $1\frac{1}{4}$ | | X | X | X | X | X | X | X | | |
| $1\frac{1}{2}$ | | X | X | X | X | X | X | X | X | |
| $1\frac{3}{4}$ | | X | X | X | X | X | X | X | X | |
| 2 | | X | X | X | X | X | X | X | X | X |
| $2\frac{1}{4}$ | | | | X | X | X | X | | | |
| $2\frac{1}{2}$ | | | | X | X | X | X | X | X | X |
| 3 | | | | | | | X | X | X | X |
| $3\frac{1}{2}$ | | | | | | | X | X | | |
| 4 | | | | | | | X | X | X | X |
| $4\frac{1}{2}$ | | | | | | | | X | X | X |
| 5 | | | | | | | | | X | X |
| $5\frac{1}{2}$ | | | | | | | | | X | X |

All dimensions are given in inches.

These pins are extensively used in the tool and machine industry and a machine reamer of nominal size may be used to produce the holes into which these pins tap or press fit. They must be straight and free from any defects that will affect their serviceability.

Reproduced from ANSI B18.8.2–1978 with the permission of the publisher, the American Society of Mechanical Engineers.

From *Drafting: Technology and Practice*, copyright 1980 by Spence. Used with permission of the publisher, Bennett Publishing Company, Peoria, IL 61615.

Suggested Shaft Diameters to Use with Taper Pins

| Pin No. | $\frac{7}{0}$ | $\frac{6}{0}$ | $\frac{5}{0}$ | $\frac{4}{0}$ | $\frac{3}{0}$ | $\frac{2}{0}$ | 0 | 1 | 2 | 3 | 4 | 5 | 6 | 7 | 8 |
|---|---|---|---|---|---|---|---|---|---|---|---|---|---|---|---|
| Suggested Shaft Dia. | | $\frac{7}{32}$ | $\frac{1}{4}$ | $\frac{5}{16}$ | $\frac{3}{8}$ | $\frac{7}{16}$ | $\frac{1}{2}$ | $\frac{9}{16}$ | $\frac{5}{8}$ | $\frac{3}{4}$ | $\frac{13}{16}$ | $\frac{7}{8}$ | 1 | $1\frac{1}{4}$ | $1\frac{1}{2}$ |

From *Drafting: Technology and Practice*, copyright 1980 by Spence. Used with permission of the publisher, Bennett Publishing Company, Peoria, IL 61615.

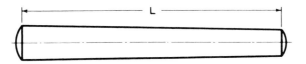

Inch Taper Pins

| Number | 7/0 | 6/0 | 5/0 | 4/0 | 3/0 | 2/0 | 0 | 1 | 2 | 3 | 4 | 5 | 6 | 7 | 8 | 9 | 10 |
|---|---|---|---|---|---|---|---|---|---|---|---|---|---|---|---|---|---|
| Size (Large End) | 0.0625 | 0.0780 | 0.0940 | 0.1090 | 0.1250 | 0.1410 | 0.1560 | 0.1720 | 0.1930 | 0.2190 | 0.2500 | 0.2890 | 0.3410 | 0.4090 | 0.4920 | 0.5910 | 0.7060 |
| Length, L | | | | | | | | | | | | | | | | | |
| 0.375 | X | X | | | | | | | | | | | | | | | |
| 0.500 | X | X | X | X | X | X | X | | | | | | | | | | |
| 0.625 | X | X | X | X | X | X | X | | | | | | | | | | |
| 0.750 | | X | X | X | X | X | X | X | X | X | | | | | | | |
| 0.875 | | | | | | X | X | X | X | X | | | | | | | |
| 1.000 | | | X | X | X | X | X | X | X | X | X | X | | | | | |
| 1.250 | | | | | | X | X | X | X | X | X | X | X | | | | |
| 1.500 | | | | | | | X | X | X | X | X | X | X | | | | |
| 1.750 | | | | | | | | X | X | X | X | X | X | | | | |
| 2.000 | | | | | | | | X | X | X | X | X | X | X | X | | |
| 2.250 | | | | | | | | | X | X | X | X | X | X | X | | |
| 2.500 | | | | | | | | | X | X | X | X | X | X | X | | |
| 2.750 | | | | | | | | | | X | X | X | X | X | X | X | |
| 3.000 | | | | | | | | | | X | X | X | X | X | X | X | |
| 3.250 | | | | | | | | | | | | | X | X | X | X | |
| 3.500 | | | | | | | | | | | | | X | X | X | X | X |
| 3.750 | | | | | | | | | | | | | X | X | X | X | X |
| 4.000 | | | | | | | | | | | | | X | X | X | X | X |
| 4.250 | | | | | | | | | | | | | | | X | X | X |
| 4.500 | | | | | | | | | | | | | | | X | X | X |
| 4.750 | | | | | | | | | | | | | | | X | X | X |
| 5.000 | | | | | | | | | | | | | | | X | X | X |
| 5.250 | | | | | | | | | | | | | | | | X | X |
| 5.500 | | | | | | | | | | | | | | | | X | X |
| 5.750 | | | | | | | | | | | | | | | | X | X |
| 6.000 | | | | | | | | | | | | | | | | X | X |

Reproduced from ANSI B18.8.2–1978 with the permission of the publisher, the American Society of Mechanical Engineers.

From *Drafting: Technology and Practice*, copyright 1980 by Spence. Used with permission of the publisher, Bennett Publishing Company, Peoria, IL 61615.

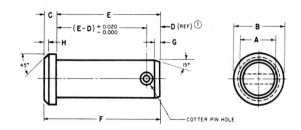

## Inch Clevis Pins

| Diameter of Pin A | | | Diameter of Head B | Height of Head C | Distance From Center of Hole to End of Pin D | Length of Pin Under Head ② E | Total Length of Pin F | Chamfer of Pin G | Chamfer of Head H | Hole Sizes +.010 −.005 | Drill Sizes |
|---|---|---|---|---|---|---|---|---|---|---|---|
| Nom | Max | Min | | | | | | | | | |
| 0.188 | 0.186 | 0.181 | 5/16 | 1/16 | 7/64 | 19/32 | 21/32 | 3/64 | 1/64 | 0.0781 | 5/64 |
| 0.250 | 0.248 | 0.243 | 3/8 | 3/32 | 1/8 | 31/64 | 57/64 | 1/16 | 1/32 | 0.0781 | 5/64 |
| 0.312 | 0.311 | 0.306 | 7/16 | 3/32 | 5/32 | 31/32 | 1 1/16 | 5/64 | 1/32 | 0.1094 | 7/64 |
| 0.375 | 0.373 | 0.368 | 1/2 | 1/8 | 5/32 | 1 3/32 | 1 7/32 | 5/64 | 1/32 | 0.1094 | 7/64 |
| 0.438 | 0.436 | 0.431 | 9/16 | 5/32 | 11/64 | 1 5/64 | 1 25/64 | 3/32 | 3/64 | 0.1094 | 7/64 |
| 0.500 | 0.496 | 0.491 | 5/8 | 5/32 | 7/32 | 1 27/64 | 1 37/64 | 7/64 | 3/64 | 0.1406 | 9/64 |
| 0.625 | 0.621 | 0.616 | 13/16 | 13/64 | 1/4 | 1 23/64 | 1 59/64 | 9/64 | 1/16 | 0.1406 | 9/64 |
| 0.750 | 0.746 | 0.741 | 15/16 | 1/4 | 19/64 | 2 3/64 | 2 19/64 | 5/32 | 5/64 | 0.1719 | 11/64 |
| 0.875 | 0.871 | 0.866 | 1 1/16 | 5/16 | 21/64 | 2 11/32 | 2 21/32 | 3/16 | 3/32 | 0.1719 | 11/64 |
| 1.000 | 0.996 | 0.991 | 1 3/16 | 11/32 | 23/64 | 2 5/8 | 2 31/32 | 7/32 | 7/64 | 0.1719 | 11/64 |

From *Drafting: Technology and Practice*, copyright 1980 by Spence. Used with permission of the publisher, Bennett Publishing Company, Peoria, IL 61615.

## Inch Cotter Pins

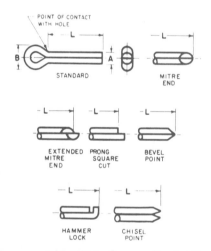

| Diameter Nominal A | Outside Eye Diameter B Min. | Hole Sizes Recommended |
|---|---|---|
| 0.031 | 1/16 | 3/64 |
| 0.047 | 3/32 | 1/16 |
| 0.062 | 1/8 | 5/64 |
| 0.078 | 5/32 | 3/32 |
| 0.094 | 3/16 | 7/64 |
| 0.109 | 7/32 | 1/8 |
| 0.125 | 1/4 | 9/64 |
| 0.141 | 9/32 | 5/32 |
| 0.156 | 5/16 | 11/64 |

| Diameter Nominal A | Outside Eye Diameter B Min. | Hole Sizes Recommended |
|---|---|---|
| 0.188 | 3/8 | 13/64 |
| 0.219 | 7/16 | 15/64 |
| 0.250 | 1/2 | 17/64 |
| 0.312 | 5/8 | 5/16 |
| 0.375 | 3/4 | 3/8 |
| 0.438 | 7/8 | 7/16 |
| 0.500 | 1 | 1/2 |
| 0.625 | 1 1/4 | 5/8 |
| 0.750 | 1 1/2 | 3/4 |

From *Drafting: Technology and Practice*, copyright 1980 by Spence. Used with permission of the publisher, Bennett Publishing Company, Peoria, IL 61615.

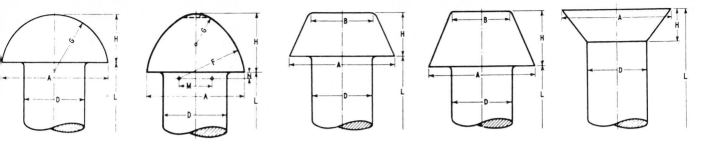

## Inch Large Rivets

| Nominal Size | Button | | | High Button | | | | Pan | | | Cone | | | Flat | |
|---|---|---|---|---|---|---|---|---|---|---|---|---|---|---|---|
| | A | H | G | A | H | F | G | A | B | H | A | B | H | A | H |
| ½ | 0.875 | 0.406 | 0.443 | 0.781 | 0.500 | 0.656 | 0.094 | 0.800 | 0.500 | 0.381 | 0.875 | 0.469 | 0.469 | 0.936 | 0.260 |
| ⅝ | 1.094 | 0.500 | 0.553 | 0.969 | 0.594 | 0.750 | 0.188 | 1.000 | 0.625 | 0.469 | 1.094 | 0.586 | 0.578 | 1.194 | 0.339 |
| ¾ | 1.312 | 0.593 | 0.664 | 1.156 | 0.688 | 0.844 | 0.282 | 1.200 | 0.750 | 0.556 | 1.312 | 0.703 | 0.687 | 1.421 | 0.400 |
| ⅞ | 1.531 | 0.687 | 0.775 | 1.344 | 0.781 | 0.937 | 0.375 | 1.400 | 0.875 | 0.643 | 1.531 | 0.820 | 0.797 | 1.647 | 0.460 |
| 1 | 1.750 | 0.781 | 0.885 | 1.531 | 0.875 | 1.031 | 0.469 | 1.600 | 1.000 | 0.731 | 1.750 | 0.938 | 0.906 | 1.873 | 0.520 |

Head dimensions are for manufactured head after driving. Large rivets are available in length increments of ⅛ inch.

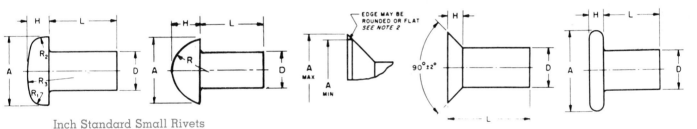

## Inch Standard Small Rivets

| Nominal Size or Basic Shank Diameter | | Diameter of Shank D | Pan Head | | | | | Button Head | | | Countersunk Head | | Flat Head | |
|---|---|---|---|---|---|---|---|---|---|---|---|---|---|---|
| | | | A | H | R₁ | R₂ | R₃ | A | H | R | A | H | A | H |
| 1/16 | 0.062 | 0.064 | 0.118 | 0.040 | 0.019 | 0.052 | 0.217 | 0.122 | 0.052 | 0.055 | 0.118 | 0.027 | 0.140 | 0.027 |
| 3/32 | 0.094 | 0.096 | 0.173 | 0.060 | 0.030 | 0.080 | 0.326 | 0.182 | 0.077 | 0.084 | 0.176 | 0.040 | 0.200 | 0.038 |
| 1/8 | 0.125 | 0.127 | 0.225 | 0.078 | 0.039 | 0.106 | 0.429 | 0.235 | 0.100 | 0.111 | 0.235 | 0.053 | 0.260 | 0.048 |
| 5/32 | 0.156 | 0.158 | 0.279 | 0.096 | 0.049 | 0.133 | 0.535 | 0.290 | 0.124 | 0.138 | 0.293 | 0.066 | 0.323 | 0.059 |
| 3/16 | 0.188 | 0.191 | 0.334 | 0.114 | 0.059 | 0.159 | 0.641 | 0.348 | 0.147 | 0.166 | 0.351 | 0.079 | 0.387 | 0.069 |
| 7/32 | 0.219 | 0.222 | 0.391 | 0.133 | 0.069 | 0.186 | 0.754 | 0.405 | 0.172 | 0.195 | 0.413 | 0.094 | 0.453 | 0.080 |
| 1/4 | 0.250 | 0.253 | 0.444 | 0.151 | 0.079 | 0.213 | 0.858 | 0.460 | 0.196 | 0.221 | 0.469 | 0.106 | 0.515 | 0.091 |
| 9/32 | 0.281 | 0.285 | 0.499 | 0.170 | 0.088 | 0.239 | 0.963 | 0.518 | 0.220 | 0.249 | 0.528 | 0.119 | 0.579 | 0.103 |
| 5/16 | 0.313 | 0.317 | 0.552 | 0.187 | 0.098 | 0.266 | 1.070 | 0.572 | 0.243 | 0.276 | 0.588 | 0.133 | 0.641 | 0.113 |
| 11/32 | 0.344 | 0.348 | 0.608 | 0.206 | 0.108 | 0.292 | 1.176 | 0.630 | 0.267 | 0.304 | 0.646 | 0.146 | 0.705 | 0.124 |
| 3/8 | 0.375 | 0.380 | 0.663 | 0.225 | 0.118 | 0.319 | 1.286 | 0.684 | 0.291 | 0.332 | 0.704 | 0.159 | 0.769 | 0.135 |
| 13/32 | 0.406 | 0.411 | 0.719 | 0.243 | 0.127 | 0.345 | 1.392 | 0.743 | 0.316 | 0.358 | 0.763 | 0.172 | 0.834 | 0.146 |
| 7/16 | 0.438 | 0.443 | 0.772 | 0.261 | 0.137 | 0.372 | 1.500 | 0.798 | 0.339 | 0.387 | 0.823 | 0.186 | 0.896 | 0.157 |

# Inch Parallel, Plain Taper, and Gib Head Key Dimensions and Tolerances

| Key | | | Nominal Key Size Width, $W$ | | Tolerance | |
|---|---|---|---|---|---|---|
| | | | Over | To (Incl) | Width, $W$ | Height, $H$ |
| Parallel | Square | Bar Stock[1] | — $^3/_4$ $1^1/_2$ | $^3/_4$ $1^1/_2$ $2^1/_2$ | +0.000 −0.002<br>+0.000 −0.003<br>+0.000 −0.004 | +0.000 −0.002<br>+0.000 −0.003<br>+0.000 −0.004 |
| | Rectangular | Bar Stock[1] | — $^3/_4$ $1^1/_2$ | $^3/_4$ $1^1/_2$ 3 | +0.000 −0.003<br>+0.000 −0.004<br>+0.000 −0.005 | +0.000 −0.003<br>+0.000 −0.004<br>+0.000 −0.005 |
| Taper | Plain or Gib Head Square or Rectangular | | — $1^1/_4$ | $1^1/_4$ 3 | +0.001 −0.000<br>+0.002 −0.000 | +0.005 −0.000<br>+0.005 −0.000 |

[1]Two types of stock are used for parallel keys. One is a bar stock with a negative tolerance. Another is a key stock with a close plus tolerance.

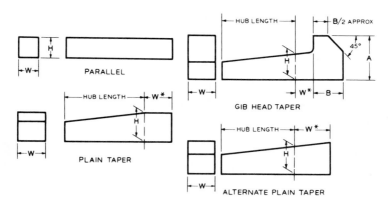

# Inch Key Sizes for Shaft Diameters for Parallel, Plain Taper, and Gib Head Keys

| Nominal Shaft Diameter | | Nominal Key Size | | | Nominal Keyseat Depth | |
|---|---|---|---|---|---|---|
| | | | Height, $H$ | | $H/2$ | |
| Over | To (Incl) | Width, $W$ | Square | Rectangular | Square | Rectangular |
| $^5/_{16}$ | $^7/_{16}$ | $^3/_{32}$ | $^3/_{32}$ | | $^3/_{64}$ | |
| $^7/_{16}$ | $^9/_{16}$ | $^1/_8$ | $^1/_8$ | $^3/_{32}$ | $^1/_{16}$ | $^3/_{64}$ |
| $^9/_{16}$ | $^7/_8$ | $^3/_{16}$ | $^3/_{16}$ | $^1/_8$ | $^3/_{32}$ | $^1/_{16}$ |
| $^7/_8$ | $1^1/_4$ | $^1/_4$ | $^1/_4$ | $^3/_{16}$ | $^1/_8$ | $^3/_{32}$ |
| $1^1/_4$ | $1^3/_8$ | $^5/_{16}$ | $^5/_{16}$ | $^1/_4$ | $^5/_{32}$ | $^1/_8$ |
| $1^3/_8$ | $1^3/_4$ | $^3/_8$ | $^3/_8$ | $^1/_4$ | $^3/_{16}$ | $^1/_8$ |
| $1^3/_4$ | $2^1/_4$ | $^1/_2$ | $^1/_2$ | $^3/_8$ | $^1/_4$ | $^3/_{16}$ |
| $2^1/_4$ | $2^3/_4$ | $^5/_8$ | $^5/_8$ | $^7/_{16}$ | $^5/_{16}$ | $^7/_{32}$ |
| $2^3/_4$ | $3^1/_4$ | $^3/_4$ | $^3/_4$ | $^1/_2$ | $^3/_8$ | $^1/_4$ |
| $3^1/_4$ | $3^3/_4$ | $^7/_8$ | $^7/_8$ | $^5/_8$ | $^7/_{16}$ | $^5/_{16}$ |
| $3^3/_4$ | $4^1/_2$ | 1 | 1 | $^3/_4$ | $^1/_2$ | $^3/_8$ |
| $4^1/_2$ | $5^1/_2$ | $1^1/_4$ | $1^1/_4$ | $^7/_8$ | $^5/_8$ | $^7/_{16}$ |
| $5^1/_2$ | $6^1/_2$ | $1^1/_2$ | $1^1/_2$ | 1 | $^3/_4$ | $^1/_2$ |
| $6^1/_2$ | $7^1/_2$ | $1^3/_4$ | $1^3/_4$ | $1^1/_2$ | $^7/_8$ | $^3/_4$ |
| $7^1/_2$ | 9 | 2 | 2 | $1^1/_2$ | 1 | $^3/_4$ |
| 9 | 11 | $2^1/_2$ | $2^1/_2$ | $1^3/_4$ | $1^1/_4$ | $^7/_8$ |
| 11 | 13 | 3 | 3 | 2 | $1^1/_2$ | 1 |

Sizes and dimension in the unshaded area are preferred.

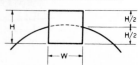

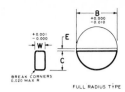
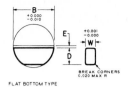

## Inch Woodruff Keys and Keyseats

| Key No.[1] | Nominal Sizes | | | | Maximum Sizes | | | | Key No.[1] | Nominal Sizes | | | | Maximum Sizes | | |
|---|---|---|---|---|---|---|---|---|---|---|---|---|---|---|---|---|
| | $W \times B$ | E | F | D | A | C | H | | $W \times B$ | E | F | D | A | C | H |
| 204 | $1/16 \times 1/2$ | $3/64$ | $1/32$ | .194 | .1718 | .203 | .0422 | 808 | $1/4 \times 1$ | $1/16$ | $1/8$ | .428 | .3130 | .438 | .1360 |
| 304 | $3/32 \times 1/2$ | $3/64$ | $3/64$ | .194 | .1561 | .203 | .0579 | 809 | $1/4 \times 1\,1/8$ | $5/64$ | $1/8$ | .475 | .3590 | .484 | .1360 |
| 305 | $3/32 \times 5/8$ | $1/16$ | $3/64$ | .240 | .2031 | .250 | .0579 | 810 | $1/4 \times 1\,1/4$ | $5/64$ | $1/8$ | .537 | .4220 | .547 | .1360 |
| 404 | $1/8 \times 1/2$ | $3/64$ | $1/16$ | .194 | .1405 | .203 | .0735 | 811 | $1/4 \times 1\,3/8$ | $3/32$ | $1/8$ | .584 | .4690 | .594 | .1360 |
| 405 | $1/8 \times 5/8$ | $1/16$ | $1/16$ | .240 | .1875 | .250 | .0735 | 812 | $1/4 \times 1\,1/2$ | $7/64$ | $1/8$ | .631 | .5160 | .641 | .1360 |
| 406 | $1/8 \times 3/4$ | $1/16$ | $1/16$ | .303 | .2505 | .313 | .0735 | 1008 | $5/16 \times 1$ | $1/16$ | $5/32$ | .428 | .2818 | .438 | .1672 |
| 505 | $5/32 \times 5/8$ | $1/16$ | $5/64$ | .240 | .1719 | .250 | .0891 | 1009 | $5/16 \times 1\,1/8$ | $5/64$ | $5/32$ | .475 | .3278 | .484 | .1672 |
| 506 | $5/32 \times 3/4$ | $1/16$ | $5/64$ | .303 | .2349 | .313 | .0891 | 1010 | $5/16 \times 1\,1/4$ | $5/64$ | $5/32$ | .537 | .3908 | .547 | .1672 |
| 507 | $5/32 \times 7/8$ | $1/16$ | $5/64$ | .365 | .2969 | .375 | .0891 | 1011 | $5/16 \times 1\,3/8$ | $3/32$ | $5/32$ | .584 | .4378 | .594 | .1672 |
| 606 | $3/16 \times 3/4$ | $1/16$ | $3/32$ | .303 | .2193 | .313 | .1047 | 1012 | $5/16 \times 1\,1/2$ | $7/64$ | $5/32$ | .631 | .4848 | .641 | .1672 |
| 607 | $3/16 \times 7/8$ | $1/16$ | $3/32$ | .365 | .2813 | .375 | .1047 | 1210 | $3/8 \times 1\,1/4$ | $5/64$ | $3/16$ | .537 | .3595 | .547 | .1985 |
| 608 | $3/16 \times 1$ | $1/16$ | $3/32$ | .428 | .3443 | .438 | .1047 | 1211 | $3/8 \times 1\,3/8$ | $3/32$ | $3/16$ | .584 | .4065 | .594 | .1985 |
| 609 | $3/16 \times 1\,1/8$ | $5/64$ | $3/32$ | .475 | .3903 | .484 | .1047 | 1212 | $3/8 \times 1\,1/2$ | $7/64$ | $3/16$ | .631 | .4535 | .641 | .1985 |
| 807 | $1/4 \times 7/8$ | $1/16$ | $1/8$ | .365 | .2500 | .375 | .1360 | ...... | ............ | ... | ... | ...... | ...... | ..... | ...... |

[1] The last two numbers of the key number indicate the diameter (B) in eighths of an inch. The other one or two numbers in front of these indicate the width of the key (A) in thirty-seconds of an inch. For example, key number 608 means the diameter is 8/8 or 1 inch and the thickness is 6/32 or 3/16 inch.

Reproduced from ANSI B17.2–1967(R1978) with the permission of the publisher, the American Society of Mechanical Engineers.

From *Drafting: Technology and Practice*, copyright 1980 by Spence. Used with permission of the publisher, Bennett Publishing Company, Peoria, IL 61615.

## Inch Woodruff Key Sizes for Shaft Diameters

| Shaft Diameter | $5/16$ to $3/8$ | $7/16$ to $1/2$ | $9/16$ to $5/8$ | $11/16$ to $3/4$ | $13/16$ |
|---|---|---|---|---|---|
| Key Numbers | 204 | 304<br>305 | 404<br>405 | 404<br>405<br>406 | 505<br>506 |
| Shaft Diameter | $7/8$ to $15/16$ | 1 | $1\,1/16$ to $1\,1/8$ | $1\,3/16$ | $1\,1/4$ to $1\,5/16$ |
| Key Numbers | 505<br>506<br>507 | 606<br>607<br>608 | 606<br>607<br>608<br>609 | 607<br>608<br>609 | 607<br>608<br>609<br>810 |
| Shaft Diameter | $1\,3/8$ to $1\,7/16$ | $1\,1/2$ to $1\,5/8$ | $1\,11/16$ to $1\,3/4$ | $1\,13/16$ to $2\,1/8$ | $2\,3/16$ to $2\,1/2$ |
| Key Numbers | 608<br>609<br>810 | 808<br>809<br>810<br>812 | 809<br>810<br>812 | 1011<br>1012 | 1211<br>1212 |

From *Drafting: Technology and Practice*, copyright 1980 by Spence. Used with permission of the publisher, Bennett Publishing Company, Peoria, IL 61615.

## Inch Slotted Head Wood Screws

| Nom-inal Size | Threads per Inch | Screw Dia. D | Flat and Oval Head | | | Round Head | |
|---|---|---|---|---|---|---|---|
| | | | A | H | O | A | H |
| 0 | 32 | 0.064 | 0.119 | 0.035 | 0.056 | 0.113 | 0.053 |
| 1 | 28 | 0.077 | 0.146 | 0.043 | 0.068 | 0.138 | 0.061 |
| 2 | 26 | 0.090 | 0.172 | 0.051 | 0.080 | 0.162 | 0.069 |
| 3 | 24 | 0.103 | 0.199 | 0.059 | 0.092 | 0.187 | 0.078 |
| 4 | 22 | 0.116 | 0.225 | 0.067 | 0.104 | 0.211 | 0.086 |
| 5 | 20 | 0.129 | 0.252 | 0.075 | 0.116 | 0.236 | 0.095 |
| 6 | 18 | 0.142 | 0.279 | 0.083 | 0.128 | 0.260 | 0.103 |
| 7 | 16 | 0.155 | 0.305 | 0.091 | 0.140 | 0.285 | 0.111 |
| 8 | 15 | 0.168 | 0.332 | 0.100 | 0.152 | 0.309 | 0.120 |
| 9 | 14 | 0.181 | 0.358 | 0.108 | 0.164 | 0.334 | 0.128 |
| 10 | 13 | 0.194 | 0.385 | 0.116 | 0.176 | 0.359 | 0.137 |
| 12 | 11 | 0.220 | 0.438 | 0.132 | 0.200 | 0.408 | 0.153 |
| 14 | 10 | 0.246 | 0.491 | 0.148 | 0.224 | 0.457 | 0.170 |
| 16 | 9 | 0.272 | 0.544 | 0.164 | 0.248 | 0.506 | 0.187 |
| 18 | 8 | 0.298 | 0.597 | 0.180 | 0.272 | 0.555 | 0.204 |
| 20 | 8 | 0.324 | 0.650 | 0.196 | 0.296 | 0.604 | 0.220 |
| 24 | 7 | 0.376 | 0.756 | 0.228 | 0.344 | 0.702 | 0.254 |

Screw lengths: 1/4 to 1 inch by 1/8 inch increments.
1 to 3 inches by 1/4 inch increments.
3 to 5 inches by 1/2 inch increments.
Thread lengths equal to 2/3 screw length.

Reproduced from ANSI B18.6.1–1972(R1977) with the permission of the publisher, the American Society of Mechanical Engineers.

From *Drafting: Technology and Practice*, copyright 1980 by Spence. Used with permission of the publisher, Bennett Publishing Company, Peoria, IL 61615.

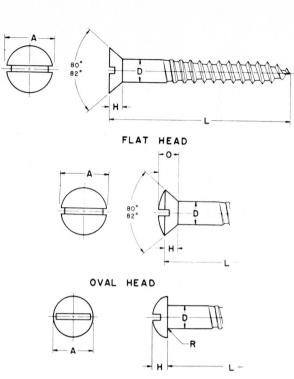

FLAT HEAD

OVAL HEAD

ROUND HEAD

## Inch Two Recessed Flat Head Wood Screws

| Nominal Size | M | T | N |
|---|---|---|---|
| 0 | 0.079 | 0.039 | 0.021 |
| 1 | 0.097 | 0.051 | 0.023 |
| 2 | 0.114 | 0.062 | 0.026 |
| 3 | 0.131 | 0.073 | 0.028 |
| 4 | 0.148 | 0.082 | 0.031 |
| 5 | 0 165 | 0.094 | 0.033 |
| 6 | 0.182 | 0.105 | 0.036 |
| 7 | 0.199 | 0.116 | 0.038 |
| 8 | 0.216 | 0.122 | 0.041 |
| 9 | 0.234 | 0.133 | 0.043 |
| 10 | 0.251 | 0.145 | 0.046 |
| 12 | 0.286 | 0.167 | 0.051 |
| 14 | 0.320 | 0.182 | 0.056 |
| 16 | 0.354 | 0.204 | 0.061 |
| 18 | 0.388 | 0.226 | 0.066 |
| 20 | 0.423 | 0.249 | 0.071 |
| 24 | 0.491 | 0.293 | 0.081 |

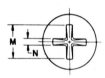

Head Dimensions are the same as those of slotted head wood screws.

Reproduced from ANSI B18.6.1–1972(R1977) with the permission of the publisher, the American Society of Mechanical Engineers.

From *Drafting: Technology and Practice*, copyright 1980 by Spence. Used with permission of the publisher, Bennett Publishing Company, Peoria, IL 61615.

# APPENDIX G: PREFERRED METRIC LIMITS AND FITS

Preferred Hole Basis Clearance Fits

Dimensions in mm.

| BASIC SIZE | | LOOSE RUNNING Hole H11 | Shaft c11 | Fit | FREE RUNNING Hole H9 | Shaft d9 | Fit | CLOSE RUNNING Hole H8 | Shaft f7 | Fit | SLIDING Hole H7 | Shaft g6 | Fit | LOCATIONAL CLEARANCE Hole H7 | Shaft h6 | Fit |
|---|---|---|---|---|---|---|---|---|---|---|---|---|---|---|---|---|
| 1 | MAX | 1.060 | 0.940 | 0.180 | 1.025 | 0.980 | 0.070 | 1.014 | 0.994 | 0.030 | 1.010 | 0.998 | 0.018 | 1.010 | 1.000 | 0.016 |
|   | MIN | 1.000 | 0.880 | 0.060 | 1.000 | 0.955 | 0.020 | 1.000 | 0.984 | 0.006 | 1.000 | 0.992 | 0.002 | 1.000 | 0.994 | 0.000 |
| 1.2 | MAX | 1.260 | 1.140 | 0.180 | 1.225 | 1.180 | 0.070 | 1.214 | 1.194 | 0.030 | 1.210 | 1.198 | 0.018 | 1.210 | 1.200 | 0.016 |
|   | MIN | 1.200 | 1.080 | 0.060 | 1.200 | 1.155 | 0.020 | 1.200 | 1.184 | 0.006 | 1.200 | 1.192 | 0.002 | 1.200 | 1.194 | 0.000 |
| 1.6 | MAX | 1.660 | 1.540 | 0.180 | 1.625 | 1.580 | 0.070 | 1.614 | 1.594 | 0.030 | 1.610 | 1.598 | 0.018 | 1.610 | 1.600 | 0.016 |
|   | MIN | 1.600 | 1.480 | 0.060 | 1.600 | 1.555 | 0.020 | 1.600 | 1.584 | 0.006 | 1.600 | 1.592 | 0.002 | 1.600 | 1.594 | 0.000 |
| 2 | MAX | 2.060 | 1.940 | 0.180 | 2.025 | 1.980 | 0.070 | 2.014 | 1.994 | 0.030 | 2.010 | 1.998 | 0.018 | 2.010 | 2.000 | 0.016 |
|   | MIN | 2.000 | 1.880 | 0.060 | 2.000 | 1.955 | 0.020 | 2.000 | 1.984 | 0.006 | 2.000 | 1.992 | 0.002 | 2.000 | 1.994 | 0.000 |
| 2.5 | MAX | 2.560 | 2.440 | 0.180 | 2.525 | 2.480 | 0.070 | 2.514 | 2.494 | 0.030 | 2.510 | 2.498 | 0.018 | 2.510 | 2.500 | 0.016 |
|   | MIN | 2.500 | 2.380 | 0.060 | 2.500 | 2.455 | 0.020 | 2.500 | 2.484 | 0.006 | 2.500 | 2.492 | 0.002 | 2.500 | 2.494 | 0.000 |
| 3 | MAX | 3.060 | 2.940 | 0.180 | 3.025 | 2.980 | 0.070 | 3.014 | 2.994 | 0.030 | 3.010 | 2.998 | 0.018 | 3.010 | 3.000 | 0.016 |
|   | MIN | 3.000 | 2.880 | 0.060 | 3.000 | 2.955 | 0.020 | 3.000 | 2.984 | 0.006 | 3.000 | 2.992 | 0.002 | 3.000 | 2.994 | 0.000 |
| 4 | MAX | 4.075 | 3.930 | 0.220 | 4.030 | 3.970 | 0.090 | 4.018 | 3.990 | 0.040 | 4.012 | 3.996 | 0.024 | 4.012 | 4.000 | 0.020 |
|   | MIN | 4.000 | 3.855 | 0.070 | 4.000 | 3.940 | 0.030 | 4.000 | 3.978 | 0.010 | 4.000 | 3.988 | 0.004 | 4.000 | 3.992 | 0.000 |
| 5 | MAX | 5.075 | 4.930 | 0.220 | 5.030 | 4.970 | 0.090 | 5.018 | 4.990 | 0.040 | 5.012 | 4.996 | 0.024 | 5.012 | 5.000 | 0.020 |
|   | MIN | 5.000 | 4.855 | 0.070 | 5.000 | 4.940 | 0.030 | 5.000 | 4.978 | 0.010 | 5.000 | 4.988 | 0.004 | 5.000 | 4.992 | 0.000 |
| 6 | MAX | 6.075 | 5.930 | 0.220 | 6.030 | 5.970 | 0.090 | 6.018 | 5.990 | 0.040 | 6.012 | 5.996 | 0.024 | 6.012 | 6.000 | 0.020 |
|   | MIN | 6.000 | 5.855 | 0.070 | 6.000 | 5.940 | 0.030 | 6.000 | 5.978 | 0.010 | 6.000 | 5.988 | 0.004 | 6.000 | 5.992 | 0.000 |
| 8 | MAX | 8.090 | 7.920 | 0.260 | 8.036 | 7.960 | 0.112 | 8.022 | 7.987 | 0.050 | 8.015 | 7.995 | 0.029 | 8.015 | 8.000 | 0.024 |
|   | MIN | 8.000 | 7.830 | 0.080 | 8.000 | 7.924 | 0.040 | 8.000 | 7.972 | 0.013 | 8.000 | 7.986 | 0.005 | 8.000 | 7.991 | 0.000 |
| 10 | MAX | 10.090 | 9.920 | 0.260 | 10.036 | 9.960 | 0.112 | 10.022 | 9.987 | 0.050 | 10.015 | 9.995 | 0.029 | 10.015 | 10.000 | 0.024 |
|   | MIN | 10.000 | 9.830 | 0.080 | 10.000 | 9.924 | 0.040 | 10.000 | 9.972 | 0.013 | 10.000 | 9.986 | 0.005 | 10.000 | 9.991 | 0.000 |
| 12 | MAX | 12.110 | 11.905 | 0.315 | 12.043 | 11.950 | 0.136 | 12.027 | 11.984 | 0.061 | 12.018 | 11.994 | 0.035 | 12.018 | 12.000 | 0.029 |
|   | MIN | 12.000 | 11.795 | 0.095 | 12.000 | 11.907 | 0.050 | 12.000 | 11.966 | 0.016 | 12.000 | 11.983 | 0.006 | 12.000 | 11.989 | 0.000 |
| 16 | MAX | 16.110 | 15.905 | 0.315 | 16.043 | 15.950 | 0.136 | 16.027 | 15.984 | 0.061 | 16.018 | 15.994 | 0.035 | 16.018 | 16.000 | 0.029 |
|   | MIN | 16.000 | 15.795 | 0.095 | 16.000 | 15.907 | 0.050 | 16.000 | 15.966 | 0.016 | 16.000 | 15.983 | 0.006 | 16.000 | 15.989 | 0.000 |
| 20 | MAX | 20.130 | 19.890 | 0.370 | 20.052 | 19.935 | 0.169 | 20.033 | 19.980 | 0.074 | 20.021 | 19.993 | 0.041 | 20.021 | 20.000 | 0.034 |
|   | MIN | 20.000 | 19.760 | 0.110 | 20.000 | 19.883 | 0.065 | 20.000 | 19.959 | 0.020 | 20.000 | 19.980 | 0.007 | 20.000 | 19.987 | 0.000 |
| 25 | MAX | 25.130 | 24.890 | 0.370 | 25.052 | 24.935 | 0.169 | 25.033 | 24.980 | 0.074 | 25.021 | 24.993 | 0.041 | 25.021 | 25.000 | 0.034 |
|   | MIN | 25.000 | 24.760 | 0.110 | 25.000 | 24.883 | 0.065 | 25.000 | 24.959 | 0.020 | 25.000 | 24.980 | 0.007 | 25.000 | 24.987 | 0.000 |
| 30 | MAX | 30.130 | 29.890 | 0.370 | 30.052 | 29.935 | 0.169 | 30.033 | 29.980 | 0.074 | 30.021 | 29.993 | 0.041 | 30.021 | 30.000 | 0.034 |
|   | MIN | 30.000 | 29.760 | 0.110 | 30.000 | 29.883 | 0.065 | 30.000 | 29.959 | 0.020 | 30.000 | 29.980 | 0.007 | 30.000 | 29.987 | 0.000 |

Reproduced from ANSI B4.1–1967(R1974) with the permission of the publisher, the American Society of Mechanical Engineers.

Preferred Hole Basis Clearance Fits (Continued)

Dimensions in mm.

| BASIC SIZE | | LOOSE RUNNING | | | FREE RUNNING | | | CLOSE RUNNING | | | SLIDING | | | LOCATIONAL CLEARANCE | | |
|---|---|---|---|---|---|---|---|---|---|---|---|---|---|---|---|---|
| | | Hole H11 | Shaft c11 | Fit | Hole H9 | Shaft d9 | Fit | Hole H8 | Shaft f7 | Fit | Hole H7 | Shaft g6 | Fit | Hole H7 | Shaft h6 | Fit |
| 40 | MAX | 40.160 | 39.880 | 0.440 | 40.062 | 39.920 | 0.204 | 40.039 | 39.975 | 0.089 | 40.025 | 39.991 | 0.050 | 40.025 | 40.000 | 0.041 |
| | MIN | 40.000 | 39.720 | 0.120 | 40.000 | 39.858 | 0.080 | 40.000 | 39.950 | 0.025 | 40.000 | 39.975 | 0.009 | 40.000 | 39.984 | 0.000 |
| 50 | MAX | 50.160 | 49.870 | 0.450 | 50.062 | 49.920 | 0.204 | 50.039 | 49.975 | 0.089 | 50.025 | 49.991 | 0.050 | 50.025 | 50.000 | 0.041 |
| | MIN | 50.000 | 49.710 | 0.130 | 50.000 | 49.858 | 0.080 | 50.000 | 49.950 | 0.025 | 50.000 | 49.975 | 0.009 | 50.000 | 49.984 | 0.000 |
| 60 | MAX | 60.190 | 59.860 | 0.520 | 60.074 | 59.900 | 0.248 | 60.046 | 59.970 | 0.106 | 60.030 | 59.990 | 0.059 | 60.030 | 60.000 | 0.049 |
| | MIN | 60.000 | 59.670 | 0.140 | 60.000 | 59.826 | 0.100 | 60.000 | 59.940 | 0.030 | 60.000 | 59.971 | 0.010 | 60.000 | 59.981 | 0.000 |
| 80 | MAX | 80.190 | 79.850 | 0.530 | 80.074 | 79.900 | 0.248 | 80.046 | 79.970 | 0.106 | 80.030 | 79.990 | 0.059 | 80.030 | 80.000 | 0.049 |
| | MIN | 80.000 | 79.660 | 0.150 | 80.000 | 79.826 | 0.100 | 80.000 | 79.940 | 0.030 | 80.000 | 79.971 | 0.010 | 80.000 | 79.981 | 0.000 |
| 100 | MAX | 100.220 | 99.830 | 0.610 | 100.087 | 99.880 | 0.294 | 100.054 | 99.964 | 0.125 | 100.035 | 99.988 | 0.069 | 100.035 | 100.000 | 0.057 |
| | MIN | 100.000 | 99.610 | 0.170 | 100.000 | 99.793 | 0.120 | 100.000 | 99.929 | 0.036 | 100.000 | 99.966 | 0.012 | 100.000 | 99.978 | 0.000 |
| 120 | MAX | 120.220 | 119.820 | 0.620 | 120.087 | 119.880 | 0.294 | 120.054 | 119.964 | 0.125 | 120.035 | 119.988 | 0.069 | 120.035 | 120.000 | 0.057 |
| | MIN | 120.000 | 119.600 | 0.180 | 120.000 | 119.793 | 0.120 | 120.000 | 119.929 | 0.036 | 120.000 | 119.966 | 0.012 | 120.000 | 119.978 | 0.000 |
| 160 | MAX | 160.250 | 159.790 | 0.710 | 160.100 | 159.855 | 0.345 | 160.063 | 159.957 | 0.146 | 160.040 | 159.986 | 0.079 | 160.040 | 160.000 | 0.065 |
| | MIN | 160.000 | 159.540 | 0.210 | 160.000 | 159.755 | 0.145 | 160.000 | 159.917 | 0.043 | 160.000 | 159.961 | 0.014 | 160.000 | 159.975 | 0.000 |
| 200 | MAX | 200.290 | 199.760 | 0.820 | 200.115 | 199.830 | 0.400 | 200.072 | 199.950 | 0.168 | 200.046 | 199.985 | 0.090 | 200.046 | 200.000 | 0.075 |
| | MIN | 200.000 | 199.470 | 0.240 | 200.000 | 199.715 | 0.170 | 200.000 | 199.904 | 0.050 | 200.000 | 199.956 | 0.015 | 200.000 | 199.971 | 0.000 |
| 250 | MAX | 250.290 | 249.720 | 0.860 | 250.115 | 249.830 | 0.400 | 250.072 | 249.950 | 0.168 | 250.046 | 249.985 | 0.090 | 250.046 | 250.000 | 0.075 |
| | MIN | 250.000 | 249.430 | 0.280 | 250.000 | 249.715 | 0.170 | 250.000 | 249.904 | 0.050 | 250.000 | 249.956 | 0.015 | 250.000 | 249.971 | 0.000 |
| 300 | MAX | 300.320 | 299.670 | 0.970 | 300.130 | 299.810 | 0.450 | 300.081 | 299.944 | 0.189 | 300.052 | 299.983 | 0.101 | 300.052 | 300.000 | 0.084 |
| | MIN | 300.000 | 299.350 | 0.330 | 300.000 | 299.680 | 0.190 | 300.000 | 299.892 | 0.056 | 300.000 | 299.951 | 0.017 | 300.000 | 299.968 | 0.000 |
| 400 | MAX | 400.360 | 399.600 | 1.120 | 400.140 | 399.790 | 0.490 | 400.089 | 399.938 | 0.208 | 400.057 | 399.982 | 0.111 | 400.057 | 400.000 | 0.093 |
| | MIN | 400.000 | 399.240 | 0.400 | 400.000 | 399.650 | 0.210 | 400.000 | 399.881 | 0.062 | 400.000 | 399.946 | 0.018 | 400.000 | 399.964 | 0.000 |
| 500 | MAX | 500.400 | 499.520 | 1.280 | 500.155 | 499.770 | 0.540 | 500.097 | 499.932 | 0.228 | 500.063 | 499.980 | 0.123 | 500.063 | 500.000 | 0.103 |
| | MIN | 500.000 | 499.120 | 0.480 | 500.000 | 499.615 | 0.230 | 500.000 | 499.869 | 0.068 | 500.000 | 499.940 | 0.020 | 500.000 | 499.960 | 0.000 |

Reproduced from ANSI B4.1–1967(R1974) with the permission of the publisher, the American Society of Mechanical Engineers.

Preferred Hole Basis Transition and Interference Fits

Dimensions in mm.

| BASIC SIZE | | LOCATIONAL TRANSN. Hole H7 | Shaft k6 | Fit | LOCATIONAL TRANSN. Hole H7 | Shaft n6 | Fit | LOCATIONAL INTERF. Hole H7 | Shaft p6 | Fit | MEDIUM DRIVE Hole H7 | Shaft s6 | Fit | FORCE Hole H7 | Shaft u6 | Fit |
|---|---|---|---|---|---|---|---|---|---|---|---|---|---|---|---|---|
| 1 | MAX | 1.010 | 1.006 | 0.010 | 1.010 | 1.010 | 0.006 | 1.010 | 1.012 | 0.004 | 1.010 | 1.020 | -0.004 | 1.010 | 1.024 | -0.008 |
|  | MIN | 1.000 | 1.000 | -0.006 | 1.000 | 1.004 | -0.010 | 1.000 | 1.006 | -0.012 | 1.000 | 1.014 | -0.020 | 1.000 | 1.018 | -0.024 |
| 1.2 | MAX | 1.210 | 1.206 | 0.010 | 1.210 | 1.210 | 0.006 | 1.210 | 1.212 | 0.004 | 1.210 | 1.220 | -0.004 | 1.210 | 1.224 | -0.008 |
|  | MIN | 1.200 | 1.200 | -0.006 | 1.200 | 1.204 | -0.010 | 1.200 | 1.206 | -0.012 | 1.200 | 1.214 | -0.020 | 1.200 | 1.218 | -0.024 |
| 1.6 | MAX | 1.610 | 1.606 | 0.010 | 1.610 | 1.610 | 0.006 | 1.610 | 1.612 | 0.004 | 1.610 | 1.620 | -0.004 | 1.610 | 1.624 | -0.008 |
|  | MIN | 1.600 | 1.600 | -0.006 | 1.600 | 1.604 | -0.010 | 1.600 | 1.606 | -0.012 | 1.600 | 1.614 | -0.020 | 1.600 | 1.618 | -0.024 |
| 2 | MAX | 2.010 | 2.006 | 0.010 | 2.010 | 2.010 | 0.006 | 2.010 | 2.012 | 0.004 | 2.010 | 2.020 | -0.004 | 2.010 | 2.024 | -0.008 |
|  | MIN | 2.000 | 2.000 | -0.006 | 2.000 | 2.004 | -0.010 | 2.000 | 2.006 | -0.012 | 2.000 | 2.014 | -0.020 | 2.000 | 2.018 | -0.024 |
| 2.5 | MAX | 2.510 | 2.506 | 0.010 | 2.510 | 2.510 | 0.006 | 2.510 | 2.512 | 0.004 | 2.510 | 2.520 | -0.004 | 2.510 | 2.524 | -0.008 |
|  | MIN | 2.500 | 2.500 | -0.006 | 2.500 | 2.504 | -0.010 | 2.500 | 2.506 | -0.012 | 2.500 | 2.514 | -0.020 | 2.500 | 2.518 | -0.024 |
| 3 | MAX | 3.010 | 3.006 | 0.010 | 3.010 | 3.010 | 0.006 | 3.010 | 3.012 | 0.004 | 3.010 | 3.020 | -0.004 | 3.010 | 3.024 | -0.008 |
|  | MIN | 3.000 | 3.000 | -0.006 | 3.000 | 3.004 | -0.010 | 3.000 | 3.006 | -0.012 | 3.000 | 3.014 | -0.020 | 3.000 | 3.018 | -0.024 |
| 4 | MAX | 4.012 | 4.009 | 0.011 | 4.012 | 4.016 | 0.004 | 4.012 | 4.020 | 0.000 | 4.012 | 4.027 | -0.007 | 4.012 | 4.031 | -0.011 |
|  | MIN | 4.000 | 4.001 | -0.009 | 4.000 | 4.008 | -0.016 | 4.000 | 4.012 | -0.020 | 4.000 | 4.019 | -0.027 | 4.000 | 4.023 | -0.031 |
| 5 | MAX | 5.012 | 5.009 | 0.011 | 5.012 | 5.016 | 0.004 | 5.012 | 5.020 | 0.000 | 5.012 | 5.027 | -0.007 | 5.012 | 5.031 | -0.011 |
|  | MIN | 5.000 | 5.001 | -0.009 | 5.000 | 5.008 | -0.016 | 5.000 | 5.012 | -0.020 | 5.000 | 5.019 | -0.027 | 5.000 | 5.023 | -0.031 |
| 6 | MAX | 6.012 | 6.009 | 0.011 | 6.012 | 6.016 | 0.004 | 6.012 | 6.020 | 0.000 | 6.012 | 6.027 | -0.007 | 6.012 | 6.031 | -0.011 |
|  | MIN | 6.000 | 6.001 | -0.009 | 6.000 | 6.008 | -0.016 | 6.000 | 6.012 | -0.020 | 6.000 | 6.019 | -0.027 | 6.000 | 6.023 | -0.031 |
| 8 | MAX | 8.015 | 8.010 | 0.014 | 8.015 | 8.019 | 0.005 | 8.015 | 8.024 | 0.000 | 8.015 | 8.032 | -0.008 | 8.015 | 8.037 | -0.013 |
|  | MIN | 8.000 | 8.001 | -0.010 | 8.000 | 8.010 | -0.019 | 8.000 | 8.015 | -0.024 | 8.000 | 8.023 | -0.032 | 8.000 | 8.028 | -0.037 |
| 10 | MAX | 10.015 | 10.010 | 0.014 | 10.015 | 10.019 | 0.005 | 10.015 | 10.024 | 0.000 | 10.015 | 10.032 | -0.008 | 10.015 | 10.037 | -0.013 |
|  | MIN | 10.000 | 10.001 | -0.010 | 10.000 | 10.010 | -0.019 | 10.000 | 10.015 | -0.024 | 10.000 | 10.023 | -0.032 | 10.000 | 10.028 | -0.037 |
| 12 | MAX | 12.018 | 12.012 | 0.017 | 12.018 | 12.023 | 0.006 | 12.018 | 12.029 | 0.000 | 12.018 | 12.039 | -0.010 | 12.018 | 12.044 | -0.015 |
|  | MIN | 12.000 | 12.001 | -0.012 | 12.000 | 12.012 | -0.023 | 12.000 | 12.018 | -0.029 | 12.000 | 12.028 | -0.039 | 12.000 | 12.033 | -0.044 |
| 16 | MAX | 16.018 | 16.012 | 0.017 | 16.018 | 16.023 | 0.006 | 16.018 | 16.029 | 0.000 | 16.018 | 16.039 | -0.010 | 16.018 | 16.044 | -0.015 |
|  | MIN | 16.000 | 16.001 | -0.012 | 16.000 | 16.012 | -0.023 | 16.000 | 16.018 | -0.029 | 16.000 | 16.028 | -0.039 | 16.000 | 16.033 | -0.044 |
| 20 | MAX | 20.021 | 20.015 | 0.019 | 20.021 | 20.028 | 0.006 | 20.021 | 20.035 | -0.001 | 20.021 | 20.048 | -0.014 | 20.021 | 20.054 | -0.020 |
|  | MIN | 20.000 | 20.002 | -0.015 | 20.000 | 20.015 | -0.028 | 20.000 | 20.022 | -0.035 | 20.000 | 20.035 | -0.048 | 20.000 | 20.041 | -0.054 |
| 25 | MAX | 25.021 | 25.015 | 0.019 | 25.021 | 25.028 | 0.006 | 25.021 | 25.035 | -0.001 | 25.021 | 25.048 | -0.014 | 25.021 | 25.061 | -0.027 |
|  | MIN | 25.000 | 25.002 | -0.015 | 25.000 | 25.015 | -0.028 | 25.000 | 25.022 | -0.035 | 25.000 | 25.035 | -0.048 | 25.000 | 25.048 | -0.061 |
| 30 | MAX | 30.021 | 30.015 | 0.019 | 30.021 | 30.028 | 0.006 | 30.021 | 30.035 | -0.001 | 30.021 | 30.048 | -0.014 | 30.021 | 30.061 | -0.027 |
|  | MIN | 30.000 | 30.002 | -0.015 | 30.000 | 30.015 | -0.028 | 30.000 | 30.022 | -0.035 | 30.000 | 30.035 | -0.048 | 30.000 | 30.048 | -0.061 |

Reproduced from ANSI B4.1-1967(R1974) with the permission of the publisher, the American Society of Mechanical Engineers.

Preferred Hole Basis Transition and Interference Fits (Continued)

Dimensions in mm.

| BASIC SIZE | | LOCATIONAL TRANSN. Hole H7 | Shaft k6 | Fit | LOCATIONAL TRANSN. Hole H7 | Shaft n6 | Fit | LOCATIONAL INTERF Hole H7 | Shaft p6 | Fit | MEDIUM DRIVE Hole H7 | Shaft s6 | Fit | FORCE Hole H7 | Shaft u6 | Fit |
|---|---|---|---|---|---|---|---|---|---|---|---|---|---|---|---|---|
| 40 | MAX | 40.025 | 40.018 | 0.023 | 40.025 | 40.033 | 0.008 | 40.025 | 40.042 | 0.001 | 40.025 | 40.059 | -0.018 | 40.025 | 40.076 | -0.035 |
|    | MIN | 40.000 | 40.002 | -0.018 | 40.000 | 40.017 | -0.033 | 40.000 | 40.026 | -0.042 | 40.000 | 40.043 | -0.059 | 40.000 | 40.060 | -0.076 |
| 50 | MAX | 50.025 | 50.018 | 0.023 | 50.025 | 50.033 | 0.008 | 50.025 | 50.042 | -0.001 | 50.025 | 50.059 | -0.018 | 50.025 | 50.086 | -0.045 |
|    | MIN | 50.000 | 50.002 | -0.018 | 50.000 | 50.017 | -0.033 | 50.000 | 50.026 | -0.042 | 50.000 | 50.043 | -0.059 | 50.000 | 50.070 | -0.086 |
| 60 | MAX | 60.030 | 60.021 | 0.028 | 60.030 | 60.039 | 0.010 | 60.030 | 60.051 | -0.002 | 60.030 | 60.072 | -0.023 | 60.030 | 60.106 | -0.057 |
|    | MIN | 60.000 | 60.002 | -0.021 | 60.000 | 60.020 | -0.039 | 60.000 | 60.032 | -0.051 | 60.000 | 60.053 | -0.072 | 60.000 | 60.087 | -0.106 |
| 80 | MAX | 80.030 | 80.021 | 0.028 | 80.030 | 80.039 | 0.010 | 80.030 | 80.051 | -0.002 | 80.030 | 80.078 | -0.029 | 80.030 | 80.121 | -0.072 |
|    | MIN | 80.000 | 80.002 | -0.021 | 80.000 | 80.020 | -0.039 | 80.000 | 80.032 | -0.051 | 80.000 | 80.059 | -0.078 | 80.000 | 80.102 | -0.121 |
| 100 | MAX | 100.035 | 100.025 | 0.032 | 100.035 | 100.045 | 0.012 | 100.035 | 100.059 | -0.002 | 100.035 | 100.093 | -0.036 | 100.035 | 100.146 | -0.089 |
|     | MIN | 100.000 | 100.003 | -0.025 | 100.000 | 100.023 | -0.045 | 100.000 | 100.037 | -0.059 | 100.000 | 100.071 | -0.093 | 100.000 | 100.124 | -0.146 |
| 120 | MAX | 120.035 | 120.025 | 0.032 | 120.035 | 120.045 | 0.012 | 120.035 | 120.059 | -0.002 | 120.035 | 120.101 | -0.044 | 120.035 | 120.166 | -0.109 |
|     | MIN | 120.000 | 120.003 | -0.025 | 120.000 | 120.023 | -0.045 | 120.000 | 120.037 | -0.059 | 120.000 | 120.079 | -0.101 | 120.000 | 120.144 | -0.166 |
| 160 | MAX | 160.040 | 160.028 | 0.037 | 160.040 | 160.052 | 0.013 | 160.040 | 160.068 | -0.003 | 160.040 | 160.125 | -0.060 | 160.040 | 160.215 | -0.150 |
|     | MIN | 160.000 | 160.003 | -0.028 | 160.000 | 160.027 | -0.052 | 160.000 | 160.043 | -0.068 | 160.000 | 160.100 | -0.125 | 160.000 | 160.190 | -0.215 |
| 200 | MAX | 200.046 | 200.033 | 0.042 | 200.046 | 200.060 | 0.015 | 200.046 | 200.079 | -0.004 | 200.046 | 200.151 | -0.076 | 200.046 | 200.265 | -0.190 |
|     | MIN | 200.000 | 200.004 | -0.033 | 200.000 | 200.031 | -0.060 | 200.000 | 200.050 | -0.079 | 200.000 | 200.122 | -0.151 | 200.000 | 200.236 | -0.265 |
| 250 | MAX | 250.046 | 250.033 | 0.042 | 250.046 | 250.060 | 0.015 | 250.046 | 250.079 | -0.004 | 250.046 | 250.169 | -0.094 | 250.046 | 250.313 | -0.238 |
|     | MIN | 250.000 | 250.004 | -0.033 | 250.000 | 250.031 | -0.060 | 250.000 | 250.050 | -0.079 | 250.000 | 250.140 | -0.169 | 250.000 | 250.284 | -0.313 |
| 300 | MAX | 300.052 | 300.036 | 0.048 | 300.052 | 300.066 | 0.018 | 300.052 | 300.088 | -0.004 | 300.052 | 300.202 | -0.118 | 300.052 | 300.382 | -0.298 |
|     | MIN | 300.000 | 300.004 | -0.036 | 300.000 | 300.034 | -0.066 | 300.000 | 300.056 | -0.088 | 300.000 | 300.170 | -0.202 | 300.000 | 300.350 | -0.382 |
| 400 | MAX | 400.057 | 400.040 | 0.053 | 400.057 | 400.073 | 0.020 | 400.057 | 400.098 | -0.005 | 400.057 | 400.244 | -0.151 | 400.057 | 400.471 | -0.378 |
|     | MIN | 400.000 | 400.004 | -0.040 | 400.000 | 400.037 | -0.073 | 400.000 | 400.062 | -0.098 | 400.000 | 400.208 | -0.244 | 400.000 | 400.435 | -0.471 |
| 500 | MAX | 500.063 | 500.045 | 0.058 | 500.063 | 500.080 | 0.023 | 500.063 | 500.108 | -0.005 | 500.063 | 500.292 | -0.189 | 500.063 | 500.580 | -0.477 |
|     | MIN | 500.000 | 500.005 | -0.045 | 500.000 | 500.040 | -0.080 | 500.000 | 500.068 | -0.108 | 500.000 | 500.252 | -0.292 | 500.000 | 500.540 | -0.580 |

Reproduced from ANSI B4.1–1967(R1974) with the permission of the publisher, the American Society of Mechanical Engineers.

Preferred Shaft Basis Clearance Fits

Dimensions in mm.

| BASIC SIZE | | LOOSE RUNNING | | | FREE RUNNING | | | CLOSE RUNNING | | | SLIDING | | | LOCATIONAL CLEARANCE | | |
|---|---|---|---|---|---|---|---|---|---|---|---|---|---|---|---|---|
| | | Hole C11 | Shaft h11 | Fit | Hole D9 | Shaft h9 | Fit | Hole F8 | Shaft h7 | Fit | Hole G7 | Shaft h6 | Fit | Hole H7 | Shaft h6 | Fit |
| 1 | MAX | 1.120 | 1.000 | 0.180 | 1.045 | 1.000 | 0.070 | 1.020 | 1.000 | 0.030 | 1.012 | 1.000 | 0.018 | 1.010 | 1.000 | 0.016 |
| | MIN | 1.060 | 0.940 | 0.060 | 1.020 | 0.975 | 0.020 | 1.006 | 0.990 | 0.006 | 1.002 | 0.994 | 0.002 | 1.000 | 0.994 | 0.000 |
| 1.2 | MAX | 1.320 | 1.200 | 0.180 | 1.245 | 1.200 | 0.070 | 1.220 | 1.200 | 0.030 | 1.212 | 1.200 | 0.018 | 1.210 | 1.200 | 0.016 |
| | MIN | 1.260 | 1.140 | 0.060 | 1.220 | 1.175 | 0.020 | 1.206 | 1.190 | 0.006 | 1.202 | 1.194 | 0.002 | 1.200 | 1.194 | 0.000 |
| 1.6 | MAX | 1.720 | 1.600 | 0.180 | 1.645 | 1.600 | 0.070 | 1.620 | 1.600 | 0.030 | 1.612 | 1.600 | 0.018 | 1.610 | 1.600 | 0.016 |
| | MIN | 1.660 | 1.540 | 0.060 | 1.620 | 1.575 | 0.020 | 1.606 | 1.590 | 0.006 | 1.602 | 1.594 | 0.002 | 1.600 | 1.594 | 0.000 |
| 2 | MAX | 2.120 | 2.000 | 0.180 | 2.045 | 2.000 | 0.070 | 2.020 | 2.000 | 0.030 | 2.012 | 2.000 | 0.018 | 2.010 | 2.000 | 0.016 |
| | MIN | 2.060 | 1.940 | 0.060 | 2.020 | 1.975 | 0.020 | 2.006 | 1.990 | 0.006 | 2.002 | 1.994 | 0.002 | 2.000 | 1.994 | 0.000 |
| 2.5 | MAX | 2.620 | 2.500 | 0.180 | 2.545 | 2.500 | 0.070 | 2.520 | 2.500 | 0.030 | 2.512 | 2.500 | 0.018 | 2.510 | 2.500 | 0.016 |
| | MIN | 2.560 | 2.440 | 0.060 | 2.520 | 2.475 | 0.020 | 2.506 | 2.490 | 0.006 | 2.502 | 2.494 | 0.002 | 2.500 | 2.494 | 0.000 |
| 3 | MAX | 3.120 | 3.000 | 0.180 | 3.045 | 3.000 | 0.070 | 3.020 | 3.000 | 0.030 | 3.012 | 3.000 | 0.018 | 3.010 | 3.000 | 0.016 |
| | MIN | 3.060 | 2.940 | 0.060 | 3.020 | 2.975 | 0.020 | 3.006 | 2.990 | 0.006 | 3.002 | 2.994 | 0.002 | 3.000 | 2.994 | 0.000 |
| 4 | MAX | 4.145 | 4.000 | 0.220 | 4.060 | 4.000 | 0.090 | 4.028 | 4.000 | 0.040 | 4.016 | 4.000 | 0.024 | 4.012 | 4.000 | 0.020 |
| | MIN | 4.070 | 3.925 | 0.070 | 4.030 | 3.970 | 0.030 | 4.010 | 3.988 | 0.010 | 4.004 | 3.992 | 0.004 | 4.000 | 3.992 | 0.000 |
| 5 | MAX | 5.145 | 5.000 | 0.220 | 5.060 | 5.000 | 0.090 | 5.028 | 5.000 | 0.040 | 5.016 | 5.000 | 0.024 | 5.012 | 5.000 | 0.020 |
| | MIN | 5.070 | 4.925 | 0.070 | 5.030 | 4.970 | 0.030 | 5.010 | 4.988 | 0.010 | 5.004 | 4.992 | 0.004 | 5.000 | 4.992 | 0.000 |
| 6 | MAX | 6.145 | 6.000 | 0.220 | 6.060 | 6.000 | 0.090 | 6.028 | 6.000 | 0.040 | 6.016 | 6.000 | 0.024 | 6.012 | 6.000 | 0.020 |
| | MIN | 6.070 | 5.925 | 0.070 | 6.030 | 5.970 | 0.030 | 6.010 | 5.988 | 0.010 | 6.004 | 5.992 | 0.004 | 6.000 | 5.992 | 0.000 |
| 8 | MAX | 8.170 | 8.000 | 0.260 | 8.076 | 8.000 | 0.112 | 8.035 | 8.000 | 0.050 | 8.020 | 8.000 | 0.029 | 8.015 | 8.000 | 0.024 |
| | MIN | 8.080 | 7.910 | 0.080 | 8.040 | 7.964 | 0.040 | 8.013 | 7.985 | 0.013 | 8.005 | 7.991 | 0.005 | 8.000 | 7.991 | 0.000 |
| 10 | MAX | 10.170 | 10.000 | 0.260 | 10.076 | 10.000 | 0.112 | 10.035 | 10.000 | 0.050 | 10.020 | 10.000 | 0.029 | 10.015 | 10.000 | 0.024 |
| | MIN | 10.080 | 9.910 | 0.080 | 10.040 | 9.964 | 0.040 | 10.013 | 9.985 | 0.013 | 10.005 | 9.991 | 0.005 | 10.000 | 9.991 | 0.000 |
| 12 | MAX | 12.205 | 12.000 | 0.315 | 12.093 | 12.000 | 0.136 | 12.043 | 12.000 | 0.061 | 12.024 | 12.000 | 0.035 | 12.018 | 12.000 | 0.029 |
| | MIN | 12.095 | 11.890 | 0.095 | 12.050 | 11.957 | 0.050 | 12.016 | 11.982 | 0.016 | 12.006 | 11.989 | 0.006 | 12.000 | 11.989 | 0.000 |
| 16 | MAX | 16.205 | 16.000 | 0.315 | 16.093 | 16.000 | 0.136 | 16.043 | 16.000 | 0.061 | 16.024 | 16.000 | 0.035 | 16.018 | 16.000 | 0.029 |
| | MIN | 16.095 | 15.890 | 0.095 | 16.050 | 15.957 | 0.050 | 16.016 | 15.982 | 0.016 | 16.006 | 15.989 | 0.006 | 16.000 | 15.989 | 0.000 |
| 20 | MAX | 20.240 | 20.000 | 0.370 | 20.117 | 20.000 | 0.169 | 20.053 | 20.000 | 0.074 | 20.028 | 20.000 | 0.041 | 20.021 | 20.000 | 0.034 |
| | MIN | 20.110 | 19.870 | 0.110 | 20.065 | 19.948 | 0.065 | 20.020 | 19.979 | 0.020 | 20.007 | 19.987 | 0.007 | 20.000 | 19.987 | 0.000 |
| 25 | MAX | 25.240 | 25.000 | 0.370 | 25.117 | 25.000 | 0.169 | 25.053 | 25.000 | 0.074 | 25.028 | 25.000 | 0.041 | 25.021 | 25.000 | 0.034 |
| | MIN | 25.110 | 24.870 | 0.110 | 25.065 | 24.948 | 0.065 | 25.020 | 24.979 | 0.020 | 25.007 | 24.987 | 0.007 | 25.000 | 24.987 | 0.000 |
| 30 | MAX | 30.240 | 30.000 | 0.370 | 30.117 | 30.000 | 0.169 | 30.053 | 30.000 | 0.074 | 30.028 | 30.000 | 0.041 | 30.021 | 30.000 | 0.034 |
| | MIN | 30.110 | 29.870 | 0.110 | 30.065 | 29.948 | 0.065 | 30.020 | 29.979 | 0.020 | 30.007 | 29.987 | 0.007 | 30.000 | 29.987 | 0.000 |

Reproduced from ANSI B4.1-1967(R1974) with the permission of the publisher, the American Society of Mechanical Engineers.

Preferred Shaft Basis Clearance Fits (Continued)

Dimensions in mm.

| BASIC SIZE | | LOOSE RUNNING | | | FREE RUNNING | | | CLOSE RUNNING | | | SLIDING | | | LOCATIONAL CLEARANCE | | |
|---|---|---|---|---|---|---|---|---|---|---|---|---|---|---|---|---|
| | | Hole C11 | Shaft h11 | Fit | Hole D9 | Shaft h9 | Fit | Hole F8 | Shaft h7 | Fit | Hole G7 | Shaft h6 | Fit | Hole H7 | Shaft h6 | Fit |
| 40 | MAX | 40.280 | 40.000 | 0.440 | 40.142 | 40.000 | 0.204 | 40.064 | 40.000 | 0.089 | 40.034 | 40.000 | 0.050 | 40.025 | 40.000 | 0.041 |
|    | MIN | 40.120 | 39.840 | 0.120 | 40.080 | 39.938 | 0.080 | 40.025 | 39.975 | 0.025 | 40.009 | 39.984 | 0.009 | 40.000 | 39.984 | 0.000 |
| 50 | MAX | 50.290 | 50.000 | 0.450 | 50.142 | 50.000 | 0.204 | 50.064 | 50.000 | 0.089 | 50.034 | 50.000 | 0.050 | 50.025 | 50.000 | 0.041 |
|    | MIN | 50.130 | 49.840 | 0.130 | 50.080 | 49.938 | 0.080 | 50.025 | 49.975 | 0.025 | 50.009 | 49.984 | 0.009 | 50.000 | 49.984 | 0.000 |
| 60 | MAX | 60.330 | 60.000 | 0.520 | 60.174 | 60.000 | 0.244 | 60.076 | 60.000 | 0.106 | 60.040 | 60.000 | 0.059 | 60.030 | 60.000 | 0.049 |
|    | MIN | 60.140 | 59.810 | 0.140 | 60.100 | 59.926 | 0.100 | 60.030 | 59.970 | 0.030 | 60.010 | 59.981 | 0.010 | 60.000 | 59.981 | 0.000 |
| 80 | MAX | 80.340 | 80.000 | 0.530 | 80.174 | 80.000 | 0.248 | 80.076 | 80.000 | 0.106 | 80.040 | 80.000 | 0.059 | 80.030 | 80.000 | 0.049 |
|    | MIN | 80.150 | 79.810 | 0.150 | 80.100 | 79.926 | 0.100 | 80.030 | 79.970 | 0.030 | 80.010 | 79.981 | 0.010 | 80.000 | 79.981 | 0.000 |
| 100 | MAX | 100.390 | 100.000 | 0.610 | 100.207 | 100.000 | 0.294 | 100.090 | 100.000 | 0.125 | 100.047 | 100.000 | 0.069 | 100.035 | 100.000 | 0.057 |
|     | MIN | 100.170 | 99.780 | 0.170 | 100.120 | 99.913 | 0.120 | 100.036 | 99.965 | 0.036 | 100.012 | 99.978 | 0.012 | 100.000 | 99.978 | 0.000 |
| 120 | MAX | 120.400 | 120.000 | 0.620 | 120.207 | 120.000 | 0.294 | 120.090 | 120.000 | 0.125 | 120.047 | 120.000 | 0.069 | 120.035 | 120.000 | 0.057 |
|     | MIN | 120.180 | 119.780 | 0.180 | 120.120 | 119.913 | 0.120 | 120.036 | 119.965 | 0.036 | 120.012 | 119.978 | 0.012 | 120.000 | 119.978 | 0.000 |
| 160 | MAX | 160.460 | 160.000 | 0.710 | 160.245 | 160.000 | 0.345 | 160.106 | 160.000 | 0.146 | 160.054 | 160.000 | 0.079 | 160.040 | 160.000 | 0.065 |
|     | MIN | 160.210 | 159.750 | 0.210 | 160.145 | 159.900 | 0.145 | 160.043 | 159.960 | 0.043 | 160.014 | 159.975 | 0.014 | 160.000 | 159.975 | 0.000 |
| 200 | MAX | 200.530 | 200.000 | 0.826 | 200.285 | 200.000 | 0.400 | 200.122 | 200.000 | 0.168 | 200.061 | 200.000 | 0.090 | 200.046 | 200.000 | 0.075 |
|     | MIN | 200.240 | 199.710 | 0.240 | 200.170 | 199.885 | 0.170 | 200.050 | 199.954 | 0.050 | 200.015 | 199.971 | 0.015 | 200.000 | 199.971 | 0.000 |
| 250 | MAX | 250.570 | 250.000 | 0.860 | 250.285 | 250.000 | 0.400 | 250.122 | 250.000 | 0.168 | 250.061 | 250.000 | 0.090 | 250.046 | 250.000 | 0.075 |
|     | MIN | 250.280 | 249.710 | 0.280 | 250.170 | 249.885 | 0.170 | 250.050 | 249.954 | 0.050 | 250.015 | 249.971 | 0.015 | 250.000 | 249.971 | 0.000 |
| 300 | MAX | 300.650 | 300.000 | 0.970 | 300.320 | 300.000 | 0.450 | 300.137 | 300.000 | 0.189 | 300.069 | 300.000 | 0.101 | 300.052 | 300.000 | 0.084 |
|     | MIN | 300.330 | 299.680 | 0.330 | 300.190 | 299.870 | 0.190 | 300.056 | 299.948 | 0.056 | 300.017 | 299.968 | 0.017 | 300.000 | 299.968 | 0.000 |
| 400 | MAX | 400.760 | 400.000 | 1.120 | 400.350 | 400.000 | 0.490 | 400.151 | 400.000 | 0.208 | 400.075 | 400.000 | 0.111 | 400.057 | 400.000 | 0.093 |
|     | MIN | 400.400 | 399.640 | 0.400 | 400.210 | 399.860 | 0.210 | 400.062 | 399.943 | 0.062 | 400.018 | 399.964 | 0.018 | 400.000 | 399.964 | 0.000 |
| 500 | MAX | 500.880 | 500.000 | 1.280 | 500.385 | 500.000 | 0.540 | 500.165 | 500.000 | 0.228 | 500.083 | 500.000 | 0.123 | 500.063 | 500.000 | 0.103 |
|     | MIN | 500.480 | 499.600 | 0.480 | 500.230 | 499.845 | 0.230 | 500.068 | 499.937 | 0.068 | 500.020 | 499.960 | 0.020 | 500.000 | 499.960 | 0.000 |

Reproduced from ANSI B4.1-1967(R1974) with the permission of the publisher, the American Society of Mechanical Engineers.

Preferred Shaft Basis Transition and Interference Fits

| BASIC SIZE | | LOCATIONAL TRANSN. Hole K7 | Shaft h6 | Fit | LOCATIONAL TRANSN. Hole N7 | Shaft h6 | Fit | LOCATIONAL INTERF. Hole P7 | Shaft h6 | Fit | MEDIUM DRIVE Hole S7 | Shaft h6 | Fit | FORCE Hole U7 | Shaft h6 | Fit |
|---|---|---|---|---|---|---|---|---|---|---|---|---|---|---|---|---|
| 1 | MAX | 1.000 | 1.000 | 0.006 | 0.996 | 1.000 | 0.002 | 0.994 | 1.000 | 0.000 | 0.986 | 1.000 | -0.008 | 0.982 | 1.000 | -0.012 |
|  | MIN | 0.990 | 0.994 | -0.010 | 0.986 | 0.994 | -0.014 | 0.984 | 0.994 | -0.016 | 0.976 | 0.994 | -0.024 | 0.972 | 0.994 | -0.028 |
| 1.2 | MAX | 1.200 | 1.200 | 0.006 | 1.196 | 1.200 | 0.002 | 1.194 | 1.200 | 0.000 | 1.186 | 1.200 | -0.008 | 1.182 | 1.200 | -0.012 |
|  | MIN | 1.190 | 1.194 | -0.010 | 1.186 | 1.194 | -0.014 | 1.184 | 1.194 | -0.016 | 1.176 | 1.194 | -0.024 | 1.172 | 1.194 | -0.028 |
| 1.6 | MAX | 1.600 | 1.600 | 0.006 | 1.596 | 1.600 | 0.002 | 1.594 | 1.600 | 0.000 | 1.586 | 1.600 | -0.008 | 1.582 | 1.600 | -0.012 |
|  | MIN | 1.590 | 1.594 | -0.010 | 1.586 | 1.594 | -0.014 | 1.584 | 1.594 | -0.016 | 1.576 | 1.594 | -0.024 | 1.572 | 1.594 | -0.028 |
| 2 | MAX | 2.000 | 2.000 | 0.006 | 1.996 | 2.000 | 0.002 | 1.994 | 2.000 | 0.000 | 1.986 | 2.000 | -0.008 | 1.982 | 2.000 | -0.012 |
|  | MIN | 1.990 | 1.994 | -0.010 | 1.986 | 1.994 | -0.014 | 1.984 | 1.994 | -0.016 | 1.976 | 1.994 | -0.024 | 1.972 | 1.994 | -0.028 |
| 2.5 | MAX | 2.500 | 2.500 | 0.006 | 2.496 | 2.500 | 0.002 | 2.494 | 2.500 | 0.000 | 2.486 | 2.500 | -0.008 | 2.482 | 2.500 | -0.012 |
|  | MIN | 2.490 | 2.494 | -0.010 | 2.486 | 2.494 | -0.014 | 2.484 | 2.494 | -0.016 | 2.476 | 2.494 | -0.024 | 2.472 | 2.494 | -0.028 |
| 3 | MAX | 3.000 | 3.000 | 0.006 | 2.996 | 3.000 | 0.002 | 2.994 | 3.000 | 0.000 | 2.986 | 3.000 | -0.008 | 2.982 | 3.000 | -0.012 |
|  | MIN | 2.990 | 2.994 | -0.010 | 2.986 | 2.994 | -0.014 | 2.984 | 2.994 | -0.016 | 2.976 | 2.994 | -0.024 | 2.972 | 2.994 | -0.028 |
| 4 | MAX | 4.003 | 4.000 | 0.011 | 3.996 | 4.000 | 0.004 | 3.992 | 4.000 | 0.000 | 3.985 | 4.000 | -0.007 | 3.981 | 4.000 | -0.011 |
|  | MIN | 3.991 | 3.992 | -0.009 | 3.984 | 3.992 | -0.016 | 3.980 | 3.992 | -0.020 | 3.973 | 3.992 | -0.027 | 3.969 | 3.992 | -0.031 |
| 5 | MAX | 5.003 | 5.000 | 0.011 | 4.996 | 5.000 | 0.004 | 4.992 | 5.000 | 0.000 | 4.985 | 5.000 | -0.007 | 4.981 | 5.000 | -0.011 |
|  | MIN | 4.991 | 4.992 | -0.009 | 4.984 | 4.992 | -0.016 | 4.980 | 4.992 | -0.020 | 4.973 | 4.992 | -0.027 | 4.969 | 4.992 | -0.031 |
| 6 | MAX | 6.003 | 6.000 | 0.011 | 5.996 | 6.000 | 0.004 | 5.992 | 6.000 | 0.000 | 5.985 | 6.000 | -0.007 | 5.981 | 6.000 | -0.011 |
|  | MIN | 5.991 | 5.992 | -0.009 | 5.984 | 5.992 | -0.016 | 5.980 | 5.992 | -0.020 | 5.973 | 5.992 | -0.027 | 5.969 | 5.992 | -0.031 |
| 8 | MAX | 8.005 | 8.000 | 0.014 | 7.996 | 8.000 | 0.005 | 7.991 | 8.000 | 0.000 | 7.983 | 8.000 | -0.008 | 7.978 | 8.000 | -0.013 |
|  | MIN | 7.990 | 7.991 | -0.010 | 7.981 | 7.991 | -0.019 | 7.976 | 7.991 | -0.024 | 7.968 | 7.991 | -0.032 | 7.963 | 7.991 | -0.037 |
| 10 | MAX | 10.005 | 10.000 | 0.014 | 9.996 | 10.000 | 0.005 | 9.991 | 10.000 | 0.000 | 9.983 | 10.000 | -0.008 | 9.978 | 10.000 | -0.013 |
|  | MIN | 9.990 | 9.991 | -0.010 | 9.981 | 9.991 | -0.019 | 9.976 | 9.991 | -0.024 | 9.968 | 9.991 | -0.032 | 9.963 | 9.991 | -0.037 |
| 12 | MAX | 12.006 | 12.000 | 0.017 | 11.995 | 12.000 | 0.006 | 11.989 | 12.000 | 0.000 | 11.979 | 12.000 | -0.010 | 11.974 | 12.000 | -0.015 |
|  | MIN | 11.988 | 11.989 | -0.012 | 11.977 | 11.989 | -0.023 | 11.971 | 11.989 | -0.029 | 11.961 | 11.989 | -0.039 | 11.956 | 11.989 | -0.044 |
| 16 | MAX | 16.006 | 16.000 | 0.017 | 15.995 | 16.000 | 0.006 | 15.989 | 16.000 | 0.000 | 15.979 | 16.000 | -0.010 | 15.974 | 16.000 | -0.015 |
|  | MIN | 15.988 | 15.989 | -0.012 | 15.977 | 15.989 | -0.023 | 15.971 | 15.989 | -0.029 | 15.961 | 15.989 | -0.039 | 15.956 | 15.989 | -0.044 |
| 20 | MAX | 20.006 | 20.000 | 0.019 | 19.993 | 20.000 | 0.006 | 19.986 | 20.000 | -0.001 | 19.973 | 20.000 | -0.014 | 19.967 | 20.000 | -0.020 |
|  | MIN | 19.985 | 19.987 | -0.015 | 19.972 | 19.987 | -0.028 | 19.965 | 19.987 | -0.035 | 19.952 | 19.987 | -0.048 | 19.946 | 19.987 | -0.054 |
| 25 | MAX | 25.006 | 25.000 | 0.019 | 24.993 | 25.000 | 0.006 | 24.986 | 25.000 | -0.001 | 24.973 | 25.000 | -0.014 | 24.960 | 25.000 | -0.027 |
|  | MIN | 24.985 | 24.987 | -0.015 | 24.972 | 24.987 | -0.028 | 24.965 | 24.987 | -0.035 | 24.952 | 24.987 | -0.048 | 24.939 | 24.987 | -0.061 |
| 30 | MAX | 30.006 | 30.000 | 0.019 | 29.993 | 30.000 | 0.006 | 29.986 | 30.000 | -0.001 | 29.973 | 30.000 | -0.014 | 29.960 | 30.000 | -0.027 |
|  | MIN | 29.985 | 29.987 | -0.015 | 29.972 | 29.987 | -0.028 | 29.965 | 29.987 | -0.035 | 29.952 | 29.987 | -0.048 | 29.939 | 29.987 | -0.061 |

Preferred Shaft Basis Transition and Interference Fits (Continued)

Dimensions In mm.

| BASIC SIZE | | LOCATIONAL TRANSN. Hole K7 | Shaft h6 | Fit | LOCATIONAL TRANSN. Hole N7 | Shaft h6 | Fit | LOCATIONAL INTERF. Hole P7 | Shaft h6 | Fit | MEDIUM DRIVE Hole S7 | Shaft h6 | Fit | FORCE Hole U7 | Shaft h6 | Fit |
|---|---|---|---|---|---|---|---|---|---|---|---|---|---|---|---|---|
| 40 | MAX | 40.007 | 40.000 | 0.023 | 39.992 | 40.000 | 0.008 | 39.983 | 40.000 | -0.001 | 39.966 | 40.000 | -0.018 | 39.949 | 40.000 | -0.035 |
|    | MIN | 39.982 | 39.984 | -0.018 | 39.967 | 39.984 | -0.033 | 39.958 | 39.984 | -0.042 | 39.941 | 39.984 | -0.059 | 39.924 | 39.984 | -0.076 |
| 50 | MAX | 50.007 | 50.000 | 0.023 | 49.992 | 50.000 | 0.008 | 49.983 | 50.000 | -0.001 | 49.966 | 50.000 | -0.018 | 49.939 | 50.000 | -0.045 |
|    | MIN | 49.982 | 49.984 | -0.018 | 49.967 | 49.984 | -0.033 | 49.958 | 49.984 | -0.042 | 49.941 | 49.984 | -0.059 | 49.914 | 49.984 | -0.086 |
| 60 | MAX | 60.009 | 60.000 | 0.028 | 59.991 | 60.000 | 0.010 | 59.979 | 60.000 | -0.002 | 59.958 | 60.000 | -0.023 | 59.924 | 60.000 | -0.057 |
|    | MIN | 59.979 | 59.981 | -0.021 | 59.961 | 59.981 | -0.039 | 59.949 | 59.981 | -0.051 | 59.928 | 59.981 | -0.072 | 59.894 | 59.981 | -0.106 |
| 80 | MAX | 80.009 | 80.000 | 0.028 | 79.991 | 80.000 | 0.010 | 79.979 | 80.000 | -0.002 | 79.952 | 80.000 | -0.029 | 79.909 | 80.000 | -0.072 |
|    | MIN | 79.979 | 79.981 | -0.021 | 79.961 | 79.981 | -0.039 | 79.949 | 79.981 | -0.051 | 79.922 | 79.981 | -0.078 | 79.879 | 79.981 | -0.121 |
| 100 | MAX | 100.010 | 100.000 | 0.032 | 99.990 | 100.000 | 0.012 | 99.976 | 100.000 | -0.002 | 99.942 | 100.000 | -0.036 | 99.889 | 100.000 | -0.089 |
|     | MIN | 99.975 | 99.978 | -0.025 | 99.955 | 99.978 | -0.045 | 99.941 | 99.978 | -0.059 | 99.907 | 99.978 | -0.093 | 99.854 | 99.978 | -0.146 |
| 120 | MAX | 120.010 | 120.000 | 0.032 | 119.990 | 120.000 | 0.012 | 119.976 | 120.000 | -0.002 | 119.934 | 120.000 | -0.044 | 119.869 | 120.000 | -0.109 |
|     | MIN | 119.975 | 119.978 | -0.025 | 119.955 | 119.978 | -0.045 | 119.941 | 119.978 | -0.059 | 119.899 | 119.978 | -0.101 | 119.834 | 119.978 | -0.166 |
| 160 | MAX | 160.012 | 160.000 | 0.037 | 159.988 | 160.000 | 0.013 | 159.972 | 160.000 | -0.003 | 159.915 | 160.000 | -0.060 | 159.825 | 160.000 | -0.150 |
|     | MIN | 159.972 | 159.975 | -0.028 | 159.948 | 159.975 | -0.052 | 159.932 | 159.975 | -0.068 | 159.875 | 159.975 | -0.125 | 159.785 | 159.975 | -0.215 |
| 200 | MAX | 200.013 | 200.000 | 0.042 | 199.986 | 200.000 | 0.015 | 199.967 | 200.000 | -0.004 | 199.895 | 200.000 | -0.076 | 199.781 | 200.000 | -0.190 |
|     | MIN | 199.967 | 199.971 | -0.033 | 199.940 | 199.971 | -0.060 | 199.921 | 199.971 | -0.079 | 199.849 | 199.971 | -0.151 | 199.735 | 199.971 | -0.265 |
| 250 | MAX | 250.013 | 250.000 | 0.042 | 249.986 | 250.000 | 0.015 | 249.967 | 250.000 | -0.004 | 249.877 | 250.000 | -0.094 | 249.733 | 250.000 | -0.238 |
|     | MIN | 249.967 | 249.971 | -0.033 | 249.940 | 249.971 | -0.060 | 249.921 | 249.971 | -0.079 | 249.831 | 249.971 | -0.169 | 249.687 | 249.971 | -0.313 |
| 300 | MAX | 300.016 | 300.000 | 0.048 | 299.986 | 300.000 | 0.018 | 299.964 | 300.000 | -0.004 | 299.850 | 300.000 | -0.118 | 299.670 | 300.000 | -0.298 |
|     | MIN | 299.964 | 299.968 | -0.036 | 299.934 | 299.968 | -0.066 | 299.912 | 299.968 | -0.088 | 299.798 | 299.968 | -0.202 | 299.618 | 299.968 | -0.382 |
| 400 | MAX | 400.017 | 400.000 | 0.053 | 399.984 | 400.000 | 0.020 | 399.959 | 400.000 | -0.005 | 399.813 | 400.000 | -0.151 | 399.586 | 400.000 | -0.378 |
|     | MIN | 399.960 | 399.964 | -0.040 | 399.927 | 399.964 | -0.073 | 399.902 | 399.964 | -0.098 | 399.756 | 399.964 | -0.244 | 399.529 | 399.964 | -0.471 |
| 500 | MAX | 500.018 | 500.000 | 0.058 | 499.983 | 500.000 | 0.023 | 499.955 | 500.003 | -0.005 | 499.771 | 500.000 | -0.189 | 499.483 | 500.000 | -0.477 |
|     | MIN | 499.955 | 499.960 | -0.045 | 499.920 | 499.960 | -0.080 | 499.892 | 499.960 | -0.108 | 499.708 | 499.960 | -0.292 | 499.420 | 499.960 | -0.580 |

Reproduced from ANSI B4.1–1967(R1974) with the permission of the publisher, the American Society of Mechanical Engineers.

# APPENDIX H: AMERICAN STANDARD LIMITS AND FITS

American Standard Running and Sliding Fits

| Nominal Size Range Inches | Class RC 1 | | | Class RC 2 | | | Class RC 3 | | | Class RC 4 | | |
|---|---|---|---|---|---|---|---|---|---|---|---|---|
| | Limits of Clearance | Standard Limits | | Limits of Clearance | Standard Limits | | Limits of Clearance | Standard Limits | | Limits of Clearance | Standard Limits | |
| Over    To | | Hole H5 | Shaft g4 | | Hole H6 | Shaft g5 | | Hole H6 | Shaft f6 | | Hole H7 | Shaft f7 |
| 0.04– 0.12 | 0.1 0.45 | +0.2 0 | −0.1 −0.25 | 0.1 0.55 | +0.25 0 | −0.1 −0.3 | 0.3 0.8 | +0.25 0 | −0.3 −0.55 | 0.3 1.1 | +0.4 0 | −0.3 −0.7 |
| 0.12– 0.24 | 0.15 0.5 | +0.2 0 | −0.15 −0.3 | 0.15 0.65 | +0.3 0 | −0.15 −0.35 | 0.4 1.0 | +0.3 0 | −0.4 −0.7 | 0.4 1.4 | +0.5 0 | −0.4 −0.9 |
| 0.24– 0.40 | 0.2 0.6 | +0.25 0 | −0.2 −0.35 | 0.2 0.85 | +0.4 0 | −0.2 −0.45 | 0.5 1.3 | +0.4 0 | −0.5 −0.9 | 0.5 1.7 | +0.6 0 | −0.5 −1.1 |
| 0.40– 0.71 | 0.25 0.75 | +0.3 0 | −0.25 −0.45 | 0.25 0.95 | +0.4 0 | −0.25 −0.55 | 0.6 1.4 | +0.4 0 | −0.6 −1.0 | 0.6 2.0 | +0.7 0 | −0.6 −1.3 |
| 0.71– 1.19 | 0.3 0.95 | +0.4 0 | −0.3 −0.55 | 0.3 1.2 | +0.5 0 | −0.3 −0.7 | 0.8 1.8 | +0.5 0 | −0.8 −1.3 | 0.8 2.4 | +0.8 0 | −0.8 −1.6 |
| 1.19– 1.97 | 0.4 1.1 | +0.4 0 | −0.4 −0.7 | 0.4 1.4 | +0.6 0 | −0.4 −0.8 | 1.0 2.2 | +0.6 0 | −1.0 −1.6 | 1.0 3.0 | +1.0 0 | −1.0 −2.0 |
| 1.97– 3.15 | 0.4 1.2 | +0.5 0 | −0.4 −0.7 | 0.4 1.6 | +0.7 0 | −0.4 −0.9 | 1.2 2.6 | +0.7 0 | −1.2 −1.9 | 1.2 3.6 | +1.2 0 | −1.2 −2.4 |
| 3.15– 4.73 | 0.5 1.5 | +0.6 0 | −0.5 −0.9 | 0.5 2.0 | +0.9 0 | −0.5 −1.1 | 1.4 3.2 | +0.9 0 | −1.4 −2.3 | 1.4 4.2 | +1.4 0 | −1.4 −2.8 |
| 4.73– 7.09 | 0.6 1.8 | +0.7 0 | −0.6 −1.1 | 0.6 2.3 | +1.0 0 | −0.6 −1.3 | 1.6 3.6 | +1.0 0 | −1.6 −2.6 | 1.6 4.8 | +1.6 0 | −1.6 −3.2 |
| 7.09– 9.85 | 0.6 2.0 | +0.8 0 | −0.6 −1.2 | 0.6 2.6 | +1.2 0 | −0.6 −1.4 | 2.0 4.4 | +1.2 0 | −2.0 −3.2 | 2.0 5.6 | +1.8 0 | −2.0 −3.8 |
| 9.85–12.41 | 0.8 2.3 | +0.9 0 | −0.8 −1.4 | 0.8 2.9 | +1.2 0 | −0.8 −1.7 | 2.5 4.9 | +1.2 0 | −2.5 −3.7 | 2.5 6.5 | +2.0 0 | −2.5 −4.5 |
| 12.41–15.75 | 1.0 2.7 | +1.0 0 | −1.0 −1.7 | 1.0 3.4 | +1.4 0 | −1.0 −2.0 | 3.0 5.8 | +1.4 0 | −3.0 −4.4 | 3.0 7.4 | +2.2 0 | −3.0 −5.2 |

Reproduced from ANSI B4.1–1967(R1974) with the permission of the publisher, the American Society of Mechanical Engineers.

From *Drafting: Technology and Practice*, copyright 1980 by Spence. Used with permission of the publisher, Bennett Publishing Company, Peoria, IL 61615.

American Standard Running and Sliding Fits

| Nominal Size Range Inches | Class RC 5 | | | Class RC 6 | | | Class RC 7 | | | Class RC 8 | | | Class RC 9 | | |
|---|---|---|---|---|---|---|---|---|---|---|---|---|---|---|---|
| | Limits of Clearance | Standard Limits | | Limits of Clearance | Standard Limits | | Limits of Clearance | Standard Limits | | Limits of Clearance | Standard Limits | | Limits of Clearance | Standard Limits | |
| Over To | | Hole H8 | Shaft e7 | | Hole H9 | Shaft e8 | | Hole H9 | Shaft d8 | | Hole H10 | Shaft c9 | | Hole H11 | Shaft |
| 0- 0.12 | 0.6 1.6 | +0.6 0 | −0.6 −1.0 | 0.6 2.2 | +1.0 0 | −0.6 −1.2 | 1.0 2.6 | +1.0 0 | − 1.0 − 1.6 | 2.5 5.1 | +1.6 0 | − 2.5 − 3.5 | 4.0 8.1 | + 2.5 0 | − 4.0 − 5.6 |
| 0.12- 0.24 | 0.8 2.0 | +0.7 0 | −0.8 −1.3 | 0.8 2.7 | +1.2 0 | −0.8 −1.5 | 1.2 3.1 | +1.2 0 | − 1.2 − 1.9 | 2.8 5.8 | +1.8 0 | − 2.8 − 4.0 | 4.5 9.0 | + 3.0 0 | − 4.5 − 6.0 |
| 0.24- 0.40 | 1.0 2.5 | +0.9 0 | −1.0 −1.6 | 1.0 3.3 | +1.4 0 | −1.0 −1.9 | 1.6 3.9 | +1.4 0 | − 1.6 − 2.5 | 3.0 6.6 | +2.2 0 | − 3.0 − 4.4 | 5.0 10.7 | + 3.5 0 | − 5.0 − 7.2 |
| 0.40- 0.71 | 1.2 2.9 | +1.0 0 | −1.2 −1.9 | 1.2 3.8 | +1.6 0 | −1.2 −2.2 | 2.0 4.6 | +1.6 0 | − 2.0 − 3.0 | 3.5 7.9 | +2.8 0 | − 3.5 − 5.1 | 6.0 12.8 | + 4.0 0 | − 6.0 − 8.8 |
| 0.71- 1.19 | 1.6 3.6 | +1.2 0 | −1.6 −2.4 | 1.6 4.8 | +2.0 0 | −1.6 −2.8 | 2.5 5.7 | +2.0 0 | − 2.5 − 3.7 | 4.5 10.0 | +3.5 0 | − 4.5 − 6.5 | 7.0 15.5 | + 5.0 0 | − 7.0 −10.5 |
| 1.19- 1.97 | 2.0 4.6 | +1.6 0 | −2.0 −3.0 | 2.0 6.1 | +2.5 0 | −2.0 −3.6 | 3.0 7.1 | +2.5 0 | − 3.0 − 4.6 | 5.0 11.5 | +4.0 0 | − 5.0 − 7.5 | 8.0 18.0 | + 6.0 0 | − 8.0 −12.0 |
| 1.97- 3.15 | 2.5 5.5 | +1.8 0 | −2.5 −3.7 | 2.5 7.3 | +3.0 0 | −2.5 −4.3 | 4.0 8.8 | +3.0 0 | − 4.0 − 5.8 | 6.0 13.5 | +4.5 0 | − 6.0 − 9.0 | 9.0 20.5 | + 7.0 0 | − 9.0 −13.5 |
| 3.15- 4.73 | 3.0 6.6 | +2.2 0 | −3.0 −4.4 | 3.0 8.7 | +3.5 0 | −3.0 −5.2 | 5.0 10.7 | +3.5 0 | − 5.0 − 7.2 | 7.0 15.5 | +5.0 0 | − 7.0 −10.5 | 10.0 24.0 | + 9.0 0 | −10.0 −15.0 |
| 4.73- 7.09 | 3.5 7.6 | +2.5 0 | −3.5 −5.1 | 3.5 10.0 | +4.0 0 | −3.5 −6.0 | 6.0 12.5 | +4.0 0 | − 6.0 − 8.5 | 8.0 18.0 | +6.0 0 | − 8.0 −12.0 | 12.0 28.0 | +10.0 0 | −12.0 −18.0 |
| 7.09- 9.85 | 4.0 8.6 | +2.8 0 | −4.0 −5.8 | 4.0 11.3 | +4.5 0 | −4.0 −6.8 | 7.0 14.3 | +4.5 0 | − 7.0 − 9.8 | 10.0 21.5 | +7.0 0 | −10.0 −14.5 | 15.0 34.0 | +12.0 0 | −15.0 −22.0 |
| 9.85- 12.41 | 5.0 10.0 | +3.0 0 | −5.0 −7.0 | 5.0 13.0 | +5.0 0 | −5.0 −8.0 | 8.0 16.0 | +5.0 0 | − 8.0 −11.0 | 12.0 25.0 | +8.0 0 | −12.0 −17.0 | 18.0 38.0 | +12.0 0 | −18.0 −26.0 |
| 12.41-15.75 | 6.0 11.7 | +3.5 0 | −6.0 −8.2 | 6.0 15.5 | +6.0 0 | −6.0 −9.5 | 10.0 19.5 | +6.0 0 | −10.0 −13.5 | 14.0 29.0 | +9.0 0 | −14.0 −20.0 | 22.0 45.0 | +14.0 0 | −22.0 −31.0 |

Reproduced from ANSI B4.1–1967(R1974) with the permission of the publisher, the American Society of Mechanical Engineers.

From *Drafting: Technology and Practice*, copyright 1980 by Spence. Used with permission of the publisher, Bennett Publishing Company, Peoria, IL 61615.

# American Standard Clearance Locational Fits

| Nominal Size Range Inches Over To | Class LC 1 Limits of Clearance | Class LC 1 Standard Limits Hole H6 | Class LC 1 Standard Limits Shaft h5 | Class LC 2 Limits of Clearance | Class LC 2 Standard Limits Hole H7 | Class LC 2 Standard Limits Shaft h6 | Class LC 3 Limits of Clearance | Class LC 3 Standard Limits Hole H8 | Class LC 3 Standard Limits Shaft h7 | Class LC 4 Limits of Clearance | Class LC 4 Standard Limits Hole H10 | Class LC 4 Standard Limits Shaft h9 | Class LC 5 Limits of Clearance | Class LC 5 Standard Limits Hole H7 | Class LC 5 Standard Limits Shaft g6 |
|---|---|---|---|---|---|---|---|---|---|---|---|---|---|---|---|
| 0.04– 0.12 | 0 / 0.45 | +0.25 / − 0 | + 0 / −0.2 | 0 / 0.65 | +0.4 / − 0 | + 0 / −0.25 | 0 / 1 | +0.6 / − 0 | + 0 / −0.4 | 0 / 2.6 | +1.6 / − 0 | + 0 / −1.0 | 0.1 / 0.75 | +0.4 / − 0 | −0.1 / −0.35 |
| 0.12– 0.24 | 0 / 0.5 | +0.3 / − 0 | + 0 / −0.2 | 0 / 0.8 | +0.5 / − 0 | + 0 / −0.3 | 0 / 1.2 | +0.7 / − 0 | + 0 / −0.5 | 0 / 3.0 | +1.8 / − 0 | + 0 / −1.2 | 0.15 / 0.95 | +0.5 / − 0 | −0.15 / −0.45 |
| 0.24– 0.40 | 0 / 0.65 | +0.4 / − 0 | + 0 / −0.25 | 0 / 1.0 | +0.6 / − 0 | + 0 / −0.4 | 0 / 1.5 | +0.9 / − 0 | + 0 / −0.6 | 0 / 3.6 | +2.2 / − 0 | + 0 / −1.4 | 0.2 / 1.2 | +0.6 / − 0 | −0.2 / −0.6 |
| 0.40– 0.71 | 0 / 0.7 | +0.4 / − 0 | + 0 / −0.3 | 0 / 1.1 | +0.7 / − 0 | + 0 / −0.4 | 0 / 1.7 | +1.0 / − 0 | + 0 / −0.7 | 0 / 4.4 | +2.8 / − 0 | + 0 / −1.6 | 0.25 / 1.35 | +0.7 / − 0 | −0.25 / −0.65 |
| 0.71– 1.19 | 0 / 0.9 | +0.5 / − 0 | + 0 / −0.4 | 0 / 1.3 | +0.8 / − 0 | + 0 / −0.5 | 0 / 2 | +1.2 / − 0 | + 0 / −0.8 | 0 / 5.5 | +3.5 / − 0 | + 0 / −2.0 | 0.3 / 1.6 | +0.8 / − 0 | −0.3 / −0.8 |
| 1.19– 1.97 | 0 / 1.0 | +0.6 / − 0 | + 0 / −0.4 | 0 / 1.6 | +1.0 / − 0 | + 0 / −0.6 | 0 / 2.6 | +1.6 / − 0 | + 0 / −1 | 0 / 6.5 | +4.0 / − 0 | + 0 / −2.5 | 0.4 / 2.0 | +1.0 / − 0 | −0.4 / −1.0 |
| 1.97– 3.15 | 0 / 1.2 | +0.7 / − 0 | + 0 / −0.5 | 0 / 1.9 | +1.2 / − 0 | + 0 / −0.7 | 0 / 3 | +1.8 / − 0 | + 0 / −1.2 | 0 / 7.5 | +4.5 / − 0 | + 0 / −3 | 0.4 / 2.3 | +1.2 / − 0 | −0.4 / −1.1 |
| 3.15– 4.73 | 0 / 1.5 | +0.9 / − 0 | + 0 / −0.6 | 0 / 2.3 | +1.4 / − 0 | + 0 / −0.9 | 0 / 3.6 | +2.2 / − 0 | + 0 / −1.4 | 0 / 8.5 | +5.0 / − 0 | + 0 / −3.5 | 0.5 / 2.8 | +1.4 / − 0 | −0.5 / −1.4 |
| 4.73– 7.09 | 0 / 1.7 | +1.0 / − 0 | + 0 / −0.7 | 0 / 2.6 | +1.6 / − 0 | + 0 / −1.0 | 0 / 4.1 | +2.5 / − 0 | + 0 / −1.6 | 0 / 10 | +6.0 / − 0 | + 0 / −4 | 0.6 / 3.2 | +1.6 / − 0 | −0.6 / −1.6 |
| 7.09– 9.85 | 0 / 2.0 | +1.2 / − 0 | + 0 / −0.8 | 0 / 3.0 | +1.8 / − 0 | + 0 / −1.2 | 0 / 4.6 | +2.8 / − 0 | + 0 / −1.8 | 0 / 11.5 | +7.0 / − 0 | + 0 / −4.5 | 0.6 / 3.6 | +1.8 / − 0 | −0.6 / −1.8 |
| 9.85–12.41 | 0 / 2.1 | +1.2 / − 0 | + 0 / −0.9 | 0 / 3.2 | +2.0 / − 0 | + 0 / −1.2 | 0 / 5 | +3.0 / − 0 | + 0 / −2.0 | 0 / 13 | +8.0 / − 0 | + 0 / −5 | 0.7 / 3.9 | +2.0 / − 0 | −0.7 / −1.9 |
| 12.41–15.75 | 0 / 2.4 | +1.4 / − 0 | + 0 / −1.0 | 0 / 3.6 | +2.2 / − 0 | + 0 / −1.4 | 0 / 5.7 | +3.5 / − 0 | + 0 / −2.2 | 0 / 15 | +9.0 / − 0 | + 0 / −6 | 0.7 / 4.3 | +2.2 / − 0 | −0.7 / −2.1 |

Reproduced from ANSI B4.1–1967(R1974) with the permission of the publisher, the American Society of Mechanical Engineers.

From *Drafting: Technology and Practice*, copyright 1980 by Spence. Used with permission of the publisher, Bennett Publishing Company, Peoria, IL 61615.

# American Standard Clearance Locational Fits

| Nominal Size Range Inches (Over – To) | LC 6 Limits of Clearance | LC 6 Hole H9 | LC 6 Shaft f8 | LC 7 Limits of Clearance | LC 7 Hole H10 | LC 7 Shaft e9 | LC 8 Limits of Clearance | LC 8 Hole H10 | LC 8 Shaft d9 | LC 9 Limits of Clearance | LC 9 Hole H11 | LC 9 Shaft c10 | LC 10 Limits of Clearance | LC 10 Hole H12 | LC 10 Shaft | LC 11 Limits of Clearance | LC 11 Hole H13 | LC 11 Shaft |
|---|---|---|---|---|---|---|---|---|---|---|---|---|---|---|---|---|---|---|
| 0– 0.12 | 0.3 / 1.9 | +1.0 / −0 | −0.3 / −0.9 | 0.6 / 3.2 | +1.6 / −0 | −0.6 / −1.6 | 1.0 / 3.6 | +1.6 / −0 | −1.0 / −2.0 | 2.5 / 6.6 | +2.5 / −0 | −2.5 / −4.1 | 4 / 12 | +4 / −0 | −4 / −8 | 5 / 17 | +6 / −0 | −5 / −11 |
| 0.12– 0.24 | 0.4 / 2.3 | +1.2 / −0 | −0.4 / −1.1 | 0.8 / 3.8 | +1.8 / −0 | −0.8 / −2.0 | 1.2 / 4.2 | +1.8 / −0 | −1.2 / −2.4 | 2.8 / 7.6 | +3.0 / −0 | −2.8 / −4.6 | 4.5 / 14.5 | +5 / −0 | −4.5 / −9.5 | 6 / 20 | +7 / −0 | −6 / −13 |
| 0.24– 0.40 | 0.5 / 2.8 | +1.4 / −0 | −0.5 / −1.4 | 1.0 / 4.6 | +2.2 / −0 | −1.0 / −2.4 | 1.6 / 5.2 | +2.2 / −0 | −1.6 / −3.0 | 3.0 / 8.7 | +3.5 / −0 | −3.0 / −5.2 | 5 / 17 | +6 / −0 | −5 / −11 | 7 / 25 | +9 / −0 | −7 / −16 |
| 0.40– 0.71 | 0.6 / 3.2 | +1.6 / −0 | −0.6 / −1.6 | 1.2 / 5.6 | +2.8 / −0 | −1.2 / −2.8 | 2.0 / 6.4 | +2.8 / −0 | −2.0 / −3.6 | 3.5 / 10.3 | +4.0 / −0 | −3.5 / −6.3 | 6 / 20 | +7 / −0 | −6 / −13 | 8 / 28 | +10 / −0 | −8 / −18 |
| 0.71– 1.19 | 0.8 / 4.0 | +2.0 / −0 | −0.8 / −2.0 | 1.6 / 7.1 | +3.5 / −0 | −1.6 / −3.6 | 2.5 / 8.0 | +3.5 / −0 | −2.5 / −4.5 | 4.5 / 13.0 | +5.0 / −0 | −4.5 / −8.0 | 7 / 23 | +8 / −0 | −7 / −15 | 10 / 34 | +12 / −0 | −10 / −22 |
| 1.19– 1.97 | 1.0 / 5.1 | +2.5 / −0 | −1.0 / −2.6 | 2.0 / 8.5 | +4.0 / −0 | −2.0 / −4.5 | 3.0 / 9.5 | +4.0 / −0 | −3.0 / −5.5 | 5 / 15 | +6 / −0 | −5 / −9 | 8 / 28 | +10 / −0 | −8 / −18 | 12 / 44 | +16 / −0 | −12 / −28 |
| 1.97– 3.15 | 1.2 / 6.0 | +3.0 / −0 | −1.2 / −3.0 | 2.5 / 10.0 | +4.5 / −0 | −2.5 / −5.5 | 4.0 / 11.5 | +4.5 / −0 | −4.0 / −7.0 | 6 / 17.5 | +7 / −0 | −6 / −10.5 | 10 / 34 | +12 / −0 | −10 / −22 | 14 / 50 | +18 / −0 | −14 / −32 |
| 3.15– 4.73 | 1.4 / 7.1 | +3.5 / −0 | −1.4 / −3.6 | 3.0 / 11.5 | +5.0 / −0 | −3.0 / −6.5 | 5.0 / 13.5 | +5.0 / −0 | −5.0 / −8.5 | 7 / 21 | +9 / −0 | −7 / −12 | 11 / 39 | +14 / −0 | −11 / −25 | 16 / 60 | +22 / −0 | −16 / −38 |
| 4.73– 7.09 | 1.6 / 8.1 | +4.0 / −0 | −1.6 / −4.1 | 3.5 / 13.5 | +6.0 / −0 | −3.5 / −7.5 | 6 / 16 | +6 / −0 | −6 / −10 | 8 / 24 | +10 / −0 | −8 / −14 | 12 / 44 | +16 / −0 | −12 / −28 | 18 / 68 | +25 / −0 | −18 / −43 |
| 7.09– 9.85 | 2.0 / 9.3 | +4.5 / −0 | −2.0 / −4.8 | 4.0 / 15.5 | +7.0 / −0 | −4.0 / −8.5 | 7 / 18.5 | +7 / −0 | −7 / −11.5 | 10 / 29 | +12 / −0 | −10 / −17 | 16 / 52 | +18 / −0 | −16 / −34 | 22 / 78 | +28 / −0 | −22 / −50 |
| 9.85– 12.41 | 2.2 / 10.2 | +5.0 / −0 | −2.2 / −5.2 | 4.5 / 17.5 | +8.0 / −0 | −4.5 / −9.5 | 7 / 20 | +8 / −0 | −7 / −12 | 12 / 32 | +12 / −0 | −12 / −20 | 20 / 60 | +20 / −0 | −20 / −40 | 28 / 88 | +30 / −0 | −28 / −58 |
| 12.41–15.75 | 2.5 / 12.0 | +6.0 / −0 | −2.5 / −6.0 | 5 / 20.0 | +9.0 / −0 | −5 / −11 | 8 / 23 | +9 / −0 | −8 / −14 | 14 / 37 | +14 / −0 | −14 / −23 | 22 / 66 | +22 / −0 | −22 / −44 | 30 / 100 | +35 / −0 | −30 / −65 |

Reproduced from ANSI B4.1–1967(R1974) with the permission of the publisher, the American Society of Mechanical Engineers.

From *Drafting: Technology and Practice*, copyright 1980 by Spence. Used with permission of the publisher, Bennett Publishing Company, Peoria, IL 61615.

American Standard Transition Locational Fits

| Nominal Size Range Inches (Over – To) | Class LT 1 Fit | Class LT 1 Standard Limits Hole H7 | Class LT 1 Standard Limits Shaft j6 | Class LT 2 Fit | Class LT 2 Standard Limits Hole H8 | Class LT 2 Standard Limits Shaft JS7 | Class LT 3 Fit | Class LT 3 Standard Limits Hole H7 | Class LT 3 Standard Limits Shaft k6 | Class LT 4 Fit | Class LT 4 Standard Limits Hole H8 | Class LT 4 Standard Limits Shaft k7 | Class LT 6 Fit | Class LT 6 Standard Limits Hole h7 | Class LT 6 Standard Limits Shaft n7 |
|---|---|---|---|---|---|---|---|---|---|---|---|---|---|---|---|
| 0– 0.12 | −0.10 +0.50 | +0.4 −0 | +0.10 −0.10 | −0.2 +0.8 | +0.6 −0 | +0.2 −0.2 | | | | | | | −0.65 +0.15 | +0.4 −0 | −0.65 +0.25 |
| 0.12– 0.24 | −0.15 +0.65 | +0.5 −0 | +0.15 −0.15 | −0.25 +0.95 | +0.7 −0 | +0.25 −0.25 | | | | | | | −0.8 +0.2 | +0.5 −0 | +0.8 +0.3 |
| 0.24– 0.40 | −0.20 +0.8 | +0.6 −0 | +0.2 −0.2 | −0.3 +1.2 | +0.9 −0 | +0.3 −0.3 | −0.5 +0.5 | +0.6 −0 | +0.5 +0.1 | −0.7 +0.8 | +0.9 −0 | +0.7 +0.1 | −1.0 +0.2 | +0.6 −0 | +1.0 +0.4 |
| 0.40– 0.71 | −0.2 +0.9 | +0.7 −0 | +0.2 −0.2 | −0.35 +1.35 | +1.0 −0 | +0.35 −0.35 | −0.5 +0.6 | +0.7 −0 | +0.5 +0.1 | −0.8 +0.9 | +1.0 −0 | +0.8 +0.1 | −1.2 +0.2 | +0.7 −0 | +1.2 +0.5 |
| 0.71– 1.19 | −0.25 +1.05 | +0.8 −0 | +0.25 −0.25 | −0.4 +1.6 | +1.2 −0 | +0.4 −0.4 | −0.6 +0.7 | +0.8 −0 | +0.6 +0.1 | −0.9 +1.1 | +1.2 −0 | +0.9 +0.1 | −1.7 +0.3 | +0.8 −0 | +1.4 +0.6 |
| 1.19– 1.97 | −0.3 +1.3 | +1.0 −0 | +0.3 −0.3 | −0.5 +2.1 | +1.6 −0 | +0.5 −0.5 | −0.7 +0.9 | +1.0 −0 | +0.7 +0.1 | −1.1 +1.5 | +1.6 −0 | +1.1 +0.1 | −2.0 +0.4 | +1.0 −0 | +1.7 +0.7 |
| 1.97– 3.15 | −0.3 +1.5 | +1.2 −0 | +0.3 −0.3 | −0.6 +2.4 | +1.8 −0 | +0.6 −0.6 | −0.8 +1.1 | +1.2 −0 | +0.8 +0.1 | −1.3 +1.7 | +1.8 −0 | +1.3 +0.1 | −2.4 +0.4 | +1.2 −0 | +2.0 +0.8 |
| 3.15– 4.73 | −0.4 +1.8 | +1.4 −0 | +0.4 −0.4 | −0.7 +2.9 | +2.2 −0 | +0.7 −0.7 | −1.0 +1.3 | +1.4 −0 | +1.0 +0.1 | −1.5 +2.1 | +2.2 −0 | +1.5 +0.1 | −2.8 +0.4 | +1.4 −0 | +2.4 +1.0 |
| 4.73– 7.09 | −0.5 +2.1 | +1.6 −0 | +0.5 −0.5 | −0.8 +3.3 | +2.5 −0 | +0.8 −0.8 | −1.1 +1.5 | +1.6 −0 | +1.1 +0.1 | −1.7 +2.4 | +2.5 −0 | +1.7 +0.1 | −3.2 +0.4 | +1.6 −0 | +2.8 +1.2 |
| 7.09– 9.85 | −0.6 +2.4 | +1.8 −0 | +0.6 −0.6 | −0.9 +3.7 | +2.8 −0 | +0.9 −0.9 | −1.4 +1.6 | +1.8 −0 | +1.4 +0.2 | −2.0 +2.6 | +2.8 −0 | +2.0 +0.2 | −3.4 +0.6 | +1.8 −0 | +3.2 +1.4 |
| 9.85– 12.41 | −0.6 +2.6 | +2.0 −0 | +0.6 −0.6 | −1.0 +4.0 | +3.0 −0 | +1.0 −1.0 | −1.4 +1.8 | +2.0 −0 | +1.4 +0.2 | −2.2 +2.8 | +3.0 −0 | +2.2 +0.2 | −3.8 +0.6 | +2.0 −0 | +3.4 +1.4 |
| 12.41– 15.75 | −0.7 +2.9 | +2.2 −0 | +0.7 −0.7 | −1.2 +4.5 | +3.5 +0 | +1.0 −1.0 | −1.6 +2.0 | +2.2 −0 | +1.6 +0.2 | −2.4 +3.3 | +3.5 −0 | +2.4 +0.2 | −4.3 +0.7 | +2.3 −0 | +3.8 +1.6 |

Reproduced from ANSI B4.1–1967(R1974) with the permission of the publisher, the American Society of Mechanical Engineers.
From *Drafting: Technology and Practice*, copyright 1980 by Spence. Used with permission of the publisher, Bennett Publishing Company, Peoria, IL 61615.

American Standard Interferance Locational Fits

| Nominal Size Range Inches Over  To | Class LN 2 | | | Class LN 3 | | | Nominal Size Range Inches Over  To | Class LN 2 | | | Class LN 3 | | |
|---|---|---|---|---|---|---|---|---|---|---|---|---|---|
| | Limits of Interference | Standard Limits | | Limits of Interference | Standard Limits | | | Limits of Interference | Standard Limits | | Limits of Interference | Standard Limits | |
| | | Hole H7 | Shaft p6 | | Hole H7 | Shaft r6 | | | Hole H7 | Shaft p6 | | Hole H7 | Shaft r6 |
| 0.04–0.12 | 0<br>0.65 | +0.4<br>— 0 | +0.65<br>+0.4 | 0.1<br>0.75 | +0.4<br>— 0 | +0.75<br>+0.5 | 1.97– 3.15 | 0.2<br>2.1 | +1.2<br>— 0 | +2.1<br>+1.4 | 0.4<br>2.3 | +1.2<br>— 0 | +2.3<br>+1.6 |
| 0.12–0.24 | 0<br>0.8 | +0.5<br>— 0 | +0.8<br>+0.5 | 0.1<br>0.9 | +0.5<br>— 0 | +0.9<br>+0.6 | 3.15– 4.73 | 0.2<br>2.5 | +1.4<br>— 0 | +2.5<br>+1.6 | 0.6<br>2.9 | +1.4<br>— 0 | +2.9<br>+2.0 |
| 0.24–0.40 | 0<br>1.0 | +0.6<br>— 0 | +1.0<br>+0.6 | 0.2<br>1.2 | +0.6<br>— 0 | +1.2<br>+0.8 | 4.73– 7.09 | 0.2<br>2.8 | +1.6<br>— 0 | +2.8<br>+1.8 | 0.9<br>3.5 | +1.6<br>— 0 | +3.5<br>+2.5 |
| 0.40–0.71 | 0<br>1.1 | +0.7<br>— 0 | +1.1<br>+0.7 | 0.3<br>1.4 | +0.7<br>— 0 | +1.4<br>+1.0 | 7.09– 9.85 | 0.2<br>3.2 | +1.8<br>— 0 | +3.2<br>+2.0 | 1.2<br>4.2 | +1.8<br>— 0 | +4.2<br>+3.0 |
| 0.71–1.19 | 0<br>1.3 | +0.8<br>— 0 | +1.3<br>+0.8 | 0.4<br>1.7 | +0.8<br>— 0 | +1.7<br>+1.2 | 9.85–12.41 | 0.2<br>3.4 | +2.0<br>— 0 | +3.4<br>+2.2 | 1.5<br>4.7 | +2.0<br>— 0 | +4.7<br>+3.5 |
| 1.19–1.97 | 0<br>1.6 | +1.0<br>— 0 | +1.6<br>+1.0 | 0.4<br>2.0 | +1.0<br>— 0 | +2.0<br>+1.4 | 12.41–15.75 | 0.3<br>3.9 | +2.2<br>— 0 | +3.9<br>+2.5 | 2.3<br>5.9 | +2.2<br>— 0 | +5.9<br>+4.5 |

Reproduced from ANSI B4.1–1967(R1974) with the permission of the publisher, the American Society of Mechanical Engineers.

From *Drafting: Technology and Practice*, copyright 1980 by Spence. Used with permission of the publisher, Bennett Publishing Company, Peoria, IL 61615.

## American Standard Force and Shrink Fits

| Nominal Size Range Inches Over | To | Class FN 1 Limits of Interference | Class FN 1 Standard Limits Hole H6 | Class FN 1 Standard Limits Shaft | Class FN 2 Limits of Interference | Class FN 2 Standard Limits Hole H7 | Class FN 2 Standard Limits Shaft s6 | Class FN 3 Limits of Interference | Class FN 3 Standard Limits Hole H7 | Class FN 3 Standard Limits Shaft t6 | Class FN 4 Limits of Interference | Class FN 4 Standard Limits Hole H7 | Class FN 4 Standard Limits Shaft u6 | Class FN 5 Limits of Interference | Class FN 5 Standard Limits Hole H8 | Class FN 5 Standard Limits Shaft x7 |
|---|---|---|---|---|---|---|---|---|---|---|---|---|---|---|---|---|
| 0.04– | 0.12 | 0.05 | +0.25 | +0.5 | 0.2 | +0.4 | +0.85 | | | | 0.3 | +0.4 | + 0.95 | 0.3 | +0.6 | + 1.3 |
| | | 0.5 | − 0 | +0.3 | 0.85 | − 0 | +0.6 | | | | 0.95 | − 0 | + 0.7 | 1.3 | − 0 | + 0.9 |
| 0.12– | 0.24 | 0.1 | +0.3 | +0.6 | 0.2 | +0.5 | +1.0 | | | | 0.4 | +0.5 | + 1.2 | 0.5 | +0.7 | + 1.7 |
| | | 0.6 | − 0 | +0.4 | 1.0 | − 0 | +0.7 | | | | 1.2 | − 0 | + 0.9 | 1.7 | − 0 | + 1.2 |
| 0.24– | 0.40 | 0.1 | +0.4 | +0.75 | 0.4 | +0.6 | +1.4 | | | | 0.6 | +0.6 | + 1.6 | 0.5 | +0.9 | + 2.0 |
| | | 0.75 | − 0 | +0.5 | 1.4 | − 0 | +1.0 | | | | 1.6 | − 0 | + 1.2 | 2.0 | − 0 | + 1.4 |
| 0.40– | 0.56 | 0.1 | +0.4 | +0.8 | 0.5 | +0.7 | +1.6 | | | | 0.7 | +0.7 | + 1.8 | 0.6 | +1.0 | + 2.3 |
| | | 0.8 | − 0 | +0.5 | 1.6 | − 0 | +1.2 | | | | 1.8 | − 0 | + 1.4 | 2.3 | − 0 | + 1.6 |
| 0.56– | 0.71 | 0.2 | +0.4 | +0.9 | 0.5 | +0.7 | +1.6 | | | | 0.7 | +0.7 | + 1.8 | 0.8 | +1.0 | + 2.5 |
| | | 0.9 | − 0 | +0.6 | 1.6 | − 0 | +1.2 | | | | 1.8 | − 0 | + 1.4 | 2.5 | − 0 | + 1.8 |
| 0.71– | 0.95 | 0.2 | +0.5 | +1.1 | 0.6 | +0.8 | +1.9 | | | | 0.8 | +0.8 | + 2.1 | 1.0 | +1.2 | + 3.0 |
| | | 1.1 | − 0 | +0.7 | 1.9 | − 0 | +1.4 | | | | 2.1 | − 0 | + 1.6 | 3.0 | − 0 | + 2.2 |
| 0.95– | 1.19 | 0.3 | +0.5 | +1.2 | 0.6 | +0.8 | +1.9 | 0.8 | +0.8 | + 2.1 | 1.0 | +0.8 | + 2.3 | 1.3 | +1.2 | + 3.3 |
| | | 1.2 | − 0 | +0.8 | 1.9 | − 0 | +1.4 | 2.1 | − 0 | + 1.6 | 2.3 | − 0 | + 1.8 | 3.3 | − 0 | + 2.5 |
| 1.19– | 1.58 | 0.3 | +0.6 | +1.3 | 0.8 | +1.0 | +2.4 | 1.0 | +1.0 | + 2.6 | 1.5 | +1.0 | + 3.1 | 1.4 | +1.6 | + 4.0 |
| | | 1.3 | − 0 | +0.9 | 2.4 | − 0 | +1.8 | 2.6 | − 0 | + 2.0 | 3.1 | − 0 | + 2.5 | 4.0 | − 0 | + 3.0 |
| 1.58– | 1.97 | 0.4 | +0.6 | +1.4 | 0.8 | +1.0 | +2.4 | 1.2 | +1.0 | + 2.8 | 1.8 | +1.0 | + 3.4 | 2.4 | +1.6 | + 5.0 |
| | | 1.4 | − 0 | +1.0 | 2.4 | − 0 | +1.8 | 2.8 | − 0 | + 2.2 | 3.4 | − 0 | + 2.8 | 5.0 | − 0 | + 4.0 |
| 1.97– | 2.56 | 0.6 | +0.7 | +1.8 | 0.8 | +1.2 | +2.7 | 1.3 | +1.2 | + 3.2 | 2.3 | +1.2 | + 4.2 | 3.2 | +1.8 | + 6.2 |
| | | 1.8 | − 0 | +1.3 | 2.7 | − 0 | +2.0 | 3.2 | − 0 | + 2.5 | 4.2 | − 0 | + 3.5 | 6.2 | − 0 | + 5.0 |
| 2.56– | 3.15 | 0.7 | +0.7 | +1.9 | 1.0 | +1.2 | +2.9 | 1.8 | +1.2 | + 3.7 | 2.8 | +1.2 | + 4.7 | 4.2 | +1.8 | + 7.2 |
| | | 1.9 | − 0 | +1.4 | 2.9 | − 0 | +2.2 | 3.7 | − 0 | + 3.0 | 4.7 | − 0 | + 4.0 | 7.2 | − 0 | + 6.0 |
| 3.15– | 3.94 | 0.9 | +0.9 | +2.4 | 1.4 | +1.4 | +3.7 | 2.1 | +1.4 | + 4.4 | 3.6 | +1.4 | + 5.9 | 4.8 | +2.2 | + 8.4 |
| | | 2.4 | − 0 | +1.8 | 3.7 | − 0 | +2.8 | 4.4 | − 0 | + 3.5 | 5.9 | − 0 | + 5.0 | 8.4 | − 0 | + 7.0 |
| 3.94– | 4.73 | 1.1 | +0.9 | +2.6 | 1.6 | +1.4 | +3.9 | 2.6 | +1.4 | + 4.9 | 4.6 | +1.4 | + 6.9 | 5.8 | +2.2 | + 9.4 |
| | | 2.6 | − 0 | +2.0 | 3.9 | − 0 | +3.0 | 4.9 | − 0 | + 4.0 | 6.9 | − 0 | + 6.0 | 9.4 | − 0 | + 8.0 |
| 4.73– | 5.52 | 1.2 | +1.0 | +2.9 | 1.9 | +1.6 | +4.5 | 3.4 | +1.6 | + 6.0 | 5.4 | +1.6 | + 8.0 | 7.6 | +2.5 | +11.6 |
| | | 2.9 | − 0 | +2.2 | 4.5 | − 0 | +3.5 | 6.0 | − 0 | + 5.0 | 8.0 | − 0 | + 7.0 | 11.6 | − 0 | +10.0 |
| 5.52– | 6.30 | 1.5 | +1.0 | +3.2 | 2.4 | +1.6 | +5.0 | 3.4 | +1.6 | + 6.0 | 5.4 | +1.6 | + 8.0 | 9.5 | +2.5 | +13.6 |
| | | 3.2 | − 0 | +2.5 | 5.0 | − 0 | +4.0 | 6.0 | − 0 | + 5.0 | 8.0 | − 0 | + 7.0 | 13.6 | − 0 | +12.0 |
| 6.30– | 7.09 | 1.8 | +1.0 | +3.5 | 2.9 | +1.6 | +5.5 | 4.4 | +1.6 | + 7.0 | 6.4 | +1.6 | + 9.0 | 9.5 | +2.5 | +13.6 |
| | | 3.5 | − 0 | +2.8 | 5.5 | − 0 | +4.5 | 7.0 | − 0 | + 6.0 | 9.0 | − 0 | + 8.0 | 13.6 | − 0 | +12.0 |
| 7.09– | 7.88 | 1.8 | +1.2 | +3.8 | 3.2 | +1.8 | +6.2 | 5.2 | +1.8 | + 8.2 | 7.2 | +1.8 | +10.2 | 11.2 | +2.8 | +15.8 |
| | | 3.8 | − 0 | +3.0 | 6.2 | − 0 | +5.0 | 8.2 | − 0 | + 7.0 | 10.2 | − 0 | + 9.0 | 15.8 | − 0 | +14.0 |
| 7.88– | 8.86 | 2.3 | +1.2 | +4.3 | 3.2 | +1.8 | +6.2 | 5.2 | +1.8 | + 8.2 | 8.2 | +1.8 | +11.2 | 13.2 | +2.8 | +17.8 |
| | | 4.3 | − 0 | +3.5 | 6.2 | − 0 | +5.0 | 8.2 | − 0 | + 7.0 | 11.2 | − 0 | +10.0 | 17.8 | − 0 | +16.0 |
| 8.86– | 9.85 | 2.3 | +1.2 | +4.3 | 4.2 | +1.8 | +7.2 | 6.2 | +1.8 | + 9.2 | 10.2 | +1.8 | +13.2 | 13.2 | +2.8 | +17.8 |
| | | 4.3 | − 0 | +3.5 | 7.2 | − 0 | +6.0 | 9.2 | − 0 | + 8.0 | 13.2 | − 0 | +12.0 | 17.8 | − 0 | +16.0 |
| 9.85– | 11.03 | 2.8 | +1.2 | +4.9 | 4.0 | +2.0 | +7.2 | 7.0 | +2.0 | +10.2 | 10.2 | +2.0 | +13.2 | 15.0 | +3.0 | +20.0 |
| | | 4.9 | − 0 | +4.0 | 7.2 | − 0 | +6.0 | 10.2 | − 0 | + 9.0 | 13.2 | − 0 | +12.0 | 20.0 | − 0 | +18.0 |
| 11.03– | 12.41 | 2.8 | +1.2 | +4.9 | 5.0 | +2.0 | +8.2 | 7.0 | +2.0 | +10.2 | 12.0 | +2.0 | +15.2 | 17.0 | +3.0 | +22.0 |
| | | 4.9 | − 0 | +4.0 | 8.2 | − 0 | +7.0 | 10.2 | − 0 | + 9.0 | 15.2 | − 0 | +14.0 | 22.0 | − 0 | +20.0 |
| 12.41– | 13.98 | 3.1 | +1.4 | +5.5 | 5.8 | +2.2 | +9.4 | 7.8 | +2.2 | +11.4 | 13.8 | +2.2 | +17.4 | 18.5 | +3.5 | +24.2 |
| | | 5.5 | − 0 | +4.5 | 9.4 | − 0 | +8.0 | 11.4 | − 0 | +10.0 | 17.4 | − 0 | +16.0 | 24.2 | − 0 | +22.0 |

Reproduced from ANSI B4.1–1967(R1974) with the permission of the publisher, the American Society of Mechanical Engineers.

From *Drafting: Technology and Practice*, copyright 1980 by Spence. Used with permission of the publisher, Bennett Publishing Company, Peoria, IL 61615.

# APPENDIX I: BEND ALLOWANCE FOR NONFERROUS METAL

Bend Allowance* For 1° of Bend in Nonferrous Metal

| Radius | | | | | Thickness of Material | | | | | | | |
|---|---|---|---|---|---|---|---|---|---|---|---|---|
| | .016 | .020 | .025 | .032 | .040 | .051 | .064 | .072 | .081 | .091 | .102 | .125 |
| 1/32 | .00067 | .00070 | .00074 | .00079 | | | | | | | | |
| 1/16 | .00121 | .00125 | .00129 | .00135 | .00140 | .00145 | .00159 | .00165 | | | | |
| 3/32 | .00176 | .00179 | .00183 | .00188 | .00195 | .00203 | .00213 | .00220 | .00226 | .00234 | .00243 | |
| 1/8 | .00230 | .00234 | .00238 | .00243 | .00249 | .00258 | .00268 | .00274 | .00281 | .00289 | .00297 | .00315 |
| 5/32 | .00285 | .00288 | .00292 | .00297 | .00304 | .00312 | .00322 | .00328 | .00335 | .00343 | .00352 | .00370 |
| 3/16 | .00339 | .00342 | .00347 | .00352 | .00358 | .00367 | .00377 | .00383 | .00390 | .00398 | .00406 | .00424 |
| 7/32 | .00394 | .00397 | .00401 | .00406 | .00412 | .00421 | .00431 | .00437 | .00444 | .00452 | .00461 | .00479 |
| 1/4 | .00448 | .00451 | .00456 | .00461 | .00467 | .00476 | .00486 | .00492 | .00499 | .00507 | .00515 | .00533 |
| 9/32 | .00501 | .00506 | .00510 | .00515 | .00521 | .00530 | .00540 | .00546 | .00553 | .00561 | .00570 | .00588 |
| 5/16 | .00557 | .00560 | .00564 | .00570 | .00576 | .00584 | .00595 | .00601 | .00608 | .00616 | .00624 | .00642 |
| 11/32 | .00612 | .00615 | .00619 | .00624 | .00630 | .00639 | .00649 | .00655 | .00662 | .00670 | .00679 | .00697 |
| 3/8 | .00666 | .00669 | .00673 | .00679 | .00685 | .00693 | .00704 | .00710 | .00717 | .00725 | .00733 | .00751 |
| 13/32 | .00721 | .00724 | .00728 | .00733 | .00739 | .00748 | .00758 | .00764 | .00771 | .00779 | .00787 | .00806 |
| 7/16 | .00775 | .00778 | .00782 | .00787 | .00794 | .00802 | .00812 | .00819 | .00826 | .00834 | .00842 | .00860 |
| 15/32 | .00829 | .00833 | .00837 | .00842 | .00848 | .00857 | .00867 | .00873 | .00880 | .00888 | .00896 | .00915 |
| 1/2 | .00884 | .00887 | .00891 | .00896 | .00903 | .00911 | .00921 | .00928 | .00935 | .00943 | .00951 | .00969 |
| 17/32 | .00938 | .00942 | .00946 | .00951 | .00957 | .00966 | .00976 | .00982 | .00989 | .00997 | .01005 | .01023 |
| 9/16 | .00993 | .00996 | .01000 | .01005 | .01012 | .01020 | .01030 | .01037 | .01043 | .01051 | .01058 | .01078 |
| 19/32 | .01047 | .01051 | .01055 | .01060 | .01065 | .01073 | .01083 | .01091 | .01098 | .01105 | .01114 | .01132 |
| 5/8 | .01102 | .01105 | .01109 | .01114 | .01121 | .01129 | .01139 | .01146 | .01152 | .01160 | .01170 | .01187 |
| 21/32 | .01156 | .01160 | .01164 | .01170 | .01175 | .01183 | .01193 | .01200 | .01207 | .01214 | .01223 | .01241 |
| 11/16 | .01211 | .01214 | .01218 | .01223 | .01230 | .01238 | .01248 | .01254 | .01261 | .01269 | .01276 | .01296 |
| 23/32 | .01265 | .01268 | .01273 | .01276 | .01283 | .01291 | .01301 | .01309 | .01316 | .01322 | .01332 | .01350 |
| 3/4 | .01320 | .01323 | .01327 | .01332 | .01338 | .01347 | .01357 | .01363 | .01370 | .01378 | .01386 | .01405 |
| 25/32 | .01374 | .01378 | .01381 | .01386 | .01392 | .01401 | .01411 | .01418 | .01425 | .01432 | .01441 | .01459 |
| 13/16 | .01429 | .01432 | .01436 | .01441 | .01447 | .01456 | .01466 | .01472 | .01479 | .01487 | .01494 | .01514 |
| 27/32 | .01483 | .01486 | .01490 | .01494 | .01501 | .01509 | .01519 | .01527 | .01534 | .01540 | .01550 | .01568 |
| 7/8 | .01538 | .01541 | .01545 | .01550 | .01556 | .01565 | .01575 | .01581 | .01588 | .01596 | .01604 | .01623 |
| 29/32 | .01592 | .01595 | .01599 | .01604 | .01611 | .01619 | .01629 | .01636 | .01643 | .01650 | .01659 | .01677 |
| 15/16 | .01646 | .01650 | .01654 | .01659 | .01665 | .01674 | .01684 | .01690 | .01697 | .01705 | .01712 | .01732 |
| 31/32 | .01701 | .01704 | .01708 | .01712 | .01718 | .01727 | .01737 | .01745 | .01752 | .01758 | .01768 | .01786 |
| 1 | .01755 | .01759 | .01763 | .01768 | .01774 | .01783 | .01793 | .01799 | .01806 | .01814 | .01823 | .01841 |

*In inches.

From *Drafting: Technology and Practice*, copyright 1980 by Spence. Used with permission of the publisher, Bennett Publishing Company, Peoria, IL 61615.

# APPENDIX J: LOGARITHMS
# OF NUMBERS

Four-Place Common Logarithms (Base 10)

| N | 0 | 1 | 2 | 3 | 4 | 5 | 6 | 7 | 8 | 9 |
|---|---|---|---|---|---|---|---|---|---|---|
| **1.0** | .0000 | .0043 | .0086 | .0128 | .0170 | .0212 | .0253 | .0294 | .0334 | .0374 |
| **1.1** | .0414 | .0453 | .0492 | .0531 | .0569 | .0607 | .0645 | .0682 | .0719 | .0755 |
| **1.2** | .0792 | .0828 | .0864 | .0899 | .0934 | .0969 | .1004 | .1038 | .1072 | .1106 |
| **1.3** | .1139 | .1173 | .1206 | .1239 | .1271 | .1303 | .1335 | .1367 | .1399 | .1430 |
| **1.4** | .1461 | .1492 | .1523 | .1553 | .1584 | .1614 | .1644 | .1673 | .1703 | .1732 |
| **1.5** | .1761 | .1790 | .1818 | .1847 | .1875 | .1903 | .1931 | .1959 | .1987 | .2014 |
| **1.6** | .2641 | .2068 | .2095 | .2122 | .2148 | .2175 | .2201 | .2227 | .2253 | .2279 |
| **1.7** | .2304 | .2330 | .2355 | .2380 | .2405 | .2430 | .2455 | .2480 | .2504 | .2529 |
| **1.8** | .2553 | .2577 | .2601 | .2625 | .2648 | .2672 | .2695 | .2718 | .2742 | .2765 |
| **1.9** | .2788 | .2810 | .2833 | .2856 | .2878 | .2900 | .2923 | .2945 | .2967 | .2989 |
| **2.0** | .3010 | .3032 | .3054 | .3075 | .3096 | .3118 | .3139 | .3160 | .3181 | .3201 |
| **2.1** | .3222 | .3243 | .3263 | .3284 | .3304 | .3324 | .3345 | .3365 | .3385 | .3404 |
| **2.2** | .3424 | .3444 | .3464 | .3483 | .3502 | .3522 | .3541 | .3560 | .3579 | .3598 |
| **2.3** | .3617 | .3636 | .3655 | .3674 | .3692 | .3711 | .3729 | .3747 | .3766 | .3784 |
| **2.4** | .3802 | .3820 | .3838 | .3856 | .3874 | .3892 | .3909 | .3927 | .3945 | .3962 |
| **2.5** | .3979 | .3997 | .4014 | .4031 | .4048 | .4065 | .4082 | .4099 | .4116 | .4133 |
| **2.6** | .4150 | .4166 | .4183 | .4200 | .4216 | .4232 | .4249 | .4265 | .4281 | .4298 |
| **2.7** | .4314 | .4330 | .4346 | .4362 | .4378 | .4393 | .4409 | .4425 | .4440 | .4456 |
| **2.8** | .4472 | .4487 | .4502 | .4518 | .4533 | .4548 | .4564 | .4579 | .4594 | .4609 |
| **2.9** | .4624 | .4639 | .4654 | .4669 | .4683 | .4698 | .4713 | .4728 | .4742 | .4757 |
| **3.0** | .4771 | .4786 | .4800 | .4814 | .4829 | .4843 | .4857 | .4871 | .4886 | .4900 |
| **3.1** | .4914 | .4928 | .4942 | .4955 | .4969 | .4983 | .4997 | .5011 | .5024 | .5038 |
| **3.2** | .5051 | .5065 | .5079 | .5092 | .5105 | .5119 | .5132 | .5145 | .5159 | .5172 |
| **3.3** | .5185 | .5198 | .5211 | .5224 | .5237 | .5250 | .5263 | .5276 | .5289 | .5302 |
| **3.4** | .5315 | .5328 | .5340 | .5353 | .5366 | .5378 | .5391 | .5403 | .5416 | .5428 |
| **3.5** | .5441 | .5453 | .5465 | .5478 | .5490 | .5502 | .5514 | .5527 | .5539 | .5551 |
| **3.6** | .5563 | .5575 | .5587 | .5599 | .5611 | .5623 | .5635 | .5647 | .5658 | .5670 |
| **3.7** | .5682 | .5694 | .5705 | .5717 | .5729 | .5740 | .5752 | .5763 | .5775 | .5786 |
| **3.8** | .5798 | .5809 | .5821 | .5832 | .5843 | .5855 | .5866 | .5877 | .5888 | .5899 |
| **3.9** | .5911 | .5922 | .5933 | .5944 | .5955 | .5966 | .5977 | .5988 | .5999 | .6010 |
| **4.0** | .6021 | .6031 | .6042 | .6053 | .6064 | .6075 | .6085 | .6096 | .6107 | .6117 |
| **4.1** | .6128 | .6138 | .6149 | .6160 | .6170 | .6180 | .6191 | .6201 | .6212 | .6222 |
| **4.2** | .6232 | .6243 | .6253 | .6263 | .6274 | .6284 | .6294 | .6304 | .6314 | .6325 |
| **4.3** | .6335 | .6345 | .6355 | .6365 | .6375 | .6385 | .6395 | .6405 | .6415 | .6425 |
| **4.4** | .6435 | .6444 | .6454 | .6464 | .6474 | .6484 | .6493 | .6503 | .6513 | .6522 |
| **4.5** | .6532 | .6542 | .6551 | .6561 | .6571 | .6580 | .6590 | .6599 | .6609 | .6618 |
| **4.6** | .6628 | .6637 | .6646 | .6656 | .6665 | .6675 | .6684 | .6693 | .6702 | .6712 |
| **4.7** | .6721 | .6730 | .6739 | .6749 | .6758 | .6767 | .6776 | .6785 | .6794 | .6803 |
| **4.8** | .6812 | .6821 | .6830 | .6839 | .6848 | .6857 | .6866 | .6875 | .6884 | .6893 |
| **4.9** | .6902 | .6911 | .6920 | .6928 | .6937 | .6946 | .6955 | .6964 | .6972 | .6981 |
| **5.0** | .6990 | .6998 | .7007 | .7016 | .7024 | .7033 | .7042 | .7050 | .7059 | .7067 |
| **5.1** | .7076 | .7084 | .7093 | .7101 | .7110 | .7118 | .7126 | .7135 | .7143 | .7152 |
| **5.2** | .7160 | .7168 | .7177 | .7185 | .7193 | .7202 | .7210 | .7218 | .7226 | .7235 |
| **5.3** | .7243 | .7251 | .7259 | .7267 | .7275 | .7284 | .7292 | .7300 | .7308 | .7316 |
| **5.4** | .7324 | .7332 | .7340 | .7348 | .7356 | .7364 | .7372 | .7380 | .7388 | .7396 |
| N | 0 | 1 | 2 | 3 | 4 | 5 | 6 | 7 | 8 | 9 |

Continued

| N | 0 | 1 | 2 | 3 | 4 | 5 | 6 | 7 | 8 | 9 |
|---|---|---|---|---|---|---|---|---|---|---|
| 5.5 | .7404 | .7412 | .7419 | .7427 | .7435 | .7443 | .7451 | .7459 | .7466 | .7474 |
| 5.6 | .7482 | .7490 | .7497 | .7505 | .7513 | .7520 | .7528 | .7536 | .7543 | .7551 |
| 5.7 | .7559 | .7566 | .7574 | .7582 | .7589 | .7597 | .7604 | .7612 | .7619 | .7627 |
| 5.8 | .7634 | .7642 | .7649 | .7657 | .7664 | .7672 | .7679 | .7686 | .7694 | .7701 |
| 5.9 | .7709 | .7716 | .7723 | .7731 | .7738 | .7745 | .7752 | .7760 | .7767 | .7774 |
| 6.0 | .7782 | .7789 | .7796 | .7803 | .7810 | .7818 | .7825 | .7832 | .7839 | .7846 |
| 6.1 | .7853 | .7860 | .7868 | .7875 | .7882 | .7889 | .7896 | .7903 | .7910 | .7917 |
| 6.2 | .7924 | .7931 | .7938 | .7945 | .7952 | .7959 | .7966 | .7973 | .7980 | .7987 |
| 6.3 | .7993 | .8000 | .8007 | .8014 | .8021 | .8028 | .8035 | .8041 | .8048 | .8055 |
| 6.4 | .8062 | .8069 | .8075 | .8082 | .8089 | .8096 | .8102 | .8109 | .8116 | .8122 |
| 6.5 | .8129 | .8136 | .8142 | .8149 | .8156 | .8162 | .8169 | .8176 | .8182 | .8189 |
| 6.6 | .8195 | .8202 | .8209 | .8215 | .8222 | .8228 | .8235 | .8241 | .8248 | .8254 |
| 6.7 | .8261 | .8267 | .8274 | .8280 | .8287 | .8293 | .8299 | .8306 | .8312 | .8319 |
| 6.8 | .8325 | .8331 | .8338 | .8344 | .8351 | .8357 | .8363 | .8370 | .8376 | .8382 |
| 6.9 | .8388 | .8395 | .8401 | .8407 | .8414 | .8420 | .8426 | .8432 | .8439 | .8445 |
| 7.0 | .8451 | .8457 | .8463 | .8470 | .8476 | .8482 | .8488 | .8494 | .8500 | .8506 |
| 7.1 | .8513 | .8519 | .8525 | .8531 | .8537 | .8543 | .8549 | .8555 | .8561 | .8567 |
| 7.2 | .8573 | .8579 | .8585 | .8591 | .8597 | .8603 | .8609 | .8615 | .8621 | .8627 |
| 7.3 | .8633 | .8639 | .8645 | .8651 | .8657 | .8663 | .8669 | .8675 | .8681 | .8686 |
| 7.4 | .8692 | .8698 | .8704 | .8710 | .8716 | .8722 | .8727 | .8733 | .8739 | .8745 |
| 7.5 | .8751 | .8756 | .8762 | .8768 | .8774 | .8779 | .8785 | .8791 | .8797 | .8802 |
| 7.6 | .8808 | .8814 | .8820 | .8825 | .8831 | .8837 | .8842 | .8848 | .8854 | .8859 |
| 7.7 | .8865 | .8871 | .8876 | .8882 | .8887 | .8893 | .8899 | .8904 | .8910 | .8915 |
| 7.8 | .8921 | .8927 | .8932 | .8938 | .8943 | .8949 | .8954 | .8960 | .8965 | .8971 |
| 7.9 | .8976 | .8982 | .8987 | .8993 | .8998 | .9004 | .9009 | .9015 | .9020 | .9025 |
| 8.0 | .9031 | .9036 | .9042 | .9047 | .9053 | .9058 | .9063 | .9069 | .9074 | .9079 |
| 8.1 | .9085 | .9090 | .9096 | .9101 | .9106 | .9112 | .9117 | .9122 | .9128 | .9133 |
| 8.2 | .9138 | .9143 | .9149 | .9154 | .9159 | .9165 | .9170 | .9175 | .9180 | .9186 |
| 8.3 | .9191 | .9196 | .9201 | .9206 | .9212 | .9217 | .9222 | .9227 | .9232 | .9238 |
| 8.4 | .9243 | .9248 | .9253 | .9258 | .9263 | .9269 | .9274 | .9279 | .9284 | .9289 |
| 8.5 | .9294 | .9299 | .9304 | .9309 | .9315 | .9320 | .9325 | .9330 | .9335 | .9340 |
| 8.6 | .9345 | .9350 | .9355 | .9360 | .9365 | .9370 | .9375 | .9380 | .9385 | .9390 |
| 8.7 | .9395 | .9400 | .9405 | .9410 | .9415 | .9420 | .9425 | .9430 | .9435 | .9440 |
| 8.8 | .9445 | .9450 | .9455 | .9460 | .9465 | .9469 | .9474 | .9479 | .9484 | .9489 |
| 8.9 | .9494 | .9499 | .9504 | .9509 | .9513 | .9518 | .9523 | .9528 | .9533 | .9538 |
| 9.0 | .9542 | .9547 | .9552 | .9557 | .9562 | .9566 | .9571 | .9576 | .9581 | .9586 |
| 9.1 | .9590 | .9595 | .9600 | .9605 | .9609 | .9614 | .9619 | .9624 | .9628 | .9633 |
| 9.2 | .9638 | .9643 | .9647 | .9652 | .9657 | .9661 | .9666 | .9671 | .9675 | .9680 |
| 9.3 | .9685 | .9689 | .9694 | .9699 | .9703 | .9708 | .9713 | .9717 | .9722 | .9727 |
| 9.4 | .9731 | .9736 | .9741 | .9745 | .9750 | .9754 | .9759 | .9763 | .9768 | .9773 |
| 9.5 | .9777 | .9782 | .9786 | .9791 | .9795 | .9800 | .9805 | .9809 | .9814 | .9818 |
| 9.6 | .9823 | .9827 | .9832 | .9836 | .9841 | .9845 | .9850 | .9854 | .9859 | .9863 |
| 9.7 | .9868 | .9872 | .9877 | .9881 | .9886 | .9890 | .9894 | .9899 | .9903 | .9908 |
| 9.8 | .9912 | .9917 | .9921 | .9926 | .9930 | .9934 | .9939 | .9943 | .9948 | .9952 |
| 9.9 | .9956 | .9961 | .9965 | .9969 | .9974 | .9978 | .9983 | .9987 | .9991 | .9996 |
| N | 0 | 1 | 2 | 3 | 4 | 5 | 6 | 7 | 8 | 9 |

# APPENDIX K: VALUES
# OF TRIGONOMETRIC FUNCTIONS

| Degrees | Radians | Sine | Tangent | Cotangent | Cosine | | |
|---------|---------|------|---------|-----------|--------|-----|-----|
| 0° 00′ | 0.0000 | 0.0000 | 0.0000 | | 1.0000 | 1.5708 | 90° 00′ |
| 10′ | 0.0029 | 0.0029 | 0.0029 | 343.77 | 1.0000 | 1.5679 | 50′ |
| 20′ | 0.0058 | 0.0058 | 0.0058 | 171.89 | 1.0000 | 1.5650 | 40′ |
| 30′ | 0.0087 | 0.0087 | 0.0087 | 114.59 | 1.0000 | 1.5621 | 30′ |
| 40′ | 0.0116 | 0.0116 | 0.0116 | 85.940 | 0.9999 | 1.5592 | 20′ |
| 50′ | 0.0145 | 0.0145 | 0.0145 | 68.750 | 0.9999 | 1.5563 | 10′ |
| 1° 00′ | 0.0175 | 0.0175 | 0.0175 | 57.290 | 0.9998 | 1.5533 | 89° 00′ |
| 10′ | 0.0204 | 0.0204 | 0.0204 | 49.104 | 0.9998 | 1.5504 | 50′ |
| 20′ | 0.0233 | 0.0233 | 0.0233 | 42.964 | 0.9997 | 1.5475 | 40′ |
| 30′ | 0.0262 | 0.0262 | 0.0262 | 38.188 | 0.9997 | 1.5446 | 30′ |
| 40′ | 0.0291 | 0.0291 | 0.0291 | 34.368 | 0.9996 | 1.5417 | 20′ |
| 50′ | 0.0320 | 0.0320 | 0.0320 | 31.242 | 0.9995 | 1.5388 | 10′ |
| 2° 00′ | 0.0349 | 0.0349 | 0.0349 | 28.636 | 0.9994 | 1.5359 | 88° 00′ |
| 10′ | 0.0378 | 0.0378 | 0.0378 | 26.432 | 0.9993 | 1.5330 | 50′ |
| 20′ | 0.0407 | 0.0407 | 0.0407 | 24.542 | 0.9992 | 1.5301 | 40′ |
| 30′ | 0.0436 | 0.0436 | 0.0437 | 22.904 | 0.9990 | 1.5272 | 30′ |
| 40′ | 0.0465 | 0.0465 | 0.0466 | 21.470 | 0.9989 | 1.5243 | 20′ |
| 50′ | 0.0495 | 0.0494 | 0.0495 | 20.206 | 0.9988 | 1.5213 | 10′ |
| 3° 00′ | 0.0524 | 0.0523 | 0.0524 | 19.081 | 0.9986 | 1.5184 | 87° 00′ |
| 10′ | 0.0553 | 0.0552 | 0.0553 | 18.075 | 0.9985 | 1.5155 | 50′ |
| 20′ | 0.0582 | 0.0581 | 0.0582 | 17.169 | 0.9983 | 1.5126 | 40′ |
| 30′ | 0.0611 | 0.0610 | 0.0612 | 16.350 | 0.9981 | 1.5097 | 30′ |
| 40′ | 0.0640 | 0.0640 | 0.0641 | 15.605 | 0.9980 | 1.5068 | 20′ |
| 50′ | 0.0669 | 0.0669 | 0.0670 | 14.924 | 0.9978 | 1.5039 | 10′ |
| 4° 00′ | 0.0698 | 0.0698 | 0.0699 | 14.301 | 0.9976 | 1.5010 | 86° 00′ |
| 10′ | 0.0727 | 0.0727 | 0.0729 | 13.727 | 0.9974 | 1.4981 | 50′ |
| 20′ | 0.0756 | 0.0756 | 0.0758 | 13.197 | 0.9971 | 1.4952 | 40′ |
| 30′ | 0.0785 | 0.0785 | 0.0787 | 12.706 | 0.9969 | 1.4923 | 30′ |
| 40′ | 0.0814 | 0.0814 | 0.0816 | 12.251 | 0.9967 | 1.4893 | 20′ |
| 50′ | 0.0844 | 0.0843 | 0.0846 | 11.826 | 0.9964 | 1.4864 | 10′ |
| 5° 00′ | 0.0873 | 0.0872 | 0.0875 | 11.430 | 0.9962 | 1.4835 | 85° 00′ |
| 10′ | 0.0902 | 0.0901 | 0.0904 | 11.059 | 0.9959 | 1.4806 | 50′ |
| 20′ | 0.0931 | 0.0929 | 0.0934 | 10.712 | 0.9957 | 1.4777 | 40′ |
| 30′ | 0.0960 | 0.0958 | 0.0963 | 10.385 | 0.9954 | 1.4748 | 30′ |
| 40′ | 0.0989 | 0.0987 | 0.0992 | 10.078 | 0.9951 | 1.4719 | 20′ |
| 50′ | 0.1018 | 0.1016 | 0.1022 | 9.7882 | 0.9948 | 1.4690 | 10′ |
| 6° 00′ | 0.1047 | 0.1045 | 0.1051 | 9.5144 | 0.9945 | 1.4661 | 84° 00′ |
| 10′ | 0.1076 | 0.1074 | 0.1080 | 9.2553 | 0.9942 | 1.4632 | 50′ |
| 20′ | 0.1105 | 0.1103 | 0.1110 | 9.0098 | 0.9939 | 1.4603 | 40′ |
| 30′ | 0.1134 | 0.1132 | 0.1139 | 8.7769 | 0.9936 | 1.4573 | 30′ |
| 40′ | 0.1164 | 0.1161 | 0.1169 | 8.5555 | 0.9932 | 1.4544 | 20′ |
| 50′ | 0.1193 | 0.1190 | 0.1198 | 8.3450 | 0.9929 | 1.4515 | 10′ |
| 7° 00′ | 0.1222 | 0.1219 | 0.1228 | 8.1443 | 0.9925 | 1.4486 | 83° 00′ |
| 10′ | 0.1251 | 0.1248 | 0.1257 | 7.9530 | 0.9922 | 1.4457 | 50′ |
| 20′ | 0.1280 | 0.1276 | 0.1287 | 7.7704 | 0.9918 | 1.4428 | 40′ |
| 30′ | 0.1309 | 0.1305 | 0.1317 | 7.5958 | 0.9914 | 1.4399 | 30′ |
| 40′ | 0.1338 | 0.1334 | 0.1346 | 7.4287 | 0.9911 | 1.4370 | 20′ |
| 50′ | 0.1367 | 0.1363 | 0.1376 | 7.2687 | 0.9907 | 1.4341 | 10′ |
| 8° 00′ | 0.1396 | 0.1392 | 0.1405 | 7.1154 | 0.9903 | 1.4312 | 82° 00′ |
| 10′ | 0.1425 | 0.1421 | 0.1435 | 6.9682 | 0.9899 | 1.4283 | 50′ |
| 20′ | 0.1454 | 0.1449 | 0.1465 | 6.8269 | 0.9894 | 1.4254 | 40′ |
| 30′ | 0.1484 | 0.1478 | 0.1495 | 6.6912 | 0.9890 | 1.4224 | 30′ |
| 40′ | 0.1513 | 0.1507 | 0.1524 | 6.5606 | 0.9886 | 1.4195 | 20′ |
| 50′ | 0.1542 | 0.1536 | 0.1554 | 6.4348 | 0.9881 | 1.4166 | 10′ |
| 9° 00′ | 0.1571 | 0.1564 | 0.1584 | 6.3138 | 0.9877 | 1.4137 | 81° 00′ |
| | | Cosine | Cotangent | Tangent | Sine | Radians | Degrees |

From *Technical Drafting*, copyright 1980 by Spence and Atkins. Used with permission of the publisher, Bennett Publishing Company, Peoria, IL 61615.

| Degrees | Radians | Sine | Tangent | Cotangent | Cosine | | |
|---|---|---|---|---|---|---|---|
| 9° 00′ | 0.1571 | 0.1564 | 0.1584 | 6.3138 | 0.9877 | 1.4137 | 81° 00′ |
| 10′ | 0.1600 | 0.1593 | 0.1614 | 6.1970 | 0.9872 | 1.4108 | 50′ |
| 20′ | 0.1629 | 0.1622 | 0.1644 | 6.0844 | 0.9868 | 1.4079 | 40′ |
| 30′ | 0.1658 | 0.1650 | 0.1673 | 5.9758 | 0.9863 | 1.4050 | 30′ |
| 40′ | 0.1687 | 0.1679 | 0.1703 | 5.8708 | 0.9858 | 1.4021 | 20′ |
| 50′ | 0.1716 | 0.1708 | 0.1733 | 5.7694 | 0.9853 | 1.3992 | 10′ |
| 10° 00′ | 0.1745 | 0.1736 | 0.1763 | 5.6713 | 0.9848 | 1.3963 | 80° 00′ |
| 10′ | 0.1774 | 0.1765 | 0.1793 | 5.5764 | 0.9843 | 1.3934 | 50′ |
| 20′ | 0.1804 | 0.1794 | 0.1823 | 5.4845 | 0.9838 | 1.3904 | 40′ |
| 30′ | 0.1833 | 0.1822 | 0.1853 | 5.3955 | 0.9833 | 1.3875 | 30′ |
| 40′ | 0.1862 | 0.1851 | 0.1883 | 5.3093 | 0.9827 | 1.3846 | 20′ |
| 50′ | 0.1891 | 0.1880 | 0.1914 | 5.2257 | 0.9822 | 1.3817 | 10′ |
| 11° 00′ | 0.1920 | 0.1908 | 0.1944 | 5.1446 | 0.9816 | 1.3788 | 79° 00′ |
| 10′ | 0.1949 | 0.1937 | 0.1974 | 5.0658 | 0.9811 | 1.3759 | 50′ |
| 20′ | 0.1978 | 0.1965 | 0.2004 | 4.9894 | 0.9805 | 1.3730 | 40′ |
| 30′ | 0.2007 | 0.1994 | 0.2035 | 4.9152 | 0.9799 | 1.3701 | 30′ |
| 40′ | 0.2036 | 0.2022 | 0.2065 | 4.8430 | 0.9793 | 1.3672 | 20′ |
| 50′ | 0.2065 | 0.2051 | 0.2095 | 4.7729 | 0.9787 | 1.3643 | 10′ |
| 12° 00′ | 0.2094 | 0.2079 | 0.2126 | 4.7046 | 0.9781 | 1.3614 | 78° 00′ |
| 10′ | 0.2123 | 0.2108 | 0.2156 | 4.6382 | 0.9775 | 1.3584 | 50′ |
| 20′ | 0.2153 | 0.2136 | 0.2186 | 4.5736 | 0.9769 | 1.3555 | 40′ |
| 30′ | 0.2182 | 0.2164 | 0.2217 | 4.5107 | 0.9763 | 1.3526 | 30′ |
| 40′ | 0.2211 | 0.2193 | 0.2247 | 4.4494 | 0.9757 | 1.3497 | 20′ |
| 50′ | 0.2240 | 0.2221 | 0.2278 | 4.3897 | 0.9750 | 1.3468 | 10′ |
| 13° 00′ | 0.2269 | 0.2250 | 0.2309 | 4.3315 | 0.9744 | 1.3439 | 77° 00′ |
| 10′ | 0.2298 | 0.2278 | 0.2339 | 4.2747 | 0.9737 | 1.3410 | 50′ |
| 20′ | 0.2327 | 0.2306 | 0.2370 | 4.2193 | 0.9730 | 1.3381 | 40′ |
| 30′ | 0.2356 | 0.2334 | 0.2401 | 4.1653 | 0.9724 | 1.3352 | 30′ |
| 40′ | 0.2385 | 0.2363 | 0.2432 | 4.1126 | 0.9717 | 1.3323 | 20′ |
| 50′ | 0.2414 | 0.2391 | 0.2462 | 4.0611 | 0.9710 | 1.3294 | 10′ |
| 14° 00′ | 0.2443 | 0.2419 | 0.2493 | 4.0108 | 0.9703 | 1.3265 | 76° 00′ |
| 10′ | 0.2473 | 0.2447 | 0.2524 | 3.9617 | 0.9696 | 1.3235 | 50′ |
| 20′ | 0.2502 | 0.2476 | 0.2555 | 3.9136 | 0.9689 | 1.3206 | 40′ |
| 30′ | 0.2531 | 0.2504 | 0.2586 | 3.8667 | 0.9681 | 1.3177 | 30′ |
| 40′ | 0.2560 | 0.2532 | 0.2617 | 3.8208 | 0.9674 | 1.3148 | 20′ |
| 50′ | 0.2589 | 0.2560 | 0.2648 | 3.7760 | 0.9667 | 1.3119 | 10′ |
| 15°00′ | 0.2618 | 0.2588 | 0.2679 | 3.7321 | 0.9659 | 1.3090 | 75° 00′ |
| 10′ | 0.2647 | 0.2616 | 0.2711 | 3.6891 | 0.9652 | 1.3061 | 50′ |
| 20′ | 0.2676 | 0.2644 | 0.2742 | 3.6470 | 0.9644 | 1.3032 | 40′ |
| 30′ | 0.2705 | 0.2672 | 0.2773 | 3.6059 | 0.9636 | 1.3003 | 30′ |
| 40′ | 0.2734 | 0.2700 | 0.2805 | 3.5656 | 0.9628 | 1.2974 | 20′ |
| 50′ | 0.2763 | 0.2728 | 0.2836 | 3.5261 | 0.9621 | 1.2945 | 10′ |
| 16° 00′ | 0.2793 | 0.2756 | 0.2867 | 3.4874 | 0.9613 | 1.2915 | 74° 00′ |
| 10′ | 0.2822 | 0.2784 | 0.2899 | 3.4495 | 0.9605 | 1.2886 | 50′ |
| 20′ | 0.2851 | 0.2812 | 0.2931 | 3.4124 | 0.9596 | 1.2857 | 40′ |
| 30′ | 0.2880 | 0.2840 | 0.2962 | 3.3759 | 0.9588 | 1.2828 | 30′ |
| 40′ | 0.2909 | 0.2868 | 0.2994 | 3.3402 | 0.9580 | 1.2799 | 20′ |
| 50′ | 0.2938 | 0.2896 | 0.3026 | 3.3052 | 0.9572 | 1.2770 | 10′ |
| 17° 00′ | 0.2967 | 0.2924 | 0.3057 | 3.2709 | 0.9563 | 1.2741 | 73° 00′ |
| 10′ | 0.2996 | 0.2952 | 0.3089 | 3.2371 | 0.9555 | 1.2712 | 50′ |
| 20′ | 0.3025 | 0.2979 | 0.3121 | 3.2041 | 0.9546 | 1.2683 | 40′ |
| 30′ | 0.3054 | 0.3007 | 0.3153 | 3.1716 | 0.9537 | 1.2654 | 30′ |
| 40′ | 0.3083 | 0.3035 | 0.3185 | 3.1397 | 0.9528 | 1.2625 | 20′ |
| 50′ | 0.3113 | 0.3062 | 0.3217 | 3.1084 | 0.9520 | 1.2595 | 10′ |
| 18° 00′ | 0.3142 | 0.3090 | 0.3249 | 3.0777 | 0.9511 | 1.2566 | 72° 00′ |
| | | Cosine | Cotangent | Tangent | Sine | Raidans | Degrees |

| Degrees | Radians | Sine | Tangent | Cotangent | Cosine | | |
|---------|---------|------|---------|-----------|--------|--------|--------|
| 18° 00′ | 0.3142 | 0.3090 | 0.3249 | 3.0777 | 0.9511 | 1.2566 | 72° 00′ |
| 10′ | 0.3171 | 0.3118 | 0.3281 | 3.0475 | 0.9502 | 1.2537 | 50′ |
| 20′ | 0.3200 | 0.3145 | 0.3314 | 3.0178 | 0.9492 | 1.2508 | 40′ |
| 30′ | 0.3229 | 0.3173 | 0.3346 | 2.9887 | 0.9483 | 1.2479 | 30′ |
| 40′ | 0.3258 | 0.3201 | 0.3378 | 2.9600 | 0.9474 | 1.2450 | 20′ |
| 50′ | 0.3287 | 0.3228 | 0.3411 | 2.9319 | 0.9465 | 1.2421 | 10′ |
| 19° 00′ | 0.3316 | 0.3256 | 0.3443 | 2.9042 | 0.9455 | 1.2392 | 71° 00′ |
| 10′ | 0.3345 | 0.3283 | 0.3476 | 2.8770 | 0.9446 | 1.2363 | 50′ |
| 20′ | 0.3374 | 0.3311 | 0.3508 | 2.8502 | 0.9436 | 1.2334 | 40′ |
| 30′ | 0.3403 | 0.3338 | 0.3541 | 2.8239 | 0.9426 | 1.2305 | 30′ |
| 40′ | 0.3432 | 0.3365 | 0.3574 | 2.7980 | 0.9417 | 1.2275 | 20′ |
| 50′ | 0.3462 | 0.3393 | 0.3607 | 2.7725 | 0.9407 | 1.2246 | 10′ |
| 20° 00′ | 0.3491 | 0.3420 | 0.3640 | 2.7475 | 0.9397 | 1.2217 | 70° 00′ |
| 10′ | 0.3520 | 0.3448 | 0.3673 | 2.7228 | 0.9387 | 1.2188 | 50′ |
| 20′ | 0.3549 | 0.3475 | 0.3706 | 2.6985 | 0.9377 | 1.2159 | 40′ |
| 30′ | 0.3578 | 0.3502 | 0.3739 | 2.6746 | 0.9367 | 1.2130 | 30′ |
| 40′ | 0.3607 | 0.3529 | 0.3772 | 2.6511 | 0.9356 | 1.2101 | 20′ |
| 50′ | 0.3636 | 0.3557 | 0.3805 | 2.6279 | 0.9346 | 1.2072 | 10′ |
| 21° 00′ | 0.3665 | 0.3584 | 0.3839 | 2.6051 | 0.9336 | 1.2043 | 69° 00′ |
| 10′ | 0.3694 | 0.3611 | 0.3872 | 2.5826 | 0.9325 | 1.2014 | 50′ |
| 20′ | 0.3723 | 0.3638 | 0.3906 | 2.5605 | 0.9315 | 1.1985 | 40′ |
| 30′ | 0.3752 | 0.3665 | 0.3939 | 2.5386 | 0.9304 | 1.1956 | 30′ |
| 40′ | 0.3782 | 0.3692 | 0.3973 | 2.5172 | 0.9293 | 1.1926 | 20′ |
| 50′ | 0.3811 | 0.3719 | 0.4006 | 2.4960 | 0.9283 | 1.1897 | 10′ |
| 22° 00′ | 0.3840 | 0.3746 | 0.4040 | 2.4751 | 0.9272 | 1.1868 | 68° 00′ |
| 10′ | 0.3869 | 0.3773 | 0.4074 | 2.4545 | 0.9261 | 1.1839 | 50′ |
| 20′ | 0.3898 | 0.3800 | 0.4108 | 2.4342 | 0.9250 | 1.1810 | 40′ |
| 30′ | 0.3927 | 0.3827 | 0.4142 | 2.4142 | 0.9239 | 1.1781 | 30′ |
| 40′ | 0.3956 | 0.3854 | 0.4176 | 2.3945 | 0.9228 | 1.1752 | 20′ |
| 50′ | 0.3985 | 0.3881 | 0.4210 | 2.3750 | 0.9216 | 1.1723 | 10′ |
| 23° 00′ | 0.4014 | 0.3907 | 0.4245 | 2.3559 | 0.9205 | 1.1694 | 67° 00′ |
| 10′ | 0.4043 | 0.3934 | 0.4279 | 2.3369 | 0.9194 | 1.1665 | 50′ |
| 20′ | 0.4072 | 0.3961 | 0.4314 | 2.3183 | 0.9182 | 1.1636 | 40′ |
| 30′ | 0.4102 | 0.3987 | 0.4348 | 2.2998 | 0.9171 | 1.1606 | 30′ |
| 40′ | 0.4131 | 0.4014 | 0.4383 | 2.2817 | 0.9159 | 1.1577 | 20′ |
| 50′ | 0.4160 | 0.4041 | 0.4417 | 2.2637 | 0.9147 | 1.1548 | 10′ |
| 24° 00′ | 0.4189 | 0.4067 | 0.4452 | 2.2460 | 0.9135 | 1.1519 | 66° 00′ |
| 10′ | 0.4218 | 0.4094 | 0.4487 | 2.2286 | 0.9124 | 1.1490 | 50′ |
| 20′ | 0.4247 | 0.4120 | 0.4522 | 2.2113 | 0.9112 | 1.1461 | 40′ |
| 30′ | 0.4276 | 0.4147 | 0.4557 | 2.1943 | 0.9100 | 1.1432 | 30′ |
| 40′ | 0.4305 | 0.4173 | 0.4592 | 2.1775 | 0.9088 | 1.1403 | 20′ |
| 50′ | 0.4334 | 0.4200 | 0.4628 | 2.1609 | 0.9075 | 1.1374 | 10′ |
| 25° 00′ | 0.4363 | 0.4226 | 0.4663 | 2.1445 | 0.9063 | 1.1345 | 65° 00′ |
| 10′ | 0.4392 | 0.4253 | 0.4699 | 2.1283 | 0.9051 | 1.1316 | 50′ |
| 20′ | 0.4422 | 0.4279 | 0.4734 | 2.1123 | 0.9038 | 1.1286 | 40′ |
| 30′ | 0.4451 | 0.4305 | 0.4770 | 2.0965 | 0.9026 | 1.1257 | 30′ |
| 40′ | 0.4480 | 0.4331 | 0.4806 | 2.0809 | 0.9013 | 1.1228 | 20′ |
| 50′ | 0.4509 | 0.4358 | 0.4841 | 2.0655 | 0.9001 | 1.1199 | 10′ |
| 26° 00′ | 0.4538 | 0.4384 | 0.4877 | 2.0503 | 0.8988 | 1.1170 | 64° 00′ |
| 10′ | 0.4567 | 0.4410 | 0.4913 | 2.0353 | 0.8975 | 1.1141 | 50′ |
| 20′ | 0.4596 | 0.4436 | 0.4950 | 2.0204 | 0.8962 | 1.1112 | 40′ |
| 30′ | 0.4625 | 0.4462 | 0.4986 | 2.0057 | 0.8949 | 1.1083 | 30′ |
| 40′ | 0.4654 | 0.4488 | 0.5022 | 1.9912 | 0.8936 | 1.1054 | 20′ |
| 50′ | 0.4683 | 0.4514 | 0.5059 | 1.9768 | 0.8923 | 1.1025 | 10′ |
| 27° 00′ | 0.4712 | 0.4540 | 0.5095 | 2.9626 | 0.8910 | 1.0996 | 63° 00′ |
| | | Cosine | Cotangent | Tangent | Sine | Radians | Degrees |

| Degrees | Radians | Sine | Tangent | Cotangent | Cosine | | |
|---|---|---|---|---|---|---|---|
| 27° 00′ | 0.4712 | 0.4540 | 0.5095 | 1.9626 | 0.8910 | 1.0996 | 63° 00′ |
| 10′ | 0.4741 | 0.4566 | 0.5132 | 1.9486 | 0.8897 | 1.0966 | 50′ |
| 20′ | 0.4771 | 0.4592 | 0.5169 | 1.9347 | 0.8884 | 1.0937 | 40′ |
| 30′ | 0.4800 | 0.4617 | 0.5206 | 1.9210 | 0.8870 | 1.0908 | 30′ |
| 40′ | 0.4829 | 0.4643 | 0.5243 | 1.9074 | 0.8857 | 1.0879 | 20′ |
| 50′ | 0.4858 | 0.4669 | 0.5280 | 1.8940 | 0.8843 | 1.0850 | 10′ |
| 28° 00′ | 0.4887 | 0.4695 | 0.5317 | 1.8807 | 0.8829 | 1.0821 | 62° 00′ |
| 10′ | 0.4916 | 0.4720 | 0.5354 | 1.8676 | 0.8816 | 1.0792 | 50′ |
| 20′ | 0.4945 | 0.4746 | 0.5392 | 1.8546 | 0.8802 | 1.0763 | 40′ |
| 30′ | 0.4974 | 0.4772 | 0.5430 | 1.8418 | 0.8788 | 1.0734 | 30′ |
| 40′ | 0.5003 | 0.4797 | 0.5467 | 1.8291 | 0.8774 | 1.0705 | 20′ |
| 50′ | 0.5032 | 0.4823 | 0.5505 | 1.8165 | 0.8760 | 1.0676 | 10′ |
| 29° 00′ | 0.5061 | 0.4848 | 0.5543 | 1.8040 | 0.8746 | 1.0647 | 61° 00′ |
| 10′ | 0.5091 | 0.4874 | 0.5581 | 1.7917 | 0.8732 | 1.0617 | 50′ |
| 20′ | 0.5120 | 0.4899 | 0.5619 | 1.7796 | 0.8718 | 1.0588 | 40′ |
| 30′ | 0.5149 | 0.4924 | 0.5658 | 1.7675 | 0.8704 | 1.0559 | 30′ |
| 40′ | 0.5178 | 0.4950 | 0.5696 | 1.7556 | 0.8689 | 1.0530 | 20′ |
| 50′ | 0.5207 | 0.4975 | 0.5735 | 1.7437 | 0.8675 | 1.0501 | 10′ |
| 30° 00′ | 0.5236 | 0.5000 | 0.5774 | 1.7321 | 0.8660 | 1.0472 | 60° 00′ |
| 10′ | 0.5265 | 0.5025 | 0.5812 | 1.7205 | 0.8646 | 1.0443 | 50′ |
| 20′ | 0.5294 | 0.5050 | 0.5851 | 1.7090 | 0.8631 | 1.0414 | 40′ |
| 30′ | 0.5323 | 0.5075 | 0.5890 | 1.6977 | 0.8616 | 1.0385 | 30′ |
| 40′ | 0.5352 | 0.5100 | 0.5930 | 1.6864 | 0.8601 | 1.0356 | 20′ |
| 50′ | 0.5381 | 0.5125 | 0.5969 | 1.6753 | 0.8587 | 1.0327 | 10′ |
| 31° 00′ | 0.5411 | 0.5150 | 0.6009 | 1.6643 | 0.8572 | 1.0297 | 59° 00′ |
| 10′ | 0.5440 | 0.5175 | 0.6048 | 1.6534 | 0.8557 | 1.0268 | 50′ |
| 20′ | 0.5469 | 0.5200 | 0.6088 | 1.6426 | 0.8542 | 1.0239 | 40′ |
| 30′ | 0.5498 | 0.5225 | 0.6128 | 1.6319 | 0.8526 | 1.0210 | 30′ |
| 40′ | 0.5527 | 0.5250 | 0.6168 | 1.6212 | 0.8511 | 1.0181 | 20′ |
| 50′ | 0.5556 | 0.5275 | 0.6208 | 1.6107 | 0.8496 | 1.0152 | 10′ |
| 32° 00′ | 0.5585 | 0.5299 | 0.6249 | 1.6003 | 0.8480 | 1.0123 | 58° 00′ |
| 10′ | 0.5614 | 0.5324 | 0.6289 | 1.5900 | 0.8465 | 1.0094 | 50′ |
| 20′ | 0.5643 | 0.5348 | 0.6330 | 1.5793 | 0.8450 | 1.0065 | 40′ |
| 30′ | 0.5672 | 0.5373 | 0.6371 | 1.5697 | 0.8434 | 1.0036 | 30′ |
| 40′ | 0.5701 | 0.5398 | 0.6412 | 1.5597 | 0.8418 | 1.0007 | 20′ |
| 50′ | 0.5730 | 0.5422 | 0.6453 | 1.5497 | 0.8403 | 0.9977 | 10′ |
| 33° 00′ | 0.5760 | 0.5446 | 0.6494 | 1.5399 | 0.8387 | 0.9948 | 57° 00′ |
| 10′ | 0.5789 | 0.5471 | 0.6536 | 1.5301 | 0.8371 | 0.9919 | 50′ |
| 20′ | 0.5818 | 0.5495 | 0.6577 | 1.5204 | 0.8355 | 0.9890 | 40′ |
| 30′ | 0.5847 | 0.5519 | 0.6619 | 1.5108 | 0.8339 | 0.9861 | 30′ |
| 40′ | 0.5876 | 0.5544 | 0.6661 | 1.5013 | 0.8323 | 0.9832 | 20′ |
| 50′ | 0.5905 | 0.5568 | 0.6703 | 1.4919 | 0.8307 | 0.9803 | 10′ |
| 34° 00′ | 0.5934 | 0.5592 | 0.6745 | 1.4826 | 0.8290 | 0.9774 | 56° 00′ |
| 10′ | 0.5963 | 0.5616 | 0.6787 | 1.4733 | 0.8274 | 0.9745 | 50′ |
| 20′ | 0.5992 | 0.5640 | 0.6830 | 1.4641 | 0.8258 | 0.9716 | 40′ |
| 30′ | 0.6021 | 0.5664 | 0.6873 | 1.4550 | 0.8241 | 0.9687 | 30′ |
| 40′ | 0.6050 | 0.5688 | 0.6916 | 1.4460 | 0.8225 | 0.9657 | 20′ |
| 50′ | 0.6080 | 0.5712 | 0.6959 | 1.4370 | 0.8208 | 0.9628 | 10′ |
| 35° 00′ | 0.6109 | 0.5736 | 0.7002 | 1.4281 | 0.8192 | 0.9599 | 55° 00′ |
| 10′ | 0.6138 | 0.5760 | 0.7046 | 1.4193 | 0.8175 | 0.9570 | 50′ |
| 20′ | 0.6167 | 0.5783 | 0.7089 | 1.4106 | 0.8158 | 0.9541 | 40′ |
| 30′ | 0.6196 | 0.5807 | 0.7133 | 1.4019 | 0.8141 | 0.9512 | 30′ |
| 40′ | 0.6225 | 0.5831 | 0.7177 | 1.3934 | 0.8124 | 0.9483 | 20′ |
| 50′ | 0.6254 | 0.5854 | 0.7221 | 1.3848 | 0.8107 | 0.9454 | 10′ |
| 36° 00′ | 0.6283 | 0.5878 | 0.7265 | 1.3764 | 0.8090 | 0.9425 | 54° 00′ |
| | | Cosine | Cotangent | Tangent | Sine | Radians | Degrees |

| Degrees | Radians | Sine | Tangent | Cotangent | Cosine | | |
|---------|---------|------|---------|-----------|--------|---|---|
| 36° 00′ | 0.6283 | 0.5878 | 0.7265 | 1.3764 | 0.8090 | 0.9425 | 54° 00′ |
| 10′ | 0.6312 | 0.5901 | 0.7310 | 1.3680 | 0.8073 | 0.9396 | 50′ |
| 20′ | 0.6341 | 0.5925 | 0.7355 | 1.3597 | 0.8056 | 0.9367 | 40′ |
| 30′ | 0.6370 | 0.5948 | 0.7400 | 1.3514 | 0.8039 | 0.9338 | 30′ |
| 40′ | 0.6400 | 0.5972 | 0.7445 | 1.3432 | 0.8021 | 0.9308 | 20′ |
| 50′ | 0.6429 | 0.5995 | 0.7490 | 1.3351 | 0.8004 | 0.9279 | 10′ |
| 37° 00′ | 0.6458 | 0.6018 | 0.7536 | 1.3270 | 0.7986 | 0.9250 | 53° 00′ |
| 10′ | 0.6487 | 0.6041 | 0.7581 | 1.3190 | 0.7969 | 0.9221 | 50′ |
| 20′ | 0.6516 | 0.6065 | 0.7627 | 1.3111 | 0.7951 | 0.9192 | 40′ |
| 30′ | 0.6545 | 0.6088 | 0.7673 | 1.3032 | 0.7934 | 0.9163 | 30′ |
| 40′ | 0.6574 | 0.6111 | 0.7720 | 1.2954 | 0.7916 | 0.9134 | 20′ |
| 50′ | 0.6603 | 0.6134 | 0.7766 | 1.2876 | 0.7898 | 0.9105 | 10′ |
| 38° 00′ | 0.6632 | 0.6157 | 0.7813 | 1.2799 | 0.7880 | 0.9076 | 52° 00′ |
| 10′ | 0.6661 | 0.6180 | 0.7860 | 1.2723 | 0.7862 | 0.9047 | 50′ |
| 20′ | 0.6690 | 0.6202 | 0.7907 | 1.2647 | 0.7844 | 0.9018 | 40′ |
| 30′ | 0.6720 | 0.6225 | 0.7954 | 1.2572 | 0.7826 | 0.8988 | 30′ |
| 40′ | 0.6749 | 0.6248 | 0.8002 | 1.2497 | 0.7808 | 0.8959 | 20′ |
| 50′ | 0.6778 | 0.6271 | 0.8050 | 1.2423 | 0.7790 | 0.8930 | 10′ |
| 39° 00′ | 0.6807 | 0.6293 | 0.8098 | 1.2349 | 0.7771 | 0.8901 | 51° 00′ |
| 10′ | 0.6836 | 0.6316 | 0.8146 | 1.2276 | 0.7753 | 0.8872 | 50′ |
| 20′ | 0.6865 | 0.6338 | 0.8195 | 1.2203 | 0.7735 | 0.8843 | 40′ |
| 30′ | 0.6894 | 0.6361 | 0.8243 | 1.2131 | 0.7716 | 0.8814 | 30′ |
| 40′ | 0.6923 | 0.6383 | 0.8292 | 1.2059 | 0.7698 | 0.8785 | 20′ |
| 50′ | 0.6952 | 0.6406 | 0.8342 | 1.1988 | 0.7679 | 0.8756 | 10′ |
| 40° 00′ | 0.6981 | 0.6428 | 0.8391 | 1.1918 | 0.7660 | 0.8727 | 50° 00′ |
| 10′ | 0.7010 | 0.6450 | 0.8441 | 1.1847 | 0.7642 | 0.8698 | 50′ |
| 20′ | 0.7039 | 0.6472 | 0.8491 | 1.1778 | 0.7623 | 0.8668 | 40′ |
| 30′ | 0.7069 | 0.6494 | 0.8541 | 1.1708 | 0.7604 | 0.8639 | 30′ |
| 40′ | 0.7098 | 0.6517 | 0.8591 | 1.1640 | 0.7585 | 0.8610 | 20′ |
| 50′ | 0.7127 | 0.6539 | 0.8642 | 1.1571 | 0.7566 | 0.8581 | 10′ |
| 41° 00′ | 0.7156 | 0.6561 | 0.8693 | 1.1504 | 0.7547 | 0.8552 | 49° 00′ |
| 10′ | 0.7185 | 0.6583 | 0.8744 | 1.1436 | 0.7528 | 0.8523 | 50′ |
| 20′ | 0.7214 | 0.6604 | 0.8796 | 1.1369 | 0.7509 | 0.8494 | 40′ |
| 30′ | 0.7243 | 0.6626 | 0.8847 | 1.1303 | 0.7490 | 0.8465 | 30′ |
| 40′ | 0.7272 | 0.6648 | 0.8899 | 1.1237 | 0.7470 | 0.8436 | 20′ |
| 50′ | 0.7301 | 0.6670 | 0.8952 | 1.1171 | 0.7451 | 0.8407 | 10′ |
| 42° 00′ | 0.7330 | 0.6691 | 0.9004 | 1.1106 | 0.7431 | 0.8378 | 48° 00′ |
| 10′ | 0.7359 | 0.6713 | 0.9057 | 1.1041 | 0.7412 | 0.8348 | 50′ |
| 20′ | 0.7389 | 0.6734 | 0.9110 | 1.0977 | 0.7392 | 0.8319 | 40′ |
| 30′ | 0.7418 | 0.6756 | 0.9163 | 1.0913 | 0.7373 | 0.8290 | 30′ |
| 40′ | 0.7447 | 0.6777 | 0.9217 | 1.0850 | 0.7353 | 0.8261 | 20′ |
| 50′ | 0.7476 | 0.6799 | 0.9271 | 1.0786 | 0.7333 | 0.8232 | 10′ |
| 43° 00′ | 0.7505 | 0.6820 | 0.9325 | 1.0724 | 0.7314 | 0.8203 | 47° 00′ |
| 10′ | 0.7534 | 0.6841 | 0.9380 | 1.0661 | 0.7294 | 0.8174 | 50′ |
| 20′ | 0.7563 | 0.6862 | 0.9435 | 1.0599 | 0.7274 | 0.8145 | 40′ |
| 30′ | 0.7592 | 0.6884 | 0.9490 | 1.0538 | 0.7254 | 0.8116 | 30′ |
| 40′ | 0.7621 | 0.6905 | 0.9545 | 1.0477 | 0.7234 | 0.8087 | 20′ |
| 50′ | 0.7650 | 0.6926 | 0.9601 | 1.0416 | 0.7214 | 0.8058 | 10′ |
| 44° 00′ | 0.7679 | 0.6947 | 0.9657 | 1.0355 | 0.7193 | 0.8029 | 46° 00′ |
| 10′ | 0.7709 | 0.6967 | 0.9713 | 1.0295 | 0.7173 | 0.7999 | 50′ |
| 20′ | 0.7738 | 0.6988 | 0.9770 | 1.0235 | 0.7153 | 0.7970 | 40′ |
| 30′ | 0.7767 | 0.7009 | 0.9827 | 1.0176 | 0.7133 | 0.7941 | 30′ |
| 40′ | 0.7796 | 0.7030 | 0.9884 | 1.0117 | 0.7112 | 0.7912 | 20′ |
| 50′ | 0.7825 | 0.7050 | 0.9942 | 1.0058 | 0.7092 | 0.7883 | 10′ |
| 45° 00′ | 0.7854 | 0.7071 | 1.0000 | 1.0000 | 0.7071 | 0.7854 | 45° 00′ |
| | | Cosine | Cotangent | Tangent | Sine | Radians | Degrees |

# APPENDIX L: STANDARD WELDING SYMBOLS

## Standard Welding Symbols

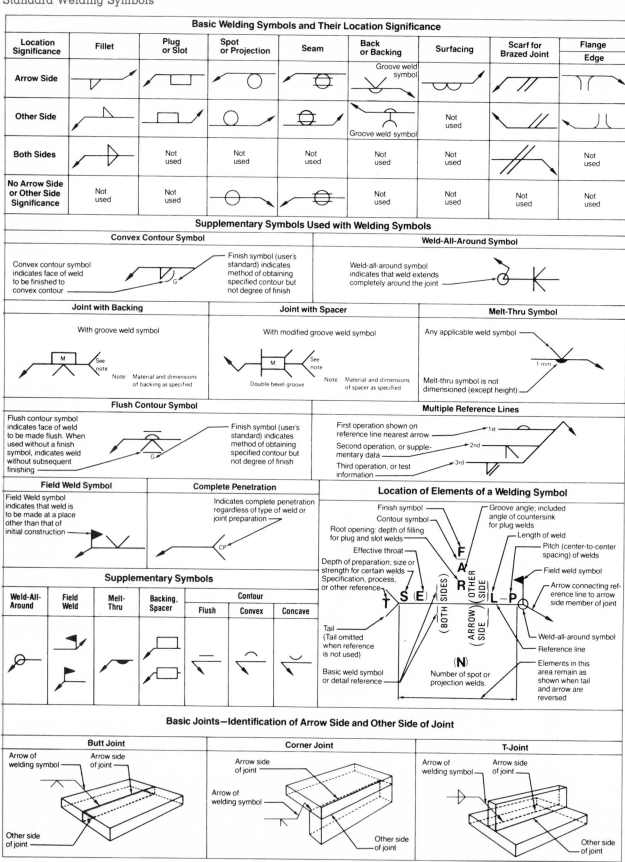

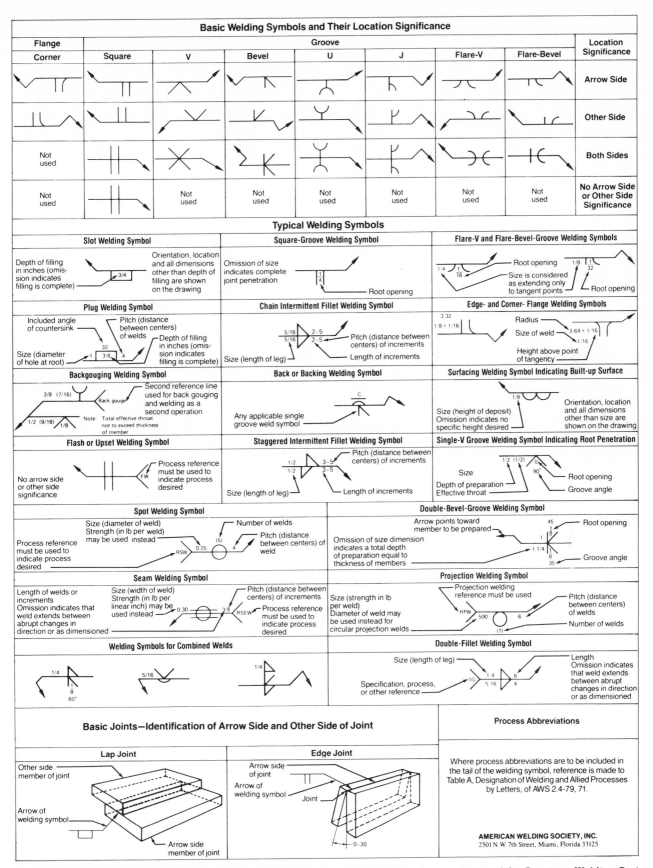

Reproduced from AWS A2.4–79, Symbols for Welding and Nondestructive Testing, by permission of the American Welding Society.

# APPENDIX M: METAL SHEET AND WIRE STANDARD

Preferred Widths for Flat Metal Products (in Millimeters)

Applicable to bar, foil, flat wire, plate, ribbon, sheet, strip, etc., only where the width falls within the 10 to 300 millimeter range

| Preferred Width | Second Preference |
|---|---|
| 10 | |
| 12 | |
| 16 | |
| 20 | |
| 25 | |
| 30 | |
| 35 | |
| 40 | |
| | 45 |
| 50 | |
| 60 | |
| 80 | |
| | 90 |
| 100 | |
| | 110 |
| 120 | |
| 140 | |
| | 150 |
| 160 | |
| 180 | |
| 200 | |
| | 250 |
| | 300 |

Reproduced from ANSI B32.3–1977 with the permission of the publisher, the American Society of Mechanical Engineers.

From *Technical Drafting*, copyright 1980 by Spence and Atkins. Used with permission of the publisher, Bennett Publishing Company, Peoria, IL 61615.

Preferred Thicknesses for All Flat Metal Products (in Millimeters)

| Preferred Thickness | Second Preference | Preferred Thickness | Second Preference |
|---|---|---|---|
| 0.050 | | 4.5 | |
| 0.060 | | | 4.8 |
| 0.080 | | 5.0 | |
| 0.10 | | | 5.5 |
| 0.12 | | 6.0 | |
| | 0.14 | 7.0 | |
| 0.16 | | 8.0 | |
| | 0.18 | | 9.0 |
| 0.20 | | 10 | |
| | 0.22 | | 11 |
| 0.25 | | 12 | |
| | 0.28 | | 14 |
| 0.30 | | 16 | |
| | 0.35 | | 18 |
| 0.40 | | 20 | |
| | 0.45 | | 22 |
| 0.50 | | 25 | |
| | 0.55 | | 28 |
| 0.60 | | 30 | |
| | 0.65 | | 32 |
| | 0.70 | 35 | |
| 0.80 | | | 38 |
| | 0.90 | 40 | |
| 1.0 | | | 45 |
| | 1.1 | 50 | |
| 1.2 | | | 55 |
| | 1.4 | 60 | |
| 1.6 | | | 70 |
| | 1.8 | 80 | |
| 2.0 | | | 90 |
| | 2.2 | 100 | |
| 2.5 | | | 110 |
| | 2.8 | 120 | |
| 3.0 | | | 130 |
| | 3.2 | 140 | |
| 3.5 | | | 150 |
| | 3.8 | 160 | |
| 4.0 | | | |

Reproduced from ANSI B32.3–1977 with the permission of the publisher, the American Society of Mechanical Engineers.

From *Technical Drafting*, copyright 1980 by Spence and Atkins. Used with permission of the publisher, Bennett Publishing Company, Peoria, IL 61615.

Standard Wire and Sheet-Metal Gages*

| Gage number | (A) Brown & Sharpe or American | (B) American Steel & Wire Co. | (C) Piano wire | (E) U.S. St'd. | Gage number |
|---|---|---|---|---|---|
| 0000000 | .6513 | .4900 | — | .5000 | 0000000 |
| 000000 | .5800 | .4615 | .004 | .4688 | 000000 |
| 00000 | .5165 | .4305 | .005 | .4375 | 00000 |
| 0000 | .4600 | .3938 | .006 | .4063 | 0000 |
| 000 | .4096 | .3625 | .007 | .3750 | 000 |
| 00 | .3648 | .3310 | .008 | .3438 | 00 |
| 0 | .3249 | .3065 | .009 | .3125 | 0 |
| 1 | .2893 | .2830 | .010 | .2813 | 1 |
| 2 | .2576 | .2625 | .011 | .2656 | 2 |
| 3 | .2294 | .2437 | .012 | .2500 | 3 |
| 4 | .2043 | .2253 | .013 | .2344 | 4 |
| 5 | .1819 | .2070 | .014 | .2188 | 5 |
| 6 | .1620 | .1920 | .016 | .2031 | 6 |
| 7 | .1443 | .1770 | .018 | .1875 | 7 |
| 8 | .1285 | .1620 | .020 | .1719 | 8 |
| 9 | .1144 | .1483 | .022 | .1563 | 9 |
| 10 | .1019 | .1350 | .024 | .1406 | 10 |
| 11 | .0907 | .1205 | .026 | .1250 | 11 |
| 12 | .0808 | .1055 | .029 | .1094 | 12 |
| 13 | .0720 | .0915 | .031 | .0938 | 13 |
| 14 | .0641 | .0800 | .033 | .0781 | 14 |
| 15 | .0571 | .0720 | .035 | .0703 | 15 |
| 16 | .0508 | .0625 | .037 | .0625 | 16 |
| 17 | .0453 | .0540 | .039 | .0563 | 17 |
| 18 | .0403 | .0475 | .041 | .0500 | 18 |
| 19 | .0359 | .0410 | .043 | .0438 | 19 |
| 20 | .0320 | .0348 | .045 | .0375 | 20 |
| 21 | .0285 | .0317 | .047 | .0344 | 21 |
| 22 | .0253 | .0286 | .049 | .0313 | 22 |
| 23 | .0226 | .0258 | .051 | .0281 | 23 |
| 24 | .0201 | .0230 | .055 | .0250 | 24 |
| 25 | .0179 | .0204 | .059 | .0219 | 25 |
| 26 | .0159 | .0181 | .063 | .0188 | 26 |
| 27 | .0142 | .0173 | .067 | .0172 | 27 |
| 28 | .0126 | .0162 | .071 | .0156 | 28 |
| 29 | .0113 | .0150 | .075 | .0141 | 29 |
| 30 | .0100 | .0140 | .080 | .0125 | 30 |
| 31 | .0089 | .0132 | .085 | .0109 | 31 |
| 32 | .0080 | .0128 | .090 | .0102 | 32 |
| 33 | .0071 | .0118 | .095 | .0094 | 33 |
| 34 | .0063 | .0104 | .100 | .0086 | 34 |
| 35 | .0056 | .0095 | .106 | .0078 | 35 |
| 36 | .0050 | .0090 | .112 | .0070 | 36 |
| 37 | .0045 | .0085 | .118 | .0066 | 37 |
| 38 | .0040 | .0080 | .124 | .0063 | 38 |
| 39 | .0035 | .0075 | .130 | — | 39 |
| 40 | .0031 | .0070 | .138 | — | 40 |

* Dimensions in decimal parts of an inch.
(A) Standard in United States for sheet metal and wire (except steel and iron).
(B) Standard for iron and steel wire (U.S. Steel Wire Gage).
(C) American Steel and Wire Company's music (or piano) wire gage sizes. Recognized by U.S. Bureau of Standards.
(E) U.S. Standard for iron and steel plate. However, plate is now generally specified by its thickness in decimals of an inch.

From *Fundamentals of Engineering Drawing: For Design, Product Development, and Numerical Control*, 8th ed., copyright 1981 by Prentice-Hall, Inc. Used with permission of the publisher, Prentice-Hall, Inc., Englewood Cliffs, NJ, 07632.

# APPENDIX N: AMERICAN NATIONAL STANDARDS

Following are a few of the hundreds of standards available from the American National Standards Institute. A complete listing of standards, "A Catalog of American National Standards", can be purchased from the American National Standards Institute, 1430 Broadway, New York, N.Y. 10018

## Abbreviations

ANSI Y1.1–1972  Abbreviations for use on Drawings and in Text

## Bearings

ANSI B3.7–1960(R1977)  Ball and Roller Bearings and Parts, Terminology and Definitions for

ANSI/AFBMA 11–1978  Roller Bearings, Load Ratings and Fatigue Life for

ANSI/AFBMA 22–1976  Spherical Plain Bearings, Joint Type

## Brick

ANSI/ASTM C62–75  Building Brick (Solid Masonry Units made from clay or shale), Specifications for

ANSI/ASTM C55–71  Concrete Building Brick, Specifications for

## Charts and Graphs

ANSI Y15.1M–1979  Illustrations for Publication and Projection

ANSI Y15.3M–1979  Materials Handling Operation and Flow Charts

ANSI Y15.3M–1979  Process Flow Charts

ANSI Y15.2M–1979  Time-Series Charts

## Computers

ANSI/NFPA 75–1976  Electronic Computer Systems

## Concrete

ANSI/ASTM C494–80  Chemical Admixtures for Concrete, Specifications for

## Conductors

ANSI/ASTM B49–70  Hot-Rolled Copper Rods for Electrical Purposes, Specification for

## Dimensioning and Surface Finish

ANSI B4.1–1967(R1979)  Preferred Limits and Fits for Cylindrical Parts

ANSI B4.2–1978  Preferred Metric Limits and Fits

ANSI B4.3–1978  General Tolerances for Metric Dimensioned Products

ANSI B46.1–1978  Surface Texture

## Drafting Manual

ANSI Y14.1–1975  Drawing Sheet Size and Format

ANSI Y14.2–1979  Line Conventions and Lettering

ANSI Y14.3–1975  Multi and Sectional View Drawings

ANSI Y14.4–1957  Pictorial Drawing

ANSI Y14.5–1973  Dimensioning and Tolerancing

ANSI Y14.6–1978  Screw Thread Representation, Engineering Drawing and Related Documentation Practice

ANSI Y14.9–1958  Forgings

ANSI Y14.15a–1971  Interconnection Diagrams

ANSI Y14.15–1966(R1973)  Electrical and Electronics Diagrams

ANSI Y14.15b–1973  Electrical and Electronics Diagrams (supplement)

ANSI Y14.15a–1970  Electrical and Electronics Diagrams (supplement)

ANSI Y14.17–1966(R1980)  Fluid Power Diagrams

ANSI Y14.26.3–1975  Dictionary of Terms for Computer-Aided Preparation of Product Definition Data

ANSI Y14.32.1–1974  Chassis Frames—Passenger Car and Light Truck—Ground Vehicle Practices

ANSI Y14.7.1–1971  Gear Drawing Standards, Part 1 for Spur, Helical, Double Helical and Rack

ANSI Y14.7.2–1978  Gear and Spline Drawing Standards, Part 2, Bevel and Hypoid Gears

ANSI Y14.36–1978  Surface Texture Symbols

ANSI Y14 Report 1  Digital Representation of Physical Object Shapes

ANSI Y14 Report 2  Documentating of Computer Systems used in Computer-Aided Preparation of Product Definition Data, Guideline for-User Instructions

ANSI Y14 Report 2  Documentating of Computer Systems used in Computer-Aided Preparation of Product Definition Data, Guideline for-Design Requirements

## Fasteners

ANSI B18.2.3.6M–1979  Bolts, Metric, Heavy Hex

ANSI B18.2.3.5M–1979  Bolts, Metric Hex

ANSI B18.2.4.5M  Hex Jam Nuts, Metric

ANSI B18.2.4.6M  Hex Nuts, Heavy, Metric

ANSI B18.2.4.3M  Hex Nuts, Slotted, Metric

ANSI B18.3.2.M  Keys and Bits, Hexagon (metric series)

ANSI B18.1.2–1972(R1977) Large Rivets (½ inch Nominal Diameter and Larger

ANSI B18.1.1–1972(R1977) Small Solid Rivets (⁷⁄₁₆ inch nominal and smaller)

ANSI B18.21.1–1972 Lock Washers

ANSI B18.23.1–1967(R1975) Beveled Washers

ANSI B18.22.1–1965(R1975) Plain Washers

ANSI B18.3.4M–1979 Screws, Hexagon Socket Button Head Cap, Metric Series

ANSI B18.6.2–1972(R1977) Slotted Head Cap Screws, Square Head Set Screws, and Slotted Headless Set Screws

ANSI B18.2.1–1972 Square and Hex Bolts and Screws, including Askew Head Bolts, Hex Cap Screws, and Lag Screws

ANSI B18.2.2–1972 Square and Hex Nuts

ANSI B18.5–1978 Round Head Bolts

ANSI B18.8.1–1972(R1977) Clevis Pins and Cotter Pins

## Gears

ANSI B6.9–1977 Design for Fine-Pitch Worm Gears

ANSI/AGMA 112.05–1976 Gear Nomenclature-Terms, Definitions, Symbols and Abbreviations

ANSI/AGMA 208.03–1979 Straight Bevel Gears, System for

## Graphic Symbols

ANSI/IEEE 315 Electrical and Electronics Diagrams (including reference designation class designation letters), Graphic Symbols for

ANSI/IEEE 200 Electrical and Electronics Parts and Equipment, Reference Designations for

ANSI Y32.9–1972 Electrical Wiring and Layout Diagrams Used in Architecture and Building Construction, Graphic Symbols for

ANSI Y32.10–1967(R1979) Fluid Power Diagrams, Graphic Symbols for

ANSI Y32.2.6–1950(R1956) Heat-Power Apparatus, Graphic Symbols for

ANSI Y32.2.4–1949(R1953) Heating, Ventilating and Air Conditioning, Graphic Symbols for

ANSI/ISA S5.1–1975 Instrumentation Symbols and Identification

ANSI/IEEE 91–1973 Logic Diagrams (two-state devices) Graphic Symbols for

ANSI Y32.18–1972(R1978) Mechanical and Acoustical Elements As Used in Schematic Diagrams, Symbols for

ANSI Y32.2.3–1949(R1953) Pipe Fittings, Valves, and Piping, Graphic Symbols for

ANSI Y32.4–1977 Plumbing, Fixtures for Diagrams Used in Architecture and Building Construction, Graphic Symbols for

ANSI Y32.11–1961 Process Flow Diagrams in the Petroleum and Chemical Industries, Graphic Symbols for

ANSI Y32.7–1972(R1979) Railroad Maps and Profiles, Graphic Symbols for

ANSI/AWS A2.4–1979 Symbols for Welding and Nondestructive Testing, including Brazing

## Keys and Keyseats

ANSI B17.1–1967(R1973) Keys and Keyseats

ANSI B17.2–1967(R1978) Woodruff Keys and Keyseats

## Metric Practice

ANSI/ASTM E 380–76/IEEE 268–1976 Metric Practice

## Nuclear

ANSI N1.1–1976 Nuclear Science and Technology, Glossary of terms in

## Petroleum

ANSI/ASTM D288–61(1978) Petroleum, Definitions of Terms Relating to

## Pipe

ANSI/ASTM B241–79 Aluminum-Alloy Seamless Pipe and Seamless Extruded Tube, Specifications for

ANSI/ASTM B42–80 Seamless Copper Pipe Standard Sizes, Specifications for

ANSI/ASTM C14–79 Concrete Sewer, Storm Drain and Culvert Pipe, Specifications for

ANSI/ASTM Poly(vinyl chloride) (PVC) Sewer Pipe and Fittings, Specification for

ANSI B36.10–1979 Welded and Seamless Wrought Steel Pipe

ANSI/ASTM A74–80 Cast Iron Soil Pipe and Fittings, Specification for

## Rubber

ANSI/ASTM D1566–80a Rubber and Rubber-like Materials, Definition of Terms Relating to

## Semiconductors

ANSI C34.2–1968(R1973) Semiconductor Power Rectifiers, Practices and Requirements for

## Shock and Vibration

ANSI S2.10–1971(R1976) Analysis and Presentation of Shock and Vibration Data, Methods for

## Solar

ANSI/ASTM B638-80 Copper and Copper-Alloy Solar Heat Absorber Panels, Specifications for

## Threads

ANSI B2.1–1968   Pipe Threads

ANSI B1.1–1974   Unified Inch Screw Threads (UN and UNR Thread Form)

ANSI B1.5–1977   ACME Screw Threads

ANSI B1.13M–1979   Metric Screw Threads—M Profile

ANSI B1.21M–1978   Metric Screw Threads—MJ Profile

ANSI B1.7–1977   Screw Threads, Nomenclature, Definitions and Letter Symbols for

ANSI B1.10–1958   Unified Miniature Screw Threads

## Wood

ANSI/ASTM D1165–76   Domestic Hardwoods and Softwoods, Nomenclature of

ANSI/Vol PROD STD PS 56–73   Structural Glued Laminated Timber

# STANDARDS OF THE INTERNATIONAL ORGANIZATION FOR STANDARDIZATION (ISO)

Information on ISO standards is available from the American National Standards Institute. It is the only source of these standards in the United States. A catalog listing of available standards can be purchased. Following are a few of the standards available.

## Drafting

ISO/R 128   Engineering Drawing—Principles of Presentation

ISO/R 129   Engineering Drawing—Dimensioning

ISO/R 3098   Lettering

ISO 2203   Conventional Representation of Gears

ISO 2162   Representation of Springs

## Tolerancing

ISO/R 1101/01   Tolerances of Form and of Position—Part 1

ISO/R 1101/11   Tolerances of Form and of Position—Part 2

ISO/R 1660   Tolerances of Form and Position—Part 3 Dimensioning and Tolerancing Profiles

ISO/R 1661   Tolerances of Form and Position—Part 4

ISO/R 406   Inscription of Linear and Angular Tolerances

## Fasteners

R 278   Hexagon Bolts and Nuts

R 724   ISO General Purpose Metric Screw Threads

R 887   Washers for Hexagon Bolts and Nuts—Metric Series

R 1478   Tapping Screw Thread

R 1479   Hexagon Head Tapping Screw—Metric Series

R 1580   Slotted Pan Head Screws

ISO 2339   Taper Pins, Unhardened, Metric Series

ISO 2343   Hexagon Socket Set Screws

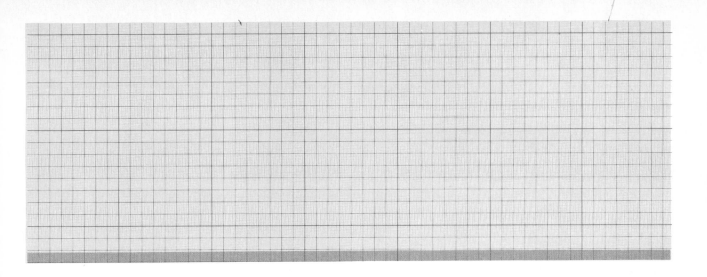

# INDEX

770